T0315084

Sourcebook in the

Mathematics of Ancient Greece and the Eastern Mediterranean

Sourcebook in the

Mathematics of Ancient Greece and the Eastern Mediterranean

Edited by

VICTOR J. KATZ and
CLEMENCY MONTELLE

PRINCETON UNIVERSITY PRESS • PRINCETON AND OXFORD

Copyright © 2024 by Princeton University Press

Princeton University Press is committed to the protection of copyright and the intellectual property our authors entrust to us. Copyright promotes the progress and integrity of knowledge. Thank you for supporting free speech and the global exchange of ideas by purchasing an authorized edition of this book. If you wish to reproduce or distribute any part of it in any form, please obtain permission.

Requests for permission to reproduce material from this work
should be sent to permissions@press.princeton.edu

Published by Princeton University Press
41 William Street, Princeton, New Jersey 08540
99 Banbury Road, Oxford OX2 6JX

press.princeton.edu

All Rights Reserved

Library of Congress Cataloging-in-Publication Data

Names: Katz, Victor J., editor. | Montelle, Clemency, editor.
Title: Sourcebook in the mathematics of Ancient Greece and the Eastern Mediterranean / edited by Victor J. Katz and Clemency Montelle.
Description: Princeton, New Jersey : Princeton University Press, [2024] | Includes bibliographical references and index.
Identifiers: LCCN 2023038765 (print) | LCCN 2023038766 (ebook) | ISBN 9780691202815 | ISBN 9780691257686 (ebook)
Subjects: LCSH: Mathematics, Greek. | Mathematics—Greece—History—to 1500. | Mathematics—Middle East—History—to 1500. | BISAC: MATHEMATICS / History & Philosophy | SCIENCE / History
Classification: LCC QA22 .S66 2024 (print) | LCC QA22 (ebook) | DDC 510.9/01—dc23/eng/20231214
LC record available at https://lccn.loc.gov/2023038765
LC ebook record available at https://lccn.loc.gov/2023038766

British Library Cataloging-in-Publication Data is available

Editorial: Diana Gillooly, Whitney Rauenhorst
Jacket: Wanda España
Production: Danielle Amatucci
Publicity: William Pagdatoon

Cover images: (From top to bottom): Medieval illustration of *Anicius Manlius Severinus Boëthius*, circa 1150. Fragment of MS Ii.3.12; f. 61v. Reproduced by kind permission of the Syndics of Cambridge University Library; Created by authors; *Table of Fractions* P.Mich.inv. 621; Recto. Courtesy of University of Michigan Library Digital Collections; Created by authors; Page from Ptolemy's *Almagest*, translated by Gerard Cremona. Image courtesy of Bibliothéque nationale de France.

Printed in the United States of America

10 9 8 7 6 5 4 3 2 1

Contents

Introduction

Why is there a need for a new sourcebook in Greek mathematics? After all, the major Greek authors—Euclid, Archimedes, Apollonius, Ptolemy, Diophantus, Pappus – all have well-regarded and easily available English editions or, in the case of Archimedes, an English edition in process. And passages from Greek mathematics are available in some of the more general sourcebooks in mathematics. For example, the 1987 *History of Mathematics: A Reader*, edited by John Fauvel and Jeremy Gray, compiles English translations of original sources pertaining to mathematics from the very beginnings of human history to the current era. A portion of the book is dedicated to Greek mathematics; original sources are included as well as passages from historians reflecting on the field and excerpts from various scholarly debates over relevant issues. However, most of the passages are quite brief, largely because Greek mathematics is but one of the cultures and time periods that feature in this compendium of historical mathematical sources.

More specifically is Ivor Thomas's *Greek Mathematical Works*, first published as two volumes in 1939/41 as part of the Loeb Classical Library. This classic sourcebook is dedicated entirely to Greek mathematics. The volumes contain excerpts from Greek mathematical sources with Greek text on the left-hand side and translations on the facing page. The passages are organized chronologically, and their translations generally derive from scholarly authorities at the time. Little mathematical explanation accompanies the passages, although there are various mathematical and historical footnotes where appropriate.

In recent decades, there has been extensive research on Greek mathematics that has considerably enlarged the scope of the topic. Namely, although most mathematicians regard "Greek mathematics" as the axiomatically based mathematics introduced to the world in the work of Euclid, Archimedes, Apollonius, and others in the first three centuries BCE, there is in fact much more mathematics that was studied and used in the eastern Mediterranean during the time it was under substantial Greek cultural influence, say, from 400 BCE to 600 CE, and in certain areas even longer.

To reflect this, in addition to well-known material from the classic Greek texts, this sourcebook also contains passages from lesser known authors, some of whose works are now hard to find, as well as material from newly uncovered texts and passages that have been left out by past scholarship, which was often based on more narrow definitions of "mathematics." Indeed, the Thomas sourcebook was written to accompany Thomas Heath's classic *History of Greek Mathematics*, first published a century ago. More recent historical surveys, such as David Fowler's *Mathematics of Plato's Academy*, Serafina Cuomo's *Ancient Mathematics*, and Reviel Netz's *New History of Greek Mathematics*, for instance, have significantly expanded the scope and richness of the field. It is thus timely to provide an up-to-date sourcebook that reflects this increased breadth, scope, and diversity.

While the majority of passages are translations from works written in ancient Greek, we have also included passages from Greek works whose originals are no longer extant but whose translations into Arabic during the eighth and ninth centuries are still available. We have also included passages from adjacent cultures of inquiry in which there are commonalities in the mathematical material, including sources originally written in Latin, Demotic, and Babylonian cuneiform.

In general, the excerpts we have included are often longer than those found in other sourcebooks to enable readers to follow the mathematical arguments in detail and understand their context. Although it is not possible to include every prerequisite to every theorem presented, we have made an effort of give the reader all the mathematical tools necessary to understand the proofs. We have used a broad scope in defining mathematical practice, so activities such as astronomy, music, and optics find a place here. The excerpts are organized into chapters by topic, enabling the reader to see how different authors treated the same mathematical ideas. While these divisions are in part inspired by early groupings of the mathematical sciences, we acknowledge that this overall organization might not always reflect how the original authors viewed the contours of their discipline. We provide a preface to each excerpt, to put the mathematical ideas into context, as well as explanations to help readers understand the arguments and, sometimes, to put the arguments into modern mathematical notation.

In addition to prose texts, other forms of mathematical activity are included here. For example, some numerical tables provide direct evidence of Greek mathematicians carrying out hand computations and recording the resulting numerical values. Instrumentation, too, embodies mathematical principles, and to this end we include discussions on the Antikythera Mechanism, the mechanical astronomical device discovered in a the remains of a ship wreck, early in the twentieth century, that is based in part on mathematically derived cycles.

Among the other special features of this book are ancient commentaries on the classic works of Euclid, Archimedes, and Apollonius. In many cases, we have interspersed these commentaries in the main body of the work, so that readers can easily understand what the commentators were trying to accomplish. We have used a different typeface to distinguish the commentaries from the original sources, as explained in each chapter.

We have also included passages about numbers that are more philosophical than technical, which were part of the effort by the so-called neo-Pythagoreans to understand fully the nature of the concept of "number." Another group of excerpts, not included in previous sourcebooks, are sourced from papyri discovered in Egypt, written during the Greek period in either Greek or Demotic. These passages are necessarily short, because only a few such papyri are extant, and many of them are fragmentary. Even though it is sometimes difficult to discern the author's intent, it does seem that mathematical ideas were being used and that these ideas are quite different from the theoretical geometry of the classic texts. Many of these papyri appear to be from teachers of mathematics compiling lists of problems to pass on much simpler ideas to students or practitioners. Although it is difficult to draw firm conclusions because of the paucity of textual material, it does seem that teachers in Egypt during the Greek period were using the same ideas as their predecessors in Seleucid Babylonia and that Roman surveyors, some years later, also used these ideas. Precisely how such ideas were passed on from era to era and from place to place remains very much in question.

In the process of compiling excerpts from such a wide variety of modern sources that have been produced according to individual translators' conventions, we have generally preserved the format in which they appear in their modern translations, rather than edit them to achieve consistency throughout the sourcebook. This is so that if the reader wishes to refer back to the original publication in which these translations appear and read more, they can consult the original relatively easily. One way that modern translations often differ is how they represent Greek lettered points. Some modern authors use the original Greek letters (e.g., *A*, *B*, Γ, Δ, . . .), others transliterate them (i.e., *A, B, G, D*, . . .), and yet others "translate" these to the modern English alphabet (i.e., *A, B, C, D*, . . .). We have not attempted to homogenize these Greek lettering practices. We have also modified some of the translations to eliminate modern symbolism, given that the Greek authors wrote in prose sentences. However, in certain cases of standard phrases, we do use symbols to shorten the reading. For example, we use sq(*AB*) to represent "the square on the line segment *AB*" and rect(*AB, BC*) to represent "the rectangle spanned by the line segments *AB*, *BC*."

Although we have given birth and death dates where known for most of the writers whose works we present, these dates are frequently very approximate. Birth and/or death dates are known precisely for very few Greek authors. In fact, for many we are not even sure of the century of their *floruit*.

We hope that students and teachers of mathematics and the history of mathematics alike will be inspired by the broad selection of sources testifying to a diverse range of mathematical activity developed and applied in the eastern Mediterranean some two millennia ago and will use this as a springboard to investigate areas that capture their interest.

Acknowledgments

Many people have contributed to producing this work, and the editors would like to extend our thanks. First, we thank the readers of the manuscript for their thorough work and their numerous valuable suggestions. We thank Colin McKinney for his work on the Apollonius material, including his translations of the Eutocius commentary and the excerpt from Serenus. He also wrote the explanations of Archimedes' *On Spirals*. Michael Fried gave us valuable suggestions as to which parts of Apollonius's *Conics* to include. Danny Otero graciously translated the Latin poem *Carmen de Ponderibus et Mensuris*, and Reviel Netz generously allowed us to use parts of his soon-to-be-published translations of several works of Archimedes. Tessa Grant contributed her expertise in drawing the majority of the figures in the book. We also want to thank our editors at Princeton University Press, Susannah Shoemaker and Diana Gillooly, for their help and support in navigating the process from contract to production. They were assisted by Kristen Hop, Kiran Pandey, and Whitney Rauenhorst. In addition, Lisa Black, the permissions manager at the Press, helped us through the sometimes difficult process of securing permissions, while Dimitri Karetnikov dealt effectively with issues surrounding the figures. Wanda España designed our cover and oversaw the text design, while Elizabeth Byrd was our liaison for production matters and Erin Davis for the team at Westchester Publishing, who did a magnificent job in copyediting a difficult manuscript. Finally, Victor thanks his wife, Phyllis, who, as always, has made valuable suggestions and provided encouragement and love. Clemency thanks her husband, Yann, for being an anchor, a support, and a creative spark throughout the writing process. We are grateful to all of you for helping this book come to fruition.

1

The Study and Status of Mathematics

In this chapter, we look at some of the earliest references to the study and status of mathematics in ancient Greek literature. Plato is one of our earliest authorities, and in the *Republic* he proposes a division of mathematics into four subjects: arithmetic, geometry, astronomy, and music. This grouping of subjects later became known as the *quadrivium*, a term first coined by Boethius, and continued into the European Middle Ages as the basis of an exemplary education. We also include selections from Aristotle, demonstrating his own knowledge of mathematics. Several of these reveal Aristotle's desire that mathematics be based on firm principles of logical reasoning. We also include a fascinating excerpt from his *Physics* in which he discusses the idea of infinity. Furthermore, we incorporate several passages that reveal various views on the status of mathematics. Archytas, associate of Plato, links mathematics to the advancement of justice in society; Philo and Nicomachus, invoking interesting analogies, explore its connection with philosophical inquiry and wisdom. Proclus, expressing both his own views and that of earlier astronomer Geminus, considers the foundational principles of mathematics along with its applications and classification. Mathematics also featured in other genres such as the comic plays of Aristophanes. Through such passages we can get insight into the broader public perceptions of mathematics in Athens.

1.1 Plato, *Republic*

Plato (429–347 BCE) was an extraordinarily original and influential philosopher. He was born into an aristocratic family in Athens but was eventually repelled by both the oligarchs and the democrats of Athens. As a result, he withdrew his active participation in politics and instead devoted himself to philosophy. In about 387, he began to gather around himself numerous scholars to discuss diverse topics of interest. Plato's discussions took place in a property he had inherited, named after the legendary Athenian hero Akademos; and after Plato's death, thinkers continued to gather in the gardens, which, as the Academy, became the intellectual center of Greek life and lasted in some form for about three hundred years. Plato is generally considered the father of Western philosophy; in fact, according to Alfred North Whitehead, "All of Western philosophy is but a footnote to Plato." Although not a mathematician, Plato was evidently well versed in mathematics and invited a steady stream of Greek mathematicians to study and teach with him at the Academy. His works were written in the form of dialogues, often with Socrates, a real historical figure and close associate of Plato, being the principal interlocutor. Since the real Socrates died in 399, however, it is believed that most of the ideas discussed in the dialogues come from Plato himself.

In the *Republic*, Plato's best-known dialogue, Socrates discusses the meaning of justice and, in particular, how to create a just city-state. This state would be ruled by philosopher-kings who had studied for many years before being able to rule. Mathematics featured prominently in his proposed curriculum; in the selection from book 7 below, Socrates, in a discussion with Glaucon, elaborates the mathematics that the would-be philosopher-kings must study. The four subjects mentioned—arithmetic, geometry (plane and solid), astronomy, and music—later became known as the *quadrivium*, a major part of the medieval European liberal arts curriculum.

Socrates: If simple unity could be adequately perceived by the sight or by any other sense, then, . . . there would be nothng to attract towards being; but when there is some contradiction always present, and one is the reverse of one and involves the conception of plurality, then thought begins to be aroused within us, and the soul perplexed and wanting to arrive at a decision asks, "What is absolute unity?" This is the way in which the study of the one has a power of drawing and converting the mind to the contemplation of true being.

Glaucon: And surely, this occurs notably in the case of one, for we see the same thing to be both one and infinite in multitude?

Socrates: Yes, and this being true of one must be equally true of all number?

Glaucon: Yes.

Socrates: And they appear to lead the mind towards truth?

Glaucon: Yes, in a very remarkable manner.

Socrates: Then this is a knowledge of the kind for which we are seeking, having a double use, military and philosophical; for the man of war must learn the art of number or he will not know how to array his troops, and the philosopher also, because he has to rise out of the sea of change and lay hold of true being, and therefore he must be an arithmetician.

Glaucon: That is true.

Socrates: And our guardian is both warrior and philosopher?

Glaucon: Certainly.

Socrates: Then this is a kind of knowledge which legislation may fitly prescribe; and we must endeavor to persuade those who are to be the principal men of our State to go and learn arithmetic, not as amateurs, but they must carry on the study until they see the nature of numbers with the mind only; nor again, like merchants or retail-traders, with a view to buying or selling, but for the sake of their military use, and of the soul herself; and because this will be the easiest way for her to pass from becoming to truth and being.

Glaucon: That is excellent.

Socrates: Yes, and now having spoken of it, I must add how charming the science is and in how many ways it conduces to our desired end, if pursued in the spirit of a philosopher, and not of a shopkeeper!

Glaucon: How do you mean?

Socrates: I mean that arithmetic has a very great and elevating effect, compelling the soul to reason about abstract number, and rebelling against the introduction of visible or tangible objects into the argument. You know how steadily the masters of the art repel and ridicule anyone who attempts to divide absolute unity when he is calculating, and if you divide, they multiply, taking care that one shall continue as one and not become lost in fractions.

Glaucon: That is very true.

Socrates: Now, suppose a person were to say to them: O my friends, what are these wonderful numbers about which you are reasoning, in which, as you say, there is a unity such as you demand, and each unit is equal, invariable, indivisible—what would they answer?

Glaucon: They would answer that they were speaking of those numbers which can only be realized in thought.

Socrates: Then you see that this knowledge may be truly called necessary, necessitating as it clearly does the use of the pure intelligence in the attainment of pure truth?

Glaucon: Yes, that is a marked characteristic of it.

Socrates: And have you further observed, that those who have a natural talent for calculation are generally quick at every other kind of knowledge; and even the dull, if they have had an arithmetical training, although they may derive no other advantage from it, always become much quicker than they would otherwise have been?

Glaucon: Very true.

Socrates: And indeed, you will not easily find a more difficult study, and not many as difficult.

Glaucon: You will not.

Socrates: And, for all these reasons, arithmetic is a kind of knowledge in which the best natures should be trained, and which must not be given up.

Glaucon: I agree.

Socrates: Let this then be made one of our subjects of education. And next, shall we inquire whether the kindred science also concerns us?

Glaucon: You mean geometry?

Socrates: Exactly so.

Glaucon: Clearly we are concerned with that part of geometry which relates to war; for in pitching a camp, or taking up a position, or closing or extending the lines of an army, or any other military maneuver, whether in actual battle or on a march, it will make all the difference whether a general is or is not a geometrician.

Socrates: Yes, but for that purpose a very little of either geometry or calculation will be enough; the question relates rather to the greater and more advanced part of geometry—whether that tends in any degree to make more easy the vision of the idea of good; and thither, as I was saying, all things tend which compel the soul to turn her gaze towards that place, where is the full perfection of being, which she ought, by all means, to behold.

Glaucon: True.

Socrates: Then if geometry compels us to view being, it concerns us; if becoming only, it does not concern us?

Glaucon: Yes, that is what we assert.

Socrates: Yet anybody who has the least acquaintance with geometry will not deny that such a conception of the science is in flat contradiction to the ordinary language of geometricians.

Glaucon: How so?

Socrates: They have in view practice only, and are always speaking, in a narrow and ridiculous manner, of squaring and extending and applying and the like—they confuse the necessities of geometry with those of daily life; whereas knowledge is the real object of the whole science.

Glaucon: Certainly.

Socrates: Then must not a further admission be made?

Glaucon: What admission?

Socrates: That the knowledge at which geometry aims is knowledge of the eternal, and not of something perishing and transient.

Glaucon: That may be readily allowed, and is true.

Socrates: Then, my noble friend, geometry will draw the soul towards truth, and create the spirit of philosophy, and raise up that which is now unhappily allowed to fall down.

Glaucon: Nothing will be more likely to have such an effect.

Socrates: Then nothing should be more sternly laid down than that the inhabitants of your fair city should by all means learn geometry. Moreover the science has indirect effects, which are not small.

Glaucon: Of what kind?

Socrates: These are the military advantages of which you spoke; and in all departments of knowledge, as experience proves, anyone who has studied geometry is infinitely quicker of apprehension than one who has not.

Glaucon: Yes indeed, there is an infinite difference between them.

Socrates: Then shall we propose this as a second branch of knowledge which our youth will study?

Glaucon: Let us do so.

Socrates: And suppose we make astronomy the third—what do you say?

Glaucon: I am strongly inclined to it; the observation of the seasons and of months and years is as essential to the general as it is to the farmer or sailor.

Socrates: I am amused at your fear of the world, which makes you guard against the appearance of insisting upon useless studies; and I quite admit the difficulty of believing that in every man there is an eye of the soul which, when by other pursuits lost and dimmed, is by these purified and rekindled; and is more precious far than ten thousand bodily eyes, for by it alone is truth seen. Now there are two classes of persons, one class of those who will agree with you and will take your words as a revelation; another class to whom they will be utterly unmeaning, and who will naturally deem them to be idle tales, for they see no sort of profit which is to be obtained from them. And therefore you had better decide at once with which of the two you are proposing to argue. You will very likely say with neither, and that your chief aim in carrying on the argument is your own improvement; at the same time you do not begrudge to others any benefit which they may receive.

Glaucon: I think that I should prefer to carry on the argument mainly on my own behalf.

Socrates: Then take a step backward, for we have gone wrong in the order of the sciences.

Glaucon: What was the mistake?

Socrates: After plane geometry, we proceeded at once to solids in revolution, instead of taking solids in themselves; whereas after the second dimension the third, which is concerned with cubes and dimensions of depth, ought to have followed.

Glaucon: That is true, Socrates, but so little seems to be known as yet about these subjects.

Socrates: Why, yes, and for two reasons. In the first place, no government values them; this leads to a want of energy in the pursuit of them, and they are difficult; in the second place, students cannot learn them unless they have a director. But then a director can hardly be found, and even if he could, as matters now stand, the students, who are very conceited, would not attend to him. That, however, would be otherwise if the whole State became the director of these studies and gave honor to them; then disciples would want to come, and there would be continuous and earnest search, and discoveries would be made. Since even now, disregarded as they are by the world and although none of their investigators can tell the use of them, still these studies force their way by their natural charm, and very likely, if they had the help of the State, they would some day emerge into light.

Glaucon: Yes, there is a remarkable charm in them. But I do not clearly understand the change in the order. First you began with a geometry of plane surfaces?

Socrates: Yes.

Glaucon: And you placed astronomy next, and then you made a step backward?

Socrates: Yes, and I have delayed you by my hurry; the ludicrous state of solid geometry, which, in natural order, should have followed, made me pass over this branch and go on to astronomy, or motion of solids.

Glaucon: True.

Socrates: Then assuming that the science now omitted would come into existence if encouraged by the State, let us go on to astronomy, which will be fourth.

Glaucon: The right order. And now, Socrates, as you rebuked the vulgar manner in which I praised astronomy before, my praise shall be given in your own spirit. For everyone, as I think, must see that astronomy compels the soul to look upwards and leads us from this world to another.

Socrates: Everyone but myself; to everyone else this may be clear, but not to me.

Glaucon: And what then would you say?

Socrates: I should rather say that those who elevate astronomy into philosophy appear to me to make us look downwards and not upwards.

Glaucon: What do you mean?

Socrates: You have in your mind a truly sublime conception of our knowledge of the things above. And I dare say that if a person were to throw his head back and study ornaments on a ceiling, you would still think that he is looking at them with his understanding and not his eyes. And you are very likely right, and I may be a simpleton; but, in my opinion, that knowledge only which is of being and of the unseen can make the soul look upwards, and whether a man gapes at the heavens or blinks on the ground, seeking to learn some particular of sense, I would deny that he can learn, for nothing of that sort is a matter of science; his soul is looking downwards, not upwards, whether his way to knowledge is by water or by land, whether he floats, or only lies on his back.

Glaucon: I acknowledge the justice of your rebuke. Still, I should like to ascertain how astronomy can be learned in any manner more conducive to that knowledge of which we are speaking?

Socrates: I will tell you. The starry heaven which we behold is wrought upon a visible ground, and therefore, although the fairest and most perfect of visible things, must necessarily be deemed inferior far to the true motions of absolute swiftness and absolute slowness, which are relative to each other, and carry with them that which is contained in them, in the true number and in every true figure. Now, these are to be apprehended by reason and intelligence, but not by sight.

Glaucon: True.

Socrates: The spangled heavens should be used as a pattern and with a view to that higher knowlege; their beauty is like the beauty of figures or pictures excellently wrought by the hand of Daedalus, or some other great artist, which we may chance to behold; any geometrician who saw them would appreciate the exquisiteness of their workmanship, but he would never dream of thinking that in them he could find the true equal or the true double, or the truth of any other proportion.

Glaucon: No, such an idea would be ridiculous.

Socrates: And will not a true astronomer have the same feeling when he looks at the movements of the stars? Will he not think that heaven and the things in heaven are framed by the Creator of them in the most perfect manner? But he will never imagine that the proportions of night and day, or of both to the month, or of the month to the year, or of the stars to these and to one another, and any other things that are material and visible can also be eternal and subject to no deviation—that would be absurd; and it is equally absurd to take so much pains in investigating their exact truth.

Glaucon: I quite agree, though I never thought of this before.

Socrates: Then in astronomy, as in geometry, we should employ problems, and let the heavens alone if we would approach the subject in the right way and so make the natural gift of reason to be of any real use.

Glaucon: That is a work infinitely beyond our present astronomers.

Socrates: Yes, and there are many other things which must also have a similar extension given to them, if our legislation is to be of any value. But can you tell me of any other suitable study?

Glaucon: No, not without thinking.

Socrates: Motion has many forms, and not one only; two of them are obvious enough even to wits no better than ours; and there are others, as I imagine, which may be left to wiser persons.

Glaucon: But where are the two?

Socrates: There is a second which is the counterpart of the one already named.

Glaucon: And what may that be?

Socrates: The second would seem relatively to the ears to be what the first is to the eyes; for I conceive that as the eyes are designed to look up at the stars, so are the ears to hear harmonious motions; and these are sister sciences—as the Pythagoreans say, and we, Glaucon, agree with them?

Glaucon: Yes.

Socrates: But this is a laborious study, and therefore we had better go and learn of them; and they will tell us whether there are any other applications of these sciences. At the same time, we must not lose sight of our own higher object.

Glaucon: What is that?

Socrates: There is a perfection which all knowledge ought to reach, and which our pupils ought also to attain, and not to fall short of, as I was saying that they did in astronomy. For in the science of harmony, as you probably know, the same thing happens. The teachers of harmony compare the sounds and consonances which are heard only, and their labor, like that of the astronomers, is in vain.

Glaucon: Yes, by heaven; and it is as good as a play to hear them talking about their condensed notes, as they call them. They put their ears close alongside of the strings like persons catching a sound from their neighbor's wall—one set of them declaring that they distinguish an intermediate note and have found the least interval which should be the unit of measurement; the others insisting that the two sounds have passed into the same—either party setting their ears before their understanding.

Socrates: You mean, those gentlemen who tease and torture the strings and stretch them on the pegs of the instrument. I might carry on the metaphor and speak after their manner of the blows which the pick gives, and make accusations against the strings that are too responsive or too unresponsive. But this would be tedious, and therefore I will only say that these are not the men, and that I am referring to the Pythagoreans, of whom I was just now proposing to inquire about harmony. For they too are in error, like the astronomers; they investigate the numbers of the harmonies which are heard, but they never attain to problems, that is to say, they never reach the natural harmonies of number, or reflect why some numbers are harmonious and others not.

Glaucon: That is a thing of more than mortal knowledge.

Socrates: A thing which I would rather call useful; that is, if sought after with a view to the beautiful and good; but if pursued in any other spirit, useless.

Glaucon: Very true.

Socrates: Now, when all these studies reach the point of intercommunion and connection with one another, and come to be considered in their mutual affinities, then, I think, but not until then, will the pursuit of them have a value for our objects; otherwise there is no profit in them.

1.2 Plato, *Gorgias*

In this dialogue of Plato, Socrates debates with several sophists, including Gorgias, seeking the true meaning of "rhetoric." We have included only the brief section where Socrates discusses the arts of arithmetic and calculation and whether these should be called "rhetoric" as well.

Socrates: As to the arts generally, they are for the most part concerned with doing, and require little or no speaking; in painting, and statuary, and many other arts, the work may proceed in silence; and of such arts I suppose you would say that they do not come within the province of rhetoric.

Gorgias: You perfectly conceive my meaning, Socrates.

Socrates: But there are other arts which work wholly through the medium of language, and require either no action or very little, as, for example, the arts of arithmetic, of calculation, of geometry, and of playing draughts; in some of these speech is pretty nearly co-extensive with action, but in most of them the verbal element is greater—they depend wholly on words for their efficacy and power; and I take your meaning to be that rhetoric is an art of this latter sort?

Gorgias: Exactly.

Socrates: And yet I do not believe that you really mean to call any of these arts rhetoric; although the precise expression which you used was, that rhetoric is an art which works and takes effect only through the medium of discourse; and an adversary who wished to be captious might say, "And so, Gorgias, you call arithmetic rhetoric." But I do not think that you really call arithmetic rhetoric any more than geometry would be so called by you.

Gorgias: You are quite right, Socrates, in your apprehension of my meaning.

Socrates: Well, then, let me now have the rest of my answer: Seeing that rhetoric is one of those arts which works mainly by the use of words, and there are other arts which also use words, tell me what is that quality in words with which rhetoric is concerned. Suppose that a person asks me about some of the arts which I was mentioning just now; he might say, "Socrates, what is arithmetic?" and I should reply to him, as you replied to me, that arithmetic is one of those arts which take effect through words. And then he would proceed to ask: "Words about what?" and I should reply, Words about odd and even numbers, and how many there are of each. And if he asked again: "What is the art of calculation?" I should say, That also is one of the arts which is concerned wholly with words. And if he further said, "Concerned with what?" I should say, like the clerks in the assembly, "as aforesaid" of arithmetic, but with a difference, the difference being that the art of calculation considers not only the quantities of odd and even numbers, but also their numerical relations to themselves and to one another. And suppose, again, I were to say that astronomy is only words, he would ask, "Words about what, Socrates?" and I should answer, that astronomy tells us about the motions of the stars and sun and moon, and their relative swiftness.

Gorgias: You would be quite right, Socrates.

1.3 Archytas, *On Things Scientific*

Archytas (428–350 BCE) was born in Tarentum, in southern Italy, became a member of the Tarentine elite, and was a friend and associate of Plato. He evidently wrote numerous works on such subjects as music theory, geometry, number theory, and optics, but none of his works survive. There are various fragments in books by others that purport to quote Archytas, and it is from these fragments and references in the works of others that we can deduce that he was a superb mathematician as well as a political leader. The following fragment dealing with the use of calculation is taken from Iamblichus (third century CE), *On General Mathematical Science II*, in which the notion that mathematics underpins justice is explored.

Wherefore Archytas says in the *On Things Scientific*:

For it is necessary to come to know those things which you did not know, either by learning from another or by discovering yourself. Learning is from another and belongs to another, while discovery is through oneself and belongs to oneself. Discovery, while not seeking, is difficult and infrequent but, while seeking, easy and frequent, but, if one does not know how to calculate, it is impossible to seek.

Once calculation was discovered, it stopped discord and increased concord. For people do not want more than their share, and equality exists, once this has come into being. For by means of calculation we will seek reconciliation in our dealings with others. Through this, then, the poor receive from the powerful, and the wealthy give to the needy, both in the confidence that they will have what is fair on account of this. It serves as a standard and a hindrance to the unjust. It stops those who know how to calculate, before they commit injustice, persuading them that they will not be able to go undetected, whenever they appeal to it. It hinders those who do not know how to calculate from committing injustice, having revealed them as unjust by means of it.

1.4 Aristotle, *Metaphysics*

Aristotle (384–322 BCE) was the most prolific of the Greek philosophers. Having studied with Plato as a young man, he later was hired by Philip II of Macedon to undertake the education of his son Alexander, who soon after he acceded to the throne in 335 began his successful conquest of the Mediterranean world. Meanwhile, Aristotle returned to Athens and founded his own school, the Lyceum, where he spent the rest of his days writing, lecturing, and holding discussions with his associates. His writings cover many subjects including physics, biology, zoology, metaphysics, logic, ethics, aesthetics, poetry, theater, music, rhetoric, psychology, linguistics, economics, politics, and government. Even though many of his scientific findings have been overhauled and updated, intellectuals in Europe and the Islamic world through early modern times began many of their inquiries by trying to support or refute his ideas. Although Aristotle was not a mathematician, he was clearly cognizant of the mathematical ideas circulating in Greece during his lifetime. We consider here excerpts of some of his writings that refer to mathematics or that have influenced the development of mathematics, in particular his ideas on methods of proof.

In the following excerpts from the *Metaphysics*, Aristotle tells us why mathematics began in Egypt, why mathematicians believe in proof, and why the Pythagoreans thought that the principles of mathematics were the principles of all things.

Book 1

Chapter 1

At first, he who invented any art whatever that went beyond the common perceptions of man was naturally admired by men, not only because there was something useful in the inventions, but because he was thought wise and superior to the rest. But as more arts were invented, and some were directed to the necessities of life, others to recreation, the inventors of the latter were naturally always regarded as wiser than the inventors of the former, because their branches of knowledge did not aim at utility. Hence when all such inventions were already established, the sciences which do not aim at giving pleasure or at the necessities of life were discovered, and first in the places where men first began to have leisure. This is why the mathematical arts were founded in Egypt; for there the priestly caste was allowed to be at leisure.

Chapter 2

For all men begin, as we said, by wondering that things are as they are, as they do about self-moving marionettes, or about the solstices or the incommensurability of the diagonal of a square with the side; for it seems wonderful to all who have not yet seen the reason, that there is a thing which cannot be measured even by the smallest unit. But we must end in the contrary and, according to the proverb, the better state, as is the case in these instances too when men learn the cause; for there is nothing which would surprise a geometer so much as if the diagonal turned out to be commensurable.

Chapter 5

Contemporaneously with these philosophers[1] and before them, the so-called Pythagoreans, who were the first to take up mathematics, not only advanced this study, but also having been brought up in it they thought its principles were the principles of all things. Since of these principles numbers are by nature the first, and in numbers they seemed to see many resemblances to the things that exist and come into being—more than in fire and earth and water (such and such a modification of numbers being justice, another being soul and reason, another being opportunity—and similarly almost all other things being numerically expressible); since, again, they saw that the modifications and the ratios of the musical scales were expressible in numbers; since, then, all other things seemed in their whole nature to be modeled on numbers, and numbers seemed to be the first things in the whole of nature, they supposed the elements of numbers to be the elements of all things, and the whole heaven to be a musical scale and a number. And all the properties of numbers and scales which they could show to agree with the attributes and parts and the whole arrangement of the heavens, they collected and fitted into their scheme; and if there was a gap anywhere, they readily made additions so as to make their whole theory coherent. For example, as the number ten is thought to be perfect and to comprise the whole nature of numbers, they say that the bodies which move through the heavens are ten, but as the visible bodies are only nine, to meet this they invent a tenth, the "counter-earth."

1.5 Aristotle, *Prior Analytics*

In these excerpts from the *Prior Analytics*, Aristotle discusses the idea of a syllogism. He then refers to the proof that the diagonal of a square is incommensurable with

[1] Anaxagoras of Clazomenae and Empedocles of Acragas, both of the fifth century BCE.

the side. He also gives hints of a proof of the theorem that the base angles of an isosceles triangle are equal. In his writings about proof, he expresses his belief that logical arguments should always be built out of syllogisms, even though later Greek writers on mathematics generally used the ideas of propositional logic to formulate such arguments. Nevertheless, the basic ideas about proofs are embedded in Aristotle's writings.

Book 1

Chapter 1

A syllogism is discourse in which, certain things being stated, something other than what is stated follows of necessity from their being so.[2] I mean by the last phrase that they produce the consequence, and by this, that no further term is required from without in order to make the consequence necessary. I call that a perfect syllogism which needs nothing other than what has been stated to make plain what necessarily follows; a syllogism is imperfect, if it needs either one or more propositions, which are indeed the necessary consequences of the terms set down, but have not been expressly stated as premises.

Chapter 23

For all who effect an argument *per impossible* infer syllogistically what is false, and prove the original conclusion hypothetically when something impossible results from the assumption of its contradictory; e.g., that the diagonal of the square is incommensurate with the side, because odd numbers are equal to evens if it is supposed to be commensurate. One infers syllogistically that odd numbers come out equal to evens, and one proves hypothetically the incommensurability of the diagonal, since a falsehood results through contradicting this.

Chapter 24

In every syllogism one of the premises must be affirmative, and universality must be present; unless one of the premises is universal either a syllogism will not be possible, or it will not refer to the subject proposed, or the original position will be begged. Suppose we have to prove that pleasure in music is good. If one should claim as a premise that pleasure is good without adding "all," no syllogism will be possible; if one should claim that some pleasure is good, then if it is different from pleasure in music, it is not relevant to the subject proposed; if it is this very pleasure, one is assuming that which was proposed at the outset to be proved. This is more obvious in geometrical proofs, e.g., that the angles at the base of an isosceles triangle are equal. Suppose the lines A and B have been drawn to the center. If then one should assume that the angle AC is equal to the angle BD, without claiming generally that angles of semicircles are equal; and again if one should assume

[2] As a simple example, we may take the following argument: All monkeys are primates, and all primates are mammals; therefore, all monkeys are mammals.

that the angle C is equal to the angle D, without the additional assumption that every angle of a segment is equal to every other angle of the same segment; and further if one should assume that when equal angles are taken from the whole angles, which are themselves equal, the remainders E and F are equal, he will beg the thing to be proved, unless he also states that when equals are taken from equals the remainders are equal.[3]

1.6 Aristotle, *Posterior Analytics*

In the *Posterior Analytics*, Aristotle discusses some of the basic ideas of a proof. The excerpts here explore the notion that syllogistic reasoning enables one to use "old knowledge" to impart new. Of course, he notes, one must begin with some basic truths that are assumed without proof.

Book 1

Chapter 1

All instruction given or received by way of argument proceeds from pre-existent knowledge. This becomes evident upon a survey of all the species of such instruction. The mathematical sciences and all other speculative disciplines are acquired in this way, and so are the two forms of dialectical reasoning, syllogistic and inductive; for each of these latter make use of old knowledge to impart new, the syllogism assuming an audience that accepts its premises, induction exhibiting the universal as implicit in the clearly known particular. Again, the persuasion exerted by rhetorical arguments is in principle the same, since they use either example, a kind of induction, or enthymeme,[4] a form of syllogism.

The pre-existent knowledge required is of two kinds. In some cases admission of the fact must be assumed, in others comprehension of the meaning of the term used, and sometimes both assumptions are essential. Thus, we assume that every predicate can be either truly affirmed or truly denied of any subject, and that "triangle" means so and so; as regards "unit" we have to make the double assumption of the meaning of the word and the existence of the thing. The reason is that these several objects are not equally obvious to us. Recognition of a truth may in some cases contain as factors both previous knowledge and also knowledge acquired simultaneously with that recognition—knowledge, this latter, of the particulars actually falling under the universal and therein already virtually known.

Chapter 2

What I now assert is that at all events we do know by demonstration. By demonstration I mean a syllogism productive of scientific knowledge, a syllogism, that is, the grasp

[3] For a possible explanation of Aristotle's proof, see Thomas L. Heath, *The Thirteen Books of Euclid's Elements with Introduction and Commentary*, vol. 1 (New York: Dover, 1956), 252–53.

[4] A syllogism where one of the premises is not explicitly stated. Such a syllogism is often used in rhetorical arguments.

of which is *eo ipso*[5] such knowledge. Assuming then that my thesis as to the nature of scientific knowing is correct, the premises of demonstrated knowledge must be true, primary, immediate, better known than and prior to the conclusion, which is further related to them as effect to cause. Unless these conditions are satisfied, the basic truths will not be "appropriate" to the conclusion. Syllogism there may indeed be without these conditions, but such syllogism, not being productive of scientific knowledge, will not be demonstration. The premises must be true, for that which is non-existent cannot be known—we cannot know, e.g., that the diagonal of a square is commensurate with its side. The premises must be primary and indemonstrable; otherwise they will require demonstration in order to be known, since to have knowledge, if it be not accidental knowledge, of things which are demonstrable, means precisely to have a demonstration of them. The premises must be the causes of the conclusion, better known than it, and prior to it; its causes, since we possess scientific knowledge of a thing only when we know its cause; prior, in order to be causes; antecedently known, this antecedent knowledge being not our mere understanding of the meaning, but knowledge of the fact as well.

Chapter 10

I call the basic truths of every genus those elements in it the existence of which cannot be proved. As regards both these primary truths and the attributes dependent on them the meaning of the name is assumed. The fact of their existence as regards the primary truths must be assumed; but it has to be proved of the remainder, the attributes. Thus we assume the meaning alike of unity, straight, and triangular; but while as regards unity and magnitude we assume also the fact of their existence, in the case of the remainder proof is required.

Of the basic truths used in the demonstrative sciences some are peculiar to each science, and some are common, but common only in the sense of analogous, being of use only in so far as they fall within the genus constituting the province of the science in question.

Peculiar truths are, e.g., the definitions of line and straight; common truths are such as "take equals from equals and equals remain." Only so much of these common truths is required as falls within the genus in question; for a truth of this kind will have the same force even if not used generally but applied by the geometer only to magnitudes, or by the arithmetician only to numbers. Also peculiar to a science are the subjects the existence as well as the meaning of which it assumes, and the essential attributes of which it investigates, e.g., in arithmetic units, in geometry points and lines. Both the existence and the meaning of the subjects are assumed by these sciences; but of their essential attributes only the meaning is assumed. For example arithmetic assumes the meaning of odd and even, square and cube, geometry that of incommensurable, or of deflection or verging of lines, whereas the existence of these attributes is demonstrated by means of the axioms and from previous conclusions as premises.

[5] A Latin phrase that means "by that very fact."

1.7 Aristotle, *Physics*

In this selection from the *Physics*, Aristotle discourses at length about the existence of the infinite, eventually concluding that although there cannot be an infinite "body," infinity does exist—mathematically—in potential. Note that Aristotle's concept of physics is not the same as ours; in general, his physics is concerned with the notion of change, including change of place (motion), quantitative change, qualitative change, and substantial change, as in coming into or out of existence.

Book 3

Chapter 4

Belief in the existence of the infinite comes mainly from five considerations:

1. From the nature of time—for it is infinite.
2. From the division of magnitudes—for the mathematicians also use the notion of the infinite.
3. If coming to be and passing away do not give out, it is only because that from which things come to be is infinite.
4. Because the limited always finds its limit in something, so that there must be no limit, if everything is always limited by something different from itself.
5. Most of all, a reason which is peculiarly appropriate and presents the difficulty that is felt by everybody—not only number but also mathematical magnitudes and what is outside the heaven are supposed to be infinite because they never give out in our thought.

The last fact (that which is outside is infinite) leads people to suppose that body also is infinite, and that there is an infinite number of worlds. Why should there be body in one part of the void rather than in another? Grant only that mass is anywhere and it follows that it must be everywhere. Also, if void and place are infinite, there must be infinite body too, for in the case of eternal things what may be must be. But the problem of the infinite is difficult; many contradictions result whether we suppose it to exist or not to exist. If it exists, we have still to ask how it exists; as a substance or as the essential attribute of some entity? Or in neither way, yet none the less is there something which is infinite or some things which are infinitely many?

The problem, however, which specially belongs to the physicist is to investigate whether there is a sensible magnitude which is infinite. We must begin by distinguishing the various senses in which the term "infinite" is used.

1. What is incapable of being gone through, because it is not in its nature to be gone through (the sense in which the voice is "invisible").
2. What admits of being gone through, the process however having no termination, or
3. What scarcely admits of being gone through.
4. What naturally admits of being gone through, but is not actually gone through or does not actually reach an end.

Further, everything that is infinite may be so in respect of addition or division or both.

In chapter 5, Aristotle endeavors to prove that an infinite body is an impossibility. He then continues:

Chapter 6

But on the other hand to suppose that the infinite does not exist in any way leads obviously to many impossible consequences: there will be a beginning and an end of time, a magnitude will not be divisible into magnitudes, number will not be infinite. If, then, in view of the above considerations, neither alternative seems possible, an arbiter must be called in; and clearly there is a sense in which the infinite exists and another in which it does not. We must keep in mind that the word "is" means either what potentially is or what fully is. Further, a thing is infinite either by addition or by division. Now, as we have seen, magnitude is not actually infinite. But by division it is infinite. (There is no difficulty in refuting the theory of indivisible lines.) The alternative then remains that the infinite has a potential existence.

But the phrase "potential existence" is ambiguous. When we speak of the potential existence of a statue we mean that there will be an actual statue. It is not so with the infinite. There will not be an actual infinite. The word "is" has many senses, and we say that the infinite "is" in the sense in which we say "it is day" or "it is the games," because one thing after another is always coming into existence. For of these things too the distinction between potential and actual existence holds. We say that there are Olympic games, both in the sense that they may occur and that they are actually occurring. . . .

In a way the infinite by addition is the same thing as the infinite by division. In a finite magnitude, the infinite by addition comes about in a way inverse to that of the other. For in proportion as we see division going on, in the same proportion we see addition being made to what is already marked off. For if we take a determinate part of a finite magnitude and add another part *determined by the same ratio* (not taking in the same amount of the original whole), and so on, we shall not traverse the given magnitude. But if we increase the ratio of the part, so as always to take in the same amount, we shall traverse the magnitude, for every finite magnitude is exhausted by means of any determinate quantity however small.

The infinite, then, exists in no other way, but in this way it does exist, potentially and by reduction. It exists fully in the sense in which we say "it is day" or "it is the games"; and potentially as matter exists, not independently as what is finite does.

By addition then, also, there is potentially an infinite, namely, what we have described as being in a sense the same as the infinite in respect of division. For it will always be possible to take something *ab extra*. Yet the sum of the parts taken will not exceed every determinate magnitude, just as in the direction of division every determinate magnitude is surpassed in smallness and there will be a smaller part. But in respect of addition there cannot be an infinite which even potentially exceeds every assignable magnitude, unless it has the attribute of being actually infinite, as the physicists hold to be true of the body which is outside the world, whose essential nature is air or something of the kind. But if there cannot be in this way a sensible body which is infinite in the full sense, evidently there can no more be a body which is potentially infinite in respect of addition, except as the inverse of the infinite by division, as we have said. It is for this reason that Plato also makes the infinites two in number, because it is supposed to be possible to exceed all limits and to proceed *ad infinitum* in the direction both of increase and of reduction. Yet though he makes the infinites two, he does not use them. For in the numbers the infinite in the direction of reduction is not present, as the monad is the smallest; nor is the infinite in the direction of increase, for the parts number only up to the decad.

The infinite turns out to be the contrary of what it is said to be. It is not what has nothing outside it that is infinite, but what always has something outside it.

Chapter 7

It is reasonable that there should not be held to be an infinite in respect of addition such as to surpass every magnitude, but that there should be thought to be such an infinite in the direction of division. For the matter and the infinite are contained inside what contains them, while it is the form which contains. It is natural too to suppose that in number there is a limit in the direction of the minimum, and that in the other direction every assigned number is surpassed. In magnitude, on the contrary, every assigned magnitude is surpassed in the direction of smallness, while in the other direction there is no infinite magnitude. The reason is that what is one is indivisible whatever it may be, e.g., a man is one man, not many. Number on the other hand is a plurality of "ones" and a certain quantity of them. Hence number must stop at the indivisible; for "two" and "three" are merely derivative terms, and so with each of the other numbers. But in the direction of largeness it is always possible to think of a larger number; for the number of times a magnitude can be bisected is infinite. Hence this infinite is potential, never actual; the number of parts that can be taken always surpasses any assigned number. But this number is not separable from the process of bisection, and its infinity is not a permanent actuality but consists in a process of coming to be, like time and the number of time.

With magnitudes the contrary holds. What is continuous is divided *ad infinitum*, but there is not infinite in the direction of increase. For the size which it can potentially be, it can also actually be. Hence since no sensible magnitude is infinite, it is impossible to exceed every assigned magnitude; for if it were possible there would be something bigger than the heavens.... Our account does not rob the mathematicians of their science, by disproving the actual existence of the infinite in the direction of increase, in the sense of the untraversible. In point of fact they do not need the infinite and do not use it. They postulate only that the finite straight line may be produced as far as they wish. It is possible to have divided in the same ratio as the largest quantity another magnitude of any size you like. Hence, for the purposes of proof, it will make no difference to them to have such an infinite instead, while its existence will be in the sphere of real magnitudes.

1.8 Philo of Alexandria, *On Mating with the Preliminary Studies*

Philo of Alexandria (fl. 10–55 CE) was one of the most important Jewish authors at the time of Roman rule in Judea. Jews had been moving to Alexandria ever since the founding of the city, but during Philo's lifetime, an estimated 200,000 Jews lived there, the largest Jewish community outside Palestine. Philo himself was from a prominent and wealthy family and was given a strong Greek education. His many works of philosophy, written in Greek, revolve around interpreting the Bible, often in allegorical terms. In *On Mating with the Preliminary Studies*, Philo presents an interpretation of Abraham's having a child with Hagar, Sarah's handmaiden. Ideally, Abraham should have a child with Sarah, who represents virtue. But before this can happen, he must partake of "intermediate instruction," learning the subjects of the Greek *trivium* and *quadrivium*, all subjects specified by Plato for the education of Greek citizens.[6]

[13] For, says she [Sarah], "The Lord has closed me up so, that I may not bear children." . . . [14] "Therefore," says she, "go thou in to my handmaiden," that is to say, to the intermediate instruction of the intermediate and encyclical branches of knowledge, "that you may first have children by her"; for hereafter you shall be able to enjoy a connection with her mistress, tending to the procreation of legitimate children. [15] For grammar, by teaching you the histories which are to be found in the works of poets and historians, will give you intelligence and abundant learning; and, moreover, will teach you to look with contempt on all the vain fables which erroneous opinions invent, on account of the ill success which history tells us that the heroes and demigods who are celebrated among those writers, meet with. [16] And music will teach what is harmonious in the way of rhythm, and what is ill arranged in harmony, and, rejecting all that is out of tune and all that is inconsistent with melody, will guide what was previously discordant to concord. And geometry, sowing the seeds of equality and just proportion in the soul, which is fond of learning, will, by means of the beauty of continued contemplation, implant in you an admiration of justice. [17] And rhetoric, having sharpened the mind for contemplation in general, and having exercised and trained the faculties of speech in interpretation and explanations, will make man really rational, taking care of that peculiar and especial duty which nature has bestowed upon it, but upon no other animal whatever. [18] And dialectic science, which is the sister, the twin sister of rhetoric, as some persons have called it, separating true from false arguments, and refuting the plausibilities of sophistical arguments, will cure the great disease of the soul, deceit. It is profitable, therefore, to aide among these and other sciences resembling them, and to devote one's especial attention to them. For perhaps, I say, as has happened to many, we shall become known to the queenly virtues by means of their subjects and handmaidens. . . .

[146] And yet even this is not unknown to anyone, namely, that philosophy has bestowed upon all the particular sciences their first principles and seeds, from which speculations respecting them appear to arise. For it is geometry which invented equilateral and scalene triangles, and circles, and polygons, and all kinds of other figures. But it was no longer geometry that discovered the nature of a point, and line, and a surface, and a solid, which are the roots and foundations of the aforementioned figures. [147] For from when could it define and pronounce that a point is that which has no parts, that a line is length without breadth; that a surface is that which has only length and breadth; that a solid is that which has the three properties, length, breadth, and depth? For these discoveries belong to philosophy, and the consideration of these definitions belongs wholly to the philosopher.

[6] The *trivium* consists of logic, grammar, and rhetoric, while the *quadrivium* consists of arithmetic, geometry, music, and astronomy. These subjects became the bases for a liberal education starting in Greece and extending into the medieval period.

1.9 Nicomachus, *Introduction to Arithmetic*

Nicomachus was probably born in Gerasa—now Jarash, Jordan—late in the first century CE. He became a leading member of the neo-Pythagorean school. The *Introduction to Arithmetic* is an elementary treatise in two books. It is written in a very informal style, and it does not give formal proofs. Often, the work just gives statements of results and examples; sometimes, in fact, the stated theorems are not generally true. Still, Nicomachus's work proved very popular and was used, in a Latin version by Boethius, as a textbook well into the medieval period. In this selection, we present his views on the study of mathematics. Note that Nicomachus quotes Archytas of Tarentum in support of his views on the study of mathematics. Excerpts from the remainder of the treatise are in chapter 2 of this *Sourcebook*.

Book 1

Chapter 1

[1] The ancients, who under the leadership of Pythagoras first made science systematic, defined philosophy as the love of wisdom. Indeed the name itself means this, and before Pythagoras all who had knowledge were called "wise" indiscriminately—a carpenter, for example, a cobbler, a helmsman, and in a word anyone who was versed in any art or handicraft. Pythagoras, however, restricting the title so as to apply to the knowledge and comprehension of reality, and calling the knowledge of the truth in this the only wisdom, naturally designated the desire and pursuit of this knowledge philosophy, as being desire for wisdom.

Chapter 2

[3] If we crave for the goal that is worthy and fitting for man, namely, happiness of life—and this is accomplished by philosophy alone and by nothing else, and philosophy, as I said, means for us desire for wisdom, and wisdom the science of the truth in things, and of things some are properly so called, others merely share the name—it is reasonable and most necessary to distinguish and systematize the accidental qualities of things. [4] Things, then, both those properly so called and those that simply have the name, are some of them unified and continuous, for example, an animal, the universe, a tree, and the like, which are properly and peculiarly called "magnitudes"; others are discontinuous, in a side-by-side arrangement, and, as it were, in heaps, which are called "multitudes," a flock, for instance, a people, a heap, a chorus, and the like. [5] Wisdom, then, must be considered to be the knowledge of these two forms. Since, however, all multitude and magnitude are by their own nature of necessity infinite—for multitude starts from a definite root and never ceases increasing; and magnitude, when division beginning with a limited whole is carried on, cannot bring the dividing process to an end, but proceeds therefore to infinity—and since sciences are always sciences of limited things, and never of infinites, it is accordingly evident that a science dealing either with magnitude, per se, or with multitude, per se, could never be formulated, for each of them is limitless in itself, multitude in the direction of the more, and magnitude in the direction of the less. A science, however, would arise to deal

with something separated from each of them, with quantity, set off from multitude, and size, set off from magnitude.

Chapter 3

[1] Again, to start afresh, since of quantity one kind is viewed by itself, having no relation to anything else, as “even,” “odd,” “perfect,” and the like, and the other is relative to something else and is conceived of together with its relationship to another thing, like “double,” “greater,” “smaller,” “half,” “one and one-half times,” “one and one-third times,” and so forth, it is clear that two scientific methods will lay hold of and deal with the whole investigation of quantity; arithmetic, absolute quantity, and music, relative quantity. [2] And once more, inasmuch as part of “size” is in a state of rest and stability, and another part in motion and revolution, two other sciences in the same way will accurately treat of “size,” geometry the part that abides and is at rest, astronomy that which moves and revolves. [3] Without the aid of these, then, it is not possible to deal accurately with the forms of being nor to discover the truth in things, knowledge of which is wisdom, and evidently not even to philosophize properly, for “just as painting contributes to the menial arts toward correctness of theory, so in truth lines, numbers, harmonic intervals, and the revolutions of circles bear aid to the learning of the doctrines of wisdom,” says the Pythagorean Androcydes.[7] [4] Likewise Archytas of Tarentum, at the beginning of his treatise *On Harmony*, says the same thing, in about these words: “It seems to me that they do well to study mathematics, and it is not at all strange that they have correct knowledge about each thing, what it is. For if they knew rightly the nature of the whole, they were also likely to see well what is the nature of the parts. About geometry, indeed, and arithmetic and astronomy, they have handed down to us a clear understanding, and not least also about music. For these seem to be sister sciences; for they deal with sister subjects, the first two forms of being.”

Chapter 4

[1] Which then of these four methods must we first learn? Evidently, the one which naturally exists before them all, is superior and takes the place of origin and root and, as it were, of mother to the others. [2] And this is arithmetic, not solely because we said that it existed before all the others in the mind of the creating God like some universal and exemplary plan, relying upon which as a design and archetypal example the creator of the universe sets in order his material creations and makes them attain to their proper ends; but also because it is naturally prior in birth, inasmuch as it abolishes other sciences with itself, but it not abolished together with them. For example, “animal” is naturally antecedent to “man,” for abolish “animal” and “man” is abolished; but if “man” be abolished, it no longer follows that “animal” is abolished at the same time. And again, “man” is antecedent to “schoolteacher”; for if “man” does not exist, neither does “schoolteacher,” but if “schoolteacher” is nonexistent, it is still possible for “man” to be. Thus since it has the property of abolishing the other ideas with itself, it is likewise the older. . . . [4] So it

[7] Androcydes was a Pythagorean philosopher whose work survives only in some fragments. He lived before the first century BCE, possibly as early as the fourth.

is with the foregoing sciences; if geometry exists, arithmetic must also needs be implied, for it is with the help of this latter that we can speak of triangle, quadrilateral, octahedron, icosahedron, double, eightfold, or one and one-half times, or anything else of the sort which is used as a term by geometry, and such things cannot be conceived of without the numbers that are implied with each one. For how can "triple" exist, or be spoken of, unless the number 3 exists beforehand, or "eightfold" without 8? But on the contrary 3, 4, and the rest might be without the figures existing to which they give names. [5] Hence arithmetic abolishes geometry along with itself, but is not abolished by it, and while it is implied by geometry, it does not itself imply geometry.

Chapter 5

[3] So then we have rightly undertaken first the systematic treatment of this, as the science naturally prior, more honorable, and more venerable, and, as it were, mother and nurse of the rest; and here we will take our start for the sake of clearness.

1.10 Iamblichus, *On the General Science of Mathematics*

Iamblichus was a neo-Pythagorean philosopher who lived in the third century CE and taught at a school in Apamea, Syria, an ancient Greek and Roman city. The treatise, *On the General Science of Mathematics*, from which these excerpts are drawn, sets out the basic ideas of Pythagoras and his school toward mathematics, at least as Iamblichus understood them some eight hundred years after Pythagoras lived. He relates the familiar anecdote of Thales, a story picked up by Proclus in the fifth century, and also notes that Pythagoras studied with teachers in Egypt and Babylonia—although he refers to these teachers as "barbarians." There is not much specific detail here on Pythagoras's mathematics, only general principles detailing the close connection of mathematics with ethics and theology. Iamblichus also emphasizes the practical relevance of mathematics for the pursuit of the good life. In addition, he sets out in a general way the Pythagorean methods of mathematical instruction.

Chapter 7

What the special object of knowledge is which is the basis of each branch of mathematics, and how it is possible by division to make a general distinction of them, so as to know both the one and the manifold in mathematics, its character, and how it should be defined.

Since one must also, for each of the mathematical sciences, determine the proper object of knowledge that underlies each, let us distinguish, taking our start from a process of division, the species of studies with which they deal. In that way we may most easily understand the unity and the plurality of mathematical science, its character, and by what sort of differences it is distinguished. Let us, then, take our start from this point.

The nature of the continuous and the discrete for all that is, that is to say for the whole structure of the cosmos, may be conceived in two ways: there is the discrete through juxtaposition and through piling up, and the continuous through unification and through conjunction. Strictly speaking, continuity and unification should be called magnitude, being adjacent, and discrete plurality. In accordance with the essence of magnitude the cosmos would be conceived as one and would be called solid, spherical, and fused together, extended and conjoined; but, again, according to the form and concept of plurality the ordering, disposition, and joining together of the whole would be thought of as, we may say, being constructed of so many oppositions and similarities of elements, spheres, stars, kinds, animals, and plants. But in the case of the unified, division from the totality is without limit, while its increase is to a limited point; while conversely, in the case of plurality, increase is unlimited, but division limited. Naturally, indeed, and conceptually, both are unlimited and therefore indefinable by sciences; "there will be no principle of object of understanding, since all things are unlimited," according to Philolaus.[8]

Since it is necessary that the nature of a science be discerned in things thus given exactitude by divine providence, certain sciences have cut off part from each and limited their content; from plurality they developed the concept of quantity, which was already familiar, while from magnitude in the same way they developed that of extension. Both of these kinds they subsumed under sciences according to their own forms—under arithmetic, quantity, under geometry, extension. But, since these were not of a single form, each of them permitted still further subdivision. Thus of quantity part is absolute, freed from any relationship to something else, as for example the even, the odd, the whole, and the like; while some is relative in some way or other, which is termed relative quantity in the strict sense, such as the equal, unequal, a multiple, the superparticular, the superpartient, and the similar; and again some quantity both is and is thought of as unchanging, other as changing and moving. For this reason it is to be expected that two other sciences share in and are attached to the objects of each of the two above-mentioned sciences. For in connection with arithmetic, to which specially belong the study of quantity as such, music is allotted a role in the technical study of relative quantity; for its study of harmonics and concords professes nothing other than the classification of relations and ratios to each other of sounds, and the quantity of their relative height and depth; while for geometry, which is concerned with persistent and static extension, spherical astronomy arose as a collaborator and as arbiter of moving quantity, clearly of that which is most perfect and exhibits orderly and uniform motion. Therefore, since the sciences are themselves concerned with twin objects, it is sensible to think of these sciences as sisters, so as not foolishly to discount the saying of Archytas: "For these studies seem to be sisters," and it is sensible to consider them as connected together like links in a chain and joined by a common bond, in the words of the most divine Plato. The uniform kinship of these studies should manifest itself to him who studies them in the right way; and him who has received all of them in the way that Plato himself proposes he calls the most truly wise, and playfully insists on it, and exhorts those eager to be philosophers that these studies, whether difficult or easy, should be pursued and chosen above all. This is entirely sensible, since a grasp of the continuous and the discrete comes about in these ways alone, while the cosmos and all that is therein contained consists of the continuous and the discrete. The accurate grasp of these is wisdom, and philosophy is the struggle for wisdom. But philosophy is the absolutely single one of all skills and sciences that is

[8] Philolaus of Croton was a fifth-century BCE Pythagorean philosopher.

concerned with man's own natural end and leads to the state of flourishing which is proper to him alone among animals, and is naturally sought after as the goal most worth his seeking.

Chapter 18

The special methods of the Pythagorean presentation of the mathematical sciences, how and in what matters they made use of them, and for whom; also that they always took due account both of the subject matter and of learners.

Indeed the special methods of Pythagorean teaching of mathematics were marvelously accurate and greatly surpassed the skill of those who were engaged in the teaching of mathematics on the technical level. So let us present an outline of it, so far as is possible to speak about it in general terms.

So let this one thing be agreed, that by starting from above from first principles they provided the first structuring of mathematical theorems, undertaking their reasoning on the basis, as it were, of the theorems' primal being itself, and finally leading the whole mathematical enterprise back up to this focus. Further, as a consequence of this, they were accustomed to exhibit first the discoveries of the theorems and to take nothing for granted, but in all cases to show how what was being demonstrated in mathematics came to be the case. They also shared another procedure, the mathematical use of symbols, e.g. the pentad as a symbol of justice, because it signifies symbolically all the forms of justice. This form was useful to them in all philosophy, since they performed the majority of their teaching through symbols, and also believed that this procedure was appropriate to the gods and befitted nature. But, indeed, it is clear from the other mathematical sciences that they provided the first principles and discoveries of mathematics, and it is plain also from their methods in number theory. For they teach clearly how each kind and species of number first comes to be and how it is discovered by us, on the principle that the study of numbers was not scientific unless one should grasp them by starting from above.

Furthermore, they always assimilated the theorems of mathematics to true beings and everything divine, to the conditions and powers of the soul, to the heavenly phenomena and the orbits of the stars, to all the elements of bodies in becoming and the things composed of them, and to matter and the things that come to be from it, likening mathematical theorems both in general, and taking from each its own points of similarity to each part of true being. They applied mathematics to things either through their having the same reason-principles in common or through some dim impression, or through a likeness either close or distant, or through some association of images, or through some dominant cause as in the role of a paradigm, or in some other way. They also join mathematics together with things in many other ways, since things can be likened to mathematics and also mathematics has a nature to be likened to things, and both can be assimilated to each other.

So they were not at all pleased by the elaboration of terminology and the wealth of procedures, since it was rather too logic-chopping and divorced from the true nature of the facts, but they particularly welcomed recognition of the problems themselves, as contributing to the knowledge and discovery of reality. They emphasized the discovery of truths, and focus on realities rather than the shrewdness and acuity of reasonings about the facts. Therefore they did not set a high value on dialectical argumentation in mathematics, but they evidently esteemed more highly that which contributed to the discovery of realities.

So these methods and the like were those they used for mathematical teaching. But they used them scientifically, in conjunction with the theoretical philosophy of true beings, and aiming at the Good. For they thought that they ought always to give precedence to and honor the limited and the most succinct, while selecting from these whatever was useful for themselves and their companions and for complete knowledge of true beings. Also, however, in their instruction they aimed in one way at the facts, their ordering and the mutual relationships; for they marked off their first and second theorems with reference to such a sequence; but in another way they also turned their attention to their pupils, their capacity and how they could be helped by them, and what should be taught to beginners and what to the more advanced, and what sciences were esoteric and what exoteric, which could be spoken of and which could not, and to whom instruction should be communicated with scientific understanding of the facts and to whom only mathematically. For accuracy in all these matters was pursued by them not without purpose, but in order that mathematical study should hold on to one thing, the fine and good, and should be aimed at one thing, knowledge of reality and assimilation to the Good itself. So then, in this way not only a bare knowledge of mathematics was handed down, but also a suitable way of life was joined to it, and an ascent to the most honorable ends was suitably provided by it. Therefore it is worthwhile to occupy oneself in accordance with the Pythagorean practice in mathematics, as excelling and preferable to all other mathematical skills.

Chapter 19

The Pythagorean division of mathematical science as a whole into genera and the most important species, which creates a common study of them.

But since we must survey not only the merit of mathematics as a whole but also its kinds and species, their number and which should be chosen, let us make our teaching about them general and capable of extending both to the whole and equally to each branch of mathematics.

The primary study of every mathematician, including the private individual, in each matter, whatsoever it be, is the theological, by harmonizing it in rank and activities with the being and the power of the gods by some fitting analogy. This, indeed, is thought to deserve the greatest concern from men, as for example, in the case of numbers, which sorts of numbers are akin to which sorts of gods and of like nature with them, and in the cases of other branches of mathematics they are accustomed to have the same conception. Now after that among these men mathematics endeavors to exercise itself concerning truly real and intelligible being, the intelligible circle and the formal number, and they pursue other similar branches of mathematics in conformity with purest being. Again they bring the practice of mathematics under the same head as the study of self-moving being and the eternal reason-principles by defining that same self-moving number and by discovering certain measures of ratios through certain mathematical symmetries.

Much mathematical science also studies the heavens and all the heavenly orbits, both the fixed and those of the planets; it also works out not only the complicated motions of the spheres but also those of them that are uniform. It also concerns itself already with reason-principles in matter and enmattered forms, what they are like and how they were initially brought about; for such is that part of mathematics which separates conceptually form and figures from bodies. Further it endeavors to give an account, in accord with principles of natural philosophy, of things in the world of becoming, studying both the simple elements and the reason-principles concerned with bodies.

So the Pythagorean school uses all these parts of their procedure in each and every branch of mathematics, and by them produces order and purification. For as in mathematics second-level principles are made known from more basic ones, so in the case of the powers of the soul the ascent to more complete lives and activities comes about through mathematics. However, they neither neglect nor omit anything intermediate which goes to complete such a science nor leave unexplored the extremes. They traverse the whole without omission, and thus this science hands down, in the case of the most important and primary topics, that division which the science of division demonstrated. From this division it is possible to find out also the subdivisions of mathematics, of which we shall make mention as we proceed in the discussion specifically devoted to this.

Chapter 21

Who were the originators of Pythagorean mathematics, and what according to him are the salient points of such a science; how, according to him, one should organize mathematics. A general survey.

But since we are especially enquiring into Pythagorean mathematics, but it is not possible to give a complete account of it unless one comes to see its earliest beginning, for this reason it is necessary to include also those who were the predecessors of Pythagoras in such a study for the purposes of our present enquiry. For our examination of it would become as complete as possible if we were to establish it by beginning from its first roots.

Now they say that Thales first discovered a good many things in geometry and passed them on to Pythagoras. So we should rightly assimilate the mathematical investigations of Thales which we have received from him to Pythagorean mathematics. After Thales, Pythagoras spent a long time with the Egyptians and he derived from them much of value for mathematical science. Therefore we should not be acting improperly in including alongside as well many things from Egyptian mathematics. But since he also later associated with the Assyrians and those among them who are called Chaldeans (for this is what their mathematicians are called), it is necessary for us to accept much from them into our approach to mathematics.

Chapter 22

The distinctive practice of mathematical science according to Pythagoras, and how many uses of it he envisaged for the soul and for mankind; also how they practiced it throughout the whole of their own life.

But this is not after all a sufficient account. Rather, since Pythagoras added much from his own store to the mathematics he had received from the barbarians, we must both include such beginnings and add also an account of the special character of his mathematics. For he studied much of mathematics in a philosophical way and adapted it to his own focus, even though it was handed down by others; also he imposed the proper order on it and made fitting lines of inquiry about it, and he always throughout provided coherence, so as nowhere to transgress the bounds of consequentiality.

So we should conform to these principles in following in the track of Pythagorean mathematics. But let us take select parts from it as being common elements, so that we may learn from them the symbolic and unfamiliar use of mathematical terms. For since he was aiming at realities and truths he therefore applied their natural names in mathematics. He made from them a starting point of his teaching, capable of guiding his students if by sufficient experience they understood the relevant terms sufficiently. Indeed, by the clarity of his demonstrations, their elegance and accuracy, he excels all study by others in the same field; he employs much clarity, and he starts from what is familiar. But the finest aspect of his work is actually the loftiness of thought which leads up to the primary causes, which pursues its studies for the sake of the facts, which seizes purely on realities, and which sometimes conjoins mathematical insight with theological. One might put forward this much for the present as select common elements of such a science.

Now as regards how one should engage in the pursuit of this science, it is worth making the following summary remarks, in conformity with what has been passed down by these men. So then, since the great bulk of this doctrine was actively present to their minds, and was therefore preserved unwritten in memories which now no longer remain, none of it can be easily vouched for or discovered either from writings or from hearsay. So something like this must be done. We must start from small clues and continually build them up and increase them; we must take them up to suitable starting points, interpolate what is missing, and speculate so far as possible as to their doctrinal positions, what they would say if it were possible for one of them to inform us. But as it is, from pursuing the logical consequences, we are able to uncover their teachings suitably from what they have indisputably handed down to us. For such methods of investigation will enable us either to arrive at the actual Pythagorean mathematical science or to come very close to it, to the highest possible degree. I have reached the conclusion that in this way I was agreeing with its practice as it was carried on according to its own originator. For it was in every way peculiar to him, and it stands out from other disciplines, since it pays attention to the soul and to the purification of the eye of the soul, provoking the discovery of the primary forms and causes of the being of mathematics and harmonizing it with the nature of things themselves. It assimilates it to the intelligible forms and teaches its kinship with the Good and the community of the branches of mathematics with each other.

Since, then, the mathematical discipline was of this character, it searched out earnestly and keenly the theorems in its province without intermission. It contributed to the soul clarity of knowledge and subtlety in reasoning, accuracy of argument, contact with the incorporeal entities involved in it, well-adjustment, harmony, and conversion to the world of being. Also for mankind it brings order into life, peace from passions, fineness of behavior, and discovery of the other things that are of advantage to human life. They practiced it throughout their own life, weaving the gain from it both into their activities and into the conduct of their souls, the organization of their cities, and the management of their homes; also into skilled work and the equipment for war or peace, and altogether they introduced mathematics into every aspect of life, suitably to the objects of activity, profitably to the agents, and with great care for both of these and correspondingly for everything else.

One must then follow in these footsteps not simply by practicing mathematics as such; for the mathematics now popular relies too much on perception and imagination, is a stranger to truth, and has unnaturally turned rather to the world of becoming. But if we should wish to study mathematics in a Pythagorean manner we should follow eagerly its divine route which raises, purifies, and perfects.

1.11 Proclus, *Commentary on the First Book of Euclid's Elements*

Proclus (411–485) was a Greek philosopher originally from Lycia, on the southern coast of Asia Minor. He studied in Alexandria and then moved to Athens, where he became a member of the Neoplatonic Academy, probably founded in the fifth century but with no direct connection to Plato's Academy. During Proclus's time in Athens, the Academy was headed by Plutarch (350–431) and later Syrianus (d. 437). Eventually, in fact, Proclus became head of the Academy. Among his many writings is his *Commentary on the First Book of Euclid's Elements*, excerpts of which are presented here and alongside the selections from the *Elements* themselves. Here, in the first part of his prologue, Proclus discusses his mathematical philosophy, much of it inspired by the works of Plato and Aristotle. He also deals with the classification of the mathematical sciences, taken, he says, from the Pythagoreans and Geminus.

PROLOGUE

Part I

Chapter 2

The Common Principles of Mathematical Being. The Limit and the Unlimited

To find the principles of mathematical being as a whole, we must ascend to those all-pervading principles that generate everything from themselves: namely, the Limit and the Unlimited. For these, the two highest principles after the indescribable and utterly incomprehensible causation of the One, give rise to everything else, including mathematical beings. From these principles proceed all other things collectively and transcendentally, but as they come forth, they appear in appropriate divisions and take their place in an ordered procession, some coming into being first, others in the middle, and others at the end. The objects of Nous, by virtue of their inherent simplicity, are the first partakers of the Limit and the Unlimited.[9] Their unity, their identity, and their stable and abiding existence they derive from the Limit; but for their variety, their generative fertility, and their divine otherness and progression they draw upon the Unlimited. Mathematicals

[9] These speculations have their source in Plato's *Philebus*. The idea is that the human mind needs these categories to understand what is real.

are the offspring of the Limit and the Unlimited, but not of the primary principles alone, nor of the hidden intelligible cause, but also of secondary principles that proceed from them and, in cooperation with one another, suffice to generate the intermediate orders of things and the variety that they display. This is why in these orders of being there are ratios proceeding to infinity, but controlled by the principle of the Limit. For number, beginning with unity, is capable of indefinite increase, yet any number you choose is finite; magnitudes likewise are divisible without end, yet the magnitudes distinguished from one another are all bounded, and the actual parts of a whole are limited. If there were no infinity, all magnitudes would be commensurable and there would be nothing inexpressible or irrational, features that are thought to distinguish geometry from arithmetic; nor could numbers exhibit the generative power of the monad, nor would they have in them all the ratios—such as multiple and superparticular[10]—that are in things. For every number that we examine has a different ratio to unity and to the number just before it. And if the Limit were absent, there would be no commensurability or identity of ratios in mathematics, no similarity and equality of figures, nor anything else that belongs in the column of the better. There would not even be any sciences dealing with such matters, nor any fixed and precise concepts. Thus mathematics needs both these principles as do the other realms of being. As for the lowest realities, those that appear in matter and are molded by nature, it is quite obvious at once that they partake of both principles, of the Unlimited as the ground that underlies their forms and of the Limit by virtue of their ratios, figures, and shapes. It is clear, then, that the principles primary in mathematics are those that preside over all things.

Chapter 3

The Common Theorems Governing Mathematical Kinds

Just as we have noted these common principles and seen that they pervade all classes of mathematical objects, so let us enumerate the simple theorems that are common to them all, that is, the theorems generated by the single science that embraces alike all forms of mathematical knowledge; and let us see how they fit into all these sciences and can be observed alike in numbers, magnitudes, and motions. Such are the theorems governing proportion, namely, the rules of compounding, dividing, converting, and alternating: likewise the theorems concerning ratios of all kinds, multiple, superparticular, superpartient, and their counterparts; and the theorems about equality and inequality in their most general and universal aspects, not equality or inequality of figures, numbers, or motions, but each of the two by itself as having a nature common to all its forms and capable of more simple apprehension. And certainly beauty and order are common to all branches of mathematics, as are the method of proceeding from things better known to things we seek to know and the reverse path from the latter to the former, the methods called analysis and synthesis. Likeness and unlikeness of ratios are not absent from any branch of mathematics, for we call some figures similar and other dissimilar, and in the same way some numbers like and others unlike. And matters pertaining to powers obviously belong to general mathematics, whether they be roots or squares. All these

[10] These ratios are discussed in the excerpt from Boethius's *On Arithmetic* in chapter 2.

Socrates in the *Republic* puts in the mouths of his loftily-speaking Muses, bringing together in determinate limits the elements common to all mathematical ratios and setting them up in specific numbers by which the periods of fruitful birth and of its opposite, unfruitfulness, can be discerned.

Chapter 8

The Utility of Mathematics

From what we have said it is clear that mathematical science makes a contribution of the greatest importance to philosophy and to its particular branches, which we must also mention. For theology, first of all, mathematics prepares our intellectual apprehension. Those truths about the gods that are difficult for imperfect minds to discover and understand, these the science of mathematics, with the help of likenesses, shows to be trustworthy, evident, and irrefutable. It proves that numbers reflect the properties of beings above being and in the objects studied by the understanding reveals the powers of the intellectual figures. Thus Plato teaches us many wonderful doctrines about the gods by means of mathematical forms, and the philosophy of the Pythagoreans clothes its secret theological teaching in such draperies. . . .

Mathematics also makes contributions of the very greatest value to physical science. It reveals the orderliness of the ratios according to which the universe is constructed and the proportion that binds things together in the cosmos, making, as the *Timaeus* somewhere says, divergent and warring factors into friends and sympathetic companions. It exhibits the simple and primal causal elements as everywhere clinging fast to one another in symmetry and equality, the properties through which the whole heaven was perfected when it took upon itself the figures appropriate to its particular region; and it discovers, furthermore, the numbers applicable to all generated things and to their periods of activity and of return to their starting points, by which it is possible to calculate the times of fruitfulness or the reverse for each of them. All these I believe the *Timaeus* sets forth, using mathematical language throughout in expounding its theory of the nature of the universe. It regulates by numbers and figures the generation of the elements, showing how their powers, characteristics, and activities are derived therefrom and tracing the causes of all change back to the acuteness or obtuseness of their angles, the uniformity or diversity of their sides, and the number or fewness of the elements involved.

How, then, can we deny that mathematics brings many remarkable benefits to what is called political philosophy? By measuring the periods of activity and the varied revolutions of the All, it finds the numbers that are appropriate for generation, that is, those that cause homogeneity or diversity in progeny, those that are fruitful and perfecting and their opposites, those that bring a harmonious life in their train and those that being discord, and in general those that are responsible for prosperity and those that occasion want. . . .

Finally, how much benefit mathematics confers on the other sciences and arts we can learn when we reflect that to the theoretical arts, such as rhetoric and all those like it that function through discourse, it contributes completeness and orderliness, by providing for them a likeness of a whole made perfect through first, intermediate, and concluding parts; that to the poetical arts it stands as a paradigm, furnishing in itself models for the speeches that the authors compose and the meters that they employ;

and that for the practical arts it defines their motion and activity through its own fixed and unchangeable forms. In general, as Socrates says in the *Philebus*, all the arts require the aid of counting, measuring, and weighing, of one or all of them; and these arts are all included in mathematical reasonings and are made definite by them, for it is mathematics that knows the divisions of numbers, the variety of measures, and the differences of weights. These considerations will make evident to the student the utility of general mathematics both to philosophy itself and to the other sciences and arts.

Chapter 12

The Pythagorean Classification of the Mathematical Sciences

We must next distinguish the species of mathematical science and determine what and how many they are; for after its generic and all-inclusive form it is necessary to consider the specific differences between the particular sciences. The Pythagoreans considered all mathematical science to be divided into four parts: one half they marked off as concerned with quantity, the other half with magnitude; and each of these they posited as twofold. A quantity can be considered in regard to its character by itself or in its relation to another quantity, magnitudes as either stationary or in motion. Arithmetic, then, studies quantity as such, music the relations between quantities, geometry magnitude at rest, spherics [that is, astronomy] magnitude inherently moving. The Pythagoreans consider quantity and magnitude not in their generality, however, but only as finite in each case. For they say that the sciences study the finite in abstraction from infinite quantities and magnitudes, since it is impossible to comprehend infinity in either of them. Since this assertion is made by men who have reached the summit of wisdom, it is not for us to demand that we be taught about quantity in sense objects or magnitude that appears in bodies. To examine these matters is, I think the province of the science of nature, not that of mathematics itself. . . .

Chapter 13

Geminus' Classfication of the Mathematical Sciences

But others, like Geminus, think that mathematics should be divided differently; they think of one part as concerned with intelligibles only and of another as working with perceptibles and in contact with them. By intelligibles, of course, they mean those objects that the soul arouses by herself and contemplates in separation from embodied forms. Of the mathematics that deals with intelligibles they posit arithmetic and geometry as the two primary and most authentic parts, while the mathematics that attends to sensibles contains six sciences: mechanics, astronomy, optics, geodesy, canonics, and calculation. Tactics they do not think it proper to call a part of mathematics, as others do, though they admit that it sometimes uses calculation, as in the enumeration of military forces, and sometimes geodesy, as in the division and measurement of encampments. Much less do they think of history and medicine as parts of mathematics, even though writers of history often bring in mathematical theorems in describing the lie of certain regions or in calculating the size, breadth, or perimeters of cities, and physicians often clarify their own doctrines by such methods, for the utility of astronomy to medicine is made clear by Hippocrates and all who speak of seasons and places. So also the master of tactics will use the theorems of mathematics, even though he is not a mathematician, if he should ever want to lay out a circular camp to make his army appear as small as possible, or a square or pentagonal or some other form of camp to make it appear very large.

These, then, are the species of general mathematics. Geometry in its turn is divided into plane geometry and stereometry. There is no special branch of study devoted to points and lines, inasmuch as no figure can be constructed from them without planes or solids; and it is always the function of geometry, whether plane or solid, either to construct figures or to compound or divide figures already constructed. In the same way arithmetic is divided into the study of linear numbers, plane numbers, and solid numbers; for it examines number as such and its various kinds as they proceed from the number one, investigating the generation of plane numbers, both similar and dissimilar, and progressions to the third dimension. Geodesy and calculation are analogous to these sciences, since they discourse not about intelligible but about sensible numbers and figures. For it is not the function of geodesy to measure cylinders or cones, but heaps of earth considered as cones and wells considered as cylinders; and it does not use intelligible straight lines, but sensible ones, sometimes more precise ones, such as rays of sunlight, sometimes coarser ones, such as a rope or a carpenter's rule. Nor does the student of calculation consider the properties of number as such, but of numbers as present in sensible objects; and hence he gives them names from the things being numbered, calling them sheep numbers or cup numbers. He does not assert, as does the arithmetician, that something is least; nevertheless with respect to any given class he assumes a least, for when he is counting a group of men, one man is his unit. Again optics and canonics [that is, music] are offshoots of geometry and arithmetic. The former science uses visual lines and the angles made by them; it is divided into a part specifically called optics, which explains the illusory appearances presented by objects seen at a distance, such as the converging of parallel lines or the rounded appearance of square towers, and general catoptrics, which is concerned with the various ways in which light is reflected. The latter is closely bound up with the art of representation and studies what is called "scene-painting," showing how objects can be represented by images that will not seem disproportionate or shapeless when seen at a distance or on an elevation. The science of canonics deals with the perceptible ratios between notes of the musical scales and discovers the divisions of the monochord, everywhere relying on sense-perception and, as Plato says, "putting the ear ahead of the mind."

In addition to these there is the science called mechanics, a part of the study of perceptible and embodied forms. Under it comes the art of making useful engines of war, like the machines that Archimedes is credited with devising for defense against the besiegers of Syracuse, and also the art of wonder-working, which invents figures moved sometimes by wind, like those written about by Ctesibius[11] and Heron, sometimes by weights, whose imbalance and balance respectively are responsible for movement and rest, as the *Timaeus* shows, and sometimes by cords and ropes in imitation of the tendons and movements of living beings. Under mechanics also falls the science of equilibrium in general and the study of the so-called center of gravity, as well as the art of making spheres imitating the revolutions of the heavens, such as was cultivated by Archimedes,

[11] Ctsebius lived in Alexandria in the third century BCE and was probably the first head of the museum there.

and in general every art concerned with the moving of material things. There remains astronomy, which inquires into the cosmic motions, the sizes and shapes of the heavenly bodies, their illuminations and distances from the earth, and all such matters. This art draws heavily on sense-perception and coincides in large measure with the science of nature. The parts of astronomy are gnomonics, which occupies itself with marking the divisions of time by the placing of sundials; meteorology, which determines the different risings of the heavenly bodies and their distances from one another and teaches many and varied details of astronomical theory; and dioptrics, which fixes the positions of the sun, moon, and stars by means of special instruments. Such are the traditions we have received from the writings of the ancients regarding the divisions of mathematical science.

1.12 Boethius, *On Arithmetic*

Boethius (477–524) was a Roman senator and consul in the years immediately after the Ostrogoths deposed the last Roman emperor in the west. His most important contributions are to philosophy, although he also wrote works on mathematical topics and translated texts of Plato and Aristotle into Latin. Boethius's *De Institutione Arithmetica*, part of whose first chapter is included here, was a loose translation of Nicomachus's *Arithmetica*. It is especially important as it was the first explicit mention of the term *quadrivium* as a name for the four branches of mathematics that served as the mathematical curriculum from the time of Plato through the middle ages. Boethius's work today survives in numerous copies, showing how influential it was for a thousand years after its composition.

Proemium: The division of mathematics

Among all the men of ancient authority who, following the lead of Pythagoras, have flourished in the purer reasoning of the mind, it is clearly obvious that hardly anyone had been able to reach the highest perfection of the disciplines of philosophy unless the nobility of such wisdom was investigated by him in a certain four-part study, the *quadrivium*, which will hardly be hidden from those properly respectful of expertness.[12] For this is the wisdom of things which are, and the perception of truth gives to these things their unchanging character.

We say those things *are* which neither grow by stretching nor diminish by crushing, nor are changed by variations, but are always in their proper force and keep themselves secure by support of their own nature. Such things are: qualities, quantities, configuration, largeness, smallness, equalities, relations, acts, dispositions, places, times, and whatever is in any way found joined to bodies. Now those things which by their nature are incorporeal, existing by reason of an immutable substance, when affected by the participation of a body and by contact with some variable thing, pass into a condition of inconstant changeableness. Such things (since as it was said, immutable substance and forces were

[12] Although Nicomachus in his book 1 mentions the four divisions of mathematics, Boethius is the first to use the term *quadrivium* to designate this "fourfold" way of study.

delegated by nature) are truly and properly said to be. Wisdom gives name to a science in terms of these things, that is, things which properly exist, whatever their essences may be.

There are two kinds of essence. One is continuous, joined together in its parts and not distributed in separate parts, as a tree, a stone, and all the bodies of this world which are properly called magnitudes. The other essence is of itself disjoined and determined by its parts as though reduced to a single collective union, such as a flock, a populace, a chorus, a heap of things, things whose parts are terminated by their own extremities and are discrete from the extremity of some other. The proper name for these is a multitude. Again, some types of multitude exist by themselves, as a three, a four, a tetragon, or whatever number which, as it is, lacks nothing. Another kind does not exist of itself but refers to some other thing, as a duplex, a dimidium, a sesquialtar, a sesquitertial, or whatever it may be which, unless it is related to anything, is not in itself able to exist. Of magnitudes, some are stable, lacking in motion, while others are always turned in mobile change and at no time are at rest. Now of these types, arithmetic considers that multitude which exists of itself as an integral whole; the measures of musical modulation understand that multitude which exists in relation to some other; geometry offers the notion of stable magnitude; the skill of astronomical discipline explains the science of movable magnitude. If a searcher is lacking knowledge of these four sciences, he is not able to find the true; without this kind of thought, nothing of truth is rightly known. This is the knowledge of those things which truly are; it is their full understanding and comprehension. He who spurns these, the paths of wisdom, does not rightly philosophize. Indeed, if philosophy is the love of wisdom, in spurning these, one has already shown contempt for philosophy.

To this I think I should add that every force of a multitude, progressing from one point, moves on to limitless increases of growth. But a magnitude, beginning with a finite quantity, does not receive a new mode of being by division; its name includes the smallest sections of its body. This infinite and unlimited ability of nature in a multitude, philosophy spontaneously rejects. For nothing which is infinite is able to be assembled by a science or to be comprehended by the mind. But reason itself takes this matter of the infinite to itself; in these matters, reason is able to exercise the searching power of truth. It delegates the boundary of finite quality to the plurality of infinite multitude, and having rejected the aspect of interminable magnitude, it demands in a defined area a cognition of these things on its own behalf.

It stands to reason that whoever puts these matters aside has lost the whole teaching of philosophy. This, therefore, is the *quadrivium* by which we bring a superior mind from knowledge offered by the senses to the more certain things of the intellect. There are various steps and certain dimensions of progressing by which the mind is able to ascend so that by means of the eye of the mind, which (as Plato says) is composed of many corporeal eyes and is of higher dignity than they, truth can be investigated and beheld. This eye, I say, submerged and surrounded by the corporal senses, is in turn illuminated by the disciplines of the *quadrivium*.

Which of these disciplines, then, is the first to be learned but that one which holds the principal place and position of a mother to the rest? This is arithmetic. It is prior to all not only because God the creator of the massive structure of the world considered this first discipline as the exemplar of his own thought and established all things in accord with it; or that through numbers of an assigned order all things exhibiting the logic of their maker found concord; but arithmetic is said to be first for this reason also, because whatever things are prior in nature, it is to these underlying elements that the posterior elements can be referred. Now if posterior things pass away, nothing concerning the status of the prior substance is disturbed—so "animal" comes before "man," If you take away "man," "animal" does not disappear.

On the other hand, those things which are posterior infer prior things in themselves, and when these prior things are stated, they do not include in them anything of the posterior, as can be seen in that same term "man." If you say "man," you also say "animal," because it is the same as man, since man is an animal. If you say "animal," you do not at the same time include the species of man, because "animal" is not the same as "man."

The same thing is seen to occur in geometry and arithmetic. If you take away numbers, in what will consist the triangle, quadrangle, or whatever else is treated in geometry? All of those things are in the domain of number. If you were to remove the triangle and the quadrangle and all of geometry, still "three" and "four" and the terminology of the other numbers would not perish. Again, when I name some geometrical form, in that term the numbers are implicit. But when I say numbers, I have not implied any geometrical form.

The logical force of numbers is also prior to music, and this can especially be demonstrated because not only are numbers prior by their nature, since they consist of themselves and are thus prior to those things which must be referred to another in order to be, but also musical modulation itself is denoted by the names of numbers. The same relationship which we remarked in geometry can be found in music. The names diatessaron, diapente, and diapason are derived from the names of antecedent numerical terms.[13] The proportion of their sounds is found only in these particular relationships and not in other number relationships. For the sound which is in a diapason harmony, the same sound is produced in the ratio of a number doubled. The interval of a diatessaron is found in an epitrita comparison; they call the harmony diapente which is joined by a hemiola interval. An epogdous in numbers is a tone in music. I cannot undertake to explain every consequence of this idea, as to how arithmetic is prior, but the rest of this work will demonstrate it without any doubt.

Arithmetic also precedes spherical and astronomical science insofar as these two remaining studies follow the third [geometry] naturally. In astronomy, "circles," "a sphere," "a center," "concentric circles," "the median," and "the axis" exist, all of which are the concern of the discipline of geometry. For this reason, I want to demonstrate the anterior logical force of geometry. This is the case because in all things, movement naturally comes after rest; the static comes first. Thus, geometry understands the doctrine of immovable things while astronomy comprehends the science of mobile things. In astronomy, the very movement of the stars is celebrated in harmonic intervals. From this it follows that the power of music logically precedes the course of the stars; and there is no doubt that

[13] A diatessaron is a relation of 4 : 3, an interval of a fourth, or an epitriata; a diapente is a 3 : 2 relation, an interval of a fifth, or a hemiola; a diapason is an interval of an octave; an epogdous is the relation of 9 : 8 or a whole-tone interval.

arithmetic precedes astronomy since it is prior to music, which comes before astronomy. All the courses of the stars and all astronomic reasoning are established exclusively by the nature of numbers. Thus we connect rising with falling, thus we keep watch on the slowness and speed of wandering stars, thus we recognize the eclipses and multiplicities of lunar variation. Since, as it is obvious, the force of arithmetic is prior, we may take up the beginning of our exposition.

1.13 Aristophanes, *The Clouds, The Frogs, The Birds*

Aristophanes (ca. 446–386 BCE) was a noted playwright from classical antiquity who authored a number of comic dramas. His satirical sketches of politics, people, and everyday life in ancient Athens were noted for both their humor and their scathing ridicule of contemporary society. Nothing was off limits to the sharp and irreverent wit of Aristophanes, not even mathematics! Indeed, mathematics and mathematicians are the subject of a comic twist or the object of ridicule in a number of his plays. The following passages, taken from *The Clouds*, *The Frogs*, and *The Birds*, poke fun at mathematicians and quantification by having the protagonists measure distances with a flea, use mathematical techniques to judge poetry, and refer to squaring the circle as a concept with which the audience would be familiar.

From *The Clouds*

Strepsiades: Then tell me without fear, for I have come to study among you.

Disciple: Very well then, but reflect, that these are mysteries. Lately, a flea bit Chaerephon on the brow and then from there sprang on to the head of Socrates. Socrates asked Chaerephon, "How many times the length of its legs does a flea jump?"

Strepsiades: And how ever did he go about measuring it?

Disciple: Oh! it was most ingenious! He melted some wax, seized the flea and dipped its two feet in the wax, which, when cooled, left them shod with true Persian slippers. These he took off and with them measured the distance.

Strepsiades: Ah! great Zeus! what a brain! what subtlety!

...

Strepsiades (pointing to a celestial globe): In the name of all the gods, what is that? Tell me.

Disciple: That is astronomy.

Strepsiades (pointing to a map): And that?

Disciple: Geometry.

Strepsiades: What is that used for?

Disciple: To measure the land.

Strepsiades: But that is apportioned by lot.

Disciple: No, no, I mean the entire earth.

Strepsiades Ah! what a funny thing! How generally useful indeed is this invention!

Disciple: There is the whole surface of the earth. Look! Here is Athens.

Strepsiades: Athens! you are mistaken; I see no courts in session.

Disciple: Nevertheless it is really and truly the Attic territory.

Strepsiades: And where are my neighbours of Cicynna?

Disciple: They live here. This is Euboea; you see this island, that is so long and narrow.

Strepsiades: I know. Because we and Pericles have stretched it by dint of squeezing it. And where is Lacedaemon?

Disciple: Lacedaemon? Why, here it is, look.

Strepsiades: How near it is to us! Think it well over, it must be removed to a greater distance.

Disciple: But, by Zeus, that is not possible.

Strepsiades: Then, woe to you! and who is this man suspended up in a basket?

Disciple: That's himself.

Strepsiades: Who's himself?

Disciple: Socrates.

Strepsiades: Socrates! Oh! I pray you, call him right loudly for me.

Disciple: Call him yourself; I have no time to waste.

From *The Frogs*

Xanthias:
And what does Pluto now propose to do?

Aeacus:
He means to hold a tournament, and bring
Their tragedies to the proof.

Xanthias:
But Sophocles,
How came not he to claim the tragic chair?

Aeacus:
Claim it? Not he! When he came down, he kissed
With reverence Aeschylus, and clasped his hand,
And yielded willingly the chair to him.
But now he's going, says Cleidemides,
To sit third-man: and then if Aeschylus win,
He'll stay content: if not, for his art's sake,
He'll fight to the death against Euripides.

Xanthias:
Will it come off?

Aeacus:
O yes, by Zeus, directly.
And then, I hear, will wonderful things be done,
The art poetic will be weighed in scales.

Xanthias:
What I weigh out tragedy, like butcher's meat?

Aeacus:
Levels they'll bring, and measuring-tapes for words,
And moulded oblongs.

Xanthias:
Is it bricks they are making?

Aeacus:
Wedges and compasses: for Euripides
Vows that he'll test the dramas, word by word.
Xanthias:
Aeschylus chafes at this, I fancy.
Aeacus:
Well, He lowered his brows, upglaring like a bull.
Xanthias:
And who's to be the judge?
Aeacus:
There came the rub.
Skilled men were hard to find: for with the Athenians
Aeschylus, somehow, did not hit it off.

From *The Birds*

Meton: I come amongst you—

Peisthetaerus: Some new misery this! Come to do what? What's your scheme's form and outline? What's your design? What buskin's on your foot?

Meton: I come to land-survey this Air of yours, and mete it out by acres.

Peisthetaerus: Heaven and Earth! Whoever are you?

Meton: Whoever am I? I'm Meton, known throughout Hellas and Colonus.

Peisthetaerus: Aye, and what are these?

Meton: They are rods for Air-surveying. I'll just explain. The Air's, in outline, like one vast extinguisher; so then, observe, applying here my flexible rod, and fixing my compass there—do you understand?

Peisthetaerus: I don't.

Meton: With the straight rod I measure out, that so the circle may be squared; and in the center a market place; and streets be leading to it straight to the very center; just as from a star, though circular, straight rays flash out in all directions.

Peisthetaerus: Why, the man's a Thales! Meton!

Meton: Yes, what?

Peisthetaerus: You know I love you, Meton, take my advice, and slip away unnoticed.

Meton: Why, what's the matter?

Peisthetaerus: As in Cacedaemon, there's stranger-hunting and a great disturbance; and blows in plenty.

Meton: What, a Revolution?

Peisthetaerus: No, no, not that.

Meton: What then?

Peisthetaerus: They've all resolved with one consent to wallop every quack.

Meton: I'd best be going.

Peisthetaerus: Faith, I'm not quite certain if you're in time; see, the blows are coming!

Meton: O, murder! help!

Peisthetaerus: I told you how it would be. Come, measure off your steps some other way.

SOURCES, CHAPTER 1

1.1 This translation of a section of the *Republic* is by Benjamin Jowett and is found in Robert Maynard Hutchins, ed., *Great Books of the Western World*, vol. 7 (Chicago: Encyclopedia Britannica, 1952), 393–97.

1.2 This translation of a part of the *Gorgias* is by Benjamin Jowett and is found in Robert Maynard Hutchins, ed., *Great Books of the Western World*, vol. 7 (Chicago: Encyclopedia Britannica, 1952), 254.

1.3 This translation is taken from Carl A. Huffman, *Archytas of Tarentum: Pythagorean, Philosopher and Mathematician King* (Cambridge: Cambridge University Press, 2005), 183.

1.4 This translation of part of the *Metaphysics* is by W. D. Ross and is found in Robert Maynard Hutchins, ed., *Great Books of the Western World*, vol. 8 (Chicago: Encyclopedia Britannica, 1952), 500–501, 503–4.

1.5 This translation of part of the *Prior Analytics* is by A. J. Jenkinson and is found in Robert Maynard Hutchins, ed., *Great Books of the Western World*, vol. 8 (Chicago: Encyclopedia Britannica, 1952), 39, 58.

1.6 This translation of part of the *Posterior Analytics* is by G. R. G. Mure and is found in Robert Maynard Hutchins, ed., *Great Books of the Western World*, vol. 8 (Chicago: Encyclopedia Britannica, 1952), 97–98, 104–5.

1.7 This translation of part of the *Physics* is by R. P. Hardie and R. K. Gaye and is found in Robert Maynard Hutchins, ed., *Great Books of the Western World*, vol. 8 (Chicago: Encyclopedia Britannica, 1971), 281–82, 284–85.

1.8 This translation is from *The Works of Philo*, trans. C. D. Yonge (Peabody, MA: Hendrickson, 1993), 5–6.

1.9 This translation is from Nicomachus of Gerasa, *Introduction to Arithmetic*, ed. and trans. Martin Luther D'Ooge (New York: Macmillan, 1926), 181–89. It is also found in Robert Maynard Hutchins, ed., *Great Books of the Western World*, vol. 11 (Chicago: Encyclopedia Britannica, 1952), 811–13.

1.10 The translation is by J. O. Urmson, John Dillon, and Sebastian Gertz in Iamblichus, *On the General Science of Mathematics* (London: Bloomsbury Academic, 2020), 5–8.

1.11 This translation is taken from Proclus, *A Commentary on the First Book of Euclid's Elements*, trans. Glenn R. Morrow (Princeton, NJ: Princeton University Press, 1970), 4–7, 18–34.

1.12 This translation is from Michael Masi, *Boethian Number Theory: A Translation of the De Institutione Arithmetica* (Amsterdam: Rodopi, 2006), 71–75.

1.13 Translations of *The Birds* are by Benjamin Bickley Rogers and are found in Robert Maynard Hutchins, ed., *Great Books of the Western World*, vol. 5 (Chicago: Encyclopedia Britannica, 1952), 555. Those of *The Clouds* and *The Frogs* are by The Athenian Society, 1912.

2

Arithmetic

INTRODUCTION

Following Plato and his fourfold division of the mathematical sciences, as well as the comments of other authors excerpted in chapter 1, arithmetic is the first branch of mathematics that should be studied. For Plato and other mathematicians of his generation, what was meant by arithmetic was not "calculation" or "logistic" but what we would call "number theory." In this chapter, we consider some of the earliest works in number theory. We begin with books 7 to 9 of Euclid's *Elements*. These books contain some of the earliest Greek ideas on number theory, including such notions as evens and odds, prime numbers, and perfect numbers.[1]

Nicomachus, writing several centuries later, included many of the same topics as Euclid, but notably did not include proofs. He also discussed the notion of means, providing definitions of numerous types beyond the standard arithmetic, geometric, and harmonic means. One of the reasons for Nicomachus's interest in arithmetic, as well as Euclid's, was its use in the study of music. These applications to music are included in chapter 8. Boethius, writing in Latin in the early sixth century CE, presented Nicomachus's ideas to a new audience. Of note is his treatment of ratios, again a critical idea in the development of music theory.

Alongside the theory of numbers, there was also a set of mystical beliefs about numbers, particularly about the positive integers. These are developed in the works of Philo of Alexandria and Pseudo-Iamblichus. We then present excerpts from Diophantus's *Arithmetica*, a work that shows how Greeks solved actual numerical problems in rational numbers, problems that are today classified as indeterminate equations. We next include several examples of actual calculation—that is, logistic—or the solution of numerical problems, including mathematical tables, most taken from fragments of papyri that have been uncovered in Egypt, as well as a method of calculating square roots that we perhaps owe to Ptolemy. We follow this with Archimedes's methods of representing very large numbers as well as Pappus's treatment of a related theme coming from Apollonius. Finally, we look at a selection of mathematical problems

[1] A perfect number is the sum of its proper divisors.

and riddles dating from the centuries near the beginning of our era that are generally solvable by simple calculations.

2.1 Euclid, *Elements*, 7–9

The *Elements* of Euclid is the most important mathematical text of Greek times and probably of all time. It has appeared in more editions than any work other than the Bible. It has been translated into countless languages and has been continuously available in manuscript form or in printed copies in one country or another since it was produced. It has long provided mathematicians with a model of how "pure mathematics" should be written, with well-thought-out axioms, precise definitions, carefully stated theorems, and logically coherent proofs. Although Proclus writes that there were earlier versions of *Elements* before that of Euclid, his is the only one to survive, perhaps because it was the first one written after both the foundations of proportion theory and the theory of irrationals had been developed and the careful distinctions always to be made between number and magnitude had been propounded by Aristotle. The *Elements* that has come down to us today was, therefore, the most complete and up-to-date version of its time. Since the mathematical community as a whole was of limited size and copying manuscripts took considerable effort, once Euclid's work was recognized for its general excellence, there was no reason to keep another inferior work in circulation.

About Euclid himself very little is known. It is generally assumed from the work of Proclus that Euclid flourished around 300 BCE in Alexandria, during the reign of Ptolemy I Soter (323–283 BCE). Most scholars believe that Euclid was one of the first scholars active at the Museum and Library at Alexandria founded by Ptolemy.

Euclid put the results of his *Elements* together in a logical sequence, making sure that each theorem was based on either the axioms or previous results. But probably his basic reason for including the results that he did was to provide mathematicians with a "toolbox" of basic mathematical truths that could be used in solving new mathematical problems. Thus, the results in the number theory books 7 to 9 excerpted here would allow others to explore further ideas in number theory, while the results in the geometrical books (see chapter 3) were used in explorations of various geometrical problems.

Euclid begins book 7 with definitions. Most of these are standard, but Euclid's definitions of even-times even and even-times odd numbers are different from their definitions in the work of Nicomachus presented later. Among the theorems we include are the so-called Euclidean algorithm, the result that in proportions between numbers, the product of the means equals the product of the extremes, and most of what is today called the fundamental theorem of arithmetic.

Please note that for all the excerpts from the *Elements*, here and in chapter 3, although we have attempted to include enough propositions for the reader to follow Euclid's arguments in his more important results, it is not possible to include every proposition that Euclid uses without including the entire book. We occasionally reference previous propositions when they are crucial to an argument and may not be entirely familiar to readers. However, readers are encouraged to consult a complete version of the *Elements* whenever they are in doubt about the reasons supporting a particular statement in a proof. Readers will note that in the books of number theory, diagrams are included, even though often they appear not to be needed. Evidently, Euclid felt the need for lettered diagrams even here; of course, in the geometry books, they are clearly necessary to enable readers to follow the proofs. Finally, note that at the end of each proof we have often included the abbreviation Q.E.D. (for *quod erat demonstrandum*, "which was to be demonstrated") as a substitute for Euclid's own words that generally restate the theorem that was to be proved.

Book 7

Definitions

1. A *unit* is that by virtue of which each of the things that exist is called one.[2]
2. A *number* is a multitude composed of units.
3. A number is a *part* of a number, the less of the greater, when it measures the greater;
4. but *parts* when it does not measure it.
5. The greater number is a *multiple* of the less when it is measured by the less.
6. An *even number* is that which is divisible into two equal parts.
7. An *odd number* is that which is not divisible into two equal parts, or that which differs by a unit from an even number.
8. An *even-times even number* is that which is measured by an even number according to an even number.[3]
9. An *even-times odd number* is that which is measured by an even number according to an odd number.[4]
10. An *odd-times odd number* is that which is measured by an odd number according to an odd number.[5]
11. A *prime number* is that which is measured by a unit alone.
12. Numbers *prime to one another* are those which are measured by a unit alone as a common measure.
15. A number is said to *multiply* a number when that which is multiplied is added to itself as many times as there are units in the other, and thus some number is produced.
18. A *square number* is equal multiplied by equal, or a number which is contained by two equal numbers.
20. Numbers are *proportional* when the first is the same multiple, or the same part, or the same parts, of the second that the third is of the fourth.
22. A *perfect number* is that which is equal to its own parts.[6]

[2] Note that Euclid most probably did not number these definitions. In fact, the numbering of definitions, axioms, and propositions was accomplished many centuries later to help readers find their way through these books.

[3] That is, $n = 2m \cdot 2q$.

[4] That is, $n = 2m \cdot (2q+1)$. Note that this definition and the previous one are not mutually exclusive. For example, 12 is both even-times even and even-times odd.

[5] That is, $n = (2m+1)(2q+1)$.

[6] That is, n is perfect if it is the sum of all its proper divisors.

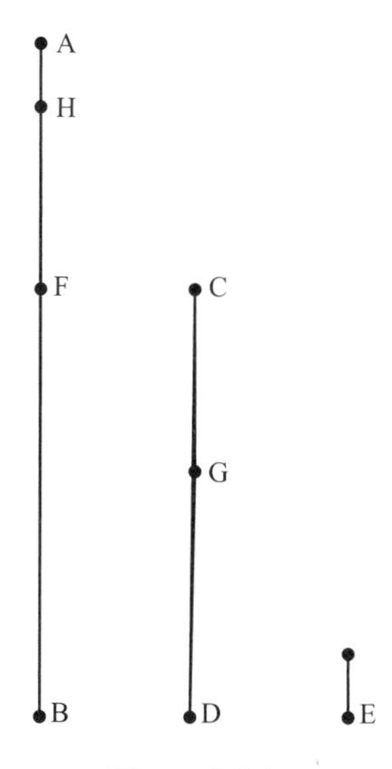

Figure 2.1.1.

Propositions

The first two propositions present the Euclidean algorithm for determining the greatest common divisor, or greatest common measure, of two numbers. These important algorithms in the arithmetic toolbox were known earlier in Greece and were also used in both Indian and Chinese mathematics after Euclid's time. The basic procedure here is that, given two numbers a, b, with $a>b$, one subtracts b from a as many times as possible. If there is a remainder c, which, of course, is less than b, one then subtracts c from b as many times as possible. Continuing in this manner, one eventually comes either to a number m which measures (divides) the one before (proposition 2) or to a unit (proposition 1). In the former case, Euclid showed that m is the greatest common measure of a and b, while in the latter case, he showed that a and b are relatively prime.

Proposition 1. *Two unequal numbers being set out, and the less being continually subtracted in turn from the greater, if the number which is left never measures the one before it until a unit is left, the original numbers will be prime to one another.*

For, the less of two unequal numbers AB, CD being continually subtracted from the greater, let the number which is left never measure the one before it until a unit is left; I say that AB, CD are prime to one another, that is, that a unit alone measures AB, CD [figure 2.1.1].

For, if AB, CD are not prime to one another, some number will measure them. Let a number measure them, and let it be E; let CD, measuring BF, leave FA less than itself, let AF, measuring DG, leave GC less than itself, and let GC, measuring FH, leave a unit HA. Since, then, E measures CD, and CD measures BF, therefore E also measures BF. But it also measures the whole BA; therefore it will also measure the remainder AF. But AF measures DG; therefore E also measures DG. But it also measures the whole DC; therefore it will also measure the remainder CG. But CG measures FH; therefore E

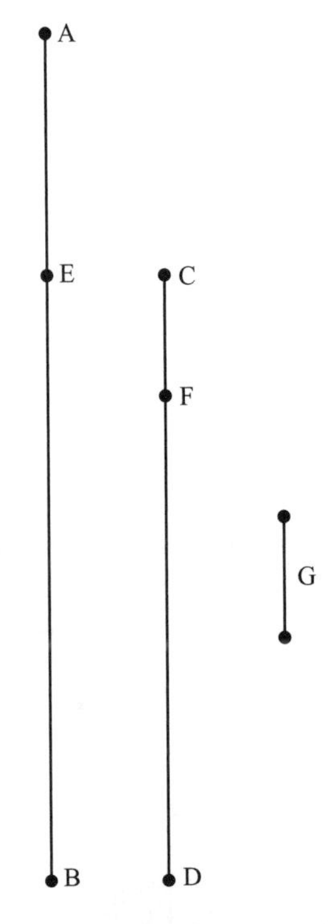

Figure 2.1.2.

also measures FH. But it also measures the whole FA; therefore it will also measure the remainder, the unit AH, though it is a number, which is impossible. Therefore no number will measure the numbers AB, CD; therefore AB, CD are prime to one another. Q.E.D.

Proposition 2. *Given two numbers not prime to one another, to find their greatest common measure.*

Let AB, CD be the two given numbers not prime to one another. Thus it is required to find the greatest common measure of AB, CD. If now CD measures AB—and it also measures itself—CD is a common measure of CD, AB. And it is manifest that it is also the greatest; for no greater number than CD will measure CD.

But, if CD does not measure AB, then, the less of the numbers AB, CD being continually subtracted from the greater, some number will be left which will measure the one before it. For a unit will not be left; otherwise AB, CD will be prime to one another, which is contrary to the hypothesis. Therefore some number will be left which will measure the one before it. Now let CD measuring BE, leave EA less than itself, let EA, measuring DF, leave FC less than itself, and let CF measure AE [figure 2.1.2].

Since, then, CF measures AE, and AE measures DF, therefore CF will also measure DF. But it also measures itself; therefore it will also measure the whole CD. But CD measures BE; therefore CF also measures BE. But it also measures EA; therefore it will also measure the whole BA. But it also measures CD; therefore CF measures AB, CD. Therefore CF is a common measure of AB, CD.

I say next that it is also the greatest. For, if CF is not the greatest common measure of AB, CD, some number which is greater than CF will measure the numbers AB, CD. Let such a number measure them, and let it be G. Now, since G measures CD, while CD measures BE, G also measures BE. But it also measures the whole BA; therefore it will also measure the remainder AE. But AE measures DF; therefore G will also measure DF. But it also measures the whole DC; therefore it will also measure the remainder CF, that is, the greater will measure the less, which is impossible. Therefore no number which is greater than CF will measure the numbers AB, CD; therefore CF is the greatest common measure of AB, CD. Q.E.D.

The following proposition gives the usual test for proportionality in numbers, that $a:b=c:d$ if and only if $ad=bc$. This is analogous to book VI, proposition 16, for line segments and rectangles, but naturally the proof is quite different.

Proposition 19. *If four numbers be proportional, the number produced from the first and fourth will be equal to the number produced from the second and third; and, if the number produced from the first and fourth be equal to that produced from the second and third, the four numbers will be proportional.*

Let A, B, C, D be four numbers in proportion, so that as A is to B, so is C to D; and let A by multiplying D make E; and let B by multiplying C make F; I say that E is equal to F. For let A by multiplying C make G. Since, then, A by multiplying C has made G, and by multiplying D has made E, the number A by multiplying the two numbers C, D has made G, E. Therefore, as C is to D, so is G to E. But, as C is to D, so is A to B; therefore also, as A is to B, so is G to E. Again, since A by multiplying C has made G, but, further, B has also by multiplying C made F, the two numbers A, B by multiplying a certain number C have made G, F. Therefore, as A is to B, so is G to F. But further, as A is to B, so is G to E also; therefore also, as G is to E, so is G to F. Therefore G has to each of the numbers E, F the same ratio; therefore E is equal to F.

Again, let E be equal to F; I say that, as A is to B, so is C to D. For, with the same construction, since E is equal to F, therefore, as G is to E, so is G to F. But, as G is to E, so is C to D, and, as G is to F, so is A to B. Therefore also, as A is to B, so is C to D. Q.E.D.

Proposition 20 shows that if a, b are the smallest numbers in the ratio $a:b$, then if $c:d=a:b$, we know that a and b each divide c and d respectively the same number of times. As consequences, Euclid proved in the following two propositions that relatively prime numbers are the least of those in the same ratio, and conversely.

Proposition 20. *The least numbers of those which have the same ratio with them measure those which have the same ratio the same number of times, the greater the greater and the less the less.*

For let CD, EF be the least numbers of those which have the same ratio with A, B; I say that CD measures A the same number of times that EF measures B [figure 2.1.3].

Now CD is not parts of A. For, if possible, let it be so; therefore EF is also the same parts of B that CD is of A. Therefore, as many parts of A as there are in CD, so many parts of B are there also in EF. Let CD be divided into the parts of A, namely, CG, GD,

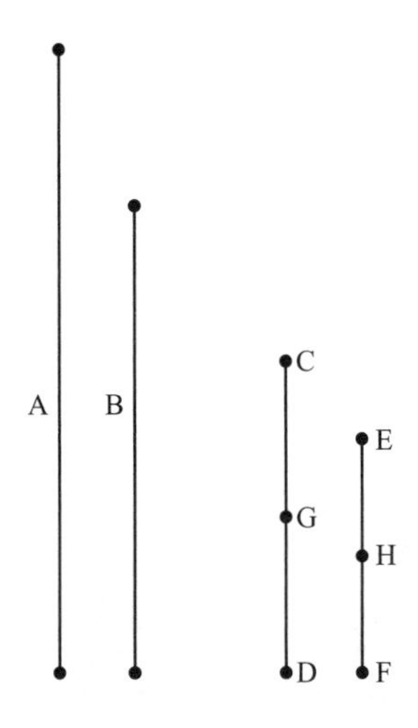

Figure 2.1.3.

and EF into the parts of B, namely, EH, HF; thus the multitude of CG, GD will be equal to the multitude of EH, HF.

Now, since the numbers CG, GD are equal to one another, and the numbers EH, HF are also equal to one another, while the multitude of CG, GD is equal to the multitude of EH, HF, therefore, as CG is to EH, so is GD to HF. Therefore also, as one of the antecedents is to one of the consequents, so will all the antecedents be to all the consequents. Therefore, as CG is to EH, so is CD to EF. Therefore CG, EH are in the same ratio with CD, EF, being less than they: which is impossible, for by hypothesis CD, EF are the least numbers of those which have the same ratio with them. Therefore CD is not parts of A; therefore it is a part of it. And EF is the same part of B that CD is of A; therefore CD measures A the same number of times that EF measures B. Q.E.D.

Proposition 21. *Numbers prime to one another are the least of those which have the same ratio with them.*

Proposition 22. *The least numbers of those which have the same ratio with them are prime to one another.*

The following three propositions, together with book 9, proposition 14, present Euclid's version of the fundamental theorem of arithmetic—that every number can be uniquely expressed as a product of prime numbers.

Proposition 30. *If two numbers by multiplying one another make some number, and any prime number measure the product, it will also measure one of the original numbers.*

Proposition 31. *Any composite number is measured by some prime number.*

Let A be a composite number; I say that A is measured by some prime number. For, since A is composite, some number will measure it. Let a number measure it, and let it be B. Now, if B is prime, what was enjoined will have been done. But if it is composite, some number will measure it. Let a number measure it, and let it be C. Then, since C

measures B, and B measures A, therefore C also measures A. And, if C is prime, what was enjoined will have been done. But if it is composite, some number will measure it. Thus, if the investigation be continued in this way, some prime number will be found which will measure the number before it, which will also measure A. For, if it is not found, an infinite series of numbers will measure the number A, each of which is less than the other, which is impossible in numbers. Therefore some prime number will be found which will measure the one before it, which will also measure A. Therefore any composite number is measured by some prime number. Q.E.D.

Proposition 32. *Any number either is prime or is measured by some prime number.*

Book 8

Book 8 is primarily concerned with numbers in continued proportion, that is, with sequences $a_1, a_2, \ldots, a_n$ such that $a_1 : a_2 = a_2 : a_3 = \ldots$. In modern terms, such a sequence is called a geometric progression. It is generally thought today that Archytas is responsible for much of the material in this book. In particular, proposition 8 is a generalization of a result due to Archytas and evidently emerged from his interest in music. The original result was that there is no mean proportional between two numbers whose ratio in lowest terms is equal to $(n+1) : n$. Recall that the ratio of the lengths of two strings whose sound is an octave apart is $2 : 1$. This ratio is the compound of the musical fifth and fourth, expressed by the ratios $4 : 3$ and $3 : 2$. Archytas's result then states that the octave cannot be divided into two equal musical intervals. Of course, in this case, the result is equivalent to the incommensurability of $\sqrt{2}$ with 1. But the result also shows that one cannot divide a whole tone, whose ratio of lengths is $9 : 8$, into two equal intervals.

Proposition 8. *If between two numbers there fall numbers in continued proportion with them, then, however many numbers fall between them in continued proportion, so many will also fall in continued proportion between the numbers which have the same ratio with the original numbers.*

Let the numbers C, D fall between the two numbers A, B in continued proportion with them, and let E be made in the same ratio to F as A is to B; I say that, as many numbers as have fallen between A, B in continued proportion, so many will also fall between E, F in continued proportion [figure 2.1.4]. For, as many as A, B, C, D are in multitude, let so many numbers G, H, K, L, the least of those which have the same ratio with A, C, D, B, be taken ; therefore the extremes of the G, L are prime to one another. Now, since A, C, D, B are in the same ratio with G, H, K, L, and the multitude of the numbers A, C, D, B is equal to the multitude of the numbers G, H, K, L, therefore, *ex aequali*,[7] as A is to B, so is G to L. But, as A is to B, so is E to F; therefore also, as G is to L, so is E to F. But G, L are prime, primes are also least, and the least numbers measure those which have the same ratio the same number of times, the greater the greater and the less the less,

[7] Definition 17 in *Elements* book 5: If $a:b=d:e$ and $b:c=e:f$, then the proportion $a:c=d:f$ is obtained *ex aequali*. This can be extended to any number of magnitude pairs.

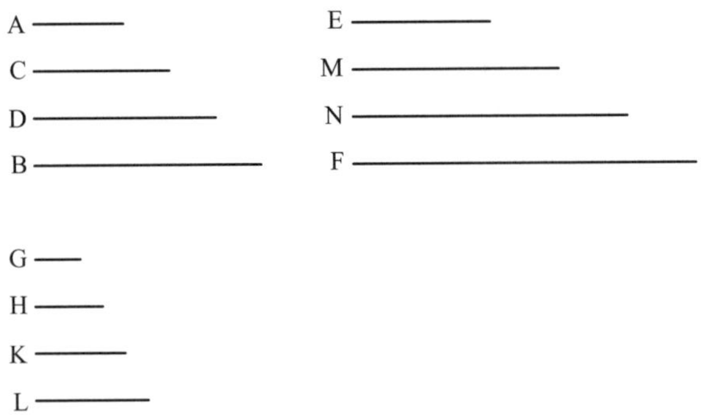

A ——— E ———
C ——— M ———
D ——— N ———
B ——— F ———

G ———
H ———
K ———
L ———

Figure 2.1.4.

that is, the antecedent the antecedent and the consequent the consequent. Therefore G measures E the same number of times as L measures F.

Next, as many times as G measures E, so many times let H, K, also measure M, N respectively; therefore G, H, K, L measure E, M, N, F the same number of times. Therefore G, H, K, L are in the same ratio with E, M, N, F. But G, H, K, L are in the same ratio with A, C, D, B; therefore A, C, D, B are also in the same ratio with E, M, N, F. But A, C, D, B are in continued proportion; therefore E, M, N, F are also in continued proportion. Therefore, as many numbers as have fallen between A, G in continued proportion with them, so many numbers have also fallen between E, F in continued proportion. Q.E.D.

The following propositions, along with several others in book 8, deal with further aspects of numbers in proportion.

Proposition 11. *Between two square numbers there is one mean proportional number, and the square has to the square the ratio duplicate of that which the side has to the side.*

Let A, B be square numbers, and C be the side of A, and D of B; I say that between A, B there is one mean proportional number, and A has to B the ratio duplicate of that which C has to D. For let C by multiplying D make E. Now, since A is a square and C is its side, therefore C by multiplying itself has made A. For the same reason also, D by multiplying itself has made B. Since, then, C by multiplying the numbers C, D has made A, E respectively, therefore, as C is to D, so is A to E. For the same reason also, as C is to D, so is E to B. Therefore also, as A is to E, so is E to B. Therefore between A, B there is one mean proportional number.

I say next that A also has to G the ratio duplicate of that which C has to D. For, since A, E, B are three numbers in proportion, therefore A has to B the ratio duplicate of that which A has to E. But, as A is to E, so is C to D. Therefore A has to B the ratio duplicate of that which the side C has to D. Q.E.D.

Proposition 12. *Between two cube numbers there are two mean proportional numbers, and the cube has to the cube the ratio triplicate of that which the side has to the side.*

Proposition 22. *If three numbers be in continued proportion, and the first be square, the third will also be square.*

Book 9

Book 9 concludes the number theory section of the *Elements* and contains several propositions of particular importance. In particular, as noted, proposition 14 completes Euclid's version of the fundamental theorem of arithmetic, while proposition 20 shows that the sequence of prime numbers has no end.

Proposition 8. *If as many numbers as we please beginning from a unit be in continued proportion, the third from the unit will be square, as will also those which successively leave out one; the fourth will be cube, as will also all those which leave out two; and the seventh will be at once cube and square, as will also those which leave out five.*

Let there be as many numbers as we please, A, B, C, D, E, F, beginning from a unit and in continued proportion; I say that B, the third from the unit, is square, as are also all those which leave out one; C, the fourth, is cube, as are also all those which leave out two; and F, the seventh, is at once cube and square, as are also all those which leave out five. For since, as the unit is to A, so is A to B, therefore the unit measures the number A the same number of times that A measures B. But the unit measures the number A according to the units in it; therefore A also measures B according to the units in A. Therefore A by multiplying itself has made B; therefore B is square. And, since B, C, D are in continued proportion, and B is square, therefore D is also square. For the same reason F is also square. Similarly we can prove that all those which leave out one are square.

I say next that C, the fourth from the unit, is cube, as are also all those which leave out two. For since, as the unit is to A, so is B to C, therefore the unit measures the number A the same number of times that B measures C. But the unit measures the number A according to the units in A; therefore B also measures C according to the units in A. Therefore A by multiplying B has made C. Since then A by multiplying itself has made B, and by multiplying B has made C, therefore C is cube. And, since C, D, E, F are in continued proportion, and C is cube, therefore F is also cube. But it was also proved square; therefore the seventh from the unit is both cube and square. Similarly we can prove that all the numbers which leave out five are also both cube and square. Q.E.D.

Proposition 14. *If a number be the least that is measured by prime numbers, it will not be measured by any other prime number except those originally measuring it.*

For let the number A be the least that is measured by the prime numbers B, C, D; I say that A will not be measured by any other prime number except B, C, D. For, if possible, let it be measured by the prime number E, and let E not be the same with any one of the numbers B, C, D. Now, since E measures A, let it measure it according to F; therefore E by multiplying F has made A. And A is measured by the prime numbers B, C, D. But, if two numbers by multiplying one another make some number, and any prime number measure the product, it will also measure one of the original numbers; therefore B, C, D will measure one of the numbers E, F. Now they will not measure E; for E is prime and not the same with any one of the numbers B, C, D. Therefore they will measure F, which is less than A, which is impossible, for A is by hypothesis the least number measured by B, C, D. Therefore no prime number will measure A except B, C, D. Q.E.D.

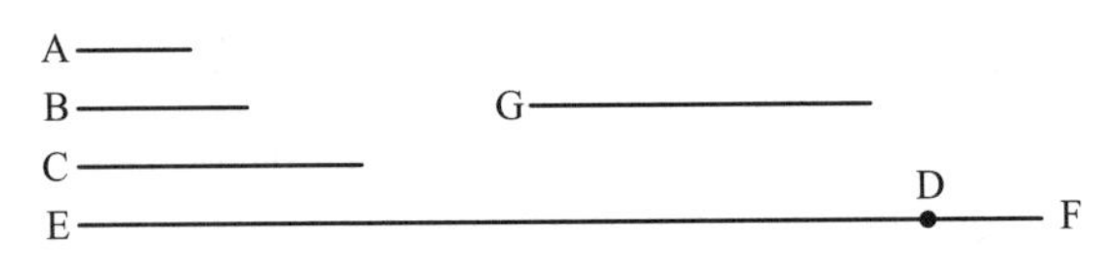

Figure 2.1.5.

Proposition 20. *Prime numbers are more than any assigned multitude of prime numbers.*

Let A, B, C be the assigned prime numbers; I say that there are more prime numbers than A, B, C. For let the least number measured by A, B, C be taken, and let it be DE; let the unit DF be added to DE [figure 2.1.5].

Then EF is either prime or not. First, let it be prime; then the prime numbers A, B, C, EF have been found which are more than A, B, C.

Next, let EF not be prime; therefore it is measured by some prime number. Let it be measured by the prime number G. I say that G is not the same with any of the numbers A, B, C. For, if possible, let it be so. Now A, B, C measure DE; therefore G also will measure DE. But it also measures EF. Therefore G, being a number, will measure the remainder, the unit DF, which is absurd. Therefore G is not the same with any one of the numbers A, B, C. And by hypothesis it is prime. Therefore the prime numbers A, B, C, G have been found which are more than the assigned multitude of A, B, C. Q.E.D.

The following three propositions, each of whose proof is straightforward, show examples of some of Euclid's very elementary material on even and odd numbers.

Proposition 32. *Each of the numbers which are continually doubled beginning from a dyad is even-times even only.*

Proposition 33. *If a number has its half odd, it is even-times odd only.*

Proposition 34. *If a number neither be one of those which are continually doubled from a dyad, not have its half odd, it is both even-times even and even-times odd.*

Proposition 35 in effect shows how to calculate the sum of a geometric progression. If we represent the sequence of numbers in "continued proportion" by $a, ar, ar^2, \ldots, ar^n$ and the sum of "all those before [the last]" by S_n, then Euclid's result states that $(ar^n - a) : S_n = (ar - a) : a$. In modern terms, this result can be written as $S_n = \frac{a(r^n-1)}{r-1}$.

Proposition 35. *If as many numbers as we please be in continued proportion, and there be subtracted, from the second and the last, numbers equal to the first, then, as the excess of the second is to the first, so will the excess of the last be to all those before it.*

Let there be as many numbers as we please in continued proportion, A, BC, D, EF, beginning from A as least, and let there be subtracted from BC and EF the numbers BG, FH, each equal to A; I say that, as GC is to A, so is EH to A, BC, D [figure 2.1.6].

For let FK be made equal to BC, and FL equal to D. Then, since FK is equal to BC, and of these the part FH is equal to the part BG, therefore the remainder HK is equal to the remainder GC. And since, as EF is to D, so is D to BC, and BC to A, while D is equal to FL, BC to FK, and A to FH, therefore, as EF is to FL, so is LF to FK, and FK

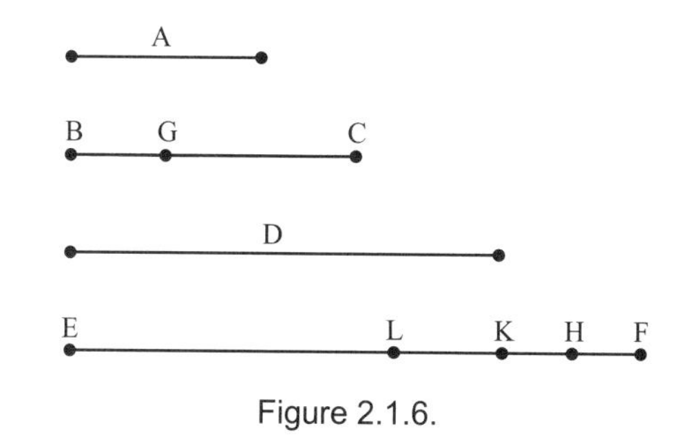

Figure 2.1.6.

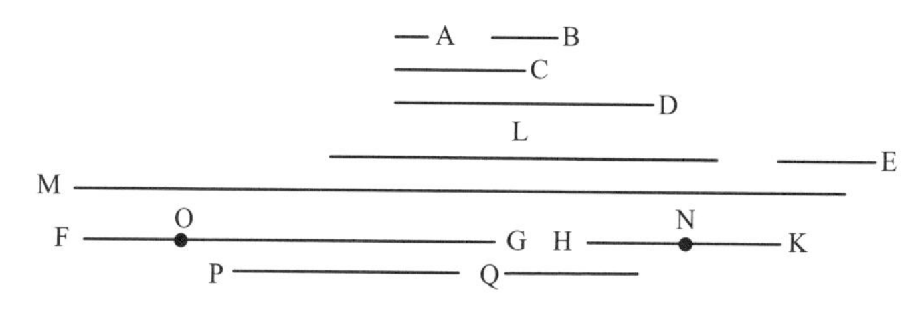

Figure 2.1.7.

to FH. *Separando*,[8] as EL is to LF, so is LK to FK, and KH to FH. Therefore also, as one of the antecedents is to one of the consequents, so are all the antecedents to all the consequents; therefore, as KH is to FH, so are LE, LK, KH to LF, FK, HF. But KH is equal to CG, FH to A, and LF, FK, HF to D, BC, A; therefore, as CG is to A, so is EH to D, BC, A. Therefore, as the excess of the second is to the first, so is the excess of the last to all those before it. Q.E.D.

The final proposition of book 9 shows how to find perfect numbers. In modern notation, the result states that if the sum of the terms of the sequence $1, 2, 2^2, \ldots, 2^n$ is prime, then the product of that sum and 2^n is perfect. For example, since $1+2+2^2=7$ is prime, we know that $7 \times 4 = 28$ is perfect. Interestingly, the Greeks knew only four perfect numbers—6, 28, 496, and 8128—corresponding to $n=1, 2, 4, 6$. It is fascinating that Euclid devoted the culminating theorem of the number theory books to the study of a class of numbers of which only four were known.

Proposition 36. *If as many numbers as we please beginning from a unit be set out continuously in double proportion, until the sum of all becomes prime, and if the sum multiplied into the last make some number, the product will be perfect.*

For let as many numbers as we please, A, B, C, D, beginning from a unit be set out in double proportion until the sum of all becomes prime, let E be equal to the sum, and let E by multiplying D make FG; I say that FG is perfect [figure 2.1.7].

[8] This is defined in book 5, definition 15: If $a:b=c:d$, then the proportion $a:a-b=c:c-d$ is obtained *separando*.

For, however many A, B, C, D are in multitude, let so many E, HK, L, M be taken in double proportion beginning from E; therefore, *ex aequali*, as A is to D, so is E to M. Therefore the product of E, D is equal to the product of A, M. And the product of E, D is FG; therefore the product of A, M is also FG. Therefore A by multiplying M has made FG; therefore M measures FG according to the units in A. And A is a dyad; therefore FG is double of M. But M, L, HK, E are continuously double of each other; therefore E, HK, L, M, FG are continuously proportional in double proportion.

Now let there be subtracted from the second HK and the last FG the numbers HN, FO, each equal to the first E; therefore, as the excess of the second is to the first, so is the excess of the last to all those before it [9:35]. Therefore, as NK is to E, so is OG to M, L, KH, E. And NK is equal to E; therefore OG is also equal to M, L, HK, E. But FO is also equal to E, and E is equal to A, B, C, D and the unit. Therefore the whole FG is equal to E, HK, L, M and A, B, C, D and the unit; and it is measured by them.

I say also that FG will not be measured by any other number except A, B, C, D, E, HK, L, M and the unit. For, if possible, let some number P measure FG, and let P not be the same with any of the numbers A, B, C, D, E, HK, L, M. And, as many times as P measures FG, so many units let there be in Q; therefore Q by multiplying P has made FG. But, further, E has also by multiplying D made FG; therefore, as E is to Q, so is P to D. And, since A, B, C, D are continuously proportional beginning from a unit, therefore D will not be measured by any other number except A, B, C. And, by hypothesis, P is not the same with any of the numbers A, B, C; therefore P will not measure D. But, as P is to D, so is E to Q; therefore neither does E measure Q. And E is prime; and any prime number is prime to any number which it does not measure. Therefore E, Q are prime to one another.

But primes are also least and the least numbers measure those which have the same ratio the same number of times, the antecedent the antecedent and the consequent the consequent; and, as E is to Q, so is P to D; therefore E measures P the same number of times that Q measures D. But D is not measured by any other number except A, B, C; therefore Q is the same with one of the numbers A, B, C. Let it be the same with B. And, however many B, C, D are in multitude, let so many E, HK, L be taken beginning from E. Now E, HK, L are in the same ratio with B, C, D; therefore, *ex aequali*, as B is to D, so is E to L. Therefore the product of B, L is equal to the product of D, E. But the product of D, E is equal to the product of Q, P; therefore the product of Q, P is also equal to the product of B, L. Therefore, as Q is to B, so is L to P. And Q is the same with B; therefore L is also the same with P; which is impossible, for by hypothesis P is not the same with any of the numbers set out. Therefore no number will measure FG except A, B, C, D, E, HK, L, M and the unit. And FG was proved equal to A, B, C, D, E, HK, L, M and the unit; and a perfect number is that which is equal to its own parts; therefore FG is perfect. Q.E.D.

2.2 Nicomachus, *Introduction to Arithmetic*

Most of book 1 of Nicomachus's *Arithmetic* is devoted to the classification of integers and their relations. Many of his concepts come from Euclid's books on number theory, but his exposition is very different; instead of proofs, he provides numerical examples.

He was clearly writing for a different audience. For example, as an aspect of his classification, he considers the even-times even, the even-times odd, and the odd-times even integers as subclasses of the even integers. He also deals with prime and composite numbers, pairs of numbers that are relatively prime, and abundant, deficient, and perfect numbers.

Book 1

Chapter 7

[1] Number is limited multitude or a combination of units or a flow of quantity made up of units; and the first division of number is even and odd. [2] The even is that which can be divided into two equal parts without a unit intervening in the middle; and the odd is that which cannot be divided into two equal parts because of the aforesaid intervention of a unit. [3] Now this is the definition after the ordinary conception; by the Pythagorean doctrine, however, the even number is that which admits of division into the greatest and the smallest parts at the same operation, greatest in size and smallest in quantity, in accordance with the natural contrariety of these two *genera*; and the odd is that which does not allow this to be done to it, but is divided into two unequal parts.

Chapter 8

[3] By subdivision of the even, there are the even-times even, the odd-times even, and the even-times odd. The even-times even and the even-times odd are opposite to one another, like extremes, and the odd-times even is common to them both like a mean term. [4] Now the even-times even is a number which is itself capable of being divided into two equal parts, in accordance with the properties of its genus, and with each of its parts similarly capable of division, and again in the same way each of their parts divisible into two equals until the division of the successive subdivisions reaches the naturally indivisible unit.[9] [5] Take for example 64; one half of this is 32, and of this 16, and of this the half is 8, and of this 4, and of this 2, and then finally unity is half of the latter, and this is naturally indivisible and will not admit of a half. [8] There is a method of producing the even-times even, so that none will escape, but all successively fall under it, if you do as follows: [9] As you proceed from unity, as from a root, by the double ratio to infinity, as many terms as there are will all be even-times even, and it is impossible to find others besides these; for instance, 1, 2, 4, 8, 16, 32, 64, 128, 256, 512, . . . [12] It is the property of all these terms when they are added together successively to be equal to the next in the series, lacking one unit, so that of necessity their summation in any way whatsoever will be an odd number, for that which fails by a unit of being equal to an even number is odd.

[9] Euclid's definition of even-times even in book 7, definition 8, is different. According to Euclid, an even-times even number is one that is measured by an even number according to an even number. So, for example, 12 is even-times even for Euclid, since it can be expressed as 2×6.

Chapter 9

[1] The even-times odd number is one which is by its genus itself even, but is specifically opposed to the aforesaid even-times even. It is a number of which, though it admits of the division into two equal halves, after the fashion of the genus common to it and even-times even, the halves are not immediately divisible into two equals, for example, 6, 10, 14, 18, 22, 26, and the like; for after these have been divided their halves are found to be indivisible.[10] [4] This number is produced from the series beginning with unity, with a difference of 2, namely, the odd numbers, set forth in proper order as far as you like and then multiplied by 2. The numbers produced would be, in order, these: 6, 10, 14, 18, 22, 26, 30, and so on, as far as you care to proceed. The greater terms always differ by 4 from the next smaller ones, the reason for which is that their original basic forms, the odd numbers, exceed one another by 2 and were multiplied by 2 to make this series, and 2 times 2 makes 4.

Chapter 10

[1] The odd-times even number is the one which displays the third form of the even, belonging in common to both the previously mentioned species like a single mean between two extremes, for in one respect it resembles the even-times even, and in another the even-times odd, and that property wherein it varies from the one it shares with the other, and by that property which it shares with the one it differs from the other.[11] [2] The odd-times even number is an even number which can be divided into two equal parts, whose parts also can so be divided, and sometimes even the parts of its parts, but it cannot carry the division of its parts as far as unity. Such numbers are 24, 28, 40; for each of these has its own half and indeed the half of the half, and sometimes one is found among them that will allow the halving to be carried even farther among its parts. There is none, however, that will have its parts divisible into halves as far as the naturally indivisible unit. [3] Now in admitting more than one division, the odd-times even is like the even-times even and unlike the even-times odd; but in that its subdivision never ends with unity, it is like the even-times odd and unlike the even-times even.

Chapter 11

[1] Again, while the odd is distinguished over against the even in classification and has nothing in common with it, since the latter is divisible into equal halves and the former is not thus divisible, nevertheless there are found three species of the odd, differing from one another, of which the first is called the prime and incomposite, that which is opposed to it the secondary and composite, and that which is midway between both of these and is viewed as a mean among extremes, namely, the variety which, in itself, is secondary and composite, but relatively is prime and incomposite. [2] Now the first species, the prime and incomposite, is found whenever an odd number admits of no other factor save the

[10] Again, Euclid's definition differs (book 7, definition 9); it says that an even-times odd number is one that is measured by an even number according to an odd number. So since $12 = 4 \times 3$, 12 is such a number. Thus, for Euclid, unlike for Nicomachus, an even number may fall into more than one class.

[11] Euclid does not have this definition at all.

one with the number itself as denominator, which is always unity; for example, 3, 5, 7, 11, 13, 17, 19, 23, 29, 31. None of these numbers will by any chance be found to have a fractional part with a denominator different from the number itself, but only the one with this as denominator, and this part will be unity in each case; for 3 has only a third part, which has the same denominator as the number and is of course unity, 5 a fifth, 7 a seventh, and 11 only an eleventh part, and in all of them these parts are unity.

Chapter 12

[1] The secondary, composite number is an odd number, indeed, because it is distinguished as a member of this same class, but it has no elementary quality, for it gets its origin by the combination of something else. For this reason it is characteristic of the secondary number to have, in addition to the fractional part with the number itself as denominator, yet another part or parts with different denominators, the former always, as in all cases, unity, the latter never unity, but always either that number or those numbers by the combination of which it was produced. For example, 9, 15, 21, 25, 27, 33, 35, 39; each one of these is measured by unity, as other numbers are, and like them has a fractional part with the same denominator as the number itself, by the nature of the class common to them all; but by exception and more peculiarly they also employ a part, or parts, with a different denominator; 9, in addition to the ninth part, has a third part besides; 15 a third and a fifth besides a fifteenth; 21 a seventh and a third besides a twenty-first, and 25, in addition to the twenty-fifth, which has as a denominator 25 itself, also a fifth, with a different denominator.

Chapter 13

[1] Now while these two species of the odd are opposed to each other, a third one is conceived of between them, deriving, as it were, its specific form from them both, namely, the number which is in itself secondary and composite, but relatively to another number is prime and incomposite. This exists when a number, in addition to the common measure, unity, is measured by some other number and is therefore able to admit of a fractional part, or parts, with denominator other than the number itself, as well as the one with itself as denominator. When this is compared with another number of similar properties, it is found that it cannot be measured by a measure common to the other, nor does it have a fractional part with the same denominator as those in the other. As an illustration, let 9 be compared with 25. Each in itself is secondary and composite, but relatively to each other they have only unity as a common measure, and no factors in them have the same denominator, for the third part in the former does not exist in the latter nor is the fifth part in the latter found in the former. . . .

[10] We shall now investigate how we may have a method of discerning whether numbers are prime and incomposite, or secondary and composite, relatively to each other, since of the former unity is the common measure, but of the latter some other number also besides unity; and what this number is. [11] Suppose there be given us two odd numbers and someone sets the problem and directs us to determine whether they are prime and incomposite relatively to each other or secondary and composite, and if they

are secondary and composite, what number is their common measure. We must compare the given numbers and subtract the smaller from the larger as many times as possible; then after this subtraction, subtract in turn from the other as many times as possible; for this changing about and subtraction from one and the other in turn will necessarily end either in unity or in some one and the same number, which will necessarily be odd. [12] Now when the subtractions terminate in unity they show that the numbers are prime and incomposite relatively to each other; and when they end in some other number, odd in quantity and twice produced, then say that they are secondary and composite relatively to each other, and that their common measure is that very number which twice appears. For example, if the given numbers were 23 and 45, subtract 23 from 45, and 22 will be the remainder; subtracting this from 23, the remainder is 1, subtracting this from 22 as many times as possible you will end with unity. Hence they are prime and incomposite to one another, and unity, which is the remainder, is their common measure. [13] But if one should propose other numbers, 21 and 49, I subtract the smaller from the larger and 28 is the remainder. Then again I subtract the same 21 from this, for it can be done, and the remainder is 7. This I subtract in turn from 21 and 14 remains; from which I subtract 7 again, for it is possible, and 7 will remain. But it is not possible to subtract 7 from 7; hence the termination of the process with a repeated 7 has been brought about, and you may declare the original numbers 21 and 49 secondary and composite relatively to each other, and 7 their common measure in addition to the universal unit.

Chapter 14

[1] To make again a fresh start, of the simple even numbers, some are superabundant, some deficient, like extremes set over against each other, and some are intermediary between them and are called perfect. . . . [3] Now the superabundant number is one which has, over and above the factors which belong to it and fall to its share, others in addition, just as if an animal should be created with too many parts or limbs, with ten tongues . . . or a hundred hands, or too many fingers on one hand. . . . If, when all the factors in a number are examined and added together in one sum, it proves upon investigation that the number's own factors exceed the number itself; this is called a superabundant number, for it oversteps the symmetry which exists between the perfect and its own parts. Such are 12, 24, and certain others, for 12 has a half, 6, a third, 4, a fourth, 3, a sixth, 2, and a twelfth, 1, which added together make 16, which is more than the original 12; its parts, therefore are greater than the whole itself. [4] And 24 has a half, a third, fourth, sixth, eighth, twelfth, and twenty-fourth, which are 12, 8, 6, 4, 3, 2, 1. Added together they make 36, which compared to the original number, 24, is found to be greater than it, although made up solely of its factors. Hence in this case also the parts are in excess of the whole.

Chapter 15

[1] The deficient number is one which has qualities the opposite of those pointed out, and whose factors added together are less in comparison than the number itself. It is as if . . . one should be one-handed, or have fewer than five fingers on one hand, or lack a tongue, or some such member. Such a one would be called deficient and so to speak

maimed, after the peculiar fashion of the number whose factors are less than itself, such as 8 or 14. For 8 has the factors half, fourth, and eighth, which are 4, 2, and 1, and added together they make 7, and less than the original number. The parts, therefore, fall short of making up the whole. [2] Again, 14 has a half, a seventh, a fourteenth, 7, 2, and 1, respectively; and all together they make 10, less than the original number. So this number also is deficient in its parts, with respect to making up the whole out of them.

Chapter 16

[1] While these two varieties are opposed after the manner of extremes, the so-called perfect number appears as a mean, which is discovered to be in the realm of equality, and neither makes its parts greater than itself, added together, nor shows itself greater than its parts, but is always equal to its own parts. For the equal is always conceived of as in the mid-ground between greater and less, and is, as it were, moderation between excess and deficiency, and that which is in tune, between pitches too high and too low. [2] Now when a number, comparing with itself the sum and combination of all the factors whose presence it will admit, neither exceeds them in multitude nor is exceeded by them, then such a number is properly said to be perfect, as one which is equal to its own parts. Such numbers are 6 and 28; for 6 has the factors half, third, and sixth, 3, 2, and 1, respectively, and these added together make 6 and are equal to the original number, and neither more nor less. Twenty-eight has the factors half, fourth, seventh, fourteenth, and twenty-eighth, which are 14, 7, 4, 2, and 1; these added together make 28, and so neither are the parts greater than the whole, nor the whole greater than the parts, but their comparison is in equality, which is the peculiar quality of the perfect number. [3] It comes about that even as fair and excellent things are few and easily enumerated, while ugly and evil ones are widespread, so also the superabundant and deficient numbers are found in great multitude and irregularly placed—for the method of their discovery is irregular—but the perfect numbers are easily enumerated and arranged with suitable order; for only one is found among the units, 6, only one other among the tens, 28, and a third in the rank of the hundreds, 496 alone, and a fourth within the limits of the thousands, that is, below ten thousand, 8128. And it is their accompanying characteristic to end alternately in 6 or 8, and always to be even.[12] [4] There is a method of producing them, neat and unfailing, which neither passes by any of the perfect numbers nor fails to differentiate any of those that are not such, which is carried out in the following way. You must set forth the even-times even numbers from unity, advancing in order in one line, as far as you please: 1, 2, 4, 8, 16, 32, 64, 128, 256, 512, 1024, 2048, 4096, ... Then you must add them together, one at a time, and each time you make a summation observe the result to see what it is. If you find that it is a prime, incomposite number, multiply it by the quantity of the last number added, and the result will always be a perfect number. If, however, the result is secondary and composite, do not multiply, but add the next and observe again what the resulting number is; if it is secondary and composite, again pass it by and do not multiply; but add the next; but if it is prime and

[12] Nicomachus was wrong in two of his apparent claims here. The fifth perfect number is 33,550,336, not a five-digit number, while the sixth is 8,589,869,056, which, like its predecessor, ends in 6. Although it has long been suspected that all perfect numbers are even, neither proof nor counterexample has yet been produced.

incomposite, multiply it by the last term added, and the result will be a perfect number; and so on to infinity. In similar fashion you will produce all the perfect numbers in succession, overlooking none. For example, to 1 I add 2, and observe the sum, and find that it is 3, a prime and incomposite number in accordance with our previous demonstrations; for it has no factor with denominator different from the number itself, but only that with denominator agreeing. Therefore I multiply it by the last number to be taken into the sum, that is, 2; I get 6 and this I declare to be the first perfect number in actuality, and to have those parts which are beheld in the numbers of which it is composed. For it will have unity as the factor with denominator the same as itself, that is, its sixth part; and 3 as the half, which is seen in 2, and conversely 2 as its third part. [5] Twenty-eight likewise is produced by the same method when another number, 4, is added to the previous ones. For the sum of the three, 1, 2, and 4, is 7, and is found to be prime and incomposite, for it admits only the factor with denominator like itself, the seventh part. Therefore I multiply it by the quantity of the term last taken into the summation, and my result is 28, equal to its own parts, and having its factors derived from the numbers already adduced, a half corresponding to 2; a fourth, to 7; a seventh, to 4; a fourteenth to offset the half; and a twenty-eighth, in accordance with its own nomenclature, which is 1 in all numbers. [6] When these have been discovered, 6 among the units and 28 in the tens, you must do the same to fashion the next. [7] Again add the next number, 8, and the sum is 15. Observing this, I find that we no longer have a prime and incomposite number, but in addition to the factor with denominator like the number itself, it has also a fifth and a third, with unlike denominators. Hence I do not multiply it by 8, but add the next number, 16, and 31 results. As this is a prime, incomposite number, of necessity it will be multiplied, in accordance with the general rule of the process, by the last number added, 16, and the result is 496, in the hundreds; and then comes 8,128 in the thousands, and so on, as far as it is convenient for one to follow.

The final seven chapters of book 1 are devoted to an elaborate tenfold classification scheme for naming ratios of unequal numbers, a scheme that probably had its origin in early music theory. The scheme was in common use in medieval and Renaissance arithmetic texts and is sometimes found in early printed editions of Euclid's *Elements*. We have included this in the version presented by Boethius below. We continue here with Nicomachus's discussion of figurate numbers, beginning in book 2, chapter 8.

Book 2

Chapter 8

[1] Now a triangular number is one which, when it is analyzed into units, shapes into triangular form the equilateral placement of its parts in a plane. 3, 6, 10, 15, 21, 28, and so on, are examples of it; for their regular formations, expressed graphically, will be at once triangular and equilateral. As you advance you will find that such a numerical series as far as you like takes the triangular form, if you put as the most elementary form the one that arises from unity, so that unity may appear to be potentially a triangle, and 3 the first actually. [2] Their sides will increase by the successive numbers, for the side of the

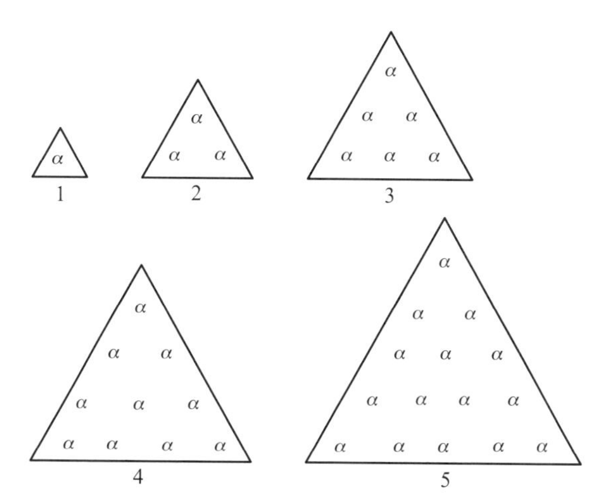

Figure 2.2.1.

one potentially first is unity; that of the one actually first, that is, 3, is 2; that of 6, which is actually second, 3; that of the third, 4; the fourth, 5; the fifth, 6; and so on. [3] The triangular number is produced from the natural series of number set forth in a line, and by the continued addition of successive terms, one by one, from the beginning; for by the successive combinations and additions of another term to the sum, the triangular numbers in regular order are completed. For example, from this natural series, 1, 2, 3, 4, 5, 6, 7, 8, 9, 10, 11, 12, 13, 14, 15, I take the first term and have the triangular number which is potentially first, 1 [figure 2.2.1, 1]; then adding the next term I get the triangle actually first, for 2 plus 1 equals 3. In its graphic representation it is thus made up: Two units, side by side, are set beneath one unit, and the number three is made a triangle [figure 2.2.1, 2]. Then when next after these the following number, 3, is added, simplified into units, and joined to the former, it gives 6, the second triangle in actuality, and furthermore, it graphically represents this number [figure 2.2.1, 3]. Again, the number that naturally follows, 4, added in and set down below the former, reduced to units, gives the one in order next after the aforesaid, 10, and takes a triangular form [figure 2.2.1, 4]. 5, after this, then 6, then 7, and all the numbers in order are added, so that regularly the sides of each triangle will consist of as many numbers as have been added from the natural series to produce it [figure 2.2.1, 5].

Chapter 9

[1] The square is the next number after this, which shows us no longer 3, like the former, but 4 angles in its graphic representation, but is none the less equilateral. Take, for example, 1, 4, 9, 16, 25, 36, 49, 64, 81, 100; for the representations of these numbers are equilateral, square figures, as here shown; and it will be similar as far as you wish to go [figure 2.2.2]. [2] It is true of these numbers, as it was also of the preceding, that the advance in their sides progresses with the natural series. The side of the square potentially first, 1, is 1; that of 4, the first in actuality, 2; that of 9, actually the second, 3; that of 16, the next, actually the third, 4; that of the fourth, 5; of the fifth, 6, and so

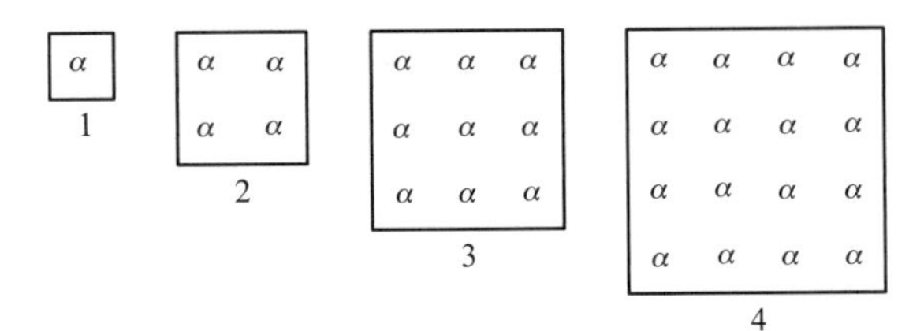

Figure 2.2.2.

on in general with all that follow. [3] This number also is produced if the natural series is extended in a line, increasing by 1, and no longer the successive numbers are added to the numbers in order, as was shown before, but rather all those in alternate places, that is, the odd numbers. For the first, 1, is potentially the first square; the second, 1 plus 3, is the first in actuality; the third, 1 plus 3 plus 5, is the second in actuality; the fourth, 1 plus 3 plus 5 plus 7, is the third in actuality; the next is produced by adding 9 to the former numbers, the next by the addition of 11, and so on. [4] In these cases, also, it is a fact that the side of each consists of as many units as there are numbers taken into the sum to produce it.

Chapter 10

[1] The pentagonal number is one which likewise upon its resolution into units and depiction as a plane figure assumes the form of an equilateral pentagon. 1, 5, 12, 22, 35, 51, 70, and analogous numbers are examples. [2] Each side of the first actual pentagon, 5, is 2, for 1 is the side of the pentagon potentially first, 1; 3 is the side of 12, the second of those listed; 4, that of the next, 22; 5, that of the next in order, 35, and 6 of the succeeding one, 51, and so on. In general the side contains as many units as are the numbers that have been added together to produce the pentagon, chosen out of the natural arithmetical series set forth in a row. For in a like and similar manner, there are added together to produce the pentagonal numbers the terms beginning with 1 to any extent whatever that are two places apart, that is, those that have a difference of 3. Unity is the first pentagon, potentially, and is thus depicted [figure 2.2.3, 1]. 5, made up of 1 plus 4, is the second, similarly represented [figure 2.2.3, 2]. 12, the third, is made up out of the two former numbers with 7 added to them, so that it may have 3 as a side, as three numbers have been added to make it. Similarly the preceding pentagon, 5, was the combination of two numbers and had 2 as its side. The graphic representation of 12 is below [figure 2.2.3, 3]. The other pentagonal numbers will be produced by adding together one after another in due order the terms after 7 that have the difference 3, as, for example, 10, 13, 16, 19, 22, 25, and so on. The pentagons will be 22, 35, 51, 70, 92, 117, and so forth.

Chapter 11

[1] The hexagonal, heptagonal, and succeeding numbers will be set forth in their series by following the same process, if from the natural series of number there be set forth series with their differences increasing by 1. For as the triangular number was produced

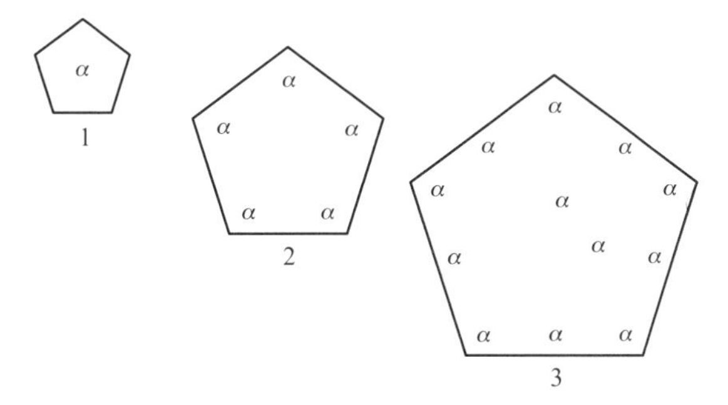

Figure 2.2.3.

by admitting into the summation the terms that differ by 1 and do not pass over any in the series; as the square was made by adding the terms that differ by 2 and are one place apart, and the pentagon similarly by adding terms with a difference of 3 and two places apart (and we have demonstrated these, by setting forth examples both of them and of the polygonal numbers made from them), so likewise the hexagons will have as their root-numbers those which differ by 4 and are three places apart in the series, which added together in succession will produce the hexagons. For example, 1, 5, 9, 13, 17, 21, and so on; so that the hexagonal numbers produced will be 1, 6, 15, 28, 45, 66, and so on, as far as one wishes to go. [2] The heptagonal, which follow these, have as their root-numbers terms differing by 5 and four places apart in the series, like 1, 6, 11, 16, 21, 26, 31, 36, and so on. The heptagons that thus arise are 1, 7, 18, 34, 55, 81, 112, 148 and so forth. [3] The octagonals increase after the same fashion, with a difference of 6 in their root-numbers and corresponding variation in the total constitution. [4] In order that, as you survey all cases, you may have a rule generally applicable, note that the root-numbers of any polygonal differ by 2 less than the number of the angles shown by the name of the polygonal—that is, by 1 in the triangle, 2 in the square, 3 in the pentagon, 4 in the hexagon, 5 in the heptagon, and so on, with similar increase.

Chapter 12

[1] Concerning the nature of plane polygonals this is sufficient for a first introduction. That, however, the doctrine of these numbers is to the highest degree in accord with their geometrical representation, and not out of harmony with it, would be evident, not only from the graphic representation in each case, but also from the following: Every square figure diagonally divided is resolved into two triangles and every square number is resolved into two consecutive triangular numbers, and hence is made up of two successive triangular numbers. For example, 1, 3, 6, 10, 15, 21, 28, 36, 45, 55, and so on, are triangular numbers and 1, 4, 9, 16, 25, 36, 49, 64, 81, 100, squares. [2] If you add any two consecutive triangles that you please, you will always make a square, and hence, whatever square you resolve, you will be able to make two triangles of it. Again, any triangle joined to any square figure makes a pentagon, for example, the triangle 1 joined with the square 4 makes the pentagon 5; the next triangle, 3, of course, with 9, the next square, makes the pentagon 12; the next, 6, with the next square, 16, gives the next pentagon, 22; 10 and 25 give 35; and so on.

[3] Similarly, if the triangles are added to the pentagons, following the same order, they will produce the hexagonals in due order, and again the same triangles with the latter will make the heptagonals in order, the octagonals after the heptagonals, and so on to infinity. [4] To remind us, let us set forth rows of the polygonals, written in parallel lines, as follows: The first row, triangles, the next squares, after them pentagonals, then hexagonals, then heptagonals, then if one wishes, the succeeding polygonals.

Triangles	1	3	6	10	15	21	28	36	45	55
Squares	1	4	9	16	25	36	49	64	81	100
Pentagonals	1	5	12	22	35	51	70	92	117	145
Hexagonals	1	6	15	28	45	66	91	120	153	190
Heptagonals	1	7	18	34	55	81	112	148	189	235

[5] In general, you will find that the squares are the sum of the triangles above those that occupy the same place in the series, plus the numbers of that same class in the next place back; for example, 4 equals 3 plus 1, 9 equals 6 plus 3, 16 equals 10 plus 6, 25 equals 15 plus 10, 36 equals 21 plus 15, and so on. The pentagons are the sum of the squares above them in the same place in the series, plus the elementary triangles that are one place further back in the series; for example, 5 equals 4 plus 1, 12 equals 9 plus 3, 22 equals 16 plus 6, 35 equals 25 plus 10, and so on. . . . [8] Naturally, then, the triangle is the element of the polygon both in figures and in numbers, and we say this because in the table, reading either up and down or across, the successive numbers in the rows are discovered to have as differences the triangles in regular order.

Chapter 13

[1] From this it is easy to see what the solid number is and how its series advances with equal sides; for the number which, in addition to the two dimensions contemplated in graphic representation in a plane, length and breadth, has a third dimension, which some call depth, others thickness, and some height, that number would be a solid number, extended in three directions and having length, depth, and breadth. . . . [6] Among numbers, each linear number increases from unity, as from a point, as for example, 1, 2, 3, 4, 5, and successive numbers to infinity; and from these same numbers, which are linear and extended in one direction, combined in no random manner, the polygonal and plane numbers are fashioned—the triangles by the combination of root-numbers immediately adjacent, the square by adding every other term, the pentagons every third term, and so on. [7] In exactly the same way, if the plane polygonal numbers are piled one upon the other and as it were built up, the pyramids that are akin to each of them are produced, the triangular pyramid from the triangles, the square pyramid from the squares, the pentagonal from the pentagons, the hexagonal from the hexagons, and so on throughout. [8] The pyramids with a triangular base, then, in their proper order, are these: 1, 4, 10, 20, 35, 56, 84, and so on; and their origin is the piling up of the triangular numbers one upon the other, first 1, then 1, 3, then 1, 3, 6, then 10 in addition to these, and next 15 together with the foregoing, then 21 besides these, next 28, and so on to infinity.

Chapter 14

[1] The next pyramids in order are those with a square base which rise in this shape to one and the same point. These are formed in the same way as the triangular pyramids of which we have just spoken. For if I extend in series the square numbers in order beginning with unity, thus, 1, 4, 9, 16, 25, 36, 49, 64, 81, 100, and again set the successive terms, as in a pile, one upon the other in the dimension height, when I put 1 on top of 4, the first actual pyramid with square base, 5, is produced, for here again unity is potentially the first. [2] Once more, I put this same pyramid entire, composed of 5 units, just as it is, upon the square 9, and there is made up for me the pyramid 14, with square base and side 3—for the former pyramid had the side 2, and the one potentially first 1 as a side. For here too each side of any pyramid whatsoever must consist of as many units as there are polygonal numbers piled together to create it. [3] Again, I place the whole pyramid 14, with the square 9 as its base, upon the square 16 and I have 30, the third actual pyramid of those that have a square base, and by the same order and procedure from a pentagonal, hexagonal, or heptagonal base, and even going on farther, we shall produce pyramids by piling upon one another the corresponding polygonal numbers, starting with unity as the smallest and going on to infinity in each case.

In the next several chapters, Nicomachus discusses solid figurate numbers, including cubes and "scalene" solids. Book 2 then concludes with a treatment of proportion, in which Nicomachus begins with three types: arithmetic, geometric, and harmonic. We will look at an earlier fragment from Archytas on this same subject in chapter 8, given that Archytas was primarily concerned with the application of proportions and means to musical theory.

Chapter 21

[1] After this it would be the proper time to incorporate the nature of proportions, a thing most essential for speculation about the nature of the universe and for the propositions of music, astronomy, and geometry, and not least for the study of the works of the ancients, and thus to bring the *Introduction to Arithmetic* to the end that is at once suitable and fitting. [2] A proportion, then, is in the proper sense, the combination of two or more *ratios*, but by the more general definition the combination of two or more *relations*, even if they are not brought under the same ratio, but rather a difference, or something else. [3] Now a ratio is the relation of two terms to one another, and the combination of such is a proportion, so that three is the smallest number of terms of which the latter is composed, although it can be a series of more, subject to the same ratio or the same difference. For example, 1:2 is one ratio, where there are two terms; but 2:4 is another similar ratio; hence 1, 2, 4 is a proportion, for it is a combination of ratios, or of three terms that are observed to be in the same ratio to one another.... [5] Now if the same term, one and unchanging, is compared to those on either side of it, to the greater as consequent and to the lesser as antecedent, such a proportion is called continued: for example 1, 2, 4 is a continued proportion as regards quality, for 4:2 equals 2:1, and conversely 1:2 equals 2:4. In quantity, 1, 2, 3, for example, is a continued proportion, for as 3 exceeds 2, so 2 exceeds 1, and conversely, as 1 is less than 2, by so much 2 is less than 3. [6] If, however,

one term answers to the lesser term, and becomes its antecedent and a greater term, and another, not the same, takes the place of consequent and lesser term with reference to the greater, such a mean and such a proportion is called no longer continued, but disjunct. For example, as regards quality, 1, 2, 4, 8, for 2:1 equals 8:4, and conversely 1:2 equals 4:8, and again 1:4 equals 2:8 or 4:1 equals 8:2; and in quantity, 1, 2, 3, 4, for as 1 is exceeded by 2, by so much 3 is exceeded by 4, or as 4 exceeds 3, so 2 exceeds 1, and by interchange, as 3 exceeds 1, so 4 exceeds 2, or as 1 is exceeded by 3, by so much 2 is exceeded by 4.

Chapter 22

[1] The first three proportions, which are acknowledged by all the ancients, Pythagoras, Plato, and Aristotle, are the arithmetic, geometric, and harmonic; and there are three others subcontrary to them, which do not have names of their own, but are called in more general terms the fourth, fifth, and sixth forms of mean; after which the moderns discover four others as well, making up the number ten, which, according to the Pythagorean view, is the most perfect possible.... [2] Now, however, we must treat from the beginning, first, that form of proportion which by quantity reconciles and binds together the comparison of the terms, which is a quantitative equality as regards the difference of the several terms to one another. This would be the arithmetic proportion....

Chapter 23

[1] It is an arithmetic proportion, then, whenever three or more terms are set forth in succession, or are so conceived, and the same quantitative difference is found to exist between the successive numbers, but not the same ratio among the terms, one to another. For example, 1, 2, 3, 4, 5, 6, 7, 8, 9, 10, 11, 12, 13; for in this natural series of numbers, examined consecutively and without any omissions, every term whatsoever is discovered to be placed between two and to preserve the arithmetic proportion to them. For its differences as compared with those ranged on either side of it are equal; the same ratio, however, is not preserved among them. [2] And we understand that in such a series there comes about both a continued and a disjunct proportion; for if the same middle term answers to those on either side as both antecedent and consequent, it would be a continued proportion, but if there is another mean along with it, a disjunct proportion comes about. [3] Now if we separate out of this series any three consecutive terms whatsoever, after the form of the continued proportion, or four or more terms after the disjunct form, and consider them, the difference of them all would be 1, but their ratios would be different throughout. If, however, again we select three or more terms, not adjacent, but separated, separated nevertheless by a constant interval, if one term was omitted in setting down each term, the difference in every case will be 2; and once more with three terms it will be a continued proportion; with more, disjunct. If two terms are omitted, the difference will always be 3 in all of them, continued or disjunct; if three, 4; if four, 5; and so on.... [5] A thing is peculiar to this proportion that does not belong to any other, namely, the mean is either half of, or equal to, the sum of the extremes, whether the proportion be viewed as continuous or disjunct or by

alternation; for either the mean term with itself, or the mean terms with one another, are equal to the sum of the extremes. [6] It has still another peculiarity; what ratio each term has to itself, this the differences have to the differences; that is, they are equal. Again, the thing which is most exact, and which has escaped the notice of the majority, the product of the extremes when compared to the square of the mean is found to be smaller than it by the product of the differences, whether they be 1, 2, 3, 4, or any number whatsoever. In the fourth place, a thing which all previous writers also have noted, the ratios between the smaller terms are larger, as compared to those between the greater terms. . . .

Chapter 24

[1] The next proportion after this one, the geometric, is the only one in the strict sense of the word to be called proportion, because the terms are seen to be in the same ratio. It exists whenever, of three or more terms, as the greatest is the next greatest, so the latter is to the one following, and if there are more terms, as this again is to the one following it, but they do not, however, differ from one another by the same quantity, but rather by the same quality of ratio, the opposite of what was seen to be the case with the arithmetic proportion. [2] For an example, set forth the numbers beginning with 1 that advance by the double ratio, 1, 2, 4, 8, 16, 32, 64, and so on, or by the triple ratio, 1, 3, 9, 27, 81, 243, and so on, or by the quadruple, or in some similar way. In each one of these series three adjacent terms, or four, or any number whatever that may be taken, will give the geometric proportion to one another; as the first is to the next smaller, so is that to the next smaller, and again that to the next smaller, and so on as far as you care to go, and also by alternation. For instance, 2, 4, 8; the ratio which 8 bears to 4 [is the same] that 4 bears to 2, and conversely; they do not, however, have the same quantitative difference. Again, 2, 4, 8, 16; for not only does 16 have the same ratio to 8 as before, though not the same difference, but also by alternation it preserves a similar relation—as 16 is to 4, so 8 is to 2; and conversely, as 2 is to 8, so 4 is to 16; and disjunctly, as 2 is to 4, so 8 is to 16; and conversely and in disjunct form, as 16 is to 8 so 4 is to 2; for it has the double ratio. [3] The geometric proportion has a peculiar property shared by none of the rest, that the differences of the terms are in the same ratio to each other as the terms to those adjacent to them, the greater to the less, and vice versa. Still another property is that the greater terms have as a differences, with respect to the lesser, the lesser terms themselves, and similarly difference differs from difference, by the smaller difference itself, if the terms are set forth in the double ratio; in the triple ratio both terms and differences will have as a difference twice the next smaller, in the quadruple ratio thrice, in the quintuple four times, and so on. . . .

Chapter 25

[1] The proportion that is placed in the third order is one called the harmonic, which exists whenever among three terms the mean on examination is observed to be neither in the same ratio to the extremes, antecedent of one and consequent of the other, as in the geometric proportion, nor with equal intervals, but an inequality of ratios, as in the

arithmetic, but on the contrary, as the greatest term is to the smallest, so the difference between greatest and mean terms is to the difference between mean and smallest term. For example, take 3, 4, 6, or 2, 3, 6. For 6 exceeds 4 by one third of itself, since 2 is one third of 6, and 3 falls short of 4 by one third of itself, for 1 is one third of 3. In the first example, the extremes are in double ratio and their differences with the mean term are again in the same double ratio to one another; but in the second they are each in the triple ratio. [2] It has a peculiar property, opposite, as we have said, to that of the arithmetic proportion; for in the latter the ratios were greater among the smaller terms, and smaller among the greater terms. Here, however, on the contrary, those among the greater terms are greater and those among the smaller terms smaller, so that in the geometric proportion, like a mean between them, there may be observed the equality of ratios on either side, a midground between greater and smaller. [3] Furthermore, in the arithmetic proportion the mean term is seen to be greater and smaller than those on either side by the same fraction of itself, but by different fractions of the terms that flank it; in the harmonic, however, it is the opposite, for the middle term is greater and less than the terms on either side by different fractions of itself, but always the same fraction of those terms at its sides, a half of them or a third; but the geometric, as if in the midground between them, shows this property neither in the mean term exclusively nor in the extremes, but in both mean and extreme. [4] Once more, the harmonic proportion has as a peculiar property the fact that when the extremes are added together and multiplied by the mean, it makes twice the product of themselves multiplied by one another. [5] The harmonic proportion was so called because the arithmetic proportion was distinguished by quantity, showing an equality in this respect with the intervals from one term to another, and the geometric by quality, giving similar qualitative relations between one term and another, but this form with reference to relativity, appears now in one form, now in another, neither in its terms exclusively nor in its differences exclusively, but partly in the terms and partly in the differences; for as the greatest term is to the smallest, so also is the difference between the greatest and the next greatest, or middle, term to the difference between the least term and the middle term, and vice versa.

Chapter 27

[1] Just as in the division of the musical canon, when a single string is stretched or one length of a pipe is used, with immovable ends, and the mid-point shifts in the pipe by means of the finger-holes, in the string by means of the bridge, and as in one way after another the aforesaid proportions, arithmetic, geometric, and harmonic, can be produced, so that the fact becomes apparent that they are logically and very properly named, since they are brought about through changing and shifting the middle term in different ways, so too it is both reasonable and possible to insert the mean term that fits each of the three proportions between two arithmetic terms, which stay fixed and do not change, whether they are both even or odd. In the arithmetic proportion this mean term is one that exceeds and is exceeded by an equal amount; in the geometric proportion it is differentiated from the extremes by the same ratio, and in the harmonic it is greater and smaller than the extremes by the same fraction of those same extremes. [2] Let there be given then, first,

two even terms, between which we must find how the three means would be inserted, and what they are. Let them be 10 and 40. [3] First, then, I fit to them the arithmetic mean. It is 25, and the attendant properties of the arithmetic proportion are all preserved; for as each term is to itself, so also is difference to difference; they are in equality, therefore. And as much as the greater exceeds the means by so much the latter exceeds the lesser term; the sum of the extremes is twice the mean; the ratio of the lesser terms is greater than that of the greater; the product of the extremes is less than the square of the mean by the amount of the square of the differences; and the middle term is greater and less than the extremes by the same fraction of itself, but by different fractions regarded as parts of the extremes. [4] If, however, I insert 20 as a mean between the given even terms, the properties of the geometrical proportion come into view and those of the arithmetic are done away with. For as the greater term is to the middle term, so is the middle term to the lesser; the product of the extremes is equal to the square of the mean; the differences are observed to be in the same ratio to one another as that of the terms; neither in the extremes alone nor in the middle term alone does there reside the sameness of the fraction concerned in the relative excess and deficiency of the terms, but in the middle term and one of the extremes by turns; and both between greater and smaller terms there is the same ratio. [5] But if I select 16 as the mean, again the properties of the two former proportions disappear and those of the harmonic are seen to remain fixed, with respect to the two even terms. For as the greatest term is to the least, so is the difference of the greater terms to that of the lesser; by what fractions, seen as fractions of the greater term, the mean is smaller than the greater term, by these the same mean term is greater than the smallest term when they are looked upon as fractions of the smallest term; the ratio between the greater terms is greater, and that of the smaller terms, smaller, a thing which is not true of any other proportion; and the sum of the extremes multiplied by the mean is double the product of the extremes. [6] If, however, the two terms that are given are not even but odd, like 5, 45, the same number, 25, will make the arithmetic proportion; and the reason for this is that the terms on either side overpass it and fail to come up to it by an equal number, keeping the same quantitative difference with respect to it. 15 substituted makes the geometric proportion, as it is the triple and subtriple of each respectively; and if 9 takes over the function of the mean term, it gives the harmonic; for by those parts of the smaller term by which it exceeds, namely, four-fifths of the smaller, it is also less than the greater, if they be regarded as parts of the greater term, for this too is four fifths, and if you try all the previously mentioned properties of the harmonic ratio you will find that they will fit. [7] And let this be your method whereby you might scientifically fashion the mean terms that are illustrated in the three proportions. For the two terms given you, whether odd or even, you will find the arithmetic mean by adding the extremes and putting down half of them as the mean, or if you divide by 2 the excess of the greater over the smaller, and add this to the smaller, you will have the mean. As for the geometric mean, if you find the square root of the product of the extremes, you will produce it, or, observing the ratio of the terms to one another, divide this by 2 and make the mean, for example, the double, in the case of a quadruple ratio. For the harmonic mean, you must multiply the difference of the extremes by the lesser term and divide the product by the sum of the extremes, then add the quotient to the lesser term, and the result will be the harmonic mean.

Chapter 28

[1] So much, then, concerning the three proportions celebrated by the ancients, which we have discussed the more clearly and at length for just this reason, that they are to be met with frequently and in various forms in the writings of those authors. The succeeding forms, however, we must only epitomize, since they do not occur frequently in the ancient writings, but are included merely for the sake of our own acquaintance with them and, so to speak, for the completeness of our reckoning. . . . [3] The fourth, and the one called subcontrary, because it is opposite to, and has opposite properties to, the harmonic proportion, exists when, in three terms, as the greatest is to the smallest, so the difference of the smaller terms is to that of the greater, for example 3, 5, 6. . . . [4] The two proportions, fifth and sixth, were both fashioned after the geometrical, and they differ from each other thus. The fifth form exists, whenever, among three terms, as the middle term is to the lesser, so their difference is to the difference between the greater and the mean, as in 2, 4, 5. . . . [5] The sixth form comes about when in a group of three terms, as the greatest is to the mean, so the excess of the mean over the lesser is to the excess of the greater over the mean, for example, 1, 4, 6. . . . [7] The seventh in the list of them all, exists when, as the greatest term is to the least, so their difference is to the difference of the lesser terms, as 6, 8, 9. . . . [8] The eighth proportion . . . comes about when, as the greatest is to the lest term, so the difference of the extremes is to the difference of the greater terms, as 6, 7, 9. . . . [9] The ninth in the complete list . . . exists when there are three terms and whatever ratio the mean bears to the least, that also the difference of the extremes has in comparison with that of the smallest terms, as 4, 6, 7. [10] The tenth, in the full list, which concludes them all . . . is seen when, among three terms, as the mean is to the lesser, so the difference of the extremes is to the difference of the greater terms, as 3, 5, 8. . . .

2.3 Boethius, *On Arithmetic*

Recall that it was Boethius's adaptation of Nicomachus's *Arithmetica* that preserved the latter into the medieval period. To show the relationship of Boethius's work to that of Nicomachus, we provide a translation of chapter 20 of book 1, dealing with perfect numbers. Compare this to the similar material in Nicomachus's *Arithmetica*, book 1, chapter 16. We also give Boethius's treatment of ratios, a treatment which is very similar to that of Nicomachus in the final seven chapters of his book 1, but which we will refer to in the context of music in chapter 8 of this *Sourcebook*.

20. Concerning the generation of the perfect number.

There is in these [the perfect numbers] a great similarity to the virtues and vices. You find the perfect numbers rarely, you may enumerate them more easily, and they are produced in a very regular order. But you find superfluous or diminished numbers to be many and infinite and not disposed in any order, but arranged randomly and illogically, not generated from a certain point. Within the first ten numbers there is only one perfect number, 6; within the first hundred, there is 28; with a thousand, 496; within ten thousand, 8128. These perfect numbers always end in one of two numbers, 6 or 8, and these numbers always provide the final term in alternating fashion for the perfect numbers. First there is 6, then

28, after this 496, which ends in 6, which is like the first number, then 8128, which ends in 8 and that is like the second number.

The generation and production of these numbers is fixed and firm and they cannot be achieved in any other way nor, if they are brought about in this way, is it at all valid to create them by another mode. Let the even-times even numbers be disposed from one, and in order as far as you wish. Then you will add together the second with the first, and if a prime and incomposite number is made from that addition, you will then multiply the sum by the second number you have added on. If from that addition, a prime number does not emerge, but a secondary and composite number, skip this number and add to it the next one which follows. If then you do not yet emerge with a prime and incomposite number, add another number and see what emerges. But if you find a prime and incomposite number, then multiply it by the number added on from the last sum. Now the even-times even numbers may be disposed in this manner: 1, 2, 4, 8, 16, 32, 64, 128. Then work in this fashion. Put down one and add two, and see what number is made from this addition; from it comes 3, which of course is prime and incomposite. After unity, you add the number two. If you then multiply three, which is the number achieved by this addition, by two, which is the last number added on to the sum, then without a doubt a perfect number is born. Twice 3 makes six, which has one part which is its factor, namely six, and 3 is its half, while 2, the second number added on, is its third and these two have been multiplied to give the product 6.

Twenty-eight is produced in the same way. If to one and two, which make three, you add the following even-times even number, that is four, you arrive at seven. Then take the last number four, which you added in sequence, multiply the total by it, and a perfect number is produced. Seven times 4 is 28, which is a number equal to the sum of its parts; it has one as its factor, which is its twenty-eighth, and a half of 14, a fourth of 7, a seventh of 4, and a fourteenth of 2, and that corresponds to the medial term.

After these numbers are found, if you want to go on to discover the others, it is necessary that you pursue them by the same reasoning. It is necessary that you put down one, and after this 2 and 4, which add up to seven. The perfect number 28 showed itself a while ago by means of this process. The even-times even which follows this number is 8; this is added on, increasing the previous number to 15. But this is not a prime and incomposite number, for it has another factor beyond the number by which it is named in itself, that is beyond the factor of one into fifteen. Because this number is secondary and incomposite, pass by it and add to the previous number the next even-times even number, that is 16, which when added to 15, makes thirty-one. This is prime and incomposite. Multiply this number by the last number added on to the total, so that 16 times 31 yields 496. Now this sum is the perfect number within the order of 1000, and it is equal to the sum of its parts.

21. Concerning a quantity that is related to another.

Whenever one quantity is related to another, there is a basic two fold way in this may occur. Any given thing in comparison to another is either equal or unequal with it. . . .

There is a double division of unequal quantity. It is divided since it can be unequal as larger or as smaller, and these relationships operate in denominations which are contrary to each other. The larger is greater than the smaller and the smaller is lesser than the larger and the two cannot be known by the same term, as we said about equality. But they are marked by diverse and separate natures accordingly as they are compared in contrary terms, as are a teacher and his student, a striker and the one struck, or whatever other contrary thing may be related to something in terms of opposites.

22. Concerning the types of major and minor quantity.

Of major inequalities, there are five kinds. One is called multiplex, another superparticular, a third superpartient, a fourth multiplex superparticular, and a fifth multiplex superpartient. Now to these five types of major inequality are contrasted another five types of minor inequality. Each major type is opposed in parallel fashion to a minor type, and these types of minor inequalities are individually related to each of the major types. The major were given above, so the minor are called by the same names, but distinguished by the prefix "sub." They are called submultiplex, subsuperparticular, subsuperpartient, submultiplex superparticular, and submultiplex superpartient.

23. Concerning the multiplex number, its types, and their generation.

The multiplex is the first kind of major inequality, prior by nature to all the others and more distinguished than they, as it will be shown a little later. The multiplex is of such a type that, when compared with another, in contrast to the one it is compared to, it contains that other number more than once.[13] This happens to the first number in the disposition of natural numbers. All those numbers which follow the number one maintain the sequence and variety of all the multiplex proportions. To the number one, that is to unity, two is a duplex, three is a triplex, four is a quadruple, and so on, until all the multiplex quantities are covered, proceeding in that order. That number is said to be more than one if it takes its beginning from the binary number and so proceeds on to infinity through the third, fourth, and on in the order and sequence of all the numbers. Contrasted with this type of proportion is distinguished that which is called the submultiplex and this is the first type of the minor inequalities. This term is of such a type that when brought into comparison with another, it numbers the sum of the larger, and beginning with its own quantity, equally is a factor for that term and proceeds equally. It is measured, I say, by the same way that it factors. If a smaller number measures a larger number twice, it is called subduplex; if it measures it three times, it is called subtriplex; if four times, subquadruple, and this process goes on *ad infinitum*. You will name these numbers always with the prefix "sub" added on, as one compared to 2 is a subduplex, one compared to three is subtriplex, one compared to four is subquadruple, and so on. . . .

24. Concerning the superparticular number, its types, and their generation.

The superparticular is a number compared to another in such a way that it has in itself the entire smaller number and a fractional part of it.[14] If it has half the smaller number, it is called sesquialter; it it has a third it is called sesquitertius; if it has a fourth, it is called sesquiquartus, and if it has a fifth it is called sesquiquintus. The order of the superparticular number proceeds with these terms carried on *ad infinitum*.[15] The larger numbers are

[13] $a{:}b$ is multiplex when $a = nb$.

[14] $A{:}B$ is superparticular if, when reduced to lowest terms, $a{:}b$, $a = b + 1$.

[15] Boethius displays a table with many examples. So $3{:}2 = 6{:}4 = 9{:}6$ are examples of a sesquialter ratio, while $4{:}3 = 8{:}6 = 12{:}9$ are examples of a sesquitertial ratio.

called after this manner and the smaller numbers which are contained in them, their entire sum plus a fractional part, are called subsesquialter, another subsesquitertius, another subsesquiquartus, another subsesquiquintus, and these are extended in this way according to the manner and multitude of the larger numbers. I call the larger numbers "leader" and the smaller numbers "followers."...

28. Concerning the third type of inequality, which is called superpartient, and its types, and their generation.

After these first two types of proportion, the multiplex and the superparticular, and the types which are under them, the submultiplex and the subsuperparticular, there is found a third type of inequality which we have already called the superpartient. This occurs when one number compared to another contains that number entirely within itself and its aliquant parts as well, either two or three or four or however many that comparison brings out.[16]... If, therefore, a number contains another number in itself and two of its fractional parts, it is called superbipartient; if it has three fractional parts, it is called supertripartient; if it has four, it is called superquadripartient. So, one may work out all of these terms and proceed *ad infinitum*.... If we were to consider 5 in comparison to the number 3, it would be a superpartient, which would be called superbipartient, for the 5 would have the entire 3 in itself and 2 fractional parts of it....

29. Concerning the multiplex superparticular.

Now these have been the first and simple types of one quantity related to another. There are two other types which are thought of as put together from other principles, such as the multiplex superparticular and the multiplex superpartient; their followers are the submultiplex superparticular and the submultiplex superpartient. In these, as in the aforesaid proportions, smaller numbers are all said to be added under a proposition whose definition can be given thus: the multiplex superparticular is one that as often as one number is compared to another number, it has that number more than once and then a part of it, that is, it has the number double, or triple, or quadruple, or however many times, and then a certain fraction of it, either a half or a third or a fourth, or whatever other part might be added to it by excess. This number therefore consists of a multiplex and a superparticular.... Here are some examples of these relationships. The duplex sesquialter is 5 to 2. Five has the two twice and then half of it, which is one. The duplex sesquitertius is 7 compared to 3. 9 compared to 4 is a duplex sesquiquartus. If we compare 11 to 5, we have a duplex sesquiquintus....

31. Concerning the multiplex superpartient.

The multiplex superpartient is that number which as often as it is compared to another number, contains the other number in itself entirely more than once and one or two or more numbers in addition, according to the form of the superpartient number.... It is not difficult according to the examples of the prior numbers given to find these numbers and the numbers outside of our examples. These will be called according to their own fractions, duplex superbipartient, or duplex supertripartient, or duplex superquadripartient

[16] $A{:}B$ is superpartient if, when reduced to lowest terms, $a{:}b$, $a = b + k$ $(1 < k < b)$.

and again, triple superbipartient, and triple supertripartient and triple superquadripartient and so on. So 8 compared to 3 makes a duplex superbipartient.

2.4 Philo of Alexandria, *On the Creation* and *Who Is the Heir of Divine Things?*

In his *On the Creation*, Philo discusses order emerging out of chaos. For Philo, order involves number. Therefore, this work contains much numerical symbolism, including wide-ranging discussions of the meaning of four, six, and seven.

(13) And [Moses] says that the world was made in six days, not because the Creator stood in need of a length of time (for it is natural that God should do everything at once, not merely by uttering a command, but by even thinking of it); but because the things created required arrangement; and, of all numbers, six is, by the laws of nature, the most productive; for of all the numbers, from the unit upwards, it is the first perfect one, being made equal to its parts, and being made complete by them; the number three being half of it, and the number two a third of it, and the unit a sixth of it, and, so to say, it is formed so as to be both male and female, and is made up of the power of both natures; for in existing things the odd number is the male, and the even number is the female; accordingly, of odd numbers the first is the number three, and of even numbers the first is two, and the two numbers multiplied together make six. (14) It was fitting therefore, that the world, being the most perfect of created things, should be made according to the perfect number, namely, six; and, as it was to have in it the causes of both, which arise from combination, that it should be formed according to a mixed number, the first combination of odd and even numbers, since it was to embrace the character both of the male who sows the seed, and of the female who receives it.

(47) . . . And next the heaven was embellished in the perfect number four, and if anyone were to pronounce this number the origin and source of the all-perfect decade he would not err. For what the decade is in actuality, that the number four, as it seems, is in potentiality, at all events if the numerals from the unit to four are placed together in order, they will make ten which is the limit of the number of immensity, around which the numbers wheel and turn as around a goal. (48) Moreover the number four also comprehends the principles of the harmonious concords in music, that in fourths, and in fifths, and the diapason, and besides the double diapason, from which sounds the most perfect system of harmony is produced. For the ratio of the sounds in fourths is as four to three; and in fifths as three to two; and in the diapason that ratio is doubled; and in the double diapason it is increased fourfold, all which ratios the number four comprehends. At all events the first, or the epistritus, is the ratio of four to three; the second, or the hemiolius, is that of three to two; the twofold ratio is that of two to one, or four to two; and the fourfold ratio is that of four to one.

(49) There is also another power of the number four which is a most wonderful one to speak of and to contemplate. For it was this number that first displayed the nature of the solid cube, the numbers before four being assigned only to incorporeal things. For it is according to the unit that that thing is reckoned which is spoken of in geometry as a point; and a line is spoken of according to the number two, because it is arranged by nature from

a point; and a line is length without breadth. But when breadth is added to it, it becomes a surface, which is arranged according to the number three. And a surface, when compared with the nature of a solid cube, wants one thing, namely depth, and when this one thing is added to the three, it becomes four. On which account it has happened that this number is a thing of great importance, inasmuch as from an incorporeal substance perceptible only by intellect, it has led us on to a comprehension of a body divisible in a threefold manner, and which by its own nature is first perceived by the external senses. (50) And he who does not comprehend what is here said may learn to understand it from a game which is very common. Those who play with nuts are accustomed when they have placed three nuts on the floor, to place one more on the top of them producing a figure like a pyramid. Accordingly the triangle stands on the floor, arranged up to the number three, and the nut which is placed upon it makes up four in number, and in figure it produces a pyramid, being now a solid body.

(51) And in addition to this there is this point also of which we should not be ignorant, the number four is the first number which is a square, being equal on all sides, the measure of justice and equality. And that it is the only number the nature of which is such that it is produced by the same numbers, whether in combination, or in power. In combination when two and two are added together; and again in power when we speak of twice two; and in this it displays an exceedingly beautiful kind of harmony, which is not the lot of any other number. If we examine the number six which is composed of two threes, if these two numbers are multiplied it is not the number six that is produced, but a different one, the number nine. (52) And the number four has many other powers also, which we must subsequently show more accurately in a separate essay appropriated to it. At present it is sufficient to add this that it was the foundation of the creation of the whole heaven and the whole world. For the four elements, out of which this universe was made, flowed from the number four as from a fountain. And in addition to the four elements the seasons of the year are also four, which are the causes of the generation of animals and plants, the year being divided into the quadruple division of winter, and spring, and summer, and autumn.

(89) But after the whole world had been completed according to the perfect nature of the number six, the Father hallowed the day following, the seventh, praising it, and calling it holy. For that day is the festival, not of one city or one country, but of all the earth; a day which alone it is right to call the day of festival for all people, and the birthday of the world. (90) And I know not if anyone would be able to celebrate the nature of the number seven in adequate terms, since it is superior to every form of expression. But it does not follow that because it is more admirable than anything that can be said of it, that on that account one ought to keep silence; but rather we ought to try, even if one cannot say everything which is proper, or even that which is most proper, at all event to utter such things as may be attainable by our capacities.

(91) The number seven is spoken of in two ways; the one within the number ten which is measured by repeating the unit alone seven times, and which consists of seven units; the other is the number outside ten, the beginning of which is altogether the unit increasing according to a twofold or threefold, or any other proportion whatever; as are the numbers sixty-four, and seven hundred and twenty-nine; the one number of which is increased by doubling on from the unit, and the other by trebling. And it is not well to examine either

species superficially, but the second has a most manifest pre-eminence. (92) For in every case the number which is combined from the unit in double or treble ratio, or any other ratio whatsoever, is the seventh number, a cube and a square, embracing both species, both that of the incorporeal and that of the corporeal essence. That of the incorporeal essence according to the surface which quadrangular figures present, and that of the corporeal essence according to the other figure which cubes make; (93) and the clearest proof of this is afforded by the numbers already spoken of. In the seventh number increasing immediately from the unit in a twofold ratio, namely, the number sixty-four, is a square formed by the multiplication of eight by eight, and it is also a cube by the multiplication of four and four and four. And again, the seventh number from the unit being increased in a threefold ratio, that is to say, the number seven hundred and twenty-nine, is a square, the number seven and twenty being multiplied by itself; and it is also a cube, by nine being multiplied by nine and nine. (94) And in every case a man making his beginning from the unit, and proceeding on to the seventh number, and increasing in the same ratio till he comes to the number seven, will at all times find the number, when increased, both a cube and a square. At all events, he who begins with the number four, and combines them in a doubling ratio, will make the seventh number four thousand and ninety-six, which is both a square and a cube, having sixty-four as its square root, and sixteen as its cube root.

(95) And we must also pass on to the other species of the number seven, which is contained in the number ten, and which displays an admirable nature, and one not inferior to the previously mentioned species. The number seven consists of one and two and four, numbers which have two most harmonious ratios, the twofold and the fourfold ratio; the former of which affects the diapason harmony, while the fourfold ratio causes that of the double diapason. It also comprehends other divisions, existing in some kind of yoke-like combination. For it is divided first of all into the number one, and the number six; then into the two and the five; and last of all, into the three and the four. (96) And the proportion of these numbers is a most musical one; for the number six bears to the number one a sixfold ratio, and the sixfold ratio causes the greatest possible difference between existing tones; the distance namely, by which the sharpest tone is separated from the flattest, as we shall show when we pass on from numbers to the discussion of harmony. Again, the ratio of four to two displays the greatest power in harmony, almost equal to that of the diapason, as is most evidently shown in the rules of that art. And the ratio of four to three effects the first harmony, that in the thirds, which is the diatessaron.

(97) The number seven displays also another beauty which it possesses, and one which is most sacred to think of. For as it consists of three and four, it displays in existing things a line which is free from all deviation and upright by nature. And in what way it does so I must show. The rectangular triangle, which is the beginning of all qualities, consists of the numbers three and four and five; and the three and the four, which are the essence of the seven, contain the right angle; for the obtuse angle and the acute angle show irregularity, and disorder, and inequality; for one may be more acute or more obtuse than another. But a right angle does not admit of comparison, nor is one right angle more a right angle than another; but one remains similar to another, never changing its peculiar nature. But if the right-angled triangle is the beginning of all figures and of all qualities, and if the essence of the number seven, that is to say, the numbers three and four together, supply the most

necessary part of this, namely, the right angle, then seven may be rightly thought to be the fountain of every figure and of every quality.

In *Who Is the Heir of Divine Things?,* Philo interprets God as being the best mathematician of all, the only one capable of perfect division as well as the creator of symmetry in the divisions of time.

(141) But since Moses not only uses the expression, "he divided," but says further, "he divided in the midst," it is necessary to say a few words on the subject of equal divisions; for that which is divided skillfully just in the middle makes two equal divisions. (142) And no man could ever possibly divide anything into two exactly equal parts; but it is inevitable that one of the divisions must fall a little short, or exceed a little, if not much, at all events by a small quantity, in every instance, which indeed escapes the perception of our outward senses which attend only to the larger and more tangible burdens of nature and custom, but which are unable to comprehend atoms and indivisible things. (143) But it is established by the incorruptible word of truth that there is nothing equal in inequality. God alone therefore seems to be exactly just, and to be the only being able to divide in the middle bodies and things, in such a manner that none of the divisions shall be greater or less than the other by the smallest and most indivisible portion, and he alone is able to attain to sublime and perfect equality.

(144) If therefore there were but one idea of perfect equality, what has been said would be quite sufficient for the purpose. But as there are many, we must not hesitate to add some considerations which are suitable. For the word "equal" is used in one sense when speaking of numbers, as when we say that two are equal to two, and three to three; and speak of other numbers in the same manner. But in another sense when speaking of magnitude, as equal in length or breadth, or depth, which are all different proportions. For wrestler compared with wrestler, or cubit with cubit are equal in magnitude but different in power, as is the case also with measures and weights. (145) But the idea of equality is a necessary one, and so is that of equality in proportion, according to which a few things are looked upon as equal to many, and small things are equal to larger ones. And their proportionate equality, cities are accustomed to use at suitable times, when they command every citizen to contribute an equal share of his property, not equal in number, but in proportion to the value of his assessment, so that in some cases he who contributes a hundred drachmas will appear to have brought an equal sum with him who contributes a talent.

(146) These things being thus previously sketched out, see now how God, dividing things in the middle, has divided them into equal portions according to all the ideas of equality which occur in the creation of the universe. He has divided the heavy things so as to make them equal in number to the light ones, two to two; that is to say, so that the earth and the water, being things of weight, are equal in number to those which are by nature light, air and fire. Again, he has made one equal to one, the driest thing to the wettest thing, the earth to the water; and the coldest thing to the hottest thing, the air to the fire. So, in the same manner, he had divided light from darkness, and day from night, and summer from winter, and autumn from spring; and so on. (147) Again, he has divided things so as to make his divisions equal in point of magnitude; such as the parallel cycles in heaven, and

those which belong to the equinoxes both of spring and autumn, and those which belong to the winter and summer solstice. And on the earth he has divided the zones, two being equal to one another, which being placed close to the poles are frozen with cold, and on this account are uninhabitable. And two he has placed on the borders between these two and the torrid zone, and these two they say are the abode of a happy temperature of the air, one of them lying towards the south and the other toward the north.

(148) Now the divisions of time are equal in point of length, the longest day being equal to the longest night, and the shortest day being equal to the shortest night, and the mean length of day to the mean length of night. And the equal magnitude of other days and nights appears to be indicated chiefly by the equinoxes. (149) From the spring equinox to the summer solstice, day receives an addition to its length, and night, on the other hand, submits to a diminution; until the longest day and the shortest night are both completed. And then after the summer solstice the sun, turning back again the same road, neither more quickly nor more slowly than he advanced, but always preserving the same difference in the same manner, having a constantly equal arrangement, proceeds on till the autumnal equinox; and then, having made day and night both equal, begins to increase the length of the night, diminishing the day until the time of the winter solstice. (150) And when it has made the night the longest night, and the day the shortest day, then returning back again and adopting the same distances as before, he again comes to the spring equinox. Thus the differences of time which appear to be unequal, do in reality possess a perfect equality in respect of magnitude, not indeed at the same seasons, but at different seasons of the year.

2.5 Pseudo-Iamblichus, *The Theology of Arithmetic*

The Theology of Arithmetic was written by an unknown author probably in the fourth century CE. It is a compilation of sources, rather than a wholly original work, with large sections taken from *The Theology of Arithmetic* by Nicomachus and from *On the Decad* by Anatolius, bishop of Laodicea (d. 283). The sections are linked by other material, which may have been lecture notes. The work has been attributed to Iamblichus, but it was evidently written after his death. It is possible, however, that some of the connecting text was based on his lectures.

The work has ten sections, each based on one of the first ten counting numbers. We have excerpted material from each of these sections.

On the Monad

The monad is the non-spatial source of number. . . . Everything has been organized by the monad, because it contains everything potentially: for even if they are not yet actual, nevertheless the monad holds seminally the principles that are within all numbers, including those that are within the dyad. For the monad is even and odd and even-odd, linear and plane and solid (cubical and spherical and in the form of pyramids from those with four angles to those with an indefinite number of angles); perfect and over-perfect and defective; proportionate and harmonic; prime and incomposite, and secondary; diagonal and side; and it is the source of every relation, whether one of equality or inequality, as has been proved in the *Introduction*.[17] Moreover, it is demonstrably both point and angle (with all forms of angle), and beginning, middle and end of all things, since, if you decrease it, it limits the infinite dissection of what is continuous, and if you increase it, it defines the increase as being the same as the dividends (and this is due to the disposition of divine, not human, nature. . . .

Nicomachus says that God coincides with the monad, since He is seminally everything which exists, just as the monad is in the case of number, and there are encompassed in it in potential things that, when actual, seem to be extremely opposed (in all the ways in which things may, generally speaking, be opposed), just as it is seen, throughout the *Introduction to Arithmetic*, to be capable, thanks to its ineffable nature, of becoming all classes of things, and to have encompassed the beginning, middle and end of all things (whether we understand them to be composed by continuity or by juxtaposition), because the monad is the beginning, middle and end of quantity, of size and moreover of every quality.

Just as without the monad there is in general no composition of anything, so also without it there is no knowledge of anything whatsoever, since it is a pure light, most authoritative over everything in general, and it is sun-like and ruling, so that in each of these respects it resembles God, and especially because it has the power of making things cohere and combine, even when they are composed of many ingredients and are very different from one another, just as He made this universe harmonious and unified out of things that are likewise opposed.

Furthermore, the monad produces itself and is produced from itself, since it is self-sufficient and has no power set over it and is everlasting, and it is evidently the cause of permanence, just as God is thought to be in the case of actual physical things, and to be the preserver and maintainer of natures.

On the Dyad

Adding dyad to dyad is equivalent to multiplying them; adding them and multiplying them have the same result, and yet in all other cases multiplication is greater than addition. . . .

The dyad would be the mid-point between plurality, which is regarded as falling under the triad, and that which is opposed to plurality, which falls under the monad. Hence it simultaneously has the properties of both. It is the property of 1, as source, to make something more by addition than by the blending power of multiplication (and that is why $1 + 1$ is more than 1×1), and it is the property of plurality, on the other hand, as product, to do the opposite; for it makes something more by multiplication than by addition. For plurality is no longer like a source, but each number is generated one out of another and by blending (and that is why 3×3 is more than $3 + 3$). And while the monad and the triad have opposite properties, the dyad is, as it were, the mean, and will admit the properties of both at once, as it occupies the mid-point between each. And we say that the mean between what is greater and what is smaller is what is equal. Therefore equality lies in

[17] The reference is probably to Nicomachus's *Introduction to Arithmetic*.

this number alone. Therefore the product of its multiplication will be equal to the sum of its addition: for $2 + 2 = 2 \times 2$. Hence they used to call it "equal."

That it also causes everything that directly relates to it to have the same property of being equal is clear not only (and this is why it is the first to express equality in a plane and solid fashion—equality of length and breadth in the plane number 4, and in the solid number eight equality of depth and height as well) in its very divisibility into two monads that are equal to each other, but also in the number that is said to be "evolved" from it (that is, 16, which is $2 \times 2 \times 2 \times 2$), which is a plane number of the so-called "color" on base 2: for 16 is 4×4. And this number is obviously in a sense a sort of mean between greater and lesser in the same way that the dyad is. For the squares before it have perimeters that are greater than their surface areas, while the squares after it, on the other hand, have perimeters that are less than their surface areas, but this square alone has perimeter equal to surface area. This is apparently why Plato in *Theaetetus* went up to 16, but stopped "for some reason" at the square whose area is 17 feet, when he was faced with the manifestation of the specific property of 16 and the appearance of a certain shared equality. . . .

The dyad is clearly formless, because the infinite sequence of polygons arise in actuality from triangularity and the triad, while as a result of the monad everything is together in potential, and no rectilinear figure consists of two straight lines or two angles. So what is indefinite and formless falls under the dyad alone. . . .

The dyad is not number, nor even, because it is not actual; at any rate, every even number is divisible into both equal and unequal parts, but the dyad alone cannot be divided into unequal parts; and also, when it is divided into equal parts, it is completely unclear to which class its parts belong, as it is like a source.

On the Triad

The triad has a special beauty and fairness beyond all numbers, primarily because it is the very first to make actual the potentialities of the monad—oddness, perfection, proportionality, unification, limit. For 3 is the first number to be actually odd, since conformity with its descriptions it is "more than equal" and has something more than the equal in another part, and it is special in respect of being successive to the two sources and a system of them both.

At any rate, it is perfect in a more particular way than the other numbers to which consecutive numbers from the monad to the tetrad are found to be equal—I mean, that is, the monad, triad, hexad and decad. The monad, as the basic number of this series, is equal to the monad; the triad is equal to monad and dyad; the hexad is equal to monad, dyad and triad; the decad is equal to monad, dyad, triad and tetrad. So the triad seems to have something extra in being successive to those to which it is also equal. . . .

The triad is pervasive in the nature of number: for there are three types of odd number—prime and incomposite, secondary and composite, and mixed, which is secondary in itself, but otherwise prime; and again, there are over-perfect, imperfect and perfect numbers; and in short, of relative quantity some is greater, some less, and some equal. The triad is very well suited to geometry; for the basic element in plane figures is the triangle, and there are three kinds of triangle—acute-angled, obtuse-angled, and scalene.

On the Tetrad

Everything in the universe turns out to be completed in the natural progression up to the tetrad, in general and in particular, as does everything numerical—in short, everything whatever its nature. The fact that the decad, which is gnomon and joiner, is consummated by the tetrad along with the numbers which precede it, is special and particularly important for the harmony that completion brings; so is the fact that it provides the limit of corporeality and three-dimensionality. For the pyramid, which is the minimal solid and the one that first appears, is obviously contained by a tetrad, either of angles or of faces, just as what is perceptible as a result of matter and form, which is a complete result in three dimensions, exists in four terms.

Moreover, it is better and less liable to error to apprehend the truth in things and to gain secure, scientific knowledge by means of the quadrivium of mathematical sciences. For since all things in general are subject to quantity when they are juxtaposed and heaped together as discrete things, and are subject to size when they are combined and continuous, and since, in terms of quantity, things are conceived as either absolute or relative, and, in terms of size, as either at rest or in motion, accordingly the four mathematical systems or sciences will make their respective apprehensions in a manner appropriate to each thing: arithmetic apprehends quantity in general, but especially absolute quantity; music apprehends quantity when it is relative; and geometry apprehends size in general, but especially static size; astronomy apprehends size when it is in motion and undergoing orderly change. If number is the form of things, and the terms up to the tetrad are the roots and elements, as it were, of number, then these terms would contain the aforementioned properties and the manifestations of the four mathematical sciences—the monad of arithmetic, the dyad of music, the triad of geometry and the tetrad of astronomy. . . .

In the first place, the association of arithmetic with the monad is reasonable; for when arithmetic is abolished, so are the other sciences, and they are generated when it is generated, but not vice versa, with the result that it is more primal than them and is their mother, just as the monad evidently is as regards the numbers that follow it. But also every specific identity and property and attribute of number is found first of all in the monad, as in a seed.

The monad is in a sense quantity regarded as absolute and as the sole agent of limit and true definition; for if anything is conjoined with anything else, it cannot be alone, but must fall under the dyad, for the dyad contains the primary conception of difference. And music obviously pertains to difference in some way, since it is a relation and a harmonious fitting together of things that are altogether dissimilar and involved in difference.

And geometry falls under the triad, not only because it is concerned with three-dimensionality and its parts and kinds, but also because it was characteristic of this teacher always to call surfaces the limiters of geometry, on the grounds that geometry concerns itself primarily with planes; but the most elementary plane is contained by a triad, either of angles or of lines; and when depth is added, from this as a base to a single point, then in turn the most elementary of solids, the pyramid, is formed, which is fitted together by virtue of three equal dimensions, and these dimensions form the limits of anything subsisting in Nature as a solid.

And astronomy—the science of the heavenly spheres—falls under the tetrad, because of all solids the most perfect and the one that particularly embraces the rest by nature, and is outstanding in thousands of other respects, is the sphere, which is a body consisting of four things—center, diameter, circumference and area. . . .

Anatolius reports that it is called "justice," since the square that is based on it is equal to the perimeter, for the perimeter of squares before it is greater than the area of those squares, and the perimeter of squares after it is less than the area, but in its case the perimeter is equal.

The tetrad is the first to display the nature of solidity: the sequence is point, line, plane, solid. Four is the first number that is even-times-even. Four is the first number that contains the sesquitertian ratio, which belongs to the primary concord, the fourth. In its case, everything is equal—area, angles, sides.

On the Pentad

The pentad is the first number to encompass the specific identity of all numbers, since it encompasses 2, the first even number, and 3, the first odd number. Hence it is called "marriage," since it is formed of male and female. It is the midpoint of the decad. When it is squared, it always encompasses itself, for $5 \times 5 = 25$; and when it is multiplied again, it both encompasses the square as a whole and terminates at itself, for $5 \times 25 = 125$. There are five solid figures with equal sides and equal angles—the tetrahedron, octahedron, icosahedron, cube and dodecahedron. And Plato says that the first is the figure of fire, the second of air, the third of water, the fourth of earth, and the fifth of the universe.

Moreover, there are five planets, not counting the sun and moon. The square on base 5 is the first to be equal to two squares—the one on base 3 and the one on base 4. A tetrachord is said to consist of the first even and the first odd number. . . . Moreover, whatever you use to add up to 10, 5 will be found to be the arithmetic mean—e.g., 9 + 1, 8 + 2, 7 + 3, 6 + 4. Each sum adds up to 10, and 5 is found to be the arithmetic mean. . . .

The pentad is the first to exhibit the best and most natural mediacy, when, in conjunction with the dyad, it is taken in disjunct proportion to both the limits of natural number—to the monad as source and the decad as end: for as 1 is to 2, so 5 is to 10, and again as 10 is to 5, so 2 is to 1; and alternately, as 10 is to 2, 5 is to 1, and as 2 is to 10, 1 is to 5. And the product of the limits is equal to the product of the means, as is the way with geometrical proportion: for $2 \times 5 = 1 \times 10$. Reciprocally, we are able to see first in the pentad, compared with the greater limit, the principle of half, just as we see this principle first in the dyad, compared with the smaller limit: for 2 is double 1, and 5 is half 10. . . .

On the Hexad

The hexad is the first perfect number, for it is counted by its own parts, as containing a sixth, a third and a half. When squared, it includes itself, for $6 \times 6 = 36$; when cubed, it no longer maintains itself as a square, for $6 \times 36 = 216$, which includes 6, but not 36.

It arises out of the first even and first odd numbers, male and female, as a product and by multiplication; hence it is called "androgynous." It is also called "marriage," in the strict sense that it arises not by addition, as the pentad does, but by multiplication. Moreover, it is called "marriage" because it is equal to its own parts, and it is the function of marriage to make offspring similar to parents.

The harmonic mean is first formed by the hexad, since the sesquitertian ratio of 8 set against 6, and the double ratio of 12 set against 6, are both gained. For by the same fraction, namely, a third, 8 both exceeds and is exceeded by the extremes. The arithmetic mean also falls under 6, since the sesquialter ratio of 9 set against it, and the double ratio of 12 set against it, are both gained. For by the same number, 3, 9 both exceeds one extreme and is exceeded by the other. Moreover, its parts (namely, 1, 2, 3) have a certain arithmetical proportion. Moreover, 6 forms a geometric mean—3, 6, 12. Moreover, there are six extensions of solid bodies. . . .

Now, inasmuch as it is relevant to the hexad, we must briefly see what is the result of forming the sequence that starts with the monad in the Pythagorean right-angled triangle: first, there is the one actual right angle in it, while there are two angles that are unequal to each other, but both together are equal to the previously mentioned angle, just as both the squares formed on each of the sides that subtend these two angles are equal to the square based on the line that subtends the right angle; three is the quantity of the smaller of the two sides that contain the right angle, four the quantity of the larger one, five the quantity of the hypotenuse, and six the quantity of the area, i.e., of half of the parallelogram, which half is defined by the diagonal of the parallelogram. . . .

On the Heptad

Seven is not born of any mother and is a virgin. The sequence from the monad to it added together totals 28, the 28 days of the moon are fulfilled hebdomad by hebdomad. Starting with the monad and making a sequence by doubling, seven numbers yield 64, the first square that is also a cube: 1, 2, 4, 8, 16, 32, 64. Doing the same, but trebling, seven numbers yield 729, the second square-and-cube: 1, 3, 9, 27, 81, 243, 729.

Moreover, the hebdomad consisting of the three dimensions (length, breadth and depth) and the four limits (point, line, surface and solid) reveals corporeality. Seven is said to be the number of the primary concord, the fourth [4:3], and of geometric proportion (1, 2, 4). It is also called "that which brings completion," for seven-month children are viable. The hebdomad is critical in illnesses. Seven encompasses the sides around the right angle of the archetypal right-angled triangle: the length of one is 4, of the other, 3. There are seven planets. . . .

Children cut their teeth at seven months, and at twice seven sit up and gain an unswaying posture, and at three times seven they begin to articulate speech and make their first efforts at talking, and at four times seven they stand without falling over and try to walk, and at five times seven they are naturally weaned and milk ceases to be their food. And at seven years they shed their natural teeth and grow ones that are suitable for hard food, and at twice seven years they come to puberty and, just as in the first hebdomad of years they acquired in an articulated manner the full range of expressed speech, consisting of as many simple words as are natural and useful for such expression, so they now begin to embark on the articulation of abstract speech, in so far as there is now a rational creature, and there being, according to most philosophers, seven senses that train the rational and are completed especially at this time: for in addition to the commonly recognized five senses, some count the faculties of speech and procreation, and the latter is completed at the time when the procreative faculty naturally changes for all humans—for males by means of seed, for females by means of menstruation. Hence they only then acquire fitness of engendering life, and among the Babylonians they do not play a part in religious ceremonies or partake in their priestly wisdom, but are debarred from all the initiations there before this time.

Since in the next period it is possible for them to have children and substitute others for themselves for the fulfillment of the universe, then the poets are being reasonable when they classify a generation as the thirty-year symmetry of the appearance of children; and because of the perfection of the triad, a complete succession consists of three—father, son and grandson. In the third hebdomad, they generally conclude growth in terms of length, and in the fourth they complete growth in terms of breadth, and there is no other bodily increase remaining to them; for 28 is a complete number.

In the fifth hebdomad, thanks to the manifestation of the harmonic 35, all increase as regards strength is checked, and after these years it is no longer possible for people to become stronger than they are. Hence, when athletes reach this age, some have already stopped winning and do not expect to achieve anything more, though others do not yet give up. And the legal codes of the best constitutions have conscription up to this hebdomad (though some have it until the next hebdomad), and after this point allow people to be officers but not to serve in the ranks any more. Finally, when the principle of the decad is blended with that of the hebdomad and ten times seven is reached, then man should be released from all tasks and dedicated to the enjoyment of happiness, as they say. . . .

On the Octad

We describe the octad as the first actual cube, and as the only number with the decad to be even-times-even, since 4 appears to combine the characteristics of being odd-even and even-times-even in admitting only two divisions up to the monad, one of itself, the other of its parts. All the ways in which it is put together are excellent and equilibrated tunings. First, it results from the only two numbers within the decad that are neither engenderers nor engendered (I mean, from 1 and 7); then, it results from the two that are even-odd, one potentially, the other actually—that is, from 2 and 6; then it results from the first two odd numbers—that is, from 3 and 5 (and this is the combination that is elementary for the generation of cubes, and is the first such sum, since the cube before it, 1, comes about without combination, while the one after it results from the next three odd numbers—7, 9 and 11—and the one after that from 4 continuous odd numbers—13, 15, 17, and 19); and fourth, it results from 4 taken twice, and four is the only number that both engenders and is engendered. The consequence is that 8 is completed by means of the first two unengendered numbers, and from their opposites (numbers that engender) and from the number that contains both characteristics. . . .

The number 8 is the source of the musical ratios, and the terms of the composition of the universe are as follows: the number 8 is in a sesquioctaval relation to 9 (9 exceeds 8 by a monad); 12 is the sesquialter of 8 and the sesquitertian of 9 (it exceeds 9 by a triad); 16 is the sesquitertian of 12 (the excess is 4); 18 is the sesquialter of 12 (the excess is a hexad); 21 is the double sesquitertian of 9 (the excess is 12); 24 is the sesquitertian of 18 (the excess is 6); 32 is the sesquitertian of 24 (the excess is 8); 36 is double 18 and the sesquialter of 24 (the excess is 12).

The 9 of the moon is in sesquioctaval relation to 8; the 12 of Mercury is the sequialter of 8; the 16 of Venus is double 8; the 18 of the sun is double 9 and the sesquioctave of 16; the 21 of Mars is the double sesquitertian of 9; the 24 of Jupiter is double 12, which is the sesquialter of 8; the 32 of Saturn is quadruple 8; the 36 of the fixed stars is quadruple 9 and the sesquioctave of 32. The excesses are: 36, by 4; 32, by 8; 24, by 3; 21, by 3; 18, by 2; 16, by 4; 12, by 3; 9, by 1. Alternatively, 9 exceeds 8 by a monad, 12 exceeds 9 by a triad, 16 exceeds 12 by a tetrad, 18 exceeds 16 by a dyad, and so on for the rest.

On the Ennead

The ennead is the greatest of the numbers within the decad and is an unsurpassable limit. At any rate, it marks the end of the formation of specific identities as follows: not only does it happen that, after the ninth pitch, there is no further superparticular musical ratio, but also addition naturally turns from the natural end to the beginning, and from both of these to the middle. At any rate, as regards the word, it is probably a riddling reference to affinity and equivalence, in the sense that it is called “ennead” as if it were the “henad” of everything within it, by derivation from “one.”. . .

The ennead is the first square based on an odd number. It too is called “that which brings completion,” and it completes nine-month children; moreover, it is called “perfect,” because it arises out of 3, which is a perfect number. The heavenly spheres revolve around the Earth, which is ninth. Nine is also said to contain the principles of the concords—4, 3 and 2; the sesquitertian is 4:3, the sesquialter is 3:2, and the double is 4:2. It is the first number to be in the sesquiocataval ratio.

On the Decad

We have often said before that the creative mind wrought the construction and composition of the universe and everything in the universe by reference to the likeness and similarity of numbers, as if to a perfect paradigm. But since the whole was an indefinite multitude and the whole substance of number was inexhaustible, it was not reasonable or scientific to employ an incomprehensible paradigm, and there was a need of commensurability, so that the Creator God, in his craftsmanship, might prevail over and overcome the terms and measures that were set before him, and might neither contract in an inferior fashion nor expand in a discordant fashion to a lesser or greater result than what was appropriate. However, a natural equilibration and commensurability and wholeness existed above all in the decad. It has encompassed seminally within itself all things, both solid and plane, even and odd and even-odd, perfect in all manners of perfection, prime and incomposite, and equality and inequality, the ten relations, and diagonal numbers, spherical numbers and circular numbers; in itself it has no special or natural variation, apart from the fact that it runs and circles back to itself. Hence it was reasonable for God to use it as a measure for things and as a gnomon and straight edge when he added things to one another and fitted them together harmoniously. And this is why, both in general and in particular, things from heaven to Earth are found to have been organized by it. . . .

Speusippus, the son of Plato's sister Potone, and head of the Academy before Xenocrates, compiled a polished little book from the Pythagorean writings that were particularly valued at any time, and especially from the writings of Philolaus; he entitled the book *On Pythagorean Numbers*. . . . He speaks in this manner about the decad:

> Ten is a perfect number, and it is both correct and in accordance with Nature that we Greeks and all men, without making any special effort, arrive at this number in all sorts of ways when we count. For it has many of the properties that are suitable for a number that is perfect in this way, and it also has many properties that are not peculiar to it, but which a perfect number ought to have. So in the first place, a perfect number ought to be even, so that it contains an equal amount of odd and even numbers, without imbalance; for since an odd number always precedes an even number, then if the final number is not even, the other sort will predominate. Secondly, it is necessary for a perfect number to contain an equal amount of prime and incomposite numbers, and secondary and composite numbers. Ten does have an equal amount, and no number less than ten has this property, though numbers more than ten might (such as 12 and others), but ten is the base number of the series. Since it is the first and smallest of those numbers that have this property, it has a kind of perfection, and this is a property peculiar to it, that it is the first in which an equal amount of incomposite and composite numbers are seen.
>
> Moreover, in addition to this property, it contains an equal amount of multiples and submultiples; for it contains as submultiples all the numbers up to and including five, while those from six to ten are multiples of the former ones. But since seven is a multiple of none of them, it must be excluded, and so must four, as a multiple of two, with the result that the amounts are again equal. Furthermore, all the ratios are contained by 10—that of the equal, and the greater and the less, and the superparticular and all the remaining kinds are in it, as are linear, plane and solid numbers. For one is a point, two a line, three a triangle and four a pyramid: these are all primary and are the sources of the things that are of the same category as each of them. In these numbers is also seen the first of the proportions, which is the one where the ratios of excess are constant and the limit is ten. The primary elements in plane and solid figures are these: point, line, triangle, pyramid. They contain the number ten and are limited by it. For there is a tetrad in the angles or bases of a pyramid, and a hexad in its sides, which makes 10. And again, there is a tetrad in the intervals and limits of a point and a line, and a hexad in the sides and angles of a triangle, which again makes 10.

2.6 Diophantus, *Arithmetica*

About Diophantus's life, little is known other than what is found in book 14, epigram 126 of the *Greek Anthology*, included in section 2.15 of this *Sourcebook*. Although it seems likely that Diophantus lived in Alexandria, when he lived is still the subject of debate. Most scholars believe, though, that he lived sometime in the middle of the third century CE, but an earlier or later date is certainly possible. Diophantus's major work is the *Arithmetica*, a book of problems determining rational solutions to equations mostly involving squares and cubes. Only six of the books have survived in

Greek, while four others have survived in an Arabic version. From internal references, it appears that the four Arabic books follow the third Greek book, while the final three Greek books come later. We will refer to the Greek books as 1–6 and the Arabic books as A, B, C, D. As will be clear from our excerpts, the style of the Arabic books is somewhat different from that of the Greek in that each step in the solution of a problem is explained in more detail. It is quite possible, therefore, that the Arabic work is not a direct translation of Diophantus's work but of a commentary on the *Arithmetica*, perhaps written by Hypatia around 400.

One major innovation in the *Arithmetica* is Diophantus's use of symbols to represent unknown quantities and their powers. In earlier works on equations from Egypt and Mesopotamia, everything was written out in words. But Diophantus introduced symbolic abbreviations for the various terms involved in equations, as we will see in his preface to book 1. Most of the problems Diophantus sets out to solve are what we would call indeterminate, in that we would write them as a set of k equations in more than k unknowns. But Diophantus is always interested in finding just one solution, so he generally specifies some of the numerical constants involved before actually demonstrating the solution. We have selected problems from each of the ten extant books, problems that demonstrate some of the techniques that Diophantus used. But we emphasize that the *Arithmetica* is by no means a systematic treatment of methods of solving equations. Although often similar problems are grouped together—for instance, one involves an addition whereas the next one involves a subtraction—in general, one never knows what type of problem will come next. Often the solutions, too, are specific to the problem at hand, so that a method used in one problem will not help in the solution of the next.

Toward the end of the preface, Diophantus states that the goal in manipulating his equation is to end up with an equation that in modern symbols can be written as $ax^m = bx^n$, which can then be solved, assuming necessary roots are rational. However, he notes that at some point he will show how to solve equations of the form, say, $ax^2 = bx + c$. In none of the extant books, however, does he present a rule for solving such a quadratic equation. However, as we will see, he does occasionally solve quadratic equations and even quadratic inequalities through the use of the standard Mesopotamian rule for such solutions. We know that these methods were known in Greece, as is witnessed in several other sources in this volume.[18] Indeed, since Diophantus claimed he would show the rule, we can only assume that this was done in one of the missing books. In what follows, we will use modern symbolism in place of Diophantus's own symbolism.

Book 1

Preface

Knowing that you are anxious, my most esteemed Dionysius,[19] to learn how to solve problems in numbers, I have tried, beginning from the foundations on which the subject is

[18] For example, see Heron's *Metrica*, section 6.3, and several of the papyri excerpted in chapter 6.
[19] It is not known who this person is.

built, to set forth the nature and power in numbers. Perhaps the subject will appear to you rather difficult, as it is not yet common knowledge, and the minds of beginners are apt to be discouraged by mistakes; but it will be easy for you to grasp, with your enthusiasm and my teaching; for keenness backed by teaching is a swift road to knowledge.

As you know, in addition to these things, that all numbers are made up of some multitude of units, it is clear that their formation has no limit. Among them are—

squares, which are formed when any number is multiplied by itself; the number itself is called the *side of the square*;

cubes, which are formed when squares are multiplied by their sides;

square-squares, which are formed when squares are multiplied by themselves;

square-cubes, which are formed when squares are multiplied by the cubes formed from the same side;

cube-cubes, which are formed when cubes are multiplied by themselves;

and it is from the addition, subtraction, or multiplication of these numbers or from the ratio which they bear one to another or to their own sides that most arithmetical problems are formed; you will be able to solve them if you follow the method shown below.

Now each of these numbers, which have been given abbreviated names, is recognized as an element in arithmetical science; the *square* [of the unknown quantity] is called *dynamis* and its sign is Δ with the index Υ, that is, Δ^{Υ}; the cube is called *kubos* and has for its sign K with the index Υ, that is, K^{Υ}; the square multiplied by itself is called *dynamo-dynamis* and its sign is two *deltas* with the index Υ, that is, $\Delta^{\Upsilon}\Delta$; the square multiplied by the cube formed from the same root is called *dynamo-kubos* and its sign is ΔK with the index Υ, that is, ΔK^{Υ}; the cube multiplied by itself is called *kubo-kubos* and its sign is two *kappas* with the index Υ, $K^{\Upsilon}K$. The number which has none of these characteristics, but merely has in it an undetermined multitude of units, is called *arithmos*, and its sign is ς.[20] There is also another sign denoting the invariable element in determinate numbers, the unit, and its sign is M with the index O, that is, $\overset{\circ}{M}$.[21]

As in the case of numbers, the corresponding fractions are called after the numbers, a *third* being called after 3 and a *fourth* after 4, so the functions named above will have reciprocals called after them:

arithmos $[x]$	*arithmoston* $\left[\frac{1}{x}\right]$
dynamis $[x^2]$	*dynamoston* $\left[\frac{1}{x^2}\right]$
kubos $[x^3]$	*kuboston* $\left[\frac{1}{x^3}\right]$
dynamodynamis $[x^4]$	*dynamodynamoston* $\left[\frac{1}{x^4}\right]$
dynamokubos $[x^5]$	*dynamokuboston* $\left[\frac{1}{x^5}\right]$
kubokubos $[x^6]$	*kubokuboston* $\left[\frac{1}{x^6}\right]$

[20] In the problems, this symbol in the manuscript will be translated as x. The symbol is possibly a contraction of the first two letters of *arithmos*. Jean Christianides and Jeffrey Oaks, authors of the new book *The Arithmetica of Diophantus: A Complete Translation and Commentary* (London: Routledge, 2023), use the word *arithmos* in their translation rather than the x that previous translators have used.

[21] Again, this is a contraction of *monas*, meaning "unit."

Figure 2.6.1. Part of a page from a manuscript of Diophantus's *Arithmetica*, dated 1296 (Vat. gr. 191, f. 388v). Image courtesy of the Vatican.

And each of these will have the same sign as the corresponding process, but with the mark χ to distinguish its nature.

Diophantus continues by demonstrating how to multiply the powers of the unknown, both the positive powers and the negative ones. He follows this with the law of signs.

A minus multiplied by a minus makes a plus; a minus multiplied by a plus makes a minus; and the sign of a minus is a truncated ⩛ turned upside down, that is ⩚.

It is well that one who is beginning this study should have acquired practice in the addition, subtraction, and multiplication of the various species. He should know how to add positive and negative terms with different coefficients to other terms, themselves either positive or likewise partly positive and partly negative, and how to subtract from a combination of positive and negative terms other terms either positive or likewise partly positive and partly negative.

Next, if a problem leads to an equation in which certain terms are equal to terms of the same species but with different coefficients, it will be necessary to subtract like from like on both sides, until one term is found equal to one term. If by chance there are on either side or on both sides any negative terms, it will be necessary to add the negative terms on both sides, until the terms on both sides are positive, and then again to subtract like from like until one term only is left on each side. This should be the object aimed at in framing the hypotheses of propositions, that is to say, to reduce the equations, if possible, until one term is left equal to one term; but I will show you later how, in the case also where two terms are left equal to one term, such a problem is solved.[22]

Now we begin with the problems themselves, which have a great deal of material to learn from. But since there are very many, and some very long, they must be studied slowly by those who read them. Therefore, I have tried, as far as possible, especially at the beginning, to provide the elements of this study and to distinguish clearly the simpler ones from the more complicated ones. For in that way they will become more accessible to beginners and the solution process will be better learned. The treatise contains thirteen books.

Diophantus did not have a symbol for addition of terms but merely put the terms together; if a "minus" was necessary, he put the symbol in front of all the terms to which it applied. However, in what follows, we translate Diophantus's symbolism into modern symbolism to make it easy for the reader to understand.

Problems

1. *To divide a given number into two numbers having a given difference.*

Let the given number be 100, the given difference 40. We want to find the two numbers. Let the smaller number be x, so the larger is therefore $x+40$. The sum of the two numbers is $2x+40$. This is equal to 100. So 100 is equal to $2x+40$. We subtract 40 from both sides, leaving $2x$ equal to 60 and therefore x is 30. So the smaller number is 30, the larger 70 and the proof is obvious.

5. *To divide a given number into two numbers such that given fractions (not the same) of each number when added together produce a given number.*

It is necessary that the latter given number must lie between the two numbers arising when the given fractions respectively are taken of the first given number. Let the first given number be 100. This is to be divided into two numbers such that $\frac{1}{3}$ of the first added to $\frac{1}{5}$ of the second produce 30. I put x equal to $\frac{1}{5}$ of the second number, so the second number is $5x$. Therefore $\frac{1}{3}$ of the first number is $30-x$ and the first number is $90-3x$. Since the sum of the two numbers is 100, and this sum of $2x+90$, the latter is equal to 100. After subtracting 90 from both sides, we get that 10 is equal to $2x$, so x is 5. Since $\frac{1}{5}$ of the second number is x, and this is 5, the second number is 25. Then $\frac{1}{3}$ of the first number is $30-x$, or 25. So the first number is 75 and thus $\frac{1}{3}$ of the first number added to $\frac{1}{5}$ of the second number is 30. We have done what was proposed.

7. *From the same number to subtract two given numbers so that the remainder will have a given ratio to one another.*

Let the numbers to be subtracted from the same number be 100 and 20, and let the larger remainder be three times that of the smaller. Let the required number be x. If I subtract from it 100, the remainder is $x-100$; if I subtract from it 20, the remainder is $x-20$. Now the larger remainder will have to be three times the smaller. Therefore three

[22] This promise is not fulfilled in any of the extant books, although Diophantus evidently uses the standard quadratic equation solving procedure in a few problems, as we will see later.

times the smaller will be equal to the larger. Now three times the smaller is $3x-300$, and this is equal to $x-20$. Let the deficiency be added in both cases. $3x$ equals $x+280$. If we subtract equals from equals, $2x$ equals 280, and x is 140. Now as to our problem. I have set the required number as x; it will therefore be 140. If I subtract from it 100, the remainder is 40; and if I subtract from it 20, the remainder is 120. And the larger remainder is three times the smaller.

9. *From two given numbers to subtract the same number and make the remainders have a given ratio to each other.*

It is necessary that the given ratio is greater than the ratio of the larger given number to the smaller. So we propose to subtract a number from 20 and 100 so that the greater remainder is six times the smaller. Let the number to be subtracted be x; then if this is subtracted from 100, the remainder is $100-x$; from 20, the remainder is $20-x$. And the greater remainder is six times the smaller. So 6 times the smaller will be equal to the greater. But 6 times the smaller is $120-6x$; this is equal to $100-x$. We add the $6x$ to both sides and then subtract 100 from both sides. There remains $5x=20$, so $x=4$. Since the number to be subtracted is $x=4$, the remainder from 100 is 96 and the remainder from 20 is 16. And the ratio of 96 to 16 is 6.

The following problem is the abstract version of a problem that appears frequently in medieval mathematics, in Hebrew and Arabic as well as in Latin, the problem of men buying a horse. For example, Fibonacci, in his *Liber Abaci* of 1202, states the problem as, "There are three men having bezants who desire to buy a horse. And as none of them can buy it, the first proposes to take from the other two men $\frac{1}{3}$ of their bezants. And the second proposes to take $\frac{1}{4}$ of the bezants of the other two men. And similarly the third proposes to take $\frac{1}{5}$ of the others, and thus each proposes to buy the horse."[23] Fibonacci's solution does not use algebra at all, so it is a bit more complicated than Diophantus's solution.

24. *To find three numbers such that, if each receives a given fraction of the sum of the remaining two, the results are all equal.*

Let the first receive $\frac{1}{3}$ of the second plus the third, the second receive $\frac{1}{4}$ of the third plus the first, and the third receive $\frac{1}{5}$ of the first plus the second. All these results will be equal. We put the first equal to x and, for convenience, let the sum of the second and third be a number of units divisible by 3, say 3. Therefore the sum of the three will be $x+3$ and the first plus $\frac{1}{3}$ the sum of the second and third will be $x+1$. Therefore, the second plus $\frac{1}{4}$ of the third plus the first will also be $x+1$. But 4 times the sum of the second plus $\frac{1}{4}$ of the sum of the other two is 3 times the second plus the sum of all three. Therefore 3 times the second plus the sum of all three will be $4x+4$. If we subtract the sum of all three, namely, $x+3$, we get that three times the second is equal to $3x+1$. Therefore, the second is equal to $x+\frac{1}{3}$. But also the third plus $\frac{1}{5}$ of the sum of the other two is $x+1$. By the same argument, we conclude that the third is equal to $x+\frac{1}{2}$. The sum of the three is $x+3$, [which is

[23] Victor J. Katz et al., eds., *Sourcebook in the Mathematics of Medieval Europe and North Africa* (Princeton, NJ: Princeton University Press, 2016), p. 83.

equal to $x+(x+\frac{1}{3})+(x+\frac{1}{2})=3x+\frac{5}{6}$], so $x=\frac{13}{12}$. Multiplying by the common denominator gives that the three numbers are 13, 17, 19, and these satisfy the conditions of the problem.

The following two problems, as well as problems 29 and 30, which we have not included, are standard problems that occur in Mesopotamian tablets from 2,000 years before Diophantus. These two problems also occur in the Seleucid tablet BM 34568 in section 6.4 of this *Sourcebook*. Diophantus's solutions are algebraic, however, while the Mesopotamian solutions appear to be based on geometry.

27. *To find two numbers whose sum and product are given numbers.*

It is necessary that the square of half the sum exceeds the product by a square. This is of the nature of a formula. Let it be required that the sum be 20 and the product be 96. I put the difference of the two numbers as $2x$. For if we take half of the given sum of 20, we get 10. If we take half of the difference, namely x, and add it to one part and subtract it from the other, then the sum will be 20 with difference $2x$. Therefore we put $x+10$ for the larger number; then the smaller will be $10-x$ and the sum will be 20 with difference $2x$. Since the product is 96, we have 96 equal to $100-x^2$. Therefore x is 2. Therefore the larger number is 12, the smaller is 8, and the conditions of the problem are satisfied.

28. *To find two numbers such that their sum and the sum of their squares are given numbers.*

It is a necessary condition that double the sum of their squares exceeds the square of their sum by a square. This is of the nature of a formula. Let it be required to make their sum 20 and the sum of their squares 208. Let their difference be $2x$, and let the greater be $x+10$ and the lesser be $10-x$. Then again their sum is 20 and their difference $2x$. It remains to make the sum of their squares 208. But the sum of their squares is $2x^2+200$. Therefore, $2x^2+200$ is equal to 208, and x is equal to 2. To return to the hypotheses, the greater is 12 and the lesser is 8. And these satisfy the conditions of the problem.

Book 2

The following problem is the one that Pierre de Fermat (1607–1665) annotated in his copy of the 1621 Latin edition of the *Arithmetica* with his famous statement that "one cannot split a cube into two cubes, nor a fourth power into two fourth powers, nor in general any power beyond the square *in infinitum* into two powers of the same name. For this I have discovered a truly wonderful proof, but the margin is too small to contain it." The problem also illustrates one of Diophantus's most common techniques of finding a square that he had expressed as a quadratic polynomial. Namely, he chose his square in the form $(ax \pm b)^2$, with a and b selected so that either the quadratic term or the constant term is eliminated from the equation. Thus, he ensured that he could find a rational solution.

8. *To divide a given square number into two squares.*

Let it be required to divide 16 into two squares. And let the first square be x^2; then the other will be $16-x^2$. It shall be required therefore to make $16-x^2$ a square. I take a square of the form $(mx-4)^2$, m being any integer and 4 the root of 16; for example, let the side be $2x-4$, and the square itself $4x^2+16-16x$. Then $4x^2+16-16x$ is equal to $16-x^2$. Add to both sides the negative terms and take like from like. Then $5x^2$ equals $16x$ and x is $\frac{16}{5}$. One number will therefore be $\frac{256}{25}$, the other $\frac{144}{25}$, and their sum is $\frac{400}{25}$ or 16, and each is a square.

10. *To find two square numbers having a given difference.*

Assume that their difference is 60. Let the side of one square be x, the side of the other x plus any number whose square is not greater than or equal to the given difference; thus one species will be equal to one species, and the problem can be solved. Let it be $x+3$. The square of the first is then x^2, of the second x^2+6x+9, and their difference is $6x+9$. If we set this equal to 60, then $x=8\frac{1}{2}$. So the first side is $8\frac{1}{2}$ and the second is $11\frac{1}{2}$. Their squares are $72\frac{1}{4}$ and $132\frac{1}{4}$, so they satisfy the requirement.

In the next problem, Diophantus introduces the method of the double equation, in which he needs to have two expressions both equal to squares. His method is to find two numbers whose product is the difference of the two expressions and then to take either the square of half the difference between these factors and set it equal to the lesser expression or the square of half the sum and set it equal to the greater. Diophantus does not mention that the factoring of the difference must be carefully chosen so that the solution is rational. As we will see in this and other examples, he does not show how he chooses an appropriate factorization.

11. *To add the same* [*required*] *number to two given numbers so as to make each of them a square.*

Let the given numbers be 2 and 3 and the required number be x. Therefore $x+2$ and $x+3$ must both be squares. This is called a double equation. We need to find two numbers whose product is the difference [of the two expressions]. These are 4 and $\frac{1}{4}$. Then take either the square of half the difference between these factors and equate it to the lesser expression or the square of half the sum and equate it to the greater. Here the square of half the difference is $\frac{225}{64}$. Setting this equal to $x+2$ gives x equal to $\frac{97}{64}$. Taking the square of half the sum gives $\frac{289}{64}$; equating this to the greater, $x+3$, also gives x equals to $\frac{97}{64}$. Therefore the required number is $\frac{97}{64}$ and this is what we wanted to find.

19. *To find three squares such that the difference between the greatest and the middle has to the difference between the middle and the least a given ratio.*

Assume that the ratio of the differences is $3:1$. Let the least square be x^2, the middle be x^2+2x+1 (whose side is $x+1$). Therefore the greatest square is x^2+8x+4; therefore x^2+8x+4 must be a square. We form a square from x and a certain number of units chosen so that in the expression for the square, the coefficient of x is less than 8 and the constant is greater than 4. This is 3. Then the square of $x+3$ is x^2+6x+9 and this is equal to x^2+8x+4. So $x=2\frac{1}{2}$. Therefore the greatest square is $30\frac{1}{4}$, the least is $6\frac{1}{4}$, the middle one is $12\frac{1}{4}$, and the problem is solved.

20. *To find two numbers such that the square of either, added to the other, shall make a square.*

Let the first be x, and the second $2x+1$, in order that the square on the first, added to the second, may make a square. There remains to be satisfied the condition that the square on the second, added to the first, shall make a square. But the square on the second, added to the first, is $4x^2+5x+1$; and therefore this must be a square. I form the square from $2x-2$; it will be $4x^2+4-8x$; and x is equal to $\frac{3}{13}$. The first number will be $\frac{3}{13}$, the second $\frac{19}{13}$, and they satisfy the conditions of the problem.

Book 3

6. *To find three numbers such that their sum is a square and such that the sum of any pair is also a square.*

Let the sum of all three be the square x^2+2x+1. If the sum of the first two is x^2, then the third is the remainder $2x+1$. But the sum of the second and third is also a square, say x^2+1-2x, whose root is $x-1$. Since the sum of the three is x^2+2x+1, it follows that the first is $4x$. But we have supposed that the sum of the first and second is x^2. Therefore, the second is x^2-4x. But the sum of the first and third is $6x+1$, and this is also a square. Suppose this square is 121. Then x is 20. So the first is 80, the second is 320, and the third is 41. These satisfy the conditions.

10. *To find three numbers such that the product of any pair of them added to a given number gives a square.*

Let the given number be 12. Since the product of the first and second numbers added to 12 gives a square, if from that square we subtract 12, we will get the product. So suppose the square is 25. If we subtract 12, the remainder, 13, will be the product of the first and second numbers. So let the first number be $13x$ and the second number $\frac{1}{x}$. Again, if we subtract 12 from another square, we will have the product of the second and third numbers. Let that square be 16. Then the remainder of 4 will be the product of the second and third numbers. But the second number is $\frac{1}{x}$, so the third number is $4x$. Finally, we know that the product of the first and third numbers added to 12 makes a square. But that product is $52x^2$, so $52x^2+12$ must be a square.

Now if 13, the coefficient of x in the first number, were a square, it would be easy to solve the equation. But it is not a square. So we must find two numbers such that their product is a square and such that if we add 12 to either, we get a square. But if each of these numbers is already a square, then the product will be a square. So we must find two squares such that, if we add 12 to either, the result is a square. This is easy, and, as we said, it makes the equation easy to solve. The numbers 4 and $\frac{1}{4}$ satisfy this condition; each of them added to 12 produces a square.

So returning to the beginning, we make the first number $4x$, the second number $\frac{1}{x}$, and the third number $\frac{1}{4}x$. We must make the product of the first and third added to 12 a square. But the product of the first and third is x^2; so x^2+12 must be a square. Let the square have the root $x+3$. Then x^2+6x+9 is the desired square and $x=\frac{1}{2}$. This produces the desired numbers.

13. *To find three numbers such that the product of any two minus the third gives a square.*

Let x be the first number and $x+4$ the second. Therefore the product of these two is x^2+4x. Since this product less the third is to be a square, if we put $4x$ as the third, then that condition is satisfied. But also the product of the second and third less the first is to be a square as is the product of the first and third less the second. But the product of the second and third less the first is $4x^2+15x$, while the product of the first and third less the second is $4x^2-x-4$. We use the method of the double equation. The difference of these two expressions is $16x+4$, so we need to find two quantities whose product is $16x+4$. These are 4 and $4x+1$. Then we either set the square of half the sum of these two quantities equal to the larger expression or the square of half the difference equal to the smaller expression. We conclude that x is $\frac{25}{20}$. Then the three desired numbers are $\frac{25}{20}$, $\frac{105}{20}$, and $\frac{100}{20}$.

Because the problems in book A and the remainder of the Arabic books involve cubes and even higher powers, Diophantus begins book A with a new introduction in which he gives the rules for multiplying such powers. But again he notes that the goal in manipulating the equation is to end up with one species equal to another species, and then to have one species equal to a number.

Book A

I have presented in detail, in the preceding part of this treatise on arithmetical problems, many problems in which we ultimately, after the restoration and the reduction, arrived at one term equal to one term, namely, those problems involving either of the two species of linear and plane numbers and also those which are composite. I have done that according to categories which beginners can memorize and grasp the nature of.

In order that you miss nothing, in treating which you would acquire ability in that science, I consider it also appropriate to write, once again, for you, in what follows, many problems of this kind, but now involving the species of number called solid alone as well as in association with one of the first two species. In it, I shall follow the same path and advance you along it from one step to another and from one kind to another for the sake of experience and skill. Then, when you are acquainted with what I have presented, you will be able to find the answer to many problems which I have not presented, since I shall have shown to you the procedure for solving a great many problems and shall have explained to you an example of each of their types.

I say the following: Every square multiplied by its side gives an x^3. When I then divide x^3 by x^2, the result is the side of x^3; if x^3 is divided by x, namely, the root of the said x^2, the result is x^2.

When I then multiply x^3 by x, the result is the same as when x^2 is multiplied by itself, and it is called x^4. If x^4 is divided by x^3, the result is x, namely, the root of x^2; if it is divided by x^2, the result is x^2; if it is divided by x, namely, the root of x^2, the result is x^3.

When x^4 is then multiplied by x, namely, the root of x^2, the result is the same as when x^3 is multiplied by x^2, and it is called x^5. If x^5 is divided by x, namely, the root of x^2, the result is x^4; if it is divided by x^2, the result is x^3; if it is divided by x^3, the result is x^2; and if it is divided by x^4, the result is x, namely, the root of x^2.

When x^5 is then multiplied by x, the result is the same as when x^3 is multiplied by itself and when x^2 is multiplied by x^4, and it is called x^6. If x^6 is divided by x, namely, the root of x^2, the result is x^5; if it is divided by x^2, the result is x^4; if it is divided by x^3, the result is x^3; if it is divided by x^4, the result is x^2; if it is divided by x^5, the result is x, namely, the root of x^2.

After the restoration and the reduction—one means by restoration the adding of what is negative to both sides of the equation and by reduction the removing of what is equal from both sides—the treatment will result for us in the equality of one of these species—the mutual multiplications and divisions of which we have explained above—with another; it will then be necessary to divide the whole by a unit of the side having the lesser degree in order to obtain one species equal to a number.

3. *We wish to find two square numbers the sum of which is a cubic number.*

We put x^2 as the smaller square and $4x^2$ as the greater square. The sum of the two squares is $5x^2$, and this must be equal to a cubic number. Let us make its side any number of x's we please, say x again, so that the cube is x^3. Therefore, $5x^2$ is equal to x^3. As the side which contains the x^2 is the lesser in degree, we divide the whole by x^2; hence x is equal to 5. Then, since we assumed the smaller square to be x^2, and since x^2 arises from the multiplication of x—which we found to be 5—by itself, x^2 is 25. And, since we put for the greater square $4x^2$, it is 100. The sum of the two squares is 125, which is a cubic number with 5 as its side. Therefore, we have found two square numbers the sum of which is a cubic number, namely, 125. This is what we intended to find.

9. *We wish to find two cubic numbers which comprise a square.*

We set $4x$ as the side of the greater cube and x as the side of the smaller cube. Then the greater cube is $64x^3$, the smaller, x^3, and the number they comprise is $64x^6$; this must be equal to a square number. We put as its side x^2's, the coefficient of which is equal to the side of the square arising from the multiplication of the 64 by the 4, namely, 256, having as its side 16. Therefore, we put as the side of the square $16x^2$, so that the square is $256x^4$. Then $64x^6$ equals $256x^4$. So we divide the whole by x^4, since the x^4's are the lower in degree of the two sides; the division of the $64x^6$ by x^4 gives $64x^2$, while we obtain 256 from the division of the $256x^4$ by x^4. Therefore, $64x^2$ equals 256, hence x^2 equals 4; x^2 being a square, as well as 4, their sides are thus equal; the side of x^2 being x, and that of 4 being 2, x is 2. Then, since we set x as the side of the smaller cube, the smaller cube is 8, and since we set $4x$, i.e., 8, as the slide of the larger cube, the larger cube is 512. When we multiply it by the smaller cube, the result is the number they comprise, namely 4096, which is a square having 64 as its side. Therefore, we have found two cubic numbers which comprise a square number, namely 8 and 512. This is what we intended to find.

11. *We wish to find a cubic number such that, when we diminish it by an arbitrary multiple of the square having the same side, the remainder is a square number.*

We set x as the side of the cube, so that the cube is x^3, and we assume 6 to be the multiplicative factor. We want the remainder of x^3 after the subtraction of the $6x^2$ to be a square. We set any number of x's we please for its side, say $2x$, so that the square is $4x^2$. Thus $x^3 - 6x^2 = 4x^2$. We restore x^3 with the $6x^2$ and add them to the $4x^2$; then x^3 equals $10x^2$. Dividing the whole by x^2 gives us x equal to 10. Then, since we assumed the side of the cube to be x, the cube is 1000. The square of the side is 100, six times which is 600, and the remainder of the 1000 after the subtraction of 600 is 400, which is a square number with 20 as its side. Therefore, we have found a cubic number such that, when we diminish it by the square of its side taken six times, the remainder is a square number; the said cube is 1000 and its side, 10.

13. *We wish to find a cubic number such that, when we diminish it by an arbitrary multiple of the square having the same side, the remainder is a cubic number.*

We put x as the side of the cube, so that the cube is x^3. We put 7 as the multiplicative factor, so that the remainder [of the subtraction] is $x^3 - 7x^2$; this, then is equal to a cubic number. We put as its side some fraction of x, say $\frac{1}{2}x$, so that the cube is one part of 8 parts of x^3; this, then, equals $x^3 - 7x^2$. We restore and reduce; hence $\frac{7}{8}x^3$ is equal to $7x^2$. Dividing then the whole by x^2 yields 7 equal to $\frac{7}{8}x$. Thus x is 8, and x^3 is 512. Then, if we subtract from the latter seven times the 64, the remainder is 64, which is a cube.

We shall now treat this problem by another method. We make the side of the first cube any number of x's, say $2x$, so that the cube is $8x^3$. Then, the difference between x^3 and $8x^3$, $7x^3$, is equal to seven times the square having the same side as the greater cube. This side being $2x$, its square, $4x^2$, and seven times that being $28x^2$, $28x^2$ is equal to $7x^3$. Dividing the whole by x^2 yields 28 equal to $7x$, so x equals 4. Thus, the smaller cube is 64, for its side was x, and the greater cube, since $2x$ was set as its side, has the side 8, while the cube is 512. Therefore, it has been found that the other cube, the larger, exceeds the smaller cube by seven times the square of the side of the larger cube; and this was the condition imposed upon us in this problem. This is what we intended to find.

22. *We wish to find a cubic number such that when we multiply it by two given numbers, the results are a cube and the side of that cube.*

It is necessary to find first the characteristic of the two given numbers. We then say: Having set x^3 for the required cube and multiplied it by the two given numbers, each of the two products is x^3's, and one of these two products is a cube having the other product as its side. Now if those x^3's of the two products which form the cube are divided by those which form the side, the resulting quotient is a number, equal to the square of the x^3's forming the side. Consequently, the number resulting from the division must be a square in order that its side may be set equal to the x^3's forming the side. Thus we shall suppose the two given numbers to be such that the division of the one by the other produces a square. Again, the number which is the side of the square number resulting from the division is equal to the x^3's, which are the side and which have their coefficient equal to that one of the two given numbers which is the divisor; so it is necessary that the division of the said number by the [coefficient of the] x^3's equal to it produces a cube, in order that x^3 be equal to a cubic number. Hence the characteristic of these two numbers is now in its complete form, which is: the division of the one by the other results in a square and the division of the side of this square by the divisor results in a cube.

We must now determine these two numbers. We assume the first to be 2 and we wish to find the second. Since the result of the division of one of these two numbers by the other is a square, the side of which, when divided by the divisor, gives a cube, we have to seek a number which, when divided by 2, gives a cube;[24] such is $6 + \frac{1}{2} + \frac{1}{4}$. Now $6 + \frac{1}{2} + \frac{1}{4}$ is the side of the square arising from the division of one of the two given numbers by the other, the square generated by the $6 + \frac{1}{2} + \frac{1}{4}$ being $45 + \frac{1}{2} + \frac{1}{2} \cdot \frac{1}{8}$, and the number from which

[24] Suppose the cube is $\left(\frac{3}{2}\right)^3$. Then the side of the square is $2 \cdot \left(\frac{3}{2}\right)^3 = 6\frac{3}{4}$, as stated.

that number arises by division of it by 2 being $91\frac{1}{8}$. The second number we were looking for is $91\frac{1}{8}$. . . .

So one of the two given numbers is 2 and the other, $91\frac{1}{8}$, and we wish to find a cubic number which when multiplied by $91\frac{1}{8}$ gives a cube and which when multiplied by 2 gives the side of that cube. We set x^3 as the cube and proceed as we did in the previous problems.[25] Then we shall find that the required cube is $3\frac{3}{8}$. The multiplication of it by $91\frac{1}{8}$ gives a cube, namely, 307 and 35 parts of 64 parts; and the same number when multiplied by 2 gives $6+\frac{1}{2}+\frac{1}{4}$, which is the side of the cube 307 and 35 parts of 64 parts. Therefore, we have found a cubic number fulfilling the condition imposed upon us. This is what we intended to find.

25. *We wish to find two numbers, one square and the other cubic, such that the sum of their squares is a square.*

We put x as the side of the cube, so that the cube is x^3, and any number of x's, say $2x$, as the side of the square, which is then $4x^2$. The square of the cube is x^6 and the square of the square, $16x^4$; their sum is x^6+16x^4, and this is equal to a square number. It is then necessary to determine the number which is the side of the square. We say then: If we put as the said side x^2's, the square equal to x^6+16x^4 is x^4's; after the subtraction of the $16x^4$, which is common, from both sides, there remain x^4's equal to x^6, and the division of the two by x^4, which constitutes the lower in degree of the two sides, gives x^2 equal to a number. This number, being equal to x^2, must be a square. But the said number is the excess of the [coefficient of the] x^4's in a square number over 16. Thus it is necessary that the coefficient of the x^4's is a square number exceeding 16 by a square number. Consequently, we are led to search for two square numbers having 16 as their difference. We then find 25 for the larger square and 9 for the smaller square. So we put $25x^4$, the side of which is $5x^2$, as the square equal to x^6+16x^4. Removing the $16x^4$, which is common, from both sides, we obtain x^6 equal to $9x^4$. Hence x^2 equals 9. As x^2 is a square with side x and 9 a square with side 3, x is 3. Since we assumed the side of the cube to be x, the side is 3 and the cube, 27. And, since we assumed the side of the square to be $2x$, the side is 6 and the square, 36. The square of 27 is 729 and the square of 36, 1296; the sum of these is 2025, which is a square with 45 as its side. Therefore, we have found two numbers, one cubic and the other square, such that the sum of their squares is a square; and these are 27 and 36. This is what we intended to find.

31. *We wish to find two numbers, one square and the other cubic, such that the excess of the square of the square over the cube of the cube is a square number.*

We set x^3 as the cube, so that its cube is x^6 [multiplied] by x^3, which is the so-called x^9. We put the side of the square $2x^2$, so that the square is $4x^4$ and its square, $16x^4$ [multiplied] by x^4, that is to say, [16 times] the so-called x^8. Thus the $16x^8$, which is the square of the square number, [must] exceed x^9 by a square number. Let us put $2x^4$ as the side of that square; as the result of the division of any square by its side equals the said side, the result of the division of $16x^8-x^9$ by $2x^4$ equals $2x^4$. But as $16x^8$ results from the multiplication of $16x^4$ by x^4, the division of it by $2x^4$ gives $8x^4$; and, as x^9 results from the

[25] Since $2x^3=\sqrt[3]{91\frac{1}{8}x^3}$, we get $2x^3=4\frac{1}{2}x$, or $x^2=\frac{9}{4}$, or, finally, $x=\frac{3}{2}$. Then $x^3=3\frac{3}{8}$.

multiplication of x^6 by x^3, while x^6 is the product of x^4 and x^2, x^9 is the product of x^4 and x^5, and, thus, the result of the division of x^9 by $2x^4$ is $\frac{1}{2}x^5$. Hence we obtain, from the division [of $16x^8 - x^9$ by $2x^4$], $8x^4 - \frac{1}{2}x^5$, and this is equal to $2x^4$. We make $\frac{1}{2}x^5$ common by adding it to both sides, so that we have $8x^4$ equal to $2x^4 + \frac{1}{2}x^5$. Let us remove the $2x^4$, which is common, from both sides, so $\frac{1}{2}x^5$ equals $6x^4$; after the division, we obtain $\frac{1}{2}x$ equal to 6, hence x is equal to 12. Since we put x as the side of the cube, the side is 12 and the cube, 1728; and, since we put $2x^2$ as the side of the square, and x^2 is 144—for x is 12—the side of the square is 288 and the square, 82,944. The cube of the cube is 5,159,780,352 and the square of the square, 6,879,707,136; the excess of the latter number over the cube of the cube is 1,719,926,784, which is a square number with 41,472 as its side. Therefore, we have found two numbers fulfilling the condition required by us, and these are 1728 and 82,944. This is what we intended to find.

34. *We wish to find two numbers, one cubic and other square, such that the cube when increased by the square gives a square number and when decreased by the square also gives a square number.*

We put x^3 as the cube and $4x^2$ as the square; then, $x^3 + 4x^2$ is equal to a square number and $x^3 - 4x^2$ is likewise equal to a square number.

We treat that by the method of the double equation. We take the difference between the two said squares, namely, $8x^2$, and seek two numbers [of x's] such that the multiplication of the one by the other give $8x^2$; such are $2x$ and $4x$. Their difference is $2x$, half of which is x. The square of x is x^2, and this equals $x^3 - 4x^2$. Adding then the $4x^2$ in common to both sides, we obtain x^3 equal to $5x^2$. Again, if we add the $2x$ to the $4x$, we obtain $6x$; half of it is $3x$, the square of which is $9x^2$, and this equals $x^3 + 4x^2$. Removing then the $4x^2$, which is common, from both sides, we obtain x^3 equal to $5x^2$. Thus the (resulting) equation for the two equations [of the proposed system] turned out to be the same, ending in each one with x^3 equal to $5x^2$. Let us divide all of this by x^2; we obtain x equal to 5. Thus the side of the cube is 5 and the cube, 125, and the side of the square is 10 and the square, 100. The addition of the 100 to the cubic number results in 225, which is a square number with side 15; and, the subtraction of the same from the cubic number gives 25, which is a square with side 5.

We also treat this [problem] by the procedure avoiding the double equation. We say: $x^3 + 4x^2$ is equal to a square number. If we put x's for its side, the square is x^2's, [which are] equal to $x^3 + 4x^2$. The subtraction of the $4x^2$, which is common, from both sides leaves x^3 equal to x^2's, and consequently, the number assumed to be x in the problem equals the coefficient of the x^2's left over. Again, [we say] $x^3 - 4x^2$ is equal to a square number. If we also put x's for its side, the square is x^2's. The addition of the $4x^2$ in common to both sides results in x^3 equal to x^2's, and, consequently, the number assumed to be x in the problem equals the coefficient of the x^2's added up. Therefore, it is necessary that the coefficient of the x^2's left over in the first equation be equal to the coefficient of the x^2's added up in the second equation. But the [coefficient of the] x^2's left over in the first equation is the remainder of a square number after subtracting 4, while the [coefficient] of the x^2's added up in the second equation is a number formed by the addition of a square number and 4. Thus we shall seek two square numbers such that the larger diminished by 4 and the smaller increased by 4 be equal. So we must look for two square numbers having 8 as their difference. Such are $12\frac{1}{4}$ and $20\frac{1}{4}$. We put for the greater square, which is equal to x^3+4x^2, $20\frac{1}{4}x^2$, and for the smaller square, which is equal to x^3-4x^2, $12\frac{1}{4}x^2$. Thus, in both equations, we shall end up with x^3 equal to $16\frac{1}{4}x^2$; hence x is equal to $16\frac{1}{4}$. Since we set x as the side of the cube, the side of the cube is $16\frac{1}{4}$ and the cube, 4291 and one part of 64 parts of 1; and since we set $2x$ as the side of the square, the said side is $32\frac{1}{2}$ and the square $1056\frac{1}{4}$. The addition of the latter to the cubic number results in 5347 and 17 parts of 64, which is a square number with side $73\frac{1}{8}$, and the subtraction of the same from the cubic number gives 3234 and 49 parts of 64 parts of 1, which is a square with side $56\frac{7}{8}$.

Therefore, we have found two numbers, one cubic and the other square, such that the cubic number when increased by the square number gives a square number, and when decreased by the square number also gives a square number.

Book B

1. *We wish to find two numbers, one square and the other cubic, such that when we add to the square of the square a given multiple of the cubic number, the result is a square number, and when we subtract from the same another given multiple of the cubic number, the remainder is a square number.*

Let the positive multiplier be 4 and the negative one, 3. We wish to find two numbers as indicated by us. We put x as the side of the square, so that the square is x^2 and the square of the square, x^4; the latter, together with four times a certain cube, is equal to a square, and minus three times the same cube, is again equal to a square. Hence the cube is equal to a certain quantity, having to x^4 a given ratio, and such that four times it when added to x^4 gives a square and three times it when subtracted from x^4 leaves a square. So we shall seek three square numbers such that the excess of the largest over the middle is to the excess of the middle over the smallest as four is to three. Let these numbers be 81, 49 and 25, x^4 being put 49 parts, the quantity given in ratio to x^4 such that four times it—i.e., 32 parts of 49 parts (of x^4)—when added to x^4 gives a square and three times it—i.e., 24 parts of 49 parts (of x^4)—when subtracted from x^4 leaves a square, is 8 parts of 49 parts of x^4. So the required cube is equal to 8 parts of 49 parts of x^4. Let us put as the side of the cube an arbitrary number of x's, say $2x$; so the cube is $8x^3$. Hence $8x^3$ is equal to 8 parts of 49 parts of x^4. Let us divide both by x^3, so 8 parts of 49 parts of x equals 8; hence x is equal to 49. Thus the side of the square is 49 and the square is 2401. Since we put $2x$ as the side of the cube, the said side is 98 and the cube, 941,192. So the square of the square is 5,764,801. When increased by four times the cubic number, that is, by 3,764,768, it results in 9,529,569, which is a square with 3087 as its side; and, when the same is decreased by three times the cubic number, that is, by 2,823,576, it results in 2,941,225, which is a square number with 1715 as its side.

Therefore, we have found two numbers fulfilling the condition required by us. This is what we intended to find.

7. *We wish to find two numbers such that their sum and the sum of their cubes are equal to two given numbers.*

It is necessary that four times that one of the two numbers which is given for the sum of the cubes of the two required numbers exceed the cube of the number given for their sum by a number which, when divided by three times the number given for the sum of the two numbers, gives a square, and which, when multiplied by three quarters of the number given for the sum of the two numbers, gives a square.[26] This [problem] belongs to the [category of] constructible problems.

Let the number given for the sum of the two numbers be 20 and the number given for the sum of their cubes be 2240. We wish to find two numbers such that their sum is 20 and the sum of their cubes, 2240. We put $2x$ as the difference of the two numbers, so that one is $10+x$ and the other, $10-x$. We form from each of them a cube. Now, whenever we wish to form a cube from [some] side made up [of the sum] of two different terms—so that a multitude of terms does not make us commit a mistake—we have to take the cubes of the two different terms, and add to them three times the results of the multiplication of the square of each term by the other; then, the result is composed of four terms, and this is the cube arising from the sum of the two different terms. But when the two terms are such that one is subtracted from the other, we take the cube of the larger, add to it three times the result of the multiplication of the square of the smaller term by the larger term, and subtract from them the cube of the smaller term and three times the result of the multiplication of the square of the larger term by the smaller; the result is then the cube arising from the difference between the two different terms. Hence the cube arising from the side $10+x$ is the sum of the cube of 10, or 1000, and of the cube of x, or x^3, plus three times the result of the multiplication of 10 by the square of x, or $30x^2$, plus, again, three times the result of the multiplication of x by the square of 10, or $300x$; thus, the cube arising from $10+x$ is $1000+x^3+300x+30x^2$. Again, the cube arising from the side $10-x$ is also equal to the cube of 10, or 1000, and to three times the result of the multiplication of 10 by the square of x, i.e., $30x^2$, minus the cube of x, or x^3, and minus three times the result of the multiplication of x by the square of 10 or $300x$; thus, the cube arising from $10-x$ is $1000+30x^2-x^3-300x$. The sum of these two cubes is $2000+60x^2$, because the subtracted x^3+300x in the one cube is cancelled by the added x^3+300x in the other. Then $2000+60x^2$ is equal to 2240. Let us subtract the 2000 which is in one side from the number which is in the other side, whence $60x^2$ equals 240; thus x^2 is 4. And, each of these being a square, their sides are also equal; but the side of x^2 is x, and the side of 4 is 2, so that x is 2. Since we put as the larger of the two required numbers $10+x$, the said number is 12; and, since we put as the smaller number $10-x$, it is 8. The cube of the larger number is 1728 and the cube of the smaller number, 512; and their sum is 2240.

Therefore, we have found two numbers such that their sum is 20 and the sum of their cubes, 2240; and these are 12 and 8. This is what we intended to find.

10. *We wish to find two numbers such that their difference is a given number and the difference of their cubes is to the square of their sum in a given ratio.*

It is necessary that the number belonging to the given ratio be greater than three quarters of the number given for the difference of the two numbers by a number comprising,

[26] If k is the sum of the two numbers and ℓ the sum of the cubes, then this condition states that $(4\ell-k^3)/3k$ is a square, or, alternatively, $(4\ell-k^3)\frac{3}{4}k$ is a square.

together with the cube of the number given for the difference of the two required numbers, a square number.[27]

Let the number given for the difference of the two required numbers be 10 and the number corresponding to the given ratio be $8\frac{1}{8}$. We wish to find two numbers such that their difference is 10 and the ratio of the difference of their cubes to the square of their sum is the ratio $8\frac{1}{8}:1$. We put $2x$ as their sum, and we set as one of the two numbers $x+5$ and as the other $x-5$ in order that their difference be 10. We take the difference between their cubes, namely $250+30x^2$. The square of the sum of the two numbers being $4x^2$, $250+30x^2$ equals $8\frac{1}{8}$ times $4x^2$, i.e., $32\frac{1}{2}x^2$. Let us remove the $30x^2$, which is common, from both sides, so 250 is equal to $2\frac{1}{2}x^2$; thus x^2 equals 100, and therefore x is 10. Since we set as the first number $x+5$, it is 15; and, since we set as the second number $x-5$, it is 5. The cube of 15 is 3375 and the cube of 5, 125, the difference of which is 3250; the square of the sum of the two numbers is 400, and the ratio of 3250 to 400 is the ratio $8\frac{1}{8}:1$.

Therefore, we have found two numbers such that their difference is 10 and the difference of their cubes is $8\frac{1}{8}$ times the square of their sum; and these are 15 and 5. This is what we intended to find.

16. *We wish to find a cubic number such that, when we subtract from a given multiple of the square of its side a given number, the result is equal to the sum of two numbers such that the subtraction of the one from the cube results in a cube, and the subtraction of the cube from the other number results in a cube.*

Let the given multiplier be 9 and the given number be 16. We wish to find a cubic number such that when we subtract 16 from nine times the square of its side the result is equal to the sum of two numbers such that the subtraction of the one from the cube results in a cube, and the subtraction of the cube from the other number results in a cube. We put x^3 as the cube, and we subtract 16 from nine times the square of its side. We form two cubes having as their sides x minus a number and a number minus x, and let the sum of the x^2's occurring in them amount to $9x^2$. Thus we form the first cube from the side $x-1$, so that it is $x^3+3x-3x^2-1$, and the second cube from the side $2-x$, so that it is $8+6x^2-x^3-12x$. Then, when the $3x^2+1-3x$ is subtracted from the required cube (or x^3), the result is a cube, which is, as already said, $x^3+3x-3x^2-1$; and, when the required cube, or x^3, is subtracted from $6x^2+8-12x$, the result is a cube, which is, likewise as already said, $8+6x^2-12x-x^3$. So let their sum be put equal to $9x^2-16$. But their sum is $9x^2+9-15x$, hence this is equal to $9x^2-16$. Let us restore and reduce that. We arrive, after the restoration and the reduction, at $15x$ equal to 25; hence x is $1\frac{2}{3}$, and this is the side of the cube, so that the cube is 4 and 17 parts of 27 parts of 1. The square of the side of the cube is 2 and 21 parts of 27 parts, and nine times that is 25. Let us subtract the 16 from it; the remainder is 9. And we had assumed that that one of the two numbers having 9 as their sum which is subtracted from the cube is $3x^2+1-3x$; as $3x^2$ is $8\frac{1}{3}$ and $3x$ is 5, the aforesaid number is $4\frac{1}{3}$, and the second number is the remainder of the 9 [after the subtraction of $4\frac{1}{3}$], namely, $4\frac{2}{3}$. If $4\frac{1}{3}$ is subtracted from the cube, that is, from 4 and 17 parts of 27 parts, the remainder is 8 parts of 27 parts of 1,

[27] If the ratio is ℓ and the difference is k, then the condition is that $(\ell-\frac{3}{4}k)k^3$ is a square.

which is a cube with side $\frac{2}{3}$; and if the second number, or $4\frac{2}{3}$, is diminished by the cube, the remainder is one part of 27 parts of 1, which is a cube with side $\frac{1}{3}$.

Therefore, we have found a cubic number fulfilling the condition stipulated by us. This is what we intended to find.

Book C

12. *We wish to find two square numbers such that the quotient of the larger divided by the lesser, when added to the larger, gives a square, and also when added to the lesser, gives a square.*

Let us put x^2 as the smaller number; we take $\frac{1}{2}x^2 + \frac{1}{2} \cdot \frac{1}{8}x^2$ as the quotient of the larger divided by the lesser. Thus, the addition of this quotient to x^2 gives a square. So the larger number is $\frac{1}{2}x^4 + \frac{1}{2} \cdot \frac{1}{8}x^4$. Then, when we increase it by $\frac{1}{2}x^2 + \frac{1}{2} \cdot \frac{1}{8}x^2$, we obtain $\frac{1}{2}x^4 + \frac{1}{2} \cdot \frac{1}{8}x^4 + \frac{1}{2}x^2 + \frac{1}{2} \cdot \frac{1}{8}x^2$, which has to be a square number. Hence, let us seek a square number which, when diminished by $\frac{1}{2} + \frac{1}{2} \cdot \frac{1}{8}$, gives a square number, and let us keep in mind that the remaining square be less than 81 parts of 256 parts of 1. Finding that is easy from what has been explained in the second Book [2:10]. The sought number is 169 parts of 256 parts of the unit, with side 13 parts of 16 parts of the unit. When we subtract $\frac{1}{2} + \frac{1}{2} \cdot \frac{1}{8}$, or 144 parts of 256 parts, from 169 parts of 256 parts of the unit, the remainder is 25 parts of 256 parts of the unit, which is a square number with side 5 parts of 16 parts. So let us put, as the side of $\frac{1}{2}x^4 + \frac{1}{2} \cdot \frac{1}{8}x^4 + \frac{1}{2}x^2 + \frac{1}{2} \cdot \frac{1}{8}x^2$, 13 parts of 16 parts of x^2; we multiply it by itself, whence 169 parts of 256 parts of x^4, which then equal $\frac{1}{2}x^4 + \frac{1}{2} \cdot \frac{1}{8}x^4 + \frac{1}{2}x^2 + \frac{1}{2} \cdot \frac{1}{8}x^2$. Let us remove the $\frac{1}{2}x^4 + \frac{1}{2} \cdot \frac{1}{8}x^4$ which is common, so 25 parts of 256 parts of x^4 equal $\frac{1}{2}x^2 + \frac{1}{2} \cdot \frac{1}{8}x^2$, and let us multiply the whole by 10 and 6 parts of 25 [i.e., $\frac{256}{25}$]; we obtain x^4 equal to $5x^2$ and 19 parts of 25 parts of x^2 [i.e., $\frac{144}{25}x^2$]. We divide the two sides by x^2, hence x^2 is equal to 5 and 19 parts of 25 parts of the unit. We had put x^2 as the smaller number, so it is 5 and 19 parts of 25 parts of the unit; let us multiply that by 25; it then becomes 144, which is part of 25 parts. And, since we set for the larger number $\frac{1}{2}x^4 + \frac{1}{2} \cdot \frac{1}{8}x^4$, it is 11,664 parts of 625 parts of the unit. Let us make the 144 parts of 25 parts, which form the smaller square, parts of 625; in other words let us multiply them by 25; then the smaller square is 3600 parts of 625. The quotient of the larger square divided by the smaller square is 3 and 6 parts of 25 parts of the unit. Let us make that parts of 625, so it becomes 2025 parts of 625. The addition of this to the larger square, that is, to 11,664 parts of 625, gives 13,689 parts of 625 parts of the unit, which is a square number with side 117 parts of 25 parts. Again, let us add the 2025 parts of 625 to the smaller square, that is, to 3600 parts of 625, so the result is 5625 parts of 625, which is a square number with side 75 parts of 25.

Therefore, we have found two numbers fulfilling the condition imposed upon us, and these are 11,664 parts of 625 parts of the unit and 3600 parts of 625 parts of the unit. This is what we intended to find.

23. *We wish to find two square numbers such that, a given square number being divided by each of them and the results of the two divisions being added, the result is a square number, and such that when the three numbers—that is to say, the two required numbers and the given number—are added, the result is a square.*

Let the given square number be 9. We wish to find two square numbers such that, 9 being divided by each of them and the results of the divisions being added, this gives a square number, and such that when the three numbers—that is to say, the two required numbers and the given 9—are added, the result is a square number. Now, whenever we divide a square number into two square parts and then divide a square number by each of the two parts, the sum of the results of the divisions is a square number. So let us take a square number, and let us divide it into two square parts. The number we take is x^2, and we divide it into two square parts, which are, say, 9 parts of 25 parts of x^2 and 16 parts of 25 parts of x^2; let these two parts be the two required numbers. We divide 9 by 9 parts of 25 parts of x^2; it becomes 25 parts of x^2. We also divide 9 by 16 parts of 25 parts of x^2, thus obtaining as a quotient 14 parts and $\frac{1}{2} \cdot \frac{1}{8}$ of a part of x^2. The addition of the results of the two divisions gives 39 parts and $\frac{1}{2} \cdot \frac{1}{8}$ of a part of x^2, which is a square number with side 6 parts and $\frac{1}{4}$ of a part of x. Now, if we add the three numbers, namely, the two required numbers and the given 9, the result is x^2+9, which has to be a square. Let us put $x+1$ as its side; we multiply it by itself and obtain x^2+2x+1, and this equals x^2+9. Remove x^2+1 from the two sides so as to have a single term equal to a single term; so $2x$ is equal to 8, hence x is 4. One of the two required numbers was 16 parts of 25 parts of x^2, and its side is $\frac{4}{5}x$, so its side is $\frac{4}{5}$ of 4, or $\frac{16}{5}$. This, when multiplied by itself, results in 256 parts of 25, which is one of the two required numbers. Again, the other number was 9 parts of 25 parts of x^2, and its side is $\frac{3}{5}x$; x being 4, the side is $\frac{12}{5}$. This, when multiplied by itself, results in 144 parts of 25 parts of 1, which is the second required number. If we divide the given number, that is, 9, or 225 parts of 25 parts, by the first number, that is, by 256 parts of 25, the result of the division is 225 parts of 256 parts; again, dividing the 9, that is, the 225 parts of 25, by the other number, that is, by the 144 parts of 25 parts, gives as a quotient 225 parts of 144 parts, or 400 parts of 256 parts. The addition of that to the result of the division of 9 by the other number, that is, to 225 parts of 256, gives 625 parts of 256, which is a square number, with side 25 parts of 16 parts of 1. Then, the addition of the three numbers, namely, 256 parts of 25 parts of the unit, 144 parts of 25, and 9, or 225 parts of 25, gives 625 parts of 25, or 25, which is a square number with side 5.

Therefore, we have found two numbers fulfilling the condition imposed upon us, and these are 256 parts of 25 parts of 1 and 144 parts of 25 parts of the unit. This is what we intended to find.

Book D

7. *We wish to divide a square number of cubic side into three parts such that the sum of any two is a square.*

Let us put x^3 as the side of the square number, so that the square number is x^6. We wish to divide x^6 into three parts such that the sum of any two is a square. Let us then seek three numbers such that any two when added give a square, and such that the number formed by the sum of the three numbers is a square. Finding that is easy on the basis of what we have expounded in the sixth problem of the third Book. So the first number is 80, the second, 320, and the third, 41; the sum of the three numbers is 441. Let us take x^4's instead of the units. Then the sum of the three numbers is $441x^4$, which is equal to

x^6. We divide the two sides by x^4; the division of x^6 by x^4 results in x^2 and the division of $441x^4$ by x^4 results in 441. So 441 is equal to x^2, thus x^2 is 441. Hence x^4 is the product of the multiplication of 441 by itself, that is, 194,481. Since we put for the first of the three parts $80x^4$, this part is 15,558,480; again, since we put for the second part $320x^4$, it is 62,233,920; again, since we put for the third part $41x^4$, it is 7,973,721. As the number which had to be divided into these three parts is the number formed by their sum, it is 85,766,121, which is a square number with side 9261, and the said side is a cubic number with side 21. Since the first of the three parts is 15,558,480 and the second part is 62,233,920, the number formed by their sum is 77,792,400, which is a square number with side 8820; again, since the second part is 62,233,920 and the third part is 7,973,721, the number formed by their sum is 70,207,641, which is a square number with side 8379; again, since the third part is 7,973,721 and the first part is 15,558,480, the number formed by their sum is 23,532,201, which is a square number with side 4851.

Therefore, we have found a number fulfilling the condition imposed upon us, and this is 85,766,121. This is what we intended to find.

11. *We wish to divide a given square number into two parts such that the addition to the said square of one of them gives a square and the subtraction from the same of the other one gives a square.*

Let the given number be 25. We wish to divide 25 into two parts such that adding the one part to 25 gives a square number and subtracting the other part from 25 gives a square. Let us aim to find a certain square which we shall divide into two parts such that adding the one to it and subtracting the other from it give, after the addition and the subtraction, a square. But if we add to x^2 twice its root plus 1, the result is x^2+2x+1, which is a square number, and if we subtract from x^2 twice its root minus 1, the remainder is x^2, plus 1, minus two roots [of x^2], which is a square number. Now, we want the sum of the added and of the subtracted number to be x^2; their sum being $4x$, $4x$ equals x^2. The division of the whole by x gives x equal to 4; and, since x is the side of x^2, x^2 is 16. The number added to x^2 was $2x+1$, which is 9, and the number subtracted from x^2, was $2x-1$, which is 7; and the addition of 9 and 7 results in 16. Hence we have attained the object of our investigation.

But the given square number was 25; so let us multiply 9 by 25, which gives 225, and divide that by 16; we obtain 225 parts of 16 parts, which is one of the two parts of 25, namely, the added part. Again, let us multiply 7 by 25, so we obtain 175, and we divide that by 16; this gives 175 parts of 16, which is the second part, namely, the one subtracted from 25. It appears that, adding the 225 parts to 25, that is, to 400 parts of 16, gives 625 parts of 16, which is a square number with side 25 parts of 4; again, subtracting the other part, namely, 175 parts of 16, from the 400 parts gives 225 parts of 16, which is a square number with side 15 parts of 4; and the sum of the two parts is 25.

Therefore, we have divided the 25 into two parts fulfilling the condition imposed upon us, and these are 225 parts of 16 and 175 parts of 16. This is what we intended to do.

And since it is not possible to find a square number such that, dividing it into two parts and increasing it by each of the parts, we obtain [in both cases] a square, we shall now present something which is possible.[28]

[28] If we could find such a number a^2, then the conditions would imply that $3a^2$ is the sum of two squares, or, equivalently, that 3 is the sum of two squares. Evidently, Diophantus knew that this was impossible.

12. *We wish to divide a given square number into two parts such that when we subtract each from the said square the remainder is* [*in both cases*] *a square.*

Let the given number be 25. We wish to divide 25 into two parts such that when we subtract each from 25 the remainder is a square. So let us seek this condition in some square. Now, for any square which is divided into two square parts, the subtraction of each of the two parts from the square gives a square, which is the other part; the way of performing that has been seen earlier in this treatise of ours [2:8]. One of the two parts is 16 and the other, 9.

Therefore, we have divided the 25 into two parts such when we subtract each from 25 the remainder is a square, and these are 9 and 16. This is what we intended to do.

13. *We wish to divide a given square number into three parts such that the addition of each to the said square gives a square.*

Let the given number be 25. We wish to divide 25 into three parts such that the addition of each to 25 gives a square. Now, the division of a square number into three parts and the addition of each one to the divided number produce three numbers such that the number formed by their sum equals four times the divided number; therefore, if we divide 25 into three parts and add each part to 25, the sum of the three [resulting] numbers is 100. Hence, let us divide 100 into three square parts and let each part be larger than 25. It has been seen earlier in this treatise of ours how to divide any square number into square parts [2:8], and we shall dispense with repetition of the treatment. So, the first part is 36, the second 30 and 370 parts of 841 parts of the unit, and the third, 33 and 471 parts of 841 parts of the unit. Since each of these three parts is composed of 25 and of one of the parts of 25, if we subtract 25 from each of these three parts, the remainder of each part is one of the parts of 25. Now, subtracting 25 from 36 gives 11, which is the first of the parts of 25. Again, let us subtract 25 from the second [found] part, that is, from 33 and 471 parts of 841; the remainder is 8 and 471 parts of 841, which is the second of the parts of 25. Again, subtracting 25 from the third [found] part, that is, from 30 and 370 parts of 841, gives 5 and 370 parts of 841, which is the third of the parts of 25. Indeed, adding these three parts together gives 25, while increasing 25 by each of them results in a square number.

Therefore, we have divided the 25 into three parts such that the addition of each to the 25 gives a square number; and these parts are the following: the first is 11; the second is 8 and 471 parts of 841 parts of the unit; the third is 5 and 370 parts of 841. This is what we intended to do.

16. *We wish to find three square numbers which are in* [*continuous*] *proportion such that the subtraction of the first from the second gives a square and the subtraction of the second from the third gives a square.*

It is the nature of [any] three square numbers which are also in proportion and are such that the subtraction of the first from the second gives a square, that the subtraction of the second from the third [also] gives a square. Let us then put as the first number 1 and as the third number x^4; thus the second number is x^2. Now the subtraction of the first number, or 1, from the second number, or x^2, gives x^2-1, which must be a square number. Let us then put as its side $x-2$, which we multiply by itself; hence we obtain x^2+4-4x. This, then, equals x^2-1. We add to the two sides $4x+1$, so x^2+4x equals x^2+5; removing the x^2, which is common, gives 5 equal to $4x$, hence x is $1\frac{1}{4}$. Since we assumed the second number to be x^2, and since the side of x^2 is x, which is $1\frac{1}{4}$, or 5 parts of 4, x^2 is 25 parts of 16 parts of the unit. And, the third number was assumed to be x^4, which is the product of the multiplication of x^2 by itself, or 625 parts of 256; thus the third number is 625 parts of 256 parts of the unit. The first number is as set by us, 1. Now, the subtraction of the first number, namely 1, from the second number, namely 25 parts of 16, gives 9 parts of 16 parts of the unit, which is a square number with 3 parts of 4 as its side. Again, the subtraction of the second number, namely 25 parts of 16 parts, or 400 parts of 256 parts, from the third number, namely 625 parts of 256, gives 225 parts of 256 parts of the unit, which is a square number with side 15 parts of 16.

Therefore, we have found three numbers fulfilling the condition imposed upon us, and these are 1,400 parts of 256, and 625 parts of 256. This is what we intended to find.

Book 4

18. *To find two numbers such that the cube of the first added to the second shall make a cube, and the square of the second added to the first shall make a square.*

Let the first number be x. Then the second will be a cube number less x^3, say $8-x^3$. And the cube of the first, added to the second, makes a cube. There remains the condition that the square on the second, added to the first, shall make a square. But the square on the second, added to the first, is $x^6+x+64-16x^3$. Let this be equal to $(x^3+8)^2$, that is, to x^6+16x^3+64. Then, by adding or subtracting like terms, $32x^3=x$; and, after dividing by x, $32x^2=1$.

Now 1 is a square, and if $32x^2$ were a square, my equation would be solvable. But $32x^2$ is formed from $2 \cdot 16x^3$, and $16x^3$ is $(2 \cdot 8)x^3$, that is, it is formed from $2 \cdot 8$. Therefore, $32x^2$ is formed from $4 \cdot 8$. My problem therefore becomes to find a cube which, when multiplied by 4, makes a square. Let the number sought be y^3. Then $4y^3$ equals a square, say $16y^2$; whence y is 4. Returning to the condition, the cube will be 64. I therefore take the second number as $64-x^3$. There remains the condition that the square on the second added to the first shall make a square. But the square on the second added to the first is $x^6+4096+x-128x^3$. To be a square, let it be $(x^3+64)^2$, or $x^6+4096+128x^3$. On taking away the common terms, $256x^3=x$, and x is $\frac{1}{16}$.

Returning to the condition, the first number is $\frac{1}{16}$ and the second number is $\frac{262143}{4096}$.

29. *To find four square numbers such that their sum added to the sum of their sides shall make a given number.*

Let the number be 12. Since any square added to its own side and $\frac{1}{4}$ makes a square, whose side minus $\frac{1}{2}$ is the number which is the side of the original square, and the four numbers added to their own sides make 12, then if we add $4 \cdot \frac{1}{4}$ they will make four squares. But $12+4 \cdot \frac{1}{4}$ is 13. Therefore it is required to divide 13 into four squares, and then, if I subtract $\frac{1}{2}$ from each of their sides, I shall have the side of the four squares. Now 13 may be divided into two squares, 4 and 9. And again, each of these may be divided into two squares, $\frac{64}{25}$ and $\frac{36}{25}$, and $\frac{144}{25}$ and $\frac{81}{25}$. I take the side of each, $\frac{8}{5}, \frac{6}{5}, \frac{12}{5}, \frac{9}{5}$, and subtract half from each side. The sides of the required squares will be $\frac{11}{10}, \frac{7}{10}, \frac{19}{10}, \frac{13}{10}$. The squares themselves are therefore respectively, $\frac{121}{100}, \frac{49}{100}, \frac{361}{100}, \frac{169}{100}$.

32. *To divide a given number into three parts such that the product of the first and second added to the third shall make a square and the third subtracted from the product of the first and second is also a square.*

Let the given number be 6. Let the third part be x and the second part any number less than 6, say 2; then the first part is $4-x$ and the two remaining conditions are that the product of the first and second added to the third is a square and the third subtracted from the product of the first and second is also a square. There results the double equation $8-x$ is a square and $8-3x$ is a square. And this does not give a rational result since the ratio of the coefficients of x is not the ratio of a square to a square. But the coefficient 1 of x is $2-1$ and the coefficient 3 of x likewise is $2+1$. Therefore my problem resolves itself into finding a number to take the place of 2 such that [the number $+1$] bears to [the number -1] the same ratio as a square to a square.

Let the number sought be y; then [the number $+1$] is $y+1$ and [the number -1] is $y-1$. We require these to have the ratio of a square to a square, say $4:1$. Now the product of $y-1$ and 4 is $4y-4$ and the product of $y+1$ and 1 is $y+1$. And these are the numbers having the ratio of a square to a square. Now I put $4y-4$ equal to $y+1$, giving $y=\frac{5}{3}$. Therefore, I make the second part $\frac{5}{3}$, for the third is x; and therefore the first is $\frac{13}{3}-x$. There remain the conditions that the product of the first and second plus the third is a square and the difference of that product and the third is also a square. But the product of the first and second plus the third is $\frac{65}{9}-\frac{2}{3}x$, a square, and the product of the first and second less the third is $\frac{65}{9}-2\frac{2}{3}x$ is also a square. Multiply throughout by 9, getting $65-6x$ equals a square and $65-24x$ equals a square. Equating the coefficients of x by multiplying the first equation by 4, I get $260-24x$ is a square and $65-24x$ is a square. Now I take their difference, which is 195 and split it into the two factors 15 and 13. Square the half of their difference and equating the result to the lesser square, I get x equals $\frac{8}{3}$. Returning to the conditions, the first part will be $\frac{5}{3}$, the second $\frac{5}{3}$, and the third $\frac{8}{3}$. And the proof is obvious.

In the following problem, Diophantus gives details for solving a quadratic equation of the form $ax^2=bx+c$. Although he is in fact solving an inequality, he goes through the basic steps, known for two thousand years, of squaring half the coefficient of x, multiplying the coefficient of x^2 by the constant term, adding those two results together, taking the square root, and then adding half the coefficient of x. Given that Diophantus had promised to show how to solve such equations, but that details are not given in any known introduction to one of the extant books, we can surmise that he provided these details in one of the missing books and, therefore, conclude that there is a book missing before this fourth Greek book.

39. *To find three numbers such that the difference of the greatest and the middle has to the difference of the middle and the least a given ratio, and further such that the sum of any two is a square.*

Let it be supposed that the difference of the greatest and the middle has to the difference of the middle and the least the ratio $3:1$. Since the sum of the middle term and the least makes a square, let it be 4. Then the middle term is greater than 2. Let it be $x+2$. Then the least term is $2-x$. And since the difference of the greatest and the middle has to the difference of the middle and the least the ratio $3:1$, and the difference of the middle and the least is $2x$, therefore the difference of the greatest and the middle is $6x$, and therefore the greatest will be $7x+2$.

There remain two conditions, that the sum of the greatest and the least make a square and the sum of the greatest and the middle make a square. And I am left with the double equation: $8x+4$ is a square and $6x+4$ is a square. And as the constants are squares, the equation is convenient to solve. I form two numbers whose product is $2x$, according to what we know about a double equation; let them be $\frac{1}{2}x$ and 4; and therefore x is 112. But, returning to the conditions, I cannot subtract x, that is, 112, from 2; I desire, then, that x be found less than 2, so that $6x+4$ is less than 16. For $2\cdot 6+4$ is 16. Then since I seek to make $8x+4$ a square and $6x+4$ a square, while $2\cdot 2$ is a square, there are three squares, $8x+4$, $6x+4$, and 4, and the difference of the greatest and the middle is one-third of the difference of the middle and least. My problem therefore resolves itself into finding three squares such that the difference of the greatest and the middle is one-third of the difference of the middle and least, and further such that the least is 4 and the middle is less than 16.

Let the least be taken as 4, and the side of the middle as $z+2$; then the square is z^2+4z+4. Then since the difference of the greatest and the middle is one-third of the difference of the middle and the least, and the difference of the middle and the least is z^2+4z, so that the difference of the greatest and the least is $\frac{1}{3}z^2+1\frac{1}{3}z$, while the middle term is z^2+4z+4, therefore the greatest term, namely, $1\frac{1}{3}z^2+5\frac{1}{3}z+4$, is a square. Multiply throughout by 9, so that $12z^2+48z+36$ is a square, and take the fourth part, so $3z^2+12z+9$ is a square. Further, I desire that the middle square be less than 16, whence clearly its side is less than 4. But the side of the middle square is $z+2$, and so $z+2$ is less than 4. Take away 2 from each side, and z is less than 2.

My equation is now $3z^2+12z+9$ equaling a square, say $(mz-3)^2$. Then $z=\frac{6m+12}{m^2-3}$, and the equation to which my problem is now reduced is that $\frac{6m+12}{m^2-3}$ is less than 2, or $\frac{2}{1}$. The inequality will be preserved when the terms are cross-multiplied, that is, $(6m+12)\cdot 1$ is less than $2\cdot(m^2-3)$, or $6m+12$ is less than $2m^2-6$. By adding 6 to both sides, we have $6m+18$ is less than $2m^2$. When we solve such an equation, we multiply half the coefficient of m into itself, getting 9; then multiply the coefficient of m^2 into the units: $2\cdot 18$, or 36; add this last number to the 9, getting 45; take the square root, which is not less than 7; add half the coefficient of m, making a number not less than 10; and divide the result by the coefficient of m^2, getting a number not less than 5.[29]

My equation is therefore $3z^2+12z+9$ equals a square on side $3-5z$, so z is $\frac{42}{22}$, or $\frac{21}{11}$. I have made the side of the middle square to be $z+2$; therefore the side will be $\frac{43}{11}$ and the square itself $\frac{1849}{121}$.

I return now to the original problem and make $\frac{1849}{121}$, which is a square, equal to $6x+4$. Multiplying by 121 throughout, I get that x is $\frac{1365}{726}$, which is less than 2. In the conditions of the original problem we made the middle term $x+2$, the least $2-x$, and the greatest $7x+2$. Therefore, the greatest is $\frac{11007}{726}$, the middle is $\frac{2817}{726}$, and the least is $\frac{87}{726}$. Since the

[29] What Diophantus means here is not that the square root of 45 is at least 7, but that, after adding 3 and dividing the sum by 2, the smallest integer to use to satisfy the quadratic inequality is 5. So that is the number he uses.

denominator, 726, is not a square, but its sixth part is, if we take 121, which is a square, and divide throughout by 6, then similarly the numbers, will be $\frac{1834\frac{1}{2}}{121}$, $\frac{469\frac{1}{2}}{121}$, $\frac{14\frac{1}{2}}{121}$. And if you prefer to use integers only, avoiding the $\frac{1}{2}$, multiply throughout by 4. Then the numbers will be $\frac{7338}{484}$, $\frac{1878}{484}$, $\frac{53}{484}$. And the proof is obvious.

Book 5

2. *To find three numbers in geometrical proportion such that each of them when added to a given number gives a square.*

Let the number be 20. Take a square which when added to 20 gives a square, say, 16. Therefore put one of the extremes as 16, and the other extreme as x^2. Therefore the middle term will be $4x$. We must make $4x+20$ a square and x^2+20 a square. Their difference is x^2-4x, which factors as x and $x-4$. The square of half the difference between these factors is 4; but equating this to the smaller expression $4x+20$ is absurd, because 4 ought not to be less than 20.[30]

But 4 is $\frac{1}{4}\cdot 16$, while the 16 is a square which when added to 20 produces a square. Therefore we must find a square whose fourth part is greater than 20 and when added to 20 gives a square. This square must therefore be greater than 80. But 81 is a square greater than 80. Therefore we can put $m+9$ for the root of the desired square; the square will then be $m^2+18m+81$, and this added to 20 must be a square. That is, $m^2+18m+101$ is a square; so assume this has the root $m-11$. Therefore $m^2+121-22m$ is equal to $m^2+18m+101$, so m is $\frac{1}{2}$ and the root is $m+9$. The square is then $[(9\frac{1}{2})^2=]90\frac{1}{4}$.

We start again and put the extremes of the proportion as $90\frac{1}{4}$ and x^2. Then the middle term is $9\frac{1}{2}x$ and we must make x^2+20 and $9\frac{1}{2}x+20$ into squares. Their difference is $x^2-9\frac{1}{2}x$, which factors as x and $x-9\frac{1}{2}$. The square of half the difference is $\frac{361}{16}$, which we equate to the smaller expression $9\frac{1}{2}x+20$. Therefore x is $\frac{41}{152}$. The three numbers are $90\frac{1}{4}$, $\frac{389\frac{1}{2}}{152}$, and $\frac{1681}{23104}$.

9. *To divide unity into two parts such that, if the same given number is added to each part, the result will be a square.*

It is necessary that the given number not be odd and the double of it plus 1 must not be divisible by any prime number, which, when 1 is added, is divisible by 4.[31] It is proposed that 6 be added to each segment and make each sum a square. Since we are cutting unity into two parts and adding 6 to each and making each sum a square, the sum of the squares is 13. Therefore, we must divide 13 into two squares each of which is greater than 6. If we divide 13 into two squares whose difference is less than one, we will solve the problem. So take half of 13, namely, $6\frac{1}{2}$, and look for a fraction which, when added to $6\frac{1}{2}$ makes a square.

[30] To get a positive value for x.

[31] Diophantus evidently realized that an integer divisible by any prime congruent to 3 modulo 4 cannot be the sum of two squares.

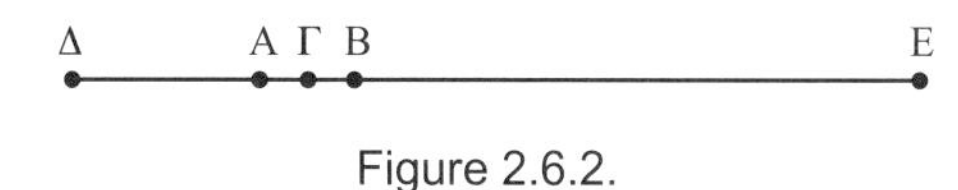

Figure 2.6.2.

Multiplying everything by 4, we must find a square fraction that when added to 26 produces a square. Let the added fraction be $\frac{1}{x^2}$; then $26+\frac{1}{x^2}$ will be a square. Multiply everything by x^2; then we need to make $26x^2+1$ a square. Let its root be $5x+1$, so x is 10. Therefore x^2 is 100, and $\frac{1}{x^2}=\frac{1}{100}$. So we must add $\frac{1}{100}$ to 26, or add $\frac{1}{400}$ to $6\frac{1}{2}$ and make a square with root $\frac{51}{20}$. Therefore, we must divide 13 into two squares, whose roots are as close as possible to $\frac{51}{20}$. That is, we need two numbers such that if we subtract the first from 3 or add the second to 2, we get $\frac{51}{20}$.

We therefore need two squares, one whose root is $11x+2$, the other whose root is $3x-9$, and whose sum will be a square. So $202x^2+13-10x$ is equal to 13 and x is equal to $\frac{5}{101}$. Therefore one of the squares has root $\frac{257}{101}$, the other $\frac{258}{101}$. If we subtract 6 from the squares of each, one segment of unity becomes $\frac{5358}{10201}$ and the other $\frac{4843}{10201}$, and manifestly each added to 6 is a square.

10. *To divide unity into two parts such that, if we add different given numbers to each, the sums will be squares.*

It is proposed that to the one part of unity we add 2 and to the other part 6 such that both sums are squares. We represent unity by AB; it is divided at Γ [figure 2.6.2].

To $A\Gamma$ we add $A\Delta$ representing 2 and to ΓB we add BE representing 6. Then $\Gamma\Delta$ and ΓE are to be squares. Since AB is 1 and $A\Delta+BE$ is 8, the entire line ΔE will be 9, which must be divided into two squares $\Gamma\Delta$ and ΓE. But one of the squares is greater than $A\Delta$, that is, greater than 2, while it is also less than ΔB, that is, less than 3. So our problem is reduced to dividing a square, 9, into two squares $\Delta\Gamma$, ΓE such that $\Gamma\Delta$ falls in the interval between 2 and 3. Then, since $A\Delta$ is 2, the remainder $A\Gamma$ is given. But AB is 1, therefore the remainder $B\Gamma$ is given. And therefore Γ, the point dividing the unit, is given. We describe the process below.

For one square, between 2 and 3, we put x^2. Then the remainder $9-x^2$ is also a square. To make this a square is easy, but we also need to have x^2 between 2 and 3. We take two squares, the smaller larger than 2 and the larger less than 3; these are $\frac{289}{144}$ and $\frac{361}{144}$. If we can make x^2 lie in the interval between these two squares, we shall solve the problem. The root x of x^2 must be greater than $\frac{17}{12}$ and less than $\frac{19}{12}$. Therefore, in making $9-x^2$ a square, we must have x greater than $\frac{17}{12}$ and less than $\frac{19}{12}$.

To make $9-x^2$ a square, we form the square root as 3 minus x with some coefficient, and we find that x must be the coefficient multiplied by 6 and divided by the square of the coefficient added to 1.[32] Therefore we must find some number such that when it is multiplied by 6 and the product divided by its square added to 1, the quotient is greater than $\frac{17}{12}$ and less than $\frac{19}{12}$. Let the sought number be m. Thus $\frac{6m}{m^2+1}$ is greater

[32] Since $9-x^2=(3-mx)^2$, we have $x=\frac{6m}{m^2+1}$.

than $\frac{17}{12}$ and less than $\frac{19}{12}$. But 17 divided by 12 produces the quotient $\frac{17}{12}$. Thus $6m : (m^2 + 1)$ must be greater than $17 : 12$. Therefore $6m \cdot 12$, or $72m$, must be greater than $(m^2 + 1) \cdot 17$, or $17m^2 + 17$. The square of half the coefficient of m is 1296; subtract the product of the coefficient of m^2 and the constant, which is 289. The remainder is 1007, the integral part of whose root is not greater than 31. Add half the coefficient of x, so the sum is not greater than 67. Divide by the coefficient of x^2. So m is not greater than $\frac{67}{17}$.[33] Similarly, from the relation $6m : (m^2 + 1)$ being less than $19 : 12$, we find that m is not less than $\frac{66}{19}$. So we choose m to be $3\frac{1}{2}$.

Therefore the root of the desired square is $3 - 3\frac{1}{2}x$. On squaring, we get $12\frac{1}{4}x^2 + 9 - 21x$ equal to $9 - x^2$. So $x = \frac{84}{63}$ and $x^2 = \frac{7056}{2809}$. We subtract 2 from the latter. Thus one segment of unity is $\frac{1438}{2809}$ and the second segment is $\frac{1371}{2809}$. The conditions of the problem are fulfilled.

11. *To divide unity into three parts such that, if we add the same number to each of the parts, the results shall all be squares.*

It is necessary that the given number be neither 2 nor any multiple of 8 increased by 2.[34] Let it be required to divide unity into three parts such that, when 3 is added to each, the results shall all be squares. Then it is required to divide 10 into three squares such that each of them is greater than 3. If then we divide 10 into three squares, according to the method of approximation, each of them will be greater than 3 and, by taking 3 from each, we shall be able to obtain the parts into which unity is to be divided.

We take, therefore, the third part of 10, which is $3\frac{1}{3}$, and try by adding some square part to $3\frac{1}{3}$ to make a square. On multiplying throughout by 9, it is required to add to 30 some square part that will make the whole a square. Let the added part be $\frac{1}{x^2}$; multiply throughout by x^2; then $30x^2 + 1$ is a square. Let the side be $5x + 1$; so, squaring, $25x^2 + 10x + 1$ is equal to $30x^2 + 1$; whence x is 2, x^2 is 4, and $\frac{1}{x^2}$ is $\frac{1}{4}$. If then we add $\frac{1}{4}$ to 30, to $3\frac{1}{3}$ there is added $\frac{1}{36}$, and the result is $\frac{121}{36}$. It is therefore required to divide 10 into three squares such that the side of each shall approximate to $\frac{11}{6}$.

But 10 is composed of two squares, 9 and 1. We divide 1 into two squares, $\frac{9}{25}$ and $\frac{16}{25}$, so that 10 is composed of three squares, 9, $\frac{9}{25}$, $\frac{16}{25}$. It is therefore required to make each of the sides approximately equal to $\frac{11}{6}$. But their sides are 3, $\frac{4}{5}$, and $\frac{3}{5}$. Multiply throughout by 30, getting 90, 24, and 18; and $\frac{11}{6}$ becomes 55. It is therefore required to make each side approximately equal to 55. [But $\frac{55}{30} = 3 - \frac{35}{30} = \frac{4}{5} + \frac{31}{30} = \frac{3}{5} + \frac{37}{30}$.] Therefore, we take the side of the first square as $3 - 35x$, of the second as $\frac{4}{5} + 31x$, and of the third as $\frac{3}{5} + 37x$. The sum of these squares is $3555x^2 + 10 - 116x$ and this equals 10; so $x = \frac{116}{3555}$. Returning to the conditions, as the sides of the squares are given, the squares themselves are also given. The rest is obvious.[35]

[33] As in problem 6:39 above, Diophantus is solving a quadratic inequality; he uses the Babylonian procedure for the quadratic equation $ax^2 + c = bx$ and then chooses a value that will satisfy the inequality.

[34] Diophantus realized that a number of the form $3(8n + 2) + 1$ cannot be the sum of three squares. As in problem 9 above, he gives no proof of this fact.

[35] The three parts of unity are $\frac{228478}{505521}$, $\frac{142381}{505521}$, $\frac{134662}{505521}$.

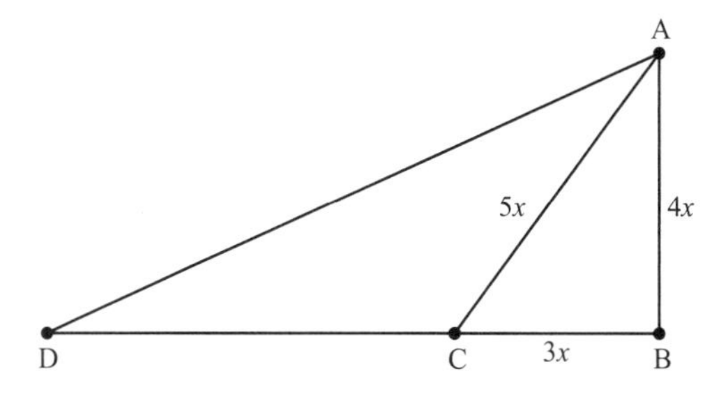

Figure 2.6.3.

Book 6

In book 6, all of the problems deal with right triangles, where Diophantus assumes that the square on the hypotenuse equals the sum of the squares on the two legs.

1. *To find a right-angled triangle such that the hypotenuse minus each of the legs gives a cube.*

Let the required triangle be formed from the two numbers x and 3.[36] So the hypotenuse is x^2+9, the altitude is $6x$, and the base is x^2-9. Then the hypotenuse minus the base x^2-9 is 18, which is not a cube. Now what is 18? It is twice the square of 3. Therefore we must find a number such that twice its square is a cube. Let the number be m. Then $2m^2$ is a cube, m^3. So m is 2. Therefore, we form the triangle not from x and 3, but from x and 2. The hypotenuse is now x^2+4, the altitude is $4x$, and the base is x^2-4, so the hypotenuse minus the base, that is, minus x^2-4, will be a cube. Now consider the hypotenuse minus the perpendicular $4x$. This is x^2+4-4x, and this must be a cube. But it is the square of $x-2$. Therefore $x-2$ must be a cube to solve the problem. Let it be 8, so x is 10. We now form the triangle from 10 and 2 and get that the hypotenuse is 104, the altitude is 40, and the base is 96, so the problem is solved.

16. *To find a right-angled triangle such that on bisecting one of the acute angles, the number representing the portion intercepted within the triangle by the bisector is rational.*

Suppose the bisector AC is $5x$ and one segment of the base, CB, is $3x$ [figure 2.6.3]. Then the perpendicular will be $4x$. Suppose the entire base DB is equal to a multiple of 3, say, 3. Then the other segment of the base CD will be $3-3x$. But since the angle is bisected, the altitude is $\frac{4}{3}$ of the adjacent segment. Therefore the hypotenuse will be $\frac{4}{3}$ of the remaining segment [*Elements* 6:3]. Since the remaining segment is $3-3x$, the hypotenuse will be $4-4x$. The square on the hypotenuse is then $16x^2+16-32x$, which is equal to the sum of the squares on the two legs, $16x^2+9$. So x is $\frac{7}{32}$. The rest is easy. If we multiply everything by 32, we find that the altitude is 28, the base is 96, the hypotenuse is 100, and the bisector is 35.

17. *To find a right-angled triangle such that its area, added to one of the perpendiculars, makes a square, while its perimeter is a cube.*

[36] The right triangle formed from two numbers p, q ($p>q$) has the legs equal to p^2-q^2, $2pq$ and the hypotenuse equal to p^2+q^2.

Let its area be x, and let its hypotenuse be some square number minus x, say $16-x$. But since we supposed the area to be x, therefore the product of the sides about the right angle is $2x$. But $2x$ can be factored into x and 2; if, then, we make one of the sides about the right angle 2, the other will be x. The perimeter then becomes 18, which is not a cube; but 18 is made up of a square plus 2. It shall be required, therefore, to find a square number which, when 2 is added, shall make a cube, so that the cube shall exceed the square by 2.

Let the side of the square be $m+1$ and that of the cube $m-1$. Then the square is m^2+2m+1 and the cube is $m^3+3m-3m^2-1$. Now I want the cube to exceed the square by 2. Therefore, by adding 2 to the square, we have that m^2+2m+3 is equal to $m^3+3m-3m^2-1$, whence m is 4.[37] Therefore, the side of the square is 5 and that of the cube is 3; hence the square is 25 and the cube 27. I now transform the right-angled triangle, and, assuming its area to be x, I make the hypotenuse $25-x$; the base remains equal to 2 and the perpendicular equal to x.

The condition is still left that the square on the hypotenuse is equal to the sum of the squares on the sides about the right angle. Thus $x^2+625-50x$ is equal to x^2+4, whence x is $\frac{621}{50}$. This satisfies the condition.

22. *To find a right-angled triangle such that its perimeter is a cube, while the perimeter plus the area is a square.*

We first must consider how, given two numbers, a right triangle can be formed such that the perimeter equals one number and the area equals the other. Let the two given numbers be 12 and 7, with 12 being the perimeter and 7 the area. Therefore the product of the two legs will be 14, so put one leg as $\frac{1}{x}$ and the other as $14x$. But the perimeter is 12. Therefore the hypotenuse is $12-\frac{1}{x}-14x$. The square of the hypotenuse is $\frac{1}{x^2}+196x^2+172-\frac{24}{x}-336x$, while the sum of the squares of the legs is $\frac{1}{x^2}+196x^2$. Setting these equal and simplifying, we get $172x$ is equal to $336x^2+24$. But this is not possible to solve, since the square of half the coefficient of x minus the product of the coefficient of x^2 and the constant term is not a square.[38]

Now the coefficient of x comes from the sum of the square of the perimeter and four times the area, while the product of the coefficients of x^2 and the constant term is eight times the product of the square of the perimeter and the area. So now let the area be m and the perimeter be both a square and a cube, say 64. Construct a triangle such that $(\frac{1}{2}(64^2+4m))^2$ less 8 times the product of x and the square of the perimeter is a square. That is, $4m^2+4194304-24576m$ must be a square. Multiply by $\frac{1}{4}$; so $m^2+1048576-6144m$ must be a square. Also, $m+64$ must be a square. Make the constant terms equal [by multiplying the second appropriately]; then take the difference, factor, and solve as usual.[39]

[37] Presumably, Diophantus realized that this equation reduced to $m^3+m=4m^2+4$, or $m(m^2+1)=4(m^2+1)$, thereby seeing the obvious solution $m=4$.

[38] Diophantus knew that if the discriminant of a quadratic equation is not a square, then the equation does not have a rational solution.

[39] By this point in the *Arithmetica*, Diophantus assumes that the reader can complete the details of the problem, although the solution is by no means obvious. First, multiply $m+64$ by 16384 to get the second expression as $16834m+1048576$, which must be a square. The difference of the two expressions is then $m^2-22528m$. The obvious

2.7 Demotic Papyri from Egypt

The next several selections are from papyri discovered in Egypt, written during the period of Greek cultural influence in the eastern Mediterranean. The papyri are mostly fragmentary, but it is clear that they come from a teaching tradition, in which problems are proposed and then solved, much like the earlier Rhind Papyrus and the Mesopotamian cuneiform tablets. Like these, the papyri are original documents that have survived for close to two thousand years, unlike the majority of our sources, which have been preserved over the centuries through a long tradition of making copies.

The first source given here, written in Egyptian Demotic script, was discovered in Egypt late in the nineteenth century. From paleographic evidence, it appears to have been written during the late Ptolemaic or early Roman period, that is, sometime between the first century BCE and the first century CE. What is included below is one of five similar problems that are the only ones readable on the fragment. Note that the scribe here wrote fractions in the Egyptian fashion, using only unit fractions except for the single fraction $\frac{2}{3}$.[40] The problem is reminiscent of problems found on both the Rhind Mathematical Papyrus and the Moscow Mathematical Papyrus of around fifteen hundred years earlier.

What is the number to whose $\frac{1}{2}$, 3 is added and to whose $\frac{1}{3}$, 3 is added and it makes 10? Here is the way of doing it: You add 3 to 3; result 6. Subtract it from 10; remainder 4. You add $\frac{1}{2}$ to $\frac{1}{3}$; result $\frac{5}{6}$. You say, "Five-sixths, what customarily goes with it to result in 1?" Result $1\frac{1}{5}$. You reckon 4 $1\frac{1}{5}$ [times]; result $4\frac{2}{3}\frac{1}{10}\frac{1}{30}$, the number described. Sum: the part $4\frac{2}{3}\frac{1}{10}\frac{1}{30}$. You take its half, amounting to $2\frac{1}{3}\frac{1}{15}$, with 3; result $5\frac{1}{3}\frac{1}{15}$. You take its third, amounting to $1\frac{1}{2}\frac{1}{10}$, with 3; result $4\frac{1}{2}\frac{1}{10}$. You add $5\frac{1}{3}\frac{1}{15}$ to $4\frac{1}{2}\frac{1}{10}$, result 10. It is correct.

The following problems are from three further Egyptian papyri that are all written in Demotic script. Altogether there are seventy-two problems on the papyri, only fifty-nine of which can be read with reasonable certainty. We present here some arithmetic problems from the papyri, saving the geometric ones for a later chapter. We note that the scribes usually check the solutions by plugging them into the original statement. The numbering of the problems is due to Richard Parker, who first translated and published these.

The first group of problems is from Papyrus Cairo J. E. 89127–30, 89137–43. It is dated to the time of Ptolemy II Philadelphus, who ruled in Egypt from 286 to 243 BCE. The problems are written on the verso of the papyrus. On the recto is the Hermopolis legal code. It is clear that the scribes were quite fluent in arithmetic operations, including the taking of square roots. And although they frequently used the old Egyptian system of unit fractions, they also could deal with other common fractions.

---

factorization of this does not work, but the factorization into $11m$, $\frac{1}{11}m-2048$ does, giving, after setting half the difference equal to the linear expression, $m=\frac{39424}{225}$. Redoing the equation for x, given this value for m, produces the quadratic equation $78848x^2+225=8432x$, which has the rational solution $x=\frac{9}{176}$. Then the legs of the desired triangle are $\frac{448}{25}$, $\frac{176}{9}$, while the hypotenuse is $\frac{5958}{225}$.

[40] For the representation of unit fractions, see section 2.11.

Problem 3. If it is said to you: "Carry $15\frac{2}{3}$ into 100," you shall subtract 6 from 100: remainder 94. You shall say: "6 customarily goes into 94 with the result $15\frac{2}{3}$." Bring it to the number $\frac{1}{3}$: result 47. You shall take the number 47 to 100. 47 to 1, 94 to 2, Remainder 6: result 6/47. Result 2 6/47. You shall reckon it 3 times: result $6\frac{18}{47}$. You shall say: "$6\frac{18}{47}$ is its number." You shall reckon $6\frac{18}{47}$, $15\frac{2}{3}$ times. $6\frac{18}{47}$ to 1; $63\frac{39}{47}$ to 10; $31\frac{1}{2}\frac{19\frac{1}{2}}{47}$ to 5. Total $95\frac{1}{2}\frac{11\frac{1}{2}}{47}$ to 15; $4\frac{12}{47}$ to $\frac{2}{3}$; Total $99\frac{1}{2}\frac{23\frac{1}{2}}{47}$ to $15\frac{2}{3}$. Its remainder is $\frac{1}{2}$, which is equivalent to $\frac{23\frac{1}{2}}{47}$. Total 100.

This problem is simply to divide 100 by $15\frac{2}{3}$. The scribe apparently began by giving an approximate answer against which to check his final result, though the statement of it is unusual. The first step in his solution is to convert the fractional number $15\frac{2}{3}$ to the integer 47 by in effect multiplying it by 3, though the text has it as giving the number of one-thirds in $15\frac{2}{3}$. The division of 100 by 47 is followed by multiplying the answer, $2\frac{6}{47}$, by 3, with the result $6\frac{18}{47}$. As a validation, this result is multiplied by $15\frac{2}{3}$ to give 100.

Problem 7. The things you should know about the articles of cloth. If it is said to you: "Have sailcloth made for the ships," and it is said to you: "Give 1,000 cloth-cubits to one sail; have the height of the sail be in the ratio 1 to $1\frac{1}{2}$ the width," here is the way of doing it. Find its half, when it happens that the ratio is 1 to $1\frac{1}{2}$: result 1,500. Cause that it reduce to its square root: result $38\frac{2}{3}\frac{1}{20}$. You shall say: "The height of the sail is $38\frac{2}{3}\frac{1}{20}$ cubits." You shall take to it $\frac{2}{3}$, since it happens that it is to $1\frac{1}{2}$ that 1 makes a ratio: result $25\frac{2}{3}\frac{1}{10}\frac{1}{90}$. It is the width.

Problem 9. A hair-cloth which is 6 cubits in height and 4 cubits in width, amounting to 24 cloth-cubits. Take one cubit off its height and add it to its width. What is that which is added to its width? The height is 6. Subtract 1: result, its $\frac{1}{6}$. You shall say: "$\frac{1}{6}$ is taken off; 5 cubits is the remainder; the height is 5 cubits." Now its taken-off area makes 4 cloth-cubits. You shall say: "4 is the what of 5?" Result, its $\frac{2}{3}\frac{1}{10}\frac{1}{30}$. Add it to 4. Result $4\frac{2}{3}\frac{1}{10}\frac{1}{30}$. The number is the width To cause that you know it, you shall reckon $4\frac{2}{3}\frac{1}{10}\frac{1}{30}$, 5 times. Result 24 cloth-cubits, which will make the given number again.

Problem 50 is from Papyrus BM 10399, which is Ptolemaic, probably somewhat later than Papyrus Cairo.

Problem 50. You are told: "$\frac{1}{9}$ is the addition; what is the subtraction?" The way of doing it: You shall add $\frac{1}{9}$ to 1: result $1\frac{1}{9}$. You shall say: "$\frac{1}{9}$ is what fraction of $1\frac{1}{9}$?" Result, its $\frac{1}{10}$. You shall say: "$\frac{1}{9}$ is the addition; $\frac{1}{10}$ is the subtraction." To cause that you know it: You shall subtract $\frac{1}{10}$ from 1: remainder $\frac{5}{6}\frac{1}{15}$. The fraction $\frac{5}{6}\frac{1}{15}$. Its $\frac{1}{9}$ is $\frac{1}{10}$. You shall add it ($\frac{1}{10}$) to it ($\frac{5}{6}\frac{1}{15}$): result 1 again.

Problems 53, 57, 61, and 62 are from Papyrus BM 10520. It is probably from the early Roman period, that is, in the one hundred years surrounding the year 1 CE.

Problem 53. 1 is filled twice up to 10. You shall say: "1 up to 10 amounts to 55." The way of doing it: You shall reckon 10, 10 times: result 100. You shall add 10 to 100: result 110.

You shall take the $\frac{1}{2}$: result 55. You shall say: "1 up to 10 amounts to 55." You shall add 2 to 10: result 12. You shall take the $\frac{1}{3}$ of 12: result 4. You shall reckon 4, 55 times: result 220. It is it. You shall say: "1 is filled twice up to 10: result 220."

The scribe here summed the integers from 1 to 10 and then the first ten triangular numbers. The final three problems are examples of calculations using Egyptian unit fractions, including the special case of $\frac{2}{3}$.

Problem 57. Reckon $\frac{1}{3}\ \frac{1}{15}$, $\frac{2}{3}\ \frac{1}{21}$ times. You shall reckon 5, 7 times: result 35. You shall bring $\frac{1}{3}\ \frac{1}{15}$ to the number 5: result 2. You shall bring $\frac{2}{3}\ \frac{1}{21}$ to the number 7: result 5. You shall reckon 2, 5 times: result 10. You shall cause that 10 makes part of 35: result $\frac{1}{4}\ \frac{1}{28}$. You shall say: "Result $\frac{1}{4}\ \frac{1}{28}$."

Problem 61. Choose $\frac{1}{3}\ \frac{1}{15}$ to $\frac{2}{3}\ \frac{1}{21}$. You shall reckon 5, 7 times: result 35. Find its $\frac{2}{3}\ \frac{1}{21}$: result 25. Find its $\frac{1}{3}\ \frac{1}{15}$: result 14. You shall add 14 to 25: result 39. You shall carry 35 into 39: result $1\frac{1}{10}\ \frac{1}{70}$. You shall say: "Result $1\frac{1}{10}\ \frac{1}{70}$."

Problem 62. Cause that 10 reduce to its square root. You shall reckon 3, 3 times: result 9, remainder 1; $\frac{1}{2}$ of 1: result $\frac{1}{2}$. You shall cause that $\frac{1}{2}$ make part of 3: result $\frac{1}{6}$. You shall add $\frac{1}{6}$ to 3: result $3\frac{1}{6}$. It is the square root. Causing knowing it: You shall reckon $3\frac{1}{6}$, $3\frac{1}{6}$ times: result $10\frac{1}{36}$. Its difference of squared root $\frac{1}{36}$.

2.8 *Papyrus Michigan 620*

The *Papyrus Michigan 620* is a Greek papyrus that was acquired in Egypt in 1921 and has been dated to no later than the second century CE. The papyrus contains portions of three algebraic problems with solutions and computations. We present here just the first one. Unfortunately, the beginning parts of this are missing on the manuscript, but the part that starts "to get the value of the fourth number" is present as well as the check with its calculation. We have included a conjecture of the beginning, where the problem's conditions are set out, but this is supported by parts of the remaining two problems as well as the last half of this one. What is especially interesting is that the symbol that is translated as x in the calculation and stands for the unknown quantity, namely ς, is the same symbol that we find in the manuscripts of Diophantus's *Arithmetica* where it also denotes the unknown quantity. From what we know of the dating of this papyrus and of Diophantus, it appears that this symbol was used to represent the unknown quantity at least a century before Diophantus.

Problem 1

[Four numbers: their sum is 9900; let the second exceed the first by one-seventh of the first; let the third exceed the sum of the first two by 300; and let the fourth exceed the sum of the first three by 300; to find the numbers.

Since the second number exceeds the first by one-seventh, let the first be seven times a particular number and the second be eight times that number. Then the third number is

15 times that number plus 300 and the fourth number is 30 times that number plus 600. Since the sum of all the numbers is 9900, we have 60 times that particular number plus 900 equals 9900. So the desired number is 150.

Then the first number is 1050 and the second number is 1200. To get the value of the third number, take 150 15 times; it gives 2250, and with the added 300, we get get 2550.] To get the value of the fourth number, again take 150 30 times; it gives 4500; and the 600 added to its value makes 5100; this is the fourth number. Add the four numbers: 1050 plus 1200 plus 1550 plus 5100 gives 9900.

Check: Since it says, "Let the second number exceed the first by one seventh," take one-seventh of the first, 1050; it is 150; add this and 1050, this gives 1200, which is the second number. Again, since it says, "Let the third exceed the first two by 300," add the first and second; it gives 2250; and add the 300 of the excess; it gives 2550, which is the third. And since it says, "Let the fourth exceed the first three by 300," add the three; it gives 4800; and the 300 of the excess; this makes 5100, which is the fourth number.

1/7		300	300	9900
$7x$	$8x$	$15x+300$	$30x+600$	
1050	1200	2550	5100	
150				

2.9 *Papyrus Mathematics*

This problem is taken from a leaf of a Greek papyrus codex found in Egypt, although the exact site is unknown. The papyrus is dated to the fourth century CE. The scribe here knows how to sum an arithmetic sequence. Namely, if the sequence is i_1, i_2, ..., i_n, with constant difference b and sum S, it is easy to see that if one takes i_0 as the term before i_1, then the sum of the terms is given by

$$S = ni_0 + (1+2+\cdots+n)b = ni_0 + \frac{n(n+1)}{2}b.$$

It follows that the first term i_1 can be calculated as

$$i_1 = \frac{1}{n}\left[S - \frac{n^2+n}{2}b\right] + b.$$

This is exactly the procedure that the scribe uses, with $n=5$, $b=8$, and $S=300$. Each remaining term is calculated by adding $b=8$ to the previous term. The scribe then checks the result. Note that the statement of the problem is missing from the papyrus, but the calculation implies that the statement must have been essentially what we have written.

[To divide 300 *artabas*[41] of wheat among five people so that each sharer exceeds the previous one by 8.] We proceed as follows. The number of sharers, 5, times itself. The result is 25. Another 5. The result is 30. Half of this is 15. Times 8. The result is 120. We subtract 120 from 300. The result is 180. I divide by the number of sharers. The result is 36. We add another 8. The result is 44. Hence the first sharer will get 44 *artabas* of wheat. We add another 8. The result is 52. Hence the second sharer will get 52 *artabas* of wheat.

[41] *Artaba* is an Egyptian dry measure.

And we add 8. The result is 60. Hence the third sharer will get 60 *artabas* of wheat. And we add 8. The result is 68. Hence the fourth sharer will get 68 *artabas* of wheat. And we add 8. The result is 76. Hence the fifth sharer will get 76 *artabas* of wheat. I add the *artabas*: 44 and 52 and 60 and 68 and 76. The result is 300. This way for similar cases.

2.10 *Papyrus Achmîm*

The Achmîm Papyrus was discovered in the necropolis of Achmîm, the ancient Panopolis, in upper Egypt, approximately midway between Cairo and Aswan. It is a well-preserved papyrus with numerous numerical tables as well as fifty arithmetical problems. There is no indication on the papyrus of its date of composition, but scholars have determined that it was written during Byzantine rule of Egypt, before the Arab conquest, that is, between the fourth century and the seventh century CE. The majority of the tables are unit fraction tables, three of which we include in section 2.11 on arithmetical tables. Here we just present seven of the arithmetical problems. The papyrus was probably written in a school context, given that the problems give procedures for solving problems without any indication of why the procedures work. The procedures are somewhat cryptic and would require the services of a teacher to justify why they work. Thus, this papyrus is similar to many others from Egypt dating back to the Rhind Mathematical Papyrus and also to many of the cuneiform tablets from Mesopotamia. The papyrus was also clearly written by a scribe fluent in the use of unit fractions, which, as other documents in this *Sourcebook* show, were in constant use in the Mediterranean basin well into the first millennium CE.

13. From a granary one person has taken $\frac{1}{13}$. From what remains, another person has taken $\frac{1}{17}$. There remains in the granary 150 units. We wish to know how much was in the granary originally.

First, we have $13 \times 17 = 221$; so $\frac{1}{13}221 = 17$; $221 - 17 = 204$; $\frac{1}{17}204 = 12$; and $204 - 12 = 192$. Then $221 \times 150 = 33150$, and we calculate $\frac{1}{192}33150$. The result is $172\frac{1}{2}\ \frac{1}{8}\ \frac{1}{48}\ \frac{1}{96}$.

In the previous problem, most of the calculation is straightforward, but it is a bit curious that the scribe used $\frac{1}{48}$ and $\frac{1}{96}$ at the end instead of the simpler $\frac{1}{32}$. In the next problem, the goal is to express $\frac{1}{22}$ as a sum of three unit fractions. The scribe first converts this to finding $\frac{5}{110}$ and then notes that since 2 is a 55th of 110, we have $\frac{5}{110} = \frac{1}{55} + \frac{3}{110}$. But $\frac{1}{10} \times 110 = 11$ and $\frac{1}{11} \times 110 = 10$, so $(\frac{1}{10} + \frac{1}{11}) \times 110 = 21$, which in turn is 3×7. Dividing everything by 7 gives the result $\frac{3}{110} = \frac{1}{70} + \frac{1}{77}$.

16. Of 1, the $\frac{1}{22}$: Decompose $\frac{1}{22}$ into 3 [unit] fractions.

Multiply 1 by 5: the result is 5. $5 \times 22 = 110$. So look for the $\frac{1}{110}$ of 5. What are the factors of 110? It is double of 55 and also ten times 11. So $5 - 2$ (a 55th of 110) = 3. Also, $10 + 11 = 21$; $21 \div 3 = 7$; $7 \times 10 = 70$; $7 \times 11 = 77$. The result is $\frac{1}{55}\ \frac{1}{70}\ \frac{1}{77}$.

In the following problem, the scribe had to perform a division of a number with unit fractions. After some initial manipulation, he needed to find the unit fraction representation of what we think of as $\frac{43}{1320}$. First, he factored 1320 as 15×88. Since $43 = 15 + 28$, he could easily decompose the original fraction into $\frac{15}{1320} + \frac{28}{1320} = \frac{1}{88} + \frac{28}{1320}$. Next, he factored 1320 as 11×120 and needed to find a multiple of $\frac{1320}{120} = 11$ that, together with $\frac{1320}{11} = 120$, summed to an integral multiple of 28. He realized, in fact, that $12 \times 11 + 120 = 252 = 9 \times 28$. Therefore, $(\frac{12}{120} + \frac{1}{11}) \cdot 1320 = 9 \times 28$, or $\frac{28}{1320} = \frac{1}{9}(\frac{12}{120} + \frac{1}{11}) = \frac{12}{1080} + \frac{1}{99}$. It remained to convert the first of these addends to a unit fraction. But since $1080 = 12 \times 90$, the scribe immediately found that to be $\frac{1}{90}$, giving the final result.

18. Of $6\frac{1}{15}\ \frac{1}{40}$, what is the 187th?

In what calculation is $\frac{1}{15}\ \frac{1}{40}$? It is $\frac{1}{120}$ of 11; multiply 120 by 6, this gives 720; with 11, it is 731. Also, $120 \times 187 = 22440$. Now $\frac{1}{17}$ of 22440 is 1320 and $\frac{1}{17}$ of 731 is 43. Now find $\frac{1}{1320}$ of 43. What are the factors of 1320? Either 15×88 or 11×120. Also, $43 - 15$ (which is $\frac{1}{88}$ of 1320) is 28.

Now $12 \times 11 = 132$; $132 + 120 = 252$; $252 \div 28 = 9$; $9 \times 11 = 99$; $9 \times 120 = 1080$, and of 12 take the $\frac{1}{1080}$. What are the factors of 1080? 12×90; subtract 12 (that is $\frac{1}{90}$). The result is $\frac{1}{88}\ \frac{1}{90}\ \frac{1}{99}$.

In the following two problems, the scribe first consulted his tables to find equivalents of some sums of unit fractions.

31. From $\frac{1}{2}\ \frac{1}{3}\ \frac{1}{42}$ subtract $\frac{1}{6}\ \frac{1}{66}$.

In what calculation is $\frac{1}{2}\ \frac{1}{3}\ \frac{1}{42}$? It is $\frac{1}{7}$ of 6. In what calculation is $\frac{1}{6}\ \frac{1}{66}$? It is $\frac{1}{11}$ of 2. Then $6 \times 11 = 66$; $2 \times 7 = 14$; $66 - 14 = 52$; $7 \times 11 = 77$ and the result is $\frac{1}{77}$ of 52.

36. 500 units produce $85\frac{2}{3}\ \frac{1}{21}$ pieces of gold as interest. How much will 100 units produce?

We calculate with $\frac{2}{3}\ \frac{1}{21}$. It is $\frac{1}{7}$ of 5. Then $7 \times 85 = 595$; $595 + 5 = 600$; $7 \times 500 = 3500$. Also, $600 \times 100 = 60{,}000$. It remains to divide 60,000 by 3500.

39. Of $3\frac{1}{2}$, find the $\frac{1}{88}$.

$3\frac{1}{2} \times 2 = 7$; $88 \times 2 = 176$. So find $\frac{1}{176}$ of 7. What are the factors of 176? 11×16. $3 \times 11 = 33$; $33 + 16 = 49$; $49 \div 7 = 7$; $7 \times 11 = 77$; $7 \times 16 = 112$. So now find $\frac{1}{112}$ of 3. What are the factors of 112? 7×16. Then $2 \times 7 = 14$; $14 + 16 = 30$; $30 \div 3 = 10$; $10 \times 7 = 70$; $10 \times 16 = 160$. So find $\frac{1}{160}$ of 2. This is $\frac{1}{80}$. The final result is $\frac{1}{70}\ \frac{1}{77}\ \frac{1}{80}$.

Here, the scribe first converts the problem to $7 \div 176$. He then notes that $\frac{1}{16} \times 176 = 11$, which multiplied by 3 gives 33. Also, $\frac{1}{11} \times 176 = 16$, while the sum of 16 and 33 is 49, of which $\frac{1}{7}$ is 7. Therefore, $(\frac{1}{11} + \frac{3}{16}) \times 176 = 49$ and, dividing by 7, we get $\frac{7}{176} = \frac{1}{77} + \frac{3}{112}$. To decompose the latter fraction, we note that $\frac{1}{16} \times 112 = 7$, double this, and add the result, 14, to $\frac{1}{7} \times 112 = 16$, giving 30, which, divided by 10 gives 3. Therefore, $\frac{1}{10}(\frac{2}{16} + \frac{1}{7}) = \frac{3}{112}$. The final result follows immediately.

The final problem we present here is a proportion problem, for which the scribe again needs his tables but then calculates the results in a familiar manner.

49. Someone had mixed the contents of three treasure boxes, with capacities of 720, 830, and 910. He then withdrew 550 units from the mixture. We want to know how much he got from each of them.

First we add and then divide: 720 + 830 + 950 = 2500; $2500 \div 550 = 4\frac{1}{2}\,\frac{1}{22}$. How do we get $\frac{1}{2}\,\frac{1}{22}$? It is $\frac{1}{11}$ of 6. Then we calculate $4 \times 11 = 44; 44 + 6 = 50$. Then $11 \times 720 = 7920; 7920 \div 50 = 158\frac{1}{3}\,\frac{1}{15}$. Also, $11 \times 830 = 9130; 9130 \div 50 = 182\frac{1}{2}\,\frac{1}{10}$. Finally, $11 \times 950 = 10450; 10450 \div 50 = 209$.

2.11 Selection of Arithmetical Tables

Multiplication Table Papyrus Michigan inv 5663

The following excerpt is from a papyrus document dating from the second or third century CE from Karanis, Egypt (see figure 2.11.1). It is a fragment from a multiplication table including the 5-, 6-, 8-, and 9-times table, aligned in rows and columns. A transcription of this incomplete set of multiplication tables and its "translation" is as follows:

...	ε	ς	λ	η	δ	...
...	ε	ζ	λε	η	ε	...
...	ε	η	μ	η	ς	...
...	ε	θ	με	η	ζ	...
...	ε	ι	ν	η	η	...
...	ς	α	ς	η	θ	...
...	ς	β	ιβ	η	ι	...
...	ς	γ	ιη	θ	α	...
...	ς	δ	κδ	θ	β	...
...	ς	ε	λ	θ	γ	...
...	ς	ς	λς	θ	δ	...
...	ς	ζ	μβ	θ	ε	...
...	ς	η	μη			...

...	5	6	30	8	4	...
...	5	7	35	8	5	...
...	5	8	40	8	6	...
...	5	9	45	8	7	...
...	5	10	50	8	8	...
...	6	1	6	8	9	...
...	6	2	12	8	10	...
...	6	3	18	9	1	...
...	6	4	24	9	2	...
...	6	5	30	9	3	...
...	6	6	36	9	4	...
...	6	7	42	9	5	...
...	6	8	48			

Table of Unit Fractions

A unit fraction table was a different sort of arithmetical table, which converts rational numbers into sums of distinct unit fractions. An example of such an arithmetical table can be seen in figure 2.11.2, which includes fragments of tables of unit fractions for 7s, 8s, and 9s. A transcription and translation of this table for the unit fractions of the form $\frac{n}{9}$ is included below. Note that the expression "of the 2 [the $\acute{9}$ is] $\acute{6}\ \acute{1}\acute{8}$" is mathematically equivalent to "$\frac{2}{9} = \frac{1}{6} + \frac{1}{18}$" or "of the 10 [the $\acute{9}$ is] 1 $\acute{9}$" is mathematically equivalent to "$\frac{10}{9} = 1 + \frac{1}{9}$."

Figure 2.11.1. An excerpt from a papyrus with a fragmentary set of multiplication tables. (P. Michigan inv 5663). Courtesy University of Michigan Library Digital Collections.

ἔνατα	ninths
τῆς α τὸ θ́ θ́	of the 1 the 9́ is 9́
τὸ θ́ χξ̔ςβ́	[of the 6000] the 9́ is 666 3̋
τῶν β ς́ίή	of the 2 [the 9́ is] 6́ 1́8́
τῶν γ γ́	of the 3 [the 9́ is] 3́
τῶν δ γ́θ́	of the 4 [the 9́ is] 3́ 9́
τῶν ε ∠ίή	of the 5 [the 9́ is] 2́ 1́8́
τῶν ς β́	of the 6 [the 9́ is] 3̋
τῶν ζ β́θ́	of the 7 [the 9́ is] 3̋ 9́
τῶν η ∠γ́ίή	of the 8 [the 9́ is] 2́ 3́ 1́8́
τῶν θ α	of the 9 [the 9́ is] 1
τῶν ι αθ́	of the 10 [the 9́ is] 1 9́
τῶν κ βς́ίή	of the 20 [the 9́ is] 2 6́ 1́8́
...	...
τῶν ϙ ι	of the 90 [the 9́ is] 10
τῶν ρ ιαθ́	of the 100 [the 9́ is] 11 9́
τῶν ϲ κβς́ίή	of the 200 [the 9́ is] 22 6́ 1́8́
...	...

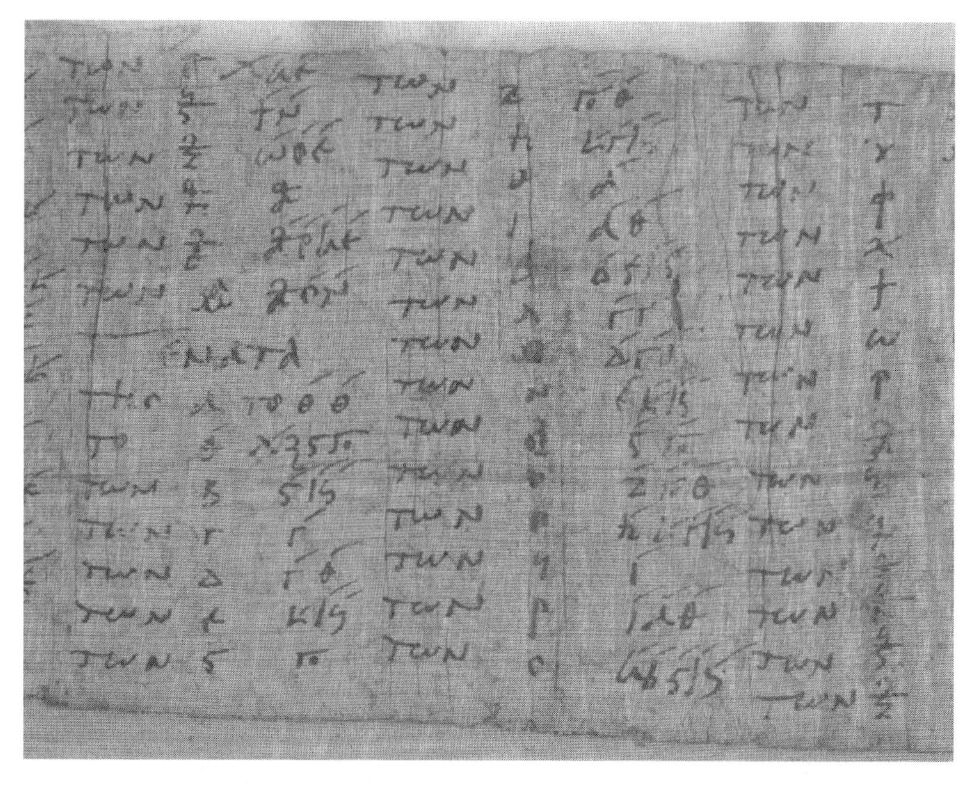

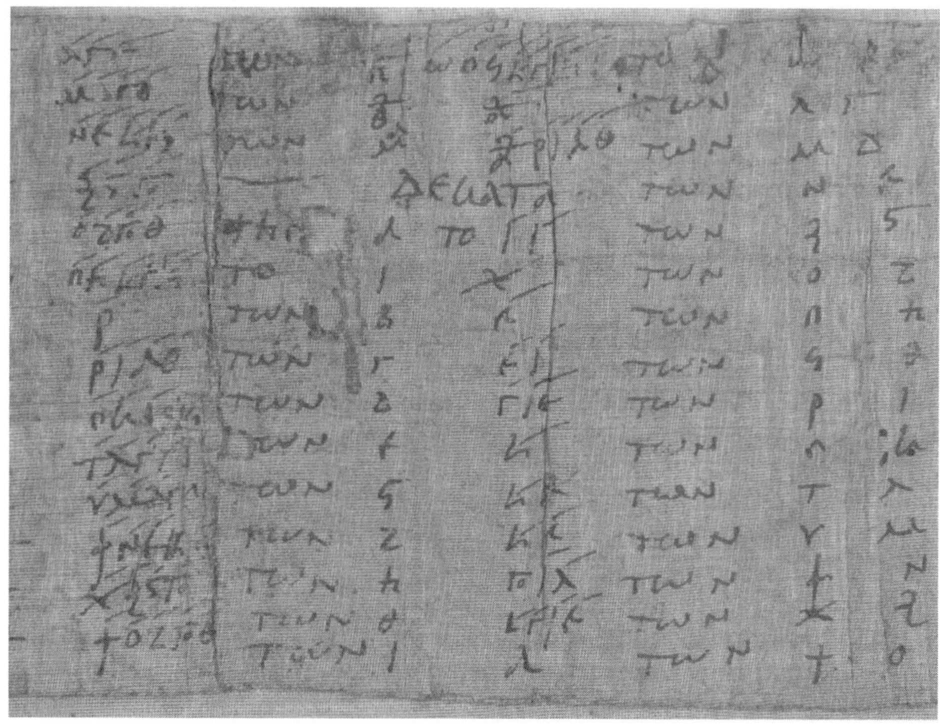

Figure 2.11.2. P. Michigan inv 621; tables of unit fractions, 9s and 10s. Courtesy University of Michigan Library Digital Collections.

τῶν ϡ ρ	of the 900 [the $\acute{9}$ is] 100
τῶν Α ριαθ́	of the 1000 [the $\acute{9}$ is] 111 $\acute{9}$
τῶν Β ϲκβϛ́ίή	of the 2000 [the $\acute{9}$ is] 222 $\acute{6}$ $\acute{18}$
...	...
τῶν Θ Α	of the 9000 [the $\acute{9}$ is] 1000
τῶν $\overset{\alpha}{\mu}$ Α ριαθ́	of the 10000 [the $\acute{9}$ is] 1111 $\acute{9}$

Papyrus Achmîm

The *Papyrus Achmîm* (perhaps seventh century CE or a bit earlier) opens with tables of divisions expressed in unit-fraction mode. The first table shows two-thirds of each number in the sequence $1, 2, 3, \ldots, 10, 20, 30, \ldots, 100, 200, 300, \ldots, 1000, 2000, 3000, \ldots, 10000$. The next table lists thirds of the same sequence, followed by tables of fourths, fifths, . . . ,tenths. These are followed by tables of elevenths through twentieths, but for each of these tables of *n*ths, the list is only for the numbers $1, 2, 3, \ldots, n$. There was a special symbol for $\frac{2}{3}$, so that value occurs in these tables; all other fractions are unit fractions. In addition, there is an extra value at the beginning of each table, namely, the given fractional value of 6000. Perhaps this is listed since there were 6000 *drachmas* in a *talent*, the basic unit of currency.

We present here the tables for sevenths, thirteenths, and nineteenths, where, unlike in the previous examples, we "translate" the tables using modern fraction notation. Recall that the Rhind Mathematical Papyrus began with a table of unit fractions, although there it was a table of the number 2 divided by every odd number up to 101. But there are other extant papyri from Egypt in the first centuries CE that also contain tables of divisions using unit fractions. Clearly, it was important for anyone doing arithmetical calculations to have at hand such a table.

7ths

of 1	$\frac{1}{7}$	of 100	$14\frac{1}{4}\,\frac{1}{28}$
of 2	$\frac{1}{4}\,\frac{1}{28}$	of 200	$28\frac{1}{2}\,\frac{1}{14}$
of 3	$\frac{1}{3}\,\frac{1}{14}\,\frac{1}{42}$	of 300	$42\frac{1}{2}\,\frac{1}{3}\,\frac{1}{42}$
of 4	$\frac{1}{2}\,\frac{1}{14}$	of 400	$57\frac{1}{7}$
of 5	$\frac{2}{3}\,\frac{1}{21}$	of 500	$71\frac{1}{3}\,\frac{1}{14}\,\frac{1}{42}$
of 6	$\frac{1}{2}\,\frac{1}{3}\,\frac{1}{42}$	of 600	$85\frac{2}{3}\,\frac{1}{21}$
of 7	1	of 700	100
of 8	$1\frac{1}{7}$	of 800	$114\frac{1}{4}\,\frac{1}{28}$
of 9	$1\frac{1}{4}\,\frac{1}{28}$	of 900	$128\frac{1}{2}\,\frac{1}{14}$
of 10	$1\frac{1}{3}\,\frac{1}{14}\,\frac{1}{42}$	of 1,000	$142\frac{1}{2}\,\frac{1}{3}\,\frac{1}{42}$

of 20	$2\frac{1}{2}\ \frac{1}{3}\ \frac{1}{42}$	
of 30	$4\frac{1}{4}\ \frac{1}{28}$	
of 40	$5\frac{2}{3}\ \frac{1}{21}$	
of 50	$7\frac{1}{7}$	
of 60	$8\frac{1}{2}\ \frac{1}{14}$	
of 70	10	
of 80	$11\frac{1}{2}\ \frac{1}{14}\ \frac{1}{42}$	
of 90	$12\frac{1}{2}\ \frac{1}{3}\ \frac{1}{42}$	

of	2,000	$285\frac{2}{3}\ \frac{1}{21}$
of	3,000	$428\frac{1}{2}\ \frac{1}{14}$
of	4,000	$571\frac{1}{2}\ \frac{1}{14}\ \frac{1}{42}$
of	5,000	$714\frac{1}{4}\ \frac{1}{28}$
of	6,000	$857\frac{1}{7}$
of	7,000	1000
of	8,000	$1142\frac{1}{2}\ \frac{1}{3}\ \frac{1}{42}$
of	9,000	$1285\frac{2}{3}\ \frac{1}{21}$
of	10,000	$1428\frac{1}{2}\ \frac{1}{14}$

13ths

of 1	$\frac{1}{13}$
of 2	$\frac{1}{7}\ \frac{1}{91}$
of 3	$\frac{1}{6}\ \frac{1}{26}\ \frac{1}{39}$
of 4	$\frac{1}{4}\ \frac{1}{26}\ \frac{1}{52}$
of 5	$\frac{1}{3}\ \frac{1}{26}\ \frac{1}{78}$
of 6	$\frac{1}{3}\ \frac{1}{13}\ \frac{1}{26}\ \frac{1}{78}$
of 7	$\frac{1}{2}\ \frac{1}{26}$
of 8	$\frac{1}{2}\ \frac{1}{13}\ \frac{1}{26}$
of 9	$\frac{2}{3}\ \frac{1}{39}$
of 10	$\frac{1}{2}\ \frac{1}{4}\ \frac{1}{52}$
of 11	$\frac{1}{2}\ \frac{1}{3}\ \frac{1}{78}$
of 12	$\frac{1}{2}\ \frac{1}{3}\ \frac{1}{13}\ \frac{1}{78}$
of 13	1

19ths

of 1	$\frac{1}{19}$
of 2	$\frac{1}{10}\ \frac{1}{190}$
of 3	$\frac{1}{15}\ \frac{1}{20}\ \frac{1}{57}\ \frac{1}{76}\ \frac{1}{95}$
of 4	$\frac{1}{5}\ \frac{1}{95}$
of 5	$\frac{1}{4}\ \frac{1}{76}$
of 6	$\frac{1}{4}\ \frac{1}{19}\ \frac{1}{76}$
of 7	$\frac{1}{3}\ \frac{1}{38}\ \frac{1}{114}$
of 8	$\frac{1}{3}\ \frac{1}{30}\ \frac{1}{38}\ \frac{1}{57}\ \frac{1}{95}$
of 9	$\frac{1}{3}\ \frac{1}{12}\ \frac{1}{38}\ \frac{1}{57}\ \frac{1}{76}$
of 10	$\frac{1}{2}\ \frac{1}{38}$
of 11	$\frac{1}{2}\ \frac{1}{19}\ \frac{1}{38}$
of 12	$\frac{1}{2}\ \frac{1}{12}\ \frac{1}{38}\ \frac{1}{76}\ \frac{1}{114}$
of 13	$\frac{2}{3}\ \frac{1}{57}$
of 14	$\frac{2}{3}\ \frac{1}{19}\ \frac{1}{57}$
of 15	$\frac{1}{2}\ \frac{1}{4}\ \frac{1}{38}\ \frac{1}{76}$
of 16	$\frac{1}{2}\ \frac{1}{4}\ \frac{1}{19}\ \frac{1}{38}\ \frac{1}{76}$
of 17	$\frac{1}{2}\ \frac{1}{3}\ \frac{1}{30}\ \frac{1}{57}\ \frac{1}{95}$
of 18	$\frac{1}{2}\ \frac{1}{3}\ \frac{1}{12}\ \frac{1}{57}\ \frac{1}{76}$
of 19	1

Figure 2.11.3. Stele from the third or second century BCE: a teacher shows a pupil a multiplication grid. Courtesy Geneva Musée d'Art et d'Histoire, Gen. MAH Inv. 27937.

A Multiplication Table

A multiplication grid (see figure 2.11.3) is included in a monumental stele (now housed in the Geneva Musée d'Art et d'Histoire) dating from the third or second century BCE. In this stele, a seated teacher (left), identified as a mathematician named

TABLE 2.1
Transcription of the 10 × 10 multiplication grid from figure 2.11.3.

A	B	Γ	Δ	E	Ϛ	Z	H	Θ	I
B	Δ	Ϛ	H	I	IB	IΔ	IϚ	IH	K
Γ	Ϛ	Θ	IB	IE	IH	KA	KΔ	KZ	Λ
Δ	H	IB	IϚ	K	KΔ	KH	ΛB	ΛϚ	M
E	I	IE	K	KE	Λ	ΛE	M	ME	N
Ϛ	IB	IH	KΔ	Λ	ΛϚ	MB	MH	NΔ	Ξ
Z	IΔ	KA	KH	ΛE	MB	MΘ	NϚ	ΞΓ	O
H	IϚ	KΔ	ΛB	M	MH	NϚ	ΞΔ	OB	Π
Θ	IH	KZ	ΛϚ	ME	NΔ	ΞΓ	OB	ΠA	Ϙ
I	K	Λ	M	N	Ξ	O	Π	Ϙ	P

Figure 2.11.4. Second-century CE papyrus fragment showing a numerical table with zero denoted (a small circle with an overbar). Lund University Library, P. Lund 35a.

Ptolemy, is teaching a student (right), pointing to a multiplication grid for the numbers one to ten.

A Papyrus with Zero

An astronomical table on a papyrus fragment P. Lund 35a dating from the second century CE includes a numerical table (left) with one of the entries denoted as zero, a small circle with a bar over the top (see figure 2.11.4). This fragmentary table inside the box is transcribed and translated as

ιθ			19		
ις	ν		16	50	
δ	λδ		4	34	
ιβ	ιζ	λ	12	17	30
ι	$\bar{o}$	ν	10	0	50

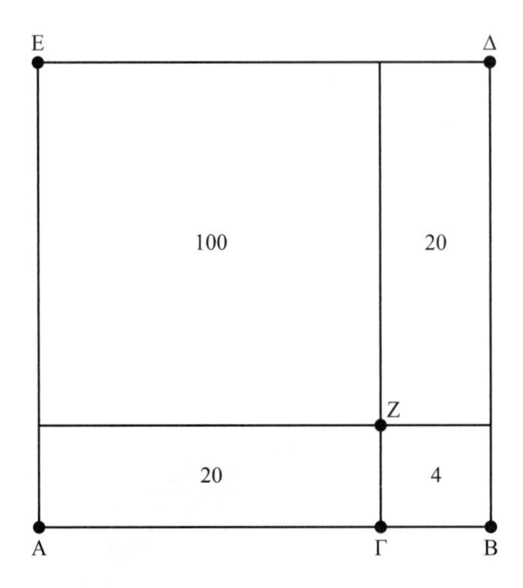

Figure 2.12.1.

2.12 Theon of Alexandria, *Commentary on Ptolemy's Almagest*

In his *Mathematikē Syntaxis* (*Mathematical Collection*), later known as the *Almagest*, Ptolemy needed to do numerous arithmetic calculations, starting with his calculation of a table of chords. Whenever he had to compute a square root, he just stated the answer without providing his working. So in his commentary on Book 1 of the *Almagest*, Theon of Alexandria (fourth century CE) gave an explicit method for calculating square roots. We do not know if this was Ptolemy's method, but it seems very likely that this was the case. This example of the method that Theon presents, the calculation of the square root of 4500, is necessary in Ptolemy's calculation of the chord of an arc of 36° in *Almagest* 1:10 (see section 7.6).

After this demonstration, the next step is to inquire in what manner, given the area of a square whose side is irrational, we may make an approximation to its side. In the case of a square with a rational side, the method is clear from the fourth theorem of the second book of the *Elements*, whose enunciation is as follows: *If a straight line be cut at random, the square on the whole is equal to the squares on the segments, and twice the rectangle contained by the segments.*

For if the given number is a square such as 144, having a rational side AB, we take the square 100, which is less than 144 and has 10 as its side, and make $A\Gamma$ equal to 10 [figure 2.12.1]. Doubling it, because the rectangle contained by $A\Gamma, \Gamma B$ is taken twice, we get 20, and by this number we divide the remainder 44, obtaining a remainder 4 as the square on ΓB, whose length will therefore be 2. Now $A\Gamma$ was 10, and therefore the whole AB is 12, which was to be proved.

In order to show visually, for one of the numbers in the *Syntaxis*, this extraction of the root by taking away the parts, we shall construct the proof for the number 4500°, whose side [Ptolemy] made $67^\circ 4' 55''$. Let $AB\Gamma\Delta$ be a square area, the square alone being rational, and let its contents be 4500°, and let it be required to calculate the side of a

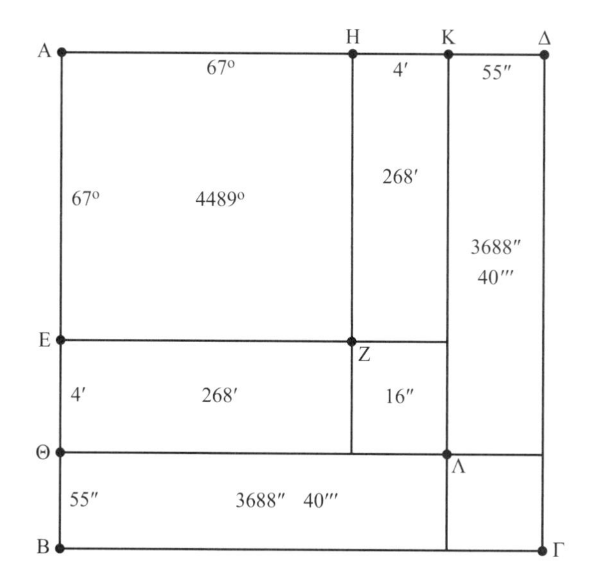

Figure 2.12.2.

square approximating to it [figure 2.12.2]. Since the square that approximates to 4500° but has a rational side and consists of a whole number of units is 4489° on a side of 67°, let the square AZ, with area 4489° and side 67°, be taken away from the square $AB\Gamma\Delta$. The remainder, the gnomon $BEZH\Delta\Gamma B$, will therefore be 11°, which we reduce to $660'$ and set out. Then we double EZ, because the rectangle on EZ has to be taken twice, as though we regarded ZH as on the straight line EZ, divide the result 134° into $660'$, and by the division get $4'$, which gives us each of $E\Theta$, HK. Completing the parallelograms ΘZ, ZK, we have for their sum $536'$, or $268'$ each.

Continuing, we reduce the reminder, $124'$, into $7440''$, and subtract from it also the complement $Z\Lambda$, which is $16''$, in order that by adding a gnomon to the original square AZ, we may have the square $A\Lambda$ on a side $67^\circ 4'$ and consisting of $4497^\circ 56' 16''$. The remainder, the gnomon $B\Theta\Lambda K\Delta\Gamma B$, consists of $2^\circ 3' 44''$, that is, $7424''$. Continuing the process, we double $\Theta\Lambda$, as though ΛK were in a straight line with $\Theta\Lambda$ and equal to it, divide the product $134^\circ 8'$ into $7424''$, and the result is approximately $55''$, which gives us an approximation to ΘB, $K\Delta$. Completing the parallelograms $B\Lambda$, $\Lambda\Delta$, we shall have for their joint area $7377''20'''$, or $3688''40'''$ each. The remainder is $46''40'''$, which approximates to the square $\Lambda\Gamma$ on a side of $55''$, and so we obtain for the side of the square $AB\Gamma\Delta$, consisting of 4500°, the approximation $67^\circ 4' 55''$.

In general, if we seek the square root of any number, we take first the side of the nearest square number, double it, divide the product into the remainder reduced to minutes, and subtract the square of the quotient; proceeding in this way we reduce the remainder to seconds, divide it by twice the quotient in degrees and minutes, and we shall have the required approximation to the size of the square area.

In his commentary on Book 13, Theon showed how to solve a problem dealing with the latitudinal motions of the planets. The problem asks to find three angles, but for our purposes it is just the numerical values of the angles that are important. In fact,

Theon works out his solution by "the process of Diophantine numbers," that is, he uses one of Diophantus's methods. This problem, aside from the numerical values, is identical to Problem 1–9 of the *Arithmetica*: To subtract the same number from two given numbers and make the remainders have to one another a given ratio. We have used the translation of Christianidis and Skoura, who used the word *arithmos* to translate the symbol used by Theon, as well as "wanting" to translate the symbol for negatives. Nevertheless, it is easy to see that the solution of the problem here is identical to Diophantus's solution given earlier.

If two numbers are given, and some equal numbers are removed from them, and the ratio of the remainders is given, those, the ratio of which was given, will also be given, and the others, that is, the equal ones, will be given too. So let in the case of the proposed numbers $4\frac{1}{3}$ and 7, and the ratio 5 to 9, work out this methodically by the process of the Diophantine numbers.

Let the number to be subtracted from each of the two, $4\frac{1}{3}$, 7 be set as 1 *arithmos* $[x]$. Now if subtracted from $4\frac{1}{3}$ units the remainder will be $4\frac{1}{3}$ units wanting 1 *arithmos* $[4\frac{1}{3}-x]$. And if from the 7, the remainder will be 7 units wanting 1 *arithmos* $[7-x]$. Therefore $4\frac{1}{3}$ units wanting 1 *arithmos* must have to 7 units wanting 1 *arithmos* the ratio that 5 has to 9. But 5 falls short of 9 by $\frac{4}{5}$ of itself. Therefore $4\frac{1}{3}$ units wanting 1 *arithmos*, likewise, falls short of 7 units wanting 1 *arithmos* by $\frac{4}{5}$ of itself. If therefore we add to $4\frac{1}{3}$ units wanting 1 *arithmos* $\frac{4}{5}$ of itself, it will be equal to 7 units wanting 1 *arithmos*. But $4\frac{1}{3}$ units wanting 1 *arithmos* when receiving the $\frac{4}{5}$ of itself is $\frac{117}{15}$ wanting $\frac{9}{5}$ of an *arithmos*, that is to say, $\frac{27}{15}$, as we will show later.[42]

Therefore, $\frac{117}{15}$ of a unit wanting $\frac{27}{15}$ of *arithmos* is equal to 7 units wanting 1 *arithmos*, that is to say, $\frac{105}{15}-\frac{15}{15}$ of *arithmos*. Let the wanting $\frac{42}{15}$ *arithmos* be added in common. Consequently, $\frac{117}{15}$ of unit and $\frac{15}{15}$ of *arithmos* is equal to $\frac{105}{15}$ unit and $\frac{27}{15}$ of *arithmos*. And we remove like from like; there remains that $\frac{12}{15}$ unit equals $\frac{12}{15}$ *arithmos*. And [multiply] all fifteen times. Therefore the *arithmos* will be 1 unit.

To the numerical values, the one of those having the given ratio $4\frac{1}{3}$ units wanting 1 *arithmos* will be $3\frac{1}{3}$ units. And the other, 7 units wanting 1 *arithmos* will be 6 units. And the rest, that is to say each of the equal ones, which are subtracted, will be 1.

2.13 Archimedes, *Sand Reckoner*

Archimedes (287–212 BCE) is generally considered the finest mathematician of the Greek period. And because he was influential in the defense of his home city of Syracuse, in Sicily, during the Second Punic War, much biographical information about him exists in the works of several Roman historians, including Polybius (see section 4.1) and Plutarch. He was the son of the astronomer Phidias and perhaps a

[42] This arithmetic calculation, $1\frac{4}{5}(4\frac{1}{3}-x)=\frac{117}{15}-\frac{9}{5}x$, is found at the end of the problem, but will not be included here.

relative of King Hiero of Syracuse, under whose rule from 270 to 215 BCE the city flourished greatly. It is probable that in his youth Archimedes studied in Alexandria, for the opening letters of his treatises refer to scholars who lived and worked there. He made enormous contributions in a number of branches in the mathematical sciences and devised many ingenious mechanical inventions.

The following excerpt is from his treatise *Psammites* (*Sand Reckoner*). In this work, Archimedes poses the ostentatious question: How many grains of sand would it take to fill the universe? Such a question is designed to motivate an approach to denoting very big numbers. Archimedes develops a systematic method for expressing these very large quantities, using multiples of the unit of the *myriad* or 10,000, the highest number for which the Greeks had a verbal expression.

Chapter 1

There are some, King Gelon,[43] who think that the number of the sand is infinite in multitude; and I mean by the sand not only that which exists about Syracuse and the rest of Sicily but also that which is found in every region whether inhabited or uninhabited. Again there are some who, without regarding it as infinite, yet think that no number has been named which is great enough to exceed its multitude. And it is clear that they who hold this view, if they imagined a mass made up of sand in other respects as large as the mass of the Earth filled up to a height equal to that of the highest of the mountains, would be many times further still from recognizing that any number could be expressed which exceeded the multitude of the sand so taken. But I will try to show you by means of geometrical proofs, which you will be able to follow, that, of the numbers named by me and given in the work which I sent to Zeuxippus,[44] some exceed not only the number of the mass of sand equal in magnitude to the Earth filled up in the way described, but also that of a mass equal in magnitude to the Universe.

Now, you are aware that "Universe" is the name given by most astronomers to the sphere whose center is the center of the Earth and whose radius is equal to the straight line between the center of the Sun and the center of the Earth. This is the common account, as you have heard from astronomers. But Aristarchus of Samos[45] brought out a book consisting of some hypotheses, in which the premises lead to the result that the Universe is many times greater than that now so called. His hypotheses are that the fixed stars and the Sun remain unmoved, that the Earth revolves about the Sun in the circumference of a circle, the Sun lying in the middle of the orbit, and that the sphere of the fixed stars, situated about the same center as the Sun, is so great that the circle in which he supposes the Earth to revolve bears such a proportion to the distance of the fixed stars as the center of the sphere bears to its surface. Now it is easy to see that this is impossible. For, since the center of the sphere has no magnitude, we cannot conceive it to bear any ratio whatever to the surface of the sphere. We must however take Aristarchus to mean this: Since we

[43] The eldest son of King Hiero of Syracuse; since he never actually ruled, the title of "king" is an honorific.

[44] Nothing is known of Zeuxippus.

[45] See section 7.2.

conceive the Earth to be, as it were, the center of the Universe, the ratio which the Earth bears to what we describe as the "Universe" is the same as the ratio which the sphere containing the circle in which he supposes the Earth to revolve bears to the sphere of the fixed stars. For he adapts the proofs of his results to a hypothesis of this kind, and in particular he appears to suppose the magnitude of the sphere in which he represents the Earth as moving to be equal to what we call the "Universe".

I say then that, even if a sphere were made up of sand as great as Aristarchus supposes the sphere of the fixed stars to be, I shall still prove that, of the numbers named in the Principles,[46] some exceed in multitude the number of the sand which is equal in magnitude to the sphere referred to, provided that the following assumptions be made:

1. The perimeter of the Earth is three hundred myriad stadia and no greater. As you know, some have tried to show that this length is thirty myriad stadia. But I, surpassing this number and setting the size of the Earth as being ten times that evaluated by my predecessors, suppose that its perimeter is three hundred myriad stadia and not greater.
2. Secondly, that the diameter of the Earth is greater than the diameter of the Moon and that the diameter of the Sun is greater than the diameter of the Earth. My hypothesis is in agreement with most earlier astronomers.
3. Third hypothesis: the diameter of the Sun is thirty times larger than that of the Moon and not greater, even though among earlier astronomers Eudoxus tried to show it as nine times larger and Phidias, my father, as twelve times larger, while Aristarchus tried to show that the diameter of the Sun lies between a length of eighteen Moon diameters and a length of twenty four Moon diameters;[47] but I, surpassing this number as well, suppose, so that my proposition may be established without dispute, that the diameter of the Sun is equal to thirty Moon diameters, and not more.
4. Finally, we state that the diameter of the Sun is greater than the side of the polygon of one thousand sides inscribed in the great circle of the Universe. I make this hypothesis because Aristarchus found that the Sun appears as the seven hundred and twentieth part of the circle of the Zodiac.

Chapter 2

These relations being given, one can also show that the diameter of the Universe is less than a line equal to a myriad diameters of the Earth and that, moreover, the diameter of the Universe is less than a line equal to one hundred myriad myriad [10^{10}] stadia. As soon as one has accepted the fact that the diameter of the Sun is not greater than thirty Moon diameters and that the diameter of the Earth is greater than the diameter of the Moon, it is clear that the diameter of the Sun is less than thirty diameters of the Earth. As we have also shown that the diameter of the Sun is greater than the side of the polygon of one thousand sides inscribed in the great circle of the Universe, it is clear that the perimeter of the indicated polygon of one thousand sides is less than one thousand diameters of the Sun. But the diameter of the Sun is less than thirty Earth diameters so it follows that

[46] This is presumably the work sent to Zeuxippus.

[47] The actual diameter of the sun is about 400 times the diameter of the moon.

the perimeter of the polygon of one thousand sides is less than thirty thousand Earth diameters. Given that the perimeter of the polygon of one thousand sides is less than thirty thousand Earth diameters and greater than three diameters of the Universe—we have shown in fact that in every circle the diameter is less than one third the perimeter of any regular polygon inscribed in the circle for which the number of sides is greater than that of the hexagon—the diameter of the Universe is less than a myriad Earth diameters. One has thus shown that the diameter of the Universe is less than a myriad Earth diameters; that the diameter of the Universe is less than one hundred myriad myriad stadia, which comes out of the following argument; since, in fact, we have supposed that the perimeter of the Earth is not greater than three hundred myriad stadia and that the perimeter of the Earth is greater than triple the diameter because in every circle the circumference is greater than triple the diameter, it is clear that the diameter of the Earth is less than one hundred myriad stadia. Given that the diameter of the Universe is less than a myriad Earth diameters, it is clear that the diameter of the world is less than one hundred myriad myriad stadia. These are my hypotheses regarding sizes and distances.

Here, now, is what I assume about the subject of sand: If one has a quantity of sand whose volume does not exceed that of a poppy-seed, the number of these grains of sand will not exceed a myriad and the diameter of the grains will not be less than a fortieth of a finger-breadth. I make these hypotheses following these observations: Poppy-seeds having been placed on a polished ruler in a straight line in such a way that each touches the next, twenty five seeds occupied a space greater than one finger-breadth. I will suppose that the diameter of the grains is smaller and to be about a fortieth of a finger-breadth for the purpose of removing any possibility of criticizing the proof of my proposition.

Chapter 3

These are thus my hypotheses; but I think it useful to explain myself about the naming of numbers so that those readers, not having been able to get hold of my book addressed to Zeuxippus, may not be thrown off by the absence in this book of any indication of the subject of this nomenclature. It so happens that tradition has given to us the name of numbers up to a myriad and we distinguish enough numbers surpassing a myriad by enumerating the number of myriads until a myriad myriad [10^8]. We will therefore call first numbers those which, after the current nomenclature, go up to a myriad myriad. We will units of second numbers the myriad myriad of first numbers and we will count among second number units and, starting with units, tens, hundreds, thousands, myriads, until a myriad myriad [$(10^8)^2 = 10^{16}$]. We will once again call third numbers a myriad myriad of second numbers and we will count among third numbers, starting with units, tens, hundreds, thousands, myriads, until the myriad myriad [$(10^8)^3) = 10^{24}$]. In the same way we will call units of fourth numbers a myriad myriad of third numbers, units of fifth numbers a myriad myriad of fourth numbers, and continuing in this way the numbers will be distinguishable until the myriad myriad of myriad myriad numbers [$(10^8)^{10^8}$].

Numbers named in this way could certainly suffice but it is possible to go still further. Let us in fact call numbers of the first period the numbers given up to this point and units of first numbers of the second period the last number of the first period. Furthermore, call the unit of second numbers of the second period the myriad myriad of first numbers of the second period. In the same way, the last of these numbers will be called the unit of third numbers of the second period, and continuing in this way, progressing through the numbers of the second period will have their names up to the myriad myriad of myriad myriad numbers. The last number of the second period will be in turn called the unit of the first numbers of the third period, and so forth until a myriad myriad units of myriad myriad numbers of the myriad myriad period.[48]

These numbers having been named, given [that] numbers are ordered by size starting from unity and if the number closest to unity is the tens, the first eight of these including the unity will belong to the numbers called first numbers, the following eight numbers called second, and the others in the same way by the distance of their octad of numbers to the first octad of numbers.[49] The eighth number of the first octad is thus one thousand myriads and the first number of the second octad, since it multiplies by ten the number preceding it, will be a myriad myriad and this number is the unit of the second numbers. The eighth number of the second octad is one thousand myriads of second numbers. The first number of the third octad will once again be, as it multiplies by ten the preceding number, a myriad myriad of second numbers, the unity of the third numbers. It is clear that the same will hold as indicated for any octad.

Chapter 4

The preceding being in part assumed and in part proved, I will now prove my proposition. As we have assumed that the diameter of a poppy-seed is not smaller than a fortieth of a finger-breadth, it is clear that the volume of the sphere having diameter one finger-breadth does not exceed that of sixty four thousand poppy-seeds; for this number indicates how many times it is the multiple of the sphere having as diameter one fortieth of a finger-breadth; it has, in fact, been shown that spheres are related to each other as the cubes of their diameters.[50] As we have also assumed that the number of grains of sand contained in one poppy-seed does not exceed a myriad, it is clear that, if the sphere having diameter one finger-breadth were filled with sand, the number of grains would not exceed sixty four thousand myriads. But this number represents six units of second numbers increased by four thousand myriad of first numbers, and is thus less than ten units of second numbers. The sphere with diameter one hundred finger-breadths is equivalent to one hundred myriad spheres of diameter one finger-breadth, since spheres are related to each other as the cubes of their diameters. If one now had a sphere filled with sand of the size of the sphere of diameter one hundred finger-breadths, it is clear that the number of grains of sand would be less than the product of ten myriad second numbers and one hundred myriads. But since ten units of second numbers make up the tenth number starting from unity in the proportional sequence of multiple ten, and the one hundred myriads of the seventh number starting from unity are in the same proportional sequence, it is clear that the number obtained will be the sixteenth starting from unity in the same proportional sequence. For we have shown that the distance of this

[48] If we let $P=(10^8)^{10^8}$, then this last number is P^{10^8}.
[49] The first octad is thus $1, 10, 10^2, 10^3, 10^4, 10^5, 10^6, 10^7$.
[50] Euclid, *Elements* 12:18.

product to unity is equal to the sum of, minus one, of the distance from unity of its two factors.

From these sixteen numbers the first eight are among, with unity, the numbers called first numbers, the following eight are part of the second numbers, and the last of these is one thousand myriad second numbers. It is now evident that the number of grains of sand whose volume is equal to one hundred finger-breadths is less than one thousand myriad second numbers. Similarly, the volume of the sphere of diameter one myriad finger-breadths is one hundred myriad times the volume of the sphere of diameter one hundred finger-breadths. If one now had a sphere, filled with sand, of the size of the sphere with diameter a myriad finger breadths, it is clear that the number of grains of sand would be less than the product of one thousand myriads of second numbers and one hundred myriads. But since one thousand myriad second numbers are the sixteenth number starting from unity in the proportional sequence and one hundred myriad are the seventh number starting from unity in the same proportional sequence, it is clear that the product will be the twenty second number starting from unity in the same proportional sequence. Of these twenty two numbers, the first eight, with unity, are among the numbers called first numbers, the following eight are among the numbers called second, and the six remaining numbers are called third numbers, the last of which being ten myriad third numbers.

It is then clear that the number of grains of sand whose volume is equal to a sphere of diameter a myriad finger-breadths is less than ten myriads of third numbers. And since the sphere with diameter one stade[51] is smaller than the sphere with diameter a myriad finger-breadths, it is also clear that the number of grains of sand contained in a volume equal to a sphere with diameter one stade is less than ten myriad third numbers. Similarly, the volume of a sphere of diameter one hundred stadia is one hundred myriads times the volume of a sphere of diameter one stade. If one now had a sphere, filled with sand, of the size of the sphere with diameter one hundred stadia, it is evident that the number of grains of sand would be less than the product of ten myriad third numbers with one hundred myriads. And since ten myriad third numbers are the twenty second numbers, starting from unity, in the proportional sequence, and one hundred myriad are the seventh number starting from unity in the same proportional sequence, it is clear that the product will be the twenty eighth number starting from unity in the proportional sequence. Of these twenty eight numbers, the first eight, with unity, are part of the numbers called first numbers, the following eight are second numbers, the following eight are third numbers, and the four remaining are called fourth, the last being one thousand units of fourth numbers. It is then evident that the number of grains of sand whose volume equals that of a sphere of diameter a hundred stadia is less than one thousand units of fourth numbers. Similarly, the volume of a sphere of diameter a myriad stadia is one hundred myriads times the volume of a sphere having diameter one hundred stadia. If one then had a sphere, filled with sand, of the size of a sphere of diameter a myriad stadia, it is clear that the number of grains of sand would be less than the product of one thousand units of fourth numbers with one hundred myriads. Just as one thousand units of fourth numbers represent the twenty-eighth number, starting from unity, in the proportional sequence, and one hundred myriad the seventh number

[51] The exact length of a stade for Archimedes is not known; estimates ranging from about 157 to 191 meters have been given by various scholars.

in the proportional sequence, starting from unity, of the same proportional sequence, it is clear that their product will be, in the same proportional sequence, with unity, the thirty-fourth number starting form unity. But of these thirty four numbers, the first eight, with unity, are among those numbers called first numbers, the following eight among second numbers, the following eight among third numbers, the following eight among fourth numbers, and the two remaining among fifth numbers, the last of these being ten units of fifth numbers. It is thus clear that the number of grains of sand whose volume is equal to that of a sphere having diameter a myriad stadia will be smaller than ten units of fifth numbers. And similarly, the volume of a sphere of diameter one hundred myriad stadia is one hundred myriads times the volume of a sphere of diameter a myriad stadia. If one had then had a sphere, filled with sand, of the size of the sphere with diameter one hundred myriad stadia, it is clear that the number of grains of sand would be smaller than the product of ten units of fifth numbers and one hundred myriads. As the ten units of fifth numbers represent the thirty fourth number starting from unity in the proportional sequence, and one hundred myriads the seventh number starting from unity in the same proportional sequence, it is clear that the product will be, in the same proportional sequence, the fortieth number starting from unity. But of these forty numbers, the first eight, with unity, are among the numbers called first numbers, the eight following are second numbers, the eight following are third numbers, the eight following are fourth numbers, the eight following are fifth numbers, the last of these being one thousand myriad fifth numbers. It is therefore clear that the number of grains of sand whose volume is equal to that of a sphere of diameter one hundred myriad stadia is less than one thousand myriad fifth numbers. But the volume of a sphere of diameter a myriad myriad stadia is one hundred myriad times the sphere of diameter one hundred myriad stadia. Thus, if one had a sphere, filled with sand, of the size of a sphere of diameter a myriad myriad stadia, it is clear that the number of grains of sand would be less than the product of one thousand myriad fifth numbers by one hundred myriads. However, since one thousand myriad fifth numbers represent the fortieth number, starting from unity, of the proportional sequence, and one hundred myriad the seventh number starting from unity in the same proportional sequence, it is clear that the product will be the forty-sixth number starting from unity. Of these forty six numbers, the first eight, with unity, are part of the numbers called first numbers, the eight following second numbers, the eight following third numbers, the eight following fourth numbers, the eight following fifth numbers, and the six left over are numbers called sixth, the last among being ten myriads of sixth numbers.

It is thus clear that the number of grains of sand whose volume is equal to a sphere of diameter a myriad myriad stadia is smaller than ten myriad sixth numbers. But the volume of a sphere of diameter one hundred myriad myriad stadia is one hundred myriad times the volume of a sphere of diameter a myriad myriad stadia. Thus, if one had a sphere, filled with sand, of the size of a sphere of diameter one hundred myriad myriad stadia, it is clear that the number of grains of sand would be smaller than the product of ten myriad sixth numbers by one hundred myriad. But, since ten myriad sixth numbers represent the forty-sixth number, starting from unity, in the proportional sequence, and one hundred myriad the seventh number starting from unity in the same proportional sequence, it is clear that the product will be the fifty-second number starting in the same proportional sequence. But of these fifty two numbers, the first forty eight, with unity, belong to numbers called first numbers, second numbers, third, fourth, fifth, and sixth, and the four remaining are among numbers called seventh numbers, the last of them being one thousand units of seventh numbers. It is thus clear that the number of grains of sand in a volume equal to a sphere whose volume is equal to that of a sphere of diameter one hundred myriad myriad stadia is smaller than one thousand units of seventh numbers.

As we have shown that the diameter of the Universe is less than one hundred myriad myriad stadia, it is clear that the number of grains of sand filling a volume equal to that of the Universe is itself less than one thousand units of seventh numbers. We have thus shown that the number of grains of sand filling a volume equal to that of the Universe, as the majority of astronomers understand it, is one thousand units of seventh numbers; we will now show that even the number of grains of sand filling a volume equal to the sphere as large as Aristarchus proposed for the fixed stars, is smaller than one thousand myriad eighth numbers. As we have assumed, in fact, that the ratio of the Earth to what we commonly call the Universe is equal to the ratio of this Universe to the sphere of fixed stars, as proposed by Aristarchus, the two spheres have the same ratio to each other. But it has been shown that the diameter of the Universe is less than a length a myriad times the multiple of the diameter of the Earth. It is thus clear that the diameter of the sphere of fixed stars is itself smaller than a length a myriad times the diameter of the Universe. But since the spheres have the ratio among themselves of their diameters, it is clear that the sphere of fixed stars, as Aristarchus proposes, is less than a volume of a myriad myriad myriad times a multiple of the volume of the Universe. But we have shown that the number of grains of sand filling a volume equal to that of the world is less than a thousand units of seventh numbers; it is therefore evident that if a sphere, as large as Aristarchus supposes that of the fixed stars to be, were to be filled with sand, the number of grains of sand would be less than the product of one thousand units [of seventh numbers] by a myriad myriad myriads. And since one thousand units of seventh numbers represent the fifty-second number in the proportional sequence starting from unity, and a myriad myriad myriads the thirteenth number starting from unity in the same proportional sequence, it is clear that the product will be the sixty-fourth number starting from unity in the same proportional sequence; but this number is the eighth of the eight numbers, which is one thousand myriads of eighth numbers.

It is therefore obvious that the number of grains of sand filling a sphere of the size that Aristarchus lends to the sphere of fixed stars is less than one thousand myriad myriads of eighth numbers [$10^{56+7} = 10^{63}$]. I conceive, King Gelon, that among men who do not have experience of mathematics, such a thing might appear incredible. On the other hand, those who know of such matters and have thought about the distances and sizes of the Earth, the Sun, the Moon, and the Universe in its entirety will accept them due to my argument, and that is why I believed that you might enjoy having brought it to your attention.

2.14 Pappus, *Collection*, 2

Very few biographical details of Pappus of Alexandra have survived. His *floruit*, established from a marginal note in a ninth-century manuscript, falls in the first few decades of the fourth century CE. Pappus was working amid various upheavals arising from the establishment of Christianity as the dominant religion in the Roman

Empire, and his ambition was to look back to the past and provide a comprehensive survey of this classical mathematics. As such, he remains a critical source for our understanding of earlier Hellenistic mathematicians and provides details for many works and contributions that have since been lost. As a legacy, Pappus's contributions also served as a key source for the transmission and reception of Greek geometry in early modern mathematics. For these reasons, he is a critical figure in the history of Greek mathematics.

Of his surviving works, the most important is the *Collection*, which covers a variety of arithmetical and geometrical topics over eight books. Unfortunately, the beginning (book 1 and the beginning of book 2) and end (the very last part of book 8) are lost. Book 2 culminates in an arithmetical challenge to be solved, which makes use of the fact that the predominant system of numeration in Greek antiquity, the Ionian system, was based on the use of letters of the Greek alphabet to denote numbers of increasing value: $1 = \alpha$, $2 = \beta$, $3 = \gamma, \ldots, 10 = \iota$, $20 = \kappa$, $30 = \lambda, \ldots$ $100 = \rho$, $200 = \sigma$, $300 = \tau$, and so on up to a "myriad," or 10,000. Pappus presents a fragment of Greek verse whose letters can be simultaneously read as numbers and seeks the product of these numbers. The result is a very large number, greater than the normal system of numeration can deal with, and thus his discussion involves not only some shortcut ways to compute the product but also how to express it using multiples of myriads. It is generally assumed that this system came originally from Apollonius, who is mentioned several times in what follows.

Book 2

[Proposition 14.] *(Let there be however so many numbers and let the numbers, being) less than one hundred, be divisible by ten, and let it be necessary to state their product without multiplying them (directly).*

Let the numbers be 50, 50, 50, 40, 40, 30. Therefore, the base-numbers (*puthmenes*) will be 5, 5, 5, 4, 4, 3. Therefore, their product will be 6,000 units (*monades*). And since the multitude of tens is 6 and dividing (that multitude) by four leaves (a remainder of) two, their product (i.e., of the tens) is a hundred straight myriads. And since the product of the tens (multiplied by) the product of the base-numbers produces the product of the original (numbers), therefore the 100 myriads (multiplied by) the 6,000 units make 60 myriads squared, so that the product of 50, 50, 50, 40, 40, 30 is 60 myriads squared.

[Proposition 16.] *Let there be two numbers A and B and let A be set as less than one thousand and divisible by a hundred, such as 500 units, and let B be set as less than a hundred and divisible by ten, such as 40 units, and let it be necessary to state the number from (the product of) these without multiplying them (directly).*

This is clear through arithmetical procedures. Indeed, 5 and 4, being the base-numbers of these (original quantities), being multiplied, produces 20 units and the number 20 a thousand times makes two myriads, being the result from (multiplying) A and B. The geometrical (*grammikos*) (approach) is made clear from those (methods) that Apollonius demonstrated.

[Proposition 25, part 2.] Indeed, it is evident from this previously considered theorem how it is possible to multiply the given line (of text) and to state the number resulting from multiplying the first number which is associated with the first of the letters by the second number which is associated with the second of the letters and (the number) resulting (from multiplying) by the third number which is associated with the third letter, and continuing in this way until the end of the line (of verse), which Apollonius gives in the beginning as:

Ἀρτέμιδος κλεῖτε κράτος ἔξοχον ἐννέα κοῦραι
Celebrate, oh nine maidens, the outstanding might of Artemis!

(He says *celebrate* instead of *memorialise*.) Indeed, there are 38 letters in the line (of verse); these contain the 10 numbers: ρ′, τ′, σ′, τ′, ρ′, τ′, σ′, χ′, υ′, ρ′, each of which is less than a thousand and divisible by a hundred and 17 numbers: μ′, ι′, ο′, κ′, λ′, ι′, κ′, ο′, ξ′, ο′, ο′, ν′, ν′, ν′, κ′, ο′,ι′, each of which is less than a hundred and divisible by ten, and the remaining 11: α′, ε′, δ′, ε′, ε′, α′, ε′, ε′, ε′, α′, α′, each of which is less than ten. If, therefore, [we double the tens (i.e., the number of those in the hundreds) and we add (that) resulting (quantity of) 20 (multitudes) to the aforementioned 17 straight numbers (of those in the tens), together the result we will have is 37 of the desired proportion of those (numbers). Or] if we substitute for the ten (numbers in the hundreds) ten equivalent numbers to the order of a hundred, and we likewise substitute the 17 (numbers in the tens) with 17 tens, it is clear from the previous theorem 12 concerning calculation (*logistic*) that the ten hundreds added to the 17 tens makes ten myriads "taken nine times" (i.e., to the power of nine). [For the resulting twice ten hundreds, that is 20, (which are added to) the 17 resulting tens, produces 37 equivalently; 37 divided by four produces 9 from the division and leaves behind (a remainder of) 1, thus ten myriads "taken nine times" (are equivalent to) ten hundreds and 17 tens.]

Indeed, the base-numbers of the numbers divisible by a hundred and divisible by ten are the following 27:[52]

1, 3, 2, 3, 1, 3, 2, 6, 4, 1;
4, 1, 7, 2, 3, 1, 2, 7, 6, 7, 7, 5, 5, 5, 2, 7, 1,

but there are 11 (numbers) less than ten, that is the numbers:

1, 5, 4, 5, 5, 1, 5, 5, 5, 1, 1.

If we multiply the (product) of the base-numbers from those 11 (numbers) and that from the 27 (numbers), the product will be 19 myriads "taken four times" and 6036 (myriads) "taken three times" and 480 (myriads) "taken two times." [The (number)[53] equal to this, (namely) the (product resulting) from the base-numbers from the line) of verse, along with the units:

[52] In the thesis by Rideout, these numbers are given in their Greek-letter equivalents, but as they are intended to be read with their numerical significance, we translate them as numbers.

[53] The following passage appears to be a later addition. In particular, the word *myriad* has been abbreviated by the Greek letter μ, and for subsequent multiplies of myriads, the printed text has superscripted the appropriated Greek-letter number to μ. We have translated this as "myr^1", "myr^2," and so on. Likewise, monade has been abbreviated to μ^{o}. We have rendered this as "mo."

Ἀρτέμιδος κλεῖτε κράτος ἔξοχον ἐννέα κοῦραι
Celebrate, oh nine maidens, the outstanding might of Artemis!

The (numbers are): 1, 1, 3, 5, 4, 1, 4, 7, 3, 3, 4, 5, 1, 3, 5, 2, 1, 1, 3, 7, 2, 5, 6, 7, 6, 7, 5, 5, 5, 5, 5, 1, 2, 7, 4, 1, 1, 1.

For, one by 1 produces 1
by 3 produces 3
by 5 produces 15
by 4 produces 60
by 1 produces 60
by 4 produces 240
by 7 produces 1,680
by 2 produces 3,360
by 2 produces 6,720
by 3 produces 2 myr^{1} and 160 mo.
by 5 produces 10 myr^{1} and 800 mo.
by 1 produces 10 myr^{1} and 800 mo.
by 3 produces 30 myr^{1} and 2,400 mo.
by 5 produces 151 myr^{1} and 2,000 mo.
by 2 produces 302 myr^{1} and 4,000 mo.
by 1 produces 302 myr^{1} and 4,000 mo.
by 1 produces 302 myr^{1} and 4,000 mo.
by 3 produces 907 myr^{1} and 2,000 mo.
by 7 produces 6,350 myr^{1} and 4,000 mo.
by 2 produces 1 myr^{2} and 2,700 myr^{1} and 8,000 mo.
by 5 produces 6 myr^{2} and 3,504 myr^{1}
by 6 produces 38 myr^{2} and 1,024 myr^{1}
by 7 produces 266 myr^{2} and 7,168 myr^{1}
by 6 produces 1,600 myr^{2} and 3,008 myr^{1}
by 7 produces 1 myr^{3} and 1,202 myr^{2} and 1,056 myr^{1}
by 5 produces 5 myr^{3} and 6,010 myr^{2} and 5,280 myr^{1}
by 5 produces 28 myr^{3} and 52 myr^{2} and 6,400 myr^{1}
by 5 produces 140 myr^{3} and 263 myr^{2} and 2,000 myr^{1}
by 5 produces 700 myr^{3} and 1,316 myr^{2}
by 5 produces 3,500 myr^{3} and 6,580 myr^{2}
by 1 produces 3,500 myr^{3} and 6,580 myr^{2}
by 2 produces 7,001 myr^{3} and 3,160 myr^{2}
by 7 produces 4 myr^{4} and 9,009 myr^{3} and 2,120 $\times$ myr^{2}
by 4 produces 19 myr^{4} and 6,036 myr^{3} and 8,480 $\times$ myr^{2}

These things multiplied together with the product of the hundreds and the tens, that is to say the previously established ten myriads "taken nine times," make 196 myriads "taken thirteen times," 368 (myriads) "taken twelve times," 4,800 (myriads) taken eleven times.

2.15 *Greek Anthology*

The *Greek Anthology* is a large collection of about 4,500 short Greek poems and other writings, by about three hundred writers. The original collection was put together by Meleager of Gedara in the first century BCE. More material was added in the first and second centuries CE and somewhat later. The material was collected and arranged in the early tenth century by Constantinus Cephalas into fifteen books and was then reedited by Maximus Planudes in the fourteenth century. It is only book 14 that contains mathematical problems and riddles, but the author of these is unknown. We present here a selection of these mathematical problems, all of which require only arithmetic to solve, although the algebra of first degree equations is the method that we would generally choose to use. The most famous of these problems is 126, which supposedly gives us virtually the only biographical information known about Diophantus.

Book 14: Arithmetical Problems, Riddles, Oracles

3. Cypris thus addressed Love, who was looking downcast: "How, my child, hath sorrow fallen on thee?" And he answered: "The Muses stole and divided among themselves, in different proportions, the apples I was bringing from Helicon, snatching them from my bosom. Clio got the fifth part, and Euterpe the twelfth, but divine Thalia the eighth. Melpemene carried off the twentieth part, and Terpsichore the fourth, and Erato the seventh; Polyhomnia robbed me of thirty apples, and Urania of a hundred and twenty, and Calliope went off with a load of three hundred apples. So I come to thee with lighter hands, bringing these fifty apples that the goddesses left me." *Solution*: 3,360.

7. I am a brazen lion; my spouts are my two eyes, my mouth, and the flat of my right foot. My right eye fills a jar in two days, my left eye in three, and my foot in four. My mouth is capable of filling it in six hours. Tell me how long all four together will take to fill it. *Solution*: $3\frac{33}{37}$ hours.

11. I desire my two sons to receive the thousand staters of which I am possessed, but let the fifth part of the legitimate one's share exceed by ten the fourth part of what falls to the illegitimate one. *Solution*: $577\frac{7}{9}$ and $422\frac{2}{9}$

51. A: I have what the second has and the third of what the third has. B: I have what the third has and the third of what the first has. C: And I have ten *minae* and the third of what the second has. *Solution*: A has 45 *minae*, B has $37\frac{1}{2}$ *minae*, C has $22\frac{1}{2}$ *minae*.

118. Myrto picked apples and divided them among her friends; she gave the fifth part to Chrysis, the fourth to Hero, the nineteenth to Psamathe, and the tenth to Cleopatra, but

she presented the twentieth part to Parthenope and gave only twelve to Evadne. Of the whole number a hundred and twenty fell to herself. *Solution*: 380.

121. From Cadiz to the city of the seven hills [Rome] the sixth of the road is to the banks of Bactis, loud with the lowing of herds, and hence a fifth to the Phocian soil of Pylades—the land is Vaccaean, its name derived from the abundance of cows. Thence to the precipitous Pyrenees is one-eighth and the twelfth part of one-tenth. Between the Pyrenees and the lofty Alps lies one-fourth of the road. Now begins Italy and straight after one-twelfth appears the amber of the Po. O blessed am I who have accomplished two thousand and five hundred stades journeying from thence! For the Palace on the Tarpeian rock is my journey's object. *Solution*: The total distance is 15,000 stades; from Cadiz to the Guadalquivir, 2,500, thence to the Vaccaci, 3,000; thence to the Pyrenees, 2,000; thence to the Alps, 3,750; thence to the Po, 1,250; thence to Rome, 2,500.

123. Take, my son, the fifth part of my inheritance, and thou, wife, receive the twelfth; and ye four sons of my departed son and my two brothers, and thou my grieving mother, take each an eleventh part of the property. But ye, my cousins, receive twelve *talents*, and let my friend Eubulus have five *talents*. To my most faithful servants I give their freedom and these recompenses in payment of their service. Let them receive as follows. Let Onesimus have twenty-five *minae* and Davus twenty *minae*, Syrus fifty, Synete ten and Tibius eight, and I give seven *minae* to the son of Syrus, Synetus. Spend thirty *talents* on adorning my tomb and sacrifice to Infernal Zeus. From two *talents* let the expense be met of my funeral pyre, the funeral cakes, and grave-clothes, and from two let my corpse receive a gift. [Note that 1 *talent* is equal to 60 *minae*.] *Solution*: 660 *talents*.

126. This tomb holds Diophantus. Ah, how great a marvel; the tomb tells scientifically the measure of his life. God granted him to be a boy for the sixth part of his life, and adding a twelfth part to this, He clothed his cheeks with down. He lit him the light of wedlock after a seventh part, and five years after his marriage He granted him a son. Alas! Late-born wretched child; after attaining the measure of half his father's life, chill Fate took him. After consoling his grief by this science of numbers for four years he ended his life. *Solution*: Diophantus lived for 84 years.

136. Brick-makers, I am in a great hurry to erect this house. Today is cloudless, and I do not require many more bricks, but I have all I want but three hundred. Thou alone in one day couldst make as many, but thy son left off working when he had finished two hundred, and thy son-in-law when he had made two hundred and fifty. Working all together, in how many hours can you make these? *Solution*: $\frac{2}{5}$ of a day.

141. Tell me the transits of the fixed stars and planets when my wife gave birth to a child yesterday. It was day, and till the sun set in the western sea it wanted six times two-sevenths of the time since dawn. *Solution*: It was $4\frac{8}{19}$ hours from sunrise.

145. A. Give me ten *minae* and I become three times as much as you. B. And if I get the same from you, I am five times as much as you. *Solution*: $A = 15\frac{5}{7}$; $B = 18\frac{4}{7}$.

SOURCES, CHAPTER 2

2.1 This translation is by Sir Thomas L. Heath in *The Thirteen Books of Euclid's Elements*, vol. 2 (New York: Dover, 1956), 296–299, 320–379, 420–422.

2.2 This translation is from Nicomachus of Gerasa, *Introduction to Arithmetic*, ed. and trans. Martin Luther D'Ooge (New York: Macmillan, 1926), 190–211, 241–281. The translation is also found in Robert Maynard Hutchins, ed., *Great Books of the Western World*, vol. 11 (Chicago: Encyclopedia Britannica, 1952).

2.3 This translation is by Michael Masi in *Boethian Number Theory: A Translation of the De Institutione Arithmetica* (Amsterdam: Rodopi, 2006), 98–113.

2.4 The translation is in *The Works of Philo*, trans. C. D. Yonge (Peabody, MA: Hendrickson 1993). Selection from *On the Creation*, 4–14; from *Who Is the Heir of Divine Things?,* 287–288.

2.5 This translation is by Robin Waterfield, in *The Theology of Arithmetic: On the Mystical, Mathematical and Cosmological Symbolism of the First Ten Numbers; Attributed to Iamblichus* (Grand Rapids, MI: Phanes, 1988), 35–113.

2.6 Some of the English translations of the problems from the Greek books have been taken from Ivor Thomas, *Greek Mathematical Works, vol. 2, Aristarchus to Pappus of Alexandria* (Cambridge, MA: Harvard University Press, 1980.) The other English translations were made by Victor Katz from the Latin in Paul Tannery, *Diophanti Alexandrini Opera Omnia* (Leibniz: Teubner, 1893). The translation of the problems from the Arabic books are found in Jacques Sesiano, ed., *Books IV to VII of Diophantus' Arithmetica in the Arabic Translation Attributed to Qustā ibn Lūqā* (New York: Springer, 1982), 87–125.

2.7 The first source comes from Richard A. Parker, "A Demotic Mathematical Papyrus Fragment," *Journal of Near Eastern Studies* 18, no. 4 (1959): 275–279. The remaining material is from Richard A. Parker, *Demotic Mathematical Papyri* (Providence, RI: Brown University Press, 1972).

2.8 This translation has been adapted from Louis C. Karpinski and Frank E. Robbins, "Michigan Papyrus 620: The Introduction of Algebraic Equations in Greece," *Science* 70, no. 1813 (1929): 311–314.

2.9 This translation is from Roger S. Bagnall and Alexander Jones, *Mathematics, Metrology, and Model Contracts: A Codex from Late Antique Business Education* (New York: New York University Press, 2019), 97.

2.10 This material was translated by Victor Katz from the French in Jules Baillet, "Le papyrus mathématique d'Akhmîm," *Mémoires publiés par les membres de la Mission Archéologique Française au Caire* 9, no. 1 (1892).

2.11 The source of the multiplication table is P. Mich. inv. 5663; University of Michigan Library Digital Collections https://quod.lib.umich.edu/a/apis/x-2527/5663ara.tif. The image of the table of unit fractions is from P. Mich. inv. 621; University of Michigan Library Digital Collections, https://quod.lib.umich.edu/a/apis/x-2701/621r.tif. The translation is found in D. H. Fowler, *The Mathematics of Plato's Academy* (Oxford: Clarendon Press, 1987). The tables from the Achmîm Papyrus are in Jules Baillet, "Le papyrus mathématique d'Akhmîm," *Mémoires publiés par les membres de la Mission Archéologique Française au Caire* 9, no. 1 (1892). The multiplication table on the stele is from Jacques Sesiano, "Greek Multiplication Tables," *SCIAMVS* 21 (2020–2021): 83–140, at 86. Finally, the papyrus with zero is found in Erik J. Knudtzon and Otto Neugebauer, "Zwei astronomische Texte," *Humanistiska Vetenskapssamfundet i Lund*, Ärsberättelse, 1946–47 2(1947): 77–78.

2.12 The first translation from Theon of Alexandria's *Commentary on Ptolemy's Almagest* is from Ivor Thomas, *Selections Illustrating the History of Greek Mathematics*, vol. 1 (Cambridge, MA: Harvard University Press, 1980), 53–61. The second section is taken from Jean Christianidis and Ioanna Skoura, "Solving Problems by Algebra in Late Antiquity: New Evidence from an Unpublished Fragment of Theon's Commentary on the Almagest," *SCIAMVS* 14 (2013): 41–57.

2.13 This translation of a section of the *Sand Reckoner* by Archimedes is by Gerard Michon and is available online at www.numericana.com/answer/archimedes.htm.

2.14 This translation of a section of Pappus's *Collection Book II* is adapted from Bronwyn Rideout, "Pappus Reborn: Pappus of Alexandria and the Changing Face of Analysis and Synthesis in Late Antiquity." MA thesis, University of Canterbury, New Zealand, 2008. (With light edits and emendments.)

2.15 This translation is from W. R. Paton, ed. and trans., *The Greek Anthology*, vol. 5 (Cambridge, MA: Harvard University Press, 1979).

3

Euclid and the Beginnings of Theoretical Geometry

The earliest organized compendium of Greek geometrical ideas still extant is the *Elements* of Euclid, written around 300 BCE. To contextualize this work, we begin this chapter with three excerpts from dialogues of Plato, in which ideas that are fully developed in Euclid's *Elements* are explored, namely, the so-called Platonic solids, the notion of incommensurability, and the doubling of a square, which is equivalent to finding a mean proportional between two lengths. We also include a selection from Aristotle's *Physics* dealing with discreteness and continuity, in which Aristotle refutes the paradoxes of Zeno. These paradoxes showed that one must think carefully about the concepts of the infinite and the infinitely small.

Given the doubling of a square, the Greeks also attempted to solve the problem of doubling a cube as well as two other interesting geometric problems: squaring the circle and trisecting an arbitrary angle. The original goal was to solve these problems using ruler and compass, procedures that came to be called "plane" techniques, that is, those based on axioms and theorems in Euclid's *Elements*, books 1–6. As it turned out, Greek mathematicians eventually realized that it was impossible to solve either of these problems with those restrictions, although formal proofs of the impossibility were not available until the nineteenth century. They therefore attempted to find solutions using other tools, some of which we consider in chapter 5. In this chapter, we see the initial attempt at squaring the circle in Hippocrates's work on the squaring of lunes. We know that Hippocrates also found that the problem of doubling the cube was equivalent to finding two mean proportionals between two given line segments. So we have the solution of this problem by Archytas, which was accomplished through nonplanar methods. Unfortunately, the originals of these two works are no longer extant; however, we have reports on them that were written several hundred years later. We have good reason to believe that these reports as written down in the works of Simplicus and Eutocius quite accurately reflect the original. Thus, the first actual mathematics in this chapter is in those two texts.

We follow this with books 1–6 of the *Elements* on plane geometry, book 10 on the classification of certain incommensurable magnitudes, and books 11–13 on solid geometry. In the fifth century CE, Proclus commented extensively on book 1, so we have interspersed his commentary in the section on book 1. We have also included Simplicius's commentary on the parallel postulate and Pappus's commentary on book 10.

We also feature two additional works of Euclid. The first is the *Data*, a work that formed the basis for the analysis of problems throughout later Greek geometry. The second is the work *On Divisions*, which is no longer extant in Greek, but is preserved in Arabic versions made in the ninth and tenth centuries. Additionally, Hypsicles wrote a work in the second century BCE that is a supplement to Euclid's book 13. In many early editions of the *Elements*, this was included as the so-called book 14. Finally, Pappus, in his fourth-century *Collection*, included considerable material related to the *Elements*. Thus, we see his description of the difference between problems and theorems, his construction inscribing the Platonic solids in a sphere, and his generalization of the Pythagorean Theorem.

3.1 Plato, *Timaeus*

The *Timaeus*, although written as a dialogue, is mainly a discourse by the main character Timaeus on the formation of the universe. In the excerpt below, Plato, speaking through Timaeus, gives one of the earliest descriptions of what are now called the Platonic solids. He associates four of the five solids to the four classical elements of which the world is made. Namely, earth was associated to the cube, fire to the tetrahedron, air to the octahedron, and water to the icosahedron. He only briefly mentions the fifth solid, the dodecahedron, as one used by God "in the delineation of the universe." It is not clear what Plato meant by this. But it does seem that Plato's contemporary, Theaetetus, was responsible for giving the first mathematical description of all five solids and showing that these were the only possible regular solids.

In the first place, then, as is evident to all, fire and earth and water and air are bodies. And every sort of body possesses solidity, and every solid must necessarily be contained in planes; and every plane rectilinear figure is composed of triangles; and all triangles are originally of two kinds, both of which are made up of one right and two acute angles; one of them has at either end of the base the half of a divided right angle, having equal sides, while in the other the right angle is divided into unequal parts, having unequal sides. These, then, proceeding by a combination of probability with demonstration, we assume to be the original elements of fire and the other bodies; but the principles which are prior to these God only knows, and he of men who is the friend of God. And next we have to determine what are the four most beautiful bodies which are unlike one another, and of which some are capable of resolution into one another; for having discovered thus much we shall know the true origin of earth and fire and of the proportionate and intermediate elements. And then we shall not be willing to allow that there are any distinct kinds of visible bodies fairer than these. Wherefore we must endeavor to construct the four forms of bodies which excel in beauty, and then we shall be able to say that we have sufficiently apprehended their nature.

Now of the two triangles, the isosceles has one form only; the scalene or unequal-sided has an infinite number. Of the infinite forms we must select the most beautiful, if we are to proceed in due order, and anyone who can point out a more beautiful form than ours for the construction of these bodies, shall carry off the palm, not as an enemy, but as a friend. Now the one which we maintain to be the most beautiful of all the many triangles (and we need not speak of the others) is that of which the double forms a third triangle which is equilateral; the reason of this would be long to tell; he who disproves what we are saying, and shows that we are mistaken, may claim a friendly victory. Then let us choose two triangles, out of which fire and the other elements have been constructed, one isosceles, the other having the square of the longer side equal to three times the square of the lesser side.

Now is the time to explain what was before obscurely said: there was an error in imagining that all the four elements might be generated by and into one another; this, I say, was an erroneous supposition, for there are generated from the triangles which we have selected four kinds—three from the one which has the sides unequal; the fourth alone is framed out of the isosceles triangle. Hence they cannot all be resolved into one another, a great number of small bodies being combined into a few large ones, or the converse. But three of them can be thus resolved and compounded, for they all spring from one, and when the greater bodies are broken up, many small bodies will spring up out of them and take their own proper figures; or, again when many small bodies are dissolved into their triangles, if they become one, they will form one large mass of another kind. So much for their passage into one another.

I have now to speak of their several kinds, and show out of what combinations of numbers each of them was formed. The first will be the simplest and smallest construction, and its element is that triangle which has its hypotenuse twice the lesser side. When two such triangles are joined at the diagonal, and this is repeated three times, and the triangles rest their diagonals and shorter sides on the same point as a center, a single equilateral triangle is formed out of six triangles; and four equilateral triangles, if put together, make out of every three plane angles one solid angle, being that which is nearest to the most obtuse of plane angles; and out of the combination of these four angles arises the first solid form which distributes into equal and similar parts the whole sphere in which it is inscribed [tetrahedron]. The second species of solid is formed out of the same triangles, which unite as eight equilateral triangles and form one solid angle out of four plane angles, and out of six such angles the second body is completed [octahedron]. And the third body is made up of 120 triangular elements, forming twelve solid angles, each of them included in five plane equilateral triangles, having altogether twenty bases, each of which is an equilateral triangle [icosahedron]. The one element [that is, the triangle which has its hypotenuse twice the lesser side] having generated these figures, generated no more; but the isosceles triangle produced the fourth elementary figure, which is compounded of four such triangles, joining their right angles in a center, and forming one equilateral quadrangle. Six of these united form eight solid angles, each of which is made by the combination of three plane right angles; the figure of the body thus composed is a cube, having six plane quadrangular equilateral bases. There was yet a fifth combination [the dodecahedron] which God used in the delineation of the universe.

Now, he who, duly reflecting on all this, inquires whether the worlds are to be regarded as indefinite or definite in number, will be of the opinion that the notion of their indefiniteness is characteristic of a sadly indefinite and ignorant mind. He, however, who raises the question whether they are to be truly regarded as one or five, takes up a more reasonable position. Arguing from probabilities, I am of the opinion that they are one; another, regarding the question from another point of view, will be of another mind. But, leaving this inquiry, let us proceed to distribute elementary forms, which have now been created in idea, among the four elements.

To earth, then, let us assign the cubical form; for earth is the most immovable of the four and the most plastic of all bodies, and that which has the most stable bases must of necessity be of such a nature. Now, of the triangles which we assumed at first, that which has two equal sides is by nature more firmly based than that which has unequal sides; and of the compound figures which are formed out of either, the plane equilateral quadrangle has necessarily a more stable basis than the equilateral triangle, both in the whole and in the parts. Wherefore, in assigning this figure to earth, we adhere to probability; and to water we assign that one of the remaining forms which is the least movable; and the most movable of them to fire; and to air that which is intermediate. Also we assign the smallest body to fire, and the greatest to water, and the intermediate in size to air; and, again, the acutest body to fire, and the next in acuteness to air, and the third to water.

Of all these elements, that which has the fewest bases must necessarily be the most movable, for it must be the acutest and most penetrating in every way, and also the lightest as being composed of the smallest number of similar particles; and the second body has similar properties in a second degree, and the third body in the third degree. Let it be agreed, then, both according to strict reason and according to probability, that the tetrahedron is the solid which is the original element and seed of fire; and let us assign the element which was next in the order of generation to air, and the third to water. We must imagine all these to be so small that no single particle of any of the four kinds is seen by us on account of their smallness; but when many of them are collected together their aggregates are seen. And the ratios of their numbers, motions, and other properties, everywhere God, as far as necessity allowed or gave consent, has exactly perfected, and harmonized in due proportion.

3.2 Plato, *Theaetetus*

In this dialogue, Socrates is discussing the nature of knowledge with Theaetetus (417–369 BCE), who was a young student of the geometer Theodorus (465–398 BCE) at the time. In fact, most of the knowledge we have of these two mathematicians comes from this dialogue. In this selection, Theaetetus discusses the idea of incommensurability and gives a classification of numbers according to whether or not their square roots are commensurable with the unit. Some of this material may well have provided the impetus for the developments of Euclid's *Elements*, book 10.

Socrates: What is knowledge? Can we answer that question? . . . We wanted to know not the subjects, nor yet the number of the arts or sciences, for we were not going to count them, but we wanted to know the nature of knowledge in the abstract. Am I not right?

Theaetetus: Perfectly right.

Socrates: Let me offer an illustration. Suppose that a person were to ask about some very trivial and obvious thing—for example, What is clay? and we were to reply, that there is a clay of potters, there is a clay of oven-makers, there is a clay of brick-makers; would not the answer be ridiculous?

Theaetetus: Truly.

Socrates: In the first place, there would be an absurdity in assuming that he who asked the question would understand from our answer the nature of "clay," merely because we added "of the image makers," or of any other workers. How can a man understand the name of anything, when he does not know the nature of it?

Theaetetus: He cannot.

Socrates: Then he who does not know what science or knowledge is, has no knowledge of the art or science of making shoes?

Theaetetus: None.

Socrates: Nor of any other science?

Theaetetus: No.

Socrates: And when a man is asked what science or knowledge is, to give in answer the name of some art of science is ridiculous; for the question is, "What is knowledge?" and he replies , "A knowledge of this or that."

Theaetetus: True.

Socrates: Moreover, he might answer shortly and simply, but he makes an enormous circuit. For example, when asked about the clay, he might have said simply, that clay is moistened earth—what sort of clay is not to the point.

Theaetetus: Yes, Socrates, there is no difficulty as you put the question. You mean, if I am not mistaken, something like what occurred to me and to my friend here, your namesake Socrates, in a recent discussion.

Socrates: What was that, Theaetetus?

Theaetetus: Theodorus was writing out for us something about roots, such as the roots of three or five, showing that they are incommensurable to the unit; he selected other examples up to seventeen—there he stopped. Now as there are innumerable roots, the notion occurred to us of attempting to include them all under one name or class.

Socrates: And did you find such a class?

Theaetetus: I think that we did; but I should like to have your opinion.

Socrates: Let me hear.

Theaetetus: We divided all numbers into two classes: those which are made up of equal factors multiplying into one another, which we compared to square figures and called square or equilateral numbers—that was one class.

Socrates: Very good.

Theaetetus: The intermediate numbers, such as three and five, and every other number which is made up of unequal factors, either of a greater multiplied by a less, or of a less multiplied by a greater, and when regarded as a figure, is contained in unequal sides—all these we compared to oblong figures, and called them oblong numbers.

Socrates: Capital; and what followed?

Theaetetus: The lines, or sides, which have for their squares the equilateral plane numbers, were called by us lengths or magnitudes; and the lines which are the roots of (or whose squares are equal to) the oblong numbers, were called powers or roots; the reason of this latter name being, that they are commensurable with the former [i.e., with the so-called lengths or magnitudes] not in linear measurement, but in the value of the superficial content of their squares; and the same about solids.

Socrates: Excellent, my boys; I think that you fully justify the praises of Theodorus, and that he will not be found guilty of false witness.

Theaetetus: But I am unable, Socrates, to give you a similar answer about knowledge, which is what you appear to want; and therefore Theodorus is a deceiver after all.

Socrates: Well, but if someone were to praise you for running, and to say that he never met your equal among boys, and afterwards you were beaten in a race by a grown-up man, who was a great runner—would the praise be any the less true?

Theaetetus: Certainly not.

Socrates: And is the discovery of the nature of knowledge so small a matter, as I just now said? Is it not one which would task the powers of men perfect in every way?

Theaetetus: By heaven, they should be the top of all perfection!

Socrates: Well, then, be of good cheer; do not say that Theodorus was mistaken about you, but do your best to ascertain the true nature of knowledge, as well as of other things.

Theaetetus: I am eager enough, Socrates, if that would bring to light the truth.

Socrates: Come, you made a good beginning just now; let your own answer about roots be your model, and as you comprehended them all in one class, try and bring the many sorts of knowledge under one definition.

3.3 Plato, *Meno*

In the *Meno*, as in earlier dialogues, Plato's spokesman is Socrates. Here, Socrates and his friend Meno consider the question about whether virtue can be taught. Socrates claims that no one can teach anyone anything. Knowledge comes from recalling what the soul has seen before birth. All that a teacher can do is help the soul to remember. To convince Meno of his ideas, Socrates proposed an experiment. By merely asking questions and not telling him anything, he will get an uneducated servant boy to understand that the square constructed on the diagonal of a given square has twice the area of that square. Whether or not Socrates's experiment is successful is up to the reader to decide. During the dialogue, Socrates refers to several different diagrams. In general, it will be easy for the reader to reconstruct these, but we have included one basic diagram that captures the main idea Socrates is getting the boy to "recall" [figure 3.3.1]. Note that in the diagram, the "4"s refer to the area of each of the quarters of the diagram, while the "8" is the area inside the square whose sides are the four diagonals.

Meno: Yes, Socrates; but what do you mean by saying that we do not learn, and that what we call learning is only a process of recollection? Can you teach me how this is?

Socrates: I told you, Meno, just now that you were a rogue, and now you ask whether I can teach you, when I am saying that there is not teaching, but only recollection; and thus you imagine that you will involve me in a contradiction.

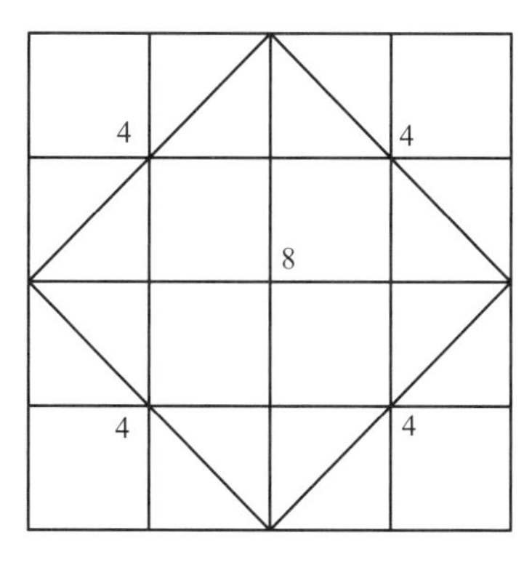

Figure 3.3.1.

Meno: Indeed, Socrates, I protest that I had no such intention. I only asked the question from habit; but if you can prove to me that what you say is true, I wish that you would.

Socrates: It will be no easy matter, but I will try to please you to the utmost of my power. Suppose that you call one of your numerous attendants, that I may demonstrate on him.

Meno: Certainly. Come hither, boy.

Socrates: He is Greek, and speaks Greek, does he not?

Meno: Yes, indeed; he was born in the house.

Socrates: Attend now to the questions which I ask him, and observe whether he learns of me or only remembers.

Meno: I will.

Socrates: Tell me, boy, do you know that a figure like this is a square?

Boy: I do.

Socrates: And you know that a square figure has these four lines equal?

Boy: Certainly.

Socrates: And if one side of the figure be of two feet, and the other side be of two feet, how much will the whole be? Let me explain: if in one direction the space was of two feet, and in the other direction of one foot, the whole would be of two feet taken once?

Boy: Yes.

Socrates: But since this side is also of two feet, there are twice two feet?

Boy: There are.

Socrates: Then the square is of twice two feet?

Boy: Yes.

Socrates: And how many are twice two feet? Count and tell me.

Boy: Four, Socrates.

Socrates: And might there not be another square twice as large as this, and having like this the lines equal?

Boy: Yes.

Socrates: And of how many feet will that be?

Boy: Of eight feet.

Socrates: And now try and tell me the length of the line which forms the side of that double square: this is two feet—what will that be?

Boy: Clearly, Socrates, it will be double.

Socrates: Do you observe, Meno, that I am not teaching the boy anything, but only asking him questions; and now he fancies that he knows how long a line is necessary in order to produce a figure of eight square feet; does he not?

Meno: Yes.

Socrates: And does he really know?

Meno: Certainly not.

Socrates: He only guesses that because the square is double, the line is double.

Meno: True.

Socrates: Observe him while he recalls the steps in regular order. Tell me, boy, do you assert that a double space comes from a double line? Remember that I am not speaking of an oblong, but of a figure equal every way, and twice the size of this—that is to say of eight feet; and I want to know whether you still say that a double square comes from a double line?

Boy: Yes.

Socrates: But does not this line become doubled if we add another such line here?

Boy: Certainly.

Socrates: And four such lines will make a space containing eight feet?

Boy: Yes

Socrates: Let us describe such a figure: Would you not say that this is the figure of eight feet?

Boy: Yes.

Socrates: And are there not these four divisions in the figure, each of which is equal to the figure of four feet?

Boy: True.

Socrates: And is not that four times four?

Boy: Certainly.

Socrates: And four times four is not double?

Boy: No, indeed.

Socrates: But how much?

Boy: Four times as much.

Socrates: Therefore the double line, boy, has given a space, not twice, but four times as much.

Boy: True.

Socrates: Four times four are sixteen—are they not?

Boy: Yes.

Socrates: What line would give you a space of eight feet, as this gives one of sixteen feet—do you see?

Boy: Yes.

Socrates: And the space of four feet is made from this half line?

Boy: Yes.

Socrates: Good; and is not a space of eight feet twice the size of this, and half the size of the other?

Boy: Certainly.

Socrates: Such a space, then, will be made out of a line greater than this one, and less than that one?

Boy: Yes, I think so.

Socrates: Very good; I like to hear you say what you think. And now tell me, is not this a line of two feet and that of four?

Boy: Yes.

Socrates: Then the line which forms the side of eight feet ought to be more than this line of two feet, and less than the other of four feet?

Boy: It ought.

Socrates: Try and see if you can tell me how much it will be.

Boy: Three feet.

Socrates: Then if we add a half to this line to two, that will be the line of three. Here are two and there is one; and on the other side, here are two also and there is one; and that makes the figure of which you speak?

Boy: Yes.

Socrates: But if there are three feet this way and three feet that way, the whole space will be three times three feet?

Boy: That is evident.

Socrates: And how much are three times three feet?

Boy: Nine.

Socrates: And how much is the double of four?

Boy: Eight.

Socrates: Then the figure of eight is not made out of a line of three?

Boy: No.

Socrates: But from what line? Tell me exactly, and if you would rather not reckon, try and show me the line.

Boy: Indeed, Socrates, I do not know.

Socrates: Do you see, Meno, what advances he has made in his power of recollection? He did not know at first, and he does not know now, what is the side of a figure of eight feet; but then he thought that he knew, and answered confidently as if he knew, and had no difficulty; now he has a difficulty, and neither knows nor fancies that he knows.

Meno: True.

Socrates: Is he not better off in knowing his ignorance?

Meno: I think that he is.

Socrates: If we have made him doubt, and given him the "torpedo's shock," have we done him any harm?

Meno: I think not.

Socrates: We have certainly, as would seem, assisted him in some degree to the discovery of the truth; and now he will wish to remedy his ignorance, but then he would have been ready to tell all the world again and again that the double space should have a double side.

Meno: True.

Socrates: But do you suppose that he would ever have enquired into or learned what he fancied that he knew, though he was really ignorant of it, until he had fallen into perplexity under the idea that he did not know, and had desired to know?

Meno: I think not, Socrates.

Socrates: Then he was the better for the torpedo's touch?

Meno: I think so.

Socrates: Mark now the farther development. I shall only ask him, and not teach him, and he shall share the enquiry with me; and do you watch and see if you find me telling or explaining anything to him, instead of eliciting his opinion. Tell me, boy, is not this a square of four feet which I have drawn?

Boy: Yes.

Socrates: And now I add another square equal to the former one?

Boy: Yes.

Socrates: And a third, which is equal to either of them?

Boy: Yes.

Socrates: Suppose that we fill up the vacant corner?

Boy: Very good.

Socrates: Here, then, there are four equal spaces?

Boy: Yes.

Socrates: And how many times larger is this space than this other?

Boy: Four times.

Socrates: But it ought to have been twice only, as you will remember.

Boy: True.

Socrates: And does not this line, reaching from corner to corner, bisect each of these spaces?

Boy: Yes.

Socrates: And are there not here four equal lines which contain this space?

Boy: There are.

Socrates: Look and see how much this space is.

Boy: I do not understand.

Socrates: Has not each interior line cut off half of the four spaces?

Boy: Yes.

Socrates: And how many spaces are there in this section?

Boy: Four.

Socrates: And how many in this?

Boy: Two.

Socrates: And four is now many times two?

Boy: Twice.

Socrates: And this space is of how many feet?

Boy: Of eight feet.

Socrates: And from what line do you get this figure?

Boy: From this.

Socrates: That is, from the line which extends from corner to corner of the figure of four feet?

Boy: Yes.

Socrates: And that is the line which the learned call the diagonal. And if this is the proper name, then you, Meno's slave, are prepared to affirm that the double space is the square of the diagonal?

Boy: Certainly, Socrates.

Socrates: What do you say of him, Meno? Were not all these answers given out of his own head?

Meno: Yes, they were all his own.

Socrates: And yet, as we were just now saying, he did not know?

Meno: True.

Socrates: But still he had in him those notions of his—had he not?

Meno: Yes.

Socrates: Then he who does not know may still have true notions of that which he does not know?

Meno: He has.

Socrates: And at present these notions have just been stirred up in him, as in a dream; but if he were frequently asked the same questions, in different forms, he would know as well as any one at last?

Meno: I dare say.

Socrates: Without any one teaching him he will recover his knowledge for himself, if he is only asked questions?

Meno: Yes.

Socrates: And this spontaneous recovery of knowledge in him is recollection?

Meno: True.

Socrates: And this knowledge which he now has must he not either have acquired or always possessed?

Meno: Yes.

Socrates: But if he always possessed this knowledge he would always have known; or if he has acquired the knowledge he could not have acquired it in this life, unless he has been taught geometry; for he may be made to do the same with all geometry and every other branch of knowledge. Now, has any one ever taught him all this? You must know about him, if, as you say, he was born and bred in your house.

Meno: And I am certain that no one ever did teach him.

Socrates: And yet he has the knowledge?

Meno: The fact, Socrates, is undeniable.

Socrates: But if he did not acquire the knowledge in this life, then he must have had and learned it at some other time?

Meno: Clearly he must.

Socrates: Which must have been the time when he was not a man?

Meno: Yes.

Socrates: And if there have been always true thoughts in him, both at the time when he was and was not a man, which only need to be awakened into knowledge by putting questions to him, his soul must have always possessed this knowledge, for he always either was or was not a man?

Meno: Obviously.

Socrates: And if the truth of all things always existed in the soul, then the soul is immortal. Wherefore be of good cheer, and try to recollect what you do not know, or rather what you do not remember.

Meno: I feel, somehow, that I like what you are saying.

Socrates: And I, Meno, like what I am saying. Some things I have said of which I am not altogether confident. But that we shall be better and braver and less helpless if we think that we ought to inquire, then we should have been if we indulged in the idle fancy that there was no knowing and no use in seeking to know what we do not know—that is a theme upon which I am ready to fight, in word and deed, to the utmost of my power.

3.4 Aristotle, *Physics*

In the following excerpts from the *Physics*, Aristotle discusses the ideas of "continuous," "in contact," and "in succession," thus exploring the difference between number and magnitude. These ideas are then developed in Aristotle's discussion of Zeno's four paradoxes, in which the philosopher shows how to refute each one.

Book 6

Chapter 1

Now if the terms "continuous," "in contact," and "in succession" are understood as defined above—things being "continuous" if their extremities are one, "in contact" if their extremities are together, and "in succession" if there is nothing of their own kind intermediate between them—nothing that is continuous can be composed of indivisibles: e.g., a line cannot be composed of points, the line being continuous and the point indivisible. For the extremities of two points can neither be one (since of an indivisible there can be no extremity as distinct from some other part) nor together (since that which has no parts can have no extremity, the extremity and the thing of which it is the extremity being distinct).

Moreover, if that which is continuous is composed of points, these points must be either continuous or in contact with one another; and the same reasoning applies in the case of all indivisibles. Now for the reason given above they cannot be continuous; and one thing can be in contact with another only if whole is in contact with whole or part with part or part with whole. But since indivisibles have no parts, they must be in contact with one another as whole with whole. And if they are in contact with one another as whole with whole, they will not be continuous; for that which is continuous has distinct parts; and these parts into which it is divisible are different in this way, i.e., spatially separate.

Nor, again, can a point be in succession to a point or a moment to a moment in such a way that length can be composed of points or time of moments; for things are in succession if there is nothing of their own kind intermediate between them, whereas that which is intermediate between points is always a line and that which is intermediate between moments is always a period of time.

Again, if length and time could thus be composed of indivisibles, they could be divided into indivisibles, since each is divisible into the parts of which it is composed. But as we saw, no continuous thing is divisible into things without parts. Nor can there be anything of any other kind intermediate between the parts or between the moments; for if there could be any such thing it is clear that it must be either indivisible or divisible, and if it is divisible, it must be divisible either into indivisibles or into divisibles that are infinitely divisible, in which case it is continuous.

Moreover, it is plain that everything continuous is divisible into divisibles that are infinitely divisible; for if it were divisible into indivisibles, we should have an indivisible in contact with an indivisible, since the extremities of things that are continuous with one another are one and are in contact. The same reasoning applies equally to magnitude, to time, and to motion: either all of these are composed of indivisibles and are divisible into indivisibles, or none. . . .

Chapter 2

Since every motion is in time and a motion may occupy any time, and the motion of everything that is in motion may be either quicker or slower, both quicker motion and slower motion may occupy any time; and this being so, it necessarily follows that time also is continuous. By continuous I mean that which is divisible into divisibles that are infinitely divisible; and if we take this as the definition of continuous, it follows necessarily that time is continuous. For since it has been shown that the quicker will pass over an equal magnitude in less time than the slower, suppose that A is quicker and B slower, and that the slower has traversed the magnitude $\Gamma\Delta$ in the time ZH. Now it is clear that the quicker will traverse the same magnitude in less time than this: let us say in the time $Z\Theta$. Again, since the quicker has passed over the whole $\Gamma\Delta$ in the time $Z\Theta$, the slower will in the same time pass over ΓK, say, which is less than $\Gamma\Delta$. And since B, the slower, has passed over ΓK in the time $Z\Theta$, the quicker will pass over it in less time; so that the time $Z\Theta$ will again be divided. And if this is divided, the magnitude ΓK will also be divided just as $\Gamma\Delta$ was; and again, if the magnitude is divided, the time will also be divided. And we can carry on this process forever, taking the slower after the quicker and the quicker after the slower alternately, and using what has been demonstrated at each stage as a new point of departure; for the quicker will divide the time and the slower will divide the length. If, then, this alternation always holds good, and at every turn involves a division, it is evident that all time must be continuous. And at the same time it is clear that all magnitude is also continuous; for the divisions of which time and magnitude respectively are susceptible are the same and equal.

Moreover, the current popular arguments make it plain that, if time is continuous, magnitude is continuous also, inasmuch as a thing passes over half a given magnitude in half the time taken to cover the whole; in fact without qualification it passes over a less magnitude in less time; for the divisions of time and of magnitude will be the same. And if either is infinite, so is the other, and the one is so in the same way as the other; i.e., if time is infinite in respect of its extremities, length is also infinite in respect of its extremities; if time is infinite in respect of divisibility, length is also infinite in respect of divisibility; and if time is infinite in both respects, magnitude is also infinite in both respects.

Zeno's argument makes a false assumption in asserting that it is impossible for a thing to pass over or severally to come in contact with infinite things in a finite time. For there are two senses in which length and time and generally anything continuous are called "infinite"; they are called so either in respect of divisibility or in respect of their extremities. So while a thing in a finite time cannot come in contact with things quantitatively infinite, it can come in contact with things infinite in respect of divisibility; for in this sense the time itself is also infinite. . . .

Chapter 9

Zeno's reasoning, however, is fallacious, when he says that everything when it occupies an equal space is at rest, and if that which is in locomotion is always occupying such a space at any moment, the flying arrow is therefore motionless. This is false, for time is not composed of indivisible moments any more than any other magnitude is composed of indivisibles.

Zeno's arguments about motion, which cause so much disquietude to those who try to solve the problems that they present, are four in number. The first asserts the non-existence of motion on the ground that that which is in locomotion must arrive at the halfway stage before it arrives at the goal. This we have discussed above.

The second is the so-called Achilles, and it amounts to this, that in a race the quickest runner can never overtake the slowest, since the pursuer must first reach the point where the pursued started, so that the slower must always hold a lead. This argument is the same in principle as that which depends on bisection, though it differs from it in that the spaces with which we successively have to deal are not divided into halves. The result of the argument is that the slower is not overtaken but it proceeds along the same lines as the bisection argument . . . so that the solution must be the same. And the axiom that that which holds a lead is never overtaken is false; it is not overtaken, it is true, while it holds a lead; but it is overtaken nevertheless if it is granted that it traverses the finite distance prescribed. These then are two of his arguments.

The third is that already given above, to the effect that the flying arrow is at rest, which result follows from the assumption that time is composed of moments; if this assumption is not granted, the conclusion will not follow.

The fourth argument is that concerning the two rows of bodies, each row being composed of an equal number of bodies of equal size, passing each other on a race-course as they proceed with equal velocity in opposite directions, the one row originally occupying the space between the goal and the middle point of the course and the other that between the middle point and the starting post. This, he thinks, involves the conclusion that half a given time is equal to double that time. The fallacy of the reasoning lies in the assumption that a body occupies an equal time in passing with equal velocity a body that is in motion and a body of equal size that is at rest; which is false. For instance (so runs the argument), let A, A, . . . be the stationary bodies of equal size, B, B . . . the bodies, equal in number and in size to A, A . . . , originally occupying the half of the course from the starting post to the middle of the A's, and C, C . . . those originally occupying the other half from the goal to the middle of the A's equal in number, size, and velocity to B, B, . . . Then three consequences follow:

```
       A A A      A A A
     B B B        B B B
          C C C   C C C
```

First, as the B's and the C's pass one another, the first B reaches the last C at the same moment as the first C reaches the last B. Secondly, at this moment the first C has passed all the A's, whereas the first B has passed only half the A's, and has consequently occupied only half the time occupied by the first C, since each of the two occupies an equal time in passing each A. Thirdly, at the same moment all the B's have passed all the C's; for the first C and the first B will simultaneously reach the opposite ends of the course, since (so says Zeno) the time occupied by the first C in passing each of the B's is equal to that occupied by it in passing each of the A's, because an equal time is occupied by both the first B and the first C in passing all the A's. This the argument, but it presupposed the aforesaid fallacious assumption.

3.5 Hippocrates, *Quadrature of Lunes*

Simplicius of Cilicia (on the southern coast of what is now Turkey) (490–560) was one of the last of the Neoplatonic philosophers and was probably active in Athens when the Roman emperor Justinian closed the pagan schools there. His extant writings are mostly commentaries on works of Aristotle. In particular, he preserved the work of Hippocrates of Chios (5th c. BCE) on quadratures of lunes in his comments on a brief passage at the beginning of Aristotle's *Physics* dealing with supposed quadratures of circles, that is, on determining the area of a circle by equating it to a rectilinear area. In fact, Simplicius claimed that he was basically reproducing an excerpt from the *History of Geometry* of Eudemus (370–300 BCE). Although Simplicius sometimes added material of his own to clarify Eudemus's remarks, modern scholars generally believe that the story he tells about Hippocrates is accurate. Hippocrates himself taught in Athens and may have been one of the first to write an *Elements of Geometry*, although the work is no longer extant. Besides calculating the area of lunes, he also showed that the problem of doubling a cube was equivalent to that of finding two mean proportionals between a line segment and its double. The excerpts about lunes that follow were, to the best of our knowledge, written by Eudemus.

The quadratures of lunes, which seemed to belong to an uncommon class of propositions by reason of the close relationship to the circle, were first investigated by Hippocrates, and seemed to be set out in correct form; therefore we shall deal with them at length and go through them. He made his starting point, and set out as the first of the theorems useful to his purpose, that similar segments of circles have the same ratios as the squares on their bases. And this he proved by showing that the squares on the diameters have the same ratios as the circles.[1]

Having first shown this he described in what way it was possible to square a lune whose outer circumference was a semicircle. He did this by circumscribing about a right-angled isosceles triangle a semicircle and about the base a segment of a circle similar to those cut off by the sides [figure 3.5.1].[2]

[1] Euclid proved this result in *Elements* 12:2 by use of the method of exhaustion. We do not know how Hippocrates proved it.

[2] That this can be done is proved in *Elements* 3:33 by showing how to construct a segment admitting a given angle. Again, we have no knowledge as to Hippocrates's solution.

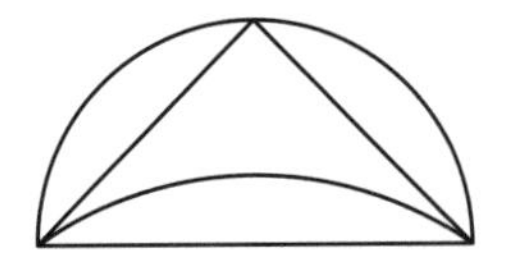

Figure 3.5.1.

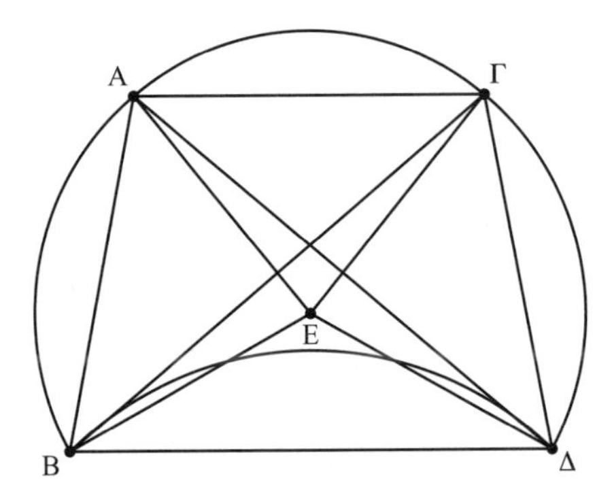

Figure 3.5.2.

Since the segment about the base is equal to the sum of those about the sides, it follows that when the part of the triangle above the segment about the base is added to both, the lune will be equal to the triangle. Therefore the lune, having been proved equal to the triangle, can be squared. In this way, taking a semicircle as the outer circumference of the lune, Hippocrates readily squared the lune.

Next in order he assumes [an outer circumference] greater than a semicircle obtained by constructing a trapezoid having three sides equal to one another while one, the greater of the parallel sides, is such that the square on it is three times the square on each of those sides, and then comprehending the trapezoid in a circle and circumscribing about its greatest side a segment similar to those cut off from the circle by the three equal sides [figure 3.5.2].[3]

That the said segment is greater than a semicircle is clear if a diagonal is drawn in the trapezoid. For this diagonal, subtending two sides of the trapezoid, must be such that the square on it is greater than double the square on one of the remaining sides. Therefore the square on $B\Gamma$ is greater than double the square on either BA, $A\Gamma$, and therefore also on $\Gamma\Delta$. Therefore the square on $B\Delta$, the greatest of the sides of the trapezoid, must be less than the sum of the squares on the diagonal and that one of the other sides which is subtended by the said [greatest] side together with the diagonal. For the squares on $B\Gamma$, $\Gamma\Delta$ are greater than three times, and the square on $B\Delta$ is equal to three times, the

[3] To show that a circle can be circumscribed about the trapezoid, let E be the intersection of the bisectors of the angles at A and Γ and draw EB and $E\Delta$. Then triangle $AE\Gamma$ is isosceles and is also congruent to triangles AEB and $\Gamma E\Delta$. It follows that $BE = AE = E\Gamma = E\Delta$, and so a circle can be drawn through the four vertices of the trapezoid.

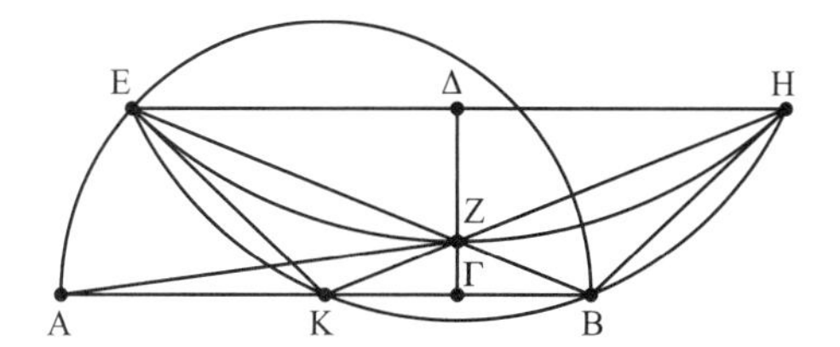

Figure 3.5.3.

square on $\Gamma\Delta$. Therefore the angle standing on the greatest side of the trapezoid is acute. Therefore the segment in which it is is greater than a semicircle. And this segment is the outer circumference of the lune.[4]

If [the outer circumference] were less than a semicircle, Hippocrates solved this also, using the following preliminary construction: Let there be a circle with diameter AB and center K. Let $\Gamma\Delta$ bisect BK at right angles, and let the straight line EXZ be placed between this and the circumference verging towards B so that the square on it is one-and-a-half times the square on one of the radii [figure 3.5.3].[5]

Let EH be drawn parallel to AB, and from K let [straight lines] be drawn joining E and Z. Let the straight line [KZ] joined to Z and produced meet EH at H, and again let [straight lines] be drawn from B joining Z and H. It is then manifest that EZ produced will pass through B—for by hypothesis EZ verges towards B—and BH will be equal to EK. This being so, I say that the trapezoid $EKBH$ can be comprehended in a circle.

Next let a segment of a circle be circumscribed about the triangle EZH; then clearly each of the segments on EZ, ZH will be similar to the segments on EK, KB, BH.[6] This being so, the lune so formed, whose outer circumference is $EKBH$, will be equal to the rectilineal figure composed of the three triangles BZH, BZK, EKZ. For the segments cut off from the rectilineal figure by the straight lines EZ, ZH are together equal to the segments outside the rectilineal figure cut off by EK, KB, BH. For each of the inner segments is one-and-a-half times each of the outer, because, by hypothesis, sq(EZ)[7] is one-and-a-half times the square on the radius, that is, the square on EK or KB or BH. Inasmuch then as the lune is made up of the three segments and the rectilineal figure less the two segments—the rectilineal figure including the two segments but not the three—while the sum of the two segments is equal to the sum of the three, it follows that the lune is equal to the rectilineal figure.

[4] Curiously, Eudemus does not explicitly square this lune, but the argument is very similar to the previous one. Namely, since the segment on $B\Delta$ equals the sum of the segments on BA, $A\Gamma$, $\Gamma\Delta$, the lune $BA\Gamma\Delta$ is equal to the trapezoid.

[5] That is, we must find Z on $\Delta\Gamma$ and E on the circumference of the circle so that $EZ^2 = \frac{3}{2}AK^2$. Since $EB \cdot BZ = AB \cdot B\Gamma = AK^2$, we have $(BZ + \sqrt{\frac{3}{2}}AK)BZ = AK^2$. Thus, BZ satisfies what amounts to a quadratic equation, one that could be solved by techniques of *Elements* 2 and 6. More precisely, it amounts to applying to a straight line a rectangle exceeding by a square figure and equal to a given rectangle.

[6] The angles in all of these segments are equal.

[7] We write "sq(EZ)" as shorthand for "the square constructed on the line segment EZ."

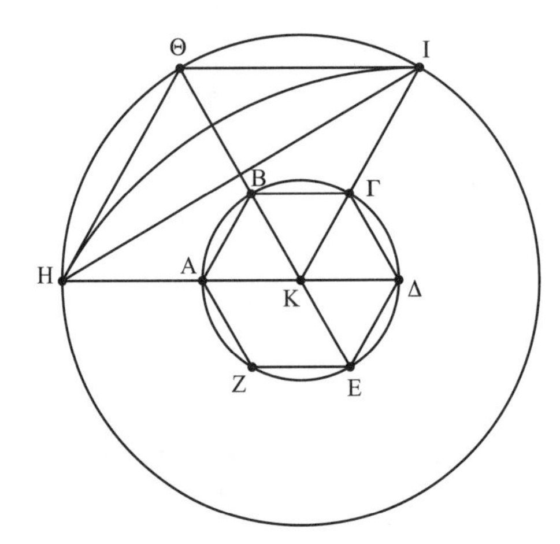

Figure 3.5.4.

That this lune has its outer circumference less than a semicircle, he proves by means of the angle EKH in the outer segment being obtuse. And that the angle EKH is obtuse, he proves thus: Since sq(EZ) is equal to half again as much as sq(EK) and sq(KB) is greater than double sq(BZ),[8] it is manifest that sq(EK) is greater than double sq(KZ). Therefore sq(EZ) is greater than the sum of sq(EK) and sq(KZ). The angle at K is therefore obtuse, so that the segment in which it is is less than a semicircle.

Thus Hippocrates squared every lune, seeing that [he squared] not only the lune that has for its outer circumference a semicircle, but also the lune in which the outer circumference is greater, and that in which it is less, than a semicircle.

Actually, as we have seen, Hippocrates has only squared three specific lunes and certainly not all possible lunes. For if, in fact, he had squared all possible lunes, the solution to the final problem would have enabled him to square the circle.

But he also squared a lune and a circle together in the following manner. Let there be two circles with K as center, such that the square on the diameter of the outer is six times the square on the diameter of the inner. Let a [regular] hexagon $AB\Gamma\Delta EZ$ be inscribed in the inner circle, and let KA, KB, $K\Gamma$ be joined from the center and produced as far as the circumference of the outer circle, and let $H\Theta$, ΘI, HI be joined [figure 3.5.4].

Then it is clear that $H\Theta$, ΘI are sides of a [regular] hexagon inscribed in the outer circle. About HI let a segment be circumscribed similar to the segment cut off by $H\Theta$. Since then sq(HI) is triple sq(ΘH) (for the square on the line subtended by two sides of the hexagon, together with the square on one other side, is equal, since they form a right angle in the semicircle, to the square on the diameter, and the square on the diameter is four times the side of the hexagon, the diameter being twice the side in length and so four times as great in square), and sq(ΘH) is six times sq(AB), it is manifest that the segment circumscribed

[8] $EB \cdot BZ = AB \cdot B\Gamma - KB^2$. Thus, $(EZ + BZ)BZ = KB^2$, or $EZ \cdot BZ + BZ^2 = KB^2 = \frac{2}{3}EZ^2$. It follows that $EZ > ZB$ and $KB^2 > 2BZ^2$.

about HI is equal to the segments cut off from the outer circle by $H\Theta$, ΘI, together with the segments cut off from the inner circle by all the sides of the hexagon. For sq(HI) is triple sq($H\Theta$), and sq(ΘI) equals sq($H\Theta$), while sq(ΘI) and sq($H\Theta$) are each equal to the sum of the squares on the six sides of the inner hexagon, since, by hypothesis, the diameter of the outer circle is six times that of the inner.

Therefore the lune $H\Theta I$ is smaller than the triangle $H\Theta I$ by the segments taken away from the inner circle by the sides of the hexagon. For the segment on HI is equal to the sum of the segments on $H\Theta$, ΘI and those taken away by the hexagon. Therefore the segments [on] $H\Theta$, ΘI are less than the segment about HI by the segments taken away by the hexagon. If to both sides there is added the part of the triangle which is above the segment about HI, out of this and the segment about HI will be formed the triangle, while out of the latter and the segments [on] $H\Theta$, ΘI will be formed the lune. Therefore the lune will be less than the triangle by the segments taken away by the hexagon. For the lune and the segments taken away by the hexagon are equal to the triangle. When the hexagon is added to both sides, this triangle and the hexagon will be equal to the aforesaid lune and to the inner circle. If then the aforementioned rectilineal figures can be squared, so also can the circle with the lune.

3.6 Archytas, *Two Mean Proportionals*

Archytas of Tarentum was apparently the first mathematician to solve the cube duplication problem, a problem that was seen to be equivalent to the construction of two mean proportionals between two line segments. He contributed greatly to other areas of mathematics, including number theory and its application to music, and was a friend and associate of Plato. Unfortunately, only fragments of his works survive (see chapter 8 for more), but the following comes from Eutocius in his commentary on Archimedes' *Sphere and Cylinder II*. Eutocius notes that this report of Archytas's solution comes from Eudemus's *History of Mathematics*, although it is unclear whether he actually had a copy of that work at hand.

It is fascinating that in one of the earliest extant geometric proofs, presumably written before Euclid's *Elements*, the author requires a rather complicated three-dimensional construction to solve a problem in the plane. And this proof also shows something about what geometric results were known before Euclid, even if we have no idea how they were proved at that time. However, we can discern the basic idea of the proof as being a generalization of the construction of one mean proportional between two line segments. In that construction, one puts the two line segments into one line, draws a semicircle with that line as the diameter, and then draws a perpendicular from the join of the two original line segments to the circumference of the semicircle. That line segment is then the mean proportional between the two original segments. If we connect the ends of the diameter with the endpoint of the perpendicular, we get a right triangle formed from two other right triangles, all similar. In the argument here, where we need to find two mean proportionals between two given magnitudes, we find Archytas constructing a right triangle with hypotenuse

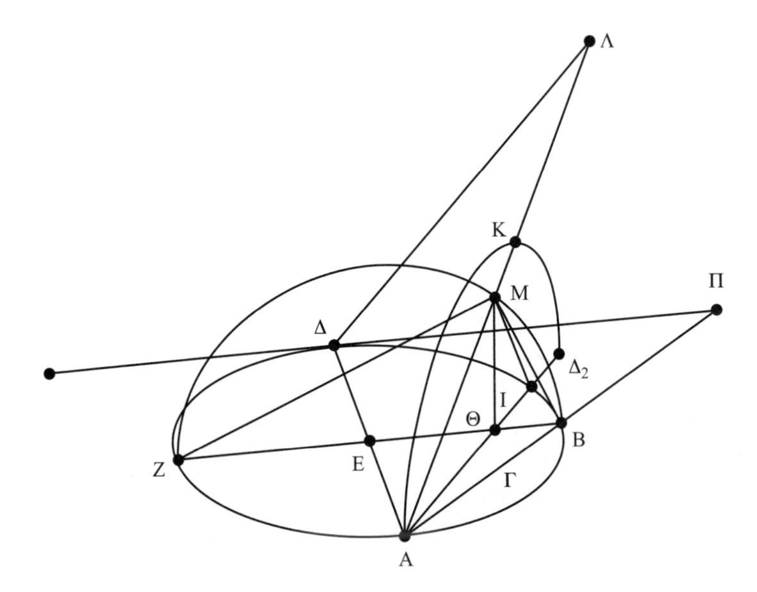

Figure 3.6.1.

equal to the larger magnitude and then looking at several other similar right triangles, with one of them having a hypotenuse equal to the smaller of the two original magnitudes. The similarity of triangles then produces the desired result. It turned out, however, that in order to construct his triangles, Archytas needed to move some of his lines into three-dimensional space; this is what caused the complexity of the proof.[9]

Let $A\Delta$ and Γ be two given straight lines: thus it is necessary to find two mean proportionals between $A\Delta$ and Γ.

For let the circle $AB\Delta Z$ be drawn around the larger segment $A\Delta$ [with $A\Delta$ as diameter], and let AB be fitted [into the circle] equal to Γ, and having extended AB, let it cut the tangent from Δ at the point Π [figure 3.6.1].

Let BEZ be drawn parallel to $\Pi\Delta$, and let a half-cylinder be imagined at right angles to the [plane of] the semicircle $AB\Delta$,[10] and let [another] semicircle $A\Delta$ be imagined at right angles to the first semicircle, this second semicircle being placed in the parallelogram of the half-cylinder.[11] Therefore this semicircle, being rotated from Δ to B, with the endpoint A of its diameter remaining fixed, will intersect the surface of the half-cylinder by its [the semicircle's] rotation, and will draw a certain curve on it.[12]

Again, when the segment $A\Delta$ is fixed, if the triangle $A\Pi\Delta$ is rotated, moving in the direction opposite of the semicircle, its motion will make a conical surface with the straight

[9] This description of the idea of the proof is taken from Reviel Netz, *A New History of Greek Mathematics* (Cambridge: Cambridge University Press, 2022).

[10] This half-cylinder is not drawn; imagine it above the the semi-circle $AB\Delta$.

[11] That is, $A\Delta$ is the diameter of this second semicircle, perpendicular to the plane of the original circle.

[12] One position of this rotated semicircle is indicated as $AK\Delta_2$.

line $A\Pi$, which, having been rotated, will meet the line on the cylinder at some point.[13] At the same time [as the rotation], the point B will also draw a semicircle on the surface of the cone. Let the rotating semicircle have its position at the place where the lines on the cylinder intersect, as $\Delta_2 KA$, and the counter-rotating triangle have as its position $\Delta\Lambda A$, and let the point of the aforementioned intersection be K, and let there be the semicircle BMZ, being drawn through the point B, and let the common section of it and the circle $B\Delta ZA$ be the line BZ.

From K, let a perpendicular be drawn to the plane of the circle $B\Delta A$: indeed it will fall on the perimeter of the circle since the cylinder was set up at right angles [and K is on the cylinder]. Let it fall as KI, and from I, extending it to A, let it [AI] meet BZ at Θ. Let $A\Lambda$ meet the semicircle BMZ at M. Let the lines $K\Delta$, MI, and $M\Theta$ be joined. So since each of the semicircles $\Delta_2 KA$ and BMZ are perpendicular to the base plane, therefore the common section of them, $M\Theta$, is also perpendicular to the plane of the circle: so that $M\Theta$ is also perpendicular to BZ. Therefore rect($B\Theta, \Theta Z$), that is, rect($A\Theta$, ΘI), equals sq($M\Theta$).[14] Therefore the triangle AMI is similar to each of the triangles $MI\Theta$ and $MA\Theta$, and the angle contained by IM, MA is right. But the angle contained by $\Delta_2 K$, KA is also right: therefore the lines $K\Delta_2$ and MI are parallel. And by similarity of the triangles, as $\Delta_2 A$ is to AK, that is, as KA is to AI, so is IA to AM.

Therefore the four segments $\Delta_2 A$, AK, AI, AM are continuously proportional. And AM is equal to Γ, since it is also equal to AB: therefore AK and AI have been found as two mean proportionals between the given lines $A\Delta$ and Γ.

3.7 Proclus, *Commentary on the First Book of Euclid's Elements*

After the general discussion of his mathematical philosophy in section 1.11 from part one of his commentary, Proclus continues in part two to consider geometry in particular, including its history. Among the ideas he develops are the the objects of geometrical inquiry and the relationship between arithmetic and geometry.

Part Two

Chapter 2: The Objects and Methods of Geometrical Science

Magnitudes, figures and their boundaries, and the ratios that are found in them, as well as their properties, their various positions and motions—these are what geometry studies, proceeding from the partless point down to solid bodies, whose many species and differences it explores, then following the reverse path from the more complex objects to the simpler ones and their principles. It makes use of synthesis and analysis, always

[13] The cone is not drawn in the figure, but it has vertex A, with the point Π sweeping out a circle of radius $\Delta\Pi$. Looking at it another way, the line $A\Pi$ is being rotated around the diameter $A\Delta$ so that the distance of Π from $A\Delta$ is constant.

[14] *Elements* 3:35 and note that BZ is the diameter of semicircle BMZ.

starting from hypotheses and first principles that it obtains from the science above it and employing all the procedures of dialectic—definition and division for establishing first principles and articulating species and genera, and demonstrations and analyses in dealing with the consequences that follow from first principles, in order to show the more complex matters both as proceeding from the simpler and also conversely as leading back to them. It treats in one part the definitions of its objects, in another the axioms and the postulates that are the starting points of its demonstrations, and in another the demonstrations of the properties that belong essentially to its objects. Each science has its own class of things that concern it and whose properties it proposes to investigate, and also its own peculiar principles that it uses in demonstration; and the essential properties likewise differ in the various sciences. The axioms are common to all sciences, although each uses them in the fashion appropriate to its own subject matter; but the genus studied and its essential properties are peculiar to each science.

Among the objects of geometrical inquiry are triangles, squares, circles, figures, and magnitudes in general and their boundaries; others are properties inherent in them, their parts, ratios, and contacts, their equalities, excesses, and deficiencies when laid alongside one another, and all such matters; still others are the postulates and axioms through which all these are demonstrated—for example, that it be permitted to draw a straight line from any point to any other, and that if equals be taken from equals the results are equal, and their consequences. Hence not every problem or question is a geometrical one, but only those that arise out of the principles of geometry; and anyone who is refuted on these principles would be refuted as a geometer; arguments not based on them are not geometrical, but ungeometrical. The latter are of two kinds: either they proceed from premises altogether unlike those of geometry, such as a question in music, which we say is ungeometrical because it arises from hypotheses quite different from the principles of geometry; or they use geometrical principles but in a perverse sense, as when it is asserted that parallel lines meet. Hence geometry also furnishes criteria whereby we can discriminate between statements that follow from its principles and those that depart from them. The various tropes for refuting fallacies when they occur have this function.

Geometrical principles yield consequences different from those that follow from arithmetical ones. . . . They are far inferior to these. For one science is more accurate than another, as Aristotle says; that is, a science that starts from simpler principles than one whose starting point is more complex, or one that states why a fact is so than one which says that it is so, or a science concerned with intelligibles than one that applies to objects in the sense world. According to these criteria of exactness, arithmetic is more precise than geometry, for its principles are simpler. A unit has no position, but a point has; and geometry includes among its principles the point with position, while arithmetic posits the unit. Likewise geometry is superior to spherics and arithmetic to music, for in general they furnish the principles of the theorems subordinate to them. And geometry is superior to mechanics and optics, for the latter discourse about objects in the sense world.

The principles of arithmetic and geometry, then, differ from those of the other sciences, yet their own hypotheses are distinct from each other, in the sense mentioned above; nevertheless they have a certain community with one another, so that some theorems demonstrated are common to the two sciences, while others are peculiar to the one or the other. The statement that every ratio is expressible belongs to arithmetic only and not at all to geometry, for geometry contains inexpressible ratios. Likewise the principle that the gnomons into which a square can be divided have a lower limit in magnitude is peculiar to arithmetic; in geometry a least magnitude has no place at all. Peculiar to geometry are the propositions regarding position (for numbers do not have position), the propositions about contacts (for contacts occur only when there are continuous magnitudes), and the propositions about irrationals (for the irrational has a place only where infinite divisibility is possible). Common to both sciences are the theorems regarding sections (such as Euclid presents in his second book), with the exception of the division of a line in extreme and mean ratio. Of these common theorems some have come to arithmetic from geometry, others from arithmetic to geometry, while others are equally at home in both because derived by them from general mathematics. The principles governing alternation, conversion, composition, and division of ratios are thus shared by both. The theory of commensurable magnitudes is developed primarily by arithmetic and then by geometry in imitation of it. This is why both sciences define commensurable magnitudes as those which have to one another the ratio of a number to a number, and this implies that commensurability exists primarily in numbers. For where there is number there also is commensurability, and where commensurability there also is number. The properties of the triangle and the square are studied primarily by geometry, but arithmetic borrows them and uses them analogically, for figures are contained in numbers as in their causes. Thus in seeking the causes of certain results we turn to numbers, both when we see precisely the same properties, such as that every polygon can be divided into triangles, and when we are content with approximation, as when, in geometry we have found a square double a given square but do not have it in numbers, we say that a square number is the double of another square number when it is short by one, like the square of seven, which is one less than the double of the square of five.

We have carried rather far this exposition of the community between the principles of these two sciences and their differences. For the geometer must understand what common first principles are required for their common theorems and what are the principles from which their special theorems are derived, so that he may distinguish between geometrical matters and those that do not belong to geometry, assigning some to one science, some to the other.

In his version of the origins of geometry, Proclus begins with the work of Thales and then mentions numerous people who have contributed to the subject before Euclid wrote his *Elements*. We know today that some of Proclus's early history is mythical, while many of the names he mentions have been lost to history. Still, Proclus's version became the standard Greek history of their own mathematics.

Chapter 4: The Origin and Development of Geometry

Next we must speak of the development of this science during the present era. The inspired Aristotle has said that the same beliefs have often recurred to men at certain regular periods in the world's history; the sciences did not arise for the first time among us or among the men of whom we know, but at countless other cycles in the past they have appeared and vanished and will do so in the future. But limiting our investigation to the origin of the arts and sciences in the present age, we say, as have most writers of history, that geometry was first discovered among the Egyptians and originated in the remeasuring of their lands. This was necessary for them because the Nile overflows and obliterates the boundary lines between their properties. It is not surprising that the discovery of this and the other sciences had its origin in necessity, since everything in the world of generation proceeds from imperfection to perfection. Thus they would naturally pass from sense perception to calculation and calculation to reason. Just as among the Phoenicians the necessities of trade and exchange gave the impetus to the accurate study of number, so also among the Egyptians the invention of geometry came about from the cause mentioned.

Thales (624–545 BCE), who had traveled to Egypt, was the first to introduce this science into Greece.[15] He made many discoveries himself and taught the principles for many others to his successors, attacking some problems in a general way and others more empirically. Next after him Mamercus,[16] brother of the poet Stesichorus, is remembered as having applied himself to the study of geometry; and Hippias of Elis (late 5th c. BCE) records that he acquired a reputation in it. Following upon these men, Pythagoras (570–495 BCE) transformed mathematical philosophy into a scheme of liberal education, surveying its principles from the highest downwards and investigating its theorems in an immaterial and intellectual manner. He it was who discovered the doctrine of proportionals and the structure of the cosmic figures.[17] After him Anaxagoras of Clazomenae (500–428 BCE) applied himself to many questions in geometry, and so did Oenopides of Chios (mid-5th c. BCE),[18] who was a little younger than Anaxagoras. Both these men are mentioned by Plato in the *Erastae* as having got a reputation in mathematics. Following them Hippocrates of Chios (5th c. BCE), who invented the method of squaring lunes, and Theodorus of Cyrene (5th c. BCE) became eminent in geometry.[19] For Hippocrates wrote a book on elements, the first of whom we have any record who did so.

Plato, who appeared after them, greatly advanced mathematics in general and geometry in particular because of his zeal for these studies. It is well known that his writings are thickly sprinkled with mathematical terms and that he everywhere tries to arouse admiration for mathematics among students of philosophy. At this time also lived Leodamas of Thasos (5th c. BCE), Archytas of Tarentum (428–347 BCE), and Theaetetus of Athens (417–369 BCE), by whom the theorems were increased in number and brought into a more scientific arrangement.[20] Younger than Leodamas were Neoclides and his pupil Leon,[21] who added many discoveries to those of their predecessors, so that Leon was able to compile a book of elements more carefully designed to take account of the

[15] Although Proclus was here repeating long-held Greek beliefs about Thales, there is no evidence that he contributed at all to mathematics.

[16] Nothing is known of Mamercus beyond this mention.

[17] As in the case of Thales, no mathematical discoveries can actually be attributed to Pythagoras.

[18] Anaxagoras is said to have been the first to explain the optical basis of eclipses, using geometrical diagrams, while Oenopides may have been the first to teach how to construct various angles in also dealing with the geometry of the heavens.

[19] The mathematical work of Theodorus is only known from Plato's description in the *Theaetetus* (see section 3.2).

[20] Nothing is known of the work of Leodamas, while Archytas's work on number theory and Theaetetus's work on incommensurables are evidently both included in the *Elements*.

[21] Nothing is known of these two other than what is written here.

number of propositions that had been proved and of their utility. He also discovered *diorismi*, whose purpose is to determine when a problem under investigation is capable of solution and when it is not. Eudoxus of Cnidus (390–337 BCE), a little later than Leon and a member of Plato's group, was the first to increase the number of the so-called general theorems;[22] to the three proportionals already known he added three more and multiplied the number of propositions concerning the "section" which had their origin in Plato, employing the method of analysis for their solution. Amyclas of Heracleia,[23] one of Plato's followers, Menaechmus (380–320 BCE),[24] a student of Eudoxus who also was associated with Plato, and his brother Dinostratus (390–320 BCE)[25] made the whole of geometry still more perfect. Theudius of Magnesia [26] had a reputation for excellence in mathematics as in the rest of philosophy, for he produced an admirable arrangement of the elements and made many partial theorems more general. There was also Athenaeus of Cyzicus, who lived about this time and became eminent in other branches of mathematics and most of all in geometry. These men lived together in the Academy, making their inquiries in common. Hermotimus of Colophon pursued further investigations already begun by Eudoxus and Theaetetus, discovered many propositions in the *Elements*, and wrote some things about locus theorems. Philippus of Mende, a pupil whom Plato had encouraged to study mathematics, also carried on his investigations according to Plato's instructions and set himself to study all the problems that he thought would contribute to Plato's philosophy.[27]

All those who have written histories bring to this point their account of the development of this science. Not long after these men came Euclid, who brought together the *Elements*, systematizing many of the theorems of Eudoxus, perfecting many of those of Theaetetus, and putting in irrefutable demonstrable form propositions that had been rather loosely established by his predecessors. He lived in the time of Ptolemy the First, for Archimedes, who lived after the time of the first Ptolemy, mentions Euclid. It is also reported that Ptolemy once asked Euclid if there was not a shorter road to geometry than through the *Elements*, and Euclid replied that there was no royal road to geometry. He was therefore later than Plato's group but earlier than Eratosthenes and Archimedes, for these two men were contemporaries, as Eratosthenes somewhere says. Euclid belonged to the persuasion of Plato and was at home in this philosophy; and this is why he thought the goal of the *Elements* as a whole to be the construction of the so-called Platonic figures.

In the final excerpts we present here, Proclus discusses the purpose of Euclid's *Elements* and, in particular, the aim of book 1, the only book for which his commentary exists.

[22] Eudoxus is credited with the theory of proportion expounded in book 5 of the *Elements* as well as the method of exhaustion dealt with in book 12. He also contributed greatly to the study of astronomy.

[23] Nothing more is known of him.

[24] Menaechmus is credited with the discovery of the conic sections.

[25] Dinostratus made use of the quadratrix to square the circle.

[26] Nothing more is known of Theudius, nor of Athenaeus or Hermotimus, mentioned later.

[27] Nothing is known of his mathematical work, although he was an editor of one of Plato's dialogues.

Chapter 6: The Purpose of the *Elements*

If now anyone should ask what the aim of this treatise is, I should reply by distinguishing between its purpose as judged by the matters investigated and its purpose with reference to the learner. Looking at its subject matter, we assert that the whole of the geometer's discourse is obviously concerned with the cosmic figures.[28] It starts from the simple figures and ends with the complexities involved in the structure of the cosmic bodies, establishing each of the figures separately but showing for all of them how they are inscribed in the sphere and the ratios that they have with respect to one another. Hence some have thought it proper to interpret with reference to the cosmos the purposes of individual books and have inscribed above each of them the utility it has for a knowledge of the universe. Of the purpose of the work with reference to the student, we shall say that it is to lay before him an elementary exposition and a method of perfecting his understanding for the whole of geometry. If we start from the elements, we shall be able to understand the other parts of this science; without the elements we cannot grasp its complexity, and the learning of the rest will be beyond us. The theorems that are simplest and most fundamental and nearest to first principles are assembled here in a suitable order, and the demonstrations of other propositions take them as the most clearly known and proceed from them. In this way also Archimedes in his book on *Sphere and Cylinder* and likewise Apollonius and all other geometers appear to use the theorems demonstrated in this very work as generally accepted starting points. This, then, is its aim: both to furnish the learner with an introduction to the science as a whole and to present the construction of the several cosmic figures.

Chapter 7: The Meaning of "Elements"

But—to inquire briefly about its title—what is the meaning of this very word *στοιχείωσις* (*stoicheiōsis*) and of the word *στοιχεῖον* (*stoicheīon*) from which it is derived? Some theorems we are accustomed to call "elements," others "elementary," and other do not qualify for either designation. We call "elements" those theorems whose understanding leads to the knowledge of the rest and by which the difficulties in them are resolved. As in written language, there are certain primal elements, simple and indivisible, to which we give the name *στοιχεῖα* and out of which every word is constructed, and every sentence, so also in geometry as a whole there are certain primary theorems that have the rank of starting points for the theorems that follow, being implicated in them all and providing demonstrations for many conjunctions of qualities; and these we call "elements." "Elementary" propositions are those that are simple and elegant and have a variety of applications but do not rank as elements because the knowledge of them is not pertinent to the whole of the science: for example, the theorem that the perpendiculars from the vertices of a triangle to the sides meet in a common point. Propositions whose understanding is not relevant to a multitude of others or which do not exhibit any grace or elegance—these do not have the force of elementary propositions. The term "element," however, can be used in two senses, as Menaechmus tells us. For what proves is called an element of what is proved by it; thus in Euclid the first theorem is an element of the second, and the fourth of the fifth.

[28] These are the five Platonic solids.

In this sense many propositions can be called elements of one another, when they can be established reciprocally. From the proposition that the exterior angles of a rectilinear figure are equal to four right angles we can prove the number of right angles to which the interior angles of the figure are equal, and vice versa. An element so regarded is a kind of lemma. But in another sense "element" means a simpler part into which a compound can be analyzed. In this sense not everything can be called an element of anything that follows from it, but only the more primary members of an argument leading to a conclusion, as postulates are elements of theorems. This is the sense of "element" that determines the arrangement of the elements in Euclid's work, some of them being elements of plane geometry, and some elements of stereometry. This also is the meaning the word has in numerous compositions in arithmetic and astronomy entitled "elementary treatises."

It is a difficult task in any science to select and arrange properly the elements out of which all other matters are produced and into which they can be resolved. Of those who have attempted it, some have brought together more theorems, some fewer; some have used rather short demonstrations, others have extended their treatment to great lengths; some have avoided the reduction to impossibility, others proportion; some have devised defenses in advance against attacks upon the starting points; and in general many ways of constructing elementary expositions have been individually invented. Such a treatise ought to be free of everything superfluous, for that is a hindrance to learning; the selections chosen must all be coherent and conducive to the end proposed, in order to be of the greatest usefulness for knowledge; it must devote great attention both to clarity and to conciseness, for what lacks these qualities confuses our understanding; it ought to aim at the comprehension of its theorems in a general form, for dividing one's subject too minutely and teaching it by bits make knowledge of it difficult to attain. Judged by all these criteria, you will find Euclid's introduction superior to others. Its usefulness contributes to the study of the primary figures; its method of proceeding from simpler to more complex matters and its laying the foundations of the science on the "common notions" produce clarity and articulateness; and by moving toward the questions under investigation by way of primary and basic theorems, it makes the demonstration general. The matters that appear to be omitted either can be learned through the same methods as those it employs, like the construction of the scalene and isosceles triangles; or they are unsuitable for a selection of elements because they lead to great and unlimited complexity, such as the material that Apollonius has elaborated at considerable length about unordered irrationals; or they can be constructed from traditional premises, such as the many species of angles and lines. These matters are passed over in this work, and though they may receive rather fuller treatment in others, they can be learned from simple premises. So much we thought it desirable to record about the general nature of this elementary introduction.

Chapter 9: The Aim of Book 1

Next we must define the aim of the first book and set forth its several divisions, and then we shall be able to begin the examination of the Definitions. What this book proposes to do is to present the principles of the study of rectilinear figures. Although the circle is naturally superior to the straight line and the study of it a higher form of being and knowledge, yet instruction in the nature of straight lines is more suitable for us who are less than perfect intelligences and are striving to convert our understanding from sensible to intelligible objects. Rectilinear figures are akin to sensibles, but the circle to intelligibles; for what is simple, uniform, and determinate accords with the nature of being, whereas to be diversified and to possess indefinitely more containing sides is a characteristic of sense objects. In this book, therefore, are presented the first and most fundamental rectilinear figures, the triangle and the parallelogram. For these are the genera that include the causal principles of the elements, the isosceles and scalene triangles and their compounds, the equilateral triangle and the square, from which the figures of the four elements are constructed. We shall therefore discover how to construct the equilateral triangle and the square, the one on a given straight line, the other from a given line. The equilateral triangle is the proximate cause of three of the elements—fire, air, water—and the square the cause of earth. Consequently the aim of the first book is dependent on the entire treatise and contributes to the full understanding of the cosmic elements. Furthermore, it introduces the learner to the science of rectilinear figures by revealing their first principles and establishing them with precision.

Chapter 10: The Divisions of Book 1

The book is divided into three major parts. The first reveals the construction of triangles and the special properties of their angles and their sides, comparing triangles with one another as well as studying each by itself. Thus it takes a single triangle and examines now the angles from the standpoint of the sides and now the sides from the standpoint of the angles, with respect to their equality or inequality; and then, assuming two triangles, it investigates the same properties in various ways. The second part develops the theory of parallelograms, beginning with the special characteristics of parallel lines and the method of constructing the parallelogram and then demonstrating the properties of parallelograms. The third part reveals the kinship between triangles and parallelograms both in their properties and in their relations to one another. Thus it proves that triangles or parallelograms on the same or equal bases have identical properties; it shows what is the relation between a triangle and a parallelogram on the same base, how to construct a parallelogram equal to a triangle, and finally, with respect to the squares on the sides of a right-angled triangle, what is the relation of the square on the side that subtends the right angle to the squares on the two sides that contain it. Something like this may be said to be the purpose of the first book of the *Elements* and the division of its contents.

3.8 Euclid, *Elements*, 1, Including Commentary by Proclus

The central parts of Euclid's *Elements*, the books that were circulated most often and have been studied most diligently, are the books on plane geometry, books 1–6. Book 1 itself begins with Euclid's definitions, postulates, and common notions (axioms), all of which are presented below. [Note that these are not numbered in the original; the numbering is due to modern editors.] We then present a selection of theorems, culminating with proposition 1:47, the Pythagorean Theorem.[29] The text of

[29] It is, of course, not possible to include every theorem to which Euclid refers, especially as we move forward in the text. The reader is invited to consult the full text of the *Elements* in any available edition.

Euclid is interspersed with excerpts from Proclus's *Commentary on the First Book of Euclid's Elements*, the introductory parts of which have been included above. Proclus's *Commentary* is distinguished from the text itself by use of a different typeface.

Book 1

Definitions

1. A *point* is that which has no part.
2. A *line* is breadthless length.
3. The extremities of a line are points.
4. A *straight line* is a line which lies evenly with the points on itself.

Plato assumes that the two simplest and most fundamental species of line are the straight and the circular and makes all other kinds mixtures of these two, both those called spiral, whether lying in planes or about solids, and the curved lines that are produced by the sections of solids. . . . Some dispute this classification, denying that there are only two simple lines and saying that there is also a third, namely, the cylindrical helix, which is traced by a point moving uniformly along a straight line that is moving around the surface of a cylinder. This moving point generates a helix any part of which coincides homeomerously with any other, as Apollonius has shown in his treatise *On the Cochlias*.[30] This characteristic belongs to this helix alone. For the segments of a spiral in a plane are dissimilar, as are those of the spirals about a cone or sphere; the cylindrical spiral alone is homeomeric, like the straight line and the circle. Are there not, then, three simple lines, instead of two only? To this difficulty we shall reply by saying that this helix is indeed homeomeric, as Apollonius has shown, but is by no means simple. . . . The very mode of generating the cylindrical helix shows that it is a mixture of simple lines, for it is produced by the movement of a straight line about the axis of a cylinder and by the movement of a point along this line. It owes its existence, then to two dissimilar simple motions, so that it is to be classed among the mixed, not the simple lines. . . .

Euclid gives the definition of the straight line that we have set forth above, making clear by it that the straight line alone covers a distance equal to that between the points that lie on it. For the interval between any two points is the length of the line that these points define, and this is what is meant by "lying evenly with the points on itself." If you take two points on a circle or any other kind of line, the length of the line between the two points taken is greater than the distance between them. This seems to be a characteristic of every line except the straight. Hence it accords with a common notion that those who go in a straight line travel only the distance they need to cover, as men say, whereas those who do not go in a straight line travel farther than is necessary. Plato, however, defines the straight line as that whose middle intercepts the view of the extremes. This is a necessary property of things lying on a straight line but need not be true of things on a circle or any other extension. . . . But Archimedes defined the straight line as the shortest of all lines having the same extremities. Because, as Euclid's definition says, it lies evenly with

[30] This work on the helix has been lost.

the points on itself, it is the shortest of all lines having the same extremities; for if there were a shorter line, this one would not lie evenly with it own extremities. In fact, all other definitions of the straight line fall back upon the same notion—that it is a line stretched to the utmost, that one part of it does not lie in a lower and another in a higher plane, that all of its parts coincide similarly with all others, that it is a line that remains fixed if its end points remain fixed, that it cannot make a figure with another line of the same nature. All these definitions express the property which the straight line has by virtue of being simple and exhibiting the single shortest route from one extremity to the other. So much for definitions of the straight line.

5. A *surface* is that which has length and breadth only.
6. The extremities of a surface are lines.
7. A *plane surface* is a surface which lies evenly with the straight lines on itself.
8. A *plane angle* is the inclination to one another of two lines in a plane which meet one another and do not lie in a straight line.

Some of the ancients put the angle in the category of relation, calling it the inclination either of lines or of planes to one another; others place it under quality, saying that, like straight and curved, it is a certain character of a surface or a solid; others refer it to quantity, asserting that it is either a surface or a solid quantity. For the angle on a surface is divided by a line, that in solids by a surface, and what is divided by them, they say, can only be a magnitude; and it is not linear magnitude, for a line is divided by a point. So it remains that it is either a surface or a solid quantity. But if it is a magnitude and all finite homogeneous magnitudes have a ratio to one another, then all homogeneous angles, at least those in planes, will have a ratio to one another, so that a horned angle[31] will have a ratio to a rectilinear. But all quantities that have a ratio to one another can exceed one another by being multiplied; a horned angle, then, may exceed a rectilinear, which is impossible, for it has been proved that a horned angle is less than any rectilinear angle.

And if it is only a quality, like heat or coldness, how can it be divided into equal parts? For equality and inequality belong no less to angles than to magnitudes, and divisibility in general is an intrinsic property of angles and magnitudes alike. But if the things to which these properties intrinsically belong are quantities and not qualities, then it is clear that angles are not qualities. . . . As to the third possibility, if the angle is an inclination and in general belongs to the class of relations, it will follow that, when the inclination is one, there is one angle and not more. For if the angle is nothing other than a relation between lines or between planes, how could there be one relation but many angles? . . . And yet it is necessary that we call the angle either a quality or a quantity or a relation. Figures are qualities, the ratios between them are relations, and so we must refer the angle also to some one of these three genera. . . .

Let us follow our head and say that the angle as such is none of the things mentioned but exists as a combination of all these categories, and this is why it

[31] A horned angle is the angle formed by a circle and a straight line tangent to it.

presents a difficulty to those who are inclined to make it any one of them.... The angle surely needs the underlying quantity implied in its size, it needs the quality by which it has something like a special shape and character of existence, and it needs also the relation of the lines that bound it or of the planes that enclose it. The angle is something that results from all of these, and is not just any one of them.... Thus one may define it as a qualified quantity, constituted by such-and-such a relation, and not quantity as such, nor quality nor relation alone.

9. And when the lines containing the angle are straight, the angle is called *rectilineal*.
10. When a straight line set up on a straight line makes the adjacent angles equal to one another, each of the equal angles is *right*, and the straight line standing on the other is called a *perpendicular* to that on which it stands.
11. An *obtuse angle* is an angle greater than a right angle.
12. An *acute angle* is an angle less than a right angle.
13. A *boundary* is that which is an extremity of anything.
14. A *figure* is that which is contained by any boundary or boundaries.
15. A *circle* is a plane figure contained by one line such that all the straight lines falling upon it from one point among those lying within the figure are equal to one another;
16. And the point is called the *center* of the circle.
17. A *diameter* of the circle is any straight line drawn through the center and terminated in both directions by the circumference of the circle, and such a straight line also bisects the circle.
18. A *semicircle* is the figure contained by the diameter and the circumference cut off by it. And the center of the semicircle is the same as that of the circle.
19. *Rectilineal* figures are those which are contained by straight lines, *trilateral* figures being those contained by three, *quadrilateral* those contained by four, and *multilateral* those contained by more than four straight lines.
20. Of trilateral figures, an *equilateral triangle* is that which has its three sides equal, an *isosceles triangle* that which has two of its sides alone equal, and a *scalene triangle* that which has its three sides unequal.
21. Further, of trilateral figures, a *right-angled triangle* is that which has a right angle, an *obtuse-angled triangle* that which has an obtuse angle, and an *acute-angled triangle* that which has its three angles acute.
22. Of quadrilateral figures, a *square* is that which is both equilateral and right-angled; an *oblong* that which is right-angled but not equilateral; a *rhombus* that which is equilateral but not right-angled; and a *rhomboid* that which has its opposite sides and angles equal to one another but is neither equilateral nor right-angled. And let quadrilaterals other than these be called *trapezia*.

Quadrilaterals ought first to be divided into two groups, one called parallelograms, the other non-parallelograms, and parallelograms into some that are both right-angled and equilateral, such as squares, others that are neither, such as rhomboids, and others either right-angled and not equilateral, such as oblongs, or equilateral and not right-angled, such as rhombi. For parallelograms necessarily

have either both equality of sides and right-angledness, or neither of them, or one of them only; and the last is possible in two ways, so that parallelograms exist in four species. Of non-parallelograms some have only two sides parallel and the other sides not, and some have no sides at all parallel; the former are called trapezia, the latter trapezoids. Of trapezia some have the sides that join the parallels equal, others have them unequal; the former are called isosceles trapezia, the latter scalene. Hence there will be seven kinds of quadrilaterals: the square, the oblong, the rhombus, the rhomboid, the isosceles trapezium, the scalene trapezium, and the trapezoid.[32]

23. *Parallel* straight lines are straight lines which, being in the same plane and being produced indefinitely in both directions, do not meet one another in either direction.

The basic propositions about parallels and the attributes by which they are recognized we shall learn later, but what parallel straight lines are is defined in the words above. They must lie in one plane, he says, and when produced in both directions do not meet but can be extended indefinitely. Lines that are not parallel may be produced to a certain distance without meeting, but what characterizes parallel lines is that they do not meet when extended indefinitely; and not simply this, they are capable of indefinite extension in both directions without meeting. Lines that are not parallel may be capable of indefinite extension on one side but not on the other; as they near each other on one side, they diverge more on the other. The reason is that two straight lines cannot enclose an area, and they would if they converged on both sides. Further, the definition rightly adds that the straight lines must be in the same plane; for if one of them should be in the given plane and the other above it, they would always be asymptotes to one another, whatever their position, but they would not for this reason be parallel. So the plane must be one, and the lines must be produced indefinitely in both directions and meet in neither. When these conditions obtain, they will be parallel straight lines.

This is the way Euclid defines parallel straight lines. But Posidonius[33] **(135–51 BCE) says that parallel lines are lines in a single plane which neither converge nor diverge but have all the perpendiculars equal that are drawn to one of them from points on the other. Those lines between which the perpendiculars become progressively longer or shorter intersect somewhere because they converge upon one another; for a perpendicular can determine both the heights of figures and the distances between lines. Hence when the perpendiculars are equal, the distances between the straight lines are equal, but when they become greater or less, the distance increases or decreases and the lines converge on the side on which the perpendiculars are shorter.**

But we must understand that absence of intersection does not always make lines parallel, for the circumferences of concentric circles do not intersect; the lines must

[32] Note that modern definitions of these terms are not the same as either Euclid's or Proclus's definitions.

[33] Posidonius was a Greek Stoic philosopher who wrote on many topics. Unfortunately, very little of his work is extant.

also be extended indefinitely. This characteristic can be found not only in straight lines but in others as well. One can think of a helix inscribed around a straight line which can be prolonged with the straight line indefinitely and never meet it. Such cases Geminus rightly distinguishes from the former ones at the outset. Some lines, he says, are finite and enclose a figure, like the circle, the perimeter of an ellipse, the cissoid, and many others; others are unlimited and can be extended indefinitely, like the straight line, the section of a right-angled or an obtuse-angled cone,[34] and the conchoid.[35] Again of those capable of being extended indefinitely some never enclose a figure, like the straight line and the above-mentioned conic sections, while others come together and, after making a figure, then extend on indefinitely. Of these some are asymptotic, namely, those which however far extended never meet, and others that do intersect are not asymptotic. Of asymptotic lines some are in the same plane with one another, others not; and of the asymptotes that are in the same plane, some are always equidistant from one another, others are constantly diminishing the distance between themselves and their straight lines, like the hyperbola and the conchoid. Although the distance between these lines constantly decreases, they remain asymptotes and, though converging upon one another, never converge completely. This is one of the most paradoxical theorems in geometry, proving as it does that some lines exhibit a non-convergent convergence. Of the lines which are always equidistant from one another those straight lines which never make the interval between them less and which lie in the same plane are parallel. . . .

Note that the first three of the postulates allow ruler and compass constructions. The fourth gives Euclid's basic measure for angles, while the fifth is the famous parallel postulate.

Postulates

Let the following be postulated:

1. To draw a straight line from any point to any point.
2. To produce a finite straight line continuously in a straight line.
3. To describe a circle with any center and distance.
4. That all right angles are equal to one another.
5. That, if a straight line falling on two straight lines makes the interior angles on the same side less than two right angles, the two straight lines, if produced indefinitely, meet on that side on which are the angles less than the two right angles.

This [postulate 5] ought to be struck from the postulates altogether. For it is a theorem—one that invites many questions, which Ptolemy proposed to resolve in one of his books—and requires for its demonstration a number of definitions as well as theorems. And the converse of it is proved by Euclid himself as a theorem.

[34] These are the old names for a parabola and a hyperbola.
[35] See the discussion of the conchoid in section 5.1.

But perhaps some persons might mistakenly think this proposition deserves to be ranked among the postulates on the ground that the angles being less than two right angles makes us at once believe in the convergence and intersection of the straight lines. To them Geminus has given the proper answer when he said that we have learned from the very founders of this science not to pay attention to plausible imaginings in determining what propositions are to be accepted in geometry. Aristotle likewise says that to accept probable reasoning from a geometer is like demanding proofs from a rhetorician. And Simmias[36] is made by Plato to say, "I am aware that those who make proofs out of probabilities are imposters." So here, although the statement that the straight lines converge when the right angles are diminished is true and necessary, yet the conclusion that because they converge more as they are extended farther they will meet at some time is plausible, but not necessary, in the absence of an argument proving that this is true of straight lines. That there are lines that approach each other indefinitely but never meet seems implausible and paradoxical, yet it is nevertheless true and has been ascertained for other species of lines. May not this, then, be possible for straight lines as for those other lines? Until we have firmly demonstrated that they meet, what is said about other lines strips our imagination of its plausibility. And although the arguments against the intersection of these lines may contain much that surprises us, should we not all the more refuse to admit into our tradition this unreasoned appeal to probability?

These considerations make it clear that we should seek a proof of the theorem that lies before us and that it lacks the special character of a postulate. But how it is to be proved, and with what arguments the objections to this proposition may be met, we can only say when the author of the *Elements* is at the point of mentioning it and using it as obvious. At that time it will be necessary to show that its obvious character does not appear independently of demonstration but is turned by proof into a matter of knowledge.

Common Notions

1. Things which are equal to the same thing are also equal to one another.
2. If equals be added to equals, the wholes are equal.
3. If equals be subtracted from equals, the remainders are equal.
4. Things which coincide with one another are equal to one another.
5. The whole is greater than the part.

Propositions

Proposition 1. *On a given finite straight line to construct an equilateral triangle.*

Let AB be the given finite straight line. Thus it is required to construct an equilateral triangle on the straight line AB. With center A and distance AB let the circle BCD be

[36] Simmias is a character in Plato's dialogue *Phaedo*.

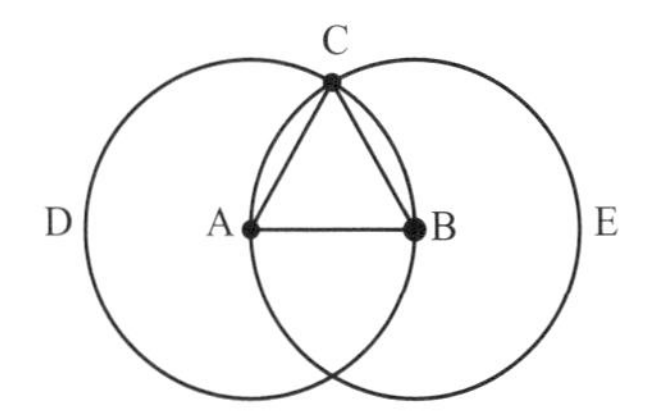

Figure 3.8.1.

described; again, with center B and distance BA let the circle ACE be described; and from the point C, in which the circles cut one another, to the points A, B let the straight lines CA, CB be joined [figure 3.8.1].

Now, since the point A is the center of the circle CDB, AC is equal to AB. Again, since the point B is the center of the circle CAE, BC is equal to BA. But CA was also proved equal to AB; therefore each of the straight lines CA, CB is equal to AB. And things which are equal to the same thing are also equal to one another; therefore CA is also equal to CB. Therefore the three straight lines CA, AB, BC are equal to one another. Therefore the triangle ABC is equilateral; and it has been constructed on the given finite straight line AB. Being what it was required to do.

Science as a whole has two parts: in one it occupies itself with immediate premises, while in the other it treats systematically the things that can be demonstrated or constructed from these first principles, or in general are consequences of them. Again this second part, in geometry, is divided into the working out of problems and the discovery of theorems. It calls "problems" those propositions whose aim is to produce, bring into view, or construct what in a sense does not exist, and "theorems" those whose purpose is to see, identify, and demonstrate the existence or nonexistence of an attribute. Problems require us to construct a figure, or set it at a place, or apply it to another, or inscribe it in or circumscribe it about another, or fit it upon or bring it into contact with another, and the like; theorems endeavor to grasp firmly and bind fast by demonstration the attributes and inherent properties belonging to the objects that are the subject matter of geometry. . . .

Every problem and every theorem that is furnished with all its parts should contain the following elements: an enunciation, an exposition, a specification, a construction, a proof, and a conclusion. Of these, the enunciation states what is given and what is being sought from it, for a perfect enunciation consists of both these parts. The exposition takes separately what is given and prepares it in advance for use in the investigation. The specification takes separately the thing that is sought and makes clear precisely what it is. The construction adds what is lacking in the given for finding what is sought. The proof draws the proposed inference by reasoning scientifically from the propositions that have been admitted. The conclusion reverts to the enunciation, confirming what has been proved. So many are the parts of a problem or a theorem. The most essential ones, and those which are always present, are enunciation, proof, and conclusion; for it is alike necessary to know in advance what is being sought, to prove it by middle terms, and to collect

what has been proved. It is impossible that any of these three should be lacking; the other parts are often brought in but are often left out when they serve no need. . . .

Let us view the things that have been said by applying them to this our first problem. Clearly it is a problem, for it bids us devise a way of constructing an equilateral triangle. In this case the enunciation consists of both what is given and what is sought. What is given is a finite straight line, and what is sought is how to construct an equilateral triangle on it. The statement of the given precedes and the statement of what is sought follows, so that we may weave them together as "If there is a finite straight line, it is possible to construct an equilateral triangle on it." If there were no straight line, no triangle could be produced, for a triangle is bounded by straight lines; nor could it if the line were not finite, for an angle can be constructed only at a definite point, and an unbounded line has no end point.

Next after the enunciation is the exposition: "Let this be the given finite line." You see that the exposition itself mentions only the given, without reference to what is sought. Upon this follows the specification: "It is required to construct an equilateral triangle on the designated finite straight line." In a sense the purpose of the specification is to fix our attention; it makes us more attentive to the proof by announcing what is to be proved, just as the exposition puts us in a better position for learning by producing the given element before our eyes. After the specification comes the construction: "Let a circle be described with center at one extremity of the line and the remainder of the line as distance; again let a circle be described with the other extremity as center and the same distance as before; and then from the point of intersection of the circles let straight lines be joined to the two extremities of the given straight line." You observe that for the construction I make use of the two postulates that a straight line may be drawn from any point to any other and that a circle may be described with any center and distance. In general, the postulates are contributory to constructions and the axioms to proofs. Next comes the proof: "Since one of the two points on the given straight line is the center of the circle enclosing it, the line drawn to the point of intersection is equal to the given straight line. For the same reason, since the other point on the given straight line is itself the center of the circle enclosing it, the line drawn from it to the point of intersection is equal to the given straight line." These inferences are suggested to us by the definition of the circle, which says that all the lines drawn from its center are equal. "Each of these lines is therefore equal to the same line; and things equal to the same thing are equal to each other" by the first axiom. "The three lines therefore are equal, and an equilateral triangle has been constructed on this given straight line." This is the first conclusion following upon the exposition. And then comes the general conclusion: "An equilateral triangle has therefore been constructed upon the given straight line." For even if you make the line double that set forth in the exposition, or triple, or of any other length greater or less than it, the same construction and proof would fit it. To these propositions he adds: "This is what it was required to do,"[37] thus showing that this is the conclusion of a problem; for in the case of a

[37] In the future, we will abbreviate this last line as Q.E.F. for *quod erat faciendum*.

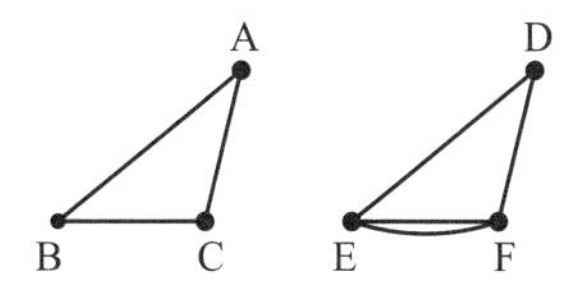

Figure 3.8.2.

theorem he adds: "This is what was to be demonstrated."[38] For problems announce that something is to be done, theorems that some truth is to be discovered and demonstrated.

Euclid made the unstated assumption here that the two circles drawn from the endpoints of AB actually intersect. Some postulate of continuity is necessary, and this was supplied in the nineteenth century. Of course, Euclid himself is basing his proof of proposition 1 on several explicit definitions and postulates.

Proposition 4. *If two triangles have the two sides equal to two sides respectively, and have the angles contained by the equal straight lines equal, they will also have the base equal to the base, the triangle will be equal to the triangle, and the remaining angles will be equal to the remaining angles respectively, namely those which the equal sides subtend.*

Let ABC, DEF be two triangles having the two sides AB, AC equal to the two sides DE, DF respectively, namely, AB to DE and AC to DF, and the angle BAC equal to the angle EDF [figure 3.8.2].

I say that the base BC is also equal to the base EF, the triangle ABC will be equal to the triangle DEF, and the remaining angles will be equal to the remaining angles respectively, namely those which the equal sides subtend, that is, the angle ABC to the angle DEF and the angle ACB to the angle DFE. For if the triangle ABC be applied to the triangle DEF, and if the point A be placed on the point D and the straight line AB on DE, then the point B will also coincide with E, because AB is equal to DE. Again, AB coinciding with DE, the straight line AC will also coincide with DF, because the angle BAC is equal to the angle EDF; hence the point D will also coincide with the point F, because AC is again equal to DF. But B also coincided with E; hence the base BC will coincide with the base EF, and will be equal to it. Thus the whole triangle ABC will coincide with the whole triangle DEF, and will be equal to it. And the remaining angles will also coincide with the remaining angles and will be equal to them, the angle ABC to the angle DEF, and the angle ACB to the angle DFE. Therefore, etc. Q.E.D.[39]

There are three things proved and two things given about these triangles. One of the given elements is the equality of two sides (really two given sides, but obviously given in ratio to one another) and the equality of the angles contained by the equal sides. And the things to be proved are three: the equality of base to base, the

[38] As above, we will abbreviate this phrase as Q.E.D. for *quod erat demonstrandum*.

[39] We will frequently abbreviate Euclid's final statement noting that the proof has been accomplished by simply writing "Therefore, etc." and "Q.E.D." or "Q.E.F."

equality of triangle to triangle, and the equality of the other angles to the other angles. Since it would be possible for the triangles to have two sides equal to two sides and yet the theorem be false because the sides are not equal one to another but one pair to the other pair he did not simply say, in his statement of the given, that the lines are equal, but that they are equal "respectively." For if it should happen that one of the triangles had one side of three units and the other of four, while the other triangle had one side of five units and another of two (the angle included between them being a right angle), the two sides of the one would be equal to the two sides of the other, since their sum is seven in each case. But this would not show the one triangle equal to the other; for the area of the former is six, of the latter five. The reason for this discrepancy is that the sides are not also equal respectively. We often fail to watch out for this in the distribution of plots of land; and many persons have taken the larger of two plots and got a reputation for justice as having chosen an equal portion, because the sum of the boundaries is the same in both cases. We must therefore take the sides as equal respectively, and whenever the author of the *Elements* adds this phrase, we should note that he does so for a reason....

Two triangles are said to be equal when their areas are equal. It can happen that two triangles with equal perimeters have unequal areas because of the inequality of their angles. "Area" I call the space itself which is cut off by the sides of the triangle, and "perimeter" the line composed of the three sides of the triangle. These are different things, and triangles with equal perimeters must also have the angles along one side equal if the areas are to be equal. It happens in some cases that, when the areas are equal, the perimeters are unequal and, when the perimeters are equal, the areas are unequal. Consider two isosceles triangles, each having its equal sides five units in length, but one having a base of eight, the other a base of six units. The person inexperienced in geometry would say that the triangle having the base of eight units is the greater, for its total perimeter is eighteen units. But the geometer would say that the area of both is twelve; and he can prove it by dropping a perpendicular from the vertex of each triangle and multiplying its length by half of the base. It is also possible, as I said, that triangles with equal perimeters have unequal areas, and some persons have wronged their associates in a distribution of lands by relying on the equality of perimeters and in fact getting a greater portion.

Although Proclus does not mention this, modern commentators note that Euclid proves this result by superposition. Namely, he imagined the first triangle being moved from its original position and placed on the second triangle. Euclid tacitly assumed that such a motion is always possible without deformation. Rather than supply such a postulate, nineteenth-century mathematicians preferred to assume this theorem itself as a postulate.

Proposition 5. *In isosceles triangles the angles at the base are equal to one another, and, if the equal straight lines be produced further, the angles under the base will be equal to one another.*

Let ABC be an isosceles triangle having the side AB equal to the side AC; and let the straight lines BD, CE be produced further in a straight line with AB, AC [figure 3.8.3].

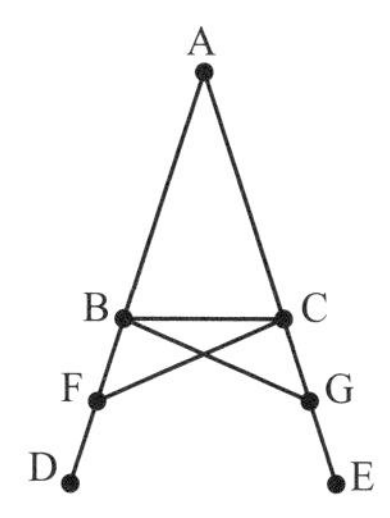

Figure 3.8.3.

I say that the angle ABC is equal to the angle ACB, and the angle CBD to the angle BCE. Let a point F be taken at random on BD; from AE the greater let AG be cut off equal to AF the less; and let the straight lines FC, GB be joined. Then, since AF is equal to AG and AB to AC, the two sides FA, AC, are equal to the two sides GA, AB, respectively; and they contain a common angle, the angle FAG. Therefore the base FC is equal to the base GB, and the triangle AFC is equal to the triangle AGB, and the remaining angles will be equal to the remaining angles respectively, namely, those which the equal sides subtend, that is, the angle ACF to the angle ABG, and the angle AFC to the angle AGB [1:4].[40] And, since the whole AF is equal to the whole AG, and in these AB is equal to AC, the remainder BF is equal to the remainder CG.

But FC was also proved equal to GB; therefore the two sides BF, FC are equal to the two sides CG, GB respectively; and the angle BFC is equal to the angle CGB, while the base BC is common to them; therefore the triangle BFC is also equal to the triangle CGB, and the remaining angles will be equal to the remaining angles respectively, namely, those which the equal sides subtend; therefore the angle FBC is equal to the angle GCB, and the angle BCF to the angle CBG. Accordingly, since the whole angle ABG was proved equal to the angle ACF, and in these the angle CBG is equal to the angle BCF, the remaining angle ABC is equal to the remaining angle ACB; and they are at the base of the triangle ABC. But the angle FBC was also proved equal to the angle GCB; and they are under the base. Therefore etc. Q.E.D.

Pappus has given a ... shorter demonstration that needs no supplementary construction, as follows. Let ABC be isosceles with side AB equal to side AC. Let us think of this triangle as two triangles and reason thus: Since AB is equal to AC and AC is equal to AB, the two sides AB and AC are equal to the two sides AC and AB, and the angle BAC is equal to the angle CAB (for they are the same); therefore all the corresponding parts are equal, BC to CB, the triangle AB to the triangle ACB, the angle ABC to the angle ACB, and angle ACB to angle ABC. For these are angles subtended by the equal sides AB and AC. Hence the angles at the base of an isosceles triangle are equal.... We are indebted to old Thales for the discovery of this and many other theorems. For he, it is said, was the first to notice

[40] We will sometimes include a reference for a given statement to a previous proposition, but for readability, we will not always do so.

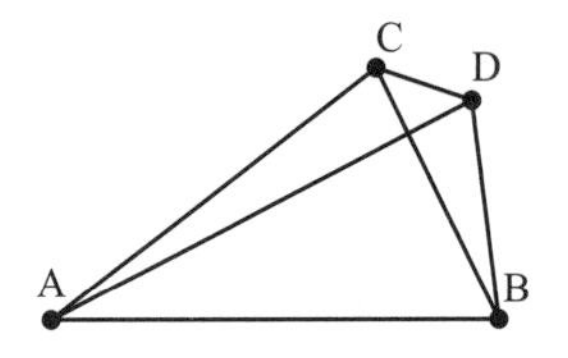

Figure 3.8.4.

and assert that in every isosceles [triangle] the angles at the base are equal, though in somewhat archaic fashion he called the equal angles similar. . . .

Proposition 7. *Given two straight lines constructed on a straight line [from its extremities] and meeting in a point, there cannot be constructed on the same straight line [from its extremities] and on the same side of it, two other straight lines meeting in another point and equal to the former two respectively, namely, each to that which has the same extremity with it.*

For, if possible, given two straight lines AC, CB constructed on the straight line AB and meeting at the point C, let two other straight lines AD, DB be constructed on the same straight line AB, on the same side of it, meeting in another point D and equal to the former two respectively, namely, each to that which has the same extremity with it, so that CA is equal to DA which has the same extremity A with it, and CB to DB which has the same extremity B with it; and let CD be joined [figure 3.8.4].

Then, since AC is equal to AD, the angle ACD is also equal to the angle ADC [1:5]; therefore the angle ADC is greater than the angle DCB; therefore the angle CDB is much greater than the angle DCB. Again, since CB is equal to DB, the angle CDB is also equal to the angle DCB. But it was also proved much greater than it: which is impossible. Therefore etc. Q.E.D.

This theorem has a character that is rare and not often found in scientific premises; for to be framed negatively and not affirmatively hardly suits their nature. At any rate the enunciations of geometrical and arithmetical theorems are usually affirmative. The reason is, as Aristotle says, that the universal affirmative proposition is best fitted for science, since it is more self-sufficient, needing no negative premise to supplement it, whereas the universal negative needs an affirmative if it is to be proved. For without an affirmative premise there is no proof nor syllogism; and this is why the sciences that demonstrate do so affirmatively for the most part and rarely make use of negative conclusions. . . .

Proposition 8. *If two triangles have two sides equal to two sides respectively, and have also the base equal to the base, they will also have the angles equal which are contained by the equal straight lines.*

Let ABC, DEF be two triangles having the two sides AB, AC equal to the two sides DE, DF respectively, namely, AB to DE, and AC to DF; and let them have the base BC equal to the base EF; I say that the angle BAC is also equal to the angle EDF [figure 3.8.5].

For, if the triangle ABC be applied to the triangle DEF, and if the point B be placed on the point E and the straight line BC on EF, the point C will also coincide with F, because

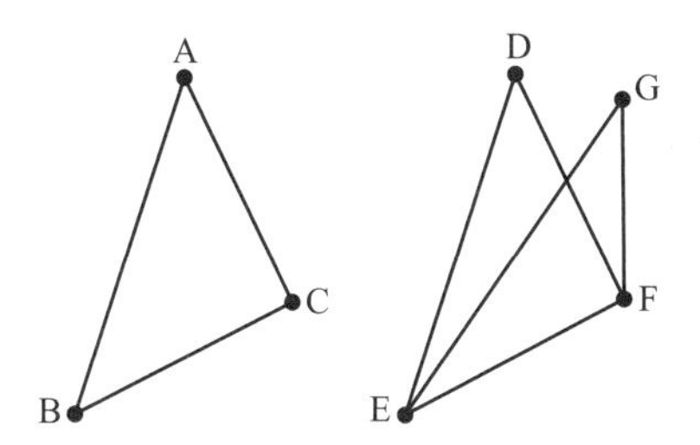

Figure 3.8.5.

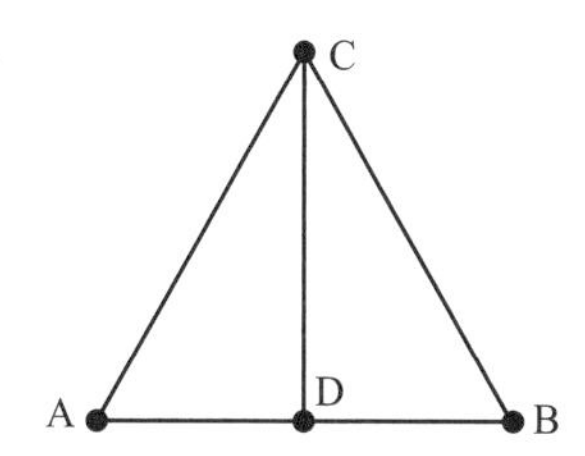

Figure 3.8.6.

BC is equal to *EF*. Then, *BC* coinciding with *EF*, *BA*, *AC* will also coincide with *ED*, *DF*; for, if the base *BC* coincides with the base *EF*, and the sides *BA*, *AC* do not coincide with *ED*, *DF* but fall beside them as *EG*, *GF*, then, given two straight lines constructed on a straight line [from its extremities] and meeting in a point, there will have been constructed on the same straight line [from its extremities], and on the same side of it, two other straight lines meeting in another point and equal to the former two respectively, namely, each to that which has the same extremity with it. But they cannot be so constructed [1:7]. Therefore it is not possible that, if the base *BC* be applied to the base *EF*, the sides *BA*, *AC* should not coincide with *ED*, *DF*; they will therefore coincide, so that the angle *BAC* will also coincide with the angle *EDF*, and will be equal to it. Therefore etc. Q.E.D.

If anyone should wonder why he did not add to the eighth the other details included in the fourth, namely, the equality of the triangles and of the remaining angles, our answer is that, when the angles at the vertex were proved to be equal, the equality of all parts to one another followed through the fourth. This, then, was the only thing it was necessary to prove independently; the rest could be inferred as consequences of it....

This proposition, like proposition 4, is also proved by superposition.

Proposition 10. *To bisect a given finite straight line.*

Let *AB* be the given finite straight line. Thus it is required to bisect the finite straight line *AB*. Let the equilateral triangle *ABC* be constructed on it, and let the angle *ACB* be bisected by the straight line *CD*; I say that the straight line *AB* has been bisected at the point *D* [figure 3.8.6].

For, since AC is equal to CB, and CD is common, the two sides AC, CD are equal to the two sides BC, CD respectively; and the angle ACD is equal to the angle BCD; therefore the base AD is equal to the base BD [1:4]. Therefore, etc. Q.E.F.

This problem may move some persons to suppose that geometers assume in advance as a hypothesis that a line does not consist of indivisible parts. For if it did, a finite line would consist of either an odd or an even number of parts. But if it has an odd number of parts, it seems that when a line is bisected the indivisible is bisected, since otherwise one segment would consist of a larger number of indivisible parts and be greater than the other. Consequently it will not be possible to bisect a given line if its magnitude consists of indivisible parts. But if it is not composed of indivisible parts, it will be divisible to infinity. This, then, they say, appears to be an agreed principle in geometry, that a magnitude consists of parts infinitely divisible. To this we shall give the reply of Geminus, that geometers do assume, in accordance with a common notion, that what is continuous is divisible. The continuous, we say, is what consists of parts that are in contact, and this can always be divided. But they do not assume that what is continuous is also divisible to infinity; rather they demonstrate it from appropriate principles. For when geometers demonstrate that there is incommensurability among magnitudes and that not all magnitudes are commensurable with one another, what else could we say they are demonstrating than that every magnitude is divisible indefinitely and that we can never reach an indivisible part which is the least common measure of magnitudes? This, then, is demonstrable, but it is an axiom that every continuum is divisible; hence a finite line, being continuous, is divisible. This is the notion that the author of the *Elements* uses in bisecting the finite straight line, not the assumption that it is divisible to infinity. That something is divisible and that it is divisible to infinity are not the same. One could use this problem also to refute the doctrine of Xenocrates[41] that asserts indivisible lines. For in general if there exists a line, it is either a straight line and can therefore be bisected, or circular and greater than some straight line—for every circular line has some straight line shorter than itself—or mixed and hence even more subject to division, since its simple components are divisible. But these matters must be reserved for study elsewhere.

Proposition 12. *To a given straight line of indefinite length, from a given point which is not on it, to draw a perpendicular straight line.*

Let AB be the given straight line, and C the given point which is not on it; thus it is required to draw to the given straight line AB, from the given point C which is not on it, a perpendicular straight line [figure 3.8.7].

For let a point D be taken at random on the other side of the straight line AB, and with center C and distance CD let the circle EFG be described; let the straight line EG be bisected at H, and let the straight lines CG, CH, CE be joined. I say that CH has been drawn perpendicular to the given infinite straight line AB from the given point C which is

[41] Xenocrates of Chalcedon (4th c. BCE) was a Greek philosopher who served for a time as head of the Platonic Academy.

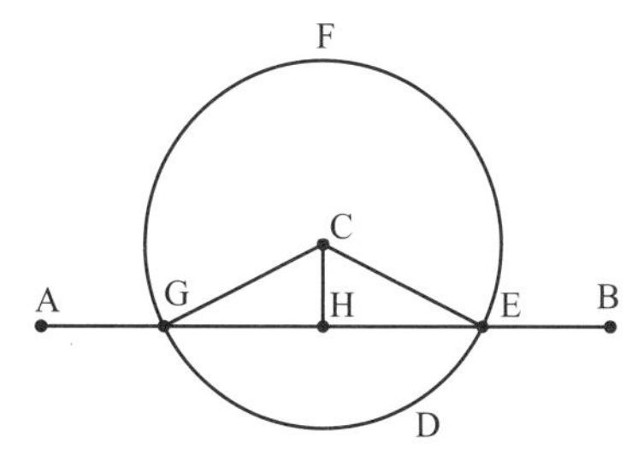

Figure 3.8.7.

not on it. For, since *GH* is equal to *HE*, and *HC* is common, the two sides *GH*, *HC* are equal to the two sides *EH*, *HC* respectively; and the base *CG* is equal to the base *CE*; therefore the angle *CHG* is equal to the angle *EHC* [1:8]. And they are adjacent angles. But, when a straight line set up on a straight line makes the adjacent angles equal to one another, each of the equal angles is right, and the straight line standing on the other is called perpendicular to that on which it stands. Therefore, etc. Q.E.F.

When considering the construction of a line at right angles, we had no need of the infinite, since the point was taken as lying on the line itself; but in the case of the perpendicular the given line is assumed to be infinite, since the point from which the perpendicular is to be dropped lies somewhere outside the line. If the line were not infinite, it would be possible so to take the point that it would lie outside the given line, but on a straight line with it, so that the line when prolonged would fall on it; and thus the problem could not be solved. For this reason he posits the straight line as infinite, so that if the point is taken only on one or the other side of the line, there will be no place left in which it can lie in a straight line with the given straight line and thus will lie outside and not on it.

This is the reason, then, why the line to which the perpendicular is to be dropped is given as infinite. But it is worth inquiring in what sense in general the infinite has existence. It is clear that, if there is an infinite line, there will also be an infinite plane, and infinite in actuality if the problem is to be a real one. That in sensible things there is not magnitude indefinitely extended in any direction has been sufficiently shown by the inspired Aristotle and by those who derive their philosophy from him. For it is not possible for the body moving in a circle to be infinite, nor any other of the simple bodies, for the place of each is determinate. But neither is it possible that there should be an infinite of this sort among separate and indivisible ideas; for if there is no extension nor magnitude in them, there can hardly be infinite magnitude.

It remains, then, that the infinite exists in the imagination, only with the imagination's knowing the infinite. For when the imagination knows, it simultaneously assigns to the object of its knowledge a form and limit, and in knowing brings to an end its movement through the imagined object; it has gone through it and comprehends it. The infinite therefore, is not the object of knowing imagination, but of imagination that is uncertain about its object, suspends further thinking, and calls infinite all that it abandons, as immeasurable and incomprehensible to

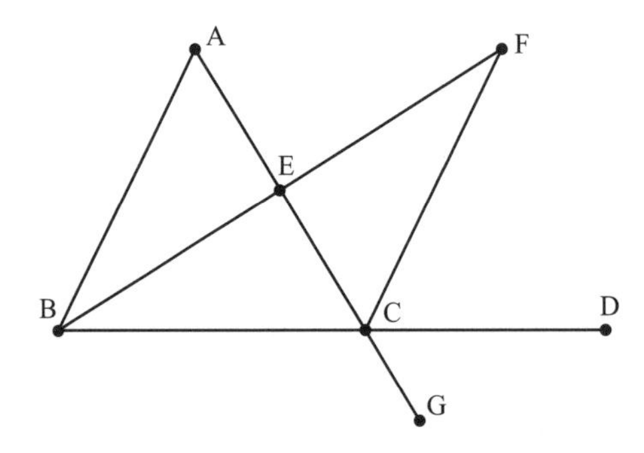

Figure 3.8.8.

thought. Just as sight recognizes darkness by the experience of not seeing, so imagination recognizes the infinite by not understanding it. It produces it indeed, because it has an indivisible power of proceeding without end, and it knows that the infinite exists because it does not know it. For whatever it dismisses as something that cannot be gone through, this it calls infinite. So if we supposed the infinite line to be given in imagination, exactly like triangles, circles, angles, lines, and all the other geometrical figures, should we not ask in wonder how a line can actually be infinite and how, being indeterminate, it associates with determinate notions? But the understanding from which our ideas and demonstrations proceed does not use the infinite for the purpose of knowing it, for the infinite is altogether incomprehensible to knowledge; rather it takes it hypothetically and uses only the finite for demonstration; that is, it assumes the infinite not for the sake of the infinite, but for the sake of the finite. If our imagination could see that the given point does not lie on the extension of the finite line and is so separated from it that no part of the line could underlie the point the demonstration would no longer need the infinite. It is therefore that it may use the finite line without risk of refutation or doubt that it posits the infinite, relying on the boundlessness of imagination as the source which generates it.

Proposition 16. *In any triangle, if one of the sides be produced, the exterior angle is greater than either of the interior and opposite angles.*

Let ABC be a triangle, and let one side of it BC be produced to D; I say that the exterior angle ACD is greater than either of the interior and opposite angles CBA, BAC [figure 3.8.8].

Let AC be bisected at E, and let BE be joined and produced in a straight line to F; let EF be made equal to BE, let FC be joined, and let AC be drawn through to G. Then, since AE is equal to EC, and BE to EF, the two sides AE, EB are equal to the two sides CE, EF respectively; and the angle AEB is equal to the angle FEC, for they are vertical angles. Therefore the base AB is equal to the base FC, and the triangle ABE is equal to the triangle CFE, and the remaining angles are equal to the remaining angles respectively, namely those which the equal sides subtend; therefore the angle BAE is equal to the angle ECF. But the angle ECD is greater than the angle ECF; therefore the angle

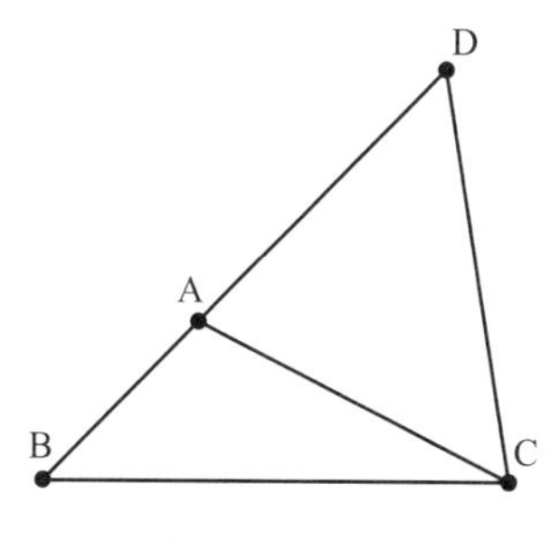

Figure 3.8.9.

ACD is greater than the angle BAE. Similarly also, if BC be bisected, the angle BCG, that is, the angle ACD, can be proved greater than the angle ABC as well. Therefore, etc. Q.E.D.

Note that Euclid assumed that EF could be made equal to BE. But there is no postulate allowing him to extend a line to any arbitrary length. A consequence of this theorem is proposition 17, that two angles of any triangle are always less than two right angles. This result, based on the faulty proof of proposition 16, was important in the developments leading to the discovery of non-Euclidean geometry.

Proposition 20. *In any triangle two sides taken together in any manner are greater than the remaining side.*

For let ABC be a triangle; I say that in the triangle ABC two sides taken together in any manner are greater than the remaining one, namely, BA, AC greater than BC; AB, BC greater than AC; BC, CA greater than AB [figure 3.8.9].

For let BA be drawn through to the point D, let DA be made equal to CA, and let DC be joined. Then, since DA is equal to AC, the angle ADC is also equal to the angle ACD; therefore the angle BCD is greater than the angle ADC. And, since DCB is a triangle having the angle BCD greater than the angle BDC, and the greater angle is subtended by the greater side, therefore DB is greater than BC. But DA is equal to AC; therefore BA, AC are greater than BC. Similarly we can prove that AB, BC are also greater than CA, and BC, CA than AB. Therefore etc. Q.E.D.

The Epicureans are wont to ridicule this theorem, saying it is evident even to an ass and needs no proof; it is as much the mark of an ignorant man, they say, to require persuasion of evident truths as to believe what is obscure without question. Now whoever lumps these things together is clearly unaware of the difference between what is and what is not demonstrated. That the present theorem is known to an ass they make out from the observation that, if straw is placed at one extremity of the sides, an ass in quest of provender will make his way along the one side and not by way of the two others. To this it should be replied that, granting the theorem is evident to sense-perception, it is still not clear for scientific thought. Many things have this character; for example, that fire warms. This is clear to perception, but it is the task of science to find out how it warms, whether by a bodiless power or by physical parts, such as spherical or pyramidal particles. Again it is clear to our

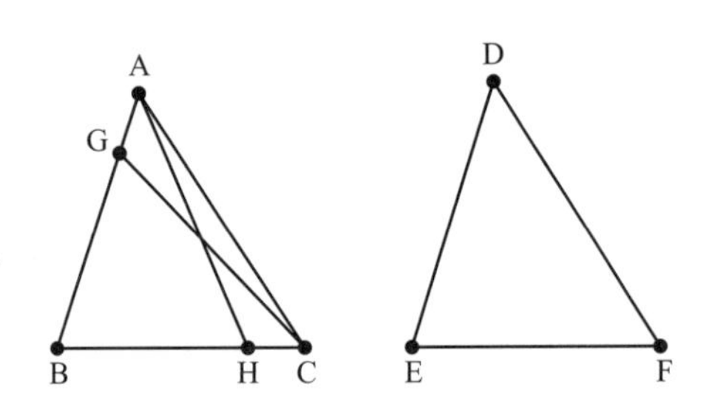

Figure 3.8.10.

senses that we move, but how we move is difficult for reason to explain, whether through a partless medium or from interval to interval, and in this case how we can traverse an infinite number of intervals, for every magnitude is divisible without end. So with respect to a triangle let it be evident to perception that two sides are greater than the third; but how this comes about it is the function of knowledge to say.

Although the next proposition has two cases, we only present the proof of the first because of the length of the demonstration and its repetitiveness.

Proposition 26. *If two triangles have the two angles equal to two angles respectively, and one side equal to one side, namely, either the side adjoining the equal angles, or that subtending one of the equal angles, they will also have the remaining sides equal to the remaining sides and the remaining angle to the remaining angle.*

Let ABC, DEF be two triangles having the two angles ABC, BCA equal to the two angles DEF, EFD respectively, namely, the angle ABC to the angle DEF, and the angle BCA to the angle EFD; and let them also have one side equal to one side, first that adjoining the equal angles, namely, BC to EF; I say that they will also have the remaining sides equal to the remaining sides respectively, namely, AB to DE and AC to DF, and the remaining angle to the remaining angle, namely, the angle BAC to the angle EDF [figure 3.8.10].

For, if AB is unequal to DE, one of them is greater. Let AB be greater, and let BG be made equal to DE; and let GC be joined. Then, since BG is equal to DE, and BC to EF, the two sides GB, BC are equal to the two sides DE, EF respectively; and the angle GBC is equal to the angle DEF; therefore the base GC is equal to the base DF, and the triangle GBC is equal to the triangle DEF, and the remaining angles will be equal to the remaining angles, namely, those which the equal sides subtend; therefore the angle GCB is equal to the angle DFE. But the angle DFE is by hypothesis equal to the angle BCA; therefore the angle BCG is equal to the angle BCA, the less to the greater, which is impossible. Therefore AB is not unequal to DE, and is therefore equal to it. But BC is also equal to EF; therefore the two sides AB, BC are equal to the two sides DE, EF respectively, and the angle ABC is equal to the angle DEF; therefore the base AC is equal to the base DF, and the remaining angle BAC is equal to the remaining angle EDF. . . . Therefore etc. Q.E.D.

Proposition 27. *If a straight line falling on two straight lines makes the alternate angles equal to one another, the straight lines will be parallel to one another.*

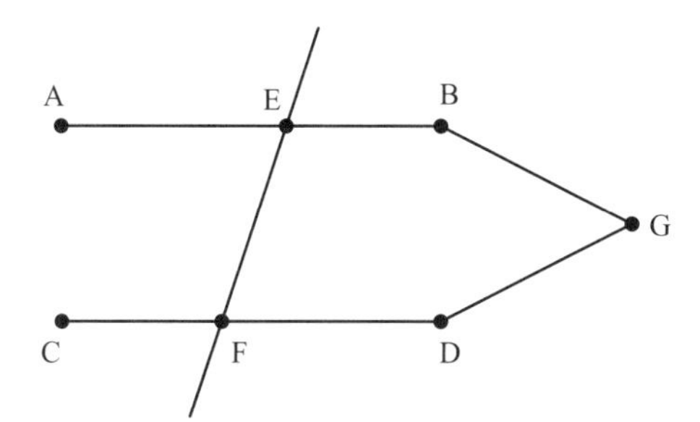

Figure 3.8.11.

For let the straight line EF falling on the two straight lines AB, CD make the alternate angles AEF, EFD equal to one another; I say that AB is parallel to CD [figure 3.8.11].

For, if not, AB, CD when produced will meet either in the direction of B, D, or towards A, C. Let them be produced and meet, in the direction of B, D, at G. Then, in the triangle GEF, the exterior angle AEF is equal to the interior and opposite angle EFG: which is impossible [1:16]. Therefore AB, CD when produced will not meet in the direction of B, D. Similarly it can be proved that neither will they meet towards A, C. But straight lines which do not meet in either direction are parallel; therefore AB is parallel to CD. Therefore etc. Q.E.D.

Proposition 28. *If a straight line falling on two straight lines makes the exterior angle equal to the interior and opposite angle on the same side, or the interior angles on the same side equal to two right angles, the straight lines will be parallel to one another.*

The proof here reduces the theorem to proposition 27.

Proposition 29. *A straight line falling on parallel straight lines makes the alternate angles equal to one another, the exterior angle equal to the interior and opposite angle, and the interior angles on the same side equal to two right angles.*

For let the straight line EF fall on the parallel straight lines AB, CD; I say that it makes the alternate angles AGH, GHD equal, the exterior angle EGB equal to the interior and opposite angle GHD, and the interior angles on the same side, namely, BGH, GHD, equal to two right angles [figure 3.8.12].

For, if the angle AGH is unequal to the angle GHD, one of them is greater. Let the angle AGH be greater. Let the angle BGH be added to each; therefore the angles AGH, BGH are greater than the angles BGH, GHD. But the angles AGH, BGH are equal to two right angles; therefore the angles BGH, GHD are less than two right angles. But straight lines produced indefinitely from angles less than two right angles meet [postulate 5]; therefore AB, CD, if produced indefinitely, will meet; but they do not meet, because they are by hypothesis parallel. Therefore the angle AGH is not unequal to the angle GHD, and is therefore equal to it. Again, the angle AGH is equal to the angle EGB; therefore the angle EGB is also equal to the angle GHD. Let the angle BGH be added to each; therefore the angles EGB, BGH are equal to the angles BGH, GHD. But the angles EGB, BGH are equal to two right angles; therefore the angles BGH, GHD are also equal to two right angles. Therefore etc. Q.E.D.

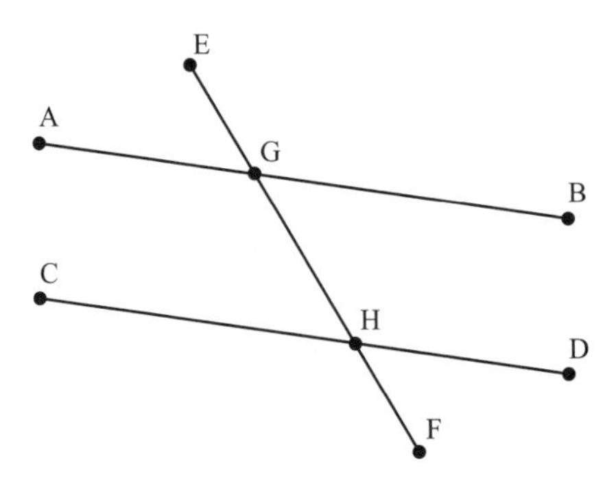

Figure 3.8.12.

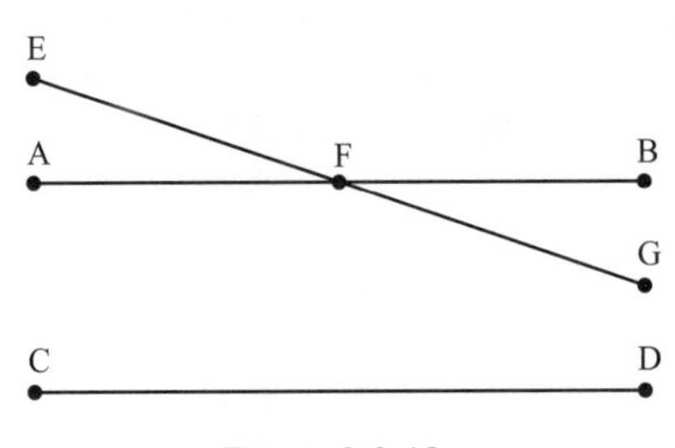

Figure 3.8.13.

In this theorem the author of the *Elements* uses for the first time the postulate, "If a straight line falling on two straight lines makes the interior angles in the same direction less than two right angles, the straight lines if produced will meet in that direction in which are the angles less than two right angles." As I said in the part of my exposition that precedes the theorems, not everyone admits that this generally accepted proposition is indemonstrable. For how could it be so when its converse is recorded among the theorems as something demonstrable? For the theorem that in every triangle any two interior angles are less than two right angles is the converse of this postulate. . . . Hence others before us have classed it among the theorems and demanded a proof of this which was taken as a postulate by the author of the *Elements*.

To anyone who wants to see this argument constructed, let us say that he must accept in advance such an axiom as Aristotle used in establishing the finiteness of the cosmos: If from a single point two straight lines making an angle are produced indefinitely, the interval between them when produced indefinitely will exceed any finite magnitude. . . . If this is laid down, I say that, if a straight line cuts one of two parallel lines, it cuts the other also. Let AB and CD be parallel lines and EFG a line cutting AB. I say that it also cuts CD [figure 3.8.13].

For since there are two straight lines through point F, when FB and FG are extended indefinitely, they will have an interval between them greater than any magnitude and hence greater than the distance between the parallel lines. And so when they are separated from each other a greater distance than that between the

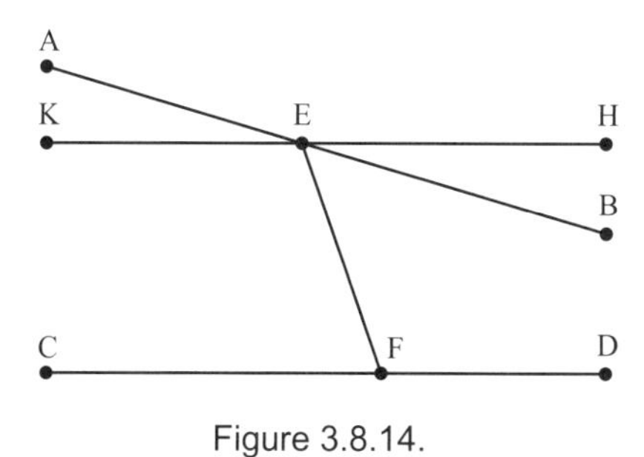

Figure 3.8.14.

parallel lines, *FG* will cut *CD*. Therefore if a straight line cuts one of two parallels, it cuts the other also.

Having proved this, we can demonstrate the proposition before us as a consequence of it. Let *AB* and *CD* be two straight lines and *EF* falling upon them and making angles *BEF* and *DFE* less than two right angles [figure 3.8.14].

I say that the straight lines will meet in that direction in which are the angles less than two right angles. For since angles *BEF* and *DFE* are less than two right angles, let angle *HEB* be equal to the excess of two right angles over them, and let *HE* be produced to *K*. Then since *EF* falls upon *KH* and *CD* and makes the interior angles equal to two right angles, namely, *HEF* and *DFE*, *HK* and *CD* are parallel straight lines. And *AB* cuts *KH*; it will therefore cut *CD*, by the proposition just demonstrated. *AB* and *CD* therefore will meet in that direction in which are the angles less than two right angles, so that the proposition before us has been demonstrated.

Of course, to accept Proclus's proof, we must also accept the axiom of Aristotle. It turns out that this axiom is equivalent to Euclid's postulate 5. Modern writers on the subject, however, usually name the result Proclus proved from Aristotle's axiom—namely, that if a straight line cuts one of two parallel lines, it cuts the other also–as Proclus's axiom. This too is equivalent to Euclid's postulate 5.

Proposition 32. *In any triangle, if one of the sides be produced, the exterior angle is equal to the two interior and opposite angles, and the three interior angles of the triangle are equal to two right angles.*

Let *ABC* be a triangle, and let one side of it *BC* be produced to *D*; I say that the exterior angle *ACD* is equal to the two interior and opposite angles *CAB*, *ABC*, and the three interior angles of the triangle *ABC*, *BCA*, *CAB* are equal to two right angles [figure 3.8.15].

For let *CE* be drawn through the point *C* parallel to the straight line *AB*. Then, since *AB* is parallel to *CE*, and *AC* has fallen upon them, the alternate angles *BAC*, *ACE* are equal to one another [1:29]. Again, since *AB* is parallel to *CE*, and the straight line *BD* has fallen upon them, the exterior angle *ECD* is equal to the interior and opposite angle *ABC* [1:29]. But the angle *ACE* was also proved equal to the angle *BAC*; therefore the whole angle *ACD* is equal to the two interior and opposite angles *BAC*, *ABC*. Let the angle *ACB* be added to each; therefore the angles *ACD*, *ACB* are equal to the three angles *ABC*, *BCA*,

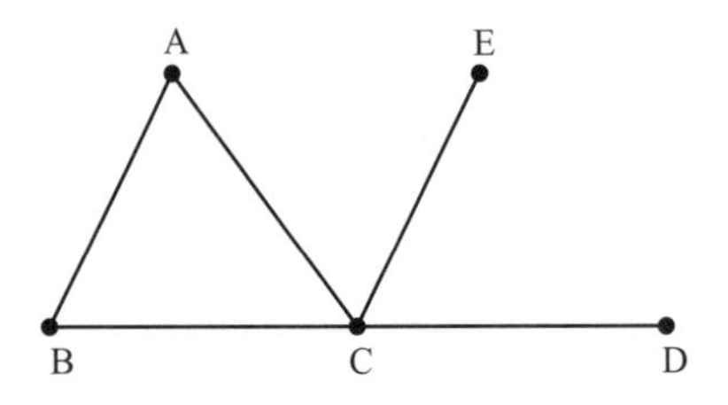

Figure 3.8.15.

CAB. But the angles *ACD*, *ACB* are equal to two right angles; therefore the angles *ABC*, *BCA*, *CAB* are also equal to two right angles. Therefore etc. Q.E.D.

We can now say that in every triangle the three angles are equal to two right angles. But we must find a method of discovering for all the other rectilinear polygonal figures—for four-angled, five-angled, and all the succeeding many-sided figures—how many right angles their angles are equal to. First of all, we should know that every rectilinear figure may be divided into triangles, for the triangle is the source from which all things are constructed. . . . Each rectilinear figure is divisible into triangles two fewer in number than the number of its sides: if it is a four-sided figure, it is divisible into two triangles; if five-sided, into three; and if six-sided, into four. For two triangles put together make at once a four-sided figure, and this difference between the number of the constituent triangles and the sides of the first figure composed of triangles is characteristic of all succeeding figures. Every many-sided figure, therefore, will have two more sides than the triangles into which it can be resolved. Now every triangle has been proved to have its angles equal to two right angles. Therefore the number which is double the number of the constituent triangles will give the number of right angles to which the angles of a many-sided figure are equal. Hence every four-sided figure has angles equal to four right angles, for it is composed of two triangles; and every five-sided figure, six right angles; and similarly for the rest.

This, then, is one inference that we can draw from this theorem with regard to all figures that are polygonal and rectilinear. Let us briefly state another that follows from it: When all the sides of a rectilinear figure are produced at one time, the exterior angles constructed are equal to four right angles. For the angles in both directions must be equal to right angles double the number of the sides, since on each of the extended sides angles are constructed equal to two right angles; and if we subtract the right angles to which the interior angles are equal, the remaining angles, the exterior ones, are equal to four right angles. For example, if the figure is a triangle and all its sides are produced at once, the interior and exterior angles produced are equal to six right angles, and of these the interior angles are equal to two, so that the remaining angles, the exterior ones, are equal to four. If it is a four-sided figure, the sum of them all will be eight right angles, double the number of sides; and of these the interior angles are equal to four, and therefore the exterior ones are equal to the other four. If it is five-sided, all the angles will equal ten right

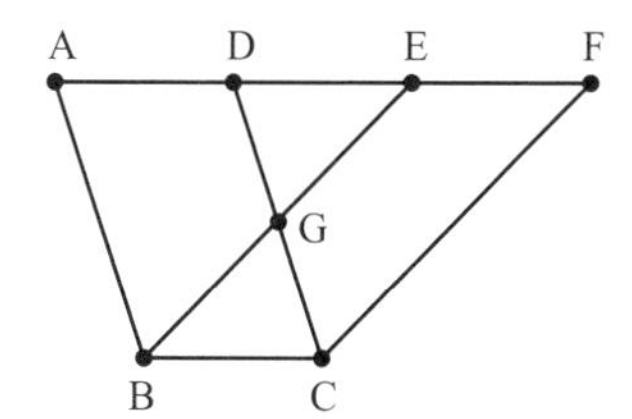

Figure 3.8.16.

angles, the interior ones being equal to six and the exterior to the other four. And so on indefinitely in the same way.

Besides these, let us list the following consequences of this theorem: that every equilateral triangle has each of its angles equal to two-thirds of a right angle; that an isosceles triangle whose vertical angle is a right angle has each of the other angles half a right angle, as in the half-square; and that the scalene half-triangle produced by dropping a perpendicular from any angle of an equilateral triangle to the side which subtends it has one of its angles right, another two-thirds of a right angle (the angle already in the equilateral triangle), and the third angle therefore one-third of a right angle, for the three must together be equal to two right angles.

Proposition 35. *Parallelograms which are on the same base and in the same parallels are equal to one another.*

Let $ABCD$, $EBCF$ be parallelograms on the same base BC and in the same parallels AF, BC. I say that $ABCD$ is equal to the parallelogram $EBCF$ [figure 3.8.16].

For, since $ABCD$ is a parallelogram, AD is equal to BC. For the same reason also EF is equal to BC, so that AD is also equal to EF; and DE is common; therefore the whole AE is equal to the whole DF. But AB is also equal to DC; therefore the two sides EA, AB are equal to the two sides FD, DC respectively, and the angle FDC is equal to the angle EAB, the exterior to the interior [1:29]; therefore the base EB is equal to the base FC, and the triangle EAB will be equal to the triangle FDC. Let DGE be subtracted from each; therefore the trapezium $ABGD$ which remains is equal to the trapezium $EGCF$ which remains. Let the triangle GBC be added to each; therefore the whole parallelogram $ABCD$ is equal to the whole parallelogram $EBCF$. Therefore etc. Q.E.D.

It may seem a great puzzle to those inexperienced in this science that the parallelograms constructed on the same base and between the same parallels should be equal to one another. For when the sides of the areas constructed on the same base can be increased indefinitely—and we can increase the length of these sides of the parallelograms as far as we can extend the parallel lines—we may well ask how the areas can remain equal when this happens. For if the breadth is the same (since the base is identical) while the side becomes greater, how could the area fail to become greater? This theorem, then, and the following one about triangles belong among what are called the "paradoxical" theorems in mathematics. The mathematicians have worked out what they call the "locus of

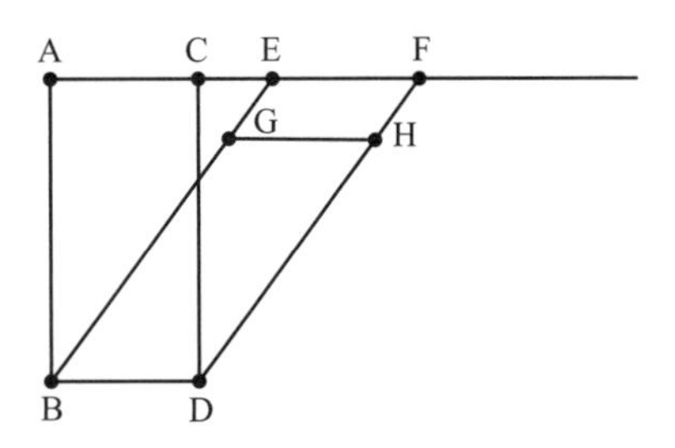

Figure 3.8.17.

paradoxes," ... and this theorem is included among them. Most people at least are immediately startled to learn that multiplying the length of the side does not destroy the equality of the areas when the base remains the same. The truth is, nevertheless, that the equality or inequality of the angles is the factor of greatest weight in determining the increase or decrease of the area. For the more unequal we make the angles, the more we decrease the area, if the side and base remain the same; hence if we are to preserve equality, we must increase the side.

Take any parallelogram, say $ABCD$, and let AC be produced indefinitely. Suppose it to be a rectangular figure, and on the base BD let another parallelogram $BEFD$ be constructed [figure 3.8.17].

Clearly the side has been increased, for BE is longer that AB, since the angle at A is a right angle. And this increase was necessary, for the angles of the parallelogram $BEFD$ have become unequal, some acute, the others obtuse; and this has happened because side BE is, as it were, folded back on BD and contracts the area. Let a line BG be taken equal to AB and GH be drawn through G parallel to BD. Then the side of parallelogram $BDGH$ is equal to the side of $ABCD$, and the breadth is the same, but its area is less, namely, less than that of $BEFD$. The inequality of the angles has clearly made the area less, and the increase in the side, by adding as much as the inequality of the angles has taken away, preserves the equality of the areas; and the limit of increase for the side is the locus of the parallels. The square is demonstrably greater than the oblong, when both are rectangular and have equal perimeters; and when both are equilateral and have equal perimeters, the rectangular figure is demonstrably greater than the non-rectangular. For the rightness of the angles and the equality of the sides are the all-important factors affecting the increase of the areas; and this is why the square is manifestly greater than all others with an equal length of boundaries, and the rhomboid is the least of all.

Proposition 44. *To a given straight line to apply, in a given rectilineal angle, a parallelogram equal to a given triangle.*

Let AB be the given straight line, C the given triangle and D the given rectilineal angle; thus it is required to apply to the given straight line AB, in an angle equal to the angle D, a parallelogram equal to the given triangle C [figure 3.8.18].

Let the parallelogram $BEFG$ be constructed equal to the triangle C, in the angle EBG which is equal to D; let it be placed so that BE is in a straight line with AB; let FG be

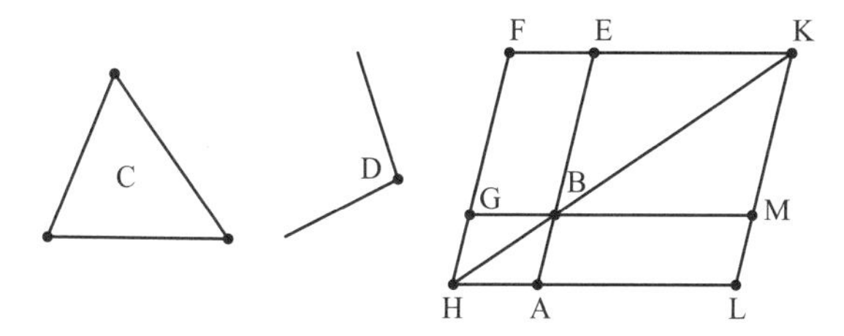

Figure 3.8.18.

drawn through to H, and let AH be drawn through A parallel to either BG or EF. Let HB be joined. Then, since the straight line HF falls upon the parallels AH, EF, the angles AHF, HFE are equal to two right angles [1:29]. Therefore the angles BHG, GFE are less than two right angles; and straight lines produced indefinitely from angles less than two right angles meet [postulate 5]; therefore HB, FE, when produced, will meet. Let them be produced and meet at K; through the point K let KL be drawn parallel to either EA or FH, and let HA, GB be produced to the points L, M. Then $HLKF$ is a parallelogram, HK is its diameter, and AG, ME are parallelograms, and LB, BF the so-called complements, about HK; therefore LB is equal to BF. But BF is equal to the triangle C; therefore LB is also equal to C. And, since the angle GBE is equal to the angle ABM, while the angle GBE is equal to D, the angle ABM is also equal to the angle D. Therefore etc. Q.E.F.

Eudemus and his school tell us that these things—the application (*paraboli*) of areas, their exceeding (*hyperboli*) and their falling short (*ellipsis*)—are ancient discoveries of the Pythagorean muse. It is from these procedures that later geometers took these terms and applied them to the so-called conic lines, calling one of them "parabola," another "hyperbola," and the third "ellipse," although those godlike men of old saw the significance of these terms in the describing of plane areas along a finite straight line. For when, given a straight line, you make the given area extend along the whole of the line, they say you "apply" the area; when you make the length of the area greater than the straight line itself, then it "exceeds"; and when less, so that there is a part of the line extending beyond the area described, then it "falls short." Euclid too in his sixth book speaks in this sense of "exceeding" and "falling short"; but here he needed "application," since he wished to apply to a given straight line an area equal to a given triangle, in order that we might be able not only to construct a parallelogram equal to a given triangle, but also to apply it to a given finite straight line. For example, when a triangle is given having an area of twelve feet and we posit a straight line whose length is four feet, we apply to the straight line an area equal to the triangle when we take its length as the whole four feet and find how many feet in breadth it must be in order that the parallelogram may be equal to the triangle. Then when we have found, let us say, a breadth of three feet and multiplied the length by the breadth, we shall have the area, that is, if the angle assumed is a right angle. Something like this is the method of "application" which has come down to us from the Pythagoreans.

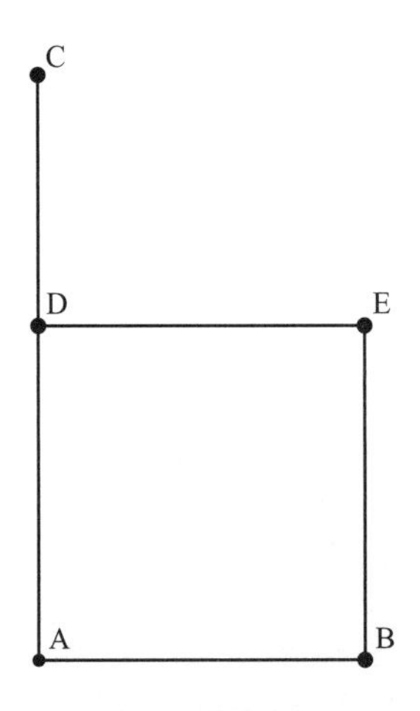

Figure 3.8.19.

Proposition 46. *On a given straight line to describe a square.*

Let AB be the given straight line; thus it is required to describe a square on the straight line AB. Let AC be drawn at right angles to the straight line AB from the point A on it, and let AD be made equal to AB; through the point D let DE be drawn parallel to AB, and through the point B let BE be drawn parallel to AD [figure 3.8.19].

Therefore $ADEB$ is a parallelogram; therefore AB is equal to DE, and AD to BE. But AB is equal to AD; therefore the four straight lines BA, AD, DE, EB are equal to one another; therefore the parallelogram $ADEB$ is equilateral. I say next that it is also right-angled. For, since the straight line AD falls upon the parallels AB, DE, the angles BAD, ADE are equal to two right angles [1:29]. But the angle BAD is right; therefore the angle ADE is also right. And in parallelogrammic areas the opposite sides and angles are equal to one another; therefore each of the opposite angles ABE, BED is also right. Therefore $ADEB$ is right-angled. And it was also proved equilateral. Therefore etc. Q.E.F.

Proposition 47, the so-called Pythagorean Theorem, is probably the most famous result in the *Elements*. This theorem was certainly known in ancient Mesopotamia, well before Pythagoras, and we indicate a possible Mesopotamian demonstration in section 6.4. However, the proof given by Euclid seems to be original to him.

Proposition 47. *In right-angled triangles the square on the side subtending the right angle is equal to the squares on the sides containing the right angle.*

Let ABC be a right-angled triangle having the angle BAC right; I say that the square on BC is equal to the squares on BA, AC. For let there be described on BC the square $BDEC$, and on BA, AC the squares GB, HC; through A let AL be drawn parallel to either BD or CE, and let AD, FC be joined [figure 3.8.20].

Then, since each of the angles BAC, BAG is right, it follows that with straight line BA, and at the point A on it, the two straight lines AC, AG, not lying on the same side, make the adjacent angles equal to two right angles; therefore CA is in a straight line with AG. For the same reason BA is also in a straight line with AH. And, since the angle DBC is equal to the angle FBA, for each is right, let the angle ABC be added to each; therefore

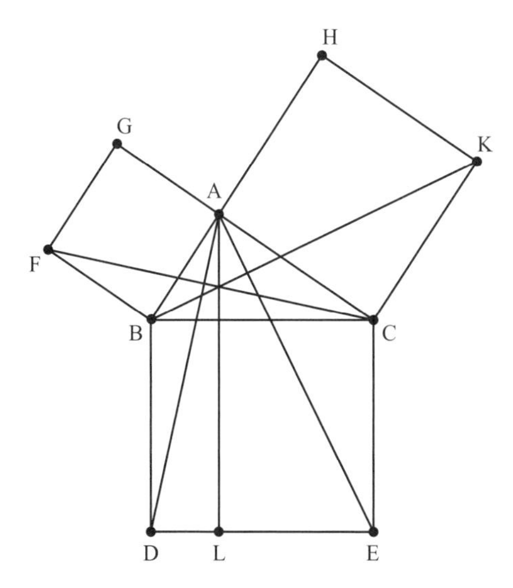

Figure 3.8.20.

the whole angle DBA is equal to the whole angle FBC. And, since DB is equal to BC, and FB to BA, the two sides AB, BD are equal to the two sides FB, BC respectively; and the angle ABD is equal to the angle FBC; therefore the base AD is equal to the base FC, and the triangle ABD is equal to the triangle FBC.

Now the parallelogram BL is double of the triangle ABD, for they have the same base BD and are in the same parallels BD, AL. And the square GB is double of the triangle FBC, for they again have the same base FB and are in the same parallels FB, GC. But the doubles of equals are equal to one another. Therefore the parallelogram BL is also equal to the square GB. Similarly, if AE, BK be joined, the parallelogram CL can also be proved equal to the square HC; therefore the whole square $BDEC$ is equal to the two squares GB, HC. And the square $BDEC$ is described on BC, and the squares GB, HC on BA, AC. Therefore the square on the side BC is equal to the squares on the sides BA, AC. Therefore etc. Q.E.D.

If we listen to those who like to record antiquities, we shall find them attributing this theorem to Pythagoras and saying that he sacrificed an ox on its discovery. For my part, though I marvel at those who first noted the truth of this theorem, I admire more the author of the *Elements*, not only for the very lucid proof by which he made it fast, but also because in the sixth book he laid hold of a theorem even more general than this and secured it by irrefutable scientific arguments. For in that book he proves generally that in right-angled triangles the figure on the side that subtends the right angle is equal to the similar and similarly drawn figures on the sides that contain the right angle.[42] Every square is of course similar to every other square, but not all similar rectilinear figures are squares, for there is similarity in triangles and in other polygonal figures. Hence the argument establishing that the figure on the side subtending the right angle, whether it be a square or any other

[42] Book 6, proposition 31.

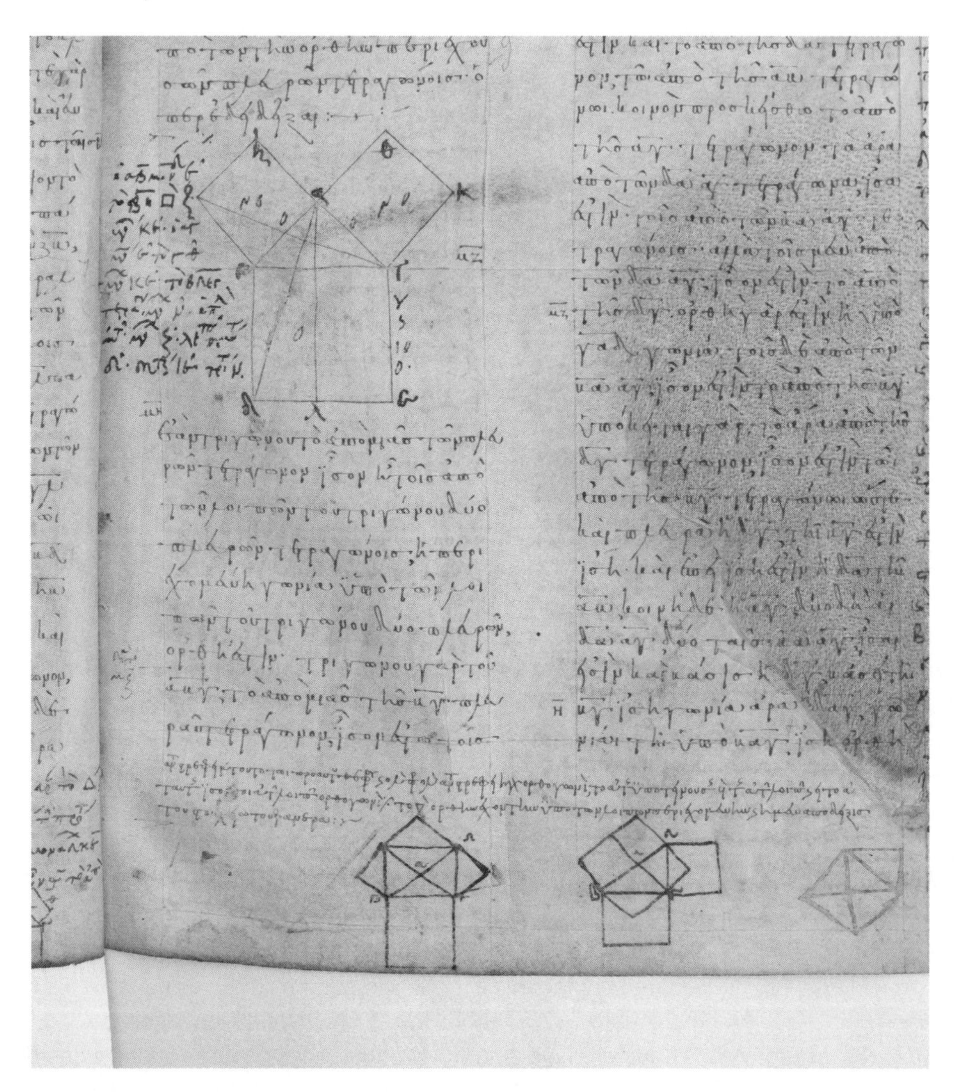

Figure 3.8.21. A part of a page containing the Pythagorean Theorem from one of the oldest extant manuscripts (10th c.) of Euclid's Elements (MS Vat. Gr. 190, pt. 1, f. 39r).

kind of figure, is equal to the similar and similarly drawn figures on the sides about the right angle, proves something more general and scientific than that which shows only that the square is equal to the squares. For there the cause of the more general proposition that is proved becomes clear: it is the rightness of the angle that makes the figure on the subtending side equal to the similar and similarly drawn figures on the containing sides, just as the obtuseness of the angle is the cause of its being greater and the acuteness of the angle the cause of its being less.

3.9 Simplicius, *Commentary on Euclid's Elements*

Simplicius's commentary on the beginning of book 1 of Euclid's *Elements* survives in an Arabic translation, which is reproduced by the ninth-century mathematician al-Faḍl ibn Ḥātim al-Nayrīzī in his own commentary on the *Elements*. We include here part of Simplicius's discussion of the nature of postulates. In particular, he refers to the five postulates in book 1, namely, the three construction postulates, the postulate on right angles, and the parallel postulate.

On Postulates

Euclid, after having stated the definitions which signify the essence of each one of the defined things, went on to enumerate the postulates. Postulates, in general are those things which are not granted, but the student is asked to grant them as a permissible hypothesis which is laid down between himself and the teacher and which is granted.

Now this hypothesis may be impossible, as for instance the postulate which Archimedes asked to be granted him, namely, to postulate that he stands outside the earth. For he promised that if this were conceded to him, he would prove that he could move the earth—saying: Young man, grant me that it is possible that I rise and stand outside the earth, and I will show you that I can move the earth. This he said when he boasted of having discovered mechanical power, demanding that that assumption, though impossible, be postulated and laid down for the sake of teaching.

Thus a postulated thing may be impossible, as we said, or it may be possible and known to the teachers but unknown to the students and required to be used at the beginning of teaching. For the things that are subject to demonstration are also known to the teachers and unknown to the student, but they are not laid down by postulation since they are not primary premises but are capable of demonstration.

But as for postulates, whoever posits them demands that they should be postulated as principles. Some are demanded to be postulated as necessary for teaching only—such as the first three postulates. Some require an easy proof in order to be assented to and accepted by themselves. The difference between these and axioms is that axioms are accepted by themselves as soon as thought takes hold of them. Whereas postulates are by nature intermediate between, on the one hand, the principles taken from the first science, whose reasons are unknown to those who use them, such as definitions, and on the other, axioms which all people equally accept. For postulates are known, not however to all people, but to the teachers in each one of the arts.

Some have thought that geometrical postulates are only intended for granting the matter inasmuch as not all operations are applicable to it. For an opponent may raise an objection on account of the matter, saying: I cannot produce a straight line on the surface of the sea, nor can I produce a straight line to infinity since the infinite does not exist. But those who say this, first, think that postulates are only for those whose geometry is material. But what would they say about the equality of right angles? How would they establish that this is postulated on account of matter? And similarly with the postulates that come after this.

It is, therefore, better to say that postulates are those things which are not accepted by the student when he first hears them, and which are required in the demonstration. Some are impossible and, therefore, are not as easily accepted as the first three, but are demanded to be granted for carrying out the process of teaching, as I said. Others are known to the teacher and accepted by him, but to the student are at the outset remote and not evident, and, accordingly, he is asked to grant them, as is the case with the postulates that come after the first three. The utility of the first three is not to allow the demonstration to be hindered by an incapacity or failing in the matter. Whereas those that follow upon the first three are required for certain demonstrations. . . .

The premise which is used in the demonstration of Proposition 29 of Book 1, namely, that every two lines drawn [from a transversal] at less than two right angles will meet, is not one of the accepted propositions. This postulate is not all that evident, but needs to be proved by means of lines, so that Abẕīnyāṭūs[43] and Diodorus[44] proved it by many different propositions. Ptolemy, too, constructed a proof and demonstration of it, using for this purpose Propositions 13, 15 and 16 of Book 1 of the *Elements*.[45] This is not objectionable, for Euclid did not use this postulate until Proposition 29 of this Book. But this notion [postulate 5] is in itself also worthy of being examined and discoursed upon and [it requires us] to prove that just as when two lines are drawn [from a transversal] at two right angles they are parallel, they meet if drawn at less than two right angles.

Although Nayrīzī seems to imply that he will provide Simplicius's proof, his commentary does not contain it. However, there is an Arabic manuscript preserved at the Bodleian Library in Oxford that apparently contains this proof. The argument that what follows comes from Simplicius is made in the article by Sabra from which the translation is drawn. The proof consists of four propositions. The central proposition is the second one, and it is here that Simplicius's proof fails. For he provides no means for constructing a chord ED such that it falls on the far side of point A. His remark that the multiples taken are unlimited and the line AB is limited unfortunately does not guarantee that such a chord will exist.

Proof of Parallel Postulate

Proposition 1. *Let ABG be an angle taken at random; then it has many chords infinite in multitude.*

About B as center let us draw any number of circles, such as DE, ZH, AG and let us draw their chords [figure 3.9.1].

Thus every angle has chords which are infinite in multitude.

Proposition 2. *Let the angle ABD be acute and the angle BAG be right; then the lines AG, BD, if produced will meet.*

Let the angle ABE be equal to the angle ZBD; let BD, BE be produced indefinitely, and from them let an unlimited number of multiples of BA be cut off; let the chords joining their extremities be drawn [figure 3.9.2].

[43] Nothing is known about this mathematician.

[44] Although Diodorus is mentioned elsewhere, nothing is known about his proof.

[45] Proclus mentions this supposed proof but notes the fallacy in it.

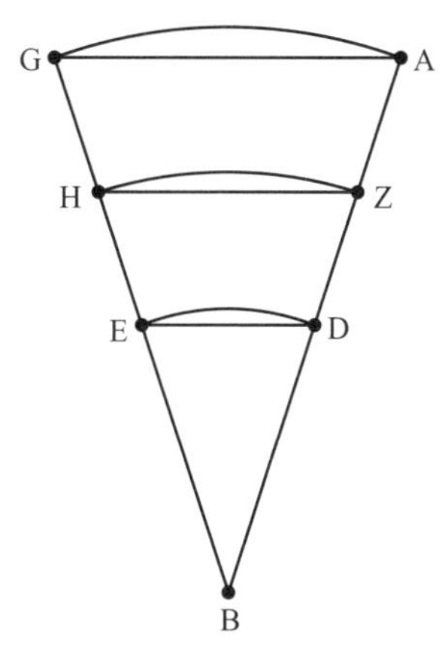

Figure 3.9.1.

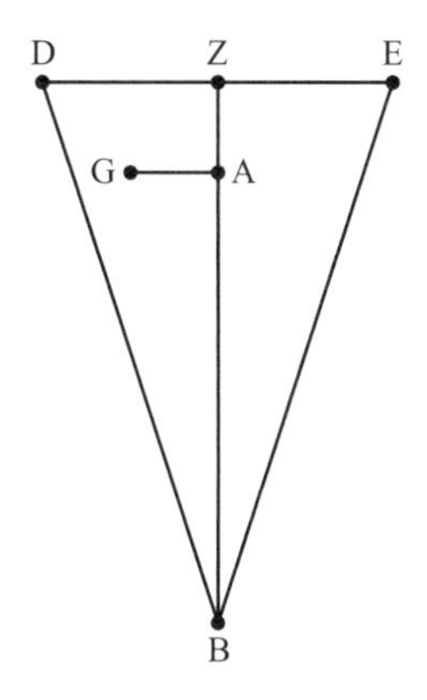

Figure 3.9.2.

Then one of these chords must fall beyond point A, towards point D. For the multiples taken are unlimited and the line AB is limited. Let this chord be as the line EZD. Then, since the angle B has been halved and BD is equal to BE, the angles at Z are right angles. Therefore, the line AG will not meet ZD, for if it did, the angle A which is exterior to the triangle made up of the lines AZ, ZD, and AG would be greater than the right angle Z; but the two angles are equal—this is impossible. Therefore, the line AG will not meet the line ZD, but as it goes out will meet the line BD. Which was to be proved.

Proposition 3. *Let each of the angles A, B be acute; then the lines AG, BD, if produced, will meet on the side of G, D.*

We draw AE perpendicular to BD. Then, the angle E being right and the angle EAG acute, the lines AG, BD will meet if produced [figure 3.9.3].

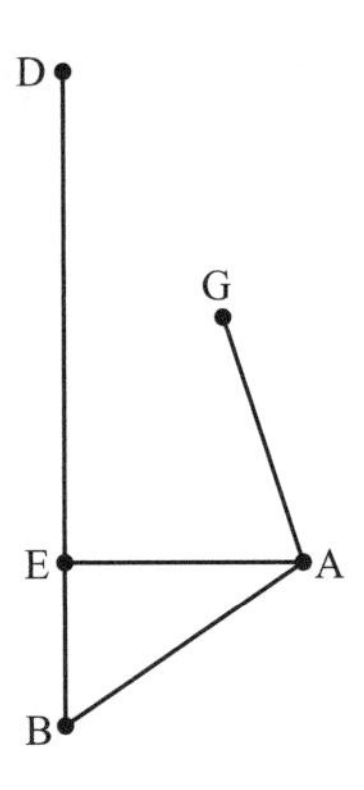

Figure 3.9.3.

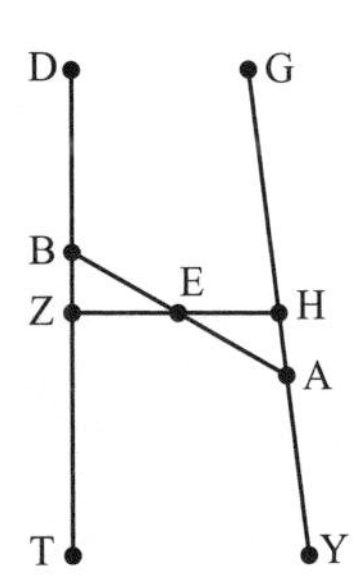

Figure 3.9.4.

Proposition 4. *The line AB falling on the lines AG, BD makes the angle ABD obtuse and the angle BAG acute, but these two angles are less than two right angles. Therefore, the two lines, when produced, will meet on the side of G, D.*

Let us divide AB into two halves at E and draw the perpendicular EZ and produce it to H [figure 3.9.4].

Then the angle ZHG is acute. For, if it were right, then the angles at E being opposite to one another, the acute angle at A would be equal to the acute angle at B, since AE, EB are equal. And since the angles at B are equal to two right angles, the angles GAB, ABD are equal to two right angles; but they are less;—this is impossible. And if the angle ZHG is obtuse, then the angle AHE is acute; but the angle HZT is a right angle; therefore, the lines GY and DT will meet when produced on the side of T, Y—which is impossible because the angles YAB, ABT are greater than two right angles. Therefore, the angle ZHG is neither obtuse nor right; then it must be acute. And the angle Z is right. Therefore, the lines AHG, BD, if produced, will meet on the side of G, D. Which was to be proved.

3.10 Euclid, *Elements*, 2–6

Book 2 deals with the relationships between various rectangles and squares. Given that most of the results can easily be translated into algebraic results by assigning

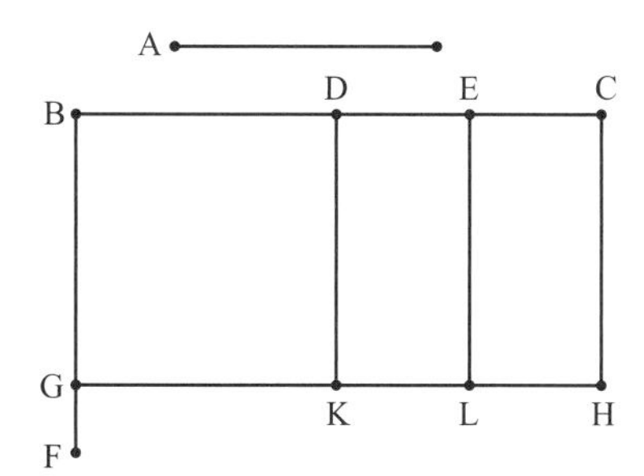

Figure 3.10.1.

lengths to each line segment, this book has often been interpreted as an introduction to "geometric algebra," particularly since propositions 5 and 6 were used by medieval Islamic mathematicians to justify their procedure for solving quadratic equations. However, most historians today believe that Euclid himself intended this book to display a relatively coherent body of geometric knowledge that could be used in the proof of further geometric theorems, even though the heritage of the book through the centuries has been algebraic. We present here selected theorems, some with proofs and others only with the statements.

Book 2

Definitions

1. Any rectangular parallelogram is said to be *contained* by the two straight lines containing the right angle.
2. And in any parallelogrammic area let any one whatever of the parallelograms about its diameter with the two complements be called a *gnomon*.

Propositions

Proposition 1. *If there be two straight lines, and one of them be cut into any number of segments whatever, the rectangle contained by the two straight lines is equal to the rectangles contained by the uncut straight line and each of its segments.*

Let A, BC be two straight lines, and let BC be cut at random at the points D, E; I say that the rectangle contained by A, BC is equal to the rectangle contained by A, BD, that contained by A, DE, and that contained by A, EC [figure 3.10.1].

For let BF be drawn from B at right angles to BC; let BG be made equal to A; through G, let GH be drawn parallel to BC, and through D, E, C let DK, EL, CH be drawn parallel to BG. Then BH is equal to BK, DL, EH. Now BH is the rectangle A, BC, for it is contained by GB, BC, and BG is equal to A; BK is the rectangle A, BD, for it is contained by GB, BD, and BG is equal to A; and DL is the rectangle A, DE, for DK, that is, BG, is equal to A. Similarly also EH is the rectangle A, EC. Therefore the rectangle A, BC is equal to the rectangle A, BD, the rectangle A, DE, and the rectangle A, EC. Therefore, etc. Q.E.D.

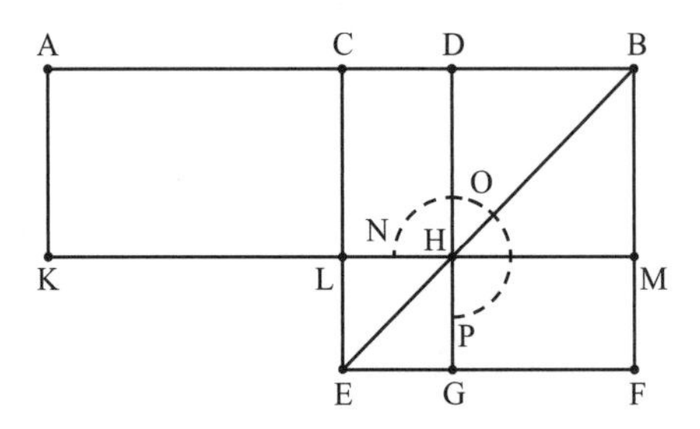

Figure 3.10.2.

This theorem can be interpreted algebraically as a version of the distributive law, $a(b+c+d+\cdots)=ab+ac+ad+\cdots$, just as we can interpret proposition 4 as simply the rule for squaring a binomial, $(a+b)^2=a^2+b^2+2ab$. One should also note that although the statement of proposition 1 mentions "any number" of segments, Euclid chose to use only three. After all, he had no way of representing "any number," so he felt his readers would understand that he used "three" just as a generalizable example, and the proof with any other number would be essentially the same.

Proposition 4. *If a straight line be cut at random, the square on the whole is equal to the squares on the segments and twice the rectangle contained by the segments.*

Proposition 5. *If a straight line be cut into equal and unequal segments, the rectangle contained by the unequal segments of the whole together with the square on the straight line between the points of section is equal to the square on the half.*

For let a straight line AB be cut into equal segments at C and into unequal segments at D; I say that the rectangle contained by AD, DB together with the square on CD is equal to the square on CB [figure 3.10.2].

For let the square $CEFB$ be described on CB, and let BE be joined; through D let DG be drawn parallel to either CE or BF, through H again let KM be drawn parallel to either AB or EF, and again through A let AK be drawn parallel to either CL or BM. Then, since the complement CH is equal to the complement HF, let DM be added to each; therefore the whole CM is equal to the whole DF. But CM is equal to AL, since AC is also equal to CB; therefore AL is also equal to DF. Let CH be added to each; therefore the whole AH is equal to the gnomon NOP. But AH is the rectangle AD, DB, for DH is equal to DB, therefore the gnomon NOP is also equal to the rectangle AD, DB. Let LG, which is equal to the square on CD, be added to each; therefore the gnomon NOP and LG are equal to the rectangle contained by AD, DB and the square on CD. But the gnomon NOP and LG are the whole square $CEFB$, which is described on CB; therefore the rectangle contained by AD, DB together with the square on CD is equal to the square on CB. Therefore, etc. Q.E.D.

Proposition 6. *If a straight line be bisected and a straight line be added to it in a straight line, the rectangle contained by the whole with the added straight line and the added straight line together with the square on the half is equal to the square on the straight line made up of the half and the added straight line* [figure 3.10.3].

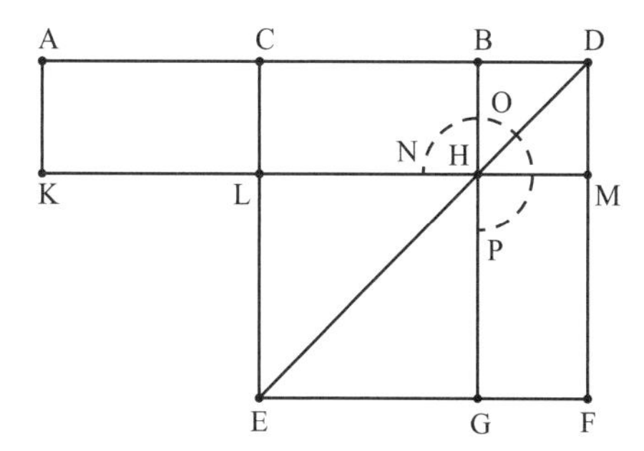

Figure 3.10.3.

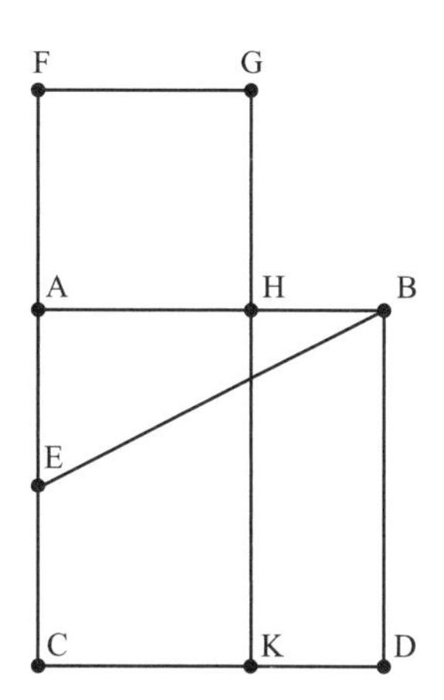

Figure 3.10.4.

Proposition 5 was used by medieval Islamic mathematicians to justify the procedure for solving the quadratic equation $bx - x^2 = c$, while proposition 6 was used to justify the procedure for solving $bx + x^2 = c$. In fact, if we translate the solution of proposition 11 into algebra, its goal is to solve the equation $a(a-x) = x^2$ or $x^2 + ax = a^2$; its procedure then seems to mirror exactly the quadratic formula as derived from proposition 6.

The following proposition shows how to cut a line segment in mean and extreme ratio, that is, to cut a segment AB at H so that $AB : AH = AH : HB$.

Proposition 11. *To cut a given straight line so that the rectangle contained by the whole and one of the segments is equal to the square on the remaining segment.*

Let AB be the given straight line; thus it is required to cut AB so that the rectangle contained by the whole and one of the segments is equal to the square on the remaining segment [figure 3.10.4].

For let the square $ABDC$ be described on AB; let AC be bisected at the point E, and let BE be joined; let CA be drawn through to F, and let EF be made equal to BE; let the square FH be described on AF, and let GH be drawn through to K. I say that AB has been

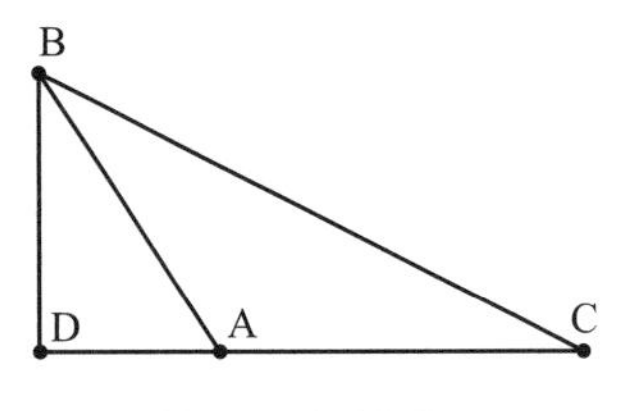

Figure 3.10.5.

cut at H so as to make the rectangle contained by AB, BH equal to the square on AH. For, since the straight line AC has been bisected at E, and FA is added to it, the rectangle contained by CF, FA together with the square on AE is equal to the square on EF [2:6]. But EF is equal to EB; therefore the rectangle CF, FA together with the square on AE is equal to the square on EB. But the squares on BA, AE are equal to the square on EB, for the angle at A is right; therefore the rectangle CF, FA together with the square on AE is equal to the squares on BA, AE. Let the square on AE be subtracted from each; therefore the rectangle CF, FA which remains is equal to the square on AB. Now the rectangle CF, FA is FK, for AF is equal to FG; and the square on AB is AD; therefore FK is equal to AD. Let AK be subtracted from each; therefore FH which remains is equal to HD. And HD is the rectangle AB, BH, for AB is equal to BD; and FH is the square on AH; therefore the rectangle contained by AB, BH is equal to the square on HA. Therefore, etc. Q.E.F.

Proposition 12. *In obtuse-angled triangles the square on the side subtending the obtuse angle is greater than the squares on the sides containing the obtuse angle by twice the rectangle contained by one of the sides about the obtuse angle, namely, that on which the perpendicular falls, and the straight line cut off outside by the perpendicular towards the obtuse angle.*

Let ABC be an obtuse-angled triangle having the angle BAC obtuse, and let BD be drawn from the point B perpendicular to CA produced; I say that the square on BC is greater than the squares on BA, AC by twice the rectangle contained by CA, AD [figure 3.10.5].

For, since the straight line CD has been cut at random at the point A, the square on DC is equal to the squares on CA, AD and twice the rectangle contained by CA, AD [2:4]. Let the square on DB be added to each; therefore the squares on CD, DB are equal to the squares on CA, AD, DB and twice the rectangle CA, AD. But the square on CB is equal to the squares on CD, DB, for the angle at D is right; and the square on AB is equal to the squares on AD, DB; therefore the square on CB is equal to the squares on CA, AB and twice the rectangle contained by CA, AD; so that the square on CB is greater than the squares on CA, AB by twice the rectangle contained by CA, AD. Therefore, etc. Q.E.D.

Proposition 13. *In acute-angled triangles the square on the side subtending the acute angle is less than the squares on the sides containing the acute angle by twice the rectangle contained by one of the sides about the acute angle, namely, that on which the perpendicular falls, and the straight line cut off within by the perpendicular towards the acute angle* [figure 3.10.6].

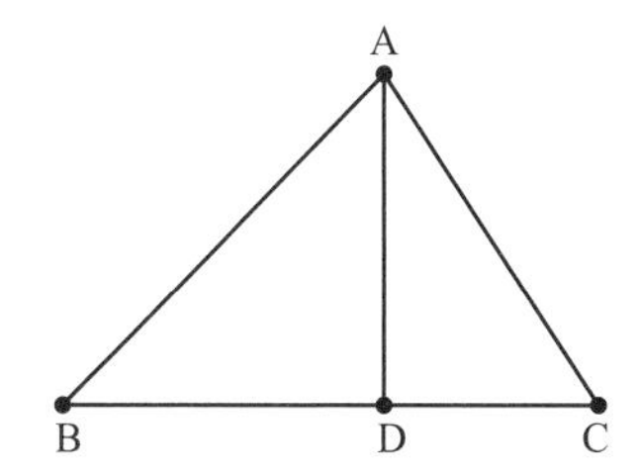

Figure 3.10.6.

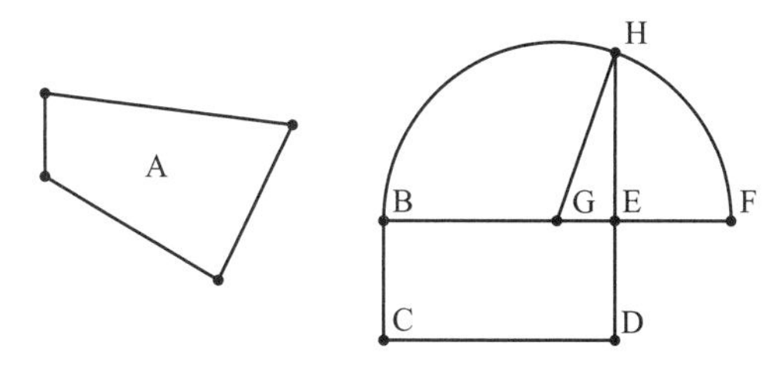

Figure 3.10.7.

In the diagram, the conclusion of this theorem is that the square on AC is less than the squares on CB, BA by twice the rectangle contained by CB, BD. In modern terms, propositions 12 and 13 together are equivalent to the trigonometric law of cosines.

Proposition 14. *To construct a square equal to a given rectilineal figure.*

Let A be the given rectilineal figure; thus it is required to construct a square equal to the rectilineal figure A [figure 3.10.7].

For let there be constructed the rectangular parallelogram BD equal to the rectilineal figure A [1:45]. Then, if BE is equal to ED, that which was enjoined will have been done; for a square BD has been constructed equal to the rectilinear figure A. But, if not, one of the straight lines BE, ED is greater. Let BE be greater, and let it be produced to F; let EF be made equal to ED, and let BF be bisected at G. With center G and distance one of the straight lines GB, GF let the semicircle BHF be described; let DE be produced to H, and let GH be joined. Then, since the straight line BF had been cut into equal segments at G, and into unequal segments at E, the rectangle contained by BE, EF together with the square on EG is equal to the square on GF [2:5]. But GF is equal to GH; therefore the rectangle BE, EF together with the square on GE is equal to the square on GH. But the squares on HE, EG are equal to the square on GH; therefore the rectangle BE, EF together with the square on GE is equal to the squares on HE, EG. Let the square on GE be subtracted from each; therefore the rectangle contained by BE, EF which remains is equal to the square on EH. But the rectangle BE, EF is BD, for EF is equal to ED; therefore the parallelogram BD is equal to the square on HE. And BD is equal to the rectilineal figure A. Therefore the rectilineal figure A is also equal to the square which can be described on EH. Therefore, etc. Q.E.F.

This proposition seems to show a geometric use of proposition 5, even though it can also be interpreted algebraically as simply solving the equation $x^2 = cd$.

After having discussed rectilinear figures in books 1 and 2, Euclid turned in book 3 to the properties of the most fundamental curved figure, the circle. The propositions here deal with various properties of the circle, but if there is any organizing principle, it is to provide for the construction, in book 4, of polygons both inscribed in and circumscribed about circles. In particular, many of the final propositions of book 3 are used in the most difficult construction in book 4, the construction of the regular pentagon. We begin with selected definitions from book 3 and continue with selected propositions, in some of which we provide the proofs.

Book 3

Definitions

2. A straight line is said to *touch a circle* which, meeting the circle and being produced, does not cut the circle.
6. A *segment of a circle* is the figure contained by a straight line and a circumference of a circle.
7. An *angle of a segment* is that contained by a straight line and a circumference of a circle.
8. An *angle in a segment* is the angle which, when a point is taken on the circumference of the segment and straight lines are joined from it to the extremities of the straight line which is the *base of the segment*, is contained by the straight lines so joined.
9. And, when the straight lines containing the angle cut off a circumference, the angle is said to *stand upon* that circumference.
10. A *sector of a circle* is the figure which, when an angle is constructed at the center of the circle, is contained by the straight lines containing the angle and the circumference cut off by them.

Propositions

Proposition 3. *If in a circle a straight line through the center bisects a straight line not through the center, it also cuts it at right angles; and if it cuts it at right angles, it also bisects it.*

Let ABC be a circle, and in it let a straight line CD through the center bisect a straight line AB not through the center at the point F; I say that it also cuts it at right angles [figure 3.10.8].

For let the center of the circle ABC be taken, and let it be E; let EA, EB be joined. Then, since AF is equal to FB, and FE is common, two sides are equal to two sides; and the base EA is equal to the base EB; therefore the angle AFE is equal to the angle BFE. But, when a straight line set up on a straight line makes the adjacent angles equal to one another, each of the equal angles is right; therefore each of the angles AFE, BFE is right. Therefore CD, which is through the center, and bisects AB which is not through the center,

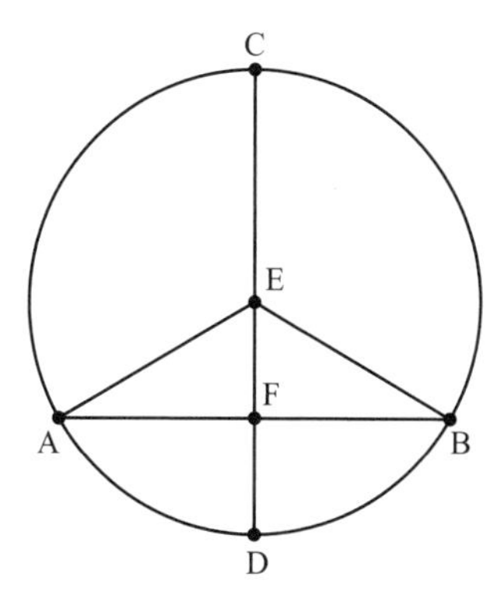

Figure 3.10.8.

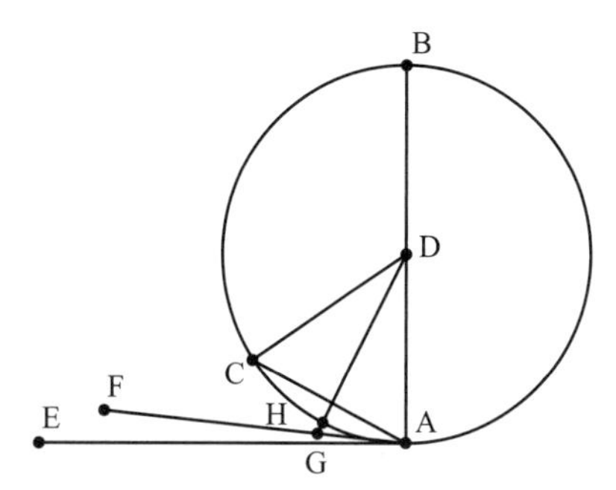

Figure 3.10.9.

also cuts it at right angles. Again, let CD cut AB at right angles; I say that it also bisects it, that is, that AF is equal to FB. For, with the same construction, since EA is equal to EB, the angle EAF is also equal to the angle EBF. But the right angle AFE is equal to the right angle BFE; therefore EAF, EBF are two triangles having two angles equal to two angles and one side equal to one side, namely, EF, which is common to them, and subtends one of the equal angles; therefore they will also have the remaining sides equal to the remaining sides; therefore AF is equal to FB. Therefore etc. Q.E.D.

Proposition 16. *The straight line drawn at right angles to the diameter of a circle from its extremity will fall outside the circle, and into the space between the straight line and the circumference another straight line cannot be interposed; further the angle of the semicircle is greater, and the remaining angle less, than any acute rectilineal angle.*

Let ABC be a circle about D as center and AB as diameter; I say that the straight line drawn from A at right angles to AB from its extremity will fall outside the circle [figure 3.10.9].

For suppose it does not, but, if possible, let it fall within as CA, and let DC be joined. Since DA is equal to DC, the angle DAC is also equal to the angle ACD. But the angle DAC is right; therefore the angle ACD is also right; thus, in the triangle ACD, the two angles DAC, ACD are equal to two right angles, which is impossible. Therefore the straight line drawn from the point A at right angles to BA will not fall within the circle. Similarly we can prove that neither will it fall on the circumference; therefore it will fall outside. Let it fall as AE; I say next that into the space between the straight line AE and the circumference CHA another straight line cannot be interposed. For, if possible, let another straight line be so interposed, as FA, and let DG be drawn from the point D perpendicular to FA. Then, since the angle AGD is right, and the angle DAG is less than a right angle, AD is greater than DG. But DA is equal to DH; therefore DH is greater than DG, the less than the greater; which is impossible. Therefore another straight line cannot be interposed into the space between the straight line and the circumference.

I say further that the angle of the semicircle contained by the straight line BA and the circumference CHA is greater than any acute rectilinear angle, and the remaining angle contained by the circumference CHA and the straight line AE is less than any acute rectilineal angle. For, if there is any rectilineal angle greater than the angle contained by the straight line BA and the circumference CHA, and any rectilineal angle less than the angle contained by the circumference CHA and the straight line AE, then into the space between the circumference and the straight line AE a straight line will be interposed such as will make an angle contained by straight lines which is greater than the angle contained by the straight line BA and the circumference CHA, and another angle contained by straight lines which is less than the angle contained by the circumference CHA and the straight line AE. But such a straight line cannot be interposed; therefore there will not be any acute angle contained by straight lines which is greater than the angle contained by the straight line BA and the circumference CHA, nor yet any acute angle contained by straight lines which is less than the angle contained by the circumference CHA and the straight line AE. Q.E.D.

Porism:[46] *From this it is manifest that the straight line drawn at right angles to the diameter of a circle from its extremity touches the circle.*

Proposition 20. *In a circle the angle at the center is double of the angle at the circumference, when the angles have the same circumference as base.*

Let ABC be a circle, let the angle BEC be an angle at the center, and the angle BAC an angle at the circumference, and let them have the same circumference BC as base; I say that the angle BEC is double of the angle BAC [figure 3.10.10].

For let AE be joined and drawn to F. Then, since EA is equal to EB, the angle EAB is also equal to the angle EBA; therefore the angles EAB, EBA are double of the angle EAB. But the angle BEF is equal to the angles EAB, EBA; therefore the angle BEF is also double of the angle EAB. For the same reason the angle FEC is also double of the angle EAC. Therefore the whole angle BEC is double of the whole angle BAC. Again, let another straight line be inflected, and let there be another angle BDC; let DE be joined and produced to G. Similarly then we can prove that the angle GEC is double of the angle EDC, of which the angle GEC is double of the angle EDB; therefore the angle BEC which remains is double of the angle BDC. Therefore, etc. Q.E.D.

Proposition 21. *In a circle the angles in the same segment are equal to one another.*

[46] A porism is a simple consequence of a particular theorem.

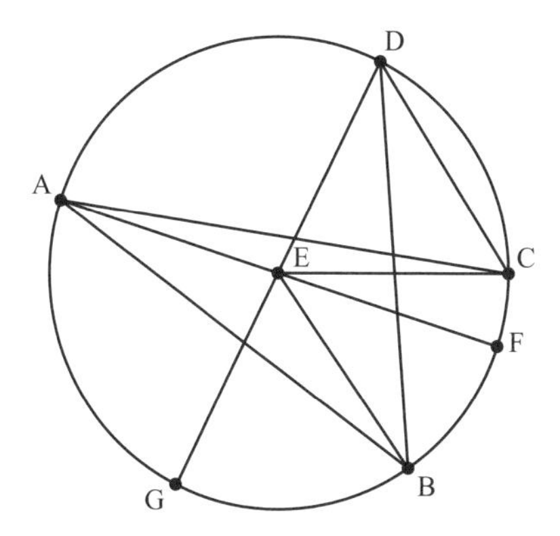

Figure 3.10.10.

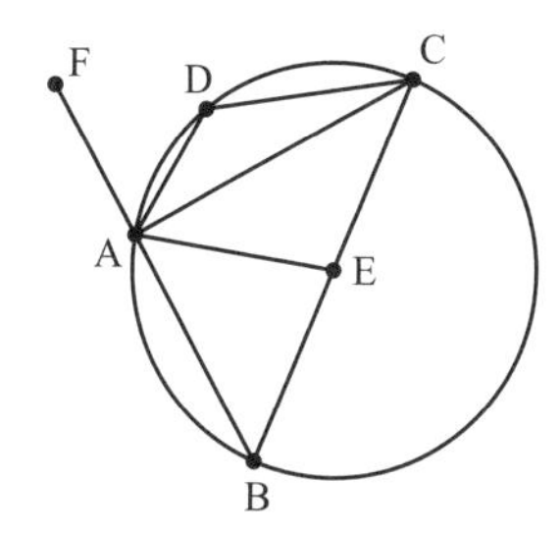

Figure 3.10.11.

Proposition 22. *The opposite angles of quadrilaterals in circles are equal to two right angles.*

Proposition 31. *In a circle the angle in the semicircle is right, that in a greater segment less than a right angle, and that in a less segment greater than a right angle; and further the angle of the greater segment is greater than a right angle, and the angle of the less segment less than a right angle.*

Let $ABCD$ be a circle, let BC be its diameter, and E its center, and let BA, AC, AD, DC be joined; I say that the angle BAC in the semicircle BAC is right, the angle ABC in the segment ABC greater than the semicircle is less than a right angle, and the angle ADC in the segment ADC less than the semicircle is greater than a right angle [figure 3.10.11].

Let AE be joined, and let BA be carried through to F. Then, since BE is equal to EA, the angle ABE is also equal to the angle BAE. Again, since CE is equal to EA, the angle ACE is also equal to the angle CAE. Therefore the whole angle BAC is equal to the two angles ABC, ACB. But the angle FAC exterior to the triangle ABC is also equal to the two angles ABC, ACB; therefore the angle BAC is also equal to the angle FAC; therefore each is right; therefore the angle BAC in the semicircle BAC is right.

Next, since in the triangle ABC the two angles ABC, BAC are less than two right angles, and the angle BAC is a right angle, the angle ABC is less than a right angle; and it is the angle in the segment ABC greater than the semicircle. Next, since $ABCD$

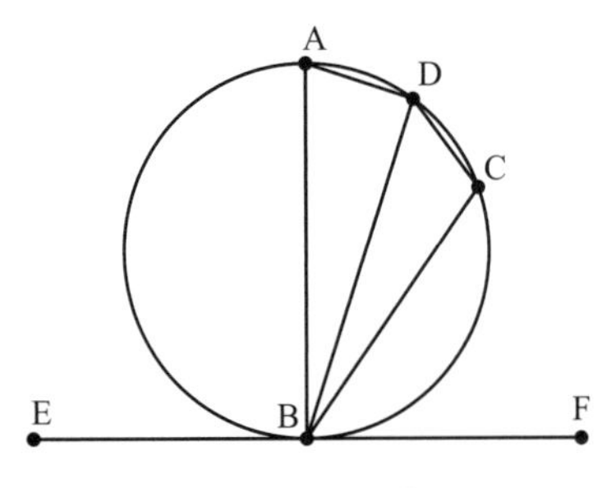

Figure 3.10.12.

is a quadrilateral in a circle, and the opposite angles of quadrilaterals in circles are equal to two right angles, while the angle *ABC* is less than a right angle, therefore the angle *ADC* which remains is greater than a right angle; and it is the angle in the segment *ADC* less than the semicircle. I say further that the angle of the greater segment, namely that contained by the circumference *ABC* and the straight line *AC*, is greater than a right angle; and the angle of the less segment, namely, that contained by the circumference *ADC* and the straight line *AC*, is less than a right angle. This is at once manifest. For, since the angle contained by the straight lines *BA*, *AC* is right, the angle contained by the circumference *ABC* and the straight line *AC* is greater than a right angle. Again, since the angle contained by the straight lines *AC*, *AF* is right, the angle contained by the straight line *CA* and the circumference *ADC* is less than a right angle. Therefore, etc. Q.E.D.

Proposition 32. *If a straight line touch a circle, and from the point of contact there be drawn across, in the circle, a straight line cutting the circle, the angles which it makes with the tangent will be equal to the angles in the alternate segments of the circle.*

Using Figure 3.10.12, this proposition proves that angle *FBD* is equal to the angle *BAD* constructed in the segment *BAD*, and the angle *EBD* is equal to the angle *DCB* constructed in the segment *DCB*.

Proposition 33. *On a given straight line to describe a segment of a circle admitting an angle equal to a given rectilineal angle.*

Let *AB* be the given straight line, and the angle at *C* the given rectilineal angle; thus it is required to describe on the given straight line *AB* a segment of a circle admitting an angle equal to the angle at *C* [figure 3.10.13]. The angle at *C* is then acute, or right, or obtuse.

First, let it be acute, and, as in the figure, on the straight line *AB* and at the point *A* let the angle *BAD* be constructed equal to the angle at *C*; therefore the angle *BAD* is also acute. Let *AE* be drawn at right angles to *DA*, let *AB* be bisected at *F*, let *FG* be drawn from the point *F* at right angles to *AB*, and let *GB* be joined. Then, since *AF* is equal to *FB* and *FG* is common, the two sides *AF*, *FG* are equal to the two sides *BF*, *FG* and the angle *AFG* is equal to the angle *BFG*; therefore the base *AG* is equal to the base *BG*. Therefore the circle described with center *G* and distance *GA* will pass through *B* also. Let it be drawn, and let it be *ABE*; let *EB* be joined.

Now, since *AD* is drawn from *A*, the extremity of the diameter *AE*, at right angles to *ADE*, therefore *AD* touches the circle *ABE* [3:16, porism]. Since then a straight line *AD*

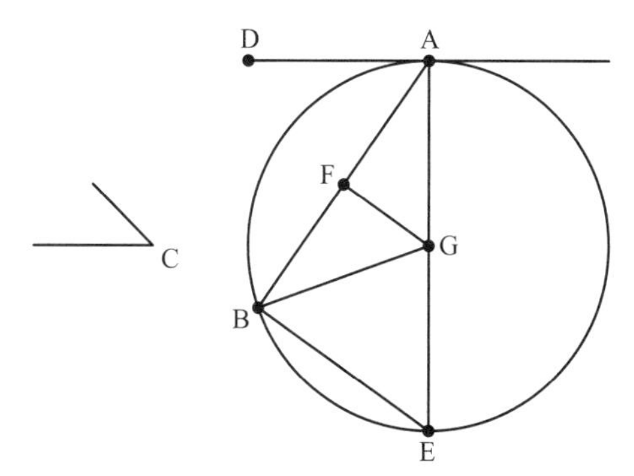

Figure 3.10.13.

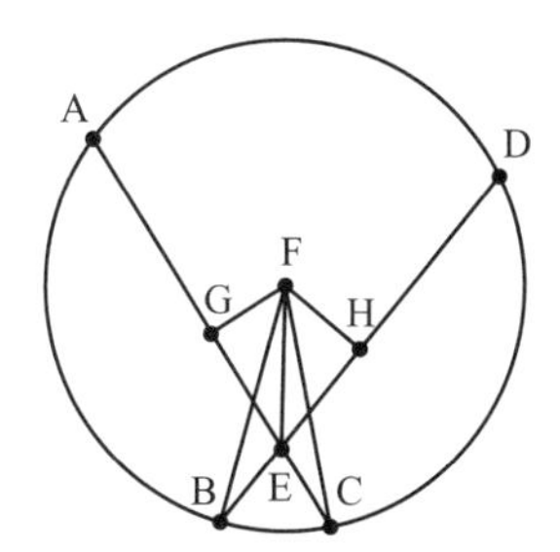

Figure 3.10.14.

touches the circle ABE, and from the point of contact at A a straight line AB is drawn across in the circle ABE, the angle DAB is equal to the angle AEB in the alternate segment of the circle [3:32]. But the angle DAB is equal to the angle at C; therefore the angle at C is also equal to the angle AEB. Therefore, on the given straight line AB the segment AEB of a circle has been described admitting the angle AEB equal to the given angle, the angle at C. Q.E.F.

We omit the proofs of the cases where the given angle is right or obtuse.

Proposition 35. *If in a circle two straight lines cut one another, the rectangle contained by the segments of the one is equal to the rectangle contained by the segments of the other.*

For in the circle $ABCD$ let the two straight lines AC, BD cut one another at the point E; I say that the rectangle contained by AE, EC is equal to the rectangle contained by DE, EB [figure 3.10.14].

If now AC, BD are through the center, so that E is the center of the circle $ABCD$, it is manifest that, AE, EC, DE, EB being equal, the rectangle contained by AE, EC is also equal to the rectangle contained by DE, EB. Next, let AC, DB not be through the center; let the center of $ABCD$ be taken, and let it be F; from F let FG, FH be drawn perpendicular to the straight lines AC, DB, and let FB, FC, FE be joined. Then since a straight line GF through the center cuts a straight line AC not through the center at right angles, it also bisects it; therefore AG is equal to GC. Since, then, the straight line AC has been

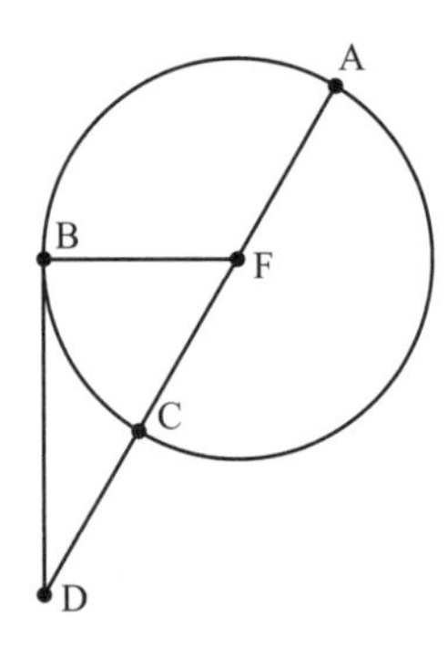

Figure 3.10.15.

cut into equal parts at G and into unequal parts at E, the rectangle contained by AE, EC together with the square on EG is equal to the square on GC [2:5]. Let the square on GF be added; therefore the rectangle AE, EC together with the squares on GE, GF is equal to the squares on CG, GF. But the square on FE is equal to the squares on EG, GF, and the square on FC is equal to the squares on CG, GF; therefore the rectangle AE, EC together with the square on FE is equal to the square on FC. And FC is equal to FB; therefore the rectangle AE, EC together with the square on EF is equal to the square on FB. For the same reason, also, the rectangle DE, EB together with the square on FE is equal to the square on FB. But the rectangle AE, EC together with the square on FE was also proved equal to the square on FB; therefore the rectangle AE, EC together with the square on FE is equal to the rectangle DE, EB together with the square on FE. Let the square on FE be subtracted from each; therefore the rectangle contained by AE, EC which remains is equal to the rectangle contained by DE, EB. Therefore, etc. Q.E.D.

Proposition 36. *If a point be taken outside a circle and from it there fall on the circle two straight lines, and if one of them cut the circle and the other touch it, the rectangle contained by the whole of the straight line which cuts the circle and the straight line intercepted on it outside between the point and the convex circumference will be equal to the square on the tangent.*

Using Figure 3.10.15, this proposition asserts, in the case that DCA goes through the center of the circle, that the rectangle contained by AD, DC is equal to the square on DB. Given that the result is reminiscent of proposition 2:6, it is not surprising that that proposition is central to the proof. The proof where DCA does not go through the center is slightly more complicated. The next proposition, proposition 37, is the converse of this one. It asserts that if two straight lines are drawn to a circle from a point outside, one a secant and one meeting the circle, and if the relationship between the rectangle and square of the previous proposition holds, then the second line is a tangent. The proof proceeds by actually drawing a tangent and then showing, using proposition 36, that the given line equals the tangent.

The propositions of book 4 demonstrate how to inscribe polygons in circles and circles in polygons and how to circumscribe polygons about circles and circles about

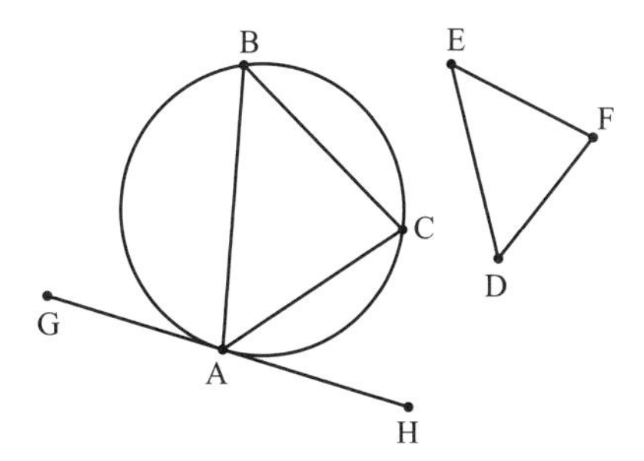

Figure 3.10.16.

polygons. The most difficult of these constructions is proposition 11, the inscription of a regular pentagon in a circle. The important idea for the construction is the prior construction of an isosceles triangle in which the two base angles are double the third angle. We include here propositions dealing with inscribing triangles, squares, pentagons, and hexagons in a given circle.

Book 4

Definitions

1. A rectilineal figure is said to be *inscribed in a rectilineal figure* when the respective angles of the inscribed figure lie on the respective sides of that in which it is inscribed.
2. Similarly a figure is said to be *circumscribed about a figure* when the respective sides of the circumscribed figure pass through the respective angles of that about which it is circumscribed.
3. A rectilineal figure is said to be *inscribed in a circle* when each angle of the inscribed figure lies on the circumference of the circle.
4. A rectilineal figure is said to be *circumscribed about a circle*, when each side of the circumscribed figure touches the circumference of the circle.
5. Similarly a circle is said to be *inscribed in a figure* when the circumference of the circle touches each side of the figure in which it is inscribed.
6. A circle is said to be *circumscribed about a figure* when the circumference of the circle passes through each angle of the figure about which it is circumscribed.
7. A straight line is said to be *fitted into a circle* when its extremities are on the circumference of the circle.

Propositions

Proposition 2. *In a given circle, to inscribe a triangle equiangular with a given triangle.*

Let ABC be the given circle and DEF the given triangle; thus it is required to inscribe in the circle ABC a triangle equiangular with the triangle DEF [figure 3.10.16].

Let GH be drawn touching the circle ABC at A. On the straight line AH, and at the point A on it, let the angle HAC be constructed equal to the angle DEF, and on the straight line AG, and at the point A on it, let the angle GAB be constructed equal to the angle DFE; let

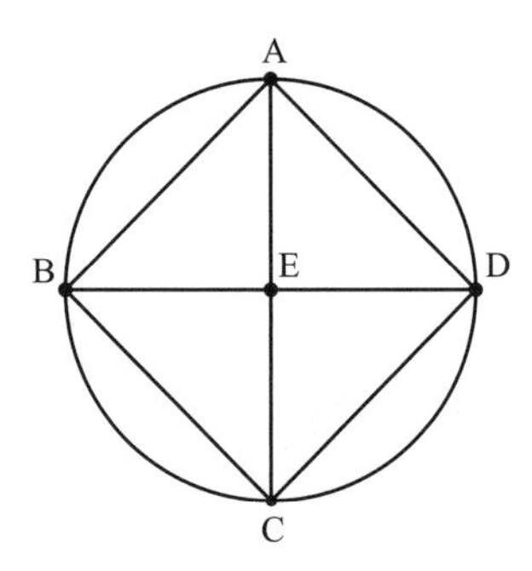

Figure 3.10.17.

BC be joined. Then, since a straight line AH touches the circle ABC, and from the point of contact at A the straight line AC is drawn across in the circle, therefore the angle HAC is equal to the angle ABC in the alternate segment of the circle [3:32]. But the angle HAC is equal to the angle DEF; therefore the angle ABC is also equal to the angle DEF. For the same reason, the angle ACB is also equal to the angle DFE; therefore the remaining angle BAC is also equal to the remaining angle EDF. Therefore, etc. Q.E.F.

Proposition 6. *In a given circle, to inscribe a square.*

Let $ABCD$ be the given circle; thus it is required to inscribe a square in the circle $ABCD$ [figure 3.10.17].

Let two diameters AC, BD of the circle $ABCD$ be drawn at right angles to one another, and let AB, BC, CD, DA be joined. Then, since BE is equal to ED, for E is the center, and EA is common and at right angles, therefore the base AB is equal to the base AD. For the same reason each of the straight lines BC, CD is also equal to each of the straight lines AB, AD; therefore the quadrilateral $ABCD$ is equilateral. I say next that it is also right-angled. For, since the straight line BD is a diameter of the circle $ABCD$, therefore BAD is a semicircle; therefore the angle BAD is right [3:31]. For the same reason, each of the angles ABC, BCD, CDA is also right; therefore the quadrilateral $ABCD$ is right-angled. But it was also proved equilateral; therefore it is a square; and it has been inscribed in the circle $ABCD$. Therefore, etc. Q.E.F.

Proposition 10. *To construct an isosceles triangle having each of the angles at the base double of the remaining one.*

Let any straight line AB be set out, and let it be cut at the point C so that the rectangle contained by AB, BC is equal to the square on CA [2:11]. With center A and distance AB let the circle BDE be described, and let there be fitted in the circle BDE the straight line BD equal to the straight line AC which is not greater than the diameter of the circle BDE [figure 3.10.18].

Let AD, DC be joined, and let the circle ACD be circumscribed about the triangle ACD. Then, since the rectangle AB, BC is equal to the square on AC, and AC is equal to BD, therefore the rectangle AB, BC is equal to the square on BD. And, since a point B has been taken outside the circle ACD, and from B the two straight lines BA, BD have fallen

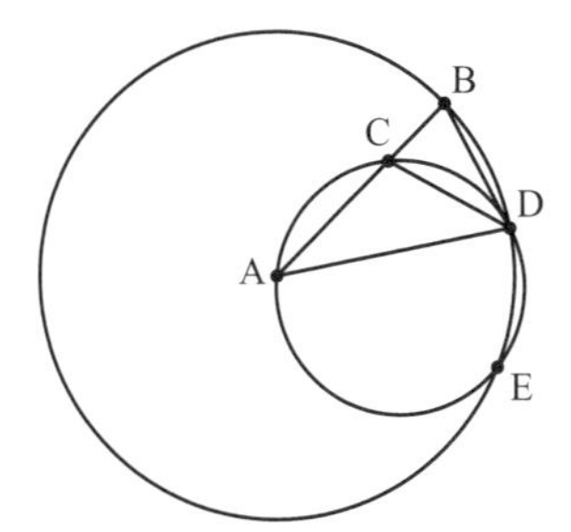

Figure 3.10.18.

on the circle ACD, and one of them cuts it, while the other falls on it, and the rectangle AB, BC is equal to the square on BD, therefore BD touches the circle ACD [3:37].

Since, then, BD touches it, and DC is drawn across from the point of contact at D, therefore the angle BDC is equal to the angle DAC in the alternate segment of the circle [3:32]. Since, then, the angle BDC is equal to the angle DAC, let the angle CDA be added to each; therefore the whole angle BDA is equal to the two angles CDA, DAC. But the exterior angle BCD is equal to the angles CDA, DAC; therefore the angle BDA is also equal to the angle BCD. But the angle BDA is equal to the angle CBD, since the side AD is also equal to AB; so that the angle DBA is also equal to the angle BCD. Therefore the three angles BDA, DBA, BCD are equal to one another. And, since the angle DBC is equal to the angle BCD, the side BD is also equal to the side DC. But BD is by hypothesis equal to CA; therefore CA is also equal to CD, so that the angle CDA is also equal to the angle DAC. Therefore the angles CDA, DAC are double of the angle DAC. But the angle BCD is equal to the angles CDA, DAC; therefore the angle BCD is also double of the angle CAD. But the angle BCD is equal to each of the angles BDA, DBA; therefore each of the angles BDA, DBA is also double of the angle DAB. Therefore, etc. Q.E.F.

To see why Euclid required the construction of an isosceles triangle with each of the base angles double the vertex angle, we perform an analysis of the construction of a regular pentagon in a circle. Namely, if the pentagon is constructed, and one connects one vertex with the two others to which it is not adjacent, the isosceles triangle so formed has base angles double the vertex angle. Thus, the construction of the pentagon is reduced to this latter construction. Furthermore, the intersection of the bisector of one of the base angles with the opposite side divides that side in extreme and mean ratio. So the triangle construction is then reduced to finding the point that divides a line in extreme and mean ratio. But that construction was accomplished in 2:11. Thus, that is where Euclid's proof of proposition 10 began.

Proposition 11. *In a given circle to inscribe an equilateral and equiangular pentagon.*

Let $ABCDE$ be the given circle; thus it is required to inscribe in the circle $ABCDE$ an equilateral and equiangular pentagon [figure 3.10.19].

Let the isosceles triangle FGH be set out having each of the angles at G, H double of the angle at F [4:10]. Let there be inscribed in the circle $ABCDE$ the triangle ACD equiangular with the triangle FGH, so that the angle CAD is equal to the angle at F and

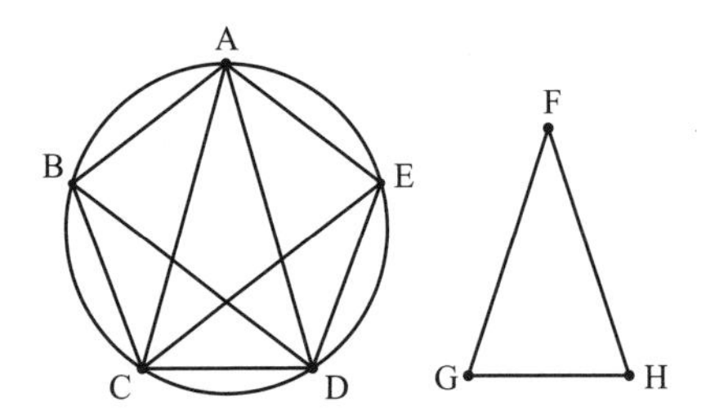

Figure 3.10.19.

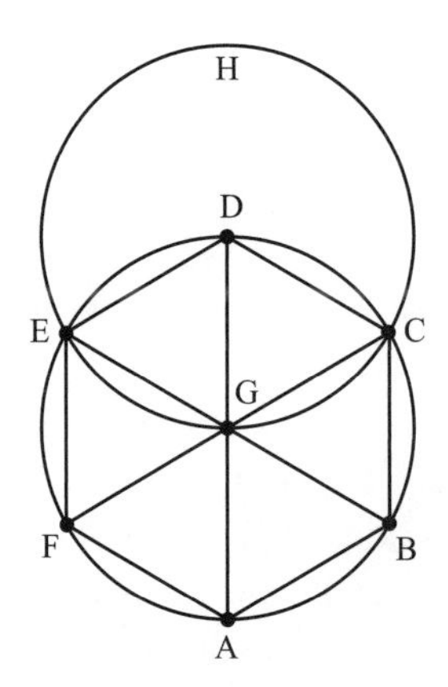

Figure 3.10.20.

the angles at G, H respectively equal to the angles ACD, CDA [4:2]; therefore each of the angles ACD, CDA is also double of the angle CAD. Now, let the angles ACD, CDA be bisected respectively by the straight lines CE, DB, and let AB, BC, DE, EA be joined. Then, since each of the angles ACD, CDA is double of the angle CAD, and they have been bisected by the straight lines CE, DB, therefore the five angles DAC, ACE, ECD, CDB, BDA are equal to one another. But equal angles stand on equal circumferences; therefore the five circumferences AB, BC, CD, DE, EA are equal to one another. But equal circumferences are subtended by equal straight lines; therefore the five straight lines AB, BC, CD, DE, EA are equal to one another; therefore the pentagon $ABCDE$ is equilateral. I say next that it is also equiangular. For, since the circumference AB is equal to the circumference DE, let BCD be added to each; therefore the whole circumference $ABCD$ is equal to the whole circumference $EDCB$. And the angle AED stands on the circumference $ABCD$, and the angle BAE on the circumference $EDCB$; therefore the angle BAD is also equal to the angle AED. For the same reason, each of the angles ABC, BCD, CDE is also equal to each of the angles BAE, AED; therefore the pentagon $ABCDE$ is equiangular. But it was also proved equilateral. Therefore, etc. Q.E.F.

Proposition 15. *In a given circle to inscribe an equilateral and equiangular hexagon.*

The basic idea of the proof of proposition 15 is that the six sides of the hexagon to be constructed are all equal to the radius of the given circle. Figure 3.10.20 illustrates the proof.

It is in book 5 that Euclid discusses the basic idea of proportionality for the case of arbitrary magnitudes. For numbers, the analogous discussion is found in book 7. It is believed that the central ideas of book 5 are due to Eudoxus, who modified earlier definitions of "same ratio," definitions which needed to be changed after Greek mathematicians understood that not all magnitudes were commensurable. We include here some of the basic definitions dealing with proportions as well as a few of the theorems.

Book 5

Definitions

1. A magnitude is a *part* of a magnitude, the less of the greater, when it measures the greater.
2. The greater is a *multiple* of the less when it is measured by the less.
3. A *ratio* is a sort of relation in respect of size between two magnitudes of the same kind.
4. Magnitudes are said to *have a ratio* to one another which are capable, when multiplied, of exceeding one another.
5. Magnitudes are said to *be in the same ratio*, the first to the second and the third to the fourth, when, if any equimultiples whatever be taken of the first and third, and any equimultiples whatever of the second and fourth, the former equimultiples alike exceed, are alike equal to, or alike fall short of the latter equimultiples respectively taken in corresponding order.
6. Let magnitudes which have the same ratio be called *proportional*.
7. When, of the equimultiples, the multiple of the first magnitude exceeds the multiple of the second, but the multiple of the third does not exceed the multiple of the fourth, then the first is said to *have a greater ratio* to the second than the third has to the fourth.
9. When three magnitudes are proportional, the first is said to have to the third the *duplicate ratio* of that which it has to the second.
10. When four magnitudes are proportional, the first is said to have to the fourth the *triplicate ratio* of that which it has to the second, and so on continually, whatever be the proportion.
12. *Alternate ratio* [*alternando*] means taking the antecedent in relation to the antecedent and the consequent in relation to the consequent.
13. *Inverse ratio* [*invertendo*] means taking the consequent as antecedent in relation to the antecedent as consequent.
14. *Composition of a ratio* [*componendo*] means taking the antecedent together with the consequent as one in relation to the consequent by itself.
15. *Separation of a ratio* [*separando*] means taking the excess by which the antecedent exceeds the consequent in relation to the consequent by itself.
16. *Conversion of a ratio* [*convertendo*] means taking the antecedent in relation to the excess by which the antecedent exceeds the consequent.
17. A ratio *ex aequali* arises when, there being several magnitudes and another set equal to them in multitude which taken two and two are in the same proportion, as

the first is to the last among the first magnitudes, so is the first to the last among the second magnitudes; or, in other words, it means taking the extreme terms by virtue of the removal of the intermediate terms.

The various ways ratios can be transformed are referred to constantly in Greek geometric works. When translating these actions, we will often use the Latin words noted above in the final six definitions, as they became standard in geometric arguments.

Propositions

Proposition 4. *If a first magnitude has to a second the same ratio as a third to a fourth, any equimultiples whatever of the first and third will also have the same ratio to any equimultiples whatever of the second and fourth respectively, taken in corresponding order.*

For let a first magnitude A have to a second B the same ratio as a third C to a fourth D; and let equimultiples E, F be taken of A, C, and G, H other, chance, equimultiples of B, D; I say that, as E is to G, so is F to H. For let equimultiples K, L be taken of E, F, and other, chance, equimultiples M, N, of G, H. Since E is the same multiple of A that F is of C, and equimultiples K, L of E, F have been taken, therefore K is the same multiple of A that L is of C. For the same reason M is the same multiple of B that N is of D. And, since, as A is to B, so is C to D, and of A, C, equimultiples K, L have been taken, and of B, D other, chance, equimultiples M, N, therefore, if K is in excess of M, L also is in excess of N, if it is equal, equal, and if less, less. And K, L are equimultiples of E, F, and M, N other, chance, equimultiples of G, H; therefore, as E is to G, so is F to H. Therefore, etc. Q.E.D.

Proposition 11. *Ratios which are the same with the same ratio are also the same with one another.*

Proposition 16. *If four magnitudes are proportional, they will also be proportional alternately.*

Let A, B, C, D be four proportional magnitudes, so that, as A is to B, so is C to D; I say that they will also be so alternately, that is, as A is to C, so is B to D. For of A, B let equimultiples E, F be taken, and of C, D other, chance equimultiples G, H. Then, since E is the same multiple of A that F is of B, and parts have the same ratio as the same multiples of them, therefore, as A is to B, so is E to F. But as A is to B, so is C to D; therefore also, as C is to D, so is E to F. Again, since G, H are equimultiples of C, D, therefore, as C is to D, so is G to H. But, as C is to D, so is E to F; therefore also, as E is to F, so is G to H. But, if four magnitudes be proportional, and the first be greater than the third, the second will also be greater than the fourth; if equal, equal; and if less, less. Therefore, if E is in excess of G, F is also in excess of H, if equal, equal, and if less, less. Now E, F are equimultiples of A, B, and G, H other, chance equimultiples of C, D; therefore, as A is to C, so is B to D. Therefore, etc. Q.E.D.

Proposition 22: *If there are any number of magnitudes whatever, and others equal to them in multitude, which taken two and two together are in the same ratio, they will also be in the same ratio* ex aequali.

Let there be any number of magnitudes A, B, C, and others D, E, F equal to them in multitude, which taken two and two together are in the same ratio, so that as A is to B, so is D to E, and, as B is to C, so is E to F; I say that they will also be in the same ratio *ex aequali*, that is, as A is to C, so is D to F. For of A, D let equimultiples G, H be taken, and of B, E other, chance, equimultiples K, L; and, further, of C, F other, chance, equimultiples M, N. Then, since, as A is to B, so is D to E, and of A, D equimultiples G, H have been taken, and of B, E other, chance, equimultiples K, L, therefore, as G is to K, so is H to L. For the same reason also, as K is to M, so is L to N. Since, then, there are three magnitudes G, K, M, and others H, L, N equal to them in multitude, which taken two and two together are in the same ratio, therefore, *ex aequali*, if G is in excess of M, H is also in excess of N; if equal, equal; and if less, less. And G, H are equimultiples of A, D, and M, N other, chance, equimultiples of C, F. Therefore, as A is to C, so is D to F. Therefore, etc. Q.E.D.

The main application of the study in book 5 of magnitudes in proportion is to the concept of similarity of geometrical objects, which Euclid treats in book 6, at least for figures in two dimensions.

Book 6

Definitions

1. *Similar rectilineal figures* are such as have their angles severally equal and the sides about the equal angles proportional.
3. A straight line is said to have been *cut in extreme and mean ratio* when, as the whole line is to the greater segment, so is the greater to the less.

Propositions

Proposition 1. *Triangles and parallelograms which are under the same height are to one another as their bases.*

This proposition is the only one in the book whose proof depends directly on the definition of proportionality given in book 5. All other propositions depend on this one.

Let ABC, ACD be triangles and EC, CF parallelograms under the same height; I say that, as the base BC is to the base CD, so is the triangle ABC to the triangle ACD, and the parallelogram EC to the parallelogram CF. For let BD be produced in both directions to the points H, L and let any number of straight lines BG, GH be made equal to the base BC, and any number of straight lines DK, KL equal to the base CD; let AG, AH, AK, AL be joined [figure 3.10.21].

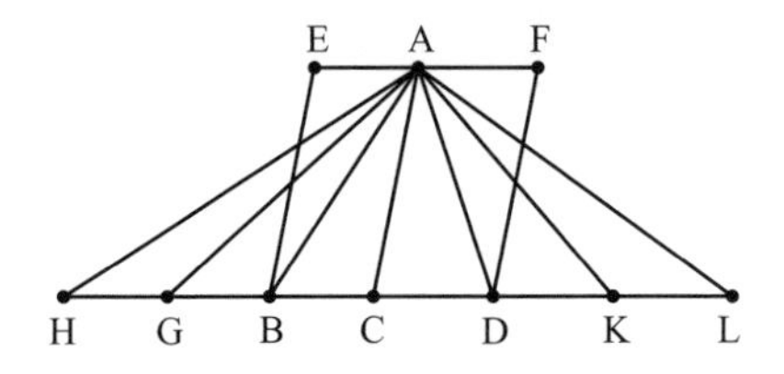

Figure 3.10.21.

Then, since *CB*, *BG*, *GH* are equal to one another, the triangles *ABC*, *ABG*, *AHG* are also equal to one another. Therefore, whatever multiple the base *HC* is of the base *BC*, that multiple also is the triangle *AHC* of the triangle *ABC*. For the same reason, whatever multiple the base *LC* of the base *CD*, that multiple also is the triangle *ALC* of the triangle *ACD*; and, if the base *HC* is equal to the base *CL*, the triangle *AHC* is also equal to the triangle *ACL*, if the base *HC* is in excess of the base *CL*, the triangle *AHC* is also in excess of the triangle *ACL*, and, if less, less.

Thus, there being four magnitudes, two bases *BC*, *CD* and two triangles *ABC*, *ACD*, equimultiples have been taken of the base *BC* and the triangle *ABC*, namely, the base *HC* and the triangle *AHC*, and of the base *CD* and the triangle *ADC* other, chance, equimultiples, namely, the base *LC* and the triangle *ALC*; and it has been proved that, if the base *HC* is in excess of the base *CL*, the triangle *AHC* is also in excess of the triangle *ALC*; if equal, equal; and, if less, less. Therefore, as the base *BC* is to the base *CD*, so is the triangle *ABC* to the triangle *ACD*.

Next, since the parallelogram *EC* is double of the triangle *ABC*, and the parallelogram *FC* is double of the triangle *ACD*, while parts have the same ratio as the same multiples of them, therefore, as the triangle *ABC* is to the triangle *ACD*, so is the parallelogram *EC* to the parallelogram *FC*. Since, then, it was proved that, as the base *BC* is to *CD*, so is the triangle *ABC* to the triangle *ACD*, and, as the triangle *ABC* is to the triangle *ACD*, so is the parallelogram *EC* to the parallelogram *CE*, therefore also, as the base *BC* is to the base *CD*, so is the parallelogram *EC* to the parallelogram *FC*. Therefore etc. Q.E.D.

Proposition 2. *If a straight line be drawn parallel to one of the sides of a triangle, it will cut the sides of the triangle proportionally; and, if the sides of the triangle be cut proportionally, the line joining the points of section will be parallel to the remaining side of the triangle.*

In Figure 3.10.22, *DE* is parallel to *BC*. The theorem then states that $BD:DA = CE:EA$.

Proposition 3. *If an angle of a triangle be bisected and the straight line cutting the angle cut the base also, the segments of the base will have the same ratio as the remaining sides of the triangle; and, if the segments of the base have the same ratio as the remaining sides of the triangle, the straight line joined from the vertex to the point of section will bisect the angle of the triangle.*

Let *ABC* be a triangle, and let the angle *BAC* be bisected by the straight line *AD*; I say that, as *BD* is to *CD*, so is *BA* to *AC* [figure 3.10.23].

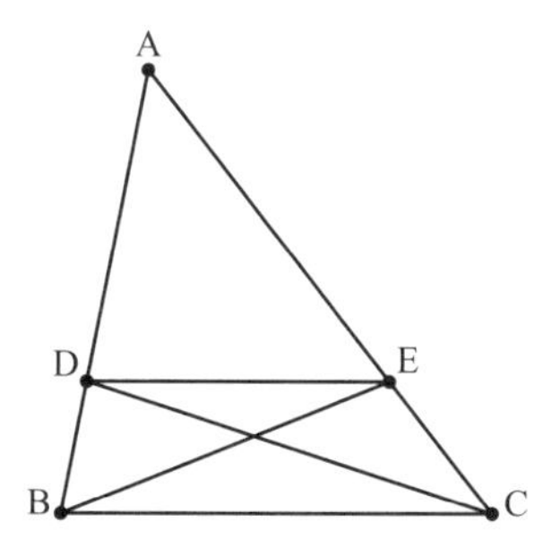

Figure 3.10.22.

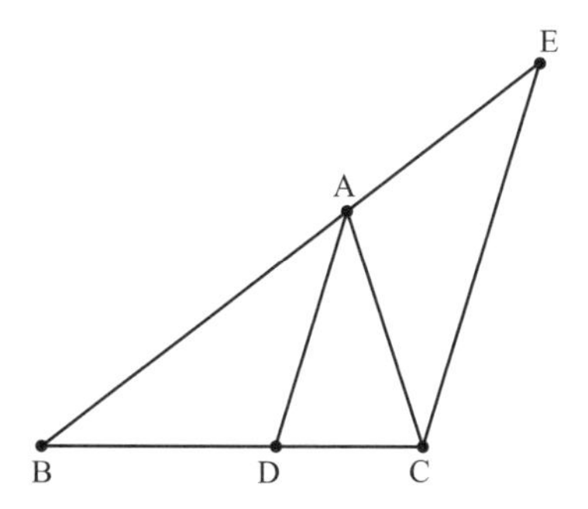

Figure 3.10.23.

For let CE be drawn through C parallel to DA, and let BA be carried through and meet it at E. Then, since the straight line AC falls upon the parallels AD, EC, the angle ACE is equal to the angle CAD. But the angle CAD is by hypothesis equal to the angle BAD; therefore the angle BAD is also equal to the angle ACE. Again, since the straight line BAE falls upon the parallels AD, EC, the exterior angle BAD is equal to the interior angle AEC. But the angle ACE was also proved equal to the angle BAD; therefore the angle ACE is also equal to the angle AEC, so that the side AE is also equal to the side AC. And, since AD has been drawn parallel to EC, one of the sides of the triangle BCE therefore, proportionally, as BD is to DC, so is BA to AE. But AE is equal to AC; therefore, as BD is to DC, so is BA to AC. [We omit the second half of the proof.] Therefore etc. Q.E.D.

Proposition 4: *In equiangular triangles the sides about the equal angles are proportional, and those are corresponding sides which subtend the equal angles.*

Let ABC, DCE be equiangular triangles having the angle ABC equal to the angle DCE, the angle BAC to the angle CDE, and further the angle ACB to the angle CED; I say that in the triangles ABC, DCE the sides about the equal angles are proportional, and those are corresponding sides which subtend the equal angles [figure 3.10.24].

For let BC be placed in a straight line with CE. Then, since the angles ABC, ACB are less than two right angles, and the angle ACB is equal to the angle DEC, therefore the angles ABC, DEC are less than two right angles; therefore BA, ED, when produced, will meet. Let them be produced and meet at F. Now, since the angle DCE is equal to

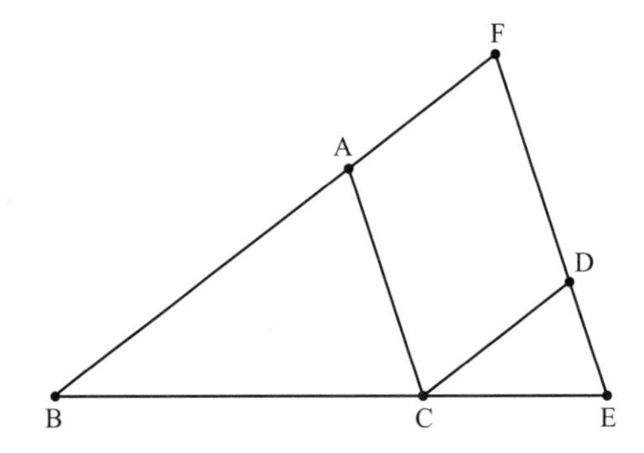

Figure 3.10.24.

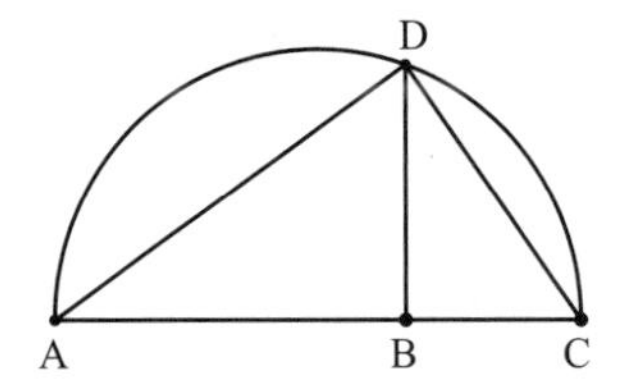

Figure 3.10.25.

the angle ABC, BF is parallel to CD. Again, since the angle ACB is equal to the angle DEC, AC is parallel to FE. Therefore $FACD$ is a parallelogram; therefore FA is equal to DC, and AC to FD. And, since AC has been drawn parallel to FE, one side of the triangle FBE, therefore, as BA is to AF, so is BC to CE. But AF is equal to CD; therefore, as BA is to CD, so is BC to CE, and alternately, as AB is to BC, so is DC to CE. Again, since CD is parallel to BF, therefore, as BC is to CE, so is FD to DE. But FD is equal to AC; therefore, as BC is to CE, so is AC to DE, and alternately, as BC is to CA, so is CE to ED. Since, then, it was proved that, as AB is to BC, so is DC to CE, and, as BC is to CA, so is CE to ED; therefore, *ex aequali* as BA is to AC, so is CD to DE. Therefore etc. Q.E.D.

Proposition 5: *If two triangles have their sides proportional, the triangles will be equiangular and will have those angles equal which the corresponding sides subtend.*

Proposition 8: *If in a right-angled triangle a perpendicular be drawn from the right angle to the base, the triangles adjoining the perpendicular are similar both to the whole and to one another.*

Proposition 13: *To two given straight lines to find a mean proportional.*

The proof is illustrated in figure 3.10.25. If AB, BC are the two given straight lines, we place them in a straight line AC and draw a semicircle with AC as the diameter. Then the line BD drawn perpendicularly from B to the circumference is the desired mean proportional.

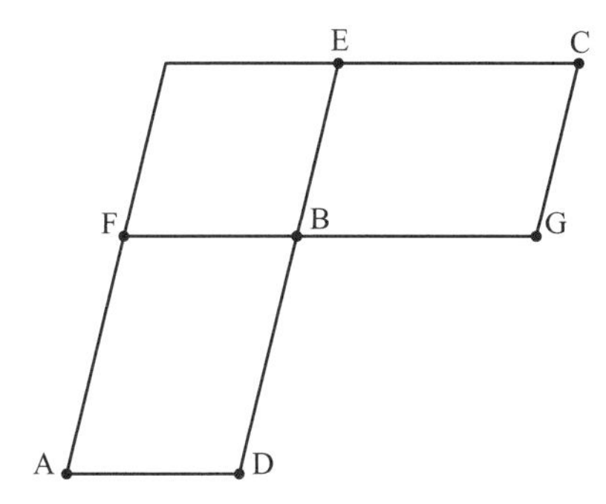

Figure 3.10.26.

Proposition 14: *In equal and equiangular parallelograms the sides about the equal angles are reciprocally proportional; and equiangular parallelograms in which the sides about the equal angles are reciprocally proportional are equal.*

Let AB, BC be equal and equiangular parallelograms having the angles at B equal, and let DB, BE be placed in a straight line; therefore FB, BG are also in a straight line. I say that, in AB, BC, the sides about the equal angles are reciprocally proportional, that is to say, that, as DB is to BE, so is GB to BF [figure 3.10.26].

For let the parallelogram FE be completed. Since, then, the parallelogram AB is equal to the parallelogram BC, and FE is another area, therefore, as AB is to FE, so is BC to FE. But, as AB is to FE, so is DB to BE, and, as BC is to FE, so is GB to BF, therefore also, as DB is to BE, so is GB to BF. Therefore in the parallelograms AB, BC the sides about the equal angles are reciprocally proportional.

Next, let GB be to BF as DB to BE; I say that the parallelogram AB is equal to the parallelogram BC. For since, as DB is to BE, so is GB to BF, while, as DB is to BE, so is the parallelogram AB to the parallelogram FE, and, as GB is to BF, so is the parallelogram BC to the parallelogram FE, therefore also, as AB is to FE, so is BC to FE; therefore the parallelogram AB is equal to the parallelogram BC. Therefore etc. Q.E.D.

Proposition 16. *If four straight lines be proportional, the rectangle contained by the extremes is equal to the rectangle contained by the means; and, if the rectangle contained by the extremes be equal to the rectangle contained by the means, the four straight lines will be proportional.*

Let the four straight lines AB, CD, E, F be proportionals, so that, as AB is to CD, so is E to F; I say that the rectangle contained by AB, F is equal to the rectangle contained by CD, E [figure 3.10.27].

Let AG, CH be drawn from the points A, C at right angles to the straight lines AB, CD, and let AG be made equal to F, and CH equal to E. Let the parallelograms BG, DH be completed. Then since, as AB is to CD, so is E to F, while E is equal to CH, and F to AG, therefore, as AB is to CD, so is CH to AG. Therefore in the parallelograms BG, DH the sides about the equal angles are reciprocally proportional. But those equiangular

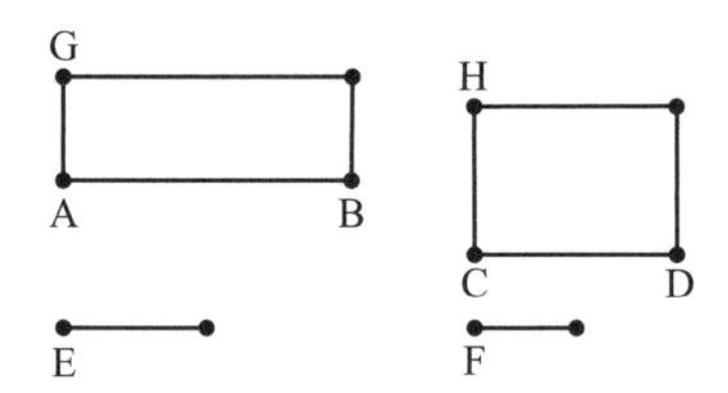

Figure 3.10.27.

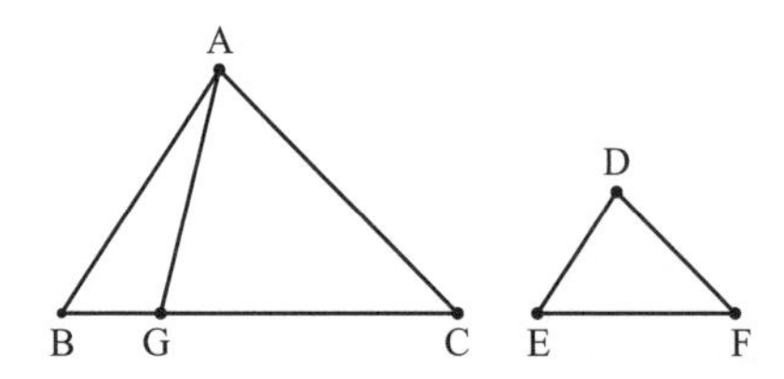

Figure 3.10.28.

parallelograms in which the sides about the equal angles are reciprocally proportional are equal [6:14]; therefore the parallelogram BG is equal to the parallelogram DH. And BG is the rectangle AB, F, for AG is equal to F; and DH is the rectangle CH, E, for E is equal to CH; therefore the rectangle contained by AB, F is equal to the rectangle contained by CD, E. [We omit the second half of the proof.] Therefore etc. Q.E.D.

Proposition 19. *Similar triangles are to one another in the duplicate ratio of the corresponding sides.*

Let ABC, DEF be similar triangles having the angle at B equal to the angle at E, and such that, as AB is to BC, so is DE to EF, so that BC corresponds to EF; I say that the triangle ABC has to the triangle DEF a ratio duplicate of that which BC has to EF [figure 3.10.28].

For let a third proportional BG be taken to BC, EF, so that, as BC is to EF, so is EF to BG; and let AG be joined. Since, then, as AB is to BC, so is DE to EF, therefore, alternately, as AB is to DE, so is BC to EF. But, as BC is to EF, so is EF to BG; therefore also, as AB is to DE, so is EF to BG. Therefore in the triangles ABG, DEF the sides about the equal angles are reciprocally proportional. But those triangles which have one angle equal to one angle, and in which the sides about the equal angles are reciprocally proportional, are equal; therefore the triangle ABG is equal to the triangle DEF. Now since, as BC is to EF, so is EF to BG, and, if three straight lines be proportional, the first has to the third a ratio duplicate of that which it has to the second, therefore BC has to BG a ratio duplicate of that which BC has to EF. But, as CB is to BG, so is the triangle ABC to the triangle ABG; therefore the triangle ABC also has to the triangle ABG a ratio duplicate of that which BC has to EF. But the triangle ABG is equal to the triangle DEF; therefore the triangle ABC also has to the triangle DEF a ratio duplicate of that which BC has to EF. Therefore etc. Q.E.D.

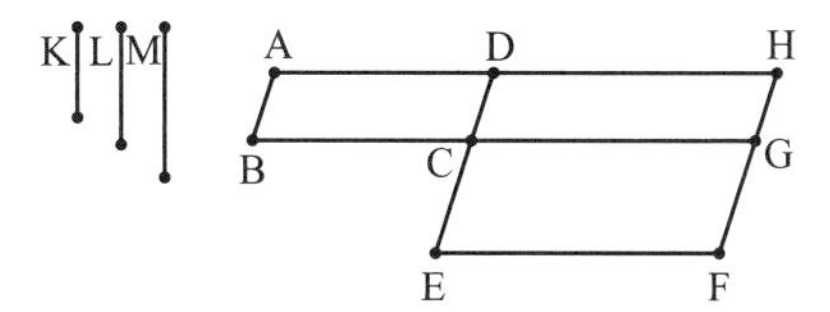

Figure 3.10.29.

Porism: *From this it is manifest that, if three straight lines be proportional, then, as the first is to the third, so is the figure described on the first to that which is similar and similarly described on the second.*

The following proposition mentions a compound ratio, a concept that Euclid did not define. Nevertheless, it is clear from the proof of the result that if one has two ratios, say $a:b$ and $c:d$, one needs to first find e such that $c:d=b:e$. Then the ratio compounded of $a:b$ and $c:d$ is the ratio $a:e$.

Proposition 23. *Equiangular parallelograms have to one another the ratio compounded of the ratios of their sides.*

Let AC, CF be equiangular parallelograms having the angle BCD equal to the angle ECG; I say that the parallelogram AC has to the parallelogram CF the ratio compounded of the ratios of the sides [figure 3.10.29].

For let them be placed so that BC is in a straight line with CG; therefore DC is also in a straight line with CE. Let the paralellogram DG be completed; let a straight line K be set out, and let it be contrived that, as BC is to CG, so is K to L, and, as DC is to CE, so is L to M. Then the ratios of K to L and of L to M are the same as the ratios of the sides, namely, of BC to CG and of DC to CE. But the ratio of K to M is compounded of the ratio of K to L and of that of L to M; so that K has also to M the ratio compounded of the ratios of the sides. Now since, as BC is to CG, so is the parallelogram AC to the parallelogram CH, while, as BC is to CG, so is K to L, therefore also, as K is to L, so is AC to CH. Again, since, as DC is to CE, so is the parallelogram CH to CF, while, as DC is to CE, so is L to M, therefore also, as L is to M, so is the parallelogram CH to the parallelogram CF. Since, then, it was proved that, as K is to L, so is the parallelogram AC to the parallelogram CH, and, as L is to M, so is the parallelogram CH to the parallelogram CF, therefore, *ex aequali*, as K is to M, so is AC to the parallelogram CF. But K has to M the ratio compounded of the ratios of the sides; therefore AC also has to CF the ratio compounded of the ratios of the sides. Therefore etc. Q.E.D.

Propositions 6:28 and 6:29 have long been used to justify the algorithm for solving quadratic equations, as it was found originally on Mesopotamian tablets. These propositions extend the idea of application of areas first mentioned in proposition 1:44 to applications with deficiency or excess. We first give the statement and proof of 6:28, along with a diagram showing the proof in full generality. We follow with just the statement of 6:29, with a diagram showing the case where the given parallelogrammic figure is a square, and therefore, the constructed parallelograms are

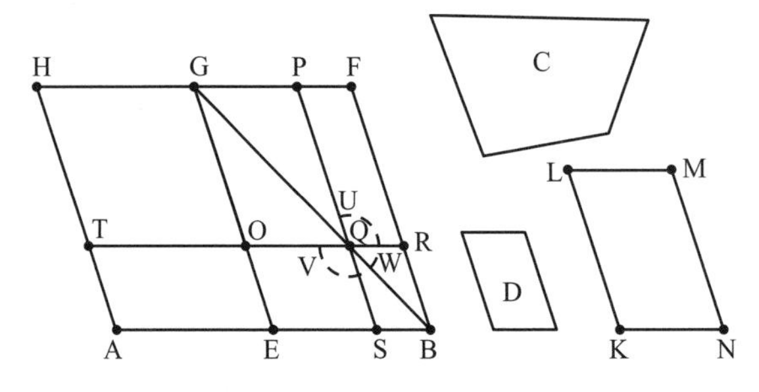

Figure 3.10.30.

all rectangles, because it is in that case that these propositions can best be interpreted algebraically.

Proposition 28. *To a given straight line to apply a parallelogram equal to a given rectilineal figure and deficient by a parallelogrammic figure similar to a given one: thus the given rectilineal figure must not be greater than the parallelogram described on the half of the straight line and similar to the defect.*

Let AB be the given straight line, C the given rectilineal figure to which the figure to be applied to AB is required to be equal, not being greater than the parallelogram described on the half of AB and similar to the defect, and D the parallelogram to which the defect is required to be similar; thus it is required to apply to the given straight line AB a parallelogram equal to the given rectilineal figure C and deficient by a parallelogrammic figure which is similar to D [figure 3.10.30].

Let AB be bisected at the point E, and on EB let $EBFG$ be described similar and similarly situated to D; let the parallelogram AG be completed. If then AG is equal to C, that which was enjoined will have been done; for there has been applied to the given straight line AB the parallelogram AG equal to the given rectilineal figure C and deficient by a parallelogrammic figure GB which is similar to D.

But, if not, let HE be greater than C. Now HE is equal to GB; therefore GB is also greater than C. Let $KLMN$ be constructed at once equal to the excess by which GB is greater than C and similar and similarly situated to D. But D is similar to GB; therefore KM is also similar to GB. Let, then, KL correspond to GE, and LM to GF. Now, since GB is equal to C, KM, therefore GB is greater than KM; therefore also GE is greater than KL, and GF than LM. Let GO be made equal to KL, and GP equal to LM; and let the parallelogram $OGPQ$ be completed; therefore it is equal and similar to KLM. Therefore GQ is also similar to GB; therefore GQ is about the same diameter with GB. Let GQB be their diameter, and let the figure be described.

Then, since BG is equal to C, KM, and in them GQ is equal to KM, therefore the remainder, the gnomon UWV, is equal to the remainder C. And, since PR is equal to OS, let QB be added to each; therefore the whole PB is equal to the whole OB. But OB is equal to TE, since the side AE is also equal to the side EB; therefore TE is also equal to PB. Let OS be added to each; therefore the whole TS is equal to the whole, the gnomon VWU. But the gnomon VWU was proved equal to C; therefore TS is also equal to C.

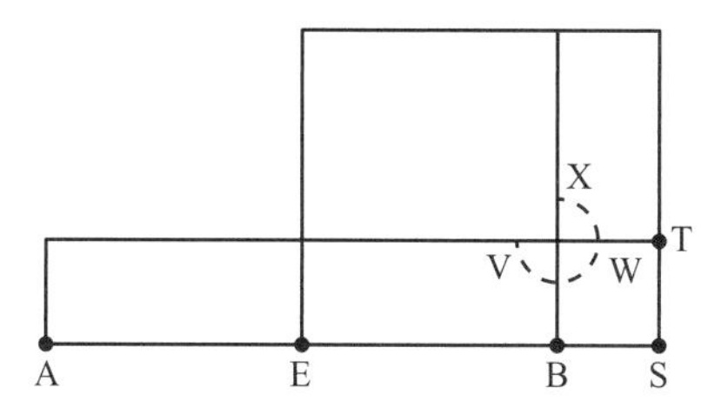

Figure 3.10.31.

Therefore to the given straight line AB there has been applied the parallelogram ST equal to the given rectilineal figure C and deficient by a parallelogrammic figure QB which is similar to D. Q.E.F.

Proposition 29. *To a given straight line to apply a parallelogram equal to a given rectilineal figure and exceeding by a parallelogrammic figure similar to a given one.*

In the special case of 6:29 we are considering, we are to apply a rectangle to the line AB of a given area exceeding by a square [figure 3.10.31].

To look at these propositions algebraically, we designate the length of AB is each case by b and the area of the given rectilineal figure by c. As noted, we also assume that all the parallelograms are rectangles. The problems then reduce to finding a point S on AB (6:28) or on AB extended (6:29) so that $x=BS$ satisfies $x(b-x)=c$ in the first case and $x(b+x)=c$ in the second. That is, in 6:28, the rectangle TS (of area $x(b-x)$) is applied to the line AB and is deficient by the square on BS (x^2). In 6:29, the rectangle AT (of area $x(b+x)$) is applied to the line AB and exceeds by the square on BS (x^2). Thus, 6:28 solves the quadratic equation $bx-x^2=c$, while 6:29 solves the equation $bx+x^2=c$. In each case, Euclid found the midpoint E of AB and constructed the square on BE, whose area is $(b/2)^2$. In the first case, S was chosen so that ES is the side of a square whose area is $(b/2)^2-c$. That is why the condition is stated in the proposition that, in effect, c cannot be greater than $(b/2)^2$. In any case, this choice for ES implies that

$$x=BS=BE-ES=\frac{b}{2}-\sqrt{\left(\frac{b}{2}\right)^2-c}.$$

In the second case, S is chosen so that ES is the side of a square whose area is $(b/2)^2+c$. Then,

$$x=BS=ES-BE=\sqrt{\left(\frac{b}{2}\right)^2+c}-\frac{b}{2}.$$

In both cases, Euclid proved that his choice was correct by showing that the desired rectangle equals the gnomon UWV (6:28) or XWV (6:29) and that the gnomon is in turn equal to the given area c. Algebraically, that result is immediate. Of course, just because these results have been interpreted algebraically is not to say that Euclid

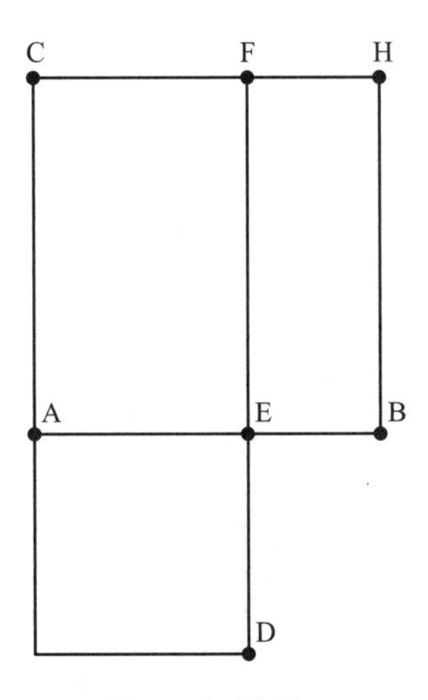

Figure 3.10.32.

conceived of them in this manner. In fact, he immediately used 6:29 to prove 6:30, a geometric theorem that he had already proved in a different way as proposition 2:11. Interestingly, in this proposition, the given parallelogrammic figure is a square.

Proposition 30: *To cut a given finite straight line in extreme and mean ratio.*

Let AB be the given finite straight line; thus it is required to cut AB in extreme and mean ratio [figure 3.10.32].

On AB let the square BC be described; and let there be applied to AC the parallelogram CD equal to BC and exceeding by the figure AD similar to BC [6:29]. Now BC is a square; therefore AD is also a square. And, since BC is equal to CD, let CE be subtracted from each; therefore the remainder BF is equal to the remainder AD. But it is also equiangular with it; therefore in BF, AD the sides about the equal angles are reciprocally proportional; therefore, as FE is to ED, so is AE to EB. But FE is equal to AB, and ED to AE. Therefore, as BA is to AE, so is AE to EB. And AB is greater than AE; therefore AE is also greater than EB. Therefore the straight line AB has been cut in extreme and mean ratio at E, and the greater segment of it is AE. Q.E.F.

The final proposition we include here is the generalization of the Pythagorean Theorem mentioned by Proclus.

Proposition 31. *In right-angled triangles the figure on the side subtending the right angle is equal to the similar and similarly described figures on the sides containing the right angle.*

Let ABC be a right-angled triangle having the angle BAC right. I say that the figure on BC is equal to the similar and similarly described figures on BA, AC [figure 3.10.33].

Let AD be drawn perpendicular. Then since, in the right-angled triangle ABC, AD has been drawn from the right angle at A perpendicular to the base BC, the triangles ABD, ADC adjoining the perpendicular are similar both to the whole ABC and to one another. And since ABC is similar to ABD, therefore, as CB is to BA, so is AB to BD. And,

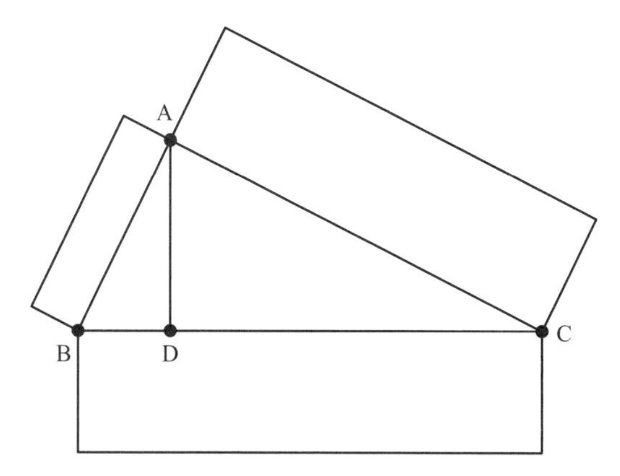

Figure 3.10.33.

since three straight lines are proportional, as the first is to the third, so is the figure on the first to the similar and similarly described figure on the second. Therefore, as CB is to BD, so is the figure on CB to the similar and similarly described figure on BA.

For the same reason also, as BC is to CD, so is the figure on BC to that on CA; so that, in addition, as BC is to BD, DC, so is the figure on BC to the similar and similarly described figures on BA, AC. But BC is equal to BD, DC; therefore the figure on BC is also equal to the similar and similarly described figures on BA, AC. Therefore etc. Q.E.D.

3.11 Euclid, *Data*

The *Data* of Euclid is, in effect, a supplement to books 1–6 of the *Elements*. The name comes from the first word of the text, $\delta\epsilon\delta o\mu \acute{\epsilon}\nu\alpha$ (*dedomena*), which, translated into Latin, is *data*, or "given." Thus, each proposition of the work takes certain parts of a geometric configuration as given, or known, and shows that certain other parts are therefore determined. In fact, in most of his proofs, Euclid explicitly shows how to determine these parts. Thus, the *Data* in essence transformed the synthetic purity of the *Elements* into a manual appropriate to one of the goals of Greek mathematics, the solutions of new geometric problems.

We present here a selection of the definitions and propositions of the *Data*. Some of the propositions are given without proof. Whenever a proof is given, we give justifications in terms of the definitions, previous propositions, or propositions from Euclid's *Elements*.

Definitions

1. Given in magnitude is said of figures and lines and angles for which we can provide equals.
2. A ratio is said to be given for which we can provide the same.
3. Rectilineal figures are said to be given in form if each angle is given and the ratios of the sides to one another are given.

4. Given in position is said of points and lines and angles which always hold the same place.
5. A circle is said to be given in magnitude if its radius is given in magnitude.
6. And a circle is said to be given in position and in magnitude if its center is given in position and its radius in magnitude.
7. Segments of circles are said to be given in magnitude if the angles in them and the bases of the segments are given in magnitude.
8. And segments of circles are said to be given in position and in magnitude if the angles in them are given in magnitude and the bases of the segments are given in position and in magnitude.

Propositions

The first several propositions are nearly axioms. One wonders if they really require proofs, but Euclid makes every effort to prove them.

Proposition 1. *The ratio of given magnitudes to one another is given.*

Let A and B be given magnitudes; I say that the ratio of A to B is given. For, since A is given, it is possible to provide a [magnitude] equal to it [definition 1]. Let it have been provided, and let it be C. Again, since B is given, it is possible to provide a [magnitude] equal to it. Let it have been provided, and let it be D. Then, since $A=C$ and $B=D$, therefore as A is to C, so is B to D. Alternately, as A is to B, so is C to D. Therefore the ratio of A to B is given, for the ratio of C to D has been provided the same as it [definition 2].

Proposition 2. *If a given magnitude has a given ratio to some other magnitude, the other is also given in magnitude.*

For, let the given magnitude A have a given ratio to some other magnitude B; I say that B is given in magnitude. For, since A is given, it is possible to provide a magnitude equal to it [definition 1]. Let it have been provided, and let it be C. And since the ratio of A to B is given, it is possible to provide the same as it [definition 2]. Let it have been provided, and let it be the ratio of C to D. And since, as A is to B, so is C to D; therefore alternately, as A is to C, so is B to D [*Elements* 5:16]. But A is equal to C; therefore B is also equal to D [5:14]; therefore B has been given in magnitude, for D has been furnished equal to it [definition 1].

Proposition 3. *If any number of given magnitudes are added together, the magnitude composed of them will also be given.*

Proposition 4. *If a given magnitude be subtracted from a given magnitude, the remainder will be given.*

Proposition 7. *If a given magnitude is divided into a given ratio, each of the segments is given.*

Proposition 8. *Magnitudes that have a given ratio to the same, will also have a given ratio to one another.*

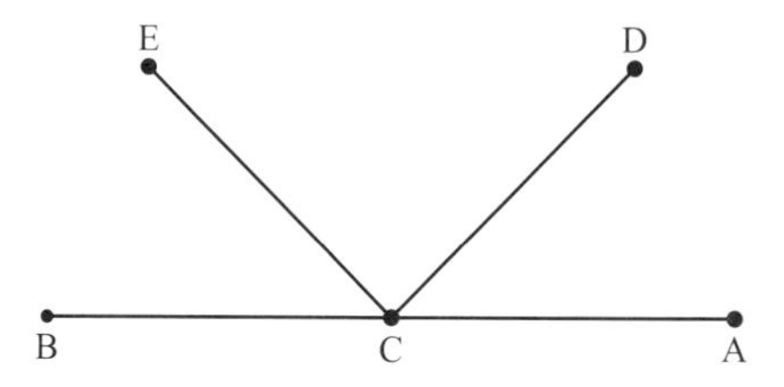

Figure 3.11.1.

In propositions 25 through 29, Euclid is dealing with the positions of points and a straight line. Notice that he makes the assumption that points and lines can change position (i.e., one can pick them up and move them) before showing that such assumptions lead to a contradiction.

Proposition 25. *If two lines given in position cut one another, their point of section is given in position.*

Let two lines given in position AB, CD cut one another at the point E. I say that the point E is given. For if not, the point E will change position. But then the position of one of the lines AB, CD will change, too. But it does not change. Therefore the point E is given [definition 4].

Proposition 26. *If the extremities of a straight line are given in position, the line is given in position and in magnitude.*

Let the extremities A, B of a straight line be given in position. I say that the straight line AB is given in position and in magnitude. For if, while A remains fixed, either the position or the magnitude of the straight line AB undergoes a change, the point B will also fall elsewhere. But it does not fall elsewhere [definition 4]. Therefore the straight line AB has been given in position and in magnitude.

Proposition 27. *If one extremity of a straight line given in position and in magnitude is given, the other will also be given.*

For, let one extremity A of the straight line given in position and in magnitude AB be given. I say that B is also given. For if, with the point A remaining fixed, the point B falls elsewhere; therefore, either the position or the magnitude of straight line AB will also undergo a change. But they do not undergo a change [definition 4]. Therefore the point B is given.

Proposition 29. *If at a straight line given in position and at a given point on it a straight line is drawn making a given angle, the straight line drawn is given in position.*

For, to the straight line given in position AB, and to the given point C on it, let the straight line CD be drawn making the given angle BCD. I say that the straight line CD is given in position. For if not, with the point C remaining fixed, the position of the straight line CD, which maintains the magnitude of the angle BCD, will fall elsewhere [Figure 3.11.1].

Let it fall elsewhere, and let it be CE. Therefore the angle DCB is equal to the angle ECB, the greater to the less, which is absurd. Therefore the position of straight

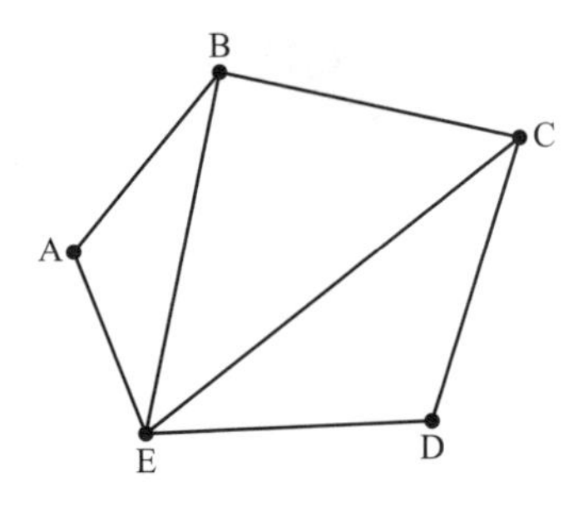

Figure 3.11.2.

line DC does not fall elsewhere. Therefore straight line CD is given in position [definition 4].

Propositions 40 through 55 deal with the concept of form for triangles and other rectilineal figures. Two such figures have the same form, in other words, if they are similar to one another. Euclid gives methods for determining this similarity.

Proposition 40. *If each of the angles of a triangle is given in magnitude, the triangle is given in form.*

Proposition 41. *If a triangle has one given angle, and the sides about the given angle have a given ratio to one another, the triangle is given in form.*

Proposition 47. *Rectilineal figures given in form can be divided into triangles given in form.*

Let $ABCDE$ be the rectilineal figure given in form. I say that the rectilineal figure $ABCDE$ can be divided into triangles given in form [figure 3.11.2].

For, let BE, EC be joined. Since the rectilineal figure $ABCDE$ is given in form, angle BAE is given, and the ratio of BA to EA is given [definition 3]. Therefore triangle BAE is given in form [proposition 41]; therefore angle ABE is given [definition 3]. And the whole angle ABC is given; therefore the remaining angle EBC is given [proposition 4]. And the ratio of AB to BE is given, and the ratio of AB to BC is given; therefore the ratio of EB to BC is given [proposition 8]. And angle CBE is given; therefore triangle BCE is given in form [proposition 41]. Then for the same reasons, triangle CDE is given in form. Therefore rectilineal figures given in form are divisible into triangles given in form.

Proposition 48. *If on the same straight line two triangles are described, given in form, they will have a given ratio to one another.*

Proposition 49. *If on the same straight line two arbitrary rectilineal figures given in form are described, they will have a given ratio to one another.*

Proposition 52. *If on a straight line given in magnitude a form given in form is described, the form described is given in magnitude.*

Proposition 55. *If a rectilineal figure is given in form and in magnitude, its sides will also be given in magnitude.*

Propositions 57 through 59 deal with the application of areas, a technique Euclid used in *Elements* 1 and 6. In fact, propositions 58 and 59 are central to the solution

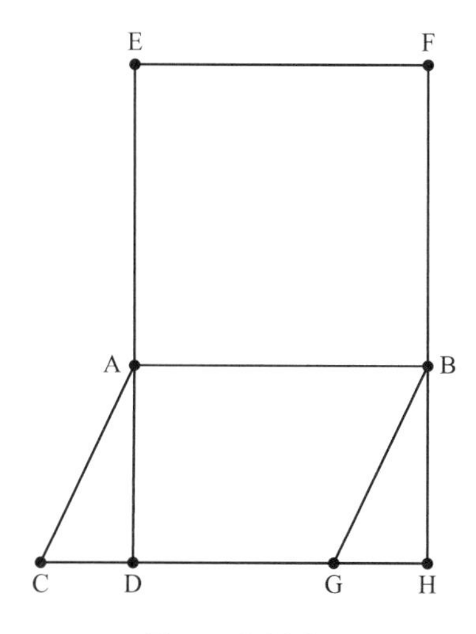

Figure 3.11.3.

of the problems posed in propositions 84 and 85. In proposition 84, Euclid shows how to find unknown lengths, given certain known ones. Although the text of this proposition speaks of two lines containing a given area in a given angle, in all of the surviving manuscripts, the angle is shown as right. Then the question could be considered as determining the width and length of a rectangle if one is given the area and the difference between the length and width. The proof of this proposition depends on proposition 59, where Euclid uses a diagram similar to that of *Elements* 6:29. One can easily show that Euclid's construction here, if translated into algebra, amounts to the modern formula for solving the system $xy=c, x-y=b$. However, that is not to say that Euclid is thinking algebraically, only that he was solving a geometric problem that dates back to ancient Mesopotamia. Similarly, proposition 85 uses proposition 58 to solve a problem that could be formulated algebraically as $xy=c, x+y=b$.

Proposition 57. *If a given [parallelogrammic] area is applied to a given straight line in a given angle, the width of the applied area is given.*

For, suppose the given [parallelogrammic] area AG has been applied to the given straight line AB in the given angle CAB [figure 3.11.3].

I say that CA is given. For suppose the square EB has been described on the straight line AB; therefore EB is given [proposition 52]. And suppose EA, FB, CG have been produced to the point D, H. And since each of EB, AG is given, therefore the ratio of EB to AG is given [proposition 1]. And GA is equal to AH [*Elements* 1:35]; therefore the ratio of EB to AH is given; so that the ratio of EA to AD is also given. But EA is equal to AB; therefore the ratio of BA to AD is also given. And since the angle CAB is given, of which the angle DAB is given, therefore the remaining angle CAD is given [proposition 4]. And the angle CDA is given because it is a right angle; therefore the remaining angle ACD is also given [propositions 3, 4]. Therefore the triangle ACD is given in form [proposition 40]; therefore the ratio of CA to AD is given [definition 3]. And the ratio of DA to AB is also given; therefore the ratio of CA to AB is also given [proposition 8]. And BA is given; therefore also AC is given [proposition 2]. And it is the width of the applied area.

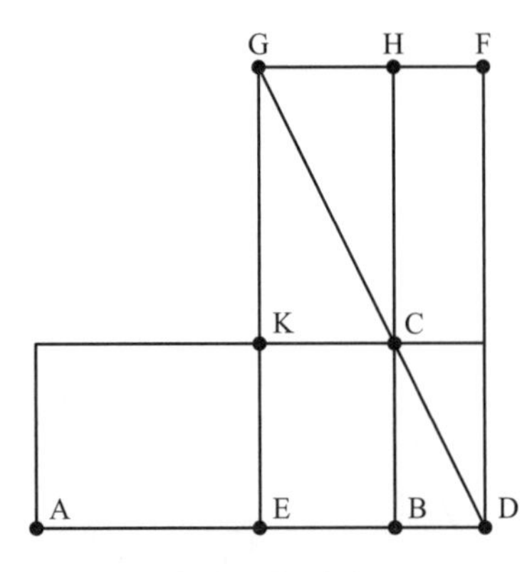

Figure 3.11.4.

Proposition 58. *If a given [parallelogrammic] area deficient by a [parallelogrammic] area given in form is applied to a given straight line, the length and width of the defect are given.*

For, suppose the given [parallelogrammic] area AC deficient by the [parallelogrammic] area given in form CD has been applied to the given straight line AD; I say that each of BC, BD is given [figure 3.11.4].

For, suppose AD has been bisected at the point E; therefore ED is given [proposition 7]. And on the straight line ED suppose the rectilineal figure EF has been described similar and similarly situated to the parallelogram CD, and suppose the figure has been completed [*Elements* 6:18][47]; therefore the figure EF has also been given in form. And since the rectilineal figure given in form EF has been described on the straight line given in magnitude ED, therefore EF has been given in magnitude [proposition 52]. And it is equal to the sum of AC, KH; therefore the sum of AC, KH is given in magnitude [proposition 4]. And AC is given in magnitude, by hypothesis; therefore the remainder KH is given in magnitude. And it is also given in form, for it is similar to CD; therefore the sides of HK are given [proposition 55]; therefore KC is given; and it is equal to EB; therefore EB is also given. And ED is also given; therefore the remainder BD is given [proposition 4]. And the ratio of BD to BC is given; therefore BC is also given [proposition 2].

Proposition 59. *If a given [parallelogrammic] area exceeding by a [parallelogrammic] area given in form is applied to a given straight line, the length and width of the excess are given.*

For, suppose the given [parallelogrammic] area AB exceeding by the [parallelogrammic] figure given in form CB has been applied to the given straight line AC; I say that each of HC, CE is given [figure 3.11.5].

For, suppose the straight line DE is bisected at the point F, and on EF suppose the figure FG has been described similar and similarly situated to CB [*Elements* 6:18]. Therefore FG is about the same diameter as CB. Suppose that HEM is the common diameter, and suppose the figure is completed. Since CB is similar to FG, and CB has

[47] On a given straight line, to describe a rectilineal figure similar and similarly situated to a given rectilineal figure.

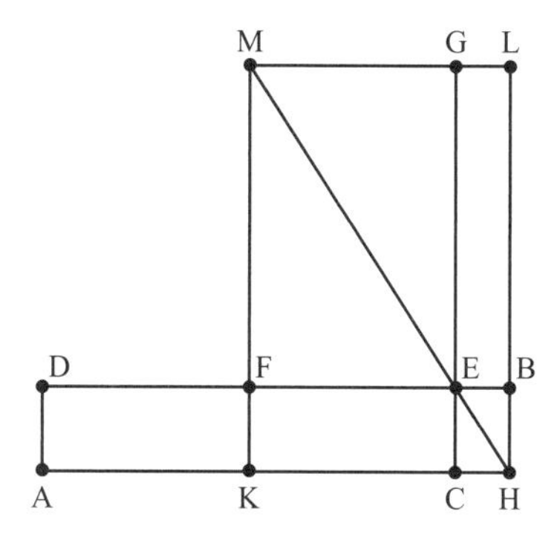

Figure 3.11.5.

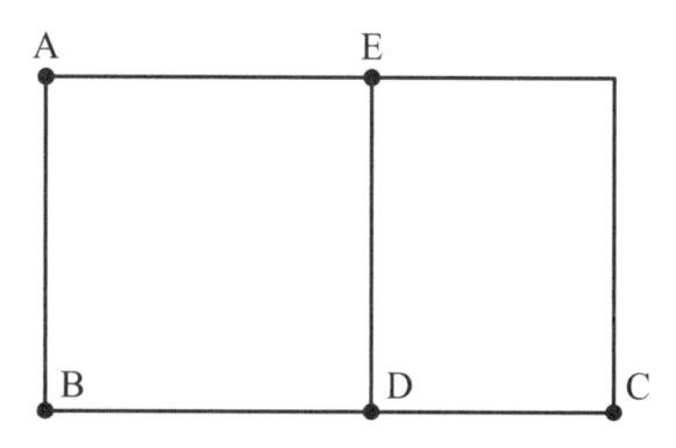

Figure 3.11.6.

been given in form, therefore FG has also been given in form; and it has been described on the given straight line FE; therefore FG is given in magnitude [proposition 52]. And AB is also given; therefore the sum of AB, FG is given [proposition 3]. And it is equal to KL. Therefore KL is given in magnitude. And it is also given in form; for it is similar to CB; therefore the sides of KL are given [proposition 55]. Therefore KH is given, of which KC is given; for it is equal to EF; therefore the remaining side CH is given [proposition 4]; and it has to HB a given ratio. Therefore HB is also given [proposition 2].

Proposition 84. *If two straight lines contain a given area in a given angle, and the one be greater than the other by a given magnitude, each of them will also be given.*

For let the two straight lines AB, BC contain the given area AC in the given angle ABC, and let CB be greater than BA by a given magnitude; I say that each of BA, BC is given [figure 3.11.6].

For, since BC is greater than BA by a given magnitude, let DC be the given magnitude; therefore the remainder DB is equal to BA [definition 9]. Let the parallelogram AD be completed. And since AB is equal to DB, therefore the ratio of AB to BD is also given; and the angle ABD is given. Therefore the parallelogram AD has been given in form. Then, since the given area AC exceeding by a figure given in form AD has been applied to the given straight line DC, therefore the width of the excess has been given; therefore BD and BA are given [proposition 59]. But DC is given; therefore the whole BC is also given [proposition 3]. And AB is given; therefore each of AB, BC is given.

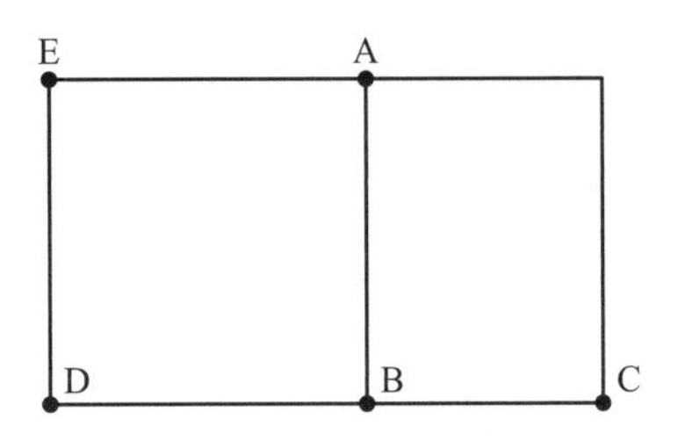

Figure 3.11.7.

Proposition 85. *If two straight lines contain a given area in a given angle, and their sum is given, each of them will be given.*

For let the two straight lines AB, BC contain the given area AC in the given angle ABC, and let the sum of AB, BC be given. I say that each of AB, BC is given [figure 3.11.7].

For, suppose CB has been produced to the point D, and let BD be made equal to AB, and through the point D suppose the straight line DE has been drawn parallel to BA, and let the parallelogram AD be completed. Since DB is equal to BA, and the angle ABD is given, since the angle adjacent to it is given, therefore the parallelogram EB has been given in form. And since the sum of AB, BC is given, and AB is equal to BD, therefore DC is given [proposition 3]. Then, since the parallelogram AC deficient by a parallelogrammic figure given in form EB has been applied to the given straight line DC, the length and width of the defect have been given; therefore AB, BD are given [proposition 58]. But the sum of AB, BC is given; therefore the remainder BC is also given [proposition 4]. Therefore each of the lines AB, BC is given.

The final propositions here deal with lines in circles and are closely related to theorems Euclid proves in *Elements* 3.

Proposition 87. *If in a circle given in magnitude a straight line is drawn cutting off a segment admitting a given angle, the line drawn is given in magnitude.*

For, in the circle given in magnitude ABC suppose the straight line AC is drawn cutting off the segment AEC admitting a given angle. I say that the straight line AC has been given in magnitude [figure 3.11.8].

For let the center D of the circle be taken, and let AD be joined and produced to E, and let CE be joined. Therefore the angle ACE is given, because it is a right angle [*Elements* 3:31]. And the angle AEC is also given; therefore the remaining angle CAE is given [*Elements* 1:32]. Therefore the triangle ACE has been given in form [proposition 40]. Therefore, the ratio of AE to AC is given. And since the circle has been given in magnitude, the straight line EA has also been given in magnitude [proposition 3]. Therefore the straight line AC is given in magnitude [proposition 2].

Proposition 88. *If in a circle given in magnitude a straight line is drawn given in magnitude, it will cut off a segment admitting a given angle.*

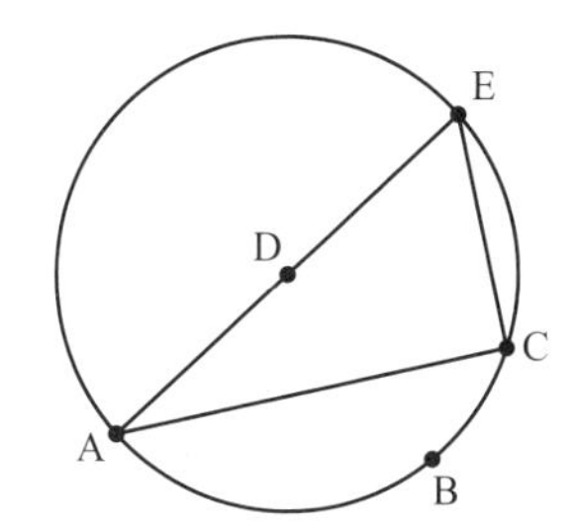

Figure 3.11.8.

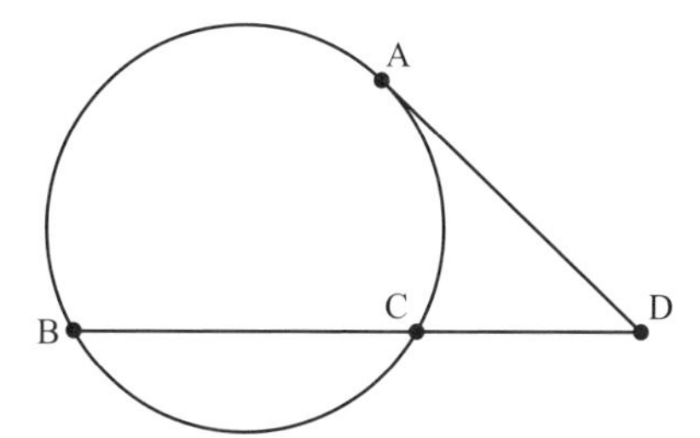

Figure 3.11.9.

Proposition 90. *If from a given point a straight line is drawn tangent to a circle given in position, the straight line drawn will be given in position and in magnitude.*

Proposition 91. *If a given point is taken outside a circle given in position, and from the point a straight line is drawn through the circle, the rectangle contained by the straight line and the straight line between the point and the convex circumference is given.*

For suppose a given point D is taken outside the circle ABC given in position, and from the point D let some straight line BD be drawn cutting the circle. I say that the rectangle BD, DC is given [figure 3.11.9].

From the point D, let the straight line AD be drawn tangent to the circle ABC [*Elements* 3:17].[48] Therefore the straight line AD is given in position and in magnitude [proposition 90]. Then, since the straight line AD is given, therefore the square on AD is also given [proposition 52]. And it is equal to the rectangle BD, DC [*Elements* 3:36]. Therefore the rectangle BD, DC is also given.

Proposition 92. *If a given point is taken inside a circle given in position, and through the point a straight line is drawn in the circle, the rectangle contained by the segments of the straight line is given.*

For let the given point A be taken inside the circle BC given in position, and through A let some straight line CB be drawn. I say that the rectangle CA, AB is given [figure 3.11.10].

[48] From a given point to draw a straight line touching a given circle.

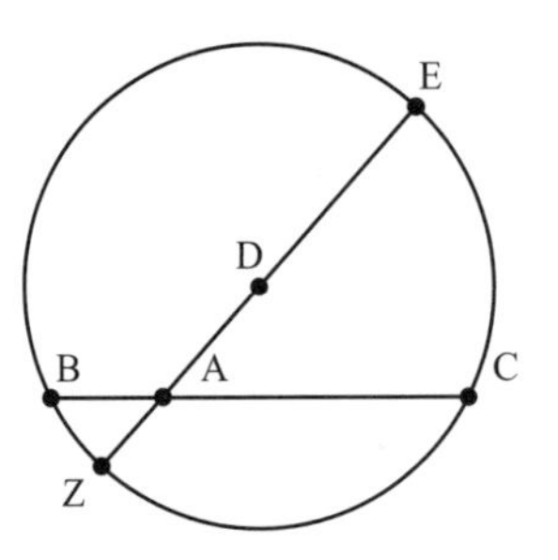

Figure 3.11.10.

For let the center D of the circle be taken, and let the joined line AD be drawn through to the points Z, E. Since each of the points D, A is given, therefore the line DA is given in position [proposition 26]. And the circle CBZ is given in position; therefore each of Z, E is given [proposition 25]. Since A is also given, therefore each of ZA, AE is given [proposition 26]. Therefore the rectangle ZA, AE is given and it is equal to the rectangle BA, AC [*Elements* 3:35]. Therefore the rectangle BA, AC is also given.

3.12 Euclid, *On Divisions*

We first learn about Euclid's *On Divisions* from Proclus's *Commentary*, where Proclus mentions the work and notes that there Euclid "divides a given figure sometimes into like and sometimes into unlike parts."[49] The Greek text of *On Divisions* disappeared centuries ago, but it was evidently known in the Islamic world since we are told that Thābit ibn Qurra (836–901) wrote an Arabic "corrected" version of it. That work has also vanished, but an abstract of *On Divisions*, written by Abū Sa'īd Aḥmad ibn Muḥammad ibn 'Abd al-Jalīl al-Sijzī (945–1020) and probably based on Thābit's work, was discovered by F. Woepcke in the mid-nineteenth century, who then published a French translation.[50] Al-Sijzī's version included all of Euclid's propositions, but only proofs of four of these, because he thought the other proofs were too easy. R. C. Archibald reconstructed the entire work of Euclid from Woepcke's translation as well as from similar propositions and proofs in *De Practica Geometrie* by Leonardo of Pisa (1170–1240).[51] We also know that Abū al-Wafā' al-Būzjānī (940–998) included many solutions of problems from *On Divisions* in his own book *On the Geometric*

[49] Proclus, *A Commentary on the First Book of Euclid's Elements*, trans. Glenn R. Morrow (Princeton, NJ: Princeton University Press, 1970), 115.

[50] Franz Woepcke, "Notice sur des traductions arabes de deux ouverages perdus d'Euclide," *Journal Asiatique* 18, no. 4 (1851): 217–47.

[51] Raymond Claire Archibald, *Euclid's Book on Divisions of Figures: With a Restoration Based on Woepcke's Text and on the Practica Geometriae of Leonardo Pisano* (Cambridge: Cambridge University Press, 1915); Barnabas Hughes, ed., *Fibonacci's De practica geometrie* (New York: Springer, 2008); Barnabas Hughes, "Leonardo of Pisa, *De practica geometrie (Practical Geometry),"* in *Sourcebook in the Mathematics of Medieval Europe and North Africa*, ed. Victor J. Katz et al. (Princeton, NJ: Princeton University Press, 2016), 130–39.

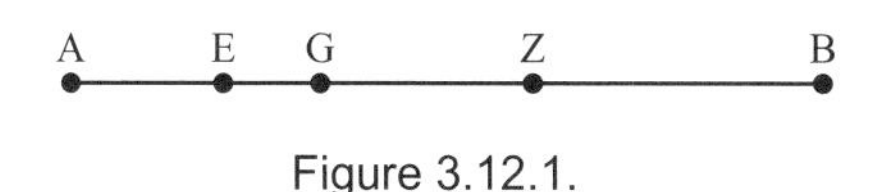

Figure 3.12.1.

Constructions Necessary for the Artisan.[52] Finally, many of these same solutions are also found in *The Treatise on Measuring Areas and Volumes* by Abraham Bar Ḥiyya (1065–1145).[53] The question, then, is how Euclid's work reached Bar Ḥiyya and then Leonardo. Since there is no record of Euclid's original Greek text, it seems most likely that both of these men had one of the Arabic versions available, presumably Thābit's version, since that did contain the proofs. The excerpts that follow are taken from Hogendijk's translation of Al-Sijzī's Arabic work. Note that constructions of 13, 14, and 15 are all found in *De Practica Geometrie*.[54]

13. We wish to demonstrate how we bisect a known quadrilateral by a straight line drawn from a known angle of it.
14. We wish to demonstrate how we cut off from a known quadrilateral some assumed part by a line drawn from a known angle of it.
15. We wish to demonstrate how we bisect a known quadrilateral by a straight line drawn from a known point which is on one of its sides.
17. We wish to apply to a straight line [AB] a rectangle equal to the area contained by lines AB, AG, and deficient from the completion of the line by a square area. After having done what was required, if someone asks, How is it possible to apply to the line AB a rectangle such that the rectangle $AE \cdot EB$ [55] is equal to the rectangle $AB \cdot AG$ and deficient by a square, we say that it is impossible, because AB is greater than BE and AG is greater than AE, so $AB \cdot AG$ is greater than $AE \cdot EB$. So, if there is applied to AB a parallelogram equal to $AB \cdot AG$, it is [like] area $AZ \cdot ZB$.[56] [figure 3.12.1]
18. We wish to demonstrate how we bisect a known triangle by a straight line which passes through a known point inside the triangle. Thus let the known triangle be triangle ABG, and let the point which is in its interior be D. We wish to let pass through D a straight line which bisects triangle ABG. Thus we draw from point D a line parallel to line BG, namely, DE. We apply to DE an area equal to half $AB \cdot BG$, let it be $TB \cdot ED$. We apply to line TB a parallelogram equal to $BT \cdot BE$, deficient from its completion by a square area; let the applied area be area $BH \cdot HT$.[57] We

[52] J. Lennart Berggren, "Abū al-Wafā' on the Geometry of Artisans," in *The Mathematics of Egypt, Mesopotamia, China, India, and Islam: A Sourcebook*, ed. Victor J. Katz (Princeton, NJ: Princeton University Press, 2007), 611–16.

[53] Roi Wagner, "Abraham bar Ḥiyya, Ḥibur Hameshiḥa Vehatishboret (The Treatise on Measuring Areas and Volumes)," in *Sourcebook in the Mathematics of Medieval Europe and North Africa*, ed. Victor J. Katz et al. (Princeton, NJ: Princeton University Press, 2016), 310–13.

[54] Hughes, *Fibonacci's De practica geometrie,* 231–35.

[55] In the Arabic version, this multiplication is indicated using a word that is usually translated as "times," so a multiplicative notation is appropriate. However, Euclid himself never multiplies lines together. So a more literal translation from the original Greek would likely be "the rectangle contained by *AE* and *EB*."

[56] That is, point Z must lie between G and B.

[57] According to 17, H must be between E and T.

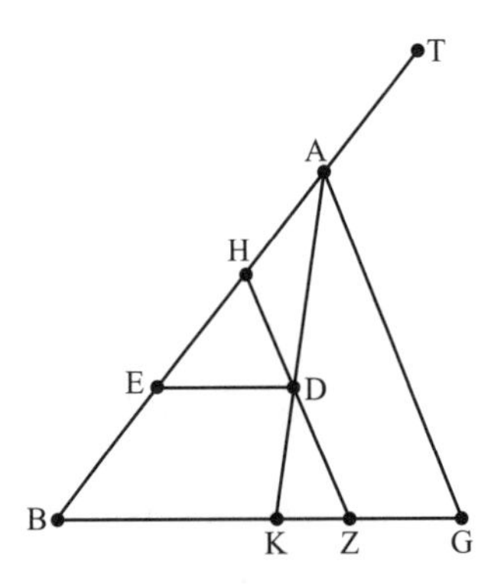

Figure 3.12.2.

join line DH and we extend it towards Z. I say that line DHZ has been drawn such as to divide triangle ABG into two equal parts, namely, HBZ, $HZGA$. [figure 3.12.2]

Proof of this: $TB \cdot BE$ is equal to $TH \cdot HB$, so the ratio of BT to TH is equal to the ratio of HB to BE. *Separando*, the ratio of TB to BH is also equal to the ratio of BH to HE. But the ratio of BH to HE is equal to the ratio of BZ to ED. Thus the ratio of TB to BH is equal to the ratio of BZ to ED. So $TB \cdot ED$ is equal to $BH \cdot BZ$. But $TB \cdot ED$ is equal to half of $AB \cdot BG$, and the ratio of $BH \cdot BZ$ to $AB \cdot BG$ is equal to the ratio of triangle HBZ to triangle ABG, because the angles at point B are common. So triangle HBZ is half of triangle ABG. So triangle ABG has been divided into two equal parts, namely, BHZ, $AHZG$. If we apply to TB a parallelogram equal to $TB \cdot BE$, deficient from its completion by a square, and if the area is $AB \cdot AT$,[58] then if we join AD and extend it towards K, we prove in the same way that triangle ABK is half of triangle ABG. That is what we wanted to demonstrate.

27. We wish to bisect a known figure contained by an arc and two straight lines which contain an angle. Thus let there be a known figure ABG, contained by arc BG and the straight lines BA, AG, containing angle BAG. We wish to draw a straight line which bisects figure ABG. Thus we join line BG, and we bisect it at point E. We draw from point E a line perpendicular to line BG, namely, EZ. We join the straight line AE. Since line BE is equal to line EG, area BZE is equal to area EZG. But triangle ABE is equal to triangle AEG, so figure $ABZE$ turns out to be equal to figure $ZGAE$. If line AE is on the rectilinear extension of line EZ, figure ABG has been divided into two equal parts, namely, $ABZE$, $GAEZ$. [figure 3.12.3]

If line AE is not on the rectilinear extension of line ZE, we join line AZ and we draw from point E a line parallel to line AZ, namely, ET. We join line TZ. I say that line TZ has been drawn so as to divide figure ABG into two equal parts, namely, $AGZT$, ZBT. Since triangles TZA, EZA are on the same base, namely, AZ, and between two parallel lines, namely, AZ, TE, triangle ZTA is equal to triangle AEZ. Let a common addition be made to them, namely, figure AZG. Then $TZGA$ turns

[58] That is, if H coincides with A.

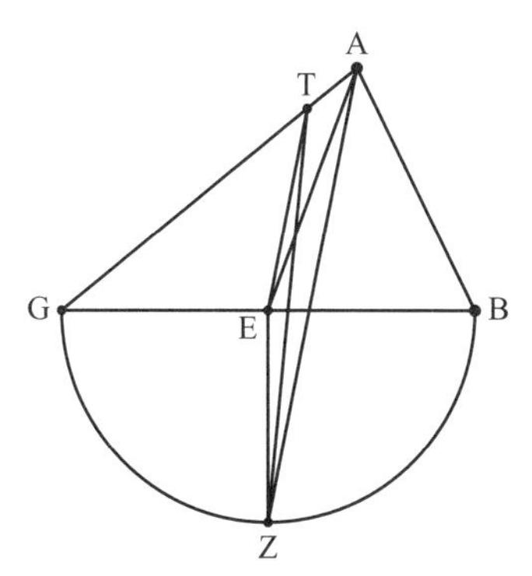

Figure 3.12.3.

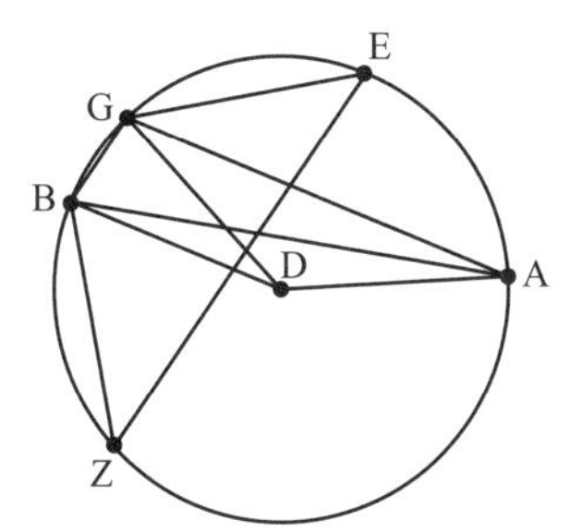

Figure 3.12.4.

out to be equal to *AGZE*, which is half of figure *ABG*. Thus the straight line *TZ* has been drawn so as to divide *BZGA* into two equal parts, namely, *AGZT*, *TZB*. That is what we wished to demonstrate.

28. We wish to draw in a known circle two parallel lines that cut off from the circle some assumed part. Thus let us make the part one-third and the circle *ABG*, and we wish to do what we have said. We make the center of circle *ABG* point *D*, and we make the side of the triangle inscribed in this circle, namely, *AG*, and we draw lines *AD*, *DG*. We let pass through point *D* a line parallel to line *AG*, namely, *DB*. We join line *GB*. We bisect arc *AG* at point *E*. We draw from point *E* a line parallel to line *BG*, namely, *EZ*. We draw line *AB*. I say that the parallel lines *EZ*, *GB* have been drawn so as to cut off from circle *ABG* one-third of it, namely, figure *ZBGE*. [figure 3.12.4]

    Proof of this: Line *AG* is parallel to line *DB*, so triangle *DAG* is equal to triangle *BAG*. Let a common addition be made to them, namely, the circular segment *AEG*. Then the whole figure *DAEG* turns out to be equal to the whole figure *BAEG*. But *DAEG* is one-third of circle *ABG*, so figure *BAEG* is one-third of this circle. Since *EZ* is parallel to *GB*, arc *EG* is equal to arc *BZ*. But *EG* is equal to *EA*, so *EA* turns out to be equal to *ZB*. We make arc *EGB* common. Then the whole arc *AB* is equal to the whole arc *EZ*, so the straight line *AB* is equal to the straight line *EZ*, and the circular segment $AEGB$ turns out to be equal to the circular segment $EGBZ$. We drop the common part, namely, the circular segment GB, then by subtraction figure $EZBG$ is equal to figure $BAEG$. But figure $BAEG$ is one-third of circle ABG. So figure $EZBG$ is one-third of circle ABG. That is what we wanted to demonstrate. If we wish to cut off from the circle one-fourth of it or one-fifth of it or another assumed part by means of two parallel lines, we draw in this circle the side of the square or the pentagon that is inscribed in it, and we draw to it from the center two lines as we have drawn them here, and we proceed as in this construction.[59]

3.13 Euclid, *Elements*, 10

The purpose of book 10 is the classification of certain incommensurable magnitudes. One of the motivations for the book was presumably the desire to characterize the edge lengths of the regular polyhedra displayed in book 13. Euclid needed a nonnumerical way of comparing the edges of the icosahedron and the dodecahedron to the diameter of the sphere in which they were inscribed. In a manner familiar in modern mathematics, this simple question was to lead to the elaborate classification scheme of book 10, far beyond its direct answer. The contents of this book are generally attributed to Theaetetus, since he is credited with some of the polyhedral constructions of book 13 and since it was in Plato's dialogue bearing his name that the question of determining which numbers have square roots incommensurable with the unit was considered. It is the answer to that question, given early in book 10, that then leads to the general classification. We will only present here a small part of Euclid's detailed classification.

Definitions

1. Those magnitudes are said to be *commensurable* which are measured by the same measure, and those *incommensurable* which cannot have any common measure.
2. Straight lines are *commensurable in square* when the squares on them are measured by the same area, and *incommensurable in square* when the squares on them cannot possibly have any area as a common measure.
3. With these hypotheses, it is proved that there exist straight lines infinite in multitude which are commensurable and incommensurable respectively, some in length only, and other in square also, with an assigned straight line. Let then the assigned straight line be called *rational*, and those straight lines which are commensurable with it, whether in length and in square or in square only, *rational*, but those which are incommensurable with it *irrational*.
4. And let the square on the assigned straight line be called *rational* and those areas which are commensurable with it *rational*, but those which are incommensurable with

[59] Although there is a Euclidean construction of the side of a square, a pentagon, or a hexagon inscribed in a circle, such a construction only exists for a limited class of regular polygons. It seems unlikely, then, that Euclid left this solution in exactly this form. Perhaps the translator decided that it was not necessary to have a Euclidean construction to begin this solution.

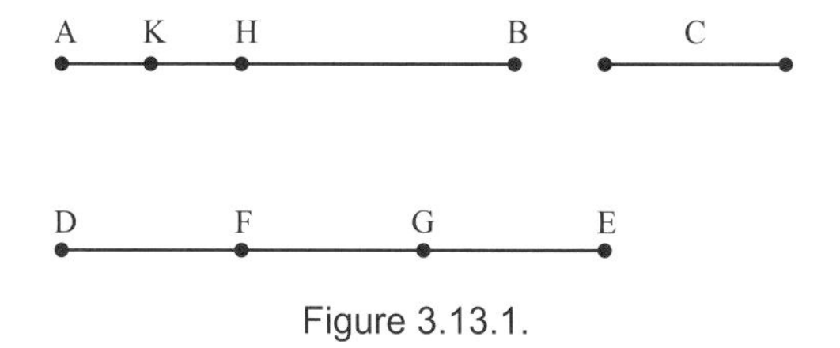

Figure 3.13.1.

it *irrational*, and the straight lines which produce them *irrational*, that is, in case the areas are squares, the sides themselves, but in case they are any other rectilineal figures, the straight lines on which are described squares equal to them.

Although the first two definitions are straightforward, the third one not only contains a theorem (which is proved later in book 10) but also presents a definition of "rational" different from our normal usage today. According to the definition, if the assigned straight line has unit length, then any line of length $\sqrt{a/b}$, where a and b are positive integers, is called "rational."

Propositions

Proposition 1. *Two unequal magnitudes being set out, if from the greater there be subtracted a magnitude greater than its half, and from that which is left a magnitude greater than its half, and if this process be repeated continually, there will be left some magnitude which will be less than the lesser magnitude set out.*

Let AB, C be two unequal magnitudes of which AB is the greater [figure 3.13.1].

I say that, if from AB there be subtracted a magnitude greater than its half, and from that which is left a magnitude greater than its half, and if this process be repeated continually, there will be left some magnitude which will be less than the magnitude C. For C if multiplied will sometime be greater than AB [5:definition 4]. Let it be multiplied, and let DE be a multiple of C, and greater than AB; let DE be divided into the parts DF, FG, GE equal to C, from AB let there be subtracted BH greater than its half, and, from AH, HK greater than its half, and let this process be repeated continually until the divisions in AB are equal in multitude with the divisions in DE.

Let, then, AK, KH, HB be divisions which are equal in multitude with DF, FG, GE. Now, since DE is greater than AB, and from DE there has been subtracted EG less than its half, and, from AB, BH greater than its half, therefore the remainder GD is greater than the remainder HA. And, since GD is greater than HA, and there has been subtracted, from GD, the half GF, and, from HA, HK greater than its half, therefore the remainder DF is greater than the remainder AK. But DF is equal to C; therefore C is also greater than AK. Therefore AK is less than C. Therefore there is left of the magnitude AB the magnitude AK which is less than the lesser magnitude set out, namely C. Q.E.D.

Porism. *And the theorem can be similarly proved even if the parts subtracted be halves.*

The following two definitions lay out what is called the "Euclidean" algorithm for magnitudes. The procedures in the two proofs are analogous to the procedures

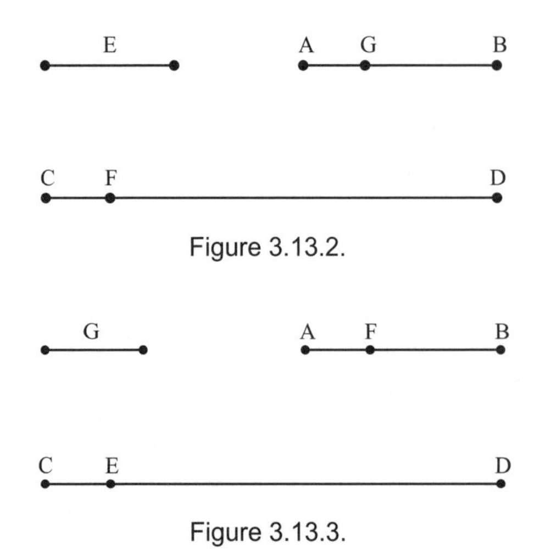

Figure 3.13.2.

Figure 3.13.3.

in the Euclidean algorithm for numbers detailed in book 7. Of course, in book 7 the algorithm always ends, while here Euclid notes that it is possible that the algorithm never ends.

Proposition 2. *If, when the less of two unequal magnitudes is continually subtracted in turn from the greater, that which is left never measures the one before it, the magnitudes will be incommensurable.*

For, there being two unequal magnitudes AB, CD, and AB being the less, when the less in continually subtracted in turn from the greater, let that which is left over never measure the one before it; I say that the magnitudes AB, CD are incommensurable [figure 3.13.2].

For, if they are commensurable, some magnitude will measure them. Let a magnitude measure them, if possible, and let it be E; let AB, measuring FD, leave CF less than itself, let CF, measuring BG, leave AG less than itself, and let this process be repeated continually, until there is left some magnitude which is less than E. Suppose this done, and let there be left AG less than E. Then, since E measures AB, while AB measured DF, therefore E will also measure FD. But it measures the whole CD also; therefore it will also measure the remainder CF. But CF measures BG; therefore E also measures BG. But it measures the whole AB also; therefore it will also measure the remainder AG, the greater the less, which is impossible. Therefore no magnitude will measure the magnitudes AB, CD; therefore the magnitudes AB, CD are incommensurable. Therefore, etc. Q.E.D.

Proposition 3. *Given two commensurable magnitudes, to find their greatest common measure.*

Let the two given commensurable magnitudes be AB, CD of which AB is the less; thus it is required to find the greatest common measure of AB, CD [figure 3.13.3].

Now the magnitude AB either measures CD or it does not. If then it measures it—and it measures itself also—AB is a common measure of AB, CD. And it is manifest that it is also the greatest; for a greater magnitude than the magnitude AB will not measure AB.

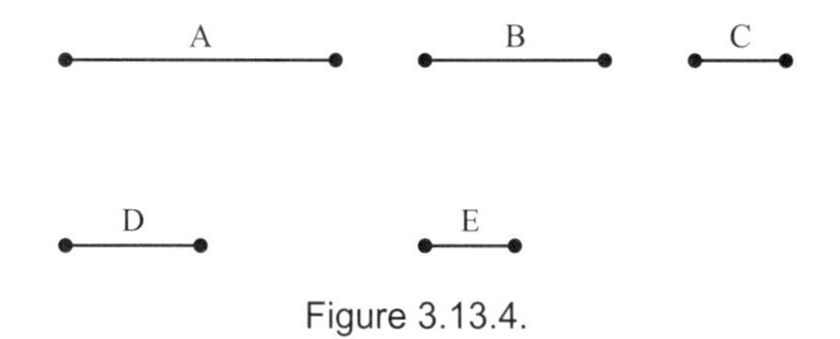

Figure 3.13.4.

Next, let AB not measure CD. Then, if the less be continually subtracted in turn from the greater, that which is left over will sometime measure the one before it, because AB, CD are not incommensurable; let AB, measuring ED leave EC less than itself, let EC, measuring FB, leave AF less than itself, and let AF measure CE. Since, then, AF measures CE, while CE measures FB, therefore AF will also measure FB. But it measures itself also; therefore AF will also measure the whole AB. But AB measures DE; therefore AF will also measure ED. But it measures CE also; therefore it also measures the whole CD. Therefore AF is a common measure of AB, CD.

I say next that it is also the greatest. For, if not, there will be some magnitude greater than AF which will measure AB, CD. Let it be G. Since then G measures AB, while AB measures ED, therefore G will also measure ED. But it measures the whole CD also; therefore G will also measure the remainder CE. But CE measures FB; therefore G will also measure FB. But it measures the whole AB also, and it will therefore measure the remainder AF, the greater the less, which is impossible. Therefore no magnitude greater than AF will measure AB, CD; therefore AF is the greatest common measure of AB, CD. Therefore, the greatest common measure of the two given commensurable magnitudes AB, CD has been found. Q.E.F.

Porism. *From this it is manifest that, if a magnitude measure two magnitudes, it will also measure their greatest common measure.*

In the following two propositions, Euclid shows that magnitudes are commensurable precisely when their ratio is that of a number to a number. So even though numbers and magnitudes are distinct notions, these results enable Euclid to apply the machinery of numerical proportion theory to commensurable magnitudes.

Proposition 5. *Commensurable magnitudes have to one another the ratio which a number has to a number.*

Let A, B be commensurable magnitudes; I say that A has to B the ratio which a number has to a number [figure 3.13.4].

For, since A, B are commensurable, some magnitude will measure them. Let it measure them, and let it be C. And, as many times as C measures A, so many units let there be in D; and, as many times as C measures B, so many units let there be in E. Since then C measures A according to the units in D, while the unit also measures D according to the units in it, therefore the unit measures the number D the same number of times as the magnitude C measures A; therefore as C is to A, so is the unit to D; therefore, inversely, as A is to C, so is D to the unit. Again, since C measures B according to the units in E, while the unit also measures E according to the units in it, therefore the unit measures

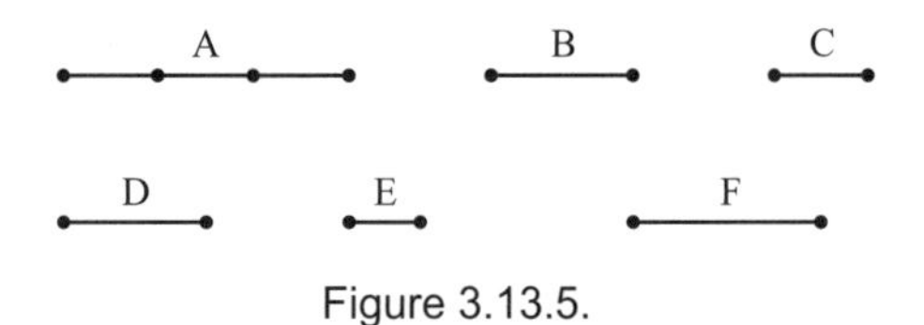

Figure 3.13.5.

E the same number of times as C measures B; therefore, as C is to B, so is the unit to E. But it was also proved that, as A is to C, so is D to the unit; therefore, *ex aequali*, as A is to B, so is the number D to E. Therefore, etc. Q.E.D.

Proposition 6. *If two magnitudes have to one another the ratio which a number has to a number, the magnitudes will be commensurable.*

For let the two magnitudes A, B have to one another the ratio which the number D has to the number E; I say that the magnitudes A, B are commensurable [figure 3.13.5].

For let A be divided into as many equal parts as there are units in D, and let C be equal to one of them; and let F be made up of as many magnitudes equal to C as there are units in E. Since then there are in A as many magnitudes equal to C as there are units in D, whatever part the unit is of D, the same part is C of A also; therefore, as C is to A, so is the unit to D. But the unit measures the number D; therefore C also measures A. And since, as C is to A, so is the unit to D, therefore, inversely, as A is to C, so is the number D to the unit. Again, since there are in F as many magnitudes equal to C as there are units in E, therefore, as C is to F, so is the unit to E. But it was also proved that, as A is to C, so is D to the unit; therefore, *ex aequali*, as A is to F, so is D to E. But, as D is to E, so is A to B; therefore also, as A is to B, so is it to F also. Therefore A has the same ratio to each of the magnitudes B, F; therefore B is equal to F. But C measures F; therefore it measures B also. Further it measures A also; therefore C measures A, B. Therefore A is commensurable with B. Therefore, etc. Q.E.D.

The following proposition provides the generalization of the early Greek discovery that the diagonal of a square is incommensurable with the side, that is, the discovery of the irrationality (in modern terms) of $\sqrt{2}$. This result, in fact, shows that the square root of every non-square integer is incommensurable with the unit.

Proposition 9. *The squares on straight lines commensurable in length have to one another the ratio which a square number has to a square number; and squares which have to one another the ratio which a square number has to a square number will also have their sides commensurable in length. But the squares on straight lines incommensurable in length have not to one another the ratio which a square number has to a square number; and squares which have not to one another the ratio which a square number has to a square number will not have their sides commensurable in length either.*

For let A, B be commensurable in length; I say that the square on A has to the square on B the ratio which a square number has to a square number [figure 3.13.6].

For, since A is commensurable in length with B, therefore A has to B the ratio which a number has to a number [10:5]. Let it have to it the ratio which C has to D. Since then, as

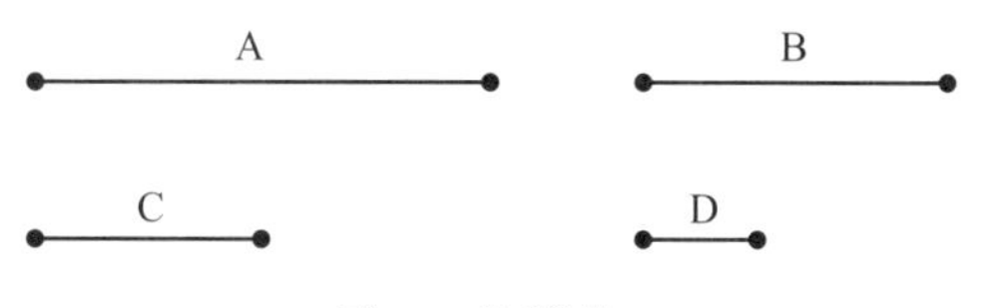

Figure 3.13.6.

A is to B, so is C to D, while the ratio of the square on A to the square on B is duplicate of the ratio of A to B, for similar figures are in the duplicate ratio of their corresponding sides; and the ratio of the square on C to the square on D is duplicate of the ratio of C to D, for between two square numbers there is one mean proportional number, and the square number has to the square number the ratio duplicate of that which the side has to the side; therefore also, as the square on A is to the square on B, so is the square on C to the square on D.

Next, as the square on A is to the square on B, so let the square on C be to the square on D; I say that A is commensurable in length with B. For since, as the square on A is to the square on B, so is the square on C to the square on D, while the ratio of the square on A to the square on B is duplicate of the ratio of A to B, and the ratio of the square on C to the square on D is duplicate of the ratio of C to D; therefore also, as A is to B, so is C to D. Therefore A has to B the ratio which the number C has to the number D; therefore A is commensurable in length with B [10:6].

Next, let A be incommensurable in length with B; I say that the square on A has not to the square on B the ratio which a square number has to a square number. For, if the square on A has to the square on B the ratio which a square number has to a square number, A will be commensurable with B. But it is not; therefore the square on A has not to the square on B the ratio which a square number has to a square number.

Again, let the square on A not have to the square on B the ratio which a square number has to a square number; I say that A is incommensurable in length with B. For, if A is commensurable with B, the square on A will have to the square on B the ratio which a square number has to a square number. But it has not; therefore A is not commensurable in length with B. Therefore, etc. Q.E.D.

The following four propositions, whose proofs we omit, are examples of Euclid's classification of irrational lines. The third and fourth results, in particular, are used in the description of the lengths of the sides of an icosahedron and a dodecahedron in book 13. In each case, we present a numerical example to help the reader understand the definition.

Proposition 21. *The rectangle contained by rational straight lines commensurable in square only is irrational, and the side of the square equal to it is irrational. Let the latter be called medial.*

The lengths 1 and $\sqrt{5}$ are commensurable in square only. Thus, the rectangle contained by these lines, of area $\sqrt{5}$, is irrational, and the side of the square equal to it, namely, $\sqrt[4]{5}$, is an example of a medial line.

Proposition 36. *If two rational straight lines commensurable in square only be added together, the whole is irrational; and let it be called binomial.*

Since 1 and $\sqrt{5}$ are two rational lengths commensurable in square only, their sum, $1+\sqrt{5}$ is an example of the irrational length called binomial.

Proposition 73. *If from a rational straight line there be subtracted a rational straight line commensurable with the whole in square only, the remainder is irrational; and let it be called an apotome.*

Since $\sqrt{15}$ and $\sqrt{3}$ are rational straight lines commensurable in square only, their difference, $\sqrt{15}-\sqrt{3}$, is an apotome.

Proposition 76. *If from a straight line there be subtracted a straight line which is incommensurable in square with the whole and which with the whole makes the squares on them added together rational, but the rectangle contained by them medial, the remainder is irrational; and let it be called minor.*

According to this proposition, a minor straight line is the difference $x-y$ between two lines such that x, y are incommensurable in square, such that x^2+y^2 is rational, and such that xy is a medial area, that is, equal to the square on a medial straight line. If we choose $x=\sqrt{5+2\sqrt{5}}$ and $y=\sqrt{5-2\sqrt{5}}$, then $x^2=5+2\sqrt{5}$ and $y^2=5-2\sqrt{5}$ are incommensurable, their sum, 10, is rational, and their product, $\sqrt{5}$, is the square on a medial straight line. Thus, $x-y=\sqrt{5+2\sqrt{5}}-\sqrt{5-2\sqrt{5}}=\sqrt{10-2\sqrt{5}}$ is a minor straight line.

3.14 Pappus, *Commentary on Euclid's Elements, Book 10*

Pappus of Alexandria was one of the last mathematicians of the classical Greek tradition. He was familiar with many of the major and minor works of the mathematicians discussed in this *Sourcebook* and is most famous for his *Collection*, excerpted in several chapters of this work. Pappus's commentary on book 10 of Euclid's *Elements*, excerpted here, is not from the *Collection* but is only preserved in Arabic, and there is only one extant manuscript. The Arab translator was Abū 'Uthmān al-Dimishqī, who lived in the tenth century in Baghdad, while the writer of the known manuscript is most probably Abū Sa'īd Aḥmad ibn Muḥammad ibn 'Abd al-Jalīl al-Sijzī, a Persian mathematician who was a correspondent of Abū l-Rāyḥan Muḥammad ibn Aḥmad al-Bīrūnī (973–1055). Pappus's aim in his commentary is to set book 10 into the general history of the Greek conception of rational and irrational quantities. Of course, he also tried to explain Euclid's definitions of the various types of irrationals and give examples to help the reader through this difficult book. In particular, he shows the relationship between three of Euclid's classes of irrational lines: the medial, the binomial, and the apotome. We present just a few excerpts from part one of this commentary.

1. The aim of Book 10 of Euclid's treatise on the *Elements* is to investigate the commensurable and incommensurable, the rational and irrational continuous quantities. This science had its origin in the sect of Pythagoras, but underwent an important development at the hands of the Athenian, Theaetetus, who had a natural aptitude for this as for other

branches of mathematics most worthy of admiration. One of the most happily endowed of men, he patiently pursued the investigation of the truth contained in these [branches of] science, as Plato bears witness for him in the book which he called after him, and was in my opinion the chief means of establishing exact distinctions and irrefragable proofs with respect to the above-mentioned quantities. For although later the great Apollonius, whose genius for mathematics was of the highest possible order, added some remarkable species of these after much laborious application, it was nevertheless Theaetetus who distinguished the squares which are commensurable in length from those which are incommensurable [in length] and who divided the more generally known irrational lines according to the different means, assigning the medial line to geometry, the binomial to arithmetic, and the apotome to harmony, as is stated by Eudemus, the Peripatetic. Euclid's object, on the other hand, was the attainment of irrefragable principles, which he established for commensurability and incommensurability in general. For rationals and irrationals he formulated definitions and (specific) differences; determined also many orders of the irrationals; and brought to light, finally, whatever of definiteness is to be found in them. Apollonius explained the species of the ordered irrationals and discovered the science of the so-called unordered, of which he produced an exceedingly large number by exact methods.

10. Since those who have been influenced by speculation concerning the knowledge of Plato, suppose that the definition of straight lines commensurable in length and square and commensurable in square only, which he gives in his book entitled *Theaetetus*, does not at all correspond with what Euclid proves concerning these lines, it seems to us that something should be said regarding this point. After, then, Theodorus had discussed with Theaetetus the proofs of the squares which are commensurable and incommensurable in length relatively to the square whose measure is a foot, the latter had recourse to a general definition of those squares, after the fashion of one who has applied himself to that knowledge which is in its nature certain. Accordingly he divided all numbers into two classes, such as are the product of equal sides, on the one hand, and on the other, such as are contained by a greater side and a less; and he represented the first [class] by a square figure and the second by an oblong, and concluded that the numbers which form into a square figure whose sides are equal are commensurable both in square and in length, but that those which form into an oblong number are incommensurable with the first [class] in the latter respect, but are commensurable occasionally with one another in one respect. Euclid, on the other hand, after he had examined this treatise carefully for some time and had determined the lines which are commensurable in length and square, those, namely, whose squares have to one another the ratio of a square number to a square number, proved that all lines of this kind are always commensurable in length. The difference between Euclid's statement and that of Theaetetus which precedes it, has not escaped us. The idea of determining these squares by means of the square numbers is a different idea altogether from that of them having to one another the ratio of a square number to a square number. For example, if there be taken, on the one hand, a square whose measure is eighteen feet, and on the other hand, another square whose measure is eight feet, it is quite clear that the one square has to the other the ratio of a square number to a square number, the numbers, namely, which these two double, notwithstanding the fact that the

two squares are determined by means of oblong numbers. Their sides, therefore, are commensurable according to the definition of Euclid, whereas according to the definition of Theaetetus they are excluded from this category. For the two squares do not form into a square figure a number whose sides are equal, but only an oblong number. So much, then, regarding what should be known concerning these things.

11. It should be observed, however, that the argument of Theaetetus does not cover every square that there is, be it commensurable in length or incommensurable, but only the squares that have ratios relative to some rational square or other, the square, namely, whose measure is a square foot. For it was with this square as basis that Theodorus began his investigation concerning the square whose measure is three [square] feet and the square whose measure is five [square] feet, and declared that they are incommensurable [in length] with the square whose measure is one [square] foot. Theaetetus explains this by saying, "We defined as lengths [the sides of the squares] whose square is a number whose factors are equal, but [the sides of the squares] whose square is an oblong number, we defined as roots, inasmuch as they are incommensurable in length with the former, the square whose measure is a [square] foot, but are commensurable with the squares that can be described upon these lengths." The argument of Euclid, on the contrary, covers every square and is not relative to some assumed rational square or line only. Moreover, it is not possible for us to prove by any theorem that the squares that we have described above are commensurable with one another in length, despite the fact that they are incommensurable in length with the square whose measure is a [square] foot, and that the unit [of measurement] which measures the lines is irrational, the lines, namely, on which these squares (i.e., the squares 18 and 8) are imagined as described. It is difficult, consequently, for those who seek to determine a recognized measure for the lines which have the power to form these squares to follow the investigation of this [problem] of irrationals, whereas whoever has carefully studied Euclid's proof, can see that the lines are undoubtedly commensurable [with one another]. For he proves that they have to one another the ratio of a number to a number. Such is the substance of our remarks concerning the uncertainty about Plato.

19. We must also consider the following fact. Having found by geometrical proportion that the medial line is a mean proportional between two rational lines commensurable in square only and, therefore, that the square upon it is equal to the area contained by these two lines—the square upon a medial line being one which is equal to the rectangle contained by the two assigned lines as its adjacent sides—Euclid always assigns the general term, medial, to a particular species of the medial line. For the medial line the square upon which is equal to the rectangle contained by two rational lines commensurable in length, is necessarily a mean proportional to these two rationals; and the line the square upon which is equal to the rectangle contained by a rational and an irrational line, is also of that type (i.e., a mean proportional); but he does not name either of these medial, but only the line the square upon which is equal to the given rectangle [i.e., that contained by two rational lines commensurable in square only]. Moreover since in every case he derives the names of the square-areas from the lines upon which they are the squares, he names the area on a rational line rational and that on a medial line medial.

21. Subsequent to his investigation and production of the medial line, Euclid began, after careful consideration, an examination of those irrational lines that are formed by addition and subtraction on the basis of the examination which he had made, of commensurability and incommensurability, commensurability and incommensurability appearing also in those lines that are formed by addition and subtraction. The first of the lines formed by addition is the binomial; for it also [like the medial with respect to all irrational lines] is the most homogeneous of such lines to the rational line, being composed of two rational lines commensurable in square. The first of the lines formed by subtraction is the apotome; for it also is produced by simply subtracting from a rational line another rational line commensurable with the whole in square. We find, therefore, the medial line by assuming a rational side and a given diagonal and taking the mean proportional between these two lines; we find the binomial by adding together the side and the diagonal, and we find the apotome by subtracting the side from the diagonal.[60] We should also recognize, however, that not only when we join together two rational lines commensurable in square, do we obtain a binomial, but three or four such lines produce the same thing. In the first case a trinomial is produced, since the whole line is irrational, in the second a quadrinomial, and so on indefinitely. The proof of the irrationality of the line composed of three rational lines commensurable in square is exactly the same as in the case of the binomial.

3.15 Euclid, *Elements*, 11–13

Book 11 is the first of three books of the *Elements* dealing mostly with solid geometry. In particular, book 11 contains the three-dimensional analogues of many of the two-dimensional results of books 1 and 6. In the introductory definitions, Euclid includes standard definitions of a pyramid and a prism. But his definition of a sphere is not the solid analogue of the definition of a circle but is in terms of the rotation of a semicircle about its diameter. Presumably, the reason for this is that Euclid did not intend to discuss the properties of a sphere as he did those of a circle in book 3 but only to calculate its volume (in book 12) and the comprehension of the regular polyhedra in spheres (in book 13). In fact, the elementary properties of a sphere were known in Euclid's time and are included in other texts in this *Sourcebook*.

Book 11

Definitions

1. A *solid* is that which has length, breadth, and depth.
2. An extremity of a solid is a surface.
11. A *solid angle* is the inclination constituted by more than two lines which meet one another and are not in the same surface, towards all the lines. Otherwise: A *solid angle* is that which is contained by more than two plane angles which are not in the same plane and are constructed to one point.

[60] For example, we can in each case consider the rectangle contained by 1 and 2, with diagonal $\sqrt{5}$.

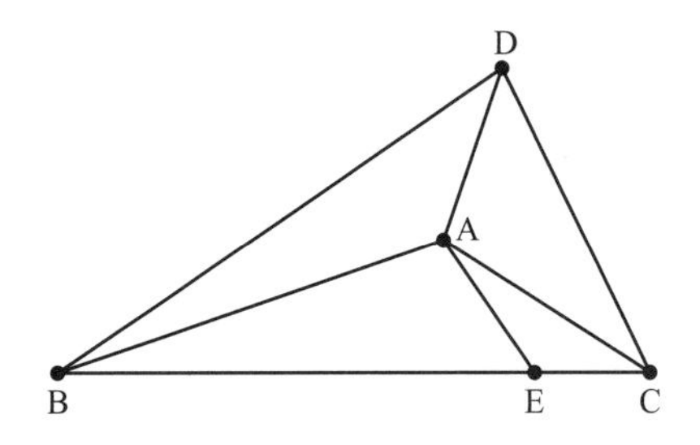

Figure 3.15.1.

12. A *pyramid* is a solid figure, contained by planes, which is constructed from one plane to one point.
13. A *prism* is a solid figure contained by planes two of which, namely, those which are opposite, are equal, similar and parallel, which the rest are parallelograms.
14. When the diameter of a semicircle remaining fixed, the semicircle is carried round and restored again to the same position from which it began to be moved, the figure so comprehended is a *sphere*.
18. When, one side of those about the right angle in a right-angled triangle remaining fixed, the triangle is carried round and restored again to the same position from which it began to be moved, the figure so comprehended is a *cone*. And, if the straight line which remains fixed be equal to the remaining side about the right angle which is carried round, the cone will be *right-angled*; if less, *obtuse-angled*; and if greater, *acute-angled*.
21. When, one side of those about the right angle in a rectangular parallelogram remaining fixed, the parallelogram is carried round and restored again to the same position from which it began to be moved, the figure so comprehended is a *cylinder*.

Propositions

Proposition 4. *If a straight line be set up at right angles to two straight lines which cut one another, at t heir common point of section, it will also be at right angles to the plane through them.*

Here and in most of the other propositions included, we do not present the proofs, as they are rather long and complicated.

Proposition 11. *From a given elevated point to draw a straight line perpendicular to a given plane.*

Proposition 20. *If a solid angle be contained by three plane angles, any two, taken together in any manner, are greater than the remaining one.*

For let the solid angle at A be contained by the three plane angles BAC, CAD, DAB; I say that any two of the angles BAC, CAD, DAB, taken together in any manner, are greater than the remaining one [figure 3.15.1].

If now the angles BAC, CAD, DAB, are equal to one another, it is manifest that any two are greater than the remaining one. But, if not, let BAC be greater, and on the straight

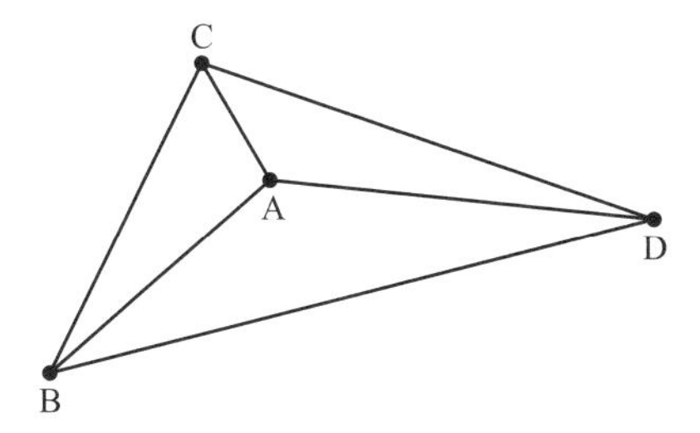

Figure 3.15.2.

line AB, and at the point A on it, let the angle BAE be constructed, in the plane through BA, AC, equal to the angle DAB; let AE be made equal to AD, and let BEC, drawn across through the point E, cut the straight lines AB, AC at the points B, C; let DB, DC be joined.

Now, since DA is equal to AE, and AB is common, two sides are equal to two sides; and the angle DAB is equal to the angle BAE; therefore the base DB is equal to the base BE. And, since the two sides BD, DC are greater than BC, and of these DB was proved equal to BE, therefore the remainder DC is greater than the remainder EC. Now, since DA is equal to AE, and AC is common, and the base DC is greater than the base EC, therefore the angle DAC is greater than the angle EAC. But the angle DAB was made equal to the angle BAE; therefore the angles DAB, DAC are greater than the angle BAC. Similarly we can prove that the remaining angles also, taken together two and two, are greater than the remaining one. Therefore, etc. Q.E.D.

The following proposition is used in the proof of the final result of book 13.

Proposition 21. *Any solid angle is contained by plane angles less than four right angles.*

Let the angle at A be a solid angle contained by the plane angles BAC, CAD, DAB; I say that the angles BAC, CAD, DAB are less than four right angles [figure 3.15.2].

For let points B, C, D be taken at random on the straight lines AB, AC, AD respectively, and let BC, CD, DB be joined. Now, since the solid angle at B is contained by the three plane angles CBA, ABD, CBD, any two are greater than the remaining one [11:20]. Therefore the angles CBA, ABD are greater than the angle CBD. For the same reason the angles BCA, ACD are also greater than the angle BCD, and the angles CDA, ADB are greater than the angle CDB; therefore the six angles CBA, ABD, BCA, ACD, CDA, ADB are greater than the three angles CBD, BCD, CDB. But the three angles CBD, BDC, BCD are equal to two right angles; therefore the six angles CBA, ABD, BCA, ACD, CDA, ADB are greater than two right angles.

And, since the three angles of each of the triangles ABC, ACD, ADB are equal to two right angles, therefore the nine angles of the three triangles, the angles CBA, ACB, BAC, ACD, CDA, CAD, ADB, DBA, BAD are equal to six right angles; and of them the six angles ABD, BCA, ACD, CDA, ADB, DBA are greater than two right angles; therefore the remaining three angles BAC, CAD, DAB containing the solid angles are less than four right angles. Therefore, etc. Q.E.D.

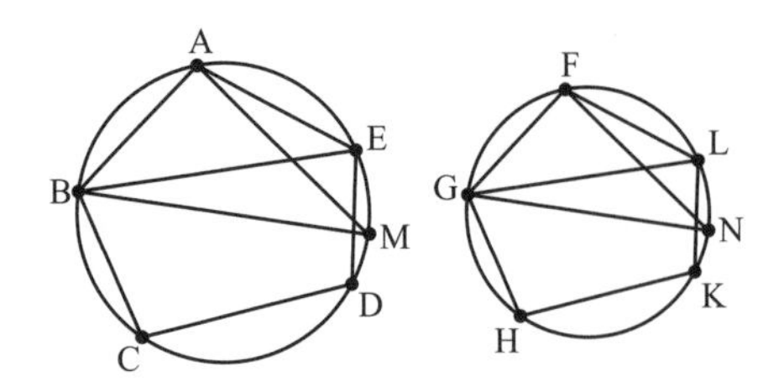

Figure 3.15.3.

The following propositions are analogous to propositions on parallelograms found in book 6. And although Euclid does not explicitly calculate volumes here, one can easily derive from these propositions the basic results on volumes of parallelepipeds.

Proposition 31. *Parallelepipedal solids which are on equal bases and of the same height are equal to one another.*

Proposition 32. *Parallelepipedal solids which are of the same height are to one another as their bases.*

Proposition 33. *Similar parallelepipedal solids are to one another in the triplicate ratio of their corresponding sides.*

Proposition 34. *In equal parallelepipedal solids the bases are reciprocally proportional to the heights; and those parallelepipedal solids in which the bases are reciprocally proportional to the heights are equal.*

The central feature of book 12 is the use of a limiting process, the so-called method of exhaustion. This process, probably developed by Eudoxus, enabled Euclid to validate the previously known results for the area of a circle and the volume of a pyramid, a cone, and a sphere, namely, that circles are to one another as the squares on their diameters, that the pyramid and cone are one-third of the surrounding prism and cylinder, respectively, and that spheres are to one another as the cubes on their diameters. However, the method of exhaustion only provided a method of proof, not a method for discovering the results in the first place.

Book 12

Propositions

Proposition 1. *Similar polygons inscribed in circles are to one another as the squares on the diameters.*

Let ABC, FGH be circles, let $ABCDE$, $FGHKL$ be similar polygons inscribed in them, and let BM, GN be diameters of the circles [figure 3.15.3].

I say that, as the square on BM is to the square on GN, so is the polygon $ABCDE$ to the polygon $FGHKL$. For let BE, AM, GL, FN be joined. Now, since the polygon $ABCDE$ is similar to the polygon $FGHKL$, the angle BAE is equal to the angle GFL, and, as BA is

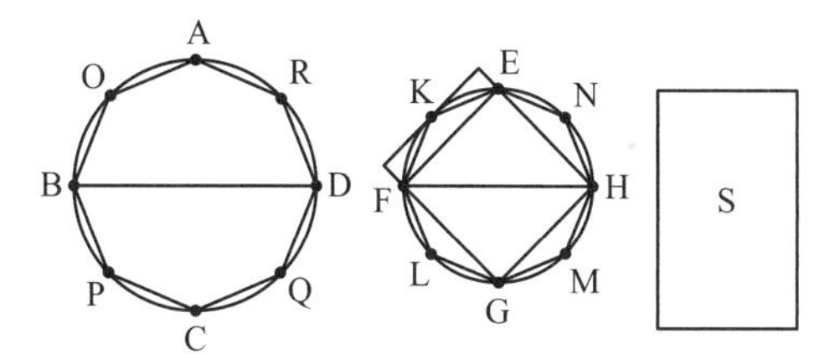

Figure 3.15.4.

to AE, so is GF to FL. Thus BAE, GFL are two triangles which have one angle equal to one angle, namely, the angle BAE to the angle GFL, and the sides about the equal angles proportional; therefore the triangle ABE is equiangular with the triangle FGL. Therefore the angle AEB is equal to the angle FLG. But the angle AEB is equal to the angle AMB, for they stand on the same circumference; and the angle FLG to the angle FNG; therefore the angle AMB is also equal to the angle FNG. But the right angle BAM is also equal to the right angle GFN; therefore the remaining angle is equal to the remaining angle. Therefore the triangle ABM is equiangular with the triangle FGN. Therefore, proportionally, as BM is to GN, so is BA to GF. But the ratio of the square on BM to the square on GN is duplicate of the ratio of BM to GN, and the ratio of the polygon $ABCDE$ to the polygon $FGHKL$ is duplicate of the ratio of BA to GF. Therefore also, as the square on BM is to the square on GN, so is the polygon $ABCDE$ to the polygon $FGHKL$. Therefore, etc. Q.E.D.

The proof of the following proposition shows Euclid's first use of the method of exhaustion. Namely, in order to prove that the area A_1 of one circle C_1 is to the area A_2 of another circle C_2 as the square on the diameter of the first, d_1^2, is to the square on the diameter of the second, d_2^2, Euclid assumes that this is not true. That is, he first assumes that the ratio of the squares on the diameters are as the area of the first circle is to an area S smaller than A_2, the area of the second circle. He then inscribes a polygon P_2 in C_2 such that $A_2 > P_2 > S$ and a similar polygon P_1 in C_1. By proposition 1, $d_1^2 : d_2^2 = P_1 : P_2 = A_1 : S$. So $P_1 : A_1 = P_2 : S$. But since $A_1 > P_1$, it follows that $S > P_2$, contradicting the assumption that $S < P_2$. Therefore, S cannot be less than A_2. Euclid then goes on to prove that S also cannot be greater than A_2 by reducing that assumption to the first case. The conclusion of the theorem then follows. As we see later in this book and in many of the works of Archimedes, this type of argument became standard in Greek mathematics in results involving areas and volumes. However, the argument was often complex and difficult to follow. Thus, ultimately, mathematicians looked for other ways to discover and prove results about areas and volumes.

Proposition 2. *Circles are to one another as the squares on the diameters.*

Let $ABCD$, $EFGH$ be circles, and BD, FH their diameters; I say that, as the circle $ABCD$ is to the circle $EFGH$, so is the square on BD to the square on FH [figure 3.15.4].

For, if the square on BD is not to the square on FH as the circle $ABCD$ is to the circle EGH, then, as the square on BD is to the square on FH, so will the circle $ABCD$ be either to some less area than the circle $EFGH$, or to a greater. First, let it be in that ratio to a less area *S*. Let the square *EFGH* be inscribed in the circle *EFGH*; then the inscribed square is greater than the half of the circle *EFGH*, inasmuch as, if through the point *E*, *F*, *G*, *H* we draw tangents to the circle, the square *EFGH* is half the square circumscribed about the circle, and the circle is less than the circumscribed square; hence the inscribed square *EFGH* is greater than the half of the circle *EFGH*.

Let the circumferences *EF*, *FG*, *GH*, *HE* be bisected at the points *K*, *L*, *M*, *N*, and let *EK*, *KF*, *FL*, *LG*, *GM*, *MH*, *HN*, *NE* be joined; therefore each of the triangles *EKF*, *FLG*, *GMH*, *HNE* is also greater than the half of the segment of the circle about it, inasmuch as, if through the points *K*, *L*, *M*, *N* we draw tangents to the circle and complete the parallelograms on the straight lines *EF*, *FG*, *GH*, *HE*, each of the triangles *EKF*, *FLG*, *GMH*, *HNE* will be half of the parallelogram about it, while the segment about it is less than the parallelogram; hence each of the triangles *EKF*, *FLG*, *GMH*, *HNE* is greater than the half of the segment of the circle about it. Thus, by bisecting the remaining circumferences and joining straight lines, and by doing this continually, we shall leave some segments of the circle which will be less than the excess by which the circle *EFGH* exceeds the area *S*. For it was proved in the first theorem of the tenth book that, if two unequal magnitudes be set out, and if from the greater there be subtracted a magnitude greater than the half, and from that which is left a greater than the half, and if this be done continually, there will be left some magnitude which will be less than the lesser magnitude set out.

Let segments be left such as described, and let the segments of the circle *EFGH*, on *EK*, *KF*, *FL*, *LG*, *GM*, *MH*, *HN*, *NE* be less than the excess by which the circle *EFGH* exceeds the area *S*. Therefore the remainder, the polygon *EKFLGMHN*, is greater than the area *S*. Let there be inscribed, also, in the circle *ABCD* the polygon *AOBPCQDR* similar to the polygon *EKFLGMHN*; therefore, as the square on *RD* is to the square on *FH*, so is the polygon *AOBPCQDR* to the polygon *EKFLGMHN* [12:1]. But, as the square on *BD* is to the square on *FH*, so also is the circle *ABCD* to the area *S*; therefore also, as the circle *ABCD* is to the area *S*, so is the polygon *AOBPCQDR* to the polygon *EKFLGMHN*. But the circle *ABCD* is greater than the polygon inscribed in it; therefore the area *S* is also greater than the polygon *EKFLGMHN*. But it is also less; which is impossible. Therefore, as the square on *BD* is to the square on *FH*, so is not the circle *ABCD* to any area less than the circle *EFGH*. Similarly we can prove that neither is the circle *ABCD* to any area less than the circle *ABCD* as the square on *FH* is to the square on *BD*.

I say next that neither is the circle *ABCD* to any area greater than the circle *EFGH* as the square on *BD* is to the square on *FH*. For, if possible, let it be in that ratio to a greater area *S*. Therefore, inversely, as the square on *FH* is to the square on *DB*, so is the area *S* to the circle *ABCD*. But, as the area *S* is to the circle *ABCD*, so is the circle *EFGH* to some area less than the circle *ABCD*; therefore also, as the square on *FH* is to the square on *BD*, so is the circle *EFGH* to some area less than the circle *ABCD*, which was proved impossible. Therefore, as the square on *BD* is to the square on *FH*, so is not the circle *ABCD* to any area greater than the circle *EFGH*. And it was proved that neither is it in that ratio to any area less than the circle *EFGH*; therefore, as the square on *BD* is to the square on *FH*, so is the circle *ABCD* to the circle *EFGH*. Therefore, etc. Q.E.D.

Euclid uses an exhaustion argument to prove results on pyramids, cones, and spheres, but we will only display the results here. For pyramids, he uses the method of exhaustion to prove the following, by dividing the pyramids into smaller similar pyramids and prisms:

Proposition 5. *Pyramids which are of the same height and have triangular bases are to one another as the bases.*

He then generalizes this result to pyramids with arbitrary polygonal bases, shows that any prism with a triangular base can be divided into three equal pyramids with triangular bases, and finally presents the volume formula for a pyramid as a porism to the last result:

Porism: *From this it is manifest that any pyramid is a third part of the prism which has the same base with it and equal height.*

Again, Euclid uses the method of exhaustion to prove the analogous result on cones, by looking at prisms that are closer and closer to the cylinder and considering pyramids enclosed in those prisms:

Proposition 10. *Any cone is a third part of the cylinder which has the same base with it and equal height.*

Euclid concludes book 12 with the result on the volume of a sphere, again proved using the method of exhaustion, here with polygonal solids inscribed in the sphere:

Proposition 18. *Spheres are to one another in the triplicate ratio of their respective diameters.*

Book 13 is devoted to the construction of the five regular polyhedra and their "comprehension" in a sphere. That is, Euclid shows how to construct each of the solids and inscribe them in a given sphere. Since many of the proofs are quite long and involved, we have chosen in several cases just to summarize the proof and/or to put it into modern terms.

Book 13

Propositions

Proposition 1. *If a straight line be cut in extreme and mean ratio, the square on the greater segment added to the half of the whole is five times the square on the half.*

For let the straight line AB be cut in extreme and mean ratio at the point C, and let AC be the greater segment; let the straight line AD be produced in a straight line with CA, and let AD be made half of AB [figure 3.15.5];

I say that the square on CD is five times the square on AD. For let the squares AE, DF be described on AB, DC, and let the figure in DF be drawn; let FC be carried through to G. Now, since AB has been cut in extreme and mean ratio at C, therefore the rectangle AB, BC is equal to the square on AC. And CE is the rectangle AB, BC, and FH the square

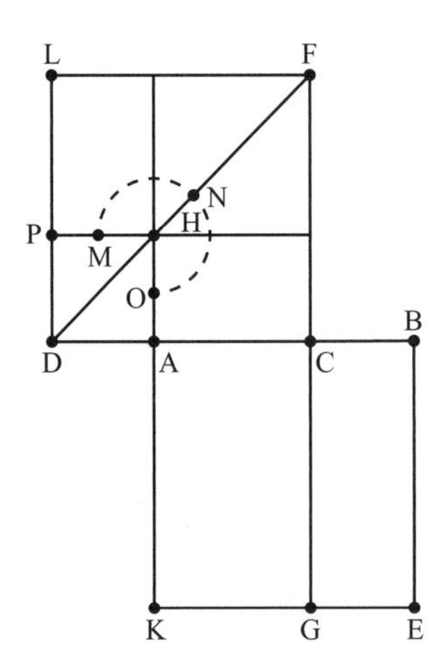

Figure 3.15.5.

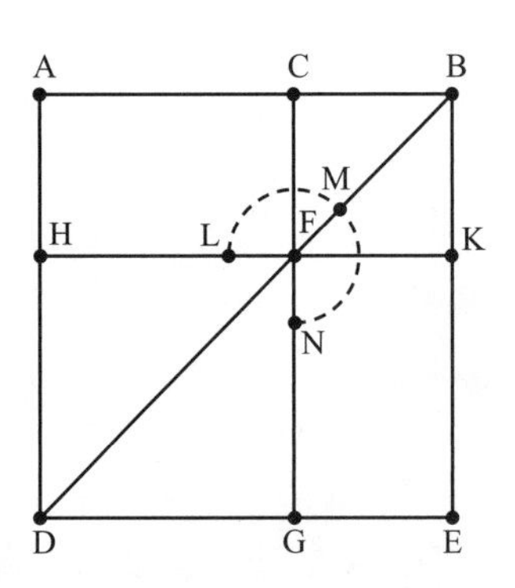

Figure 3.15.6.

on AC; therefore CE is equal to FH. And, since BA is double of AD, while BA is equal to KA, and AD to AH, therefore KA is also double of AH. But as KA is to AH, so is CK to CH; therefore CK is double of CH. But LH, HC are also double of CH. Therefore KC is equal to LH, HC. But CE was also proved equal to HF; therefore the whole square AE is equal to the gnomon MNO. And, since BA is double of AD, the square on BA is quadruple of the square on AD, that is, AE is quadruple of DH. But AE is equal to the gnomon MNO; therefore the gnomon MNO is also quadruple of AP; therefore the whole DF is five times AP. And DF is the square on DC, and AP the square on DA; therefore the square on CD is five times the square on DA. Therefore, etc. Q.E.D.

Proposition 4. *If a straight line be cut in extreme and mean ratio, the square on the whole and the square on the lesser segment together are triple of the square on the greater segment.*

Let AB be a straight line, let it be cut in extreme and mean ratio at C, and let AC be the greater segment [figure 3.15.6];

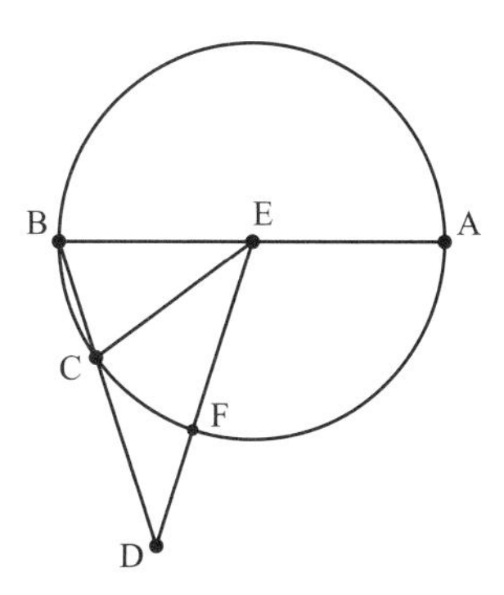

Figure 3.15.7.

I say that the square on AB, BC are triple of the square on CA. For let the square $ADEB$ be described on AB, and let the figure be drawn. Since, then, AB has been cut in extreme and mean ratio at C, and AC is the greater segment, therefore the rectangle AB, BC is equal to the square on AC. And AK is the rectangle AB, BC, and HG the square on AC; therefore AK is equal to HG. And, since AF is equal to FE, let CK be added to each; therefore the whole AK is equal to the whole CE; therefore AK, CE are double of AK. But AK, CE are the gnomon LMN and the square CK; therefore the gnomon LMN and the square CK are double of AK. But, further, CK was also proved equal to HG; therefore the gnomon LMN and the squares CK, HG are triple of the square HG. And the gnomon LMN and the squares CK, HG are the whole square AE and CK, which are the squares on AB, BC, while HG is the square on AC. Therefore the squares on AB, BC are triple of the square on AC. Therefore, etc. Q.E.D.

Propositions 1 and 4 are easily proved algebraically. If a line of length a is cut in extreme and mean ratio, then the lengths of the two segments are $\frac{\sqrt{5}-1}{2}a$ and $\frac{3-\sqrt{5}}{2}a$. Therefore, the square on the greater segment added to the half of the whole is $[(\frac{\sqrt{5}-1}{2})a + \frac{1}{2}a]^2 = \frac{5}{4}a^2$, which is clearly five times the square of $\frac{a}{2}$. Similarly, the square on the whole together with the square on the lesser segment is $a^2 + (\frac{3-\sqrt{5}}{2}a)^2 = \frac{9-3\sqrt{5}}{2}a^2$, clearly triple the square on the greater segment: $(\frac{\sqrt{5}-1}{2}a)^2 = \frac{3-\sqrt{5}}{2}a^2$.

Proposition 9. *If the side of the hexagon and that of the decagon inscribed in the same circle be added together, the whole straight line has been cut in extreme and mean ratio, and its greater segment is the side of the hexagon.*

Let ABC be a circle; of the figures inscribed in the circle ABC let BC be the side of a decagon, CD that of a hexagon, and let them be in a straight line; I say that the whole straight line BD has been cut in extreme and mean ratio, and CD is its greater segment [figure 3.15.7].

For let the center of the circle, the point E be taken, let EB, EC, ED be joined, and let BE be carried through to A. Since BC is the side of an equilateral decagon, therefore the

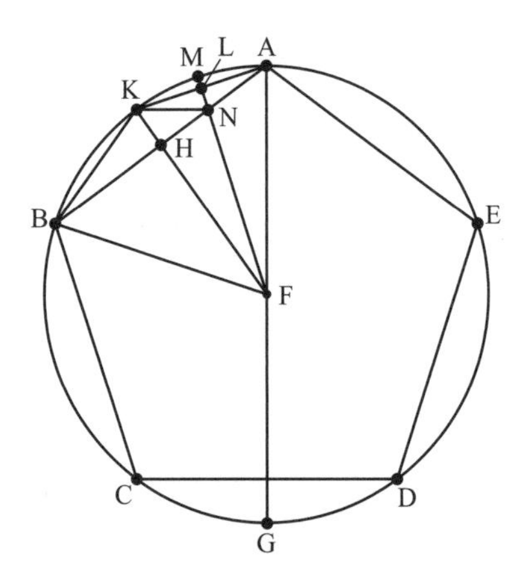

Figure 3.15.8.

circumference ACB is five times the circumference BC; therefore the circumference AC is quadruple of CB. But as the circumference AC is to CB, so is the angle AEC to the angle CEB; therefore the angle AEC is quadruple of the angle CEB. And, since the angle EBC is equal to the angle ECB, therefore the angle AEC is double of the angle ECB. And, since the straight line EC is equal to CD, for each of them is equal to the side of the hexagon inscribed in the circle ABC, the angle CED is also equal to the angle CDE; therefore the angle ECB is double of the angle EDB. But the angle AEC was proved double of the angle ECB; therefore the angle AEC is quadruple of the angle EDC. But the angle AEC was also proved quadruple of the angle BEC; therefore the angle EDC is equal to the angle BEC. But the angle EBD is common to the two triangles BEC and BED; therefore the remaining angle BED is also equal to the remaining angle ECB; therefore the triangle EBD is equiangular with the triangle EBC. Therefore, proportionally, as DB is to BE, so is EB to BC. But EB is equal to CD. Therefore, as BD is to DC, so is DC to CB. And BD is greater than DC; therefore DC is also greater than CB. Therefore the straight line BD has been cut in extreme and mean ratio, and DC is its greater segment. Q.E.D.

Proposition 10. *If an equilateral pentagon be inscribed in a circle, the square on the side of the pentagon is equal to the squares on the side of the hexagon and on that of the decagon inscribed in the same circle.*

We sketch Euclid's proof. Let F be the center of the circle and AB one side of an inscribed equilateral pentagon. Let FH be the perpendicular bisector of AB and let it be extended to K. Connect AK and KB. Then AK is the side of a decagon. Also, let FL be drawn perpendicular to AK and let it be extended to M; let N be the intersection of FM with AB and connect KN [figure 3.15.8].

Then triangle ABF is similar to triangle BFN, since both are $72° - 54° - 54°$ triangles. So $AB : BF = BF : BN$ and $AB \cdot BN = BF^2$. Also, triangle KBA is similar to triangle KNA, since both are $144° - 18° - 18°$ triangles. So $AB : AK = AK : AN$ and $AB \cdot AN = AK^2$. If we add the two equations, we get $BF^2 + AK^2 = AB \cdot BN + AB \cdot AN =$

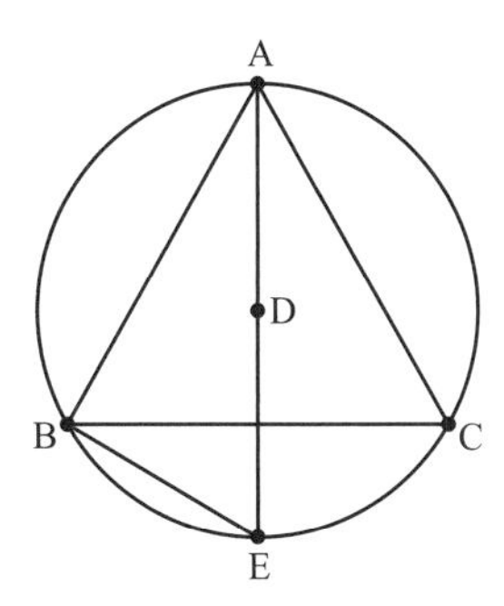

Figure 3.15.9.

$AB \cdot (BN+AN) = AB^2$. Since AB is the side of a pentagon, AK the side of a decagon, and BF, the radius, the side of a hexagon, the claim in proved. In modern terms, $\frac{AB}{BF} = 2\sin 36°$ and $\frac{AK}{BF} = 2\sin 18°$, so the result states that $4\sin^2 36° = 1 + 4\sin^2 18°$. We can calculate that $\sin 36° = \frac{\sqrt{10-2\sqrt{5}}}{4}$ while $\sin 18° = \frac{\sqrt{5}-1}{4}$. A short calculation then verifies the previous equation.

Proposition 12. *If an equilateral triangle be inscribed in a circle, the square on the side of the triangle is triple of the square on the radius of the circle.*

Let ABC be a circle, and let the equilateral triangle ABC be inscribed in it; I say that the square on one side of the triangle ABC is triple of the square on the radius of the circle [figure 3.15.9].

For let the center D of the circle ABC be taken, let AD be joined and carried through to E, and let BE be joined. Then, since the triangle ABD is equilateral, therefore the circumference BEC is a third part of the circumference of the circle ABC. Therefore the circumference BE is a sixth part of the circumference of the circle; therefore the straight line BE belongs to a hexagon; therefore it is equal to the radius DE. And, since AE is double of DE, the square on AE is quadruple of the square on ED, that is, of the square on BE. But the square on AE is equal to the squares on AB, BE; therefore the squares on AB, BE are quadruple of the square on BE. Therefore, *separando*, the square on AB is triple of the square on BE. But BE is equal to DE; therefore the square on AB is triple of the square on DE. Therefore, etc. Q.E.D.

Proposition 13. *To construct a tetrahedron, to comprehend it in a given sphere, and to prove that the square on the diameter of the sphere is one and a half times the square on the side of the tetrahedron.*

Euclid shows that if one draws a diameter of the sphere through one vertex of the tetrahedron, then the center of the opposite base is located at a point that divides the diameter in the ratio of 2 to 1. Therefore, we have the relation $a^2 = (\frac{2}{3}d)^2 + r^2$, where d is the diameter of the circumscribing sphere, a is the length of a side of the tetrahedron, and r is the radius of the circle circumscribing the triangle of the base of the tetrahedron. By proposition 13:12, the square on the side of the base is triple the

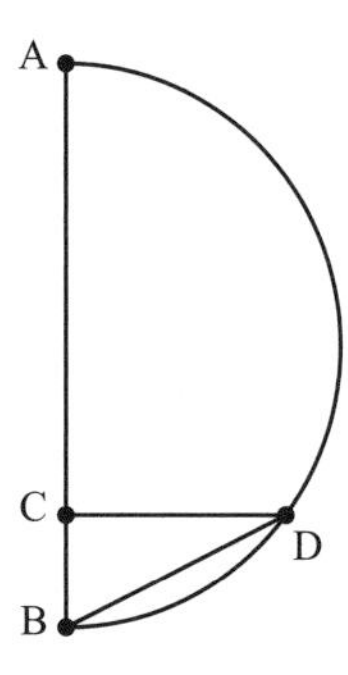

Figure 3.15.10.

square on the radius of the circle circumscribing it. Thus, $a^2 = \frac{4d^2}{9} + \frac{a^2}{3}$, so $\frac{2a^2}{3} = \frac{4d^2}{9}$, and $d^2 = \frac{3}{2}a^2$ as claimed.

Proposition 14. *To construct an octahedron and comprehend it in a sphere, as in the preceding case, and to prove that the square on the diameter of the sphere is double of the square on the side of the octahedron.*

Since an octahedron is composed of two joined square pyramids, the diameter d of the circumscribing sphere satisfies the relation $a^2 = (\frac{d}{2})^2 + \frac{a^2}{2}$, where a is the side of the octahedron. Thus, $\frac{a^2}{2} = \frac{d^2}{4}$ and $d^2 = 2a^2$ as claimed.

Proposition 15. *To construct a cube and comprehend it in a sphere, like the tetrahedron, and to prove that the square on the diameter of the sphere is triple of the square on the side of the cube.*

If a cube of side a is inscribed in a sphere of diameter d, then a diameter can be drawn from one vertex of the cube to the vertex diagonally opposite. It follows that $d^2 = 3a^2$ as claimed.

Proposition 16. *To construct an icosahedron and comprehend it in a sphere, like the aforesaid figures, and to prove that the side of the icosahedron is the irrational straight line called minor.*

Porism. *The square on the diameter of the sphere is five times the square on the radius of the circle from which the icosahedron has been described, and the diameter of the sphere is composed of the side of the hexagon and two of the sides of the decagon inscribed in the same circle.*

Euclid shows that if one draws a diameter AB of the sphere, divides it at C so that $AC = 4BC$, draws a semicircle about the diameter, draws CD perpendicular to AB, and connects DB, then DB is the radius of the circle circumscribing the pentagon formed by the bases of five of the equilateral triangles of the icosahedron [figure 3.15.10].

Now, since DC is the mean proportional between AC and CB, we see that $DC = \frac{2}{5}d$, where $d = AB$ is the diameter of the sphere. Then, $DB^2 = DC^2 + BC^2 = \frac{4}{25}d^2 +$

$\frac{1}{25}d^2 = \frac{1}{5}d^2$, and the square on the diameter of the sphere is five times the square on the radius of the circle, as claimed in the porism. Also, the side of a pentagon inscribed in a circle of radius r is $a = 2r \sin 36°$, and $\sin 36° = \frac{\sqrt{10-2\sqrt{5}}}{4}$, and, of course, a is a side of the icosahedron. Therefore,

$$a^2 = 4r^2 \sin^2 36° = \frac{4d^2}{5}\left(\frac{10-2\sqrt{5}}{16}\right) = \frac{d^2}{20}(10-2\sqrt{5}) = d^2\left(\frac{5-\sqrt{5}}{10}\right).$$

Finally, we have $a = \frac{d}{10}\sqrt{50-10\sqrt{5}}$. This is a minor straight line.

Proposition 17. *To construct a dodecahedron and comprehend it in a sphere, like the aforesaid figures, and to prove that the side of the dodecahedron is the irrational straight line called apotome.*

Porism. *When the side of the cube is cut in extreme and mean ratio, the greater segment is the side of the dodecahedron.*

Euclid begins by dividing half the edge of the cube inscribed in the same sphere in extreme and mean ratio. If q denotes half the edge of the cube and x denotes the greater segment, then by 13:4 we have the relation $q^2 + (q-x)^2 = 3x^2$, or $q^2 - qx = x^2$. Thus, $x = \frac{q(\sqrt{5}-1)}{2}$. From 13:14, we know that the square on the diameter d of the sphere is triple of the square on the side of the cube. Therefore, $d^2 = 3(2q)^2 = 12q^2$. Euclid then demonstrates that the edge of the dodecahedron is double the greater segment x that we have calculated. Since $q = \frac{\sqrt{3}}{6}d$, we get

$$x = \frac{\sqrt{3}d(\sqrt{5}-1)}{12} = \frac{(\sqrt{15}-\sqrt{3})d}{12},$$

and therefore, the edge of the dodecahedron is $a = \frac{\sqrt{15}-\sqrt{3}}{6}d$. This value is an apotome. The porism follows immediately.

Proposition 18. *To set out the sides of the five figures and to compare them with one another.*

Let AB, the diameter of the given sphere, be set out, and let it be cut at C so that AC is equal to CB, and at D so that AD is double of DB; let the semicircle AEB be described on AB, from C, D let CE, DF be drawn at right angles to AB, and let AF, FB, EB be joined [figure 3.15.11].

Then, since AD is double of DB, therefore AB is triple of BD. *Convertendo*, therefore, BA is one and a half times AD. But, as BA is to AD, so is the square on BA to the square on AF, for the triangle AFB is equiangular with the triangle AFD; therefore the square on BA is one and a half times the square on AF. But the square on the diameter of the sphere is also one and a half times the square on the side of the pyramid [13:13]. And AB is the diameter of the sphere; therefore AF is equal to the side of the tetrahedron.

Again, since AD is double of DB, therefore AB is triple of BD. But, as AB is to BD, so is the square on AB to the square on BF; therefore the square on AB is triple of the

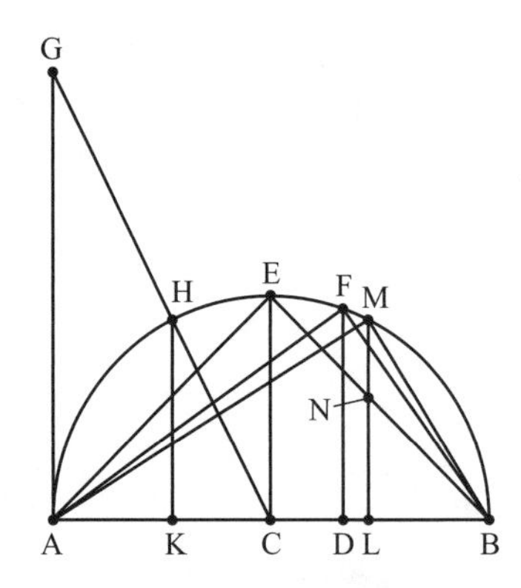

Figure 3.15.11.

square on BF. But the square on the diameter of the sphere is also triple of the square on the side of the cube [13:15]. And AB is the diameter of the sphere; therefore BF is the side of the cube.

And, since AC is equal to CB, therefore AB is double of BC. But, as AB is to BC, so is the square on AB to the square on BE; therefore the square on AB is double of the square on BE. But the square on the diameter of the sphere is also double of the square on the side of the octahedron [13:14]. And AB is the diameter of the given sphere; therefore BE is the side of the octahedron.

We sketch the remainder of the proof. Let AG be drawn perpendicular to AB and let $AG = AB$. Join GC and let HK be drawn perpendicular to AB. Since $AG = 2AC$, by similarity, $HK = 2KC$, so $HK^2 = 4KC^2$ and $HC^2 = 5KC^2$. But then also $CB^2 = 5CK^2$. Since $AB = 2CB$ and $AD = 2DB$, also $BD = 2DC$. Thus, $BC = 3DC$ and $BC^2 = 9CD^2$, so $CK > CD$. Let $CL = CK$, draw LM perpendicular to AB, and join MB. Since $KL = 2CK$, $AB = 2BC$, and $BC^2 = 5CK^2$, we have $AB^2 = 5KL^2$. By the porism to 13:16, KL is the radius of the circle from which the icosahedron is described. Therefore, KL is the side of a hexagon in that circle. But $AB = KL + AK + BL$ and $BL = AK$, so AK and BL are both sides of decagons in that circle. Since $MB^2 = ML^2 + BL^2$, it follows from 13:10 that MB is the side of a pentagon in that circle. But the side of a pentagon in that circle is also a side of the icosahedron. Therefore, MB is the side of the icosahedron.

Finally, we know that FB is the side of the cube. Divide it into extreme and mean ratio at N, where NB is the greater segment. Then, by the porism to 13:17, NB is the side of the dodecahedron.

I say next that *no other figure, besides the said five figures, can be constructed which is contained by equilateral and equiangular figures equal to one another*. For a solid angle cannot be constructed with two triangles, or indeed planes. With three triangles the angle of the tetrahedron is constructed, with four the angle of the octahedron, and with five the angle of the icosahedron; but a solid angle cannot be formed by six equilateral triangles placed together at one point, for, the angle of the equilateral triangle being two-thirds of a right angle, the six will be equal to four right angles; which is impossible, for any solid angle is contained by angles less than four right angles [11:21]. For the same reason, neither can a solid angle be constructed by more than six plane angles. By three squares the angle of the cube is contained, but by four it is impossible for a solid angle to be contained, for they will again be four right angles. By three equilateral and equiangular pentagons the angle of the dodecahedron is contained; but by four such it is impossible for any solid angle to be contained, for the angle of the equilateral pentagon being a right angle and a fifth, the four angles will be greater than four right angles, which is impossible. Neither again will a solid angle be contained by other polygonal figures by reason of the same absurdity. Therefore, etc. Q.E.D.

3.16 Hypsicles, *Elements*, 14

The Alexandrian mathematician Hypsicles (fl. ca. 190 BCE) wrote a work which came to be called book 14 of Euclid's *Elements*. This book contains a detailed study on the comparison between a dodecahedron and an icosahedron inscribed in the same sphere in eight propositions and various lemmas. In many respects, this work is a natural continuation of the mathematical content of the *Elements* and is unmistakably Euclidean in style. Furthermore, the demonstrations rely on many results established in book 13. These features, along with others, are no doubt why the book has been considered a natural complement and included as part of the tradition.

Hypsicles refers to many historical persons in his introduction, including Apollonius, Basileides of Tyre, a well-known figure in antiquity, and his own father. It is these references that help establish his dates.

When Basileides of Tyre, O Protarchus, arrived in Alexandria and met my father, he passed the majority of the visit with him because of [their] association with mathematics. And at one time, while investigating the composition by Apollonius concerning the comparison between the dodecahedron and icosahedron inscribed in the same sphere (and) what ratio they have with respect to one another, they reckoned that Apollonius had described those things incorrectly, and they themselves, amending those things, (re)wrote (it), so far as I heard from my father. Later, encountering another book published by Apollonius containing a certain demonstration of the proposed, I was greatly allured by the investigation of the problem. Indeed, the publication by Apollonius is available for all to consider. For it circulates after, it seems, it was written with due diligence. So far as I consider it necessary, I, having written a treatise, decided to dedicate it to you, on the one hand because of your progress in all of mathematics, especially in geometry, you can judge with expertise the things that will be mentioned, and on the other hand, because of your intimacy with my father and your goodwill towards me by listening graciously to my treatment. But it might be timely to stop the introduction and begin the composition.

Proposition 1. *The perpendicular drawn from the center of an arbitrary circle to the side of a pentagon inscribed in that same circle is half of the sum of the side of both the hexagon and the [side] of the decagon inscribed in the same circle.*

Let ABG be the circle, and in the circle ABG let BG be the side of a pentagon, and let the center D of the circle have been found, and let the perpendicular DE have been

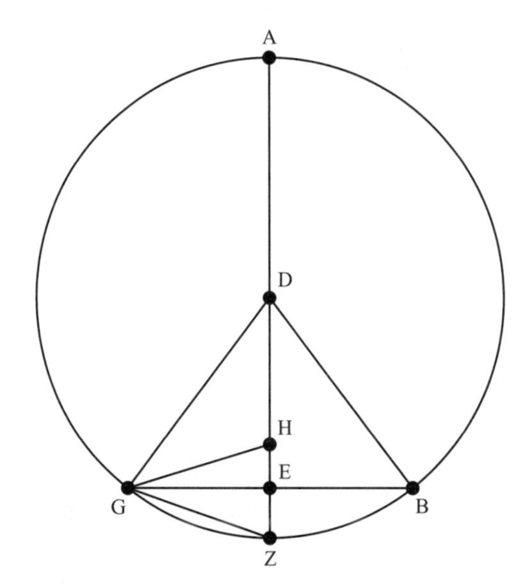

Figure 3.16.1.

drawn from D to BG, and let the straight lines EZ and DA have been extended out along the straight line DE [figure 3.16.1]. I say that DE is half [the sum of the side] of the hexagon and the [side] of the decagon inscribed in the same circle.

For let DG and GZ have been joined up, and let HE lie equal to EZ, and let HG have been joined up from the point H to G. Since therefore the arc of the whole of the circle (i.e., the circumference) is five times the arc BZG, and AGZ is half of the arc of the whole of the circle, and ZG is half of BZG, therefore the arc AGZ is five times the arc ZG. Therefore, AG is four times ZG. Just as AG is to ZG, so too the [angle] ADG is to the angle ZDG. Therefore, the [angle] ADG is four times the [angle] ZDG. And the [angle] ADG is two times the [angle] EZG. Therefore, the [angle] EZG is two times the [angle] HDG. And the [angle] EZG is equal to the [angle] EHG. Therefore, the [angle] EHG is two times the [angle] HDG. Therefore, DH is equal to HG. But HG is equal to ZG. Therefore, DH is equal to ZG. And HE is equal to EZ. Therefore, DE is equal to the sum EZ and ZG (lit., EZG). Let DE be added in common. Therefore, the sum of DZ and ZG is two times DE. And DZ is equal to the side of the hexagon and ZG is equal to the [side] of the decagon: therefore, DE is half of the [sum of the side] of both the hexagon and of the [side] of the decagon inscribed in the same circle.

Proposition 6. *With this being clear, it must be demonstrated that just as the surface of the dodecahedron is to the surface of the icosahedron, so too the side of the cube to the side of the icosahedron.*

Let a circle containing both the pentagon of the dodecahedron and the triangle of the icosahedron inscribed in the same sphere be taken as ABG, and let a side of an icosahedron GD have been inscribed the circle ABG, and [also] AG [the side] of a dodecahedron. Therefore, a side of an equilateral triangle is GD, and [the side] of a pentagon [is] AG. Let the center of the circle E have been found, and let the perpendiculars

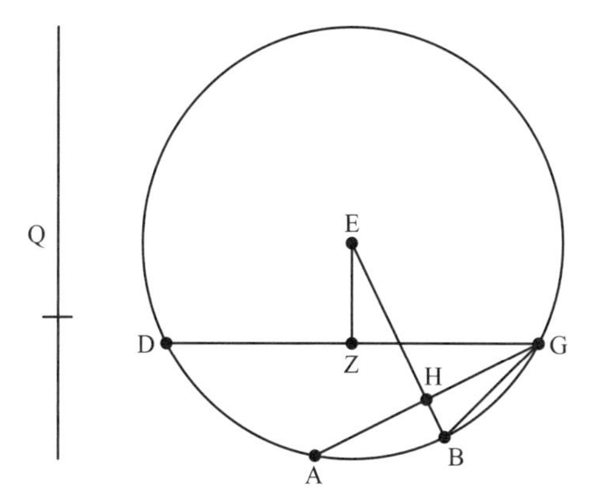

Figure 3.16.2.

EZ and EH have been drawn from E to DG and GA [respectively] and let a straight line HB have been extended to the straight line EH, and let BG have been joined, and let the side of a cube have been taken as Q [figure 3.16.2]. I say that just as the surface of the dodecahedron is to the surface of the icosahedron so too is Q to GD.

For since, when the sum of BE and BG has been cut in mean and extreme ratio, the greater segment is BE, and half of the sum of EB and BG is EH, and half of BE is EZ, therefore, when EH has been cut in mean and extreme ratio, the greater segment is EZ. Moreover, when Q has been cut in mean and extreme ratio, the greater segment is GA. Therefore, just as Q is to GA, so too EH is to EZ. Therefore, the [rectangle] under ZE and Q is equal to the [rectangle] under GA and EH. And since just as Q is to GD, so too the [rectangle] under ZE and Q is to GD and ZE, [and] the [rectangle] under GA and EH is equal to the [rectangle] under ZE and Q, therefore, just as Q is to GD, so too the [rectangle] under GA and HE is to the [rectangle] under GD and ZE, that is, the surface of the dodecahedron is to the surface of the icosahedron. Therefore, just as the surface of the dodecahedron is to the surface of the icosahedron, so too Q is to GD.

And [it is possible] to demonstrate in another way that just as the surface of the dodecahedron is to the surface of the icosahedron, so too the side of the cube is to the side of the icosahedron, with the previously written:

Let ABG be a circle, and let the sides of an equilateral pentagon AB and AG have been drawn in the circle ABG, and let BG have been joined up, and let the center of the circle D have been found, and let AD have been joined up from A to D and let a straight line DE have been extended along the straight line AD, and let DZ be half AD and let HG be a third of GQ. I say that the [rectangle] under AZ and BQ is equal to a pentagon [figure 3.16.3].

For let BD have been joined up from B to D. Since AD is two times DZ, therefore AZ is one and a half times AD. Again, since HG is three times GQ, HQ is double QG. Therefore, HG is one and a half times QH. Therefore, just as ZA is to AD, so too GH is to HQ. Therefore, the [rectangle] under AZ and HQ is equal to the [rectangle] under DA and GH. GH is equal to BH. Therefore, the [rectangle] under AD and BH is equal to the [rectangle] under ZA and HQ. But the [rectangle] under AD and BH is two ABD triangles.

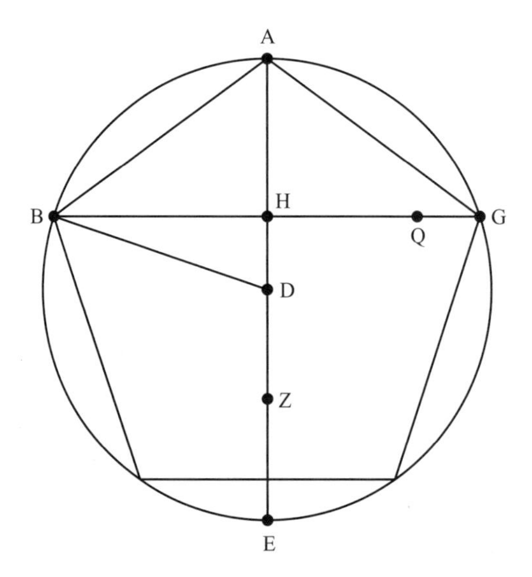

Figure 3.16.3.

And the [rectangle] under AZ and HQ is therefore two ABD triangles. Consequently, therefore, five [rectangles] under AZ and HQ are ten triangles. And ten triangles are two pentagons. Therefore, five [rectangles] under AZ and HQ are equal to two pentagons. Since, therefore, HQ is two times QG, and the [rectangle] under AZ and HQ is two times the [rectangle] under AZ and QG. Therefore, two [rectangles] under AZ and QG are equal to the [rectangle] under AZ and HQ. Therefore, ten [rectangles] under AZ and QG are equal to five [rectangles] under AZ and HQ, that is, two pentagons. Consequently five [rectangles] under AZ and QG are equal to one pentagon. Five times the [rectangle] under AZ and QG is equal to the [rectangle] under AZ and QB, since QB is five times QG, and AZ is the altitude in common. Therefore, the [rectangle] under AZ and BQ is equal to one pentagon.

With this being clear, now let there be set out the circle ABG containing both the pentagon of the dodecahedron and the triangle of the icosahedron inscribed in the same sphere, and let sides BA and AG of an equilateral pentagon have been inscribed in the circle ABG, and let BG have been joined up, and let the center of the circle E have been found, and let AE have been joined up from A to E and let it have been extended to Z, and let AH be double EH and KG be triple GQ, and let HM have been drawn from H at right angles to AZ, and let HD have been extended in a straight line from HM [figure 3.16.4]. Therefore, DM is [the side of] an equilateral triangle. Let AD and AM have been joined up. Therefore, the triangle ADM is equilateral. And since the [rectangle] under AH and QB is equal to the pentagon and the [rectangle] under AH and HD [is equal] to the triangle ADM, therefore, just as the [rectangle] under AH and QB is to the [rectangle] under DHA, so too the pentagon is to the triangle. And just as the [rectangle] under BQ and AH is to the [rectangle] under DHA, so too BQ is to DH. And therefore, as twelve QB is to twenty DH, so too twelve pentagons is to twenty triangles, that is, the surface of a pentagon to

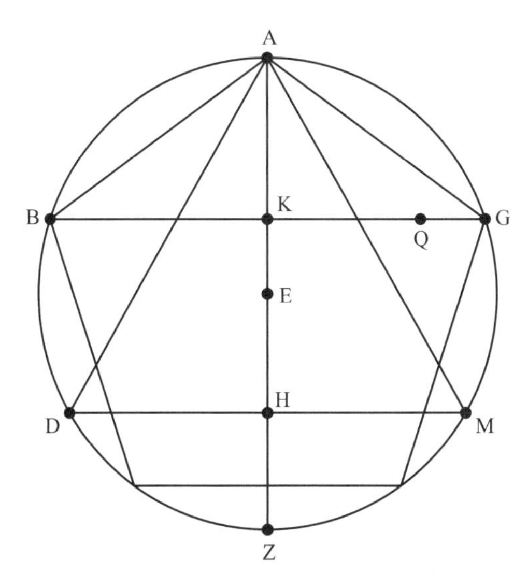

Figure 3.16.4.

the [surface] of an icosahedron. And twelve BQ are ten BG. For BQ is five times QG, and BG is six times QG. Therefore, six BQ are equal to five BG. And take the doubles. Twenty DH are ten DM: For DM is twice DH. Therefore, just as ten BG are to ten DM, thus the surface of the dodecahedron is to the surface of the icosahedron. And BG is the side of the cube, and DM is the [side] of the icosahedron. And just as the surface of a dodecahedron is to the surface of an icosahedron, so too BG is to DM, that is, the side of the cube to the side of the icosahedron.

[Conclusion] With all these things being well known to us, it is clear that if one should inscribe both a dodecahedron and an icosahedron in the same sphere, the dodecahedron will have the ratio to the icosahedron when an arbitrary straight line is cut in mean and extreme ratio, just as the square of the whole and of the greater segment is to the square of the whole and the lesser segment. For since just as the dodecahedron is to the icosahedron, so too the surface of the dodecahedron is to the surface of the icosahedron, that is, the side of the cube is to the side of the icosahedron, and just as the side of the cube is to the side of the icosahedron, so too when an arbitrary straight line is cut in mean and extreme ratio, the square on the whole and the greater segment is to the square on the whole and the lesser segment, therefore, just as the dodecahedron is to the icosahedron [those] having been inscribed in the same sphere, when an arbitrary straight line is drawn in mean and extreme ratio the square of the whole and the greater segment is to the square of the whole and the lesser segment.

3.17 Pappus, *Collection*, 3, Problems and Theorems

Like Proclus, Pappus, in the introduction to book 3 of his *Collection*, describes the difference between theorems to be proved and problems to be solved. But, more interestingly, this note is addressed to Pandrosion, a female teacher of geometry, although Pappus criticized her teaching.

Those wishing to discriminate more precisely between the [things] sought in geometry, my dear Pandrosion, think it fit to call a problem that in which one puts forward something to do or to construct, and a theorem that in which, when certain things are being supposed, one observes what follows from [these presupposed things] and in general everything that accompanies them; among the Ancients some said that all things are problems, while others said that all things are theorems.

Now, the person who puts forward the theorem, supervising in one way or another what follows, thinks it fit to search in this way and would not sanely put it forward in another way. On the other hand, he who puts forward a problem [in case he is ignorant and totally inexpert], even if he prescribes something which is somehow impossible to construct, is understandable and should not be blamed. Indeed, the task of the person who is searching is to determine also this: the possible and the impossible, and, if possible, when and how and in how many ways possible. But when a man professing to know mathematics sets an investigation wrongly, he is not free from censure.

For example, some persons professing to have learned mathematics from you lately gave me a wrong enunciation of problems. It is desirable that I should state some of the proofs of these, and of matters akin to them, for the benefit both of yourself and of other lovers of this science, in the third book of the *Collection*. Now the first of these problems was set wrongly by a person who was thought to be a great geometer. For, given two straight lines, he claimed to know how to find by plane methods two means in continuous proportion, and he even asked that I should look into the matter and comment on his construction, which is after this manner.

We have already seen one method of finding two means between two straight lines in the work of Archytas, and more methods are discussed later in this volume. But Pappus at this point describes a method of solving the problem by successive approximations, a method that is, naturally, not exact. We do not include that method here.

The second of the problems was this: A certain other [geometer] set the problem of exhibiting the three means in a semicircle. Describing a semicircle ABC, with center E, and taking any point D on AC, and from it drawing DB perpendicular to EC, and joining EB, and from D drawing DZ perpendicular to it, he claimed simply that the three means had been set out in the semicircle, EC being the arithmetic mean, DB the geometric mean and BZ the harmonic mean [figure 3.17.1].

That BD is a mean between AD, DC in geometrical proportion, and EC between AD, DC in arithmetical proportion, is clear. For AD is to DB as DB is to DC and the excess of AD over AE, that is, the excess of AD over EC, is equal to the excess of EC over CD.

But how ZB is a harmonic mean, or between what kind of lines, he did not say, but only that it is a third proportional to EB, BD, not knowing that from EB, BD, BZ, which are in geometrical proportion, the harmonic mean is formed. For it will be proved by me later that

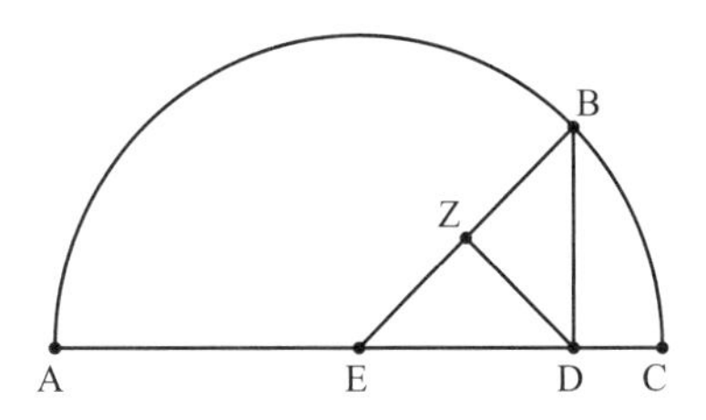

Figure 3.17.1.

a harmonic proportion can thus be formed with the greater extreme being $2EB + 3DB + BZ$, the mean term being $2BD + BZ$, and the lesser extreme being $BD + BZ$.

The quantity B is the harmonic mean between two quantities A, C if $C : A = (C - B) : (B - A)$. So, despite Pappus's criticism, ZB is, in fact, a harmonic mean between AD and DC. First, since BDE is a right triangle in which DZ is perpendicular to BE, $ZB : BD = BD : BE$, so $BZ \cdot BE = BD^2 = AD \cdot DC$. But $BE = \frac{1}{2}(AD + DC)$. So $BZ(AD + DC) = 2\,AD \cdot DC$. It follows that $AD(BZ - DC) = DC(AD - BZ)$. Therefore, $AD : DC = (AD - BZ) : (BZ - DC)$ and ZB is the harmonic mean between AD and DC. However, it is not difficult to show that Pappus's statement about finding a harmonic proportion is also valid. For since $EB \cdot BZ = BD^2$, we have $2\,EB \cdot BZ + BD^2 = 3\,BD^2$. Therefore, $2\,EB \cdot BD + 2\,EB \cdot BZ + BD^2 + BD \cdot BZ = 2\,EB \cdot BD + 3\,BD^2 + BZ \cdot BD$, or $(2\,EB + BD)(BD + BZ) = (2\,EB + 3\,DB + BZ)(BD)$, which is the same as $(2\,EB + 3\,DB + BZ) : (BD + BZ) = (2\,EB + BD) : BD$. And this ratio shows that $2\,BD + BZ$ is the mean term between $2\,EB + 3\,DB + BZ$ and $BD + BZ$ in a harmonic proportion.

The third of the problems mentioned by Pappus in his criticism of Pandrosion deals with certain constructions in a right triangle and will not be given here.

3.18 Pappus, *Collection*, 3, Inscribing Polyhedra in a Sphere

In this section of book 3 of the *Collection*, Pappus deals with the question of inscribing a regular polyhedron in a given sphere. His method is somewhat different from that of Euclid, described in *Elements* 13, propositions 13–17. Euclid's aim was, first, to construct a polyhedron that a sphere equal to the given sphere would circumscribe, and, second, to actually construct the sphere around the polyhedron. Pappus, on the other hand, begins with a given sphere and shows directly how to inscribe a polyhedron inside that sphere. For both of them, there were five different results, corresponding to the five different regular polyhedra. We present here some preliminary results, mostly without proof, and then give details of Pappus's constructions for the tetrahedron and the cube. Note that most of the preliminary results are geometrically obvious, although Pappus gives proofs of all of them.

In order to inscribe the five polyhedra in a sphere, these things are premised.

Proposition 43. *Let ABG be a circle in a sphere, with diameter AG and center D, and let it be proposed to insert in the circle a chord parallel to the diameter AG and equal to a given straight line not greater than the diameter AG.*

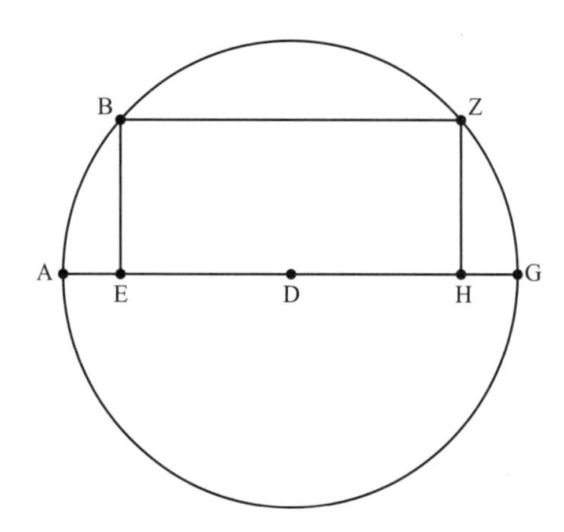

Figure 3.18.1.

Let ED be placed equal to half of the given straight line, and let EB be drawn perpendicular to the diameter AG, and let BZ be drawn parallel to AG [figure 3.18.1]; then shall this line be equal to the given straight line. For it is double of ED, inasmuch as ZH, when drawn, is parallel to BE, and it is therefore equal to EH.

Proposition 44. *Let there be two parallel circles AKD, $BEZG$ in a sphere such the line drawn through B and G is a diameter in one circle; it is proposed to draw a diameter of the circle AKD that is parallel to the given diameter BG of the first circle.*

Proposition 49. *Suppose in a sphere there are equal and parallel circles AB and GD and there are equal and parallel line segments [also designated as AB, GD] drawn in the circles situated on the same side of the centers; I say that the lines AG, BD connecting the endpoints of the two given line segments are equal, parallel, and at right angles to the planes of the circles.*

Proposition 50. *With the same equal and parallel circles in a sphere as before, suppose that the equal and parallel line segments AB, GD are not situated on the same sides of the centers; I say that the straight lines AD, BG joining the opposite ends of these line segments are each equal to a diameter of the sphere.*

Proposition 53. *Suppose that in a plane, the lines AB, BG form equal angles with the line DBE situated in the same plane, and suppose we elevate a line BZ from the plane forming equal angles with each of the lines AB, BG; I say that this line is perpendicular to the line DE.*

Proposition 54. *To inscribe a [regular] tetrahedron in a given sphere.*

Suppose it is inscribed and let A, B, G, D be the vertices in the surface of the sphere [figure 3.18.2]. Draw through the point A the line EZ parallel to the line GD; this will make equal angles with the lines AG, AD.[61] And the line AB also makes equal angles with

[61] Angles DAE and ADG are equal, as are angles GAZ and AGD. But triangle DAG, being a face of the tetrahedron, is equilateral, so angle ADG equals angle AGD. Therefore, angle DAE equals angle GAZ.

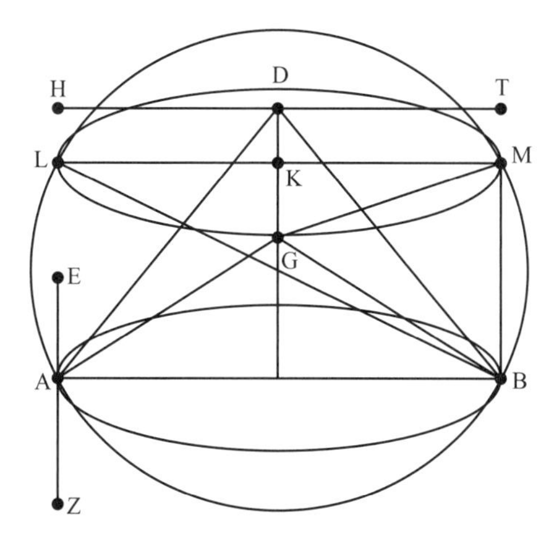

Figure 3.18.2.

lines AG, AD. Therefore, as has been demonstrated earlier [proposition 53], the line EZ is perpendicular to the line AB and tangent to the sphere. In fact, if one extends the plane that passes through the lines DA, AG, it determines a circle in which is inscribed the equilateral triangle ADG, and the line EZ is parallel to the line GD. As a consequence, the line EZ is tangent to the circle and, it follows, to the sphere. Therefore, the plane determined by the lines EZ, AB determines, as a section of the sphere, a circle of which the line AB will be the diameter, because it is perpendicular to the line EZ, which is similarly tangent. And if one draws through the point D a line HT parallel to the line AB, it will be tangent to the sphere and the line GD will be perpendicular to it. So if one extends the plane passing through the lines HT, GD, it determines a circle having as diameter the line GD. This circle is equal and parallel to the one with diameter line AB, so the lines EZ, GD and the lines AB, HT are parallel.

Now draw through the center K [of the circle with diameter GD] the line LM perpendicular to the line GD; this is therefore parallel to the line AB. If we join BL, BM, the line BM will be perpendicular to each of the lines AB, LM and thus to the plane of the circles, and the line BL will be a diameter of the sphere [propositions 49, 50]. And also, if one joins the line MG, sq(LM) is 2 sq(MG) and also sq(BG) is 2 sq(GM).[62] Also, the angle formed by the lines BM, MG is right. Therefore the line BM is equal to the line MG and so sq(LM) is 2 sq(MB).[63] Consequently, sq(BL) is $\frac{3}{2}$ sq(LM).[64] But the diameter BL of the sphere is given; therefore, so is the diameter LM of the circle. So the circles are given in position and the points A, B, G, D are given.

The synthesis is now manifest. In fact, if one draws in the sphere two equal and parallel circles, such that the square of the diameter of the sphere is one and a half times the square of each of their diameters, and then draws the two parallel diameters AB, LM as we have done earlier [proposition 44] and, through the center the line GD perpendicular

[62] Since $LM^2 = 2\,MG^2$ and $BG = AB = LM$, then $BG^2 = 2\,MG^2$.
[63] $BG^2 = BM^2 + MG^2$; thus, $MG^2 = BM^2$, or $MG = BM$.
[64] $BL^2 = LM^2 + BM^2$, but $BM^2 = \frac{1}{2}LM^2$, so $BL^2 = \frac{3}{2}LM^2$.

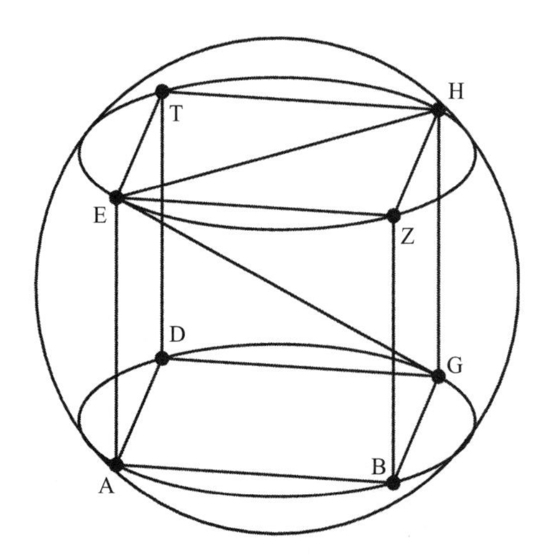

Figure 3.18.3.

to the line LM, we obtain the points A, B, G, D as the vertices of the tetrahedron. The demonstration will be the inverse of the analysis, and it will demonstrate at the same time that the square of the diameter of the sphere is one and a half times the square of a side of the tetrahedron.

Proposition 55. *To inscribe a cube in a given sphere.*

Suppose it is inscribed. Let A, B, G, D, E, Z, H, T be the vertices of the cube on the surface of the sphere, and draw planes through A, B, G, D and E, Z, H, T [figure 3.18.3]. These will determine, as sections, two equal and parallel circles. The squares inscribed in these circles are also equal and parallel. The line joining GE will be a diameter of the sphere. Now join the line EH. Since sq(EH) is 2 sq(ET), that is, 2 sq(HG), and since the angle between the lines GH and HE is right, therefore sq(GE) is $\frac{3}{2}$ sq(EH). But sq(GE) is given; thus so is sq(EH). Further, this line is a diameter of the circle $EZHT$; thus the circle is given. Therefore the circle $ABGD$ is given also, since the squares in these circles are given and the vertices of the cube are also given.

The synthesis is now manifest. If fact, one describes in the sphere two parallel circles, such that the square of the diameter of the sphere is one and a half times the squares on the equal diameters of these circles. Then one inscribes, in one of these, the square $ABGD$. Then one draws in the other circle the line ZH, equal and parallel to line BG, as we have shown earlier [proposition 43]. Complete on this line the square $EZHT$ and one obtains the inscribed cube. One then can demonstrate, following the analysis, that $BZHG$ is a square, and one can demonstrate at the same time that the square on the diameter of the sphere is triple the square on a side of the cube and that the same circles circumscribe the vertices of the tetrahedron and the cube. For in the tetrahedron, the square on the diameter of the sphere is one and a half times the square on the diameter of each of these circles.

3.19 Pappus, *Collection*, 4, Generalization of the Pythagorean Theorem

The first proposition in book 4 of Pappus's *Collection* is a generalization of the Pythagorean Theorem (*Elements* 1:47). The proof Pappus gives is in true Euclidean

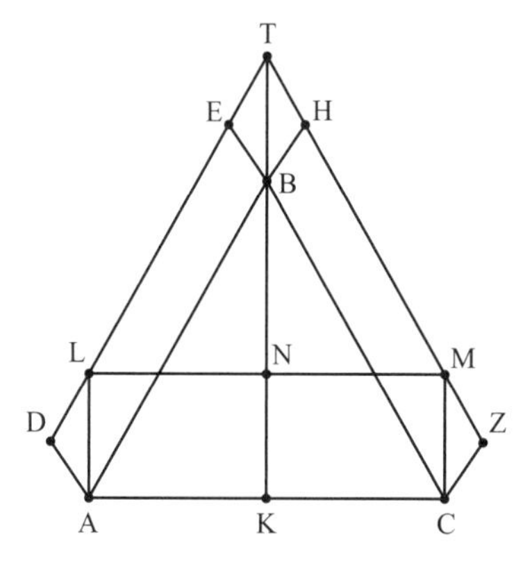

Figure 3.19.1.

style and makes no use of proportion theory: the proof relies only on ideas from *Elements* 1. To show that this is indeed a generalization of the Pythagorean Theorem, or that the Pythagorean Theorem is a special case of this one, simply take triangle *ABC* to be a right triangle with right angle at *B*, take the two parallelograms *ABED* and *BCZH* as squares, and note that the parallelogram comprised by *AC* and *TB* is then a square.

Proposition 1. *When ABC is a triangle, and over AB and BC any parallelograms ABED and BCZH are described, and DE and ZH are produced to T, and TB is joined, then the parallelograms ABED and BCZH taken together turn out to be equal to the parallelogram comprised by AC and TB, with an angle at A that is equal to the sum of the angles BAC and DTB.*

For: Produce *TB* to *K*, and through *A* and *C* draw the parallels *AL* and *CM* to *TK*, and join *LM* [figure 3.19.1]. Since *ALTB* is a parallelogram, *AL* and *TB* are equal and parallel. Similarly, *MC* and *TB* are both equal and parallel as well, so that *LA* and *MC*, also, are both equal and parallel. Therefore, *LM* and *AC* are both equal and parallel as well. Therefore, *ALMC* is a parallelogram with angle *LAC*, i.e.: with an angle that is the sum of angle *BAC* and angle *DTB*. For the angle *DTB* is equal to the angle *LAB*.

And since the parallelogram *DABE* is equal to the parallelogram *LABT* (for they are both erected over the same base *AB*, and lie within the same parallels *AB* and *DT*), but *LABT* is equal to *LAKN* (for they are both erected over the same base *LA*, and lie within the same parallels *LA* and *TK*), *ADEB* is therefore equal to *LAKN* as well. For the same reason, *BHZC* is equal to *NKCM* as well. Therefore, the parallelograms *DABE* and *BHZC* taken together are equal to *LACM*, i.e.: to the parallelogram spanned by *AC* and *TB*, with the angle *LAC*, which is equal to the sum of the angles *BAC* and *BTD*.

And this is much more general than what was proved in the *Elements* about right-angled triangles concerning the squares.

SOURCES, CHAPTER 3

3.1 This translation of a part of the *Timaeus*, a dialogue of Plato, is by Benjamin Jowett and is found in Robert Maynard Hutchins, ed., *Great Books of the Western World*, vol. 7 (Chicago: Encyclopedia Britannica, 1952), 458–59.

3.2 This translation of a section of the *Theaetetus* is by Benjamin Jowett and is found in Robert Maynard Hutchins, ed., *Great Books of the Western World*, vol. 7 (Chicago: Encyclopedia Britannica, 1952), 514–15.

3.3 This translation of part of the *Meno* is by Benjamin Jowett and is found in Robert Maynard Hutchins, ed., *Great Books of the Western World*, vol. 7 (Chicago: Encyclopedia Britannica, 1952), 180–83.

3.4 This excerpt is taken from the edition of Aristotle's works found in *Great Books of the Western World*, vol. 8 (Chicago: Encyclopedia Britannica, 1971), 312–24. The translator is G. R. G. Mure.

3.5 This translation is taken from Ivor Thomas, *Selections Illustrating the History of Greek Mathematics*, vol. 1 (Cambridge, MA: Harvard University Press, 1980), 239–53.

3.6 The translation was made by Colin McKinney from the Greek text of Eutocius in his commentary to Archimedes, *On the Sphere and Cylinder II* and was first published in *Convergence*, 2016, as part of the article "The Duplicators, Part I: Eutocius' Collection of Cube Duplications." The Greek text is in Johan Ludvig Heiberg, ed., *Archimedis Opera*, 2nd ed. (Leipzig: Teubner, 1910–1915).

3.7 This translation is taken from Proclus, *A Commentary on the First Book of Euclid's Elements*, trans. Glenn. R. Morrow (Princeton, NJ: Princeton University Press, 1970).

3.8 The translation of the *Elements* is taken from Sir Thomas Heath, ed. and trans., *The Thirteen Books of Euclid's Elements* (New York: Dover, 1956). The translation of the commentary is taken from Proclus, *A Commentary on the First Book of Euclid's Elements*, trans. Glenn R. Morrow (Princeton, NJ: Princeton University Press, 1970).

3.9 This translation from the Arabic is by A. I. Sabra in "Simplicius's Proof of Euclid's Parallels Postulate," *Journal of the Warburg and Courtauld Institutes* 32 (1969): 1–24.

3.10 The translation of the *Elements* is taken from Sir Thomas Heath, ed. and trans., *The Thirteen Books of Euclid's Elements* (New York: Dover, 1956).

3.11 The translation here is adapted from Christian Marinus Taisbak, *Euclid's Data or The Importance of Being Given* (Copenhagen: Museum Tusculanum Press, 2003) and George L. McDowell and Merle A. Sokolik, trans., *The Data of Euclid* (Baltimore, MD: Union Square, 1993).

3.12 This translation is from Jan P. Hogendijk, "The Arabic Version of Euclid's *On Divisions*," in *Vestigia Mathematica: Studies in Medieval and Early Modern Mathematics in Honour of H. L. L. Busard*, ed. M. Folkerts and J. P. Hogendijk (Amsterdam: Rodopi, 1993), 143–62.

3.13 This translation is by Sir Thomas L. Heath in *The Thirteen Books of Euclid's Elements*, vol. 3 (New York: Dover, 1956).

3.14 This translation is by William Thomson and is found in *The Commentary of Pappus on Book X of Euclid's Elements* (Cambridge, MA: Harvard University Press, 1930), 63–85.

3.15 The translation of the *Elements* is taken from Sir Thomas Heath, ed. and trans., *The Thirteen Books of Euclid's Elements* (New York: Dover, 1956).

3.16 This translation was made by Clemency Montelle from the Greek original in Johan Ludvig Heiberg, ed., *Euclid's Elements*, vol. 5 (Leipzig: Teubner, 1888).

3.17 The translation of this section of book 3 of Pappus's *Collection* is taken from Ivor Thomas, *Greek Mathematical Works*, vol. 2, *Aristarchus to Pappus of Alexandria* (Cambridge, MA: Harvard University Press, 1980), 567–71.

3.18 The translation of this section of book 3 of Pappus's *Collection* was made by Victor Katz from the French in Paul Ver Eecke, ed. and trans., *Pappus D'Alexandrie: Collection Mathématique*, vol. 1 (Paris: Desclée de Brouwer, 1933), 97–110.

3.19 The translation of this section of book 4 of Pappus's *Collection* is taken from Pappus of Alexandria, *Book 4 of the Collection*, ed. and trans. Heike Sefrin-Weis (New York: Springer, 2010), 83–84.

4

Archimedes and Related Authors

Archimedes's mathematical treatises are generally considered the finest mathematical works from antiquity. Due to the extensive foundational results set out by his predecessors, he was able to focus on detailed solutions of very specific problems. In particular, where Euclid never dealt with numerical results, Archimedes often calculated important numerical values and put bounds on his answers. We present here excerpts from many of Archimedes's works, emphasizing the most important results. We have attempted to display the significant parts of Archimedes's arguments leading up to his major theorems, but not all the subsidiary results could be included. On the other hand, where Archimedes's arguments are difficult, we have explained his strategies. Surrounding the excerpts from Archimedes, we first have a selection from Polybius's *Histories* giving some biographical material about Archimedes and then at the end a translation of a Latin poem describing Archimedes's method of solving King Hiero's crown problem using the law of hydrostatics. The chapter concludes with three excerpts from Pappus's *Collection* giving a new proof of Archimedes's result on the area bounded by one loop of a spiral, a theorem about touching circles, and a report on Archimedes's discovery of thirteen semiregular solids.

In terms of the three classical problems mentioned in chapter 3, we see in this chapter that in *Measurement of a Circle* Archimedes solved the problem of squaring the circle by showing that the area of a circle was the same as that of a right triangle with one leg equal to the radius and the other equal to the circumference. Of course, Archimedes could not find the exact circumference by Euclidean means, only by bounding the value within certain limits. Additionally, in his *On Spirals*, Archimedes found the length of the circumference of a circle by use of the spiral, a curve also constructed by nonplanar means.

Archimedes's works are extant today partly because of the edition and commentary prepared by Eutocius of Ascalon (480–540 CE) in the sixth century CE. In what follows, we often include material written by Eutocius in his commentaries.

4.1 Polybius, *The Histories*

Polybius (c. 180–118 BCE) was a Greek historian of the Hellenistic period. His most famous work, *The Histories*, covered the period from 264 to 146 BCE, with particular focus on the Punic wars between Rome and Carthage, during the second of which Archimedes was killed. Not only does Polybius describe in detail numerous historical events in that period, but he also emphasizes the aspects of Roman civilization that enabled Rome to become dominant in the Mediterranean at that time. The first five books of *The Histories* survive in their entirety; only fragments of the numerous further volumes remain. In this excerpt from book 8, Polybius describes in detail his belief that Archimedes single-handedly prevented the Roman army and navy from a quick conquest of Syracuse in 212 BCE.

Book 8

At the time that Epicydes and Hippocrates[1] seized on Syracuse, alienating themselves and the rest of the citizens from the friendship of Rome, the Romans, who had already heard of the fate of Hieronymus, tyrant of Syracuse, appointed Appius Claudius as *propraetor*, entrusting him with the command of the land forces, while they put their fleet under that of Marcus Claudius Marcellus. These commanders took up a position not far from the city and decided to attack it with their land forces in the neighborhood of the Hexapyli, and with their fleet at the Stoa Scytice in Achradina, where the wall reaches down to the very edge of the sea. Having got ready their blindages, missiles, and other siege material, they were in high hopes owing to their large numbers that in five days their works would be much more advanced than those of the enemy, but instead they did not reckon with the ability of Archimedes, or foresee that in some cases the genius of one man accomplishes much more than any number of hands. However, now they learned the truth of this saying by experience.

The strength of Syracuse lies in the fact that the wall extends in a circle along a chain of hills with overhanging brows, which are, except in a limited number of places, by no means easy of approach even with no one to hinder it. Archimedes now made such extensive preparations, both within the city and also to guard against an attack from the sea, that there would be no chance of the defenders being employed in meeting emergencies, but that every move of the enemy could be replied to instantly by a counter move. Appius, however, with his blindages and ladders attempted to use these for attacking the portion of the wall which abuts on the Hexapylus to the east. Meanwhile Marcellus was attacking Achradina from the sea with sixty *quinqueremes*,[2] each of which was full of men armed with bows, slings, and javelins, meant to repulse those fighting from the battlements. He had also eight *quinqueremes* from which the oars had been removed, the starboard oars from some and the larboard ones from others. These were lashed together two and two,

[1] Epicydes and Hippocrates were brothers from Syracuse who served as Carthaginian generals during the Second Punic War. After Hieronymus (the grandson of Hiero II, who ruled from 270 to 216 BCE) was killed in 214 BCE, they became the de facto rulers of Syracuse.

[2] A *quinquereme* is a Roman galley propelled by five banks of oars.

on their dismantled sides, and pulling with the oars on their outer sides they brought up to the wall the so-called *sambucae*. These engines are constructed as follows. A ladder was made four feet broad and of a height equal to that of the wall when planted at the proper distance. Each side was furnished with a breastwork, and it was covered by a screen at a considerable height. It was then laid flat upon those sides of the ships which were in contact and protruding a considerable distance beyond the prow. At the top of the masts there are pulleys with ropes, and when they are about to use it, they attach the ropes to the top of the ladder, and men standing at the stern pull them by means of the pulleys, while others stand on the prow, and supporting the engine with props, assure its being safely raised. After this the towers on both the outer sides of the ships bring them close to shore, and they now endeavor to set the engine I have described up against the wall. At the summit of the ladder there is a platform protected on three sides by wicker screens, on which four men mount and face the enemy, resisting the efforts of those who from the battlements try to prevent the *sambuca* from being set up against the wall. As soon as they have set it up and are on a higher level than the wall, these men pull down the wicker screens on each side of the platform and mount the battlements or towers, while the rest follow them through the *sambuca* which is held firm by the ropes attached to both ships. The construction was appropriately called a *sambuca*, for when it is raised the shape of the ship and ladder together is just like the musical instrument. Such were the contrivances with which the Romans intended to attack the towers.

But Archimedes, who had prepared engines constructed to carry to any distance, so damaged the assailants at long range, as they sailed up, with his more powerful *mangonels*[3] and heavier missiles as to throw them into much difficulty and distress; and as soon as these engines shot too high he continued using smaller and smaller ones as the range became shorter, and, finally, so thoroughly shook their courage that he put a complete stop to their advance, until Marcellus was so hard put to it that he was compelled to bring up his ships secretly while it was still night. But when they were close in to shore and too near to be struck by the mangonels, Archimedes had hit upon another contrivance for attacking the men who were fighting from the decks. He had pierced in the wall at short distances a series of loopholes of the height of a man and of about a palm's breadth on the outer side. Stationing archers and "small scorpions" opposite these inside the wall and shooting through them, he disabled the soldiers. So that he not only made the efforts of the enemy ineffective whether they were at a distance or close at hand, but destroyed the greater number of them. And when they tried to raise the *sambucae* he had engines ready all along the wall, which while invisible at other times, reared themselves when required from inside above the wall, their beams projecting far beyond the battlements, some of them carrying stones weighing as much as ten talents and others large lumps of lead. Whenever the *sambucae* approached, these beams were swung round on their axis, and by means of a rope running through a pulley dropped the stones on the *sambuca*, the consequence being that not only was the engine smashed, but the ship and those on board were in the utmost peril. There were some machines again which were directed against parties advancing under the cover of blinds and thus protected from injury by missiles shot through the wall. These machines, on the one hand, discharged stones large enough to chase the

[3] A *mangonel* is a military device for throwing stones.

assailants from the prow, and at the same time let down an iron hand attached to a chain with which the man who piloted the beam would clutch at the ship, and when he had got hold of her by the prow, would press down the opposite end of the machine which was inside the wall. Then when he had thus by lifting up the ship's prow made her stand upright on her stern, he made fast the opposite end of the machine, and by means of a rope and pulley let the chain and hand suddenly drop from it. The result was that some of the vessels fell on their sides, some entirely capsized, while the greater number, when their prows were thus dropped from a height, went under water and filled, throwing all into confusion.

Marcellus was hard put to it by the resourcefulness of Archimedes, and seeing that the garrison thus baffled his attacks not only with much loss to himself but with derision he was deeply vexed, but still made fun of his own performances, saying "Archimedes uses my ships to ladle sea-water into his wine cups, but my *sambuca* band is flogged out of the banquet in disgrace." Such was the result of the siege from the sea. And Appius, too, found himself in similar difficulties and abandoned his attempt. For his men while at a distance were mowed down by the shots from the mangonels and catapults, the supply of artillery and ammunition being admirable both as regards quantity and force, as indeed was to be expected where Hiero had furnished the means and Archimedes had designed and constructed the various contrivances. And when they did get near the wall they were so severely punished by the continuous volleys of arrows from the loopholes of which I spoke above that their advance was checked or, if they attacked under the cover of mantelets, they were destroyed by the stones and beams dropped upon their heads. The besieged also inflicted no little damage by the above-mentioned hands hanging from cranes, for they lifted up men, armor and all, and then let them drop. At last Appius retired to his camp and called a council of his military tribunes, at which it was unanimously decided to resort to any means rather than attempt to take Syracuse by storm. And to this resolution they adhered; for during their eight months' investment of the city, while leaving no stratagem or daring design untried, they never once ventured again upon an assault. Such a great and marvellous thing does the genius of one man show itself to be when properly applied to certain matters. The Romans at least, strong as they were both by sea and land, had every hope of capturing the town at once if one old man of Syracuse were removed; but as long as he was present, they did not venture even to attempt to attack in that fashion in which the ability of Archimedes could be used in the defense. On the contrary, thinking that owing to the large population of the town the best way to reduce it was by famine, they placed their hope in this, cutting off supplies from the sea by their fleet and those from the land by their army. Wishing not to spend in idleness the time during which they besieged Syracuse, but to attain some useful results outside, the commanders divided themselves and their forces, so that Appius with two-thirds of their army invested the town while Marcus took the other third and made raids on the parts of Sicily which favored the Carthaginians.

4.2 Archimedes, *Measurement of a Circle*

Measurement of a Circle is a short treatise in which Archimedes first shows by exhaustion that a circle is equal to a right triangle with one leg equal to the circumference and the other equal to the radius. In modern terms, that means the area

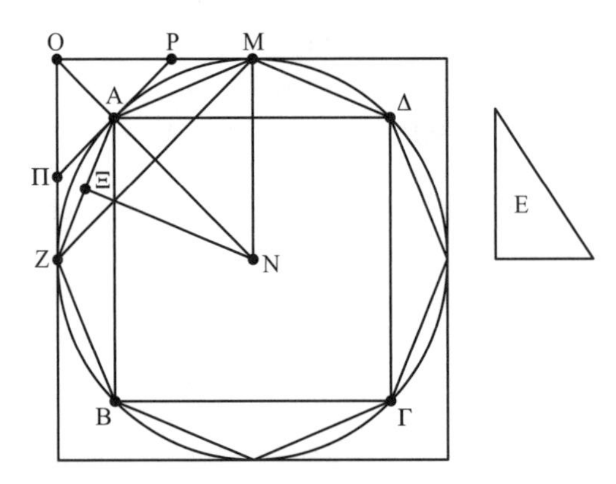

Figure 4.2.1.

of the circle is $\frac{1}{2}rC$, equivalent to our modern formula $A = \pi r^2$. But Archimedes in this case does not hesitate to do actual calculations, so in proposition 3 he calculates an approximation to the ratio of circumference to the diameter, that is, to what we call π. Proposition 2, which we have omitted, is clearly in the wrong place, since it depends on proposition 3. It uses the approximation $\frac{22}{7}$ to π to show that the ratio of the circle to the square on the diameter is 11:14.

Proposition 1. *Any circle is equal to a right-angled triangle in which one of the sides about the right angle is equal to the radius, and the base is equal to the circumference.*

Let the circle $AB\Gamma\Delta$ have to the triangle E the stated relation: I say that it is equal [figure 4.2.1]. For, if possible, let the circle be greater, and let the square $A\Gamma$ be inscribed, and let the arcs be divided into equal parts and let $BZ, ZA, AM, M\Delta, \ldots$ be drawn and let the segments be less than the excess by which the circle exceeds the triangle. The rectilineal figure is therefore greater than the triangle. Let N be the center, and $N\Xi$ perpendicular to ZA. $N\Xi$ is then less than the side of the triangle. But the perimeter of the rectilineal figure is also less than the other side, since it is less than the perimeter of the circle. The rectilineal figure is therefore less than the triangle E, which is absurd.

Let the circle be, if possible, less than the triangle E, and let the square be circumscribed, and let the arcs be divided into equal parts, and through the points of division let tangents be drawn; the angle OAP is therefore right. Therefore OP is greater than MP; for PM is equal to PA, and the triangle $PO\Pi$ is greater than half the figure $OZAM$. Let the spaces left between the circle and the circumscribed polygon, such as the figure ΠZA, be less than the excess by which E exceeds the circle $AB\Gamma\Delta$. Therefore, the circumscribed rectilineal figure is now less than E, which is absurd. For it is greater, because NA is equal to the perpendicular of the triangle, while the perimeter is greater than the base of the triangle. The circle is therefore equal to the triangle E.

. . .

Proposition 3. *The circumference of any circle is greater than three times the diameter and exceeds it by a quantity less than the seventh part of the diameter but greater than ten seventy-first parts.*

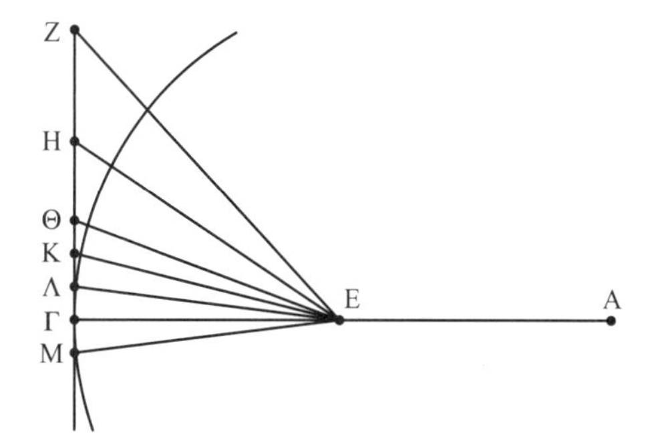

Figure 4.2.2.

In Archimedes's text of proposition 3, he leaves out most of the steps of his calculation and just presents the numerical answers. Eutocius, in his commentary, fills in many of the details, including even working out explicitly the squares of certain large numbers. But rather than put in the details, we describe the methods Archimedes uses to calculate his approximation to π, first by using polygons circumscribed around the circle and then by using polygons inscribed in the circle. He begins with a hexagon circumscribing the circle and then proceeds step by step to find the perimeters of polygons with 12, 24, 48, and 96 sides. He uses what amounts to a recursive algorithm, which we can represent in modern notation as

$$t_{i+1} = \frac{rt_i}{u_i + r}, \quad u_{i+1} = \sqrt{r^2 + t_{i+1}^2},$$

where r is the radius of the circle, t_i is half of one side of a circumscribed polygon of $3 \cdot 2^i$ sides, and u_i is the length of the line from the center of the circle to a vertex of that polygon. What is not known is exactly how Archimedes calculated the various numerical approximations to the square roots that are part of his calculation.

Let there be a circle with diameter $A\Gamma$ and center E, and let $\Gamma\Lambda Z$ be a tangent and the angle $ZE\Gamma$ one-third of a right angle [figure 4.2.2]. Then $E\Gamma$ has to ΓZ a ratio greater than 265 to 153 and EZ is to $Z\Gamma$ as 306 is to 153. Now let the angle $ZE\Gamma$ be bisected by EH. It follows that ZE is to $E\Gamma$ as ZH is to $H\Gamma$ so that $ZE + E\Gamma$ is to $Z\Gamma$ as $E\Gamma$ is to $H\Gamma$.[4] Therefore ΓE has to ΓH a ratio greater than 571 to 153. Hence sq(EH) has to sq($H\Gamma$) a ratio greater than 349450 to 23409. so that the ratio of EH to $H\Gamma$ is greater than $591\frac{1}{8}$ to 153.

Again, let angle $HE\Gamma$ be bisected by $E\Theta$; then by the same reasoning the ratio of $E\Gamma$ to $\Gamma\Theta$ is greater than $1162\frac{1}{8}$ to 153, so that the ratio of ΘE to $\Theta\Gamma$ is greater than $1172\frac{1}{8}$ to 153. Again, let angle $\Theta E\Gamma$ be bisected by EK. Then the ratio of $E\Gamma$ to ΓK is greater than $2334\frac{1}{4}$ to 153, so that the ratio of EK to ΓK is greater than $2339\frac{1}{4}$ to 153. Again, let angle $KE\Gamma$ be bisected by ΛE. Then the ratio of $E\Gamma$ to $\Lambda\Gamma$ is greater than $4673\frac{1}{2}$ to 153.

Now since angle $ZE\Gamma$, which is the third part of a right angle, has been bisected four times, angle $\Lambda E\Gamma$ is one forty-eighth of a right angle. Let angle ΓEM be placed at E

[4] By *Elements* 6:3, $ZE : E\Gamma = ZH : H\Gamma$, so $(ZE + E\Gamma) : E\Gamma = (ZH + H\Gamma) : H\Gamma = Z\Gamma : H\Gamma$, and therefore $(ZE + E\Gamma) : Z\Gamma = E\Gamma : H\Gamma$.

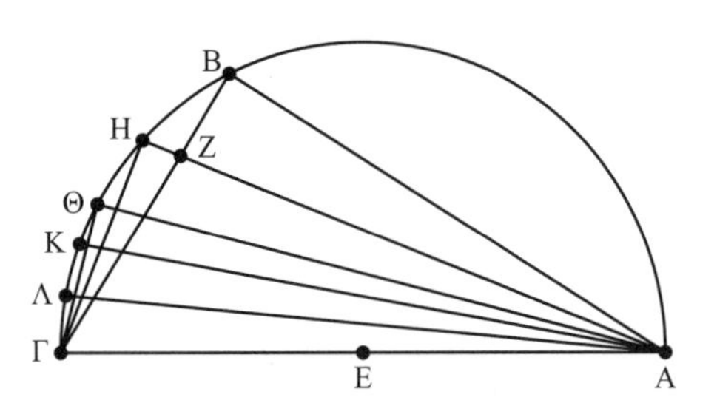

Figure 4.2.3.

equal to it. So angle ΛEM is therefore one twenty-fourth of a right angle. And ΛM is therefore the side of a polygon circumscribing the circle and having ninety-six sides. Since the ratio of $E\Gamma$ to $\Gamma\Lambda$ was proved to be greater than $4673\frac{1}{2}$ to 153 and $A\Gamma$ equals twice $E\Gamma$ while ΛM equals twice $\Gamma\Lambda$, the ratio of $A\Gamma$ to the perimeter of the 96-sided polygon is greater than $4673\frac{1}{2}$ to 14688. And the ratio of 14688 to $4673\frac{1}{2}$ is greater than 3, being in excess by $667\frac{1}{2}$, which is less than the seventh part of $4673\frac{1}{2}$; so that the perimeter of the circumscribed polygon is greater than three times the diameter by less than the seventh part; *a fortiori* therefore the circumference of the circle is less than $3\frac{1}{7}$ times the diameter.

In the second half of the proof, Archimedes deals with inscribed polygons. He begins with the side of an inscribed hexagon, presented as one leg of a right triangle with hypotenuse the diameter of the circle. He then shows how to calculate an approximation to the sides of, consecutively, a polygon of 12, 24, 48, and 96 sides. As above, the method uses elementary geometry. If we translate the algorithm into algebra, we can represent it as follows, where d is the diameter of the circle, t_i is the length of one side of the regular inscribed polygon of $3 \cdot 2^i$ sides, and u_i the length of the other leg of the right triangle formed from the diameter and the side of the polygon. Then

$$t_{i+1} = \frac{dt_i}{\sqrt{t_i^2 + (d+u_i)^2}}, \qquad u_{i+1} = \sqrt{d^2 - t_{i+1}^2}.$$

Let there be a circle with diameter $A\Gamma$ and angle $BA\Gamma$ one-third of a right angle [figure 4.2.3]. Then the ratio of AB to $B\Gamma$ is less than 1351 to 780. Let angle $BA\Gamma$ be bisected by AH. Now since angle BAH equals angle $H\Gamma B$ and angle BAH equals angle $HA\Gamma$, therefore angle $H\Gamma B$ equals angle $HA\Gamma$. And the right angle $AH\Gamma$ is common. Therefore the third angle $HZ\Gamma$ equals the third angle $A\Gamma H$. The triangle $AH\Gamma$ is therefore equiangular with the triangle ΓHZ. Therefore AH is to $H\Gamma$ as ΓH is to HZ as $A\Gamma$ is to ΓZ. But $A\Gamma$ is to ΓZ as ΓA, AB is to $B\Gamma$. Therefore BA, $A\Gamma$ is to $B\Gamma$ as AH is to $H\Gamma$. Therefore the ratio of AH to $H\Gamma$ is less than 2911 to 780. Hence the ratio of $A\Gamma$ to ΓH is less than $3013\frac{3}{4}$ to 780.

Let angle ΓAH be bisected by $A\Theta$. By the same reasoning the ratio of $A\Theta$ to $\Theta\Gamma$ is less than $5924\frac{3}{4}$ to 780, or $\frac{4}{13} \cdot 5924\frac{3}{4}$ to $\frac{4}{13} \cdot 780$, or, finally, 1823 to 240. Therefore the ratio of $A\Gamma$ to $\Gamma\Theta$ is less than $1838\frac{9}{11}$ to 240. Further, let angle $\Theta A\Gamma$ be bisected by KA. Then the ratio of AK to $K\Gamma$ is less than $\frac{11}{40} \cdot 3661\frac{9}{11}$ is to $\frac{11}{40} \cdot 240$, or less than 1007 to 66.

Therefore the ratio of $A\Gamma$ to $K\Gamma$ is less than $1009\frac{1}{6}$ to 66. Further, let angle $KA\Gamma$ be bisected by ΛA. Then the ratio of $A\Lambda$ to $\Lambda\Gamma$ is less than $2016\frac{1}{6}$ to 66. Therefore the ratio of $A\Gamma$: to $\Gamma\Lambda$ is less than $2017\frac{1}{4}$ to 66 and *invertendo*.

But $\Gamma\Lambda$ is the side of a polygon of 96 sides; and accordingly the perimeter of the polygon bears to the diameter a ratio greater than 6336 to $2017\frac{1}{4}$, which is greater than $3\frac{10}{71}$. Therefore the perimeter of the 96-sided polygon is greater than $3\frac{10}{71}$ times the diameter, so that *a fortiori* the circle is greater than $3\frac{10}{71}$ times the diameter.

The perimeter of the circle is therefore more than three times the diameter, exceeding by a quantity less than the seventh part but greater than ten seventy-first parts.

4.3 Archimedes, *Quadrature of the Parabola*

Quadrature of the Parabola contains Archimedes's "geometrical" proof for the area of a parabolic segment. But the first half of the treatise is devoted to a different proof, one using mechanical arguments based on the law of the lever. As shown in section 4.4, Archimedes demonstrates in the first proposition of his *Method* a mechanical method that led him to the solution of the problem, but of which he noted that a purely geometrical proof would be necessary to confirm the result. So it is a bit curious that in this treatise, although he provided a purely geometrical proof, he also gave another mechanical method of deriving the result. As in most of his works, this treatise is prefaced by a letter written to Dositheus, a mathematician and astronomer based in Alexandria.

Archimedes to Dositheus, best wishes.

When I heard that Conon[5]—who never failed to show me friendship—was dead, but that you were known to him and familiar with geometry, while I was in grief for the dead—for a man who was a friend and for a marvelous mathematician—I set myself to write down and send you (as I thought to write to Conon) geometrical theorems, which were not studied before but have now been studied by us, first found through mechanical means but then also proved through geometrical means. Indeed, some of those who earlier engaged with geometry tried to prove that it is possible to find a rectilinear area equal to a given circle or to a given segment of a circle, and further, they tried to square the area contained by, in general, a section of the cone and a line, making inadmissible assumptions, which is why most considered that they did not find these results. But we do not know of any previous mathematician having tried to square the segment contained by both: a line as well as a section of a right-angled cone [parabola][6]—which is what we now found. For it is proved that every segment contained by a line and a parabola is a third as much again of the triangle having the same base and a height equal to the segment, with the following assumption taken for the proof: It is possible for the difference, by which the greater of unequal areas exceeds the smaller, added itself to itself, to exceed any given bounded area.

[5] Conon was a mathematician well known to Archimedes in his youth, but one who died before he had the opportunity to comment on many of Archimedes's treatises.

[6] Archimedes uses the term "section of a right-angled cone" to mean what we call a parabola; the name of this curve was first proposed by Apollonius. In what follows, we use the modern word rather than Archimedes's older terminology.

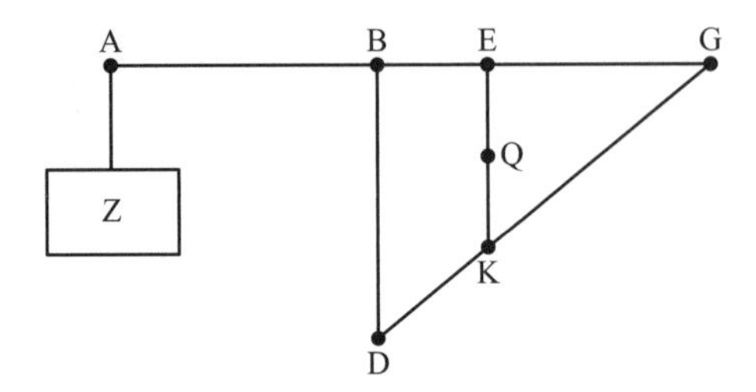

Figure 4.3.1.

Earlier geometers, too, have used this assumption. For they proved using this very assumption, both that circles have to each other the duplicate ratio of the diameters and that spheres have to each other the triplicate ratio of the diameters; and yet again also, that every pyramid is a third part of the prism having the same base as the pyramid and an equal height, and that every cone is a third part of the cylinder having the same base as the cone and an equal height, they proved using a certain assumption similar to the aforementioned. It turns out that each of the aforementioned theorems has been believed no less than those proved without this assumption, and it is sufficient to have brought the arguments now published by me to a similar credibility. Well then, having written down the proofs of the theorem, I send it: first, as studied through mechanical arguments and following that, also as proved through geometrized arguments. And first are written the conic elements useful for the proof. Farewell.

Archimedes begins the treatise by stating three results about parabolas that he assumes as known and then proving two other elementary results about these curves. Then in proposition 6, he begins the mechanical proof that he promised by showing that if an area Z is hung at one end A of a lever with fulcrum at B, then it will balance a right triangle BDG hung at a point one-third of the way from the fulcrum to the other end G of the lever provided that Z is one-third of the triangle.

Proposition 6. *And let the plane set forth in view be imagined right to the horizon, and let some part of it be imagined on the same side of the line AB as D, below, and the other part above* [figure 4.3.1]. *And let there be the triangle BDG, right-angled, having the angle at B right and the side BG be equal to half of a balance. And let the triangle hang from the points B, G, and let some other area Z hang from the other part of the balance at A, and let the area Z, hung at A, balance with triangle BDG being such as it is set now. I claim that the area Z is a third part of the triangle BDG.*

For since the balance is supposed to be in balance, the line AG shall be parallel to the horizon, and the lines drawn in the plane which is perpendicular to the horizon, at right angles to AG, shall be perpendicular to the horizon. So, let the line BG be cut at E in such a way so that GE is twice EB, and let KE be drawn parallel to DB, and let it be bisected at Q. So the point Q is center of the weight of the triangle BDG. (For this has been proved.[7]) Now, if the hanging of the triangle at B, G shall be freed, but it be hung at E instead, the triangle shall remain as it is now. For each of the things hanged shall remain at the point

[7] In proposition 12 of *Planes in Equilibrium*, Book 1.

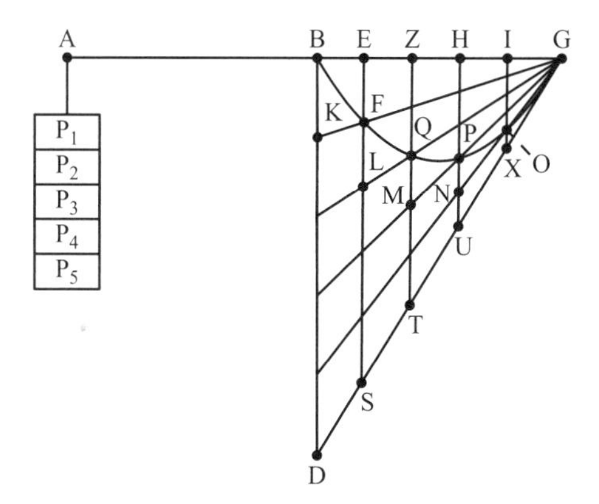

Figure 4.3.2.

from which it is set up, so that both the point of the hanging as well as the center of the weight of the things hanged, are on a perpendicular. For this too has been proved. Now, since the triangle BGD has the same setup to the balance as it had before, it shall similarly balance with the area Z. But since they are in balance, Z hung at A while the triangle BDG is hung at E, it is clear that the lengths reciprocate to the weights, and it is as AB is to BE, so the triangle BDG is to the area Z. And AB is three times BE. Therefore the triangle BDG, too, is three times the area Z. And it is also obvious that if the triangle BDG shall be three times the area Z, they shall balance.

Archimedes continues with a series of propositions using the law of the lever that involve balancing given areas against triangles, both right and obtuse, and against trapezia, also both right and obtuse. The conclusion of each proposition shows the relationship of the given areas to the areas of the triangles and trapezia involved. Finally, in propositions 14 and 15 he puts together the previous results to find a relationship between a right or an obtuse triangle that contains a parabolic segment and a set of trapezia and triangles that are drawn in the triangle. We present the statement of proposition 14 along with the relevant diagram and an outline of the proof. This result then allows Archimedes in propositions 16 and 17 to calculate the area of the parabolic segment, using an exhaustion argument similar to those in Euclid's *Elements* 12.

Proposition 14. *Let there be a segment BQG contained by a line and by a parabola* [figure 4.3.2]. *Let BG be at right angles to the diameter, and let BD be drawn from the point B parallel to the diameter, while GD is drawn tangent to the parabola. So, the triangle BGD shall be right-angled. So, let BG be divided into however many equal segments, BE, EZ, ZH, HI, IG, and, from the section points, let ES, ZT, HU, IX be drawn parallel to the diameter, and, from the points at which they cut the parabola, let them be joined to G, and produced. I claim that the triangle BDG is smaller than three times the trapezia KE, LZ, MH, NI and the triangle XIG, but greater than three times the trapezia ZF, HG, IP and the triangle IOG.*

We call the set of trapezia KE, LZ, MH, NI together with the triangle XIG the circumscribed figure to the parabolic segment, and the set ZF, HG, IP with the triangle IOG, the inscribed figure. So the theorem states that triangle BDG is smaller than

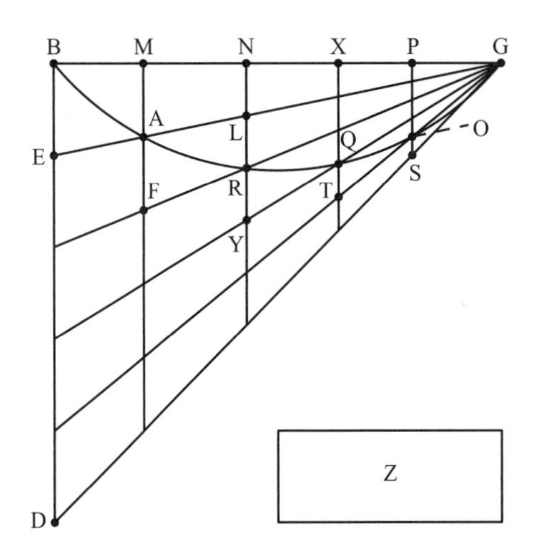

Figure 4.3.3.

three times the circumscribed figure but larger than three times the inscribed figure. Archimedes begins his argument by using the lever AG with fulcrum at B. If the trapezium DE is balanced by an area P_1 suspended at A, trapezium SZ balanced by an area P_2 suspended at A, trapezium TH by area P_3, trapezium UI by area P_4, and triangle XIG by P_5, then $P_1+P_2+P_3+P_4+P_5$ will balance the whole triangle DBG, so, by proposition 6, $P_1+P_2+P_3+P_4+P_5=\frac{1}{3}DBG$. But Archimedes has shown by proposition 5, a property of the parabola, and by propositions 8, 10, and 12 dealing with balancing areas on a lever, that $P_1<$ trapezium KE, that $FZ<P_2<LZ$, that $QH<P_3<MH$, that $PI<P_4<NI$, and finally that $IOG<P_5<XIG$. It follows by addition that the sum of the P_i, that is, $\frac{1}{3}DBG$, is simultaneously greater than the inscribed figure and smaller than the circumscribed figure or, that alternatively, that triangle BDG is greater than three times the inscribed figure and smaller than three times the circumscribed figure. (We note that propositions 7, 9, 11, and 13 all deal with obtuse rather than right triangles and trapezia, and these are used in the proof of proposition 15, in which BG is not at right angles to the diameter of the parabola.)

Proposition 16. *Again, let there be a segment BQG contained by a line and parabola, and let BD be drawn through B parallel to the diameter, and let GD be drawn from G tangent to the parabola at G, and let the area Z be a third part of the triangle BDG* [figure 4.3.3]. *So, I claim that the segment BQG is equal to the area Z.*

For if it is not equal it is either greater or smaller. So first, let it be, if possible, greater. So, the difference by which the segment BQG is greater than the area Z, added itself to itself, shall be greater than the triangle BDG, and it is possible to take some area smaller than the difference, which shall be a part of the triangle BDG. So let it be the triangle BGE, both smaller than the said difference as well as a part of the triangle BDG. And BE shall be the same part of BD. Now, let BD be divided into the parts, and let the points of the divisions be H, I, K, and let lines be joined from the points H, I, K to G. So these cut the parabola, since GD is a tangent to it at G, and let the lines MA, NR, XQ, PO be drawn from the points at which the lines cut the parabola, parallel to the diameter. So they shall be parallel to BD as well. Now, since the triangle BGE is smaller than the difference by which the segment BQG exceeds the area Z, it is clear that both taken together—the area Z and the triangle BGE—are smaller than the segment. And the triangle BGE is equal to the trapezia, through which the parabola goes—the trapezia ME, FL, QR, OQ. and the triangle GOS. For the trapezium ME is common, while the trapezium ML is equal to the trapezium FL and the trapezium LX is equal to the trapezium QR and the trapezium CX is equal to the trapezium OQ and the triangle GCP is equal to the triangle GOS. So the area Z is smaller than the trapezia ML, XR, PQ and the triangle POG, and the triangle BDG is three times the area Z. So the triangle BDG is smaller than three times the trapezia ML, RX, QP and the triangle POG, which is impossible; for it was proved to be greater than three times. Well then, the segment BQG is not greater than the area Z.

So I say, that neither is it smaller. For let it be, if possible, smaller. Therefore, again, the difference by which the area Z exceeds the segment BQG, added itself to itself, shall also exceed the triangle BDG. So, it is possible to take an area smaller than the difference, which shall be a part of the triangle BDG. Now, let the triangle BGE be smaller than the difference and a part of the triangle BDG, and let the rest be constructed the same as before. Now, since the triangle BGE is smaller than the difference by which the area Z exceeds the segment BQG, the triangle BEG and the segment BQG taken together are smaller than the area Z. But the area Z is also smaller than the quadrilaterals EM, FN, YX, PT and the triangle GPS, for the triangle BDG is three times the area Z, while it is less than three times the said areas, as was shown in the preceding propositions. Therefore the triangle BGE and the segment BQG are smaller than the quadrilaterals EM, FN, XY, PT and the triangle GPS. So that, taking away the segment as common, the triangle GBE will be greater than the remaining areas, which is impossible. For the triangle BEG was shown to be equal to the trapezia EM, FL, QR, QO and the triangle GOS, which are greater than the remaining areas. Therefore, the segment BQG is not smaller than the area Z, and it was shown that neither is it greater. Therefore the segment is equal to the area Z.

The basic argument in proposition 16 is the following: As before, we call the circumscribed figure the rectilinear figure composed of trapezia and triangles with base the line BG and having vertices at E, F, Y, T, S. So we make the intuitive assumption that the circumscribed figure is larger than the segment. We call the "difference set" the set of trapezia EM, FL, QR, OQ together with the triangle GOS. Then since the circumscribed figure is greater than the parabolic segment, it is also true that the circumscribed figure minus the difference set is greater than the segment minus the difference set. But the circumscribed figure minus the difference set is the inscribed figure, namely, the rectilinear figure with base MG and vertices A, R, Q, O. And the difference set is equal to the triangle BGE. So the inscribed figure is greater than the segment minus triangle BGE. But the segment is greater than Z plus triangle BGE, or the segment less triangle BGE is greater than Z. Therefore the inscribed figure is greater than Z. But Z is one-third triangle BDG, so the inscribed figure is greater than one-third triangle BDG, or triangle BDG is less than three times the inscribed figure. But this directly contradicts the last statement of proposition 14. A similar argument gives the second part of the proof by contradiction.

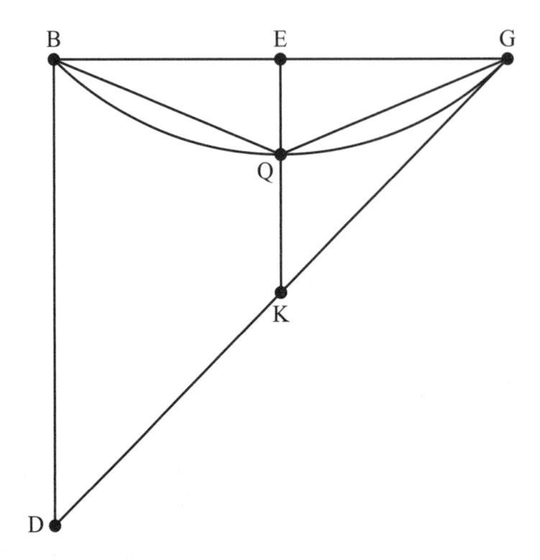

Figure 4.3.4.

Proposition 17. *This being proved it is obvious that every segment contained by both a line as well as a parabola is a third as much again as the triangle having the same base as the segment and an equal height.*

For let there be a segment contained by both a line as well as a parabola, and let its vertex be the point Q, and let a triangle BQG be inscribed inside it, having the same base as the segment and an equal height [figure 4.3.4]. Now, since the point Q is the vertex of the segment, the line drawn from Q parallel to the diameter bisects BG, and BG is parallel to the tangent to the parabola at Q. Let EQ be drawn parallel to the diameter; from B, let BD be drawn parallel to the diameter as well, and let GD be drawn from G, tangent to the parabola at G. Now, since KQ is parallel to the diameter while GD is a tangent to the parabola at G and EG is parallel to the tangent to the parabola at Q, the triangle BDG is four times the triangle BQG.[8] And since the triangle BDG is three times the segment BQG and four times the triangle BQG, it is clear that the segment BQG is a third as much as the triangle BQG.

Archimedes now turns to a new, more geometric proof of the result on the area of a parabolic segment. He begins with defining a number of concepts he uses in his proof. However, we note that he used some of these in the previous proposition.

Definitions: I call the *base* of figures contained by both a straight line as well as a curved line, the straight line, and I call *height* the greatest perpendicular drawn from the curved line on the base of the figure, and *vertex*, the point from which the greatest perpendicular is drawn.

In proposition 18, Archimedes shows that in a parabolic segment, a line drawn from the middle of the base, parallel to the diameter, cuts the parabola at a vertex of

[8] We know from the properties of the parabola that $EQ = QK$. Also, since E is the midpoint of BG, we have $BD = 2EK$. The result follows.

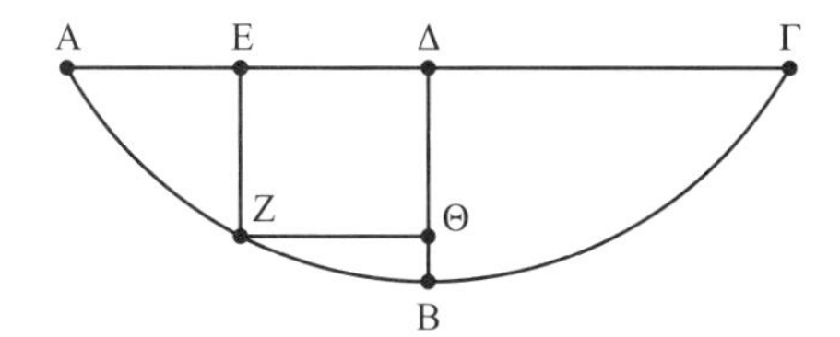

Figure 4.3.5.

the segment. He then proves a series of propositions that together give a proof, again by exhaustion, that the area of a parabolic segment is four-thirds of the triangle with the same base and equal height.

Proposition 19. *The line drawn from the middle of the base in a figure contained by a line and a parabola shall be a third as much again, in length, as the line drawn from the middle of the half.*

For let there be the segment $AB\Gamma$, contained by a line and a parabola, and let there be drawn parallel to the diameter, $B\Delta$, from the middle of $A\Gamma$, and EZ, from the middle of $A\Delta$ [figure 4.3.5]. And let $Z\Theta$, too, be drawn parallel to $A\Gamma$. Now, since $B\Delta$ has been drawn in a parabola parallel to the diameter, and $A\Delta$, $Z\Theta$ are parallel to the tangent at B, it is clear that $B\Delta$ has to $B\Theta$ the same ratio, in length, which $A\Delta$ has to $Z\Theta$, in square. Therefore, $B\Delta$, too, is four times $B\Theta$ in length. Now, it is obvious that $B\Delta$ is a third again as much as EZ in length.

Proposition 21. *If a triangle is inscribed inside a segment contained by a line and a parabola, having the same base as the segment and the same height, and other triangles are inscribed inside the remaining segments, having the same base as the segments and the same height, the triangle inscribed inside the whole segment shall be eight times either of the triangles inscribed inside the remaining segments.*

Let there be the segment $AB\Gamma$, as has been said; let $A\Gamma$ be bisected by Δ, and let $B\Delta$ be drawn parallel to the diameter [figure 4.3.6]. Therefore, the point B is vertex of the segment. Therefore, the triangle $AB\Gamma$ has the same base as the segment and the same height. Again, let $A\Delta$ be bisected by E, and let EZ be drawn parallel to the diameter, and let AB be cut at Θ. Therefore, the point Z is vertex of the segment AZB. So the triangle AZB has the same base as segment AZB, and the same height. It is to be proved that the triangle $AB\Gamma$ is eight times the triangle AZB.

Now, $B\Delta$ is a third again as much as EZ, and twice $E\Theta$. Therefore $E\Theta$ is twice ΘZ. So the triangle AEB, too, is twice the triangle ZBA. (For the triangle $AE\Theta$ is twice the triangle $A\Theta Z$, while the triangle ΘBE is twice the triangle $Z\Theta B$.) So the triangle $AB\Gamma$ is eight times the triangle AZB. The triangle $AB\Gamma$ shall be similarly proved to be eight times the triangle similarly inscribed inside the segment $BH\Gamma$.

The next result is essentially an algebraic one. It simply asserts that if one has a finite geometric series a, $\frac{1}{4}a$, $(\frac{1}{4})^2a, \ldots, (\frac{1}{4})^na$, then its sum added to the quantity $\frac{1}{3}(\frac{1}{4})^na$ is equal to $\frac{4}{3}a$. Since Archimedes then shows that the parabolic segment can

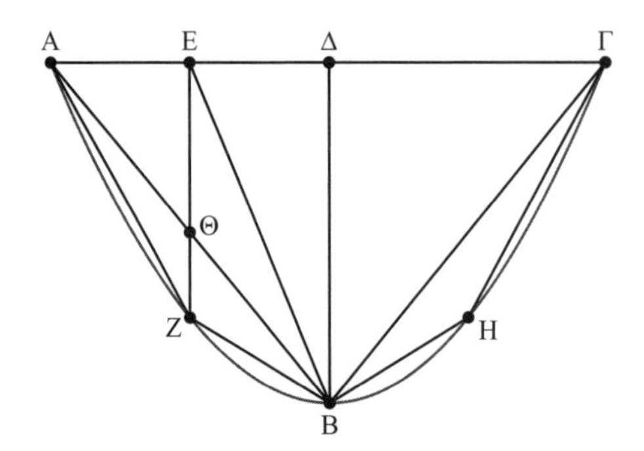

Figure 4.3.6.

be "exhausted" by a series of sets of triangles, where each set has area one-fourth of the previous set, the result for the area of the parabolic segment follows.

Proposition 23. *If magnitudes are set in sequence in the four times ratio, all the magnitudes and yet again the third part of the smallest magnitude, composed into one and the same, shall be a third again as much as the greatest magnitude.*

Indeed, let there be however many magnitudes set in sequence, A, B, Γ, Δ, E, each four times the following one, and let A be greatest, and let Z be a third of B, and let H be a third of Γ, and let Θ be a third of Δ, and let I be a third of E. Now, since Z is a third part of B, while B is a fourth part of A, both magnitudes B, Z are a third part of A. So, through the same argument, the magnitudes H, Γ, too, are a third part of B, and the magnitudes Θ, Δ are a third part of Γ, and the magnitudes I, E are a third part of Δ. So the magnitudes B, Γ, Δ, E, Z, H, Θ, I all together are a third part of the magnitudes A, B, Γ, Δ all together, while the magnitudes themselves Z, H, Θ are also a third part of the magnitudes themselves B, Γ, Δ. Therefore, the remaining magnitudes, too, B, Γ, Δ, E, I are a third part of the remaining magnitude A. Now, it is clear that the magnitudes A, B, Γ, Δ, E all together, as well as the magnitude I, that is, the third of the magnitude E, are a third again as much as the magnitude A.

Proposition 24. *Every segment contained by a line and a parabola is a third as much again as a triangle having the same base and the same height.*

For let there be the segment $A\Delta BE\Gamma$, contained by a line and a parabola, and let there be the triangle $AB\Gamma$, having the same base as the segment and the same height, and let the area K be a third again as much as the triangle $AB\Gamma$. It is to be proved that the area K is equal to the segment $A\Delta BE\Gamma$ [Fig 4.3.7].

For if it is not equal, it is either greater or smaller. Let the segment $A\Delta BE\Gamma$ be first, if possible, greater than the area K. So, I inscribed the triangles $A\Delta B$, $BE\Gamma$, as has been said, and I also inscribed other triangles inside the remaining segments having the same base as the segments and the same height, and always, I inscribe two triangles inside the segments that arise later, having the same base as the segment and the same height. So, the segments that end up remaining shall be smaller than the difference by which the segment $A\Delta BE\Gamma$ exceeds the area K so that the inscribed polygon shall be greater than the area K, which is impossible. For since there are areas set in sequence in the

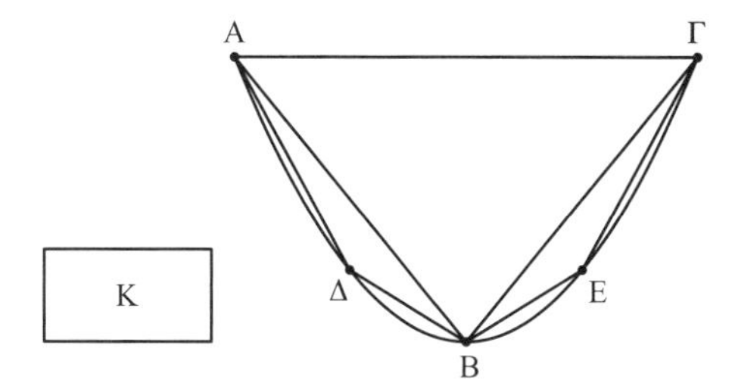

Figure 4.3.7.

four-times ratio, first the triangle $AB\Gamma$ being four times the triangles $A\Delta B$, $BE\Gamma$, then these triangles themselves, being four times the triangles inscribed inside the following segment, and always in this manner, it is clear that the areas, all together, are smaller than a third as much again as the greatest area, while the area K is a third again as much as the greatest area. Therefore the segment $A\Delta BE\Gamma$ is not greater than the area K.

But let the segment be, if possible, smaller than the area. So, let the triangle $AB\Gamma$ be set equal to the area Z, and let the area H be set a fourth of the area Z, and similarly, let the area Θ be set a fourth of the area H, and let it always be set, until the last area comes to be smaller than the difference by which the area K exceeds the segment, and let the area I be smaller than the difference. And the areas Z, H, Θ, I, and the third of the area I are a third as much again as the area Z. And the area K, too, is a third again as much as the area Z. Therefore the area K is equal to the areas Z, H, Θ, I and the third part of the area I. Now, since the area K exceeds the areas Z, H, Θ, I by a difference smaller than the area I, while the area K exceeds the segment by a difference greater than I, it is clear that the areas Z, H, Θ, I are greater than the segment, which is impossible. For it has been proved that if however many areas set in sequence in the four-times ratio, and the greatest is equal to the triangle inscribed inside the segment, the areas, all together, shall be smaller than the segment. Therefore, the segment $A\Delta BE\Gamma$ is not smaller than the area K. But it was proved that neither is it greater. Therefore, it is equal to the area K. But the area K is a third as much again as the triangle $AB\Gamma$. Therefore, the segment $A\Delta BE\Gamma$, too, is a third again as much as the triangle $AB\Gamma$.

4.4 Archimedes, *Sphere and Cylinder I*

In *On the Sphere and Cylinder I*, Archimedes's goal is to determine the surface area and the volume of a sphere and then generalize these results to spherical segments and sectors. We present here Archimedes's introductory letter to Dositheus, then some definitions and postulates, followed by the most important results, along with some of the proofs and some discussion of the ideas behind the proofs. Interspersed are the pertinent observations made by Eutocius as he prepared the edition of Archimedes's works that survived into the medieval period and, ultimately, into the Renaissance. These comments are in a **different typeface** from Archimedes's own words.

Archimedes to Dositheus: Greetings!

Earlier, I have sent you some of what we had already investigated then, writing it with a proof: that every segment contained by a straight line and by a section of the right-angled cone is a third again as much as a triangle having the same base as the segment and an equal height. Later, theorems worthy of mention suggested themselves to us, and we took the trouble of preparing their proofs. They are these: first that the surface of every sphere is four times the greatest circle of the circles in it. Further, that the surface of every segment of a sphere is equal to a circle whose radius is equal to the line drawn from the vertex of the segment to the circumference of the circle which is the base of the segment. Next to these, that, in every sphere, the cylinder having a base equal to the greatest circle of the circles in the sphere, and a height equal to the diameter of the sphere, is, itself, half as large again as the sphere; and its surface is half as large again as the surface of the sphere.

In nature, these properties always held for the figures mentioned above. But these properties were unknown to those who have engaged in geometry before us—none of them realizing that there is a common measure to those figures. Therefore I would not hesitate to compare them to the properties investigated by any other geometer, indeed to those which are considered to be by far the best among Eudoxus's investigations concerning solids: that every pyramid is a third part of a prism having the same base as the pyramid and an equal height, and that every cone is a third part of the cylinder having the base the same as the cylinder and an equal height. For even though these properties, too, always held, naturally, for those figures, and even though there were many geometers worthy of mention before Eudoxus, they all did not know it; none perceived it.

But now it shall become possible—for those who will be able—to examine those theorems.

They should have come out while Conon was still alive. For we suppose that he was probably the one most able to understand them and to pass the appropriate judgment. But we think it is the right thing, to share with those who are friendly towards mathematics, and so, having composed the proofs, we send them to you, and it shall be possible—for those who are engaged in mathematics—to examine them. Farewell.

As I found that no one before us had written down a proper treatise on the books of Archimedes *On Sphere and Cylinder*, and seeing that this has not been overlooked because of the ease of the propositions (for they require, as you know, precise attention as well as intelligent insight), I desired, as best I could, to set out clearly those things in it which are difficult to understand; and I was more led to do this by the fact that no one had yet taken up this project, than I was deterred by the difficulty; as I was also reasoning in the Socratic manner that, with god's support, most probably we shall reach the end of my efforts. And third, I thought that, even if, through my youth, something will strike out of tune, this will be made right by your scientific comprehension of philosophy in general, and especially of mathematics; and so I dedicate it to you, Ammonius, the best of philosophers.[9] It would be fitting

[9] Ammonius (ca. 480–540 CE) was a neo-Platonist philosopher, a pupil of Proclus at Athens, and a teacher at Alexandria who wrote commentaries on Aristotle.

that you help my effort. And if the book seems to you slight, then do not allow it to go from yourself to anyone else, but if it has not strayed completely off the mark, make your view upon it clear for, if it comes to be established by your own judgment, I shall try to explicate some other of the Archimedean treatises.

Definitions

1. There are in a plane some limited curved lines, which are either wholly on the same side as the straight lines joining their limits or have nothing on the other side.
2. So I call *concave in the same direction* such a line, in which, if any two points whatever being taken, the straight lines between the two points either all fall on the same side of the line, or some fall on the same side, and some on the line itself, but none on the other side.
3. Next, similarly, there are also some limited surfaces, which, while not themselves in a plane, do have the limits in a plane; and they shall either be wholly on the same side of the plane in which they have the limits, or have nothing on the other side.
4. So I call *concave in the same direction* such surfaces, in which, suppose two points being taken, the straight lines between the points either all fall on the same side of the surface, or some on the same side, and some on the surface itself, but none on the other side.
5. And, when a cone cuts a sphere, having a vertex at the center of the sphere, I call the figure internally contained by the surface of the cone, and by the surface of the sphere inside the cone, a *solid sector*.
6. And when two cones having the same base have the vertices on each of the sides of the plane of the base, so that their axes lie on a line, I call the solid figure composed of both cones a *solid rhombus*.

Postulates

1. That among lines which have the same limits, the straight line is the smallest.
2. And, among the other lines (if, being in a plane, they have the same limits); that such lines are unequal, when they are both concave in the same direction and either one of them is wholly contained by the other and by the straight line having the same limits as itself, or some is contained, and some it has as common; and the contained is smaller.
3. And similarly, that among surfaces, too, which have the same limits (if they have the limits in a plane) the plane is the smallest.
4. And that among the other surfaces that also have the same limits (if the limits are in a plane): such surfaces are unequal, when they are both concave in the same direction, and either one is wholly contained by the other surface and by the plane which has the same limits as itself, or some is contained, and some it has as common; and the contained is smaller.
5. Further, that among unequal lines, as well as unequal surfaces and unequal solids, the greater exceeds the smaller by such a difference that is capable, added itself to itself, of exceeding everything set forth (of those which are in a ratio to one another).

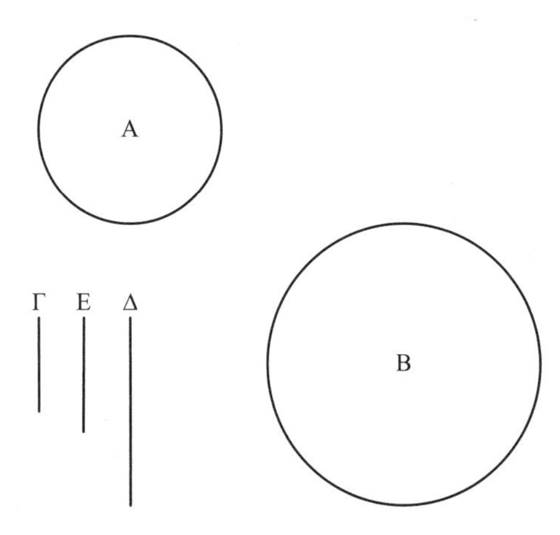

Figure 4.4.1.

Assuming these it is manifest that if a polygon is inscribed inside a circle, the perimeter of the inscribed polygon is smaller than the circumference of the circle; for each of the sides of the polygon is smaller than the circumference of the circle which is cut by it.

Propositions

Proposition 14. *The surface of every isosceles cone without the base, is equal to a circle whose radius has a mean ratio between the side of the cone and the radius of the circle which is the base of the cone.*[10]

Let there be an isosceles cone, whose base is the circle A and let the radius be Γ, and let Δ be equal to the side of the cone, and let E be a mean proportional between Γ, Δ, and let the circle B have the radius equal to E; I say that the circle B is equal to the surface of the cone without the base [figure 4.4.1].

For if it is not equal, it is either greater or smaller. Let it first be smaller. So there are two unequal magnitudes, the surface of the cone and the circle B, and the surface of the cone is greater; therefore it is possible to inscribe inside the circle B an equilateral polygon and circumscribe another similar to the inscribed, so that the circumscribed has to the inscribed a smaller ratio than that which the surface of the cone has to the circle B.[11] So let a circumscribed polygon be imagined around the circle A, too, similar to the polygon circumscribed around the circle B, and let a pyramid be set up on the polygon circumscribed around the circle A—constructed having the same vertex as the cone. Now, since the polygons circumscribed around the circles A, B are similar, they have the same ratio to each other, as the radii in square to each other. That is, the ratio that Γ has to E in square, that is Γ to Δ in length. But that ratio that Γ has to Δ in length, the circumscribed polygon around the circle A has to the surface of the pyramid circumscribed around the cone, for Γ is equal to the perpendicular drawn from the center on one side of the polygon,

[10] In modern terms, if r is the radius of the base circle and s is the slant height (side) of the cone, then the surface area is πsr.

[11] See *Sphere and Cylinder I*, proposition 5.

while Δ is equal to the side of the cone, and the perimeter of the polygon is a common height to the halves of the surfaces. Therefore the rectilinear figure around the circle A to the rectilinear figure around the circle B, and the same rectilinear figure to the surface of the pyramid circumscribed around the cone have the same ratio; so that the surface of the pyramid is equal to the rectilinear figure circumscribed around the circle B. Now, since the rectilinear figure circumscribed around the circle B has to the inscribed a smaller ratio than the surface of the cone to the circle B, the surface of the pyramid circumscribed around the cone will have to the rectilinear figure inscribed in the circle B a smaller ratio than the surface of the cone to the circle B, which is impossible, for the surface of the pyramid has been proved to be greater than the surface of the cone, while the rectilinear figure inscribed in the circle B will be smaller than the circle B. Therefore, the circle B will not be smaller than the surface of the cone.

But I say that it will not be greater, either. For if it is possible, let it be greater. So again let a polygon inscribed inside the circle B be imagined and another circumscribed, so that the circumscribed has to the inscribed a smaller ratio than that which the circle B has to the surface of the cone, and inside the circle A let an inscribed polygon be imagined, similar to the polygon inscribed inside the circle B, and let a pyramid be set up on it having the same vertex as the cone. Now, since the inscribed polygons in A, B are similar, they will have to each other the same ratio as the radii have to each other in square; therefore the polygon to the polygon, and Γ to Δ in length, have the same ratio. But Γ has to Δ a greater ratio than the polygon inscribed in the circle A to the surface of the pyramid inscribed inside the cone, for the radius of the circle A has to the side of the cone a greater ratio than the perpendicular drawn from the center on one side of the polygon to the perpendicular drawn on the side of the polygon from the vertex of the cone. Therefore the polygon inscribed in the circle A has to the polygon inscribed in the circle B a greater ratio than the same polygon to the surface of the pyramid; therefore the surface of the pyramid is greater than the polygon inscribed in the circle B. But the polygon circumscribed around the circle B has to the inscribed a smaller ratio than the circle B to the surface of the cone; much more, therefore, the polygon circumscribed around the circle B has to the surface of the pyramid inscribed in the cone a smaller ratio than the circle B to the surface of the cone; which is impossible for the circumscribed polygon is greater than the circle B, while the surface of the pyramid in the cone is smaller than the surface of the cone. Therefore neither is the circle greater than the surface of the cone. But it was proved, that neither is it smaller; therefore it is equal.

Proposition 16. *If an isosceles cone is cut by a plane parallel to the base, then a circle whose radius is the mean ratio between the side of the cone between the parallel planes and the line equal to both radii of the circles in the parallel planes, is equal to the surface of the cone between the parallel planes.*[12]

Proposition 21. *If an even-sided and equilateral polygon is inscribed inside a circle, and lines are drawn through, joining the vertices of the polygon (so that they are parallel to one of the lines subtended by two sides of the polygon), all the joined lines have to the diameter*

[12] If the two parallel circles have radii r_1, r_2 and if s is the slant height of the frustum, then the surface area of the frustum is $\pi s(r_1 + r_2)$.

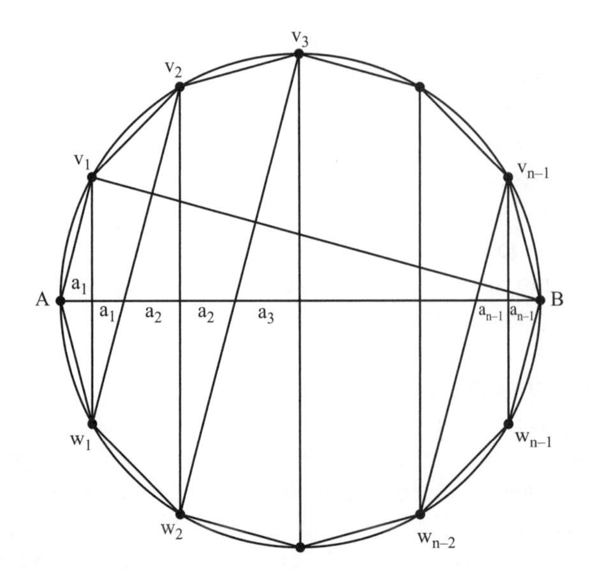

Figure 4.4.2.

of the circle that ratio, which the line subtending the sides, whose number is smaller by one than half the sides, has to the side of the polygon.

This proposition deals with an equilateral polygon of evenly many sides of length s, say $2n$ sides, inscribed inside a circle with diameter AB [figure 4.4.2]. It asserts that $(v_1w_1 + v_2w_2 + \cdots + v_{n-1}w_{n-1}) : AB = v_1B : s$. To prove this, first join v_2w_1, v_3w_2, ..., $v_{n-1}w_{n-2}$. Then the angles Av_1w_1, $v_1w_1v_2$, $w_1v_2w_2$, ... are all equal since they each subtend equal arcs on the circumference of the circle. We label the segments along the diameter AB as $a_1, a_1, a_2, a_2, \ldots, a_{n-1}, a_{n-1}$, and the halves of the lines v_1w_1, v_2w_2, ..., $v_{n-1}w_{n-1}$ as $r_1, r_2, \ldots, r_{n-1}$. Similarity of the right triangles then implies that $r_1 : a_1 = r_2 : a_2 = \cdots = r_{n-1} : a_{n-1}$. Also, each of these ratios is equal to $v_1B : Av_1$. So the ratio of the sum of the r_i to the sum of the a_i is also equal to $v_1B : Av_1 = v_1B : s$. But the sum of the r_i is half the sum $(v_1w_1 + v_2w_2 + \cdots + v_{n-1}w_{n-1})$, while the sum of the a_i is half the diameter AB. The desired result follows.

Many of the remaining propositions we include begin with this figure, although more specifically a polygon with "the number of its sides measured by 4," that is, of $4n$ sides, inscribed in a great circle in a sphere. The entire system is rotated about a diameter of the circle (AB) to form a solid inscribed in the sphere [figure 4.4.3]. This solid is composed of two cones, one on each end of the diameter, and $2n - 2$ frustums of cones. Archimedes generally refers to this as "the figure." He concerns himself both with the surface area and with the volume of this figure.

In proposition 23, Archimedes shows that the entire surface area of the figure is smaller than the surface area of the sphere in which it is inscribed. But then in proposition 24 he makes a stronger claim by actually calculating the surface area of this figure composed of conical surfaces. Using the notation of the previous paragraph,

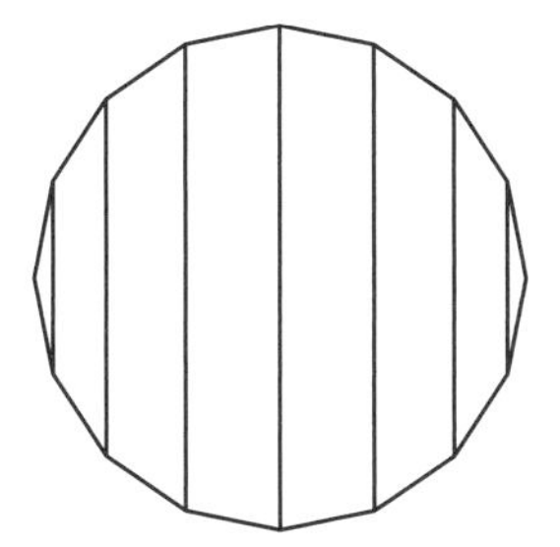

Figure 4.4.3.

the surface area of the figure is $\pi r_1 s + \pi(r_1 + r_2)s + \pi(r_2 + r_3)s + \cdots + \pi(r_{n-2} + r_{n-1})s + \pi r_{n-1}s = 2\pi(r_1 + r_2 + \cdots + r_{n-1})s = \pi(v_1w_1 + v_2w_2 + \cdots + v_{n-1}w_{n-1})s$, and proposition 24 is proved. Then by proposition 21, this last expression is equal to $\pi AB \cdot v_1B$. But AB is the diameter d of the circle and v_1B is less than the diameter of the circle. Therefore, the surface area of the figure is less than $\pi d^2 = 4\pi r^2$, where r is the radius of a great circle on the sphere. That is, the surface area is less than four times the area of a great circle on the sphere. So we have now proved proposition 25 as well. We also note that the closer the value of v_1B is to the diameter of the circle or, equivalently, the more sides the polygon has, the closer the area of the figure will be to exactly four times the area of a great circle. Archimedes, however, does not consider this idea directly. Instead, he gives a further demonstration that the area of a new figure circumscribed about the sphere is greater than four times a great circle and then shows, using the familiar *reductio ad absurdum* argument by exhaustion, that the surface area of the sphere is exactly four times a great circle. We present the theorems and some of the proofs below, as well as the analogous theorems dealing with the volume of the sphere.

[Archimedes] wants that the sides of the polygon be measured by four because it will be of use to him in the propositions following this one, that all the sides are carried along conical surfaces. For if the sides of the polygon are not measured by four, then it is possible—even if it is an even-sided polygon—that not all sides are carried along conical surfaces—which can be understood in the case of the sides of the hexagon; for two opposite parallel sides of it are in fact carried along a cylindrical surface. Which, as was said, is not of use to him in the following.

Proposition 24. *The surface of the figure inscribed inside the sphere is equal to a circle, whose radius is, in square, the rectangle contained by the side of the figure*[13] *and by the line equal to all the lines joining the sides of the polygon [which are parallel to the line connecting the two vertices on either side of one end of the diameter].*

Let there be $AB\Gamma\Delta$, greatest circle in a sphere, and let an equilateral polygon whose sides are measured by four be inscribed in it, and, on the inscribed polygon, let some

[13] Here Archimedes means the inscribed polygon.

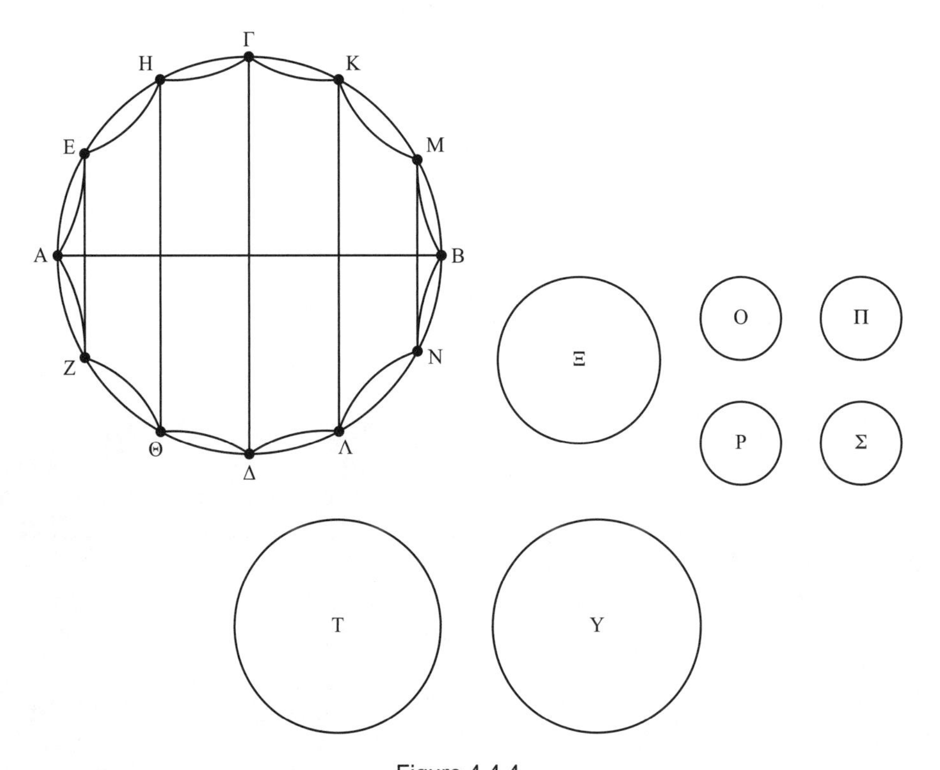

Figure 4.4.4.

figure inscribed inside the sphere be imagined, and let EZ, $H\Theta$, $\Gamma\Delta$, $K\Lambda$, MN be joined, being parallel to the line subtended by two sides; so let some circle be set out, Ξ, whose radius is, in square, the rectangle contained by AE and by the line equal to EZ, $H\Theta$, $\Gamma\Delta$, $K\Lambda$, MN; I say that this circle is equal to the surface of the figure inscribed inside the sphere [figure 4.4.4].[14]

For let there be set out circles, O, Π, P, Σ, T, Υ, and let the radius of O be in square, the rectangle contained by EA and by the half of EZ, and let the radius of Π be, in square, the rectangle contained by EA and by the half of EZ, $H\Theta$, and let the radius of P be, in square, the rectangle contained by EA and by the half of $H\Theta$, $\Gamma\Delta$, and let the radius of Σ be, in square, the rectangle contained by EA and by the half of $\Gamma\Delta$, $K\Lambda$, and let the radius of T be, in square, the rectangle contained by AE and by the half of $K\Lambda$, MN, and let the radius of Υ be in square the rectangle contained by AE and by the half of MN. So through these, the circle O is equal to the surface of the cone AEZ [proposition 14], and Π is equal to the surface of the cone between EZ, $H\Theta$ [proposition 16], and P is equal to the surface between $H\Theta$, $\Gamma\Delta$, and Σ is equal to the surface between $\Delta\Gamma$, $K\Lambda$, and yet again, T is equal to the surface of the cone between $K\Lambda$, MN, and Υ is equal to the surface of the cone MBN. Therefore all the circles are equal to the surface of the inscribed figure. And it is obvious that the radii of the circles O, Π, P, Σ, T, Υ are,

[14] In this figure, we use curved lines to represent the sides of the polygon, in accordance with some of the manuscripts.

in square, the rectangle contained by AE and by twice the halves of EZ, $H\Theta$, $\Gamma\Delta$, $K\Lambda$, MN, which are as wholes EZ, $H\Theta$, $\Gamma\Delta$, $K\Lambda$, MN; therefore the radii of the circles O, Π, P, Σ, T, Υ are, in square, the rectangle contained by AE and by all the lines EZ, $H\Theta$, $\Gamma\Delta$, $K\Lambda$, MN. But the radius of the circle Ξ, too, is, in square, the rectangle contained by AE and by the line composed of all the lines EZ, $H\Theta$, $\Gamma\Delta$, $K\Lambda$, MN; therefore the radius of the circle Ξ is, in square, the radii of the circles O, Π, P, Σ, T, Υ in square; and therefore the circle Ξ is equal to the circles O, Π, P, Σ, T, Υ. But the circles O, Π, P, Σ, T, Υ were proved equal to the surface of the said figure; therefore the circle Ξ, too, will be equal to the surface of the figure.

Proposition 25. *The surface of the figure inscribed inside the sphere, contained by the conical surfaces, is smaller than four times the greatest circle of the circles in the sphere.*

The next two propositions, presented without proof, deal with the volume of "the figure." In essence, Archimedes is thinking of a sphere as equivalent to a cone with base the surface of the sphere and vertex the center of the sphere, making the height of the cone equal to the radius. So he begins by approximating the surface area of the sphere by using the figure already described.

Proposition 26. *The figure inscribed in the sphere, contained by the conical surfaces, is equal to a cone having, as base, the circle equal to the surface of the figure inscribed in the sphere, and height equal to the perpendicular drawn from the center of the sphere on one side of the polygon.*

Proposition 27. *The figure inscribed in the sphere, contained by the conical surfaces, is smaller than four times the cone having a base equal to the greatest circle of the sphere and a height equal to the radius of the sphere.*

In proposition 28, Archimedes describes a new situation. Here, a regular polygon of $4n$ sides is circumscribed about a great circle in a sphere, and then a new circle is circumscribed about the polygon. When the entire system is rotated about the diameter of the larger circle, we get the figure of the previous propositions now circumscribing one sphere and being inscribed in a larger sphere. It follows that the surface of the circumscribed figure is greater than the surface of the smaller sphere.

Proposition 29. *The surface of the figure circumscribed around the sphere is equal to a circle, whose radius is, in square, equal to the rectangle contained by one side of the polygon and by the line equal to all the lines joining the angles of the polygon, the lines being parallel to some line among the lines subtended by two sides of the polygon.*

This result is clear, since the circumscribed figure is also an inscribed figure, so the result of proposition 24 holds.

Proposition 30. *The surface of the figure circumscribed around the sphere is greater than four times the greatest circle of the circles in the sphere.*

For let there be the sphere and the circle and the rest the same as set out before, and let the circle Λ be equal to the surface of the figure circumscribed around the smaller sphere which was set out before [figure 4.4.5]. Now, since an equilateral and even-angled

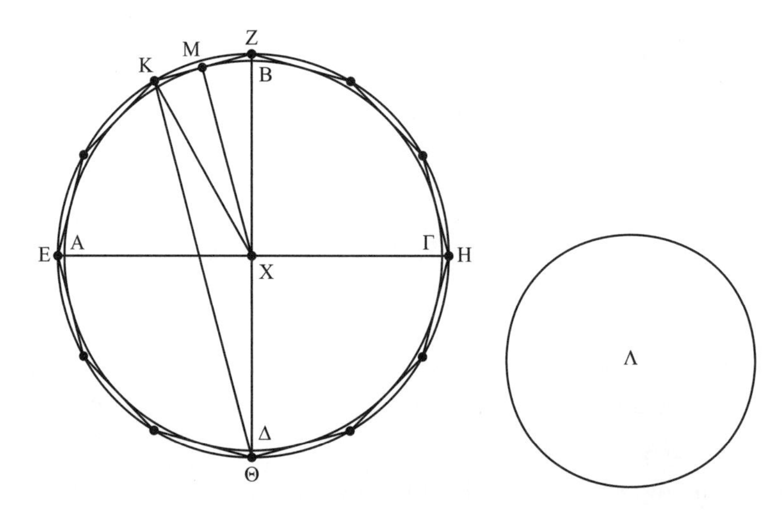

Figure 4.4.5.

polygon has been inscribed in the circle $EZH\Theta$, the lines joining the sides of the polygon, being parallel to $Z\Theta$, have to $Z\Theta$ the same ratio which ΘK has to KZ [proposition 21]. Therefore, the rectangle contained by one side of the polygon and by the line equal to all the lines joining the angles of the polygon is equal to the rectangle contained by $Z\Theta$ and ΘK. So the radius of the circle Λ is equal, in square, to the rectangle contained by $Z\Theta$, ΘK. Therefore the radius of the circle Λ is greater than ΘK. But ΘK is equal to the diameter of the circle $AB\Gamma\Delta$, for it is twice XM, which is a radius of the circle $AB\Gamma\Delta$. So it is clear that the circle Λ, that is, the surface of the figure circumscribed around the smaller sphere, is greater than four times the greatest circle of the circles in the sphere.

"But ΘK is equal to the diameter of the circle $AB\Gamma\Delta$." For if we join X to the point M at which KZ touches the circle $AB\Gamma\Delta$, and similarly join XK, then since XK is equal to XZ, and, also, the angles at M are right, KM will then be equal to MZ as well. But then ZX is equal to $X\Theta$, as well. Therefore XM is parallel to $K\Theta$ and through this it will be: as ΘZ is to ZX, so is $K\Theta$ to XM. But ΘZ is double XZ; therefore $K\Theta$ is also double XM, which is the radius of the circle $AB\Gamma\Delta$.

Proposition 31. *The figure circumscribed around the smaller sphere is equal to a cone having as base the circle equal to the surface of the figure and a height equal to the radius of the sphere.*

Corollary. *And from this it is obvious that the figure circumscribed around the smaller sphere is greater than four times a cone having, as base, the greatest circle of the circles in the sphere, and, as height, the radius of the sphere.*

Proposition 32. *If there is a figure inscribed in a sphere and another circumscribed, constructed of similar polygons, in the same manner as the above, the surface of the circumscribed figure has to the surface of the inscribed figure a ratio duplicate of the side of the polygon circumscribed around the great circle to the side of the polygon inscribed*

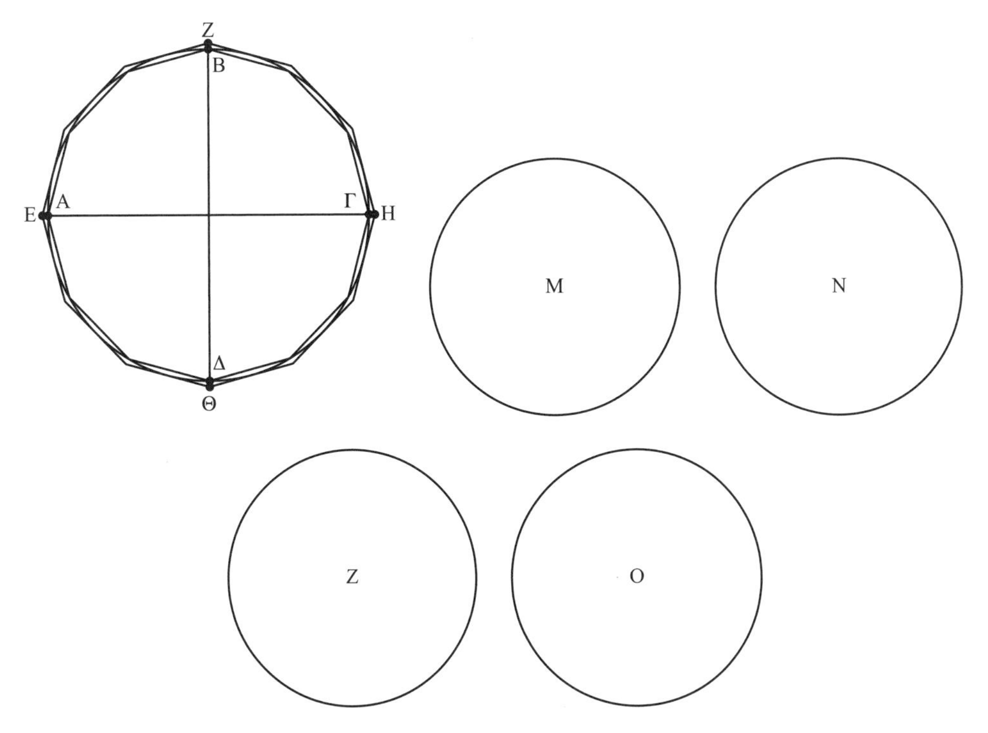

Figure 4.4.6.

in the same circle, and the circumscribed figure itself has to the figure a ratio triplicate of the same ratio.

Let there be a great circle in a sphere, $AB\Gamma\Delta$, and let an equilateral figure be inscribed inside it, and let the number of its sides be measured by four, and let another figure, similar to the inscribed, be circumscribed around the circle, and yet again, let the sides of the circumscribed polygon touch the circle at the middles of the circumferences cut by the sides of the inscribed polygon, and let EH, $Z\Theta$, at right angles to each other, be diameters of the circle containing the circumscribed polygon, and similarly placed as the diameters $A\Gamma$, $B\Delta$, and let lines be imagined joined to the opposite angles of the polygon (which will then be parallel to each other and to $ZB\Delta\Theta$) [figure 4.4.6]. So, as the diameter EH remains fixed, and the perimeters of the polygons are carried in a circular motion, one perimeter will be a figure inscribed in the sphere, the other will be a circumscribed figure; now, it is to be proved that the surface of the circumscribed figure has to the surface of the inscribed a ratio duplicate of $E\Lambda$ to AK, while the circumscribed figure has to the inscribed a ratio triplicate of the same ratio.

For let there be the circle M equal to the surface of the figure circumscribed around the sphere, and the circle N, equal to the surface of the inscribed figure; therefore the radius of the circle M is, in square, the rectangle contained by $E\Lambda$ and by the line equal to all the lines joining the angles of the circumscribed polygon, while the radius of the circle N is, in square, the rectangle contained by AK, and by the line equal to all the lines joining the angles. And since the polygons are similar, the areas contained by the said lines will also be similar, so that the areas have the same ratio to each other which the sides of the polygons have in square. But also, the areas contained by the said lines have that ratio which the radii of the circles M, N have to each other in square so that the diameters of the circles M, N, too, have the same ratio as the sides of the polygons. But the circles have to each other a ratio duplicate of the diameters, circles which, moreover, are equal to the surfaces of the circumscribed figure and the inscribed figure; so it is clear that the surface of the figure circumscribed around the sphere has to the surface of the figure inscribed inside the sphere a ratio duplicate of $E\Lambda$ to AK.

So let two cones be taken O, Ξ, and let the cone Ξ have as base the circle Ξ equal to the circle M, and let O be a cone having as base the circle O equal to the circle N, and let the cone Ξ have as height the radius of the sphere, and let the cone O have as height the perpendicular drawn from the center on AK. Therefore the cone Ξ is equal to the figure circumscribed around the sphere, and the cone O is equal to the inscribed. And since the polygons are similar, $E\Lambda$ has to AK the same ratio which the radius of the sphere has to the perpendicular drawn from the center of the sphere on AK; therefore the height of the cone Ξ has to the height of the cone O the same ratio which $E\Lambda$ has to AK. But the diameter of the circle M, also, has to the diameter of the circle N a ratio which $E\Lambda$ has to AK. Therefore the diameters of the bases of the cones Ξ, O have the same ratio as the heights, and, through this, the cone Ξ will have to the cone O a ratio triplicate of the diameter of the circle M to the diameter of the circle N. So it is clear that the circumscribed figure, too, will have to the inscribed a ratio triplicate of $E\Lambda$ to AK.

Proposition 33. *The surface of every sphere is four times the greatest circle of the circles in it.*

For let there be some sphere, and let the circle A be four times the great circle; I say that A is equal to the surface of the sphere.

For if not, it is either greater or smaller. First let the surface of the sphere be greater than the circle A. So there are two unequal magnitudes, the surface of the sphere, and the circle A; therefore it is possible to take two unequal lines, so that the greater has to the smaller a ratio smaller than that which the surface of the sphere has to the circle. Let the lines B, Γ be taken, and let Δ be a mean proportional between B, Γ, and also, let the sphere be imagined cut by a plane passing through the center, at the circle $EZH\Theta$, and also, let a polygon be imagined inscribed inside the circle, and circumscribed, so that the circumscribed is similar to the inscribed polygon, and the side of the circumscribed has to that of the inscribed a smaller ratio than that which B has to Δ [figure 4.4.7]. Therefore the duplicate ratio, too, is smaller than the duplicate ratio. And the duplicate ratio of B to Δ is the ratio of B to Γ, while, the duplicate ratio of the side of the circumscribed polygon to the side of the inscribed is the ratio of the surface of the circumscribed solid to the surface of the inscribed. Therefore, the surface of the figure circumscribed around the sphere has to the surface of the inscribed figure a smaller ratio than the surface of the sphere to the circle A, which is absurd, for the surface of the circumscribed is greater than the surface of the sphere, while the surface of the inscribed figure is smaller than the circle A, for the surface of the inscribed has been proved to be smaller than four times the greatest circle of the circles in the sphere, and the circle A is four times the greatest circle. Therefore the surface of the sphere is not greater than the circle A.

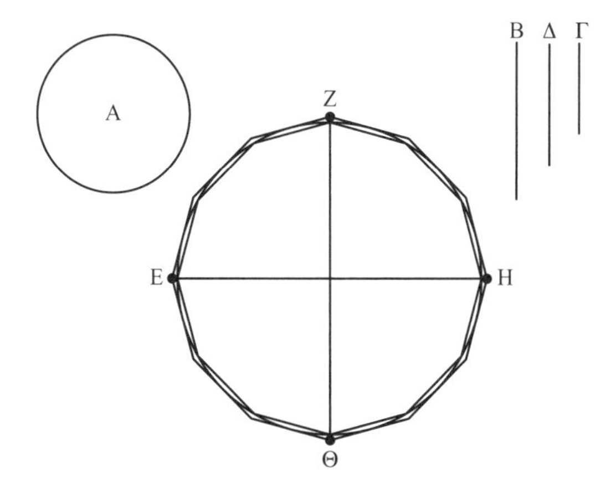

Figure 4.4.7.

So I say that neither is it smaller. For if possible, let it be smaller. And similarly let the lines B, Γ be found, so that B has to Γ a smaller ratio than that which the circle A has to the surface of the sphere, and let Δ be a mean proportional between B, Γ, and again let it be inscribed and circumscribed so that the side of the circumscribed has to the inscribed a smaller ratio than the ratio of B to Δ; therefore the duplicates, too. Therefore the surface of the circumscribed has to the surface of the inscribed a smaller ratio than B to Γ. But B has to Γ a smaller ratio than the circle A to the surface of the sphere, which is absurd, for the surface of the circumscribed is greater than the circle A, while the surface of the inscribed is smaller than the surface of the sphere.

Therefore neither is the surface of the sphere smaller than the circle A. And it was proved that neither is it greater. Therefore the surface of the sphere is equal to the circle A, that is, to four times the great circle.

Proposition 34. *Every sphere is four times a cone having a base equal to the greatest circle of the circles in the sphere and as height the radius of the sphere.*

For let there be some sphere and in it a great circle $AB\Gamma\Delta$. Now, if the sphere is not four times the said cone, let it be, if possible, greater than four times; and let there be the cone Ξ, having a base four times the circle $AB\Gamma\Delta$, and a height equal to the radius of the sphere. Now, the sphere is greater than the cone Ξ. So there will be two unequal magnitudes, the sphere and the cone. Now, it is possible to take two unequal lines, so that the greater has to the smaller a smaller ratio than that which the sphere has to the cone Ξ. Now let them be K, H, and I, Θ taken so that they exceed each other, K exceeding I, and I exceeding Θ and Θ exceeding H, by an equal difference. Let also a polygon be imagined inscribed inside the circle $AB\Gamma\Delta$ with the number of its sides measured by four, and another, circumscribed, similar to the inscribed, as in the earlier constructions [figure 4.4.8]. Let the side of the circumscribed polygon have to the side of the inscribed a smaller ratio than that which K has to I, and let $A\Gamma B\Delta$ be diameters at right angles to each other. Now, if the plane in which are the polygons is carried in a circular motion around diameter $A\Gamma$, there will be figures, the one inscribed in the sphere, the other circumscribed, and the circumscribed will have to the inscribed a ratio triplicate of the side of the circumscribed

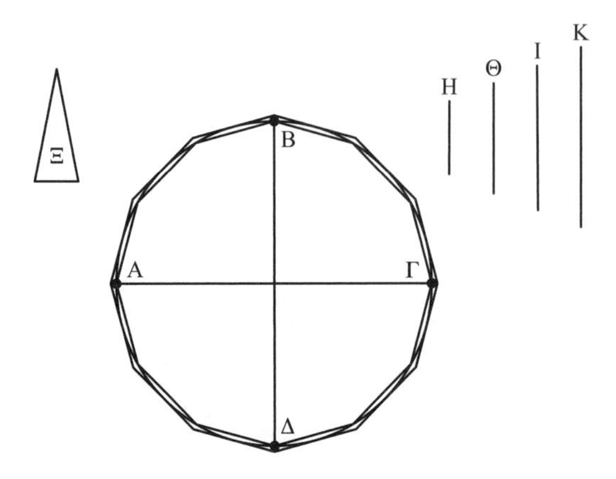

Figure 4.4.8.

polygon to the side of the polygon inscribed inside the circle $AB\Gamma\Delta$. But the side has to the side a smaller ratio than K to I; so that the circumscribed figure has a smaller ratio than triplicate of the ratio of K to I. But, also, K has to H a greater ratio than triplicate that which K has to I; much more, therefore, that which was circumscribed has to the inscribed a smaller ratio than that which K has to H. But K has to H a smaller ratio than the sphere to the cone Ξ; and alternately, which is impossible, for the circumscribed figure is greater than the sphere, while the inscribed is smaller than the cone Ξ. Therefore the sphere is not greater than four times the said cone.

Let it be, if possible, smaller than four times; so that the sphere is smaller than the cone Ξ. So let the lines K, H be taken so that K is greater than H and has to it a smaller ratio than that which the cone Ξ has to the sphere, and let Θ, I be set out, as before, and let a polygon be imagined inscribed inside the circle $AB\Gamma\Delta$, and another circumscribed, so that the side of the circumscribed has to the side of the inscribed a smaller ratio than K to I, and the rest constructed in the same way as before; therefore, the circumscribed solid figure wil also have to the inscribed a ratio triplicate of the ratio of the side of the polygon circumscribed around the circle $AB\Gamma\Delta$ to the side of the inscribed. But the side has to the side a smaller ratio than K to I; so the circumscribed figure will have to the inscribed a smaller ratio than triplicate that which K has to I. And K has to H a greater ratio than triplicate that which K has to I; so that the circumscribed figure has to the inscribed a smaller ratio than K to H. But K has to H a smaller ratio than the cone Ξ to the sphere, which is impossible, for the inscribed is smaller than the sphere, while the circumscribed is greater than the cone Ξ. Therefore neither is the sphere smaller than four times the cone having the base equal to the circle $AB\Gamma\Delta$ and, as height, the line equal to the radius of the sphere. And it was proved that neither is it greater. Therefore, it is four times.

Corollary. *And these being proved, it is obvious that every cylinder having as base the greatest circle of the circles in the sphere and a height equal to the diameter of the sphere is half as large again as the sphere and its surface with the bases is half as large again as the surface of the sphere.*

For the cylinder mentioned above is six times the cone having the same base and a height equal to the radius of the sphere, and the sphere has been proved to be four times the same cone; so it is clear that the cylinder is half as large again as the sphere. Again, since the surface of the cylinder (without the bases) has been proved equal to a circle whose radius is a mean proportional between the side of the cylinder and the diameter of the base, and the side of the said cylinder (which is around the sphere) is equal to the diameter of the base, it is clear that their mean proportional will then be equal to the diameter of the base; and the circle having the radius equal to the diameter of the base is four times the base, that is, four times the greatest circle of the circles in the sphere; therefore the surface of the cylinder without the bases, too, will be four times the great circle; therefore the whole surface of the cylinder, with the bases, will be six times the great circle. And also, the surface of the sphere is four times the great circle. Therefore the whole surface of the cylinder is half as large again as the surface of the sphere.

4.5 Archimedes, *Sphere and Cylinder II*

In *Sphere and Cylinder II*, Archimedes sets out and solves several problems involving spheres and segments of spheres, using some of the basic results of the first book. The problem we consider in detail is one in which, by a modern interpretation, Archimedes solves a cubic equation. Archimedes himself, however, was solving an interesting geometrical problem on cutting a sphere by a plane so that the volumes of the segments are in a given ratio. We present, in footnotes, the relationship of Archimedes's own work to our modern algebraic notion. As before, Archimedes begins with a letter to Dositheus, the recipient of many of his works.

Archimedes to Dositheus: Greetings!

Earlier you sent me a request to write the proofs of the problems, whose proposals I had myself sent to Conon; and for the most part they happen to be proved through the theorems whose proofs I had sent you earlier: that the surface of every sphere is four times the greatest circle of the circles in it, and that the surface of every segment of a sphere is equal to a circle, whose radius is equal to the line drawn from the vertex of the segment to the circumference of the base, and through the theorem that, in every sphere, the cylinder having, as base, the greatest circle of the circles in the sphere, and a height equal to the diameter of the sphere, is both: itself, in magnitude, half as large again as the sphere; and, in surface, half as large again as the surface of the sphere, and through the theorem that every solid sector is equal to the cone having, as base, the circle equal to the surface of the segment of the sphere contained in the sector, and a height equal to the radius of the sphere. Now, I have sent you those theorems and problems that are proved through these theorems above, having proved them in this book. And as for those that are found through some other theory, namely, those concerning spirals, and those concerning conoids, I shall try to send quickly.

Of the problems, the first was this: Given a sphere, to find a plane area equal to the surface of the sphere. And this is obviously proved from the theorems mentioned already; for the quadruple of the greatest circle of the circles in the sphere is both a plane area,

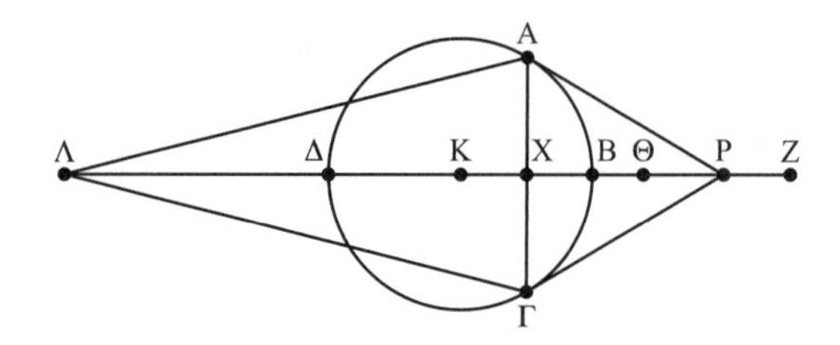

Figure 4.5.1.

and equal to the surface of the sphere. The second was: Given a cone or a cylinder, to find a sphere equal to the cone or to the cylinder [in volume].

Archimedes proves this second result at the beginning of this book and then takes up several other problems. Proposition 2 we present without proof. We then move to proposition 4, which is subject to extensive commentary from Eutocius.

Proposition 2. *Every segment of the sphere is equal to a cone having a base the same as the segment and, as height, a line which has to the height of the segment the same ratio which both the radius of the sphere and the height of the remaining segment, taken together, have to the height of the remaining segment.*

Note that every plane cutting a sphere divides it into two segments, one of which is called *the segment* and the other, *the remaining segment*. The theorem states that given a segment of a sphere of height h_1, the volume of the segment is the same as that of a cone whose base is the same as the base of the segment and whose height H satisfies the relationship $H : h_1 = (R + h_2) : h_2$, where R is the radius of the sphere and $h_2 = 2R - h_1$ is the height of the remaining segment.

Proposition 4. *To cut the given sphere so that the [volumes of the] segments of the sphere have to each other the same ratio as a given ratio.*

Let there be the given sphere, $AB\Gamma\Delta$; so it is required to cut it by a plane so that the segments of the sphere have to each other the given ratio.

As in many problems, as contrasted to theorems, Archimedes begins with the analysis, assuming that the problem has been solved and then reaching a condition that must be satisfied, before ultimately working through the synthesis that solves the problem.

Let it be cut by the plane $A\Gamma$. Therefore the ratio of the segment of the sphere $A\Delta\Gamma$ to the segment of the sphere $AB\Gamma$ is given. And let the sphere be cut through the center, and let the section be a great circle $AB\Gamma\Delta$, and let its center be K, and its diameter ΔB, and let it be made: as $K\Delta$, ΔX taken together is to ΔX, so PX is to XB,[15] and as KB, BX taken together to BX, so is ΛX to $X\Delta$,[16] and let $A\Lambda$, $\Lambda\Gamma$, AP, $P\Gamma$ be joined [figure 4.5.1]. Therefore the cone $A\Lambda\Gamma$ is equal to the segment of the sphere $A\Delta\Gamma$, while the

[15] This condition defines the point P.
[16] Again, this condition defines the point Λ.

cone $AP\Gamma$ is equal to the segment $AB\Gamma$ [proposition 2, above]. Therefore the ratio of the cone $A\Lambda\Gamma$ to the cone $AP\Gamma$ is given, too. And as the cone is to the cone, so ΛX is to XP [since, indeed, they have the same base, the circle around the diameter $A\Gamma$]. Therefore the ratio of ΛX to XP is given too. And through the same arguments as before, through the construction, as $\Lambda\Delta$ is to $K\Delta$, so is KB to BP and ΔX to XB.

For in the proposition preceding this one, it was thus concluded: Since it is, as $K\Delta$, ΔX taken together is to ΔX, so is PX to XB. Dividedly: as $K\Delta$ is to ΔX, so is PB to BX. Alternately: as $K\Delta$, that is KB, is to BP, ΔX is to XB. Again, since it is: as ΛX is to $X\Delta$, so is KB, BX taken together to XB. Dividedly and alternately: as $\Lambda\Delta$ is to ΔK, so is ΔX to XB. And it was also: as ΔX is to XB, KB is to BP. Therefore, as $\Lambda\Delta$ is to ΔK, so is ΔX to XB and KB to BP.

And since it is: as PB is to BK, so is $K\Delta$ to $\Lambda\Delta$, compoundly, as PK is to KB, that is, to $K\Delta$, so is $K\Lambda$ to $\Lambda\Delta$. Therefore the whole $P\Lambda$ is to the whole $K\Lambda$ as $K\Lambda$ is to $\Lambda\Delta$; therefore rect($P\Lambda$, $\Lambda\Delta$) is equal to sq(ΛK). Therefore as $P\Lambda$ is to $\Lambda\Delta$, sq($K\Lambda$) is to sq($\Lambda\Delta$).

For since it is: as $P\Lambda$ is to ΛK, so is $K\Lambda$ to $\Lambda\Delta$; therefore also: as the first is to the second, so is the square on the first to the square on the second; therefore it is: as $P\Lambda$ is to $\Lambda\Delta$, so sq($P\Lambda$) is to sq(ΛK). But as sq($P\Lambda$) is to sq(ΛK), so is sq(ΛK) to sq($\Lambda\Delta$); for they are proportional; therefore, as $P\Lambda$ is to $\Lambda\Delta$, so is sq(ΛK) to sq($\Lambda\Delta$).

And since it is: as $\Lambda\Delta$ to ΔK, so is ΔX to XB, it will be, inversely and compoundly: as $K\Lambda$ is to $\Lambda\Delta$, so is $B\Delta$ to ΔX [and therefore as sq($K\Lambda$) is to sq($\Lambda\Delta$), so is sq($B\Delta$) to sq(ΔX). Again, since it is: as ΛX is to ΔX, KB, BX taken together is to BX, dividedly, as $\Lambda\Delta$ is to ΔX, so is KB to BX].

And let BZ be set equal to KB; for it is clear that it will fall beyond P.

For since it is: as $X\Delta$ is to XB, so is KB to BP and ΔX is greater than SB, therefore KB, as well, is greater than BP. Therefore Z falls beyond P.

And it will be: as $\Lambda\Delta$ is to ΔX, so is ZB to BX; so that also: as $\Delta\Lambda$ is to ΛX, so is BZ to ZX. And since the ratio of $\Delta\Lambda$ to ΛX is given, as well as the ratio of $P\Lambda$ to ΛX, therefore the ratio of $P\Lambda$ to $\Lambda\Delta$, too, is given.

For since it is: as KB, BX taken together is to BX, that is ZX to XB, so is ΛX to $X\Delta$. Convertedly: as XZ is to ZB, so is $X\Lambda$ to $\Lambda\Delta$; also inversely: as BZ is to ZX, so is $\Lambda\Delta$ to ΛX. And the ratio of BZ to ZX is given, since ZB is equal to the radius of the given sphere, while BX is given because its limits B, X are given by hypothesis, the sphere being cut by the plane $A\Gamma$ and by the line ΔB being at right angles to the line $A\Gamma$, and through this the whole XZ, too, is given, as well as the ratio of XZ to ZB; so that the ratio of $X\Lambda$ to $\Lambda\Delta$ is given as well. Again, since the ratio of the segments is given, the ratio of the cone $\Lambda A\Gamma$ to the cone $AP\Gamma$ is given, as well. So that the ratio of ΛX to XP is given, too; for they are to each other as the heights; therefore the ratio of the whole $P\Lambda$ to ΛX is given. Now since the ratio of each of $P\Lambda$, $\Lambda\Delta$ to ΛX is given, therefore the ratio of $P\Lambda$ to $\Lambda\Delta$ is given as well. For the magnitudes which have a given ratio to the same, have also a given ratio to each other.

Now, since the ratio of $P\Lambda$ to ΛX is composed of both: the ratio which $P\Lambda$ has to $\Lambda\Delta$, and that which $\Delta\Lambda$ has to ΛX, but as $P\Lambda$ is to $\Lambda\Delta$, sq(ΔB) is to sq(ΔX), while as $\Delta\Lambda$ is to ΛX, so BZ is to ZX, therefore the ratio of $P\Lambda$ to ΛX is composed of both: the ratio which sq($B\Delta$) has to sq(ΔX) and the ratio which BZ has to ZX.

And let it be made: as $P\Lambda$ is to ΛX, BZ is to $Z\Theta$. And the ratio of $P\Lambda$ to ΛX is given; therefore the ratio of ZB to $Z\Theta$ is given as well. And BZ is given; for it is equal to the radius; therefore, $Z\Theta$ is given as well. Also, therefore, the ratio of BZ to $Z\Theta$ is composed of both: the ratio which sq($B\Delta$) has to sq(ΔX) and that which BZ has to ZX. But the ratio of BZ to $Z\Theta$ is composed of both: the ratio of BZ to ZX and the ratio of ZX to $Z\Theta$. Let the ratio of BZ to ZX be taken away as common; remaining, therefore, it is: as sq($B\Delta$), that is, a given, is to sq(ΔX), so XZ is to $Z\Theta$, that is, to a given. And the line $Z\Delta$ is given. Therefore it is required to cut a given line, ΔZ, at the point X and to produce: as XZ is to a given line [namely, $Z\Theta$], so the given square [namely, sq($B\Delta$)] is to sq(ΔX).

When the problem is stated in this general form, it is necessary to investigate the limits of possibility, but under the conditions of the present case, no such investigation is necessary. In the present case the problem will be as follows: Given two lines $B\Delta$, BZ, with $B\Delta$ being twice BZ, and given a point Θ on BZ, to cut ΔB at X, and to produce: as sq($B\Delta$) is to sq(ΔX), so is XZ to $Z\Theta$. And these problems will be, each, both analyzed and constructed at the end.

Although the received manuscripts of Archimedes did not contain the promised analyses and syntheses, Eutocius managed to find at least the analysis and synthesis of the general problem after a long search. We reproduce what he believed Archimedes wrote below. But first, we continue with the original text of Archimedes as he constructs the synthesis of this problem.

So the problem will be constructed like this: Let there be the given ratio, the ratio of Π to Σ, and let some sphere be given and let it be cut by a plane passing through the center, and let there be a section of the sphere and the plane, namely the circle $AB\Gamma\Delta$, and let $B\Delta$ be diameter and K center; and let BZ be set equal to KB, and let BZ be cut at Θ, so that it is: as ΘZ is to ΘB, so is Π to Σ, and yet again let $B\Delta$ be cut at X, so that it is: as XZ is to ΘZ, so is sq($B\Delta$) to sq(ΔX), and through X let a plane be produced, perpendicular to the line $B\Delta$ [figure 4.5.1]. I say that this plane cuts the sphere so that it is: as the greater segment to the smaller, so Π is to Σ.

For let it be made, first as KB, BX taken together is to BX, so is ΛX to ΔX, second as $K\Delta$, ΔX taken together is to $X\Delta$, PX is to XB, and let $A\Lambda$, $\Lambda\Gamma$, AP, $P\Gamma$ be joined;[17] so through the construction (as we proved in the analysis), rect($P\Lambda, \Lambda\Delta$) will be equal to sq(ΛK), and as $K\Lambda$ is to $\Lambda\Delta$, so is $B\Delta$ to ΔX; so that, also, as sq($K\Lambda$) is to sq($\Lambda\Delta$), so sq($B\Delta$) is to sq(ΔX). And since rect($P\Lambda, \Lambda\Delta$) is equal to sq(ΛK) [it is: as $P\Lambda$ to $\Lambda\Delta$, so sq(ΛK) is to sq($\Lambda\Delta$)], therefore it will also be: as $P\Lambda$ is to $\Lambda\Delta$, so is sq($B\Delta$) to sq(ΔX), that is, as XZ is to $Z\Theta$. And since it is: as KB, BX taken together is to BX, so ΛX is to $X\Delta$, and KB is equal to BZ. Therefore it will also be: as ZX is to XB, so is ΛX to $X\Delta$;

[17] P is defined by the property given; the stated proportions are proved in the analysis.

convertedly, as XZ is to ZB, so is $X\Lambda$ to $\Lambda\Delta$; so that also, as $\Lambda\Delta$ is to ΛX, so is BZ to ZX. And since it is: as $P\Lambda$ is to $\Lambda\Delta$, so is XZ to $Z\Theta$, and as $\Delta\Lambda$ is to ΛX, so is BZ to ZX, and through the equality in the perturbed proportion, as $P\Lambda$ is to ΛX, so is BZ to $Z\Theta$; therefore also: as ΛX is to XP, so is $Z\Theta$ to ΘB. And as $Z\Theta$ is to ΘB, so is Π to Σ; therefore also: as ΛX is to XP, that is, as the cone $A\Gamma\Lambda$ is to the cone $AP\Gamma$, that is, as the segment of the sphere $A\Delta\Gamma$ is to the segment of the sphere $AB\Gamma$, so is Π to Σ.

"And these problems will be, each, both analyzed and constructed at the end." While he [Archimedes] promised to prove the aforementioned claim at the end, it is impossible to find the promised thing in any of the manuscripts.... But, in a certain old book (for we did not cease from the search for many books), we have read theorems written very unclearly (because of the errors), and in many ways mistaken about the diagrams. But they had to do with the subject matter we were looking for, and they preserved in part the Doric language Archimedes liked using, written with the ancient names of things: the parabola called "section of a right-angled cone," the hyperbola "section of an obtuse-angled cone." From which things we began to suspect, whether these may not in fact be the things promised to be written at the end. So we read more carefully the content itself (since we have found—as had been said—that it has been an uneasy piece of writing, because of the great number of mistakes), taking apart the ideas one by one. We write it down, as far as possible, word-for-word (but in a language that is more widely used, and clearer). The first theorem is proved for the general case, so that his claim, concerning the limits on the solution, becomes clearer. Then it is also applied to the results of the analysis in the problem.

We begin, then, with Eutocius's reconstruction of Archimedes's analysis of the general problem stated earlier.

Given a line, AB, and another, $A\Gamma$, and an area, Δ: let it first be put forth: to take a point on AB, such as E, so that it is: as AE is to $A\Gamma$, so the area Δ is to the square on EB.[18]

Let it come to be, and let $A\Gamma$ be set a right angles to AB, and, having joined ΓE, let it be drawn through to Z, and let ΓH be drawn through Γ, parallel to AB, and let ZBH be drawn through B, parallel to $A\Gamma$, meeting each of the lines ΓE, ΓH, and let the parallelogram $H\Theta$ be filled in, and let $KE\Lambda$ be drawn through E parallel to either $\Gamma\Theta$ or HZ, and let rect($\Gamma H, HM$) be equal to the area Δ [figure 4.5.2].

Now since it is: as EA is to $A\Gamma$, so the area Δ is to sq(EB), but as EA is to $A\Gamma$, so is ΓH to HZ, and as ΓH is to HZ, so sq(ΓH) is to rect($\Gamma H, HZ$); therefore as sq(ΓH) is to rect($\Gamma H, HZ$), so the area Δ is to sq(EB), that is, to sq(KZ). Alternately also: as sq(ΓH) is to the area Δ, that is, to rect($\Gamma H, HM$), so rect($\Gamma H, HZ$) is to sq(ZK). But as sq(ΓH) is to rect($\Gamma H, HM$), so is ΓH to HM; therefore also: as ΓH is to HM, so rect($\Gamma H, HZ$) is to sq(ZK). But as ΓH is to HM, so (HZ taken as a common

[18] If we set $AB=a$, $A\Gamma=b$, $\Delta=c^2$, and $BE=x$, then, in modern terms, Archimedes is proposing to solve the cubic equation $\frac{a-x}{b}=\frac{c^2}{x^2}$, or $x^2(a-x)=bc^2$. He does so by finding the same method that Persian mathematician Omar Khayyam used in the twelfth century in his detailed classification and treatment of cubic equations.

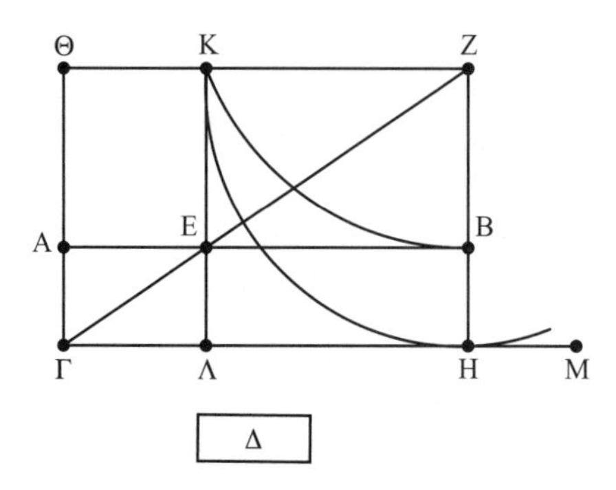

Figure 4.5.2.

height) rect($\Gamma H, HZ$) is to rect(MH, HZ). Therefore as rect($\Gamma H, HZ$) is to rect(MH, HZ), so rect($\Gamma H, HZ$) is to sq(ZK). Therefore, rect(MH, HZ) is equal to sq(ZK). Therefore if a parabola is drawn through H around the axis ZH, so that the lines drawn down to the axis are in square the rectangle applied along HM, it shall pass through K. And the parabola shall be given in position, through HM being given in magnitude as it contains, together with the given $H\Gamma$, the given area Δ. Therefore K touches a parabola given in position.[19]

Now let the parabola be drawn as has been said, and let it be as HK. Again, since the area $\Theta\Lambda$ is equal to the area ΓB, that is, rect($\Theta K, K\Lambda$) is equal to rect(AB, BH), if a hyperbola is drawn through B, around the asymptotes $\Theta\Gamma, \Gamma H$, it shall pass through K, **through the converse of the eighth theorem of the second book of the *Conic Elements* of Apollonius**,[20] and it shall be given in position through the fact that each of $\Theta\Gamma, \Gamma H$ is given in position, as well, further yet, through the fact that B is given in position, too.[21] Let it be drawn, as has been said, and let it be as KB; therefore K touches a hyperbola given in position; and it also touches a parabola given in position. Therefore K is given. And KE is a perpendicular drawn from it to a line given in position, namely, to AB; therefore E is given. Now since it is: as EA is to the given line $A\Gamma$, so the given area Δ is to sq(EB), two solids, whose bases are sq(EB) and the area Δ, and whose heights are EA, $A\Gamma$, have the bases reciprocal to the heights; so the solids are equal. Therefore the solid produced by sq(EB) on EA as the solid's height is equal to the solid produced by the given area Δ on the given line ΓA as the solid's height. But the solid produced by sq(BE) on EA as the solid's height is the greatest of all similarly taken solids on BA, when BE is twice EA, as shall be proved. Therefore the solid produced by the given area on the given line as the solid's height must be not greater than the solid produced by sq(BE) on EA as the solid's height.[22]

[19] Since $\Delta = \Gamma H \cdot HM = c^2$, we have $HM = \frac{c^2}{a}$. If we also set H as the origin of a coordinate system, the equation of the parabola is $c^2 y = ax^2$.

[20] Clearly, this is a reference added by Eutocius. In modern editions of the *Conics*, the relevant theorem is 2:12.

[21] The equation of the hyperbola is $(a-x)y = ab$.

[22] Since AE, BE are segments of AB, the volume of the solid $AE \cdot BE^2$ cannot be arbitrarily large. In fact, as will be proved, the maximum of this product occurs when $BE = 2AE$. In modern terms, the maximum of $x^2(a-x)$ occurs

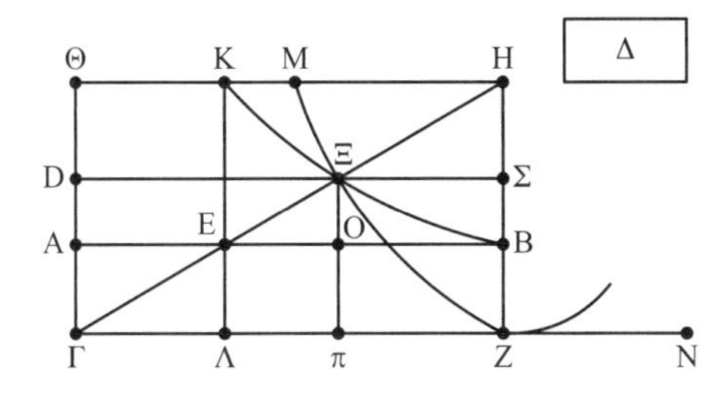

Figure 4.5.3.

The analysis is followed by the synthesis, that is, the construction of the solution to the problem.

And it will be constructed like this: let the given line be AB, and some other given line $A\Gamma$, and the given area Δ, and let it be required to cut AB, so that it is: as one segment is to the given AB, so the given area Δ is to the square on the remaining segment.

Let AE be taken, a third part of AB; therefore the area Δ on the line $A\Gamma$ is either greater than sq(BE) on EA, or equal, or smaller. Now then, if it is greater, the problem may not be constructed, as has been proved in the analysis; and if it is equal, the point E produces the problem. For, the solids being equal, the bases are reciprocal to the heights, and it is: as the line EA is to the line $A\Gamma$, so the area Δ is to sq(BE).

And if the area Δ, on $A\Gamma$, is smaller than sq(BE), on EA, it shall be constructed like this: Let $A\Gamma$ be set at right angles to AB, and let ΓZ be drawn through Γ parallel to AB, and let BZ be drawn through B parallel to the line $A\Gamma$, and let it meet ΓE produced at H, and let the parallelogram $Z\Theta$ be filled in, and let $KE\Lambda$ be drawn through E parallel to ZH [figure 4.5.3]. Now, since the area Δ, on $A\Gamma$, is smaller than sq(BE), on EA, it is: as EA is to $A\Gamma$, so the area Δ is to some area smaller than sq(BE), that is, smaller than sq(HK). So let it be: as EA is to $A\Gamma$, so the area Δ is to sq(HM), and let rect($\Gamma Z, ZN$) be equal to the area Δ. Now since it is: as EA is to $A\Gamma$, so the area Δ, that is, rect($\Gamma Z, ZN$), is to sq(HM), but as EA is to $A\Gamma$, so is ΓZ to ZH, and as ΓZ is to ZH, so is sq(ΓZ) to rect($\Gamma Z, ZH$), therefore also: as sq(ΓZ) is to rect($\Gamma Z, ZH$), so rect($\Gamma Z, ZN$) is to sq(HM); alternately also: as sq(ΓZ) is to rect($\Gamma Z, ZN$), so rect($\Gamma Z, ZH$) is to sq(HM). But as sq(ΓZ) is to rect($\Gamma Z, ZN$), so is ΓZ to ZN, and as ΓZ is to ZN (taking ZH as a common height), so is rect($\Gamma Z, ZH$) to rect(NZ, ZH); therefore also: as rect($\Gamma Z, ZH$) is to rect(NZ, ZH), so is rect($\Gamma Z, ZH$) to sq(HM); therefore sq(HM) is equal to rect(HZ, ZN).

Therefore, if we draw, through Z, a parabola around the axis ZH, so that the lines drawn down to the axis are, in square, the rectangle applied along ZN, it shall pass through M. Let it be drawn, and let it be as the parabola $M\Xi Z$. And since the area $\Theta\Lambda$ is equal to the area AZ, that is, rect($\Theta K, K\Lambda$) to rect(AB, BZ), if we draw, through B, a hyperbola around the asymptotes $\Theta\Gamma, \Gamma Z$, it shall pass through K (**through the converse of the eighth theorem of the second book of Apollonius's *Conic Elements***).[23] Let it be drawn, and

when $x = \frac{2}{3}a$. Thus, the cubic equation can be solved only when $bc^2 \leq \frac{4}{27}a^3$. As Eutocius shows below, in the "less than" case, there are two (positive) solutions, one greater than $\frac{2}{3}a$ and one smaller.

[23] As above, this is theorem 2:12.

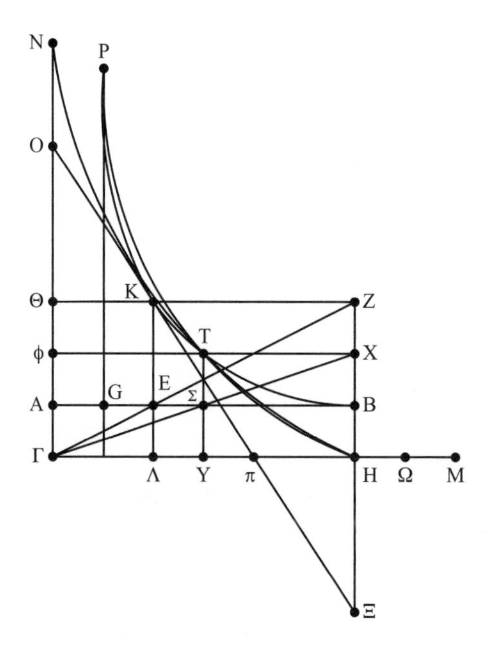

Figure 4.5.4.

let it be as the hyperbola BK, cutting the parabola at Ξ, and let a perpendicular be drawn from Ξ on AB, namely, $\Xi O\Pi$, and let the line $P\Xi\Sigma$ be drawn through Ξ parallel to AB. Now, since $B\Xi K$ is a hyperbola and $\Theta\Gamma, \Gamma Z$ are asymptotes, and $P\Xi, \Xi\Pi$ are drawn parallel to AB, BZ, rect($P\Xi, \Xi\Pi$) is equal to rect(AB, BZ); so that the area PO, too, is equal to the area OZ. Therefore if a line is joined from Γ to Σ, it shall pass through O. Let it pass and let it be as $\Gamma O\Sigma$. Now, since it is: as OA is to $A\Gamma$, so is OB to $B\Sigma$, that is, as ΓZ to $Z\Sigma$, and as ΓZ is to $Z\Sigma$ (taking ZN as a common height), rect($\Gamma Z, ZN$) is to rect($\Sigma Z, ZN$), therefore as OA is to $A\Gamma$, too, so rect($\Gamma Z, ZN$) is to rect($\Sigma Z, ZN$). And rect($\Gamma Z, ZN$) is equal to the area Δ, while rect($\Sigma Z, ZN$) is equal to sq($\Sigma\Xi$), that is, to sq(BO), through the parabola. Therefore as OA is to $A\Gamma$, so the area Δ is to sq(BO). Therefore the point O has been taken, producing the problem.

And it will be proved like this that, BE being twice EA, sq(BE), on EA, is the greatest of all magnitudes similarly taken on BA.

For let there be, as in the analysis, again: a given line, at right angles to AB, namely. $A\Gamma$, and having joined ΓE, let it be produced and let it meet at Z the line drawn through B parallel to $A\Gamma$, and through the points Γ, Z, let $\Theta Z, \Gamma H$ be drawn parallel to AB, and let ΓA be produced to Θ, and, parallel to it, let $KE\Lambda$ be drawn through E [figure 4.5.4], and let it come to be: as EA is to $A\Gamma$, so rect($\Gamma H, HM$) is to sq(EB); therefore, sq(EB), on EA, is equal to rect($\Gamma H, HM$), on $A\Gamma$, through the fact that the bases of the two solids are reciprocal to the heights. Now I say that rect($\Gamma H, HM$), on $A\Gamma$, is the greatest of all magnitudes similarly taken on BA.

For let a parabola be drawn through H, around the axis ZH, so that the lines drawn down to the axis are in square the rectangle applied along HM. So it will pass through K, as has been proved in the analysis and, produced, it will meet $\Theta\Gamma$ since it is parallel to the diameter of the section (**through the twenty-seventh theorem of the first book of Apollonius's *Conic Elements***).[24] Let the parabola be produced and let it meet the line $\Gamma\Theta$ produced at N, and let a hyperbola be drawn through B, around the asymptotes $N\Gamma$, ΓH. Therefore it will pass through K, as was said in the analysis. So let it pass, as the hyperbola BK, and, ZH being produced, let $H\Xi$ be set equal to it, and let ΞK be joined, and let it be produced to O; therefore it is obvious that ΞO will touch the parabola (**through the converse of the thirty-fourth theorem of the first book of Apollonius's *Conic Elements***).[25] Now since BE is double EA, for so it is assumed, that is, ZK is twice $K\Theta$, and the triangle $O\Theta K$ is similar to the triangle ΞZK, ΞK, too, is twice KO. And ΞK is double $K\Pi$, as well, through the facts that ΞZ, too, is double ZH, and that ΠH is parallel to KZ; therefore OK is equal to $K\Pi$.

Therefore $OK\Pi$, being in contact with the hyperbola, and lying between the asymptotes, is bisected at the point of contact with the hyperbola; therefore it touches the hyperbola (**through the converse of the third theorem of the second book of Apollonius's *Conic Elements***). And it touches the parabola, too, as the same point K. Therefore the parabola touches the hyperbola at K. So let the hyperbola, produced, as towards P, be imagined as well, and let a chance point be taken on EB, namely, Σ, and let $T\Sigma Y$ be drawn through Σ parallel to $K\Lambda$, and let it meet the hyperbola at T, and let ΦTX be drawn through T parallel to ΓH. Now since (through the hyperbola and the asymptotes) the area ΦY is equal to the area ΓB; taking the area $\Gamma\Sigma$ away as common, the area $\Phi\Sigma$ is then equal to the area ΣH, and through this, the line joined from Γ to X will pass through Σ. Let it pass, and let it be as $\Gamma\Sigma X$. And since sq(ΨX) is equal to rect(XH, HM) through the parabola, sq(TX) is smaller than rect(XH, HM). So let rect($XH, H\Omega$) come to be equal to sq(TX). Now since it is: as ΣA is to $A\Gamma$, so is ΓH to HX, but as ΓH is to HX (taking $H\Omega$ as a common height), so rect($\Gamma H, H\Omega$) is to rect($XH, H\Omega$), and rect($\Gamma H, H\Omega$) is to sq(XT) (which is equal to it), that is, to sq($B\Sigma$), therefore sq($B\Sigma$), on ΣA, is equal to rect($\Gamma H, H\Omega$), on ΓA. But rect($\Gamma H, H\Omega$), on ΓA, is smaller than rect($\Gamma H, HM$), on ΓA; therefore sq($B\Sigma$), on ΣA, is smaller than sq(BE), on EA. So it will be proved similarly also in all the points taken between E and B.

Eutocius next completes the argument by showing that if a point ζ is taken between the points E and A, than sq(BE), on EA, is also greater than the square on $B\zeta$, on ζA. In fact, it has been speculated that this part of the argument was written by Eutocius and not by Archimedes.[26] In particular, he shows that this part is unnecessary, since the point ζ is determined again below when Eutocius shows that it also produces a solution to the problem. It is clear, then, that what follows is all by Eutocius.

So it shall be proved similarly in all the points taken between the points E, A, as well. And it was also proved for all the points between the points E, B; therefore, of

[24] In modern editions, this is theorem 1:26.

[25] This is theorem 1:33.

[26] Netz, *The Works of Archimedes*, Vol. 1, *The Two Books On the Sphere and the Cylinder*, ed. and trans. Reviel Netz, vol. 1 (Cambridge: Cambridge University Press, 2004), p. 327.

all the magnitudes taken similarly on AB, the greatest is sq(BE), on EA, when BE is twice EA.

Now one must understand also the consequences of the diagram above. For since it has been proved that sq($B\Sigma$), on ΣA, and sq($B\zeta$), on ζA, are smaller than sq(BE), on EA, therefore it is possible to produce the task assigned by the original problem, by cutting the line AB at two points (when the given area on the given line is smaller than sq(BE), on EA).

And this comes to be, if we imagine a parabola drawn around the diameter XH, so that the lines drawn down to the diameter are in square the rectangle applied along $H\Omega$; for such a parabola certainly passes through the point T. And since it must meet ΓN (being parallel to the diameter), it is clear that it cuts the hyperbola at some point above K (as here, at P), and it is clear that a perpendicular drawn from P on AB (as here, $P\zeta$), cuts AB at ζ, so that the point ζ produces the task assigned by the problem, and so that sq($B\Sigma$) on ΣA is then equal to sq($B\zeta$) on ζA as is self-evident from the preceding proofs.

So that—it being possible to take two points on BA, producing the required task—one may take whichever one wishes, either the point between the points E, B, or the point between the points E, A. For if one takes the point between the points E, B, then, as has been said, one draws a parabola through the points H, T, which cuts the hyperbola at two points. The point closer to H, that is to the axis of the parabola, will procure the point between the points E, B (as here T has procured Σ), while the point more distant from the diameter will procure the point between the points E, A (as here P procures ζ).

Now, generally, the problem is analyzed and constructed like this. But in order that it may also be applied to Archimedes's text, let the diameter of the sphere ΔB be imagined (just as in the diagram of the text), and the radius BZ, and the given line $Z\Theta$. Therefore he says the problem comes down to: "To cut ΔZ at X, so that it is: as XZ is to the given line, so the given square is to sq(ΔZ). This said in this way—without qualification—is soluble only given certain conditions."

For if the given area, on the given line, turns out to be greater than sq(ΔB), on BZ, the problem would be impossible, as has been proved. And if it is equal, the point B would produce the task assigned by the problem, and in this way, too, the solution would have no relevance to what Archimedes originally put forward; for the sphere would not be cut according to the given ratio. Therefore, said in this way, without qualification, it was only solvable given certain added conditions. "But with the added qualification of the specific characteristics of the problem as hand" (that is, both that ΔB is twice BZ and that BZ is greater than $Z\Theta$), "it is always solvable." For the given sq(ΔB), on $Z\Theta$, is smaller than sq(ΔB), on BZ (through BZ being greater than $Z\Theta$), and, when this is the case, we have shown that the problem is possible, and how it then unfolds.[27]

[27] Thus the specific cubic equation to be solved for the original problem is $x^2(a-x)=gd^2$, where $\Delta Z=a$, $Z\Theta=g$, and $\Delta B=d$, the diameter of the sphere. Since we know that $d=\frac{2}{3}a$, we can rewrite the equation as $x^2(a-x)=\frac{4}{9}a^2g$, and this is solvable as long as the constant is less than $\frac{4}{27}a^3$. But this is the same as $g<\frac{1}{3}a$, or $Z\Theta<BZ$, a condition we know to be true.

It should also be noticed that Archimedes's words fit with our analysis. For previously (following his analysis) he stated, in general terms, that which the problem came down to, saying: "It is required to cut a given line, ΔZ, at the point X, and to produce: as XZ is to a given line, so the given square is to sq(ΔX). When the problem is stated in this general form, it is necessary to investigate the limits of possibility." But with the addition of specific characteristics of the problem that he has obtained (that ΔB is twice BZ, and that BZ is greater than $Z\Theta$), "no such investigation is necessary." And so he takes this problem in particular, and says this: "And the problem will be as follows: given two lines ΔB, BZ (and ΔB being twice BZ), and given a point on BZ, namely, Θ; to cut ΔB at X . . ."—and no longer saying, as previously, that it is required to cut ΔZ, but to cut ΔB, instead—because he knew (as we ourselves have proved above) that there are two points which, taken on ΔZ, produce the task assigned by the problem, one between the points Δ, B and another between the points B, Z. Of these, only the point between the points Δ, B would be of use for what Archimedes put forward originally.

So we have copied this down, in conformity with Archimedes's work, as clearly as possible.

4.6 Archimedes, *Planes in Equilibrium*

In book 1 of *Planes in Equilibrium*, Archimedes first proves the law of the lever and then proceeds to find the center of weight of parallelograms, triangles, and trapezoids. In book 2, he finds the center of weight of a parabolic segment and of a portion of a parabola intercepted between two parallel chords bisected by the diameter. Although Archimedes uses the phrase "center of weight" frequently in this work, he never gives a definition, so we are forced to assume that either the concept was so well known that a definition was unneeded or that he had given a definition in an earlier work that no longer exists. Intuitively, of course, the center of weight of a body is a point within it so that if the body is conceived to be suspended from that point, the body will remain at rest. We present here only selected propositions from book 1, some with proofs. Archimedes begins *Planes in Equilibrium* with a set of assumptions:

Introduction

1. We demand that equal weights balance at equal distances, and that equal weights do not balance at unequal distances, but are inclined to the weight which is at the greater distance.
2. If, there being weights balancing at certain distances and a weight is added to one of the weights, they shall not balance, but shall incline to that weight to which it was added.
3. Similarly, if some weight is taken away from one of the weights, they shall not balance, but shall incline to that weight from which it was not taken away.
4. The centers of weight of mutually congruent, equal and similar figures, are mutually congruent as well.

5. The centers of weight of unequal but similar figures shall be similarly situated. (We say that they are "similarly situated" in respect of similar figures, of points, from which lines drawn to the equal angles of the similar figures make equal angles with the corresponding sides.)
6. If magnitudes balance at certain distances, the magnitudes equal to them will also balance at the same distances.
7. In every figure where the perimeter is concave in the same direction, the center of weight must be inside the figure.
8. With these assumed, weights, balancing at equal distances, are equal, for if they shall be unequal, then, the excess taken away from the greater, the remaining weights shall not balance.
9. Unequal weights, at equal distances, do not balance, but incline to the greater, for the excess taken away, they shall balance; so adding that which was taken away, they shall incline to the greater.

Propositions

Proposition 1.[28] *Unequal weights shall balance at unequal distances, the greater weight being at the smaller distance.*

Proposition 2. *If two equal magnitudes do not have the same center of weight, the center of weight of the magnitude composed of both magnitudes shall be the midpoint of the line joining the centers of weight of the magnitudes.*

Proposition 3. *If the centers of weight of three magnitudes are situated on a line and the magnitudes have an equal weight and the lines between the centers are equal, the point, which is also itself the center of weight of the middle magnitude, as well is the center of weight of the magnitude composed of all the magnitudes.*

Corollary 1. *So through this it is obvious that, however many odd in number centers of weight be situated on a line, if both the magnitudes that are equally removed from the middle have equal weight and also the lines between their centers are equal, then the point that is also the center of weight of the middle magnitude shall be the center of weight of the magnitude composed of all the magnitudes.*

Corollary 2. *If the magnitudes are even in number and their centers of weight are situated on a line and their middle magnitudes and the magnitudes equally removed from them shall have an equal weight, and the lines between the centers are equal, then, too, the middle of the line joining the centers of weight of the [two middle magnitudes] shall be the center of weight of the magnitude composed of all the magnitudes.*

Proposition 4. *Commensurable magnitudes balance at distances having the same ratio reciprocally as the weights.*

[28] J. L. Heiberg, in *Archimedis Opera*, and Thomas Heath, in *The Works of Archimedes*, consider the last two items in the above list (8 and 9) as the first two propositions of this work. Netz believes, in a volume to be published soon, that those items belong with the introduction and therefore numbers the propositions beginning here. Note, of course, that Archimedes did not number either the assumptions or the propositions.

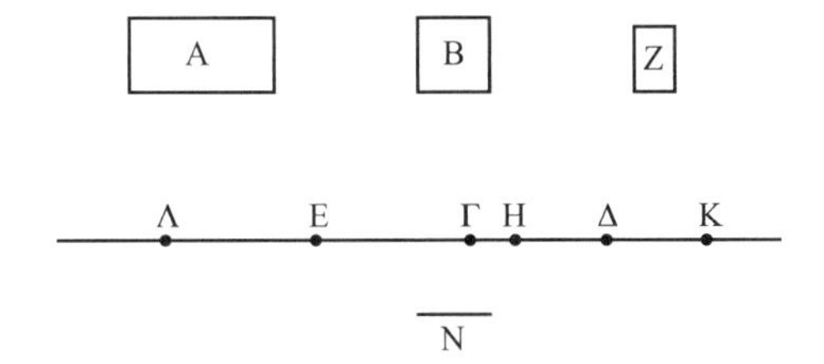

Figure 4.6.1.

Let there be commensurable magnitudes, A, B, whose centers are A, B, and let there be a certain distance, $E\Delta$, and let it be: as A is to B, so the distance $\Delta\Gamma$ is to the distance ΓE [figure 4.6.1]. It is to be proved that Γ is the center of weight of the magnitude composed of both magnitudes A and B.

For since it is: as the magnitude A is to the magnitude B, so is the distance $\Delta\Gamma$ to the distance ΓE, and the magnitude A is commensurable with the magnitude B. Therefore the distance $\Gamma\Delta$, too, is commensurable with the distance ΓE, that is, a line with a line; so that there is a common measure to the lines $E\Gamma$, $\Gamma\Delta$. So let it be the measure N, and let each of ΔH, ΔK be set equal to $E\Gamma$, and let $E\Lambda$ be set equal to $\Delta\Gamma$. And since ΔH is equal to ΓE, $\Delta\Gamma$ is equal to EH too. So that ΛE, too, is equal to EH. Therefore ΛH is twice $\Delta\Gamma$, while HK is twice ΓE, so that N measures each of ΛH, HK, too, since it also measures their halves. And since as A is to B, so is $\Delta\Gamma$ to ΓE, and as $\Delta\Gamma$ is to ΓE, so is ΛH to HK, for each is twice each, therefore also as A is to B, so is ΛH to HK. And as many times ΛH is of N, let the magnitude A, as well, be that many times the magnitude Z; therefore, as ΛH is to N, so is A to Z. And also, as KH is to ΛH, so is B to A. Therefore, through the equality, as KH is to N, so is B to Z; therefore KH is equally an equimultiple of N as B is of Z. And A, too, was proved to be a multiple of Z; so that Z is a common measure to A, B.

Now with line ΛH divided into lines equal to N, and the magnitude A divided into magnitudes equal to Z, the segments of line ΛH, which are equal in magnitude to the line N, shall be equal in multitude to the segments in the magnitude A, which are equal to Z. So that, if on each of the segments in line ΛH is put a magnitude equal to Z, having the center of weight on the middle of the segment of the line, then, both all the magnitudes are equal to A, and also E shall be the center of weight of the magnitude composed of all the magnitudes; for all the magnitudes are even in number (through ΛE being equal to HE), and they are all equal. And it shall be proved similarly also that if on each of the segments in the line KH a magnitude equal to Z is set, having the center of weight on the middle of the segment, both all the magnitudes shall be equal to B, and Δ shall be the center of weight of the magnitude composed of all the magnitudes. Now, A shall be put at E and B at Δ.

So, there shall be magnitudes equal to each other situated on a line, whose centers of weight are equally distant from each other, being even in multitude; now, it is clear that the midpoint of the line holding the centers of the middle magnitudes is the center of weight of the magnitude composed of all the magnitudes. And since ΛE is equal to $\Gamma\Delta$, while

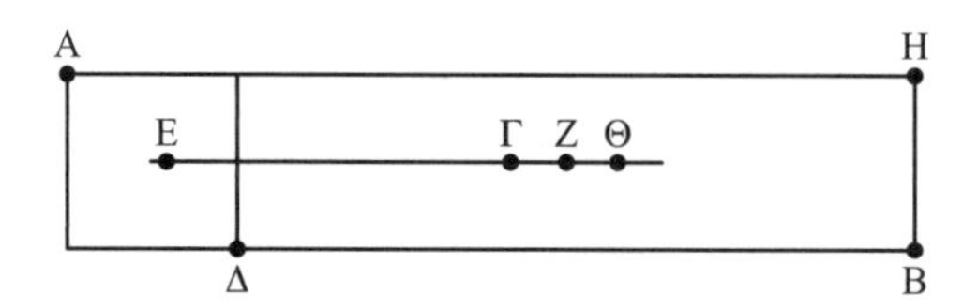

Figure 4.6.2.

$E\Gamma$ is equal to ΔK, therefore the whole $\Lambda\Gamma$, too, is equal to ΓK; so that the point Γ is the center of weight of the magnitude composed of all the magnitudes. Therefore, when A is situated at E while B is at Δ, they shall balance at Γ.

Proposition 5. *And even if the magnitudes be incommensurable, they shall similarly balance at distances having the same ratio reciprocally to the magnitudes.*

The following proposition is used in the proof of proposition 11 below and of proposition 8 of book 1 of *On Floating Bodies*.

Proposition 6. *If, from a certain magnitude, a certain magnitude (not having the same center of weight as the whole) is taken away, the center of weight of the remaining magnitude is such that its distance from the center of weight of the whole magnitude has to the line between the centers of weight of the original two magnitudes the same ratio that the weight of the magnitude taken away has to the remaining magnitude.*

Let the center of weight of a certain magnitude, the magnitude AB, be Γ, and, from magnitude AB let the magnitude $A\Delta$ (whose center of weight is E) be taken away, and, with $E\Gamma$ joined and produced, let ΓZ be taken, having to ΓE the same ratio which the magnitude $A\Delta$ has to the magnitude ΔH. It is to be proved that the point Z is the center of weight of magnitude ΔH.

For let it not be, but, if possible, let the point Θ be the center of weight. Now, since E is the center of weight of magnitude $A\Delta$, while Θ is of magnitude ΔH, the center of weight of the magnitude composed of both $A\Delta$, ΔH shall be on the line $E\Theta$, cut so that its segments reciprocate at the same ratio as the magnitudes. So the point Γ shall not be at the cut proportional as the cut mentioned. Therefore Γ is not the center of weight of the magnitude combined of the magnitudes $A\Delta$, ΔH, that is, of magnitude AB. But it is, for it was assumed. Therefore Θ is not the center of weight of magnitude ΔH.

Proposition 7. *The center of weight of every parallelogram is on the line joining the midpoints of the opposite sides of the parallelogram.*

Proposition 8. *The center of weight of every parallelogram is the point at which the diameters intersect.*

Let there be a parallelogram $AB\Gamma\Delta$ and, in it, the line EZ bisecting $AB, \Gamma\Delta$, and $K\Lambda$, bisecting $A\Gamma, B\Delta$ [figure 4.6.3]. So the center of weight of the parallelogram $AB\Gamma\Delta$ is on EZ, for this has been proved. But, through the same, it is also on $K\Lambda$. Therefore the point Θ is the center of weight. And the diameters of the parallelogram intersect at Θ, so that the claim in proved.

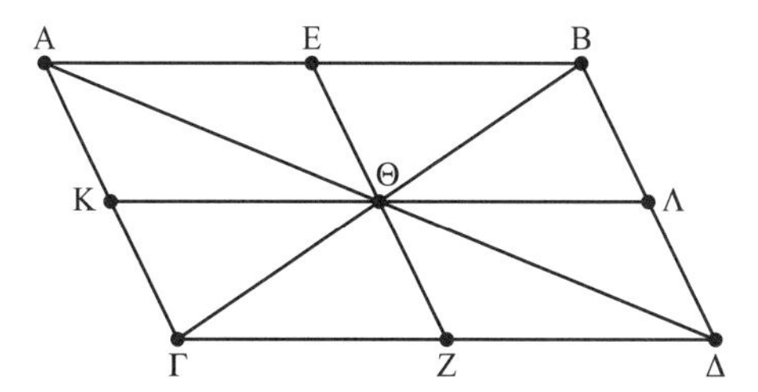

Figure 4.6.3.

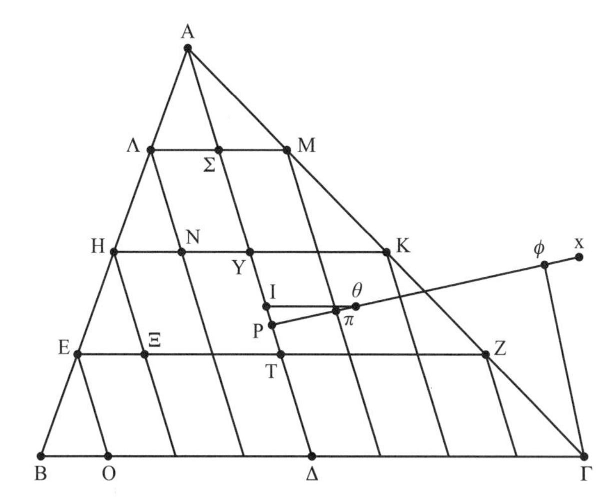

Figure 4.6.4.

Proposition 11. *The center of weight of every triangle is on the line that is drawn from the angle to the midpoint of the base.*

Let there be a triangle $AB\Gamma$ and in it $A\Delta$, drawn to the midpoint of the base $B\Gamma$; it is to be proved that the center of weight of triangle $AB\Gamma$ is on $A\Delta$ [figure 4.6.4].

For let it not be but, if possible, let Θ be the center of weight and, through it, let ΘI be drawn parallel to $B\Gamma$. So, $\Delta\Gamma$ being continually bisected, at some time the remainder shall be less than ΘI. Let each of $B\Delta$, $\Delta\Gamma$ be divided into parts equal [to that remainder], and let lines be drawn through the cuts, parallel to $A\Delta$, and let $EZ, HK, \Lambda M$ be joined. So, they themselves shall be parallel to $B\Gamma$. So, the center of the weight of parallelogram MN is on $Y\Sigma$, while the center of weight of parallelogram $K\Xi$ is on TY, and the center of weight of parallelogram ZO is on $T\Delta$. Therefore the center of weight of the magnitude composed of all is on line $\Sigma\Delta$. So, let it be P, and let $P\Theta$ be joined and produced, and let $\Gamma\Phi$ be drawn parallel to $A\Delta$. So, triangle $A\Delta\Gamma$ has to all the triangles set up on $AM, MK, KZ, Z\Gamma$, similar to $A\Delta\Gamma$, that ratio which ΓA has to AM, since $AM, MK, Z\Gamma, KZ$ are equal to each other. And since triangle $A\Delta B$, too, has to the similar triangles set up on $A\Lambda, \Lambda H, HE, EB$ the same ratio, which BA has to $A\Lambda$, therefore triangle $AB\Gamma$ has to all the said triangles that ratio which ΓA has to AM.

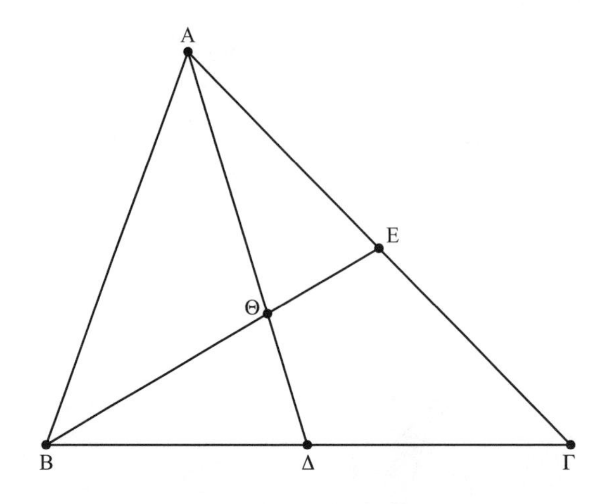

Figure 4.6.5.

But ΓA has to AM a greater ratio than that of ΦP to $P\Theta$, for the ratio of ΓA to AM is the same as that of the whole ΦP to ΠP, through the triangles being similar; therefore triangle $AB\Gamma$, too, has to the said triangles a greater ratio than ΦP to $P\Theta$, so that *dividendo* the parallelograms $MN, K\Xi, ZO$ have to the remaining triangles a greater ratio than that of $\Phi\Theta$ to ΘP. Now, let the ratio of $X\Theta$ to ΘP come to be in the ratio of the parallelograms to the triangles. Now, since there is a certain magnitude, triangle $AB\Gamma$, of which Θ is the center of weight, and a magnitude has been taken away from it, namely, the magnitude composed of the parallelograms $MN, K\Xi, ZO$, and the point P is the center of weight of the magnitude taken away, therefore the center of weight of the remaining magnitude composed of the triangles remaining around the perimeter, is on the line $P\Theta$ produced and taken to be having to ΘP the ratio which the magnitude taken away has to the remainder [proposition 6]. Therefore, the point X is the center of weight of the magnitude composed of the triangles remaining around, which is impossible. For all the triangles are on the same side in the plane of the line drawn through X parallel to $A\Delta$. So the claim is clear.

Proposition 12. *The center of weight of every triangle is the point at which the lines drawn from the angles to the midpoints of the sides meet each other.*

Let there be a triangle $AB\Gamma$ and let $A\Delta$ be drawn to the midpoint of $B\Gamma$, while BE is drawn to the midpoint of $A\Gamma$ [figure 4.6.5]. So, the center of weight of triangle $AB\Gamma$ shall be on each of the lines $A\Delta, BE$. For this has been proved. So the point Θ is the center of weight.

Archimedes seems to imply with this result that all three medians meet at the same point. And given that result, it is straightforward to show that the center of weight is one-third of the distance along the median from the base to the opposite vertex. That is, in figure 4.6.5, $A\Theta = 2\Theta\Delta$, with the same result applying to the other medians. In fact, this result is referred to in proposition 1 of *The Method*, although the result is not actually stated in *Planes in Equilibrium*.

4.7 Archimedes, *On Spirals*

Archimedes's *On Spirals* contains twenty-eight propositions, along with seven definitions. We can divide the work into four parts, as follows:

1. Propositions 1–11: Preliminary results, none of which explicitly involve spirals.
2. Definitions: Here Archimedes defines the spiral and related terms.
3. Propositions 12–20: Basic results about spirals, and results about tangents to a spiral.
4. Propositions 21–28: Results related to the area contained by spirals.

The two main results are proposition 18, in which Archimedes shows that a circumference of a circle is equal to a certain line constructed from the spiral, and proposition 24, in which he calculates the area bounded by one turn of the spiral. The preliminary propositions that enable Archimedes to prove these main results are separated into two groups. That is, these two results are essentially independent of each other, with their only commonality being that they both deal with a spiral. We have chosen to include full translations of propositions 10, 12, 14, 16, 18, 21, and 24, with just the statements of a few of the others. As in many of Archimedes's works, *On Spirals* begins with a letter. This letter describes many of the results Archimedes has proved in other books.

Archimedes to Dositheus: Greetings!

Of those theorems dispatched to Conon, about which you keep sending me letters asking that I write down the proofs—many you have, written down in the books conveyed by Heracleides,[29] while some I send you, having written them down as well in this book.

Now should you wonder why I took such a long time publishing their proofs; this came about because I wanted to allow those who busy themselves with mathematics to take up studying these theorems first. For how many of the theorems in geometry appear not to go along the right lines at first, to the one who eventually perfects them? Conon passed away without taking sufficient time for their study; otherwise he would have made them all clear, discovering them as well as many others, while advancing geometry a great deal, for we know that his mathematical understanding was extraordinary, his diligence unsurpassed.

With many years now having passed since Conon's death, we are not aware of even a single problem being set in motion, not by a single person.

I also wish to set out each of them, one by one. For it happens that there are a certain two of the theorems in the letter to Conon, not distinguished apart but added at the end, so that whose who claim to find all of them, but publish none of their proofs, would be refuted by promising to find solutions to impossible theorems. So I now find it appropriate to make clear which are those problems, and which of them are those whose proofs you have (which were sent to you already), and which I convey in this book.

So, the first of these problems was: given a sphere, to find a plane area equal to the surface of the sphere. This, indeed, was also the first to become clear following the

[29] There is no scholarly consensus on the identity of this man. It is possible that he is the person mentioned by Pappus as solving a certain geometrical problem or a biographer of Archimedes mentioned by Eutocius.

publication of the book about the sphere. For once it is proved that the surface of every sphere is four times the greatest circle of the circles in the sphere, it is obvious that it is possible to find a plane area equal to the surface of the sphere.[30]

Second, given a cone or a cylinder, to find a sphere equal to the cone or cylinder.

Third, to cut the given sphere with a plane, so that its segments have to each other the ratio assigned.

Fourth, to cut the given sphere with a plane, so that the segments of the surface have to each other the ratio assigned.

Fifth, to make the given segment of a sphere similar to another given segment of a sphere.

Sixth, given two segments of a sphere, whether of the same sphere or of another, to find a certain segment of a sphere, which will itself be similar to one of the given segments, while it will have a surface equal to the surface of the other given segment.

Seventh, to cut off a segment from a given sphere with a plane, so that the segment has to the cone having the same base as the segment and an equal height, an assigned ratio greater than that which 3 has to 2.

The proofs of all these theorems mentioned, then, Heracleides conveyed. But the one positioned, separately, following them was false. It is: if a sphere is cut by a plane into unequal segments, the greater segment has to the smaller a ratio duplicate that of the greater surface to the smaller. That this is false is clear through what was sent before, for it has been set out among those as follows: if a sphere is cut by a plane into two unequal segments at right angles to some diameter of those in the sphere, the greater segment of the surface shall have to the smaller the same ratio which the greater segment of the diameter has to the smaller; indeed, the greater segment of the sphere has to the smaller a ratio smaller than the duplicate of that which the greater surface has to the smaller, but greater than half-as-much again.

And the last one separated from among the problems was a falsehood, too: that if the diameter of a certain sphere is cut so that the square on the greater segment is three times the square on the smaller segment, and through the point where the diameter was cut the plane drawn at right angles to the diameter cuts the sphere, the figure of this kind, viz., the greater segment of the sphere, is the greatest among all the other segments which have a surface equal to it. That this is false is clear through the theorems sent before; for it has been shown that the hemisphere is greatest among the segments of a sphere, contained by an equal surface.

Following these, the problems concerning the cone are these:

If a section of a right-angled cone should be rotated, the diameter remaining fixed, so that the diameter is the axis, let the figure drawn by rotation by the section of the right-angled cone be called a *conoid*; and if a plane touches the conoid figure, and another plane, drawn parallel to the touching plane, cuts a certain figure of the conoid, let the cutting plane be called *base* of the segment cut off, and let the point at which the other plane touches the conoid be called *vertex*. So, if the figure mentioned is cut by a plane at right angles to the axis, it is clear that the section shall be a circle, but it is required to

[30] This problem and the following ones are solved in *On the Sphere and the Cylinder*.

prove that the segment cut off shall be half as much again of the cone having a base the same as the segment and an equal height. And if two segments of the conoid should be cut by any planes drawn in whichever way, it is clear that the sections shall be sections of acute-angled cones if the cutting planes are not perpendicular to the axis, but it is required to prove that the segments shall have to each other the ratio which the lines, drawn from their vertices to the cutting planes and parallel to the axis, have to each other, in square.[31]

The proofs of these have not yet been sent to you.

Following these, the problems offered concerning the spiral were these—and they are, as it were, a certain other class of problems, having nothing in common with those mentioned above; the proofs concerning which we have written for you, in this book—They are as follows:

If a straight line, being rotated in a plane in uniform speed, with one of its ends remaining fixed, should be returned again to where it started from, while at the same time, even as the line is rotated, a certain point is carried along the line, in uniform speed with itself, starting at the fixed end, the point shall draw a spiral in the plane. So I claim that (1) the area contained by the spiral and by the line that has been returned to where it started from is a third part of the circle drawn, with the fixed point as center and with the line traversed by the point in one rotation of the line as radius. And (2) if some line touches the spiral at the end of the spiral which is the last, while some other line is drawn from its fixed end at right angles to the line rotated and returned to position, so that it meets the tangent, I claim that the line produced towards the tangent is equal to the circumference of the circle. And (3) if the rotated line and the point carried along it are carried around for many rotations and are returned to where they started from, I claim that—compared to the area added by the spiral in the second rotation—the area added in the third rotation shall be double, the area added in the fourth shall be triple, the area added in the fifth shall be quadruple, and always, the areas added in the last rotations shall be multiples, according to the numbers in sequence, of the area added in the second rotation; while the area contained in the first rotation is a sixth part of the area added in the second rotation. And (4) if two points should be taken on the spiral drawn in a single rotation, and lines are joined from them to the fixed end of the rotated line, and two circles are drawn with the fixed point as center, and the lines joined to fixed end as radii, and the smaller of the joined lines is produced, I claim that the area contained by (i) the circumference of the greater circle which is on the side of the spiral between the two lines and (ii) the spiral itself and (iii) the produced line, has to the area contained by (i) the circumference of the smaller circle and (ii) the spiral itself and (iii) the line joining their ends, the ratio which the radius of the smaller circle, together with two thirds the excess by which the radius of the greater circle exceeds the radius of the smaller circle, has to the radius of the smaller circle, together with one-third part of the mentioned excess.

So, the proofs of these and other theorems concerning the spiral are written by me in this book, and preceding them (as is also the case with any other books provided in a geometrical way) are the theorems required for their proofs. And I also adopt, in these

[31] These results appear in *On Conoids and Spheroids*.

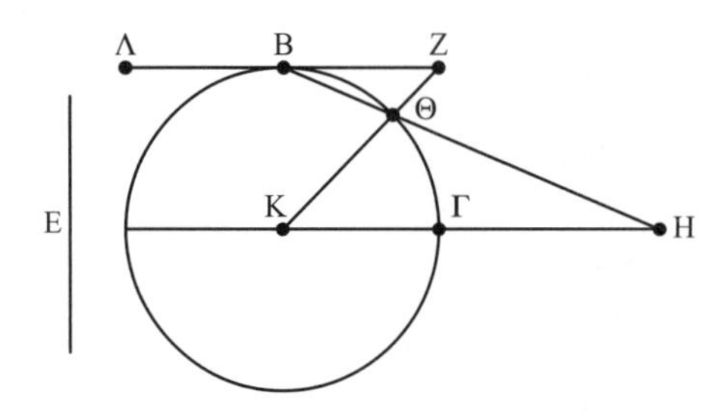

Figure 4.7.1.

theorems this lemma, which is also among the lemmas in the books sent out before: that, among unequal lines and unequal areas, the excess by which the greater exceeds the smaller, itself added onto itself, is capable of exceeding every given magnitude, of those which are said to be in a ratio to each other.

Propositions

Proposition 5. *Given a circle and a line touching the circle, it is possible to draw a straight line from the center of the circle to the tangent, so that the straight line between the tangent and the circumference of the circle has to the radius a smaller ratio than the circumference of the circle which is between the touching point and the line drawn through to the given circumference (however big) of a circle.*

This proposition is used in the proof of proposition 16. To prove it, assume that the circle has center K with a tangent at the point B [figure 4.7.1]. We take a straight line E equal to the given circumference and through K draw KH parallel to the tangent. We then draw line $B\Theta H$, meeting the circle at Θ, such that ΘH is equal to E. Finally, we draw line $K\Theta Z$. Then $Z\Theta$ is to $K\Theta$ as $B\Theta$ is to ΘH, or as $B\Theta$ is to E. And this ratio is less than that of arc $B\Theta$ to E, as claimed.

The next two propositions are used in the proof of proposition 18. We omit the proofs here. It is interesting that Archimedes gives no motivation for these rather technical results at this time, just saving them for their later use.

Proposition 7. *Given a circle and, in the circle, a line smaller than the diameter, and with that line being produced, it is possible to extend a line from the center towards the produced line so that the line between the circumference and the produced line has a given ratio to the line joined from the end of the line taken off inside to the end of the produced line, if the given ratio is greater than the ratio which the half of the line given in the circle has to the perpendicular drawn on it from the center.*

Proposition 8. *Given a circle and a line in the circle smaller than the diameter and another touching the circle at the end of the line given in the circle, it is possible to extend a certain line from the center of the circle towards the given line so that the line taken off from it between the circumference of the circle and the line given in the circle has to the line taken off from the tangent a given ratio, if the given ratio is smaller than the ratio which the half of the line given in the circle has to the perpendicular drawn on it from the center of the circle.*

In modern terms, the next result is essentially an algebraic one, giving a way to approximate the sum of the squares of the integers. But without any algebraic machinery, Archimedes must prove this geometrically using actual squares.

Proposition 10. *If however many lines are set in order, exceeding each other by an equal difference, and the excess is equal to the smallest line, and other lines are set equal to them in number while, in magnitude, each is equal to the greatest line among the lines exceeding each other, the squares on the lines equal to the greatest line adding on both the square on the greatest line, and the rectangle contained by both the smallest line among the lines exceeding each other, and the line equal to all the lines exceeding each other by an equal difference shall be three times all the squares that are on the lines exceeding each other by an equal difference.*

Let there be however many lines set in order, exceeding each other by an equal difference, viz., the lines $A, B, \Gamma, \Delta, E, Z, H, \Theta$, and let Θ be equal to the difference, and let I, equal to Θ, be added to B; K, equal to H, be added to Γ; L, equal to Z, to Δ; M equal to E, to E; N, equal to Δ, to Z; Ξ, equal to Γ, to H; O, equal to B, to Θ—and the lines that come to be are equal to each other and to the greatest, A. Now, it is to be proved that the squares on all the lines A as well as the lines coming to be, adding on both: the square on A and the rectangle contained by both: Θ and the line equal to all the lines $A, B, \Gamma, \Delta, E, Z, H, \Theta$, are three times all the squares on A, B, Γ, Δ, E, Z, H, Θ.

Archimedes chooses to prove proposition 10 using eight unequal lines, since he had no way of denoting "however many lines." But this is simply an example of a proof by generalized example, a technique that Euclid used often, and in no way affects the completeness of the proof. Archimedes explains that he will be proving what is equivalent to the following:

$$A^2+(B+I)^2+\cdots+(\Theta+O)^2+A^2+\Theta(A+B+\cdots+\Theta)=3(A^2+B^2+\cdots+\Theta^2)$$

Since $I=\Theta$, $K=H$, ..., $O=B$ and there are eight segments A, B, ..., Θ, this statement is equivalent to $9A^2+\Theta(A+B+\cdots+\Theta)=3(A^2+B^2+\cdots+\Theta^2)$. Of course, this result is true for any number n of segments, with the 9 being replaced by $n+1$. In more modern notation, then, we can write the result as

$$(n+1)n^2+(1+2+\cdots+n)=3(1^2+2^2+\cdots+n^2).$$

From figures 4.7.2, 4.7.3, 4.7.4, and 4.7.5, it is clear, as Archimedes explains below, that twice the sum of the squares of $A, B, \ldots, \Theta$ is included in the left-hand side of the displayed equation, so he needs to show that the remainder, as shown in figure 4.7.5, gives a third sum.

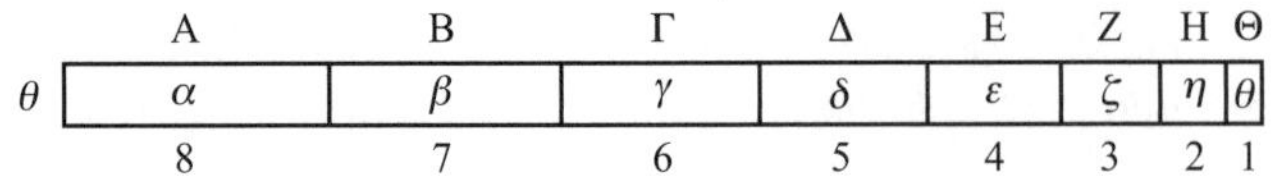

	A	B	Γ	Δ	E	Z	H	Θ
θ	α	β	γ	δ	ε	ζ	η	θ
	8	7	6	5	4	3	2	1

Figure 4.7.2.

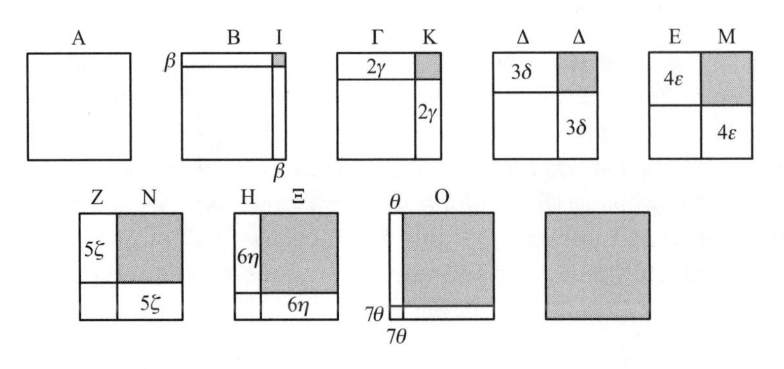

Figure 4.7.3.

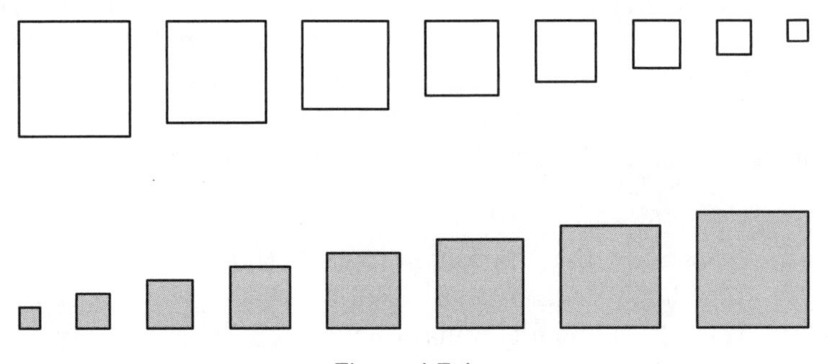

Figure 4.7.4.

So, the square on BI[32] is equal to the squares on I, B and to two rectangles contained by B, I, while the square on $K\Gamma$ is equal to the squares on K, Γ and to two rectangles contained by K, Γ [figure 4.7.3]. And also similarly, the squares on the other lines equal to the line A, are equal to the square on the segments and to two rectangles contained by the segments. Now, the squares on $A, B, \Gamma, \Delta, E, Z, H, \Theta$ and the squares on $I, K, \Lambda, M, N, \Xi, O$, adding on the square on A, are twice the squares on $A, B, \Gamma, \Delta, E, Z, H, \Theta$; we will prove what remains, that twice the rectangles contained by the segments in each line of those equal to A, adding on the rectangle contained by both: Θ and the line equal to all the lines $A, B, \Gamma, \Delta, E, Z, H, \Theta$, is equal to the squares on $A, B, \Gamma, \Delta, E, Z, H, \Theta$.

Now two rectangles, the rectangles contained by B, I are equal to two rectangles, the rectangles contained by B, Θ, while two rectangles, the rectangles contained by K, Γ are

[32] BI means the line segment composed of the two line segments B and I.

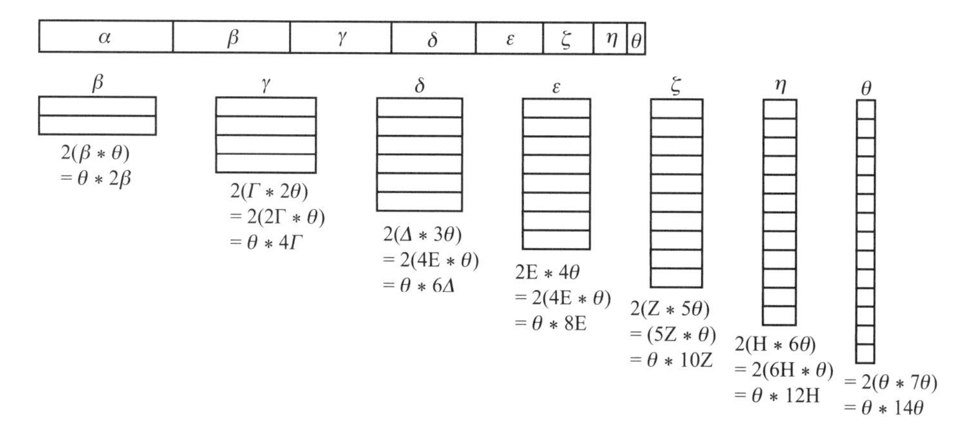

Figure 4.7.5.

equal to the rectangle contained by both Θ and four times Γ, through K being twice Θ, and two rectangles, the rectangles contained by Δ, Λ are equal to the rectangle contained by Θ and six times Δ, through Λ being three times Θ, and similarly also: the others, twice the rectangles contained by the segments are equal to the rectangle contained by both: Θ and the multiple, ever ascending according to the even numbers in sequence, of the following line. Now, all the double rectangles taken together, adding on the rectangle contained by both Θ and the line equal to all the lines $A, B, \Gamma, \Delta, E, Z, H, \Theta$, shall be equal to the rectangle contained by both Θ and the line equal to all: A, as well as three times B, and five times Γ, and the odd multiple, ever ascending according to the odd numbers in sequence, of the following line.

Here, Archimedes first rearranges every rectangle so that each has one side length equal to Θ, the smallest segment of those with which we began. Consider one of these, the rectangle on Δ, Λ. Recall that $\Lambda = Z = 3\Theta$, so the rectangle of Δ, Λ is equal to the rectangle on $3\Theta, \Delta$, which in turn is equal to three times the rectangle on Θ, Δ and the rectangle on $\Theta, 3\Delta$, because if we have three equal rectangles, we may place them end to end from either side to obtain only one rectangle with one side being a multiple of its previous length in each of the initial rectangles. So the two rectangles on Δ, Λ are equal, as Archimedes writes, to the rectangle on Θ and 6Δ, with similar results for the other rectangles. Therefore, in modern notation, $2B\Theta + 2\Gamma H + \cdots + 2H\Gamma + 2\Theta B + \Theta(A + B + \cdots + \Theta) = \Theta \cdot 2B + \Theta \cdot 4\Gamma + \cdots + \Theta \cdot 12H + \Theta \cdot 14\Theta + \Theta(A + B + \cdots + \Theta)$. Moreover, this sum of rectangles is equal to one long rectangle: $\Theta(A + 3B + 5\Gamma + \cdots + 13H + 15\Theta)$, because every rectangle has a side length of Θ to which the next can be joined. Thus we must prove that this rectangle equals the sum of the squares of $A, B, \ldots, \Theta$. That is, the bottom parts of figure 4.7.3 must be shown equal to the sum of the squares.

And the squares on $A, B, \Gamma, \Delta, E, Z, H, \Theta$, too, are equal to the rectangle contained by the same lines. For the square on A is equal to the rectangle contained by both Θ and the line equal to all: to both A, and the line equal to the remaining, of which each is equal to A. For they measure equally, both Θ measuring A and A measuring all

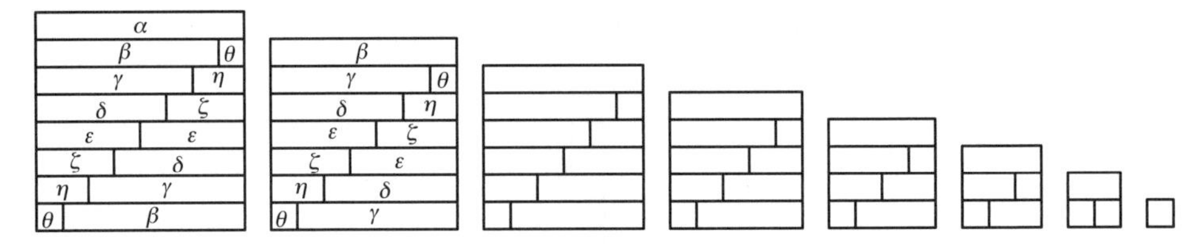

Figure 4.7.6.

the lines equal to it, with A— so that the square on A is equal to the rectangle contained by both Θ and the line equal to A, and the double of $B, \Gamma, \Delta, E, Z, H, \Theta$. For the lines equal to A, all besides A, are doubles of $B, \Gamma, \Delta, E, Z, H, \Theta$. And similarly also: the square on B is equal to the rectangle contained by both Θ and the line equal to both B and the double of $\Gamma, \Delta, E, Z, H, \Theta$, and again: the square on Γ is equal to the rectangle contained by both Θ and the line equal to both Γ and the double of Δ, E, Z, H, Θ. And similarly also the squares on the other lines are equal to the rectangles contained by both Θ and the line equal to both the line itself on which the square was formed and the double of the remaining lines. Now, it is clear that the squares on all lines are equal to the rectangle contained by both Θ and the line equal to both A, as well as three times B, and five times Γ, and the multiple, according to the odd numbers in sequence, of the following line.

In this final step, Archimedes breaks up the long rectangle he made and arranges the pieces into the squares on A through Θ. He claims that $A^2 = \Theta[A + (B + \Theta) + (\Gamma + H) + \cdots + (\Theta + B)]$, where $B + \Theta, \Gamma + H, \ldots$ are "the remaining, of which each is equal to A." Now Θ divides A into the same quantity that A divides the sum of all the lines equal to it, namely, $A + (B + \Theta) + (\Gamma + H) + \cdots + (\Theta + B)$; A and the sum of all the lines are equimultiples of Θ and A, respectively; that is, just multiply both by A to obtain the larger magnitudes. This must be true, for the first magnitude in each term of the sum is Θ less than the one that came before it, and we begin the series with the magnitude A. Since there are a number of rectangles of length A with side length Θ equal to the number of Θs in A, we may stack them to obtain a square with side length A [figure 4.7.6].

Since $\Theta[A + (B + \Theta) + (\Gamma + H) + \cdots + (\Theta + B)] = \Theta(A + 2B + 2\Gamma + \cdots + 2\Theta)$, after Archimedes cuts this rectangle off of his initial rectangle on Θ and $A + 3B + 5\Gamma + \cdots + 13H + 15\Theta$ to form the square on A, he is left with the rectangle on Θ and $B + 3\Gamma + 5\Delta + \cdots + 13\Theta$. Archimedes similarly repeats this cutting procedure to create squares on B through Θ, as illustrated in figure 4.7.6. That this process will exactly "use up" the rectangle is inductively evident.

Proposition 10 is both visually interesting and readily applicable. The unequal segments, each exceeding the last by the smallest, can be thought of as a sequence of integers, beginning with the unit. This proposition is immediately followed by a corollary, which restates the result in a more usable form:

$$3[1^2 + 2^2 + \cdots + (n-1)^2] < n^3 < 3(1^2 + 2^2 + \cdots + n^2)$$

It is this latter result that provides the key to the proof of proposition 24.

Corollary. *Now, from this it is obvious that all the squares on the lines equal to the greatest line are smaller than three times the squares on the lines exceeding each other by an equal difference—since by adding on certain magnitudes they are three times—while they are greater than three times the remaining squares, without the square on the greatest line. And, furthermore, if similar figures are set up on all the lines, on both the lines exceeding each other by an equal difference as well as the lines equal to the greatest line, the figures on the lines equal to the greatest line shall be smaller than three times the figures on the lines exceeding each other by an equal difference, while being greater than three times the remaining figures, without the figure on the greatest line. For similar figures have the same ratio as the squares.*

It is only after proposition 10 that Archimedes defines the central notion of this book, that of a spiral. The definition of the spiral is the first of seven definitions he gives at this point.

Definitions

1. If a straight line is joined in a plane, and, being rotated at uniform speed however many times, with one of its ends remaining fixed, is returned again to where it started from, while at the same time, even as the line is rotated, a certain point is carried along the line, at uniform speed with itself, starting at the fixed end, the point shall draw a *spiral* in the plane.[33]
2. Now, let the fixed end of the line, the line which is itself being moved around, be called the *start of the spiral*.
3. And let the position of the line, from which the straight line started to rotate, be called the *start of the rotation*.
4. Let a straight line, through which the point carried along the line passes during the first rotation, be called *first*; and a line which the same point completes during the second rotation, *second*; and let the other lines similarly to these be called by the same name as the rotations.
5. And let the area taken by both the spiral drawn during the first rotation, as well as by the line, which is first, be called *first*, and the area taken by both the spiral drawn during the second rotation as well as by the second line, *second*, and let the others be called in this manner in sequence.
6. And if from the point, which is the start of the spiral, some straight line is drawn, let those things which are at the same side of that line, at which the rotation is made, be called *preceding*, and those at the other side, *following*.
7. Let the circle drawn with the point, which is the start of the spiral, as center, and the line, which is first, as radius, be called *first*, and let the circle be drawn with the same center and with the double line as radius be called *second*, and let the other circles be called in sequence with these in the same manner.

[33] In modern polar coordinates, the equation of this Archimedean spiral is $r = a\theta$, for some positive number a.

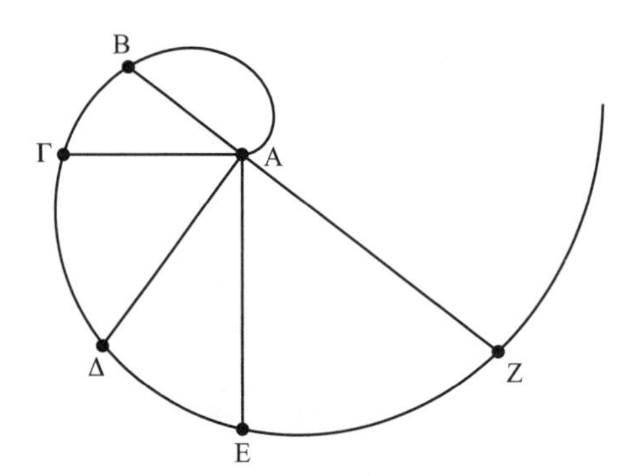

Figure 4.7.7.

Proposition 12. *If however many lines, drawn from the start of the spiral, fall on the spiral during a single rotation, making the angles equal to each other, they exceed each other by an equal difference.*

Let there be a spiral, on which are lines AB, $A\Gamma$, $A\Delta$, AE, AZ making equal angles to each other [figure 4.7.7]. It is to be proved that $A\Gamma$ exceeds AB, and $A\Delta$ exceeds $A\Gamma$ by an equal difference, and the others similarly.

For in the time in which the rotated line reaches from AB to $A\Gamma$, in that time the point being carried along the straight line passes through the difference, by which ΓA exceeds AB, and, in which time it reaches from $A\Gamma$ to $A\Delta$, in that time the point passes through the difference, by which $A\Delta$ exceeds $A\Gamma$. And the rotated line reaches both from AB to $A\Gamma$ and from $A\Gamma$ to $A\Delta$ in an equal time, since the angles are equal. Therefore the point carried along the straight line passes through the difference, by which ΓA exceeds AB, and through the difference, by which $A\Delta$ exceeds $A\Gamma$, in an equal time. Therefore $A\Gamma$ exceeds AB and $A\Delta$ exceeds $A\Gamma$ by an equal difference, as well as the rest of the lines, accordingly.

Proposition 14. *If, from the point that is the start of the spiral, two lines fall on the spiral drawn during the first rotation, and are produced to the circumference of the first circle, the lines falling on the spiral shall have to each other the same ratio that the circumferences of the circle between the end of the spiral and the ends of the lines which were produced so as to come to be on the circumferences have to each other (the circumferences being taken in the preceding direction, from the end of the spiral).*

Let there be a spiral drawn in the first rotation, $AB\Gamma\Delta E\Theta$, and let the point A be the start of the spiral, and let the line ΘA be the start of the rotation, and let ΘKH be the first circle, with the lines $AE, A\Delta$ falling on the spiral from the point A, and further falling on the circumference of the circle on the points Z, H [figure 4.7.8]. It is to be proved that they have the same ratio: AE to $A\Delta$, the same which the circumference ΘKZ has to the circumference ΘKH.

For, the line $A\Theta$ being rotated, it is clear that the point Θ is carried at a uniform speed along the circumference of the circle ΘKH while A, being carried along the line, passes through the line $A\Theta$, and the point Θ, being carried along the circumference of the circle,

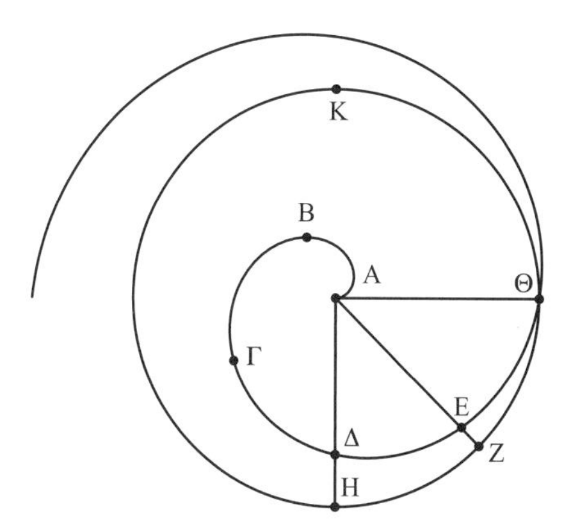

Figure 4.7.8.

passes through the circumference ΘKZ, while the point A passes through the line AE. And again both the point A passes through the line $A\Delta$ and Θ through the circumference ΘKH, each being carried itself at uniform speed with itself. Now, it is clear that they have the same ratio: AE to $A\Delta$, which the circumference ΘKZ has to the circumference ΘKH. And similarly it shall be proved that even if one of the falling lines should fall on the end of the spiral, the same thing happens.

Proposition 15. *If lines fall from the start of the spiral on the spiral drawn during the second rotation, the lines shall have to each other the same ratio, which the said circumferences [as in the previous proposition] together with an entire circumference of a circle taken have to each other.*

The proof of this proposition is similar to that of proposition 14. Both are used in the proof of proposition 18.

Proposition 16. *If a straight line touches the spiral drawn in the first rotation, and a line should be joined from the touching point to the point which is the start of the spiral, the angles which the tangent makes with the joined line will be unequal, and that in the preceding lines will be obtuse, while in the following lines acute.*

Let there be a spiral on which $AB\Gamma\Delta\Theta$, drawn in the first rotation, and let the point A be the start of the spiral, and the line $A\Theta$ be the start of the rotation, and the circle ΘKH be the first circle, and let some straight line $E\Delta Z$ touch the spiral at Δ, and let ΔA be joined from Δ to A. It is to be proved that ΔZ makes an obtuse angle with $A\Delta$ [figure 4.7.9].

Let a circle, $TN\Delta$, be drawn, with A as center and $A\Delta$ as radius. So, it is necessary that the circumference of the circle in the preceding lines falls inside the spiral, but in the following lines outside, through the fact that, among the lines falling on the spiral from A, the lines in the preceding are smaller than $A\Delta$, but those in the following are greater. Now, that the angle contained by the lines $A\Delta$, ΔZ, is not acute, is clear, since it is greater than the angle of a semicircle.[34] But it is to be proved as follows that it is not a right angle.

[34] This is the angle between the radius and the circumference of a circle. According to *Elements* 3:16, it is greater than any acute angle.

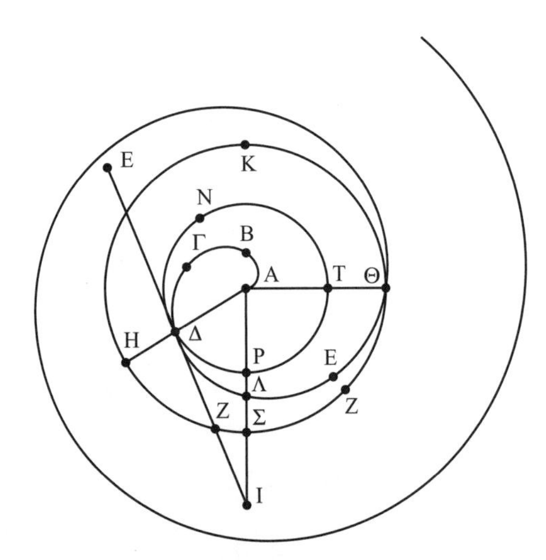

Figure 4.7.9.

For let it be, if possible, a right angle; therefore $E\Delta Z$ touches the circle ΔTN.[35] So, it is possible to insert a line from A to the tangent, so that the line between the tangent and the circumference of the circle has to the radius of the circle a smaller ratio than the circumference between the touching point and the falling line has to the given circumference [proposition 5]. So, let it fall as the line AI. So, it cuts the spiral at Λ, and the circumference of the circle $TN\Delta$ at P. And let the line PI have to the line AP a smaller ratio than the ratio which the circumference $P\Delta$ has to the circumference $TN\Delta$. Therefore IA in its entirety, too, has to AP a smaller ratio than the circumference $TN\Delta P$ to the circumference $TN\Delta$, that is, than the ratio which the circumference $\Theta KH\Sigma$ has to the circumference ΘKH. But that ratio which the circumference $\Theta KH\Sigma$ has to the circumference ΘKH, the line $A\Lambda$ has to $A\Delta$; for this has been proved [proposition 14]. Therefore AI has to ΔP a smaller ratio than, indeed, ΛA has to $A\Delta$, which indeed is impossible, for PA is equal to $A\Delta$. Therefore the angle contained by the lines $A\Delta$, ΔZ is not a right angle. And it was proved that neither is it acute. Therefore it is obtuse. Thus the remainder is acute.

And it shall be proved similarly that even if the tangent touches the spiral at the end, the same thing shall happen.

Proposition 18. *If a straight line should touch the spiral drawn in the first rotation at the end of the spiral, and a certain line is drawn from the point, which is the start of the spiral, at right angles to the start of the rotation, the drawn line shall meet the tangent, and the line between the tangent and the start of the spiral shall be equal to the circumference of the first circle.*

Let there be a spiral $AB\Gamma\Delta\Theta$, and let the point A be the start of the spiral, the line ΘA the start of the rotation, and the circle ΘHK the first circle. And let some line, say ΘZ,

35 Again, *Elements* 3:16.

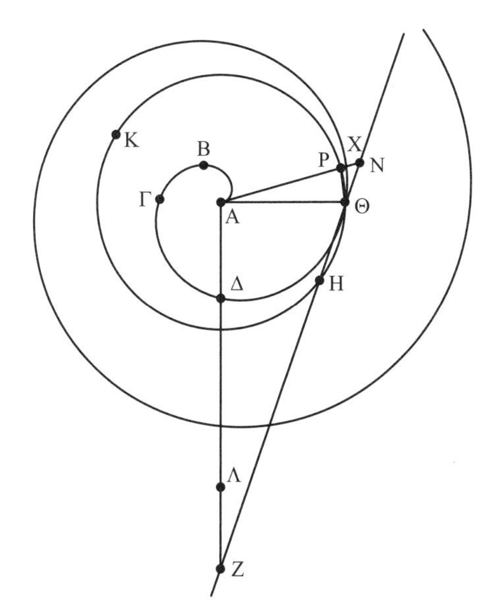

Figure 4.7.10.

touch the spiral at Θ, and let AZ be drawn from A at right angles to ΘA. So, that line shall meet ΘZ, since $Z\Theta, \Theta A$ contain an acute angle [proposition 16]. Let it meet at the point Z. It is to be proved that ZA is equal to the circumference of the circle ΘKH.

For if not, it is either greater or smaller. Let it first be, if possible, greater. So, I took a certain line ΛA, smaller than the line ZA but greater than the circumference of the circle ΘHK [figure 4.7.10]. So, there is a certain circle, ΘHK, and a line in the circle, smaller than the diameter, ΘH, and a ratio, which ΘA has to $A\Lambda$, greater than the ratio that half of $H\Theta$ has to the perpendicular to it drawn from A, because it is also greater than the ratio that ΘA has to AZ. Now, it is possible to extend a line AN from A towards the produced line, so that the line between the circumference and the produced line, NP, has to ΘP the same ratio that ΘA has to $A\Lambda$ [proposition 7]. Now, NP shall have to PA a ratio which the line ΘP has to $A\Lambda$.[36] And ΘP has to $A\Lambda$ a smaller ratio than the circumference ΘP to the circumference of the circle ΘHK; for the line ΘP is smaller than the circumference ΘP, while the line $A\Lambda$ is greater than the circumference of the circle ΘHK. Now, NP, too, shall have to PA a smaller ratio than the circumference ΘP to the circumference of the circle ΘHK. Now, NA, in its entirety, too, has to AP a smaller ratio than, indeed, the circumference ΘP together with the circumference of the circle in its entirety to the circumference of the circle ΘHK. And the ratio that the circumference ΘP with the circumference of the circle ΘHK in its entirety has to the circumference of the circle ΘHK—that ratio XA has to $A\Theta$, for this has been proved [proposition 15]; therefore NA has to AP a smaller ratio than, indeed, XA has to $A\Theta$, which indeed is impossible; for NA is greater than AZ, while AP is equal to ΘA. Therefore ZA is not greater than the circumference of the circle ΘHK.

[36] Since $NP : \Theta A = \Theta P : A\Lambda$ and $\Theta A = PA$, because both are radii of the same circle.

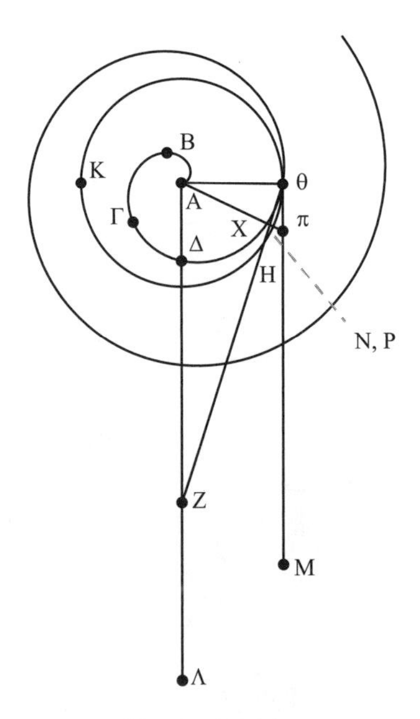

Figure 4.7.11.

So, again, let ZA be, if possible, smaller than the circumference of the circle ΘHK. So, I took a certain line, again, $A\Lambda$, greater than AZ but smaller than the circumference of the circle ΘHK [figure 4.7.11]. And, from Θ, I draw ΘM parallel to AZ. Now, again, there is a circle, ΘHK, and in it a line smaller than the diameter, ΘH, and another line, touching the circle at Θ, and a ratio, which $A\Theta$ has to $A\Lambda$, smaller than the ratio which the half of $H\Theta$ has to the perpendicular to it drawn from A, since it is also smaller than the ratio which ΘA has to AZ. Now, it is possible to draw the line $A\Pi$ from A to the tangent, so that PN, the line between the line in the circle and the circumference, has to $\Theta\Pi$, the line taken off from the tangent, the ratio which ΘA has to $A\Lambda$ [proposition 8]. So, $A\Pi$ shall cut the circle at P, and the spiral at X. And also, alternately, NP shall have to PA the same ratio which $\Theta\Pi$ has to $A\Lambda$. But $\Theta\Pi$ has to $A\Lambda$ a greater ratio than the circumference ΘP has to the circumference of the circle ΘHK; for the line $\Theta\Pi$ is greater than the circumference ΘP, while $A\Lambda$ is smaller than the circumference of the circle ΘHK. Therefore NP has a greater ratio to AP than the circumference ΘP has to the circumference of the circle ΘHK; so that PA, too, has to AN a greater ratio than the circumference of the circle ΘHK has to the circumference ΘKP. But the ratio which the circumference of the circle ΘHK has to the circumference ΘKP—that ratio the line ΘA has to AX; for this has been proved [proposition 14]. Therefore PA has to AN a greater ratio than ΘA to AX; which indeed is impossible. Therefore ZA is neither greater no smaller than the circumference of the circle ΘHK; therefore it is equal.

In modern terms, the polar equation of the spiral is $r=a\theta$. The slope of the tangent to the spiral at any point (r, θ) is given by

$$\frac{dy}{dx}=\frac{\sin\theta+\theta\cos\theta}{\cos\theta-\theta\sin\theta}.$$

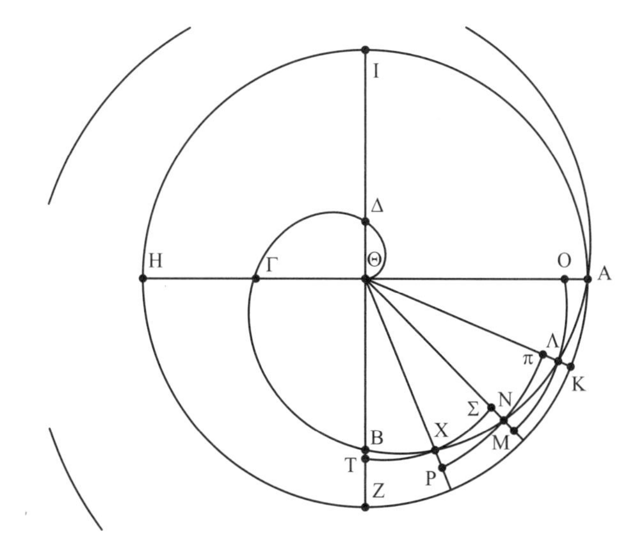

Figure 4.7.12.

Therefore, the slope of the tangent at the end of the first rotation, that is, when $\theta = 2\pi$, is $m = 2\pi$. It follows that the tangent line at that point, namely, at $(2\pi a, 0)$, is given by $y = 2\pi x - 4\pi^2 a$. Therefore the y intercept of that line is at $(0, -4\pi^2 a)$, whose distance from the start of the spiral is $4\pi^2 a$, which is, as claimed, equal to the circumference of the first circle, whose radius is $2\pi a$.

Proposition 21. *Taking the area contained by both the spiral drawn in the first rotation and the first line in the start of the rotation, it is possible to circumscribe a plane figure around it and inscribe another, composed of similar sectors, so that the circumscribed is greater than the inscribed by a magnitude smaller than any given area.*

Let there be a spiral drawn during the first rotation, which is the line $AB\Gamma\Delta$, and let the point Θ be the start of the spiral, ΘA the start of the rotation, the circle $ZHIA$ the first circle, and the diameters AH, ZI its diameters, at right angles to each other [figure 4.7.12]. So, the right angle ever again being bisected, and the sector containing the right angle, the remainder of the sector shall be smaller than the given; and let the sector have come to be, as the sector $A\Theta K$, smaller than the given area. So, let the four right angles be divided into the angles equal to the angle contained by $A\Theta$, ΘK, and let the lines making the angles be drawn as far as the spiral. So, let the point at which ΘK cuts the spiral be Λ, and let a circle be drawn with Θ as center and $\Theta\Lambda$ as radius; its circumference shall fall, towards the preceding circumference, inside the spiral, and towards the following circumference, outside. So, let the circumference OM be drawn, as far as it extends to fall on ΘA, at O, and as far as it extends to fall on the line falling on the spiral beyond the line ΘK. So, again, let the point at which ΘM cuts the spiral be N, and let a circle be drawn with Θ as center and ΘN as radius, as far as the circumference of the circle extends to fall on ΘK and on the line falling on the spiral beyond ΘM, and, similarly, let circles be drawn through all the other points at which the lines making the equal angles cut the spiral, with Θ as center, as far as each circumference extends to fall on the preceding line and on the following; so there shall be a certain figure composed of similar sectors, circumscribed around the taken area, and another inscribed.

And it shall be proved that the circumscribed figure is greater than the inscribed by a magnitude smaller than the given area. For the sector $\Theta\Lambda O$ is equal to the sector $\Theta M\Lambda$, and the sector $\Theta N\Pi$ to the sector ΘNP, and the sector $\Theta X\Sigma$ to the sector ΘXT, and also: each of the other sectors in the inscribed figure is equal to the sector having a common side, among the sectors in the circumscribed figure. Now, it is clear that all the sectors shall be equal to all the sectors; therefore the inscribed figure is equal to the figure in the area circumscribed around the area without the sector ΘAK; for this alone is not taken among the sectors in the circumscribed figure. Now, it is clear that the circumscribed figure is greater than the inscribed, by the sector $AK\Theta$, which is smaller than the given.

Corollary. *And from this it is obvious that it is possible to draw a figure around the said area, as was said, so that the circumscribed figure is greater than the area by a magnitude smaller than any given area, and again: to inscribe, so that the area, similarly, is greater than the inscribed figure by a magnitude smaller than any given area.*

In the final proposition we consider here, Archimedes proves that the area bounded by the first rotation of the spiral is one-third of the area of the first circle. As usual, Archimedes proves this by contradiction, first showing that the assumption that the area is smaller than a third of the first circle is absurd and then showing that the opposite assumption is also absurd. We present only the first part of the argument, since the second is similar. A modern calculation of the area bounded by the first rotation of the spiral $r = a\theta$ is given by

$$\frac{1}{2}\int_0^{2\pi} r^2\, d\theta = \frac{1}{2}\int_0^{2\pi} a^2\theta^2\, d\theta = \frac{4}{3}a^2\pi^3.$$

Since the radius of the first circle is $2\pi a$, one-third of its area is $\frac{1}{3}\pi(2\pi a)^2 = \frac{4}{3}a^2\pi^3$, as stated.

Proposition 24. *The area contained by both the spiral drawn in the first rotation, as well as the first line among the lines at the start of the rotation, is a third part of the first circle.*

Let there be a spiral drawn during the first rotation, on which is the line $AB\Gamma\Delta E\Theta$, and let the point Θ be the start of the spiral, the line ΘA first among the lines at the start of the rotation, and the circle $AKZHI$, first circle, of which let the circle Q be a third part. It is to be proved that the mentioned area is equal to the circle Q.

For if not, it is either greater or smaller. Let it first be, if possible, smaller. So, it is possible to circumscribe around the area contained by both the spiral $AB\Gamma\Delta E\Theta$, and the line $A\Theta$ a plane figure composed of similar sectors, so that the circumscribed figure is greater than the area by a magnitude smaller than the difference by which the circle Q exceeds the mentioned area [proposition 21]. So, let it be circumscribed, and let the greatest of the sectors, of which the mentioned figure is composed, be the sector ΘAK, and the smallest of the sectors, ΘEO [figure 4.7.13]. Now, it is clear that the circumscribed figure is smaller

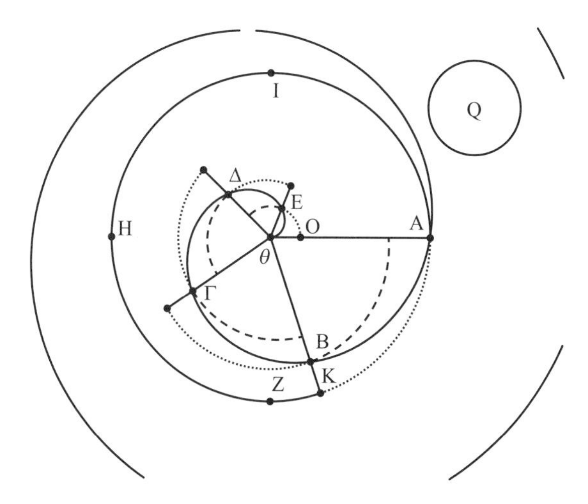

Figure 4.7.13.

than the circle Q. So, let the lines making angles at Θ be produced as far as they extend to fall on the circumference of the circle. So, there are certain lines—those falling on the spiral from Θ—exceeding each other by an equal difference [proposition 12], of which ΘA is the greatest, while ΘE is the smallest, and the smallest is equal to the difference. And there are also certain other lines, those falling on the circumference of the circle from Θ, equal to those lines falling on the spiral in multitude, while each is equal in magnitude to the greatest among the lines falling on the spiral, and similar sectors have been set up on all the lines, both on the lines exceeding each other by an equal difference as well as on the lines equal both to each other as well as to the greatest; therefore the sectors on the lines equal to the greatest are smaller than triple the sectors on the lines exceeding each other by an equal difference, for this has been proved [proposition 10, corollary]. But the sectors on the lines equal both to each other as well as to the greatest are equal to the circle $AZHI$, while the sectors on the lines exceeding each other by an equal difference are equal to the circumscribed figure; therefore the circle $AZHI$ is smaller than triple the circumscribed figure. And it is three times the circle Q; therefore the circle Q is smaller than the circumscribed figure. But it is not smaller, but greater. Therefore the area contained by both the spiral $AB\Gamma\Delta E\Theta$ as well as $A\Theta$ is not smaller than the area Q.

Archimedes then shows that the assumption that the area is greater than the circle Q also leads to a contradiction, and thus completes the proof.

4.8 Archimedes, *The Method*

The manuscript of Archimedes's *Method* was discovered in 1899 in a Greek monastery library in Constantinople. The manuscript contains several other works of Archimedes as well and is now the oldest extant manuscript of Archimedes, dating from the tenth century. The writing was, however, partially washed out in the thirteenth century and the parchment reused for a religious work. Such a reused parchment is called a *palimpsest*. Johan Ludvig Heiberg was able to inspect and photograph the manuscript in 1906 and was able to read much of the old writing, which he published in Greek in 1907. However, the palimpsest disappeared during the chaos following World War I, only to reappear in an auction in 1998. Evidently, it had been owned for much of the twentieth century by a French family, who finally decided to sell. Despite some legal challenges to the sale, it was sold by Christie's in New York for about $2 million to an anonymous buyer, who then contracted with the Walters Art Gallery in Baltimore to stabilize and preserve it. Several scholars, including Reviel Netz, have now been able to inspect the manuscript with the latest imaging devices. It turned out that Heiberg's original reading of the manuscript was relatively accurate, but a few new discoveries have been made, including some parts of the proofs of propositions 6 and 10 below, as well as some of the original diagrams.

Archimedes's aim in *The Method* was to record his method of discovery by mechanics of several important results on area and volume, most of which are rigorously proved elsewhere. The essential features of *The Method* are, first, the assumption that figures are "composed" of their indivisible cross sections; second, the balancing of cross sections of a particular figure against corresponding cross sections of a known figure, using the law of the lever; and third, the generalization of some results known in a finite case to analogous results about infinite cases. Archimedes knew that this method did not provide a rigorous proof, but, as he noted in his opening letter, it was easier for him to provide a rigorous proof once he had "acquired, by the method, some knowledge of the questions." Archimedes wrote in the introductory letter to Eratosthenes[37] that he was mainly interested in providing the proofs of two theorems on volumes. The first result, to find the volume of a segment of a cylinder cut off by a plane going through the diameter of the base circle and a point on the upper circle, is developed in proposition 10 below (as well as in propositions 9 and 11). The second result, to find the volume of the intersection of two perpendicular cylinders, is unfortunately in a part of the manuscript that has been irretrievably lost. We present here, besides the proof mentioned in proposition 10, proofs of two other results given as propositions 1 and 6, as well as the statements of several other results that are demonstrated in *The Method*. Note that Archimedes did not number the propositions, and often just continued discursively from one to the other. The numbering here is due to Netz and differs from Heiberg's original numbering, since Netz merged a few of Heiberg's separate propositions into one.

Archimedes used the phrase "section of a right-angled cone" to designate what we call a parabola. Euclid defined a cone in definition 18 of book 11 of the *Elements* as a solid generated by rotating a right triangle about one of its legs. He then classified the cones in terms of their vertex angles as right-angled (when the generating triangle is isosceles), acute-angled, or obtuse-angled. A section of such a cone is formed by cutting the cone by a plane at right angles to the generating line, the hypotenuse of the right triangle. If the cone is right-angled, the section is what is today called a parabola; if acute-angled, we get an ellipse; and if obtuse-angled, we get a hyperbola. In what

[37] Eratosthenes (ca. 276–194 BCE) was a Greek polymath who became chief librarian at the Library of Alexandria. See section 7.11.

follows, we sometimes use Archimedes's phrase, but often we replace that phrase with the modern term.

The manuscript begins with a letter to Eratosthenes, the librarian at the famous library of Alexandria.

Archimedes to Eratosthenes: Greetings!

I sent you before some of the theorems I have discovered, writing down the claims themselves, telling you to find the proofs, which I did not tell at the time. And the claims of the sent theorems were these:

1. Of the first: if a cylinder is inscribed inside a right prism having a parallelogram as a base, having the bases in the opposing parallelograms and the sides touching the remaining four planes, and, through the center of the circle which is a base of the cylinder, and through one side of the square which is in the opposite plane, a plane is drawn, the drawn plane shall cut a segment off the cylinder which is contained by two planes and a surface of a cylinder: one plane, the drawn plane, the other, that in which is the base of the cylinder, and by the surface of the cylinder between the said planes; and the said segment, cut off from the cylinder, is a sixth part of the whole prism.
2. And the claim of the other theorem is this: that if a cylinder is inscribed inside a cube, having the bases at the opposite parallelograms while touching, with its surface, the remaining four planes, and another cylinder is inscribed inside the same cube, having the bases in other parallelograms, touching, with its surface, the remaining four planes, the figure contained by the surfaces of the cylinders, which is in both cylinders, is two thirds of the whole cube.

As it happens, those theorems are different from those found before. For we have compared those figures—the conoids and the spheroids and the segments—both the figures themselves to each other, as well as with some of the cones and cylinders; but none of them was found to be equal to a solid figure contained by planes. Yet each of these figures—the one contained by two planes and the surface of a cylinder, the other by surfaces of cylinders—has been found to be equal to a solid figure contained by planes. So, writing down the proofs of these theorems, I send them to you in this roll.

And seeing that you are, as I say, serious, preeminent in a praiseworthy way in the love of wisdom, while also appreciating the contemplation of theorems in mathematics, as they may come your way, I thought it appropriate to write for you—even in the same roll—indicating the special character of a certain method, proceeding along which it shall be possible to possess starting points to enable one to see, through ideas of mechanics, some of the theorems in mathematics. And I am convinced that this is no less useful for the proof, too, of those theorems. For theorems that were first made visible to me mechanically were later proved geometrically (as seeing the theorem in the method is without a proof). For it is more feasible, having already in one's possession, through the method, a knowledge of some sort of the matters under investigation, to provide the proof, rather than investigating it, knowing nothing. For this reason one should assign not a small share of the discovery of those theorems, of which Eudoxus was the first to publish the proofs—the proof of both the cone and the pyramid, that it is a third part, the cone of the cylinder, the pyramid of the prism (having the same base and an equal height)—to Democritus, being the first to state the claim, without a proof, about the said figure.[38] And for us, the discovery of the theorem sent out now, too, came about the same as those before.

And I wanted to publish the method, writing it down, because I have made this claim about it, so that I will not appear to some to be uttering vain words, but also, at the same time, because I am convinced that it will contribute not a small service to mathematics. For I suppose that some, of those who are now or of those who will come, shall discover, through the proved method, other theorems, too, never suspected by us.

Well then, we provide first in proof form the theorem which was also the first of those made visible through mechanics that

Each segment of the section of a right-angled cone is a third again as much as a triangle having the same base and an equal height.

Then, following that, we provide in proof form each of the theorems studied through the same method. At the end of the roll we provide in proof form the proofs, put in geometrical form, of the theorems whose claims we sent you. Farewell.

Archimedes next provides a list of propositions whose results he will assume in the remainder of this work. Most of these propositions are found in Archimedes's *Planes in Equilibrium*, but a few are presumably found in works that no longer exist. These results deal with what is known today as the center of gravity, but a more accurate translation of Archimedes's words is "center of weight." Note that Archimedes does not number these results but just writes them one after the other.

1. If a magnitude is taken away from a magnitude, and the same point is the center of weight of both the whole, as well as the magnitude taken away, the same point is the center of weight of the remaining magnitude.
2. If a magnitude is taken away from a magnitude, and the same point is not the center of weight of both the whole as well as the magnitude taken away, the center of weight of the remaining magnitude is on the line joining the centers of weight of both the whole magnitude as well as the magnitude taken away, produced and taken away from it, having the line so taken to the line between the said centers of weight that ratio which the weight of the magnitude taken away has to the remaining weight of the remaining magnitude.
3. If the center of weight of however many magnitudes is on the same line, the center of weight of the magnitude composed of all magnitudes shall be on the same line, as well.
4. The center of weight of every line is the midpoint of the line.
5. The center of weight of every triangle is the point at which the lines drawn from the angles of the triangles to the midpoints of the sides cut each other.

[38] Archimedes was in error in assigning the first statement of this claim to Democritus. It seems clear that Babylonian scribes knew how to calculate the volume of a pyramid, and it is possible that Egyptian scribes did as well.

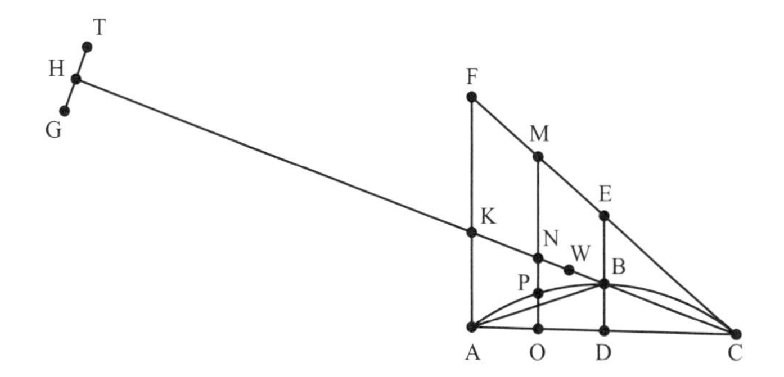

Figure 4.8.1.

6. The center of weight of every parallelogram is the point at which the diameters meet.
7. The center of weight of every circle is that point which is also the center of the circle.
8. The center of weight of every cylinder is the midpoint of the axis.
9. The center of weight of every prism is the midpoint of the axis.
10. The center of weight of every cone is on the axis, divided in such a way so that the segment towards the vertex is three times the remaining segment.
11. And we shall also make use of this theorem [in the previously written *On Conoids and Spheroids*, proposition 1]: if however many magnitudes, equal in multitude to other magnitudes, having—similarly ordered two by two—the same ratio, and the first magnitudes are in whichever ratios, whether all or some of them, to some other magnitudes, and the latter magnitudes are in the same ratios—all the first magnitudes shall have the same ratio to all the consequents, which all the latter magnitudes have to all the consequents.

Proposition 1. *Each segment of the section of a right-angled cone*[39] *is a third again as much as a triangle having the same base and an equal height.*

Let there be a segment ABC contained by a line AC and by a section of a right-angled cone ABC, and let AC be bisected at D, and let DBE be drawn parallel to the diameter, and let AB, BC be joined [figure 4.8.1]. I say that the segment ABC is a third again as much as the triangle ABC.

From the points A, C let AF be drawn parallel to DBE, while CF is drawn touching the section, and let CB be produced, and let it cut AF at K. And let KH be set equal to CK, and let CH be imagined as a balance and K its midpoint, and let MO be a chance parallel to BD. Now, since CBA is a parabola, and CF touches it and CD is drawn ordinate-wise, EB is equal to BD. For this is proved in the *Elements*.[40] So, through this and since FA, MO are parallel to ED, MN, too, is equal to NO while FK is equal to KA. And since as CA

[39] That is, a parabola.

[40] Archimedes is presumably referencing an early work on conic sections, probably by either Euclid or his contemporary, Aristaeus.

is to AO, so MO is to OP, for this lemma is proved,[41] while as CA is to AO, so CK is to KN, and CK is equal to KH, therefore as HK is to KN, so is MO to OP. And the point N is the center of weight of the line MO, since indeed MN is equal to NO, therefore if we set TG equal to OP and set H as its center of weight, TG shall balance MO, remaining in the same place, through HN being cut reciprocally to the weights TG, MO, as HK is to KN, so is MO to GT. So that K is the center of weight composed of both weights MO, GT.

And similarly also: if however many parallels to ED be drawn in the triangle FAC, they shall, remaining in the same place, balance the lines taken off them by the section, transferred around the center of weight H. And K shall be the center of weight of the weight composed of both. And since, of the lines in the triangle FAC, the triangle FAC is made up, while, of the lines in the section, taken similarly to PO, the segment ABC is made up, therefore the FAC triangle, remaining in the same place, shall balance, at the point K, the segment of the section, set around H as the center of weight, so that K is the center of weight of the weight composed of both.

Let CK be cut at W, so that CK is three times KW. Therefore the point W shall be the center of weight of the triangle AFC. For this has been proved in *Planes in Equilibrium*.[42] Now, the triangle FAC, remaining in place, shall balance, at the point K, the segment BAC set around H as the center of weight. And W is the center of weight of the triangle FAC. Therefore, as the triangle AFC is to the segment ABC, set around the center H, so is HK to WK. And HK is three times WK. Therefore the triangle AFC, too, is three times the segment ABC. And also: the triangle FAC is four times the triangle ABC, through FK being equal to KA, while AD is equal to DC. Therefore the segment ABC is a third again as much as the triangle ABC. This indeed is obvious.

So, while this has not been proved by what was now said, it did provide a certain suggestion that the conclusion is true. Hence we—seeing that it was not proved, but suspecting that the conclusion is true—discovered ourselves and set out the geometrized proof—the one published before [*Quadrature of the Parabola*, proposition 24].

We give only the statements of the next four propositions. Note that Archimedes actually says that these propositions are "studied" by his method, and not "proved." He says the same about proposition 6, but here we present his entire argument.

Proposition 2. *And that every sphere is four times the cone having the base equal to the greatest circles of the circles in the sphere, and a height equal to the radius of the sphere, and that the cylinder having the base equal to the greatest circle of the circles in the sphere, and a height equal to the diameter of the sphere is one-and-a-half times the sphere, will be studied through this method.*

Proposition 3. *And it shall be studied through this method also that, of every spheroid,*[43] *the cylinder having a base equal to the greatest circle of the circles in the spheroid, and a height equal to the axis of the spheroid, is half as much again. With this studied it is*

[41] This is straightforward to see analytically. Suppose the equation of the parabola is $y=-x^2+1$; then the tangent line at $C=(1,0)$ has the equation $y=-2x+2$. Suppose O has coordinates $(-a,0)$. Then $MO=2a+2$, $OP=-a^2+1$, $CA=2$, $AO=-a+1$. So $MO:OP=(2a+2):(1-a^2)=2:(1-a)=CA:AO$.

[42] See the discussion at the end of *Planes in Equilibrium*, 1:12.

[43] A spheroid is defined by Archimedes as the solid formed by rotating an ellipse around one of its axes.

clear that, in every spheroid cut by a plane drawn through the center, perpendicular to the axis, the half of the spheroid is twice the cone having a base the same as the segment (the half-spheroid) and the same axis.

Proposition 4. *And that every segment of a right-angled conoid*[44] *cut by a plane perpendicular to the axis is half as much again as the cone having the same base as the segment and the same axis, will be studied by the method.*

Proposition 5. *And that the center of weight of the segment of the right-angled conoid, cut by a plane perpendicular to the axis, is on the line, which is the axis of the segment, and cuts the said line in such a way that its part towards the vertex is twice the remaining segment, shall be studied.*

Proposition 6. *And it shall be studied similarly that the center of weight of a hemisphere is on a line which is an axis of the hemisphere, cut in such a way that its segment towards the vertex of the hemisphere has to the remaining segment that ratio which five has to three.*

Let there be a sphere, and let it be cut by a plane through the center, and let a segment come to be in the surface, namely, the circle $ABCD$, and let AC, BD be diameters of the circle at right angles to one another [figure 4.8.2]. Let a plane be set up from BD, perpendicular to AC, and let there be a cone having the circle around the diameter BD as base and the point A as vertex. Let BA, AD be sides of the cone, and let CA be produced. Let AH be set equal to CA, and let the line HC be imagined a balance, its center A. Let a certain line TO be drawn in the semicircle BAD, being parallel to BD, and let it cut the circumference of the semicircle at T, O and the sides of the cone at the points R, P, and AC at E. Let a plane be set up from TO perpendicular to AE. It shall make a cut in the hemisphere, a circle whose diameter is TO, and, in the cone, a cut, a circle whose diameter is RP.

And since it is: as AC is to AE, so the square on TA is to the square on AE,[45] while the squares on AE, ET are equal to the square on TA, and ER is equal to AE, therefore as AC is to AE, so are the squares on TE, ER to the square on ER. But as the squares on TE, ER are to the square on ER, so the circle with diameter TO and the circle with diameter RP are to the circle with diameter RP. And CA is equal to AH. Therefore, as HA is to AE, so the circle with diameter TO and the circle with diameter RP are to the circle with diameter RP. Therefore both circles with diameters TO, RP, remaining in their own position, balance at the point A with the circle whose diameter is RP, transposed and positioned at the point H in such a way that its center of gravity is H. Now, since the center of gravity of both circles TO, RP, remaining in their own position, is E, the circles on diameters TO, RP are to the circle with diameter RP as AH is to AE. For any other arbitrary parallel such as TO, the two circles balance the single circle moved to H, around the point A.

Since the hemisphere and the cone are filled up by the circles arising through all the parallel lines such as TO, all the circles, both in the hemisphere, and in the cone, shall balance around the point A with all the circles in the cone, transposed and positioned

[44] A right-angled conoid is defined as the solid formed by rotating a parabolic segment around its axis.

[45] By similarity, $AC:AT=AT:AE$, and the result follows.

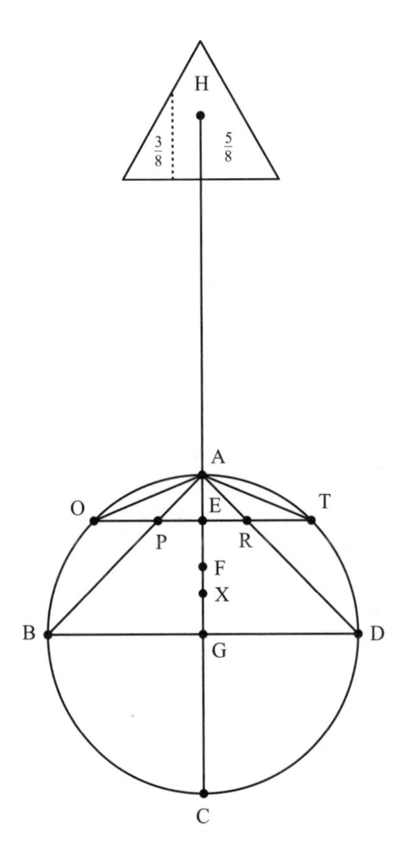

Figure 4.8.2.

at the balance at H so that the center of gravity of each of them is H. Therefore, taken together, both the hemisphere and the cone will balance around the point A with the cone transposed and positioned at the balance at H so that its center is the point H. So, let the cone be divided into two unequal parts, so that the greater has to the smaller that ratio which five has to three.[46] Now, position the smaller part of the cone so that its center of gravity is at the point H and find the point X at one fourth the axis AG of the original cone (X is one-fourth of the way from the base of the cone to its vertex; so AX is three-eighths of AC or of AH). Now the center of gravity of a cone is at one-fourth the axis.[47] Thus X is the center of gravity of the cone itself, remaining in its own position, while H is the center of gravity of the three-eighths part of it positioned at H. And it is: as HA is to AX, so the cone whose axis is AX is to the three-eighths part of the cone itself. For their ratio is the same, which eight has to three. Therefore the hemisphere, too, remaining in its own position, is in balance around the point A with the five-eighths part of the cone positioned at H.

[46] This sentence and the next two are conjectured by Netz, based on a few words he was able to read but Heiberg was not.

[47] The result that the center of gravity of a cone is on the axis at a point one-fourth of the way from the base to the vertex was assumed known by Archimedes, although it is not proved in any extant work of his.

And since the sphere is four times the cone, whose base is the circle around the diameter BD, and its vertex the point A, and the hemisphere is twice the cone, and the cone itself has to the five-eighths part of itself a ratio which eight has to five, therefore the hemisphere is to the five-eighths part of the cone as 16 is to 5. Now, let the ratio of AC to AF, too, be 16 to 5. Therefore, the center of gravity of the hemisphere shall be F.[48] And AH has to AF the ratio which 16 has to 5 and the same ratio, the ratio which 16 has to 5, the hemisphere has to the five-eighths part of the cone itself. And the axis is divided at F so that AF has to the remaining part of the axis the ratio which five has to three.

And it shall be proved similarly that the center of gravity of a spheroid cut by a plane perpendicular to the axis of the spheroid, is on the line which is the axis of the segment, the said line cut so that the part at the vertex of the spheroid has to the remaining segment of the axis that ratio which five has to three.

The statement of the following proposition is given by Netz as proposition 9. But Archimedes proceeds to give three proofs of this result, with the first using the same techniques as in the previous propositions; the second, here labeled proposition 10 and the one we present, using a related but slightly different method; and the third, labeled by Netz as proposition 11, being a rigorous proof through exhaustion, a proof we do not include here.

Proposition 10.[49] *If in a right prism with a square base a cylinder be inscribed which has its bases in the opposite squares, and its sides on the remaining planes of the prism, and if through the center of the circle which is the base of the cylinder and through one side of the square in the plane opposite to it a plane be drawn, the plane so drawn will cut off from the cylinder a segment which is bounded by two planes and the surface of the cylinder, one of the two planes being the plane which has been drawn and the other the plane in which the base of the cylinder is, and the surface being that which is between the said planes; and the segment cut off from the cylinder is one-sixth part of the whole prism* [figure 4.8.3].

Let there be a right prism having square base, and let one of its bases be the square $ABGD$, and let a cylinder be inscribed inside the prism, and let the base of the cylinder be the circle $EZHQ$, touching the square [figure 4.8.4]. Let a plane be drawn through the circle's center, and through the side, above GD, of the square plane opposite the square $ABGD$. So it shall cut, of the whole prism, another prism, which shall be a fourth part of the whole prism. This prism shall be contained by three parallelograms and two triangles opposite each other. So, let a section of a right-angled cone [i.e., a parabola] be drawn in the semicircle EZH, and let its diameter be ZK, and let the same line ZK also be that, applied on which, the lines drawn in the section are equal in square, and let some line, say MN, be drawn in the parallelogram DH, being parallel to KZ. So it shall cut the circumference of the semicircle at S and the section of the cone at L.

[48] Since the hemisphere balances at A the $\frac{5}{8}$ of the cone at H of which it is $\frac{16}{5}$, so the distance of its center of gravity from A must be $\frac{5}{16}$ of AH, which is the point F.

[49] This is proposition 13 in Heiberg's numbering, but Netz believes it is more correct to call it proposition 10.

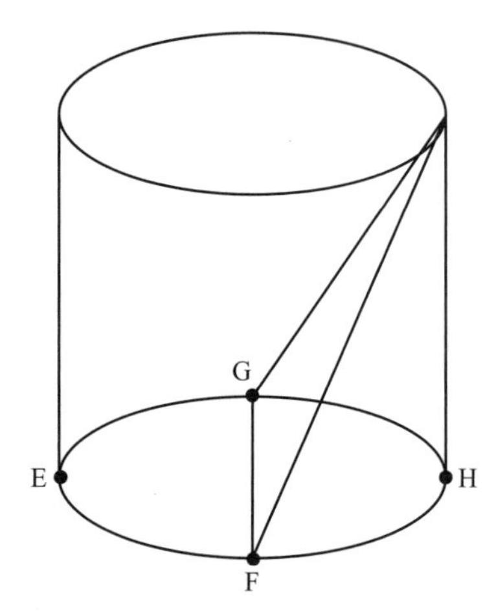

Figure 4.8.3.

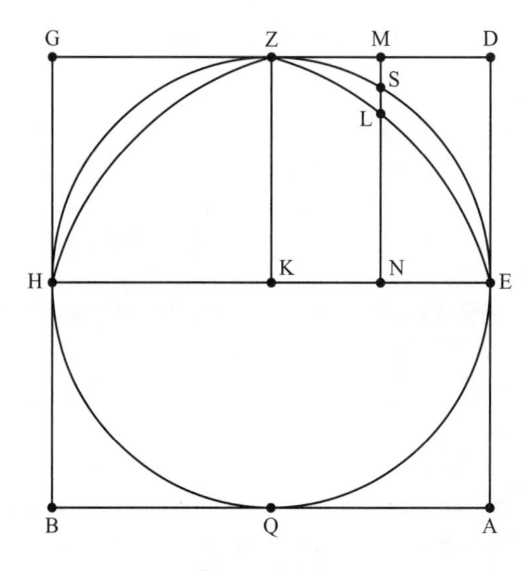

Figure 4.8.4.

And the rectangle contained by the lines MN, NL is equal to the square on NS. For this is clear.[50]

So, through this, it shall be: as MN is to NL, so the square on MN is to the square on NS.[51] And let a plane be set up on MN, perpendicular to the line EH. So the plane shall make a right angled triangle in the prism cut off from the whole prism, of which one of the sides around the right angle shall be MN, while the other shall be in the line drawn up

[50] This, in fact, is not so clear. By the defining property of the parabola, $KN^2 = ZK \cdot ML = MN \cdot ML$. Also, $MN \cdot NL + MN \cdot ML = MN^2 = KS^2 = KN^2 + NS^2$. By subtraction, $MN \cdot NL = NS^2$.

[51] Since $MN : NS = NS : NL$, we have $MN : NL = (MN : NS)(NS : NL) = (MN : NS)(MN : NS) = MN^2 : NS^2$.

from M in the plane on GD, perpendicular to the line GD, equal to the axis of the cylinder, and the hypotenuse shall be in the cutting plane itself; so it shall also make a right-angled triangle in the segment cut off from the cylinder by the plane that was drawn through EH and through the side of the square opposite GD, of which one of the sides around the right angle shall be NS, and the other shall be in the surface of the cylinder drawn up from S, perpendicular to the plane DH, and the hypotenuse shall be in the cutting plane. And the triangles are similar. And since the rectangle contained by MN, NL is equal to the square on NS (for this is obvious, as has been said), it shall be: as MN is to NL, so the square on MN is to the square on NS. But as the square on MN is to the square on NS, so the triangle on MN in the whole prism cut off is to the triangle on NS, taken in the segment cut off by the cylinder.[52] Therefore as MN is to NL, so the triangle is to the triangle. And similarly it shall be proved also that if any other line is drawn in the parallelogram DH, parallel to KZ, and a plane is set up on the drawn parallel line, perpendicular to the line EH, it shall be: as the triangle made in the prism is to the triangle in the segment cut off from the cylinder, so the line drawn in the parallelogram DH, being parallel to KZ, is to the line taken in the section of the right-angled cone HZ to the diameter EH.

Now, this parallelogram DH being filled by the lines drawn parallel to KZ, and the segment contained both by the section of the right-angled cone, and by the diameter EH, being filled by the lines in the segment, and also the prism being filled by the triangles that come to be in it, as well as the segment cut off from the cylinder, there are certain magnitudes equal to each other—the triangles in the prism; and there are other magnitudes, which are lines in the parallelogram DH, being parallel to KZ, which are both equal to each other and equal in multitude to the triangles in the prism; and those triangles, in the segment cut off, shall also be equal in multitude to the triangles that come about in the prism, and the lines drawn parallel to KZ between the section of the right-angled cone and EH, shall be equal in multitude to the lines drawn parallel to KZ in the parallelogram DH; it shall be, as well: as all the triangles in the prism are to all the triangles taken away in the segment cut off from the cylinder, so all the lines in the parallelogram DH are to all the lines between the section of the right-angled cone and the line EH.[53] And, from the triangles in the prism, is composed the prism; while, from the triangles in the segment cut off from the cylinder, is composed the segment; and, from the lines in the parallelogram DH, parallel to KZ, is composed the parallelogram DH; and, from the lines between the section of the right-angled cone and EH, is composed the segment of the parabola;[54] therefore as the prism is to the segment cut off from the cylinder, so the parallelogram DH is to the segment EZH contained by the section of the right-angled cone and by the line EH.

[52] Since the triangles are similar, the areas are as the squares on corresponding sides.

[53] For finite sets, this result is a special case of the proposition stated at the end of the introductory letter. Namely, if there are four finite sets each with k elements $A=\{a_i\}, B=\{b_i\}, C=\{c_i\}, D=\{d_i\}$. such that $a_1=a_2=\cdots=a_k$, $b_1=b_2=\cdots=b_k$, and $a_i:c_i=b_i:d_i$ for all i, then $\sum a_i:\sum c_i=\sum b_i:\sum d_i$. Archimedes here has, however, applied the result to the four infinite sets of triangles and line segments.

[54] Archimedes is here claiming that the sums of the relevant triangles are equal respectively to the prism and the segment cut off from the cylinder, while the sums of the relevant lines are equal respectively to the parallelogram and the segment of the parabola.

But the parallelogram DH is half as much again as the segment so contained by the section of the right-angled cone and by the line EH (for this has been proved in the treatises sent out previously);[55] therefore the prism, too, is half as large again as the segment taken away from the cylinder; therefore, of such parts that the segment of the cylinder is 2, the prism is three, but, of such parts that the prism is three, the whole prism around the whole cylinder is 12, through the one being 4 times the other. Therefore, of such parts that the segment of the cylinder is 2, the whole prism is 12; so that the segment cut off from the whole cylinder is a sixth part of the whole prism.

4.9 Archimedes, *On Floating Bodies*

In *On Floating Bodies*, Archimedes states and proves the law of hydrostatics and then proceeds to determine the stability of segments of a sphere and a paraboloid of revolution when they are immersed in a fluid. We first consider proposition 7, which gives the law Archimedes uses to solve the crown problem for King Hiero of Syracuse. Namely, the king had given a goldsmith a certain amount of gold with which to fashion a crown, but later suspected that the smith had stolen some of the gold and replaced it with silver. He asked Archimedes to determine whether this was in fact the case. The story is told that, after pondering the problem in his bath, Archimedes jumped out of the bath and ran through the streets shouting "Eureka!" ("I have found it" in ancient Greek). What Archimedes found was presumably proposition 7 below, with which he was able to answer the king's question. Archimedes begins his treatment of the subject with a single postulate:

Let the nature of a fluid be assumed to be such that, of its parts which lie evenly and are continuous, that which is under the lesser pressure is driven along by that under the greater pressure, and each of its parts is under pressure from the fluid which is perpendicularly above it, except when the fluid is enclosed in something and is under pressure from something else.

After showing that the surface of any fluid at rest is part of the surface of the spherical earth, Archimedes presents several theorems dealing with what happens to solids immersed in a fluid. In proposition 3, he shows that if the solid has the same density as the fluid, when immersed it will neither project above the fluid nor sink lower than the fluid's surface. On the other hand, if the solid is less dense than the fluid, then proposition 4 shows that some of the solid will project above the surface, while proposition 5 states that it will be so far immersed that the weight of the solid will be equal to the weight of the fluid displaced. The basic idea of the proof of this proposition is that since the fluid is at rest, the pressure on that part of it from the solid must be equal to the pressure exerted by the fluid displaced. Then proposition 6 states that if a solid lighter than a fluid is forcibly immersed in it, the solid will be driven upwards by a force equal to the difference between its weight and the weight of

[55] *Quadrature of the Parabola*, proposition 24.

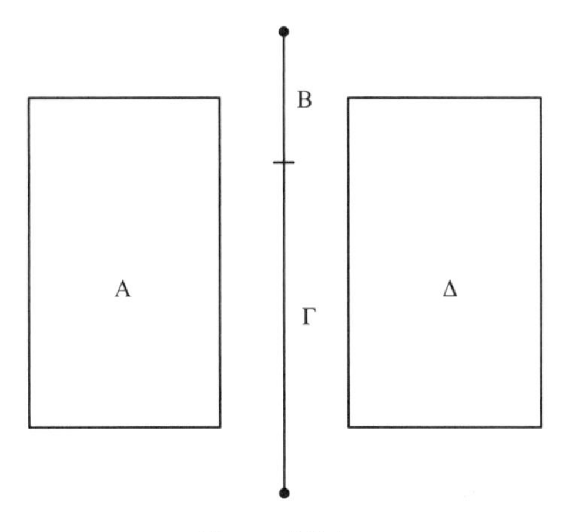

Figure 4.9.1.

the fluid displaced. Archimedes finally presents the statement and proof of one of the central theorems of the book, the result known today as the law of hydrostatics.

Proposition 7. *The magnitudes heavier than the fluid, let into the fluid, shall be carried down until they are bottomed, and they shall be lighter in the fluid by as much weight as has the part of the fluid having that much bulk, as much as is the bulk of the solid magnitude.*

Now, that it shall be carried downwards, until it is bottomed, is clear. For the parts of the fluid underneath it are pressed more than the parts lying equally with them, since the solid magnitude is assumed to be heavier than the fluid. But that they shall be lighter, as has been said, shall be proved.

Let there be a certain magnitude, the magnitude A, which is heavier than the fluid, and let the weight $B\Gamma$ be the weight of the magnitude in which is A, and let the weight B be the weight of the fluid having an equal bulk to A [figure 4.9.1]. It is to be proved that the magnitude, being in the fluid, shall have a weight equal to the weight Γ.

For let a certain magnitude be taken, the magnitude in which is the letter Δ, lighter than the fluid. Let the magnitude in which is the letter Δ be equal in weight to the weight B, and let the weight of the fluid having an equal bulk to the magnitude Δ be equal to the weight $B\Gamma$. So, with magnitudes in which are the letters $A\Delta$ being composed into the same magnitude, the magnitude composed of both taken together shall be of equal weight to the fluid; for the weight of the magnitudes taken together is equal to both weights: $B\Gamma$ as well as B, while the weight of the fluid having an equal bulk to both magnitudes is equal in weight, to the same weights. Now, the magnitudes being let into the fluid, they shall balance the fluid and be carried neither downwards nor upwards. Hence, the magnitude in which is A will be carried downwards by that much force, as much as it is pulled upwards by the magnitude in which is Δ, while the magnitude in which is Δ, since it is lighter than the fluid, will be carried upwards by that much force, as much as is the weight Γ. For it has been proved that the solid magnitudes lighter than the fluid, forced into the fluid, are carried upwards by that much force, as much as is the weight, by which the fluid having a bulk equal to the magnitude Δ is heavier than the magnitude, and the fluid having a bulk equal to the fluid Δ is heavier than the magnitude Δ by the weight Γ. Now, it is clear that the magnitude in which is A will be carried downwards by that much weight, as much as is the weight Γ.

Evangelista Torricelli in 1644 realized that Archimedes's "fluid" could also be the air. As he wrote in a letter to Michaelangelo Ricci, "We live immersed at the bottom of a sea of elemental air, which by experiment undoubtedly has weight."[56] What this implies, Torricelli showed, was that if one inverted a tube entirely full of mercury (of length about a yard) in a bowl of mercury, leaving the bottom open, the mercury in the tube will go down and then come to rest at a height of approximately 30 inches above the surface of the mercury in the bowl. That is, the weight of the column of air above the bowl balances the weight of 30 inches of mercury. Thus, air pressure is an important factor in many physical processes. In addition, Torricelli noted that when the mercury dropped down the tube, there must be a vacuum above it, given that originally there was nothing but mercury in the tube. Thus, vacuums were possible, contradicting an idea believed for centuries.

That gases have weight and that production of a vacuum is a practical possibility were two extremely important ideas that, after numerous other experiments and inventions, ultimately led to Thomas Newcomen's invention of the Newcomen engine and then James Watt's invention of the steam engine. And, of course, it was the steam engine that launched Great Britain, and ultimately the world, into the industrial revolution. According to Reviel Netz, it was the translation of Archimedes's *Floating Bodies* into Latin in 1565 that ultimately led to Torricelli, Newcomen, Watt, and the industrial revolution. And without Archimedes, this train of events would not have happened. Could there have been another route to our modern world? That is a question that we cannot answer.

The next proposition is the first of those dealing with the stability of objects in fluids. Preceding that is the following postulate:

Let it be assumed that each of the solid magnitudes carried upwards in the fluid is carried along the perpendicular drawn through its center of weight.

Proposition 8. *If a certain solid magnitude, lighter than the fluid, having a figure of a segment of a sphere, is let into the fluid in such a way that the base of the segment does not touch the fluid, the figure comes to stand upright in such a way that the axis of the segment is along a perpendicular.*

For let a certain magnitude be imagined let into the fluid as has been said, and let a plane be imagined produced through the axis of the segment and the center of the earth, and let the circle $AB\Gamma\Delta$ be the section of the surface of the fluid, while the circumference

[56] Quoted in Reviel Netz, *A New History of Greek Mathematics* (Cambridge: Cambridge University Press, 2022), 503.

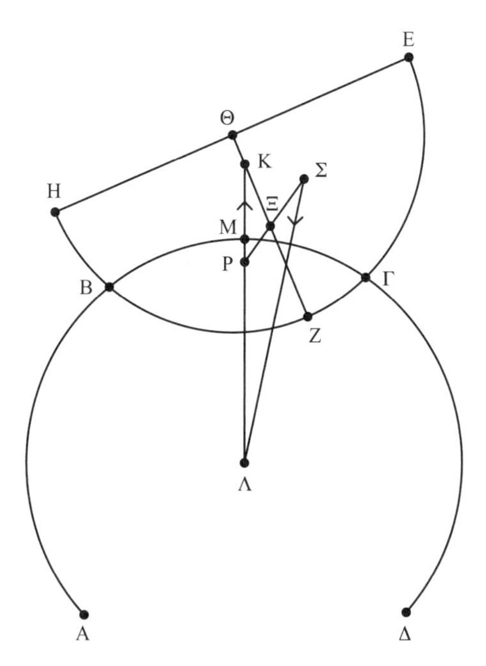

Figure 4.9.2.

$EZH\Theta$ is the section of the figure let into the fluid, and let ΘZ be the axis of the figure [figure 4.9.2]. So, the center of the sphere is on ΘZ. Let it be K and let the figure, if possible, be inclined [either pressed by something or as itself]. [Assume first that the segment is greater than a hemisphere.]

Now, it is to be proved that it will not rest but will stand up to the upright position, so that $Z\Theta$ is along a perpendicular. For since the figure is assumed to be inclined, the points Z, Θ are not along a perpendicular. So let $K\Lambda$ be drawn through K and Λ, and let Λ be assumed to be the center of the earth; so, the figure taken in the fluid by the surface of the fluid has the axis on $K\Lambda$, for if two surfaces of spheres cut each other, the segment is a circle, perpendicular to the line joining the centers of the spheres. Now, the center of weight of the figure at the circumference $BM\Gamma$, taken in the fluid, is on $K\Lambda$—let it be P—while the center of weight of the whole segment at the circumference EHZ is on $Z\Theta$—let it be Ξ. Therefore the center of weight of the remaining figure which is outside the surface of the fluid is on $P\Xi$ produced, and with a certain line taken, such as $\Sigma\Xi$, having to ΞP the same ratio which the weight of the part of the segment at the circumference $BM\Gamma$ has to the weight of the part outside the liquid. For these have been proved. So, Σ shall be the center of weight of the said figure. Now, since the weight of the figure which is outside the fluid is carried downwards along the line $\Lambda\Sigma$, while the weight in the fluid is carried upwards along the line ΛPK, it is clear that the figure shall not rest, but its parts at E will be carried downwards, those at H upwards, and will always be carried in the same direction, until $Z\Theta$ comes to be along a perpendicular. But as $Z\Theta$ comes to be along a perpendicular, the centers of weight shall be—of the part in the fluid and of the part outside—on the same perpendicular. For they shall be on $Z\Theta$. Now, they will press against each other by the same force along the same perpendicular, the one carried downwards and the other carried upwards. So that the figure shall rest. For neither shall be pushed out by the other. And the same things shall be even if the figure is a hemisphere or smaller than a hemisphere.

In book 2, Archimedes discusses the stability of various types of segments of a paraboloid of revolution when they are immersed in a fluid, where the answer depends partly on the relationship of the density of the segment to that of the fluid. We refer the interested reader to a fuller treatment in volume 3 of Netz's *The Works of Archimedes.*

4.10 *Carmen de Ponderibus et Mensuris*

This poem, written probably in the fifth century CE and now attributed to Rem(m)ius F(l)avin(n)us, is a Latin didactic poem in 208 hexameter verses. It sets out the several systems of weights and measures used in ancient Greece and Rome and then in verses 124–79 suggests a method Archimedes could have used for solving the problem of Hiero's crown. We present a translation of these verses of the poem here. Note that Vitruvius suggests a different method for solving the problem. (see section 6.14).

Now, by our talents, let us recount another tale.
Suppose one mixes gold with finest silver;
How much of it is there, and by what means can you detect it?
The first of the Syracusans[57] revealed the lofty intellect of the Master,
For they say that the Sicilian king, who once promised
To the heavenly gods that he would make them a crown of gold,
Having then discovered it stolen—for a part had been retained
When the craftsman mixed the same amount of silver with the gold,
He appealed to the talent of his countryman, one of keen mind,
Who discovered that someone had hidden the gold in finest silver,
in what had previously been pure, as it had been dedicated to the gods.
I shall teach you in a few words (listen up!) what this is all about.
With balance scales its quality can be assessed,
Silver and gold that the greedy one has purged by fire.
Place pounds of each on opposite sides to weigh them.
But then lower them into water so that the pure liquid may receive them,
Immediately, that part will incline which bears the gold;
For it is more dense than air, and at the same time thicker than water.
Mark you the yoke and fulcrum on the center hinge,
Note the interval, how far it deviates one from the other,
How many units separate the weights on the suspended wire.

[57] This description refers to Archimedes.

Set aside three drachmas. For we know
The differences between silver and gold; indeed,
A pound of the latter exceeds by three drachmas a pound of the former when submerged in water.
Then take the gold that was partly mixed with silver
And some unadulterated silver of equal weight when underwater.
Examine what has been placed on the pans: whatever of the gold is heavier,
That substance will be under the water and indicate the theft.
For if six, each will be thrice defeated by the other drachmas,
And we may say that there are only six pounds of gold;
The rest is silver, because there does not differ in weight
Silver from silver, when immersed in the liquid.
We can detect these same things in pure gold:
But if the pair be corrupted, the latter part will gain in weight.
For as many times as three unadulterated parts weigh,
You will see that there is as much corruption from the coins that are underwater
As there are pounds of silver, which fraud had mixed with gold.
Also any portion on the scale, if it remains there,
Let these coins also be pointed out to you in a similar way.
No, not without the waters to detect this theft
Does this art teach you how to test it with me.
Fashion a little bit of gold by weight,
And an equal amount of silver; then
The weight of the two equal amounts will be different, since gold is the more dense.
After this, reduce the weight you require on the pan
Of the silver, for it is already clear what we have said about the gold,
And make it a third of the heavier silver that has been found.
Then, for the gold whose corruption and theft are what you seek,
Imagine an equal amount and weight:
The other weighs less and in mass,
Is half the load: you can tell from this how much
May be hidden in gold mixed with finest silver.
For since the half we have is three-sixths,
There will be three pounds of gold; moreover, this is why,
As it happens, that treachery mixes it in with the gold.
And the reason why is so easy to learn the truth of it.

The basic idea here is that, since gold is denser than silver, a piece of gold of equal weight with a piece of silver will take up less space. Thus, when immersed, the gold will displace less water than the silver, and according to proposition 7 of *On Floating Bodies*, it will weigh more than the silver. Hence, if one takes a piece of gold of equal weight with the crown and then weighs both under water, one can easily detect if the crown has been adulterated with the silver and, in fact, the ratio of gold to silver in the crown.

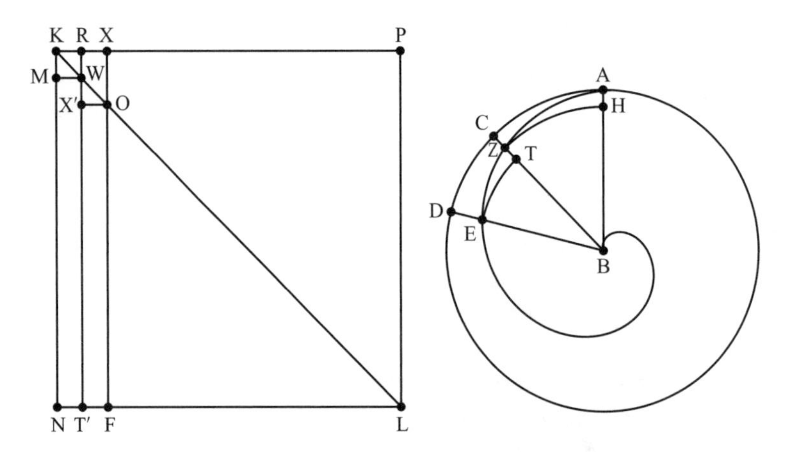

Figure 4.11.1.

4.11 Pappus, *Collection*, 4, Spiral Area

After defining the Archimedean spiral just as Archimedes did, Pappus presents as proposition 21 of book 4 a proof of the theorem giving the area bounded by one turn of the spiral that is quite different from that of Archimedes. However, some scholars believe that a version of Pappus's method was what led Archimedes to the discovery of this result in the first place.

Proposition 21. *It is shown, however, that the figure contained by the spiral and the straight line at the starting point of the rotation is the third part of the circle comprising it.*

For: Let there be given both the circle and the above-mentioned [spiral] line, and set out a rectangle $KNLP$, and cut off, on the one hand, the arc AC as a certain part of the circumference of the circle, and, on the other hand, the straight line KR as the same part of KP, and join both BC and KL, and draw the parallel RT' to KN, and the parallel WM to KP, and finally, describe the arc ZH around center B [figure 4.11.1].

Now, since as the straight line AB is to AH, i.e., as BC is to CZ, so is the whole circumference of the circle to the arc CA (for this is the principal *symptoma* of the spiral), whereas as the circumference of the circle is to the arc CA, so is PK to KR, and as PK is to KR, so is LK to KW, i.e., RT' to RW; therefore $T'R$ is to RW as BC is to CZ, also.

And *convertendo*, therefore, as the square over BC is to the square over BZ, so is the square over RT' to the square over $T'W$, also. But, on the one hand, as the square over BC is to the square over BZ, so is the sector ABC to the sector ZBH. On the other hand, as the square over RT' is to the square over $T'W$, so is the cylinder over the rectangle KT' around the axis NT' to the cylinder over the rectangle MT' around the same axis. And therefore, as the sector CBA is to the sector ZBH, so is the cylinder over the rectangle KT' around the axis NT' to the cylinder over the rectangle MT' around the same axis.

Similarly, however, when we set down, on the one hand, an arc CD equal to the arc AC, and on the other hand, RX equal to KR, and go through the same constructions,

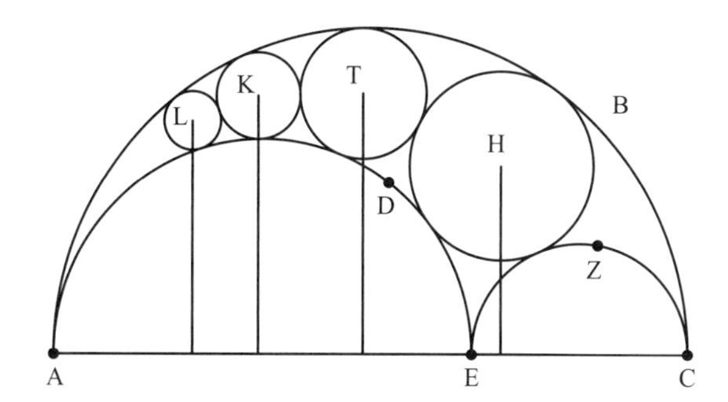

Figure 4.12.1.

the cylinder over the rectangle RF around the axis $T'F$ will be to the cylinder over the rectangle $X'F$ around the same axis as the sector DBC is to the sector EBT. Proceeding in the same manner, however, we will show that as the whole circle is to all the figures constituted out of sectors that are inscribed in the spiral taken together, so is the cylinder over the rectangle NP around the axis NL to all the figures constituted out of cylinders that are inscribed in the cone over the triangle KNL around the axis LN taken together.

And again: as the circle is to all the figures constituted out of sectors circumscribed around the spiral taken together, so is the cylinder to all the figures constituted out of cylinders circumscribed around the same cone taken together. From this result it is obvious that, as the circle is to the figure between the spiral and the straight line AB, so is the cylinder to the cone. However, the cylinder is three times the cone. Therefore, the circle is three times the said figure, also.

4.12 Pappus, *Collection*, 4, Touching Circles

In the second section of book 4, Pappus states and proves a theorem about the relative sizes of successive circles inscribed in a figure known as the *arbelos*, or "shoemaker's knife." The proof contains several lemmas and is relatively complicated, so we include only the statement of the theorem and a diagram. As is the case for the section above, many scholars believe that this result was originally discovered and proved by Archimedes.

In certain books an ancient proposition of the following sort is reported. Posit three semicircles ABC, ADE, and EZC, touching each other, and into the space between their circumferences, which is in fact called "arbelos," describe any number of circles, touching both the semicircles and each other, like the ones around the centers H, T, K, and L [figure 4.12.1].

The task is to show that the perpendicular from the center H onto AC is equal to the diameter of the circle around H, whereas the perpendicular from T is double the diameter of the circle around T, and the perpendicular from K is three times the diameter of its circle, and the perpendiculars in sequence are multiples of their respective diameters according

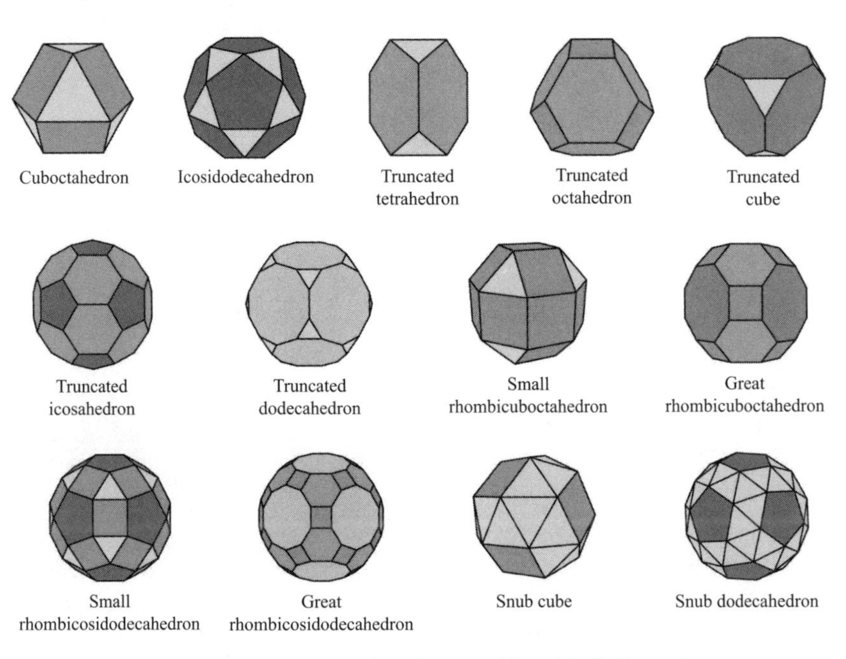

Figure 4.13.1. The thirteen Archimedean solids, with their modern names.

to the sequence of numbers exceeding one another by a unit, when the inscription of circles continues indefinitely. However, the lemmata will be proved before.

4.13 Pappus, *Collection*, 5, Archimedean Solids

In this section Pappus describes the thirteen semiregular polyhedra, whose discovery he attributes to Archimedes. The Archimedean solids are the only thirteen polyhedra that are convex, have identical vertices, and have all their faces regular polygons (although not equal as in the Platonic solids). In what follows, we give a Greek name for each solid, indicating its number of faces, but also give in brackets the name given today. The solids are pictured in figure 4.13.1.

The philosophers say that it is right that the first of the gods clothed the world with the spherical figure, chosen as the most beautiful of those that exist. They mention the natural properties of the sphere and further add that the sphere is the largest of all figures of the same surface. All that they declare to belong to the sphere is, moreover, manifest and scarcely demands demonstrations; but, as for declaring that it is larger than other figures, the philosophers do not demonstrate it and limit themselves to affirming it; and it is not easy to be convinced of this without further examination. Therefore, just as, in what precedes, we have found that the circle is the largest of the polygonal figures having the same perimeter as it, we will now try to demonstrate that the sphere is, consequently, the largest of the regular solid figures having the same area as it. We are first going to discuss a little about the very solids to which it is a question of comparing the sphere. It is indeed possible to imagine a large number of solid figures having surfaces of all kinds; but we shall have regard rather to those which appear regular. Now, these are not only the five figures which one encounters in the divine Plato, namely: the tetrahedron and the cube, the octahedron and the dodecahedron and, in the fifth place, the icosahedron, but also those thirteen which were discovered by Archimedes and are formed by equilateral and equiangular polygons, but not all identical.

First, there is the octahedron [truncated tetrahedron] composed of 4 triangles and 4 hexagons.

In addition to this one, there are three decatetrahedrons [i.e., with fourteen faces], of which the first is composed of 8 triangles and 6 squares [cuboctahedron], the second of 6 squares and 8 hexagons [truncated octahedron], and the third of 8 triangles and 6 octagons [truncated cube].

In addition to the last ones, there are two icohexahedrons [i.e., with 26 faces] of which the first is composed of 8 triangles and 18 squares [small rhombicuboctahedron], and the second of 12 squares, 8 hexagons, and 6 octagons [great rhombicuboctahedron].

In addition to these, there are three triacontadohedrons [i.e., with 32 faces], of which the first is composed of 20 triangles and 12 pentagons [icosidodecahedron], the second of 12 pentagons and 20 hexagons [truncated icosahedron], and the third of 20 triangles and 12 decagons [truncated dodecahedron].

In addition to these there is the triacontaoctahedron [i.e., with 38 faces] composed of 32 triangles and 6 squares [snub cube].

In addition to these, there are two hexacontadohedrons [i.e., with 62 faces], of which the first is composed of 20 triangles, 30 squares, and 12 pentagons [small rhombicosidodecahedron] and the second of 30 squares, 20 hexagons, and 12 decagons [great rhombicosidodecahedron].

In addition to these, the last one is the ennecontadohedron [i.e., with 92 faces] composed of 80 triangles and 12 pentagons [snub dodecahedron].

We will recognize in the following way how many angles and edges each of these thirteen polyhedral figures has.

Indeed, if, for the polyhedra whose solid angles are formed by three plane angles, we simply count the plane angles that all the bases of the polyhedron have, it is obvious that the number of solid angles is one-third of the number obtained; while for polyhedra whose solid angles are formed by four plane angles, if we count all the plane angles possessed by the bases of the polyhedron, the number of solid angles of this polyhedron is a quarter of the number obtained. Finally, for polyhedra whose solid angles are formed by five plane angles, the number of solid angles is a fifth the number of plane angles.

On the other hand, we find in the following way the number of edges that each of the polyhedra has: If we count all the edges possessed by the plane polygons that bound the polyhedra, their number will obviously be equal to the number of plane angles. But, as each of the edges is common to two faces of the polygon, it is obvious that the number of edges of the polyhedron is half this number.

For example, since the first of the thirteen non-homogeneous polyhedra is composed of 4 triangles and 4 hexagons, it has 12 solid angles and 8 faces; since the angles of the four triangles total 12 and the edges [of these triangles] are also 12, and since the angles of the four hexagons total 24 and the edges also 24, therefore the total number of edges is 36, the number of solid angles [vertices] is necessarily one-third of that number, since each of these solid angles of the polyhedron is formed from three plane angles, and since the number of edges is half of the total number, that is, of 36, there will be 18 edges.

Next, the first of the decatetrahedrons is formed from 8 triangles and 6 squares; thus there are 12 vertices (since each of the solid angles is formed from four plane angles), and 24 edges. The second of the decatetrahedrons is composed of 8 hexagons and 6 squares, so it has 24 vertices (since each of the solid angles is formed from three plane angles) and there are 36 edges. Finally, the third of the decatetrahedrons is composed of 8 triangles and 6 octagons, so it has 24 vertices and 36 edges.

Next, the first of the icohexahedrons is composed of 8 triangles and 18 squares, and has 24 vertices and 48 edges; the second of the icohexahedrons is composed of 12 squares, 8 hexagons and 6 octagons, so has 48 vertices and 72 edges. Next, the first of the triacontadohedrons, being composed of 20 triangles and 12 pentagons, has 30 vertices and 60 edges; the second of these triacontadohedrons, being composed of 12 pentagons and 20 hexagons, has 60 vertices and 90 edges; and the third of the triacontadohedrons, being composed of 20 triangles and 12 decagons, has 60 vertices and 90 edges.

Next, the triacontaoctahedron, being composed of 32 triangles and 6 squares, has 24 vertices and 60 edges. Then, the first of the hexacontadohedrons, being composed of 20 triangles, 30 squares, and 12 pentagons, has 60 vertices and 120 edges; the other one of these polyhedra is composed of 30 squares, 20 hexagons, and 12 decagons, has 120 vertices and 180 edges. Finally, the ennecontadohedron, being composed of 80 triangles and 12 pentagons, has 60 vertices and 150 edges.

At the end of this section Pappus shows that the volume of the sphere is greater than that of any regular polyhedron of the same surface area.

We will neglect for the moment these thirteen figures included under the unequal and dissimilar polygons because they are less regular, and it is appropriate to compare the sphere with the five figures which we have named; for these being formed with equal and similar planes, they are the only ones to have equal solid angles and are, for that very reason, more regular than the others. Besides, it has been shown by Euclid and others that it is impossible to find other figures besides these five that are formed from equilateral and similar polygons.

We will therefore compare these latter polyhedra with the sphere.

Proposition 18: *Let a sphere with its center be given, and let one of these five figures have surface area equal to that of the sphere. I say that the sphere is greater [in volume].*

If fact, imagine a sphere inscribed in the polyhedron so that it is tangent to the planes that form the polyhedron; it follows that the surface of the polyhedron is greater than the surface of the inscribed sphere, since it surrounds it. But, the surface of the polyhedron is equal to the surface of a sphere A; it follows that the surface of the sphere A is also greater than the surface of the sphere inscribed in the polyhedron, and the radius of sphere A is therefore greater than the radius of the inscribed sphere. But the surface of sphere A is equal to the surface of the polyhedron. Thus the cone whose base is a circle equal to the surface of sphere A [and whose height is equal to the radius of sphere A] is greater than the pyramid whose base is the rectilinear figure equal to the surface of the polyhedron, and whose height is equal to the radius of the sphere inscribed in the polyhedron. But, this cone is equal to the sphere A [in volume], in virtue of results demonstrated by Archimedes in the book *On the Sphere and the Cylinder* and other lemmas that we have stated earlier. Therefore this pyramid is equal to the polyhedron. As a consequence, the sphere A is greater than the polyhedron.

SOURCES, CHAPTER 4

4.1 This translation is from Polybius, *The Complete Histories of Polybius*, ed. and trans. W. R. Paton (Digireads.com, 2014).

4.2 This translation is taken from Ivor Thomas, *Greek Mathematical Works*, Vol. 1, *Thales to Euclid* (Cambridge, MA: Harvard University Press, 1980), pp. 317–33.

4.3 The translation is adapted from an as yet unpublished translation of *Quadrature of the Parabola* by Reviel Netz and is used with thanks.

4.4 The translation of *On the Sphere and Cylinder I* has been taken, in general, from Netz, *The Works of Archimedes*, Vol. 1, *The Two Books On the Sphere and the Cylinder*, ed. and trans. Reviel Netz (Cambridge: Cambridge University Press, 1980). In certain passages, though, we have used Ivor Thomas, *Greek Mathematical Works, Vol. 2, Aristarchus to Pappus of Alexandria* (Cambridge, MA: Harvard University Press, 1980).

4.5 The translation of *On the Sphere and Cylinder II* has been taken, in general, from Archimedes, *The Works of Archimedes*, vol. 1. In certain passages, though, we have used Ivor Thomas, *Greek Mathematical Works*, vol. 2. For more information on problem 4 of *On the Sphere and Cylinder II*, especially on how the geometrical problem of Archimedes was eventually transformed into an algebraic equation in medieval Islam, see Reviel Netz, *The Transformation of Mathematics in the Early Mediterranean World: From Problems to Equations* (Cambridge: Cambridge University Press, 2004).

4.6 This translation is adapted from an as yet unpublished translation of *Planes in Equilibrium* by Reviel Netz and is used with thanks.

4.7 This translation is taken from Netz, *The Works of Archimedes*, vol. 2.

4.8 This translation of *The Method* is adapted from an as yet unpublished translation by Reviel Netz and is used with thanks.

4.9 The translation from *On Floating Bodies* is adapted from an as yet unpublished translation by Reviel Netz and is used with thanks.

4.10 The translation of the *Carmen de Ponderibus et Mensuris* is by Daniel Otero from the original Latin in Frederick Hultsch, ed., *Metrologicorum Scriptorum Reliquae*, vol. 2 (Leipzig: Teubner, 1866), and is used with thanks.

4.11 The translation of this section of Pappus's *Collection*, book 4, is by Heike Sefrin-Weis in Pappus, *Pappus of Alexandria: Book 4 of the Collection* (London: Springer, 2010), 119–23.

4.12 The translation of this section of Pappus's *Collection*, book 4, is from Pappus, *Pappus of Alexandria*, 103–4.

4.13 The translation of this section of Pappus's *Collection*, book 5, is by Victor Katz from the French in Paul Ver Eecke, *Pappus D'Alexandrie: Collection Mathématique*, vol. 1 (Paris: Desclée de Brouwer, 1933), 272–76.

5

Apollonius and Other Responses to Archimedes

This chapter deals with the work of Apollonius and others on conic sections and also with the construction of other nonplanar curves. Apollonius was not the first to consider conic sections—he was preceded by Euclid and Aristaeus in writing texts on the subject—but his work was considered the most comprehensive text, so earlier ones were no longer copied and today are no longer extant. Apollonius's *Conics* was a work in eight books, but only four were preserved in Greek. Three others exist today only in an Arabic translation, while the eighth book seems to have disappeared entirely. We present here excerpts from the first four books, along with the commentary on these books by Eutocius (shown in a different typeface). We present excerpts from the Arabic books without commentary, as Eutocius himself evidently had no access to those books.

This chapter also contains excerpts from works dealing with other special curves invented by Greek mathematicians, including the conchoid of Nicomedes, the quadratrix of Hippias and Dinostratus, and the cissoid of Diocles. These curves were used to solve the problems of finding two mean proportionals between two given line segments and of trisecting an angle. We also include how conic sections were used to accomplish these tasks, especially as described in several selections from Pappus's *Collection*. And Pappus further explains how Greek mathematicians used analysis and synthesis to solve geometrical problems. Apollonius's *On Cutting Off a Ratio* provides a good example of the use of these techniques. We conclude with the work of Serenus, who showed that a section of a cylinder was in fact a conic section.

5.1 Apollonius, *Conics*, 1

Not much is known about Apollonius (240–190 BCE) other than what can be gleaned from the letters opening most of the books of the *Conics*, his masterpiece in eight books. He was born in Perga, a town in southern Asia Minor, and presumably went to Alexandria as a young man to study with successors of Euclid. He then remained in Alexandria for most of his life. His only works that have survived, besides seven of the eight books of the *Conics*, are *On the Cutting Off of a Ratio*, excerpted in section 5.4, and a work on expressing large numbers, which is excerpted from Pappus's *Collection* in section 2.14 of this volume. But many more works are attested, mainly in Pappus's *Treasury of Analysis*, book 7 of his *Collection*. In particular, we do know that Apollonius was famous for work in astronomy, including the development of the hypotheses of epicycles and eccentrics that formed the basis of Ptolemy's astronomical models (see section 7.6, book 3:3).

Besides the commentaries of Eutocius, who edited the first four books of the *Conics* and thereby preserved them in Greek, we also include the preface to the *Conics* from the Arabic translation by the Banū Mūsā in the ninth century.[1] These scholars were instrumental in preserving the fifth through the seventh books in Arabic. The preface, in fact, details aspects of the history of the manuscripts after they were originally prepared by Apollonius.

Book 1

Apollonius's Opening Letter

Apollonius to Eudemus:[2] Greetings!

If you are restored in body, and other things go with you to your mind, well and good; and we too fare pretty well. At the time I was with you in Pergamum, I observed you were quite eager to be kept informed of the work I was doing in conics. And so I have sent you this first book revised, and we shall dispatch the others when we are satisfied with them. For I don't believe you have forgotten hearing from me how I worked out the plan for these conics at the request of Naucrates,[3] the geometer, at the time he was with us in Alexandria lecturing, and how on arranging them in eight books we immediately communicated them in great haste because of his near departure, not revising them but putting down whatever came to us with the intention of a final going over. And so finding now the occasion of correcting them, one book after another, we publish them. And since it happened that some others among those frequenting us got acquainted with the first and second books before the revision, don't be surprised if you come upon them in a different form.

Of the eight books the first four belong to a course in the elements. The first book contains the generation of the three sections and of the opposite branches, and the principal properties in them worked out more fully and universally than in the writings of others. The second book contains properties having to do with the diameters and axes and also the asymptotes, and other things of a general and necessary use for limits of possibility. And what I call diameters and what I call axes you will know from this book. The third book contains many incredible theorems for use of the construction of solid loci and for limits of possibility of which the greatest part and the most beautiful are new. And when we had grasped these, we know that the three-line and four-line locus had not been

[1] The Banū Mūsā were the three sons of Mūsā ibn Shākir who lived in the ninth century and were associated with the House of Wisdom in Baghdad.

[2] Eudemus of Pergamum lived in the third century BCE and was the teacher of Philonides. He is not the Eudemus credited with writing a history of geometry a century earlier.

[3] Nothing more is known of Naucrates.

constructed by Euclid, but only a chance part of it and that not very happily. For it was not possible for this construction to be completed without the additional things found by us. The fourth book shows in how many ways the sections of a cone intersect with each other and with the circumference of a circle, and contains other things in addition, none of which has been written up by our predecessors, that is, in how many points the section of a cone or the circumference of a circle and the opposite branches meet the opposite branches. The rest of the books are fuller in treatment. For there is one dealing more fully with maxima and minima, and one with equal and similar sections of a cone, and one with limiting theorems, and one with determinate conic problems. And so indeed, with all of them published, those happening upon them can judge them as they see fit. Good-bye.

Eutocius: Introduction to the Commentary

Apollonius the geometer, my dear friend Anthemius,[4] was born in Perga in Pamphylia during the reign of Ptolemy Euergetes,[5] as chronicled by Heraklius, who wrote *The Life of Archimedes*. He also says that Archimedes first thought of the conic theorems; but that Apollonius, having found them unpublished by Archimedes, made them his own. But he is not correct, in my opinion, at least. For both Archimedes seems to recall in many passages the *Elements of Conics* as more ancient, and Apollonius does not write his own thoughts: for he would not have said that he had worked these things out in full and more generally than the writings of others. But the very thing which Geminus[6] says is true: that the ancients, defining a cone as the revolution of the right triangle, with one side about the right angle remaining fixed, naturally assumed that all cones are right, and that one section occurs in each: in the right-angled[7] cone, what is now called the parabola, in the obtuse-angled, the hyperbola, and in the acute-angled, the ellipse: and it is possible among them[8] to find the so-called sections. So just as the ancients theorized on the fact that in any triangle there are two right angles—first in the equilateral, in turn the isosceles, and last the scalene—their descendants proved a general theorem as follows: in every triangle, the three internal angles are equal to two right angles; likewise in the case of the sections of the cone. For the thing called a section of a right-angled cone they viewed only by means of a plane cutting orthogonal to one side of the cone, and they showed that the section of the obtuse cone occurs in an obtuse cone, and that of the acute cone in an acute one: likewise in all cones bringing planes orthogonal to one side of the cone: and he [Geminus] shows also the ancient names of the curves themselves. But later, Apollonius of

[4] Anthemius of Tralles (474–533) was a geometer and architect who worked in Constantinople. With Isidore of Miletus, he designed the Hagia Sophia church for Justinian I.

[5] Ptolemy Euergetes was the third king in the Ptolemaic dynasty in Egypt, reigning from 246 to 222 BCE.

[6] Geminus lived in the first century BCE. His mathematical works do not survive, except in extracts in the works of later mathematicians.

[7] It is important that we distinguish, as does Apollonius, that a cone is "right" when it is formed from the revolution of a right triangle. The angle at the vertex of the cone, which subtends the rotating leg, determines if the cone is right-angled, obtuse-angled, or acute-angled.

[8] That is, the different cones.

Perga theorized somewhat generally, that in every cone, both the right and the oblique, all the sections are according to a different application of the cutting plane to the cone; and his contemporaries, having marveled at the wonder of the conic theorems shown by him, called him a great geometer. So Geminus says these things in the sixth book of *The Theory of Mathematics.* But what he says, we will make clear in the diagrams below.[9]

Preface to the Conics of the Banū Mūsā

In the name of God, the merciful, the forgiving. I have success except through God. The first book of the treatise of Apollonius on Conics [in] the revision of the Banū Mūsā and the version of Hilāl b. Abī Hilāl al-Ḥimṣī.

Truly the position of the science of the sections occurring in cones and of the figures and lines occurring in them is in the highest rank in the science of geometry. The ancients used to call the propositions on conic sections "the amazing propositions" and they were of the opinion that whoever has reached the point in the science of geometry where he has mastered the understanding of this science has attained the highest rank in the science of geometry. The ancient students of the science of geometry never ceased to be interested in discovering this science, and to labor in the study of it, and to write down what they understood of it, little by little, in their books, until this process reached Apollonius. This man was an Alexandrian: he was interested in this science, and was a man outstanding in the science of geometry, and a master of it. So he composed on that subject a treatise in eight books in which he collected the advances in this science made by his predecessors and added what he himself was responsible for discovering. But then this treatise was corrupted and the mistakes in it multiplied in the course of time through the succession of people copying it, one from another. There were two causes for its corruption: one was the cause common to all books which go through a succession of hands in copying, due to the negligence of those who copy them in correcting the copies and comparing them [with the original], and to the differences between books and the obliteration of what is in them before they are renewed by copying. The other cause is peculiar to this treatise and treatises like it, but to no others: for this treatise is an obscure one, the understanding of which is difficult, and only a few people have control of it. But ease of understanding a book is helpful in emending it when the need for that arises. Furthermore it is a work the copying of which is long and difficult, and the correction of which is a labor.

So for the reasons we have described corruption took place in this work after Apollonius until there appeared in Ascalon a man, one of the geometers, called Eutocius. He was outstanding in the science of geometry, and there are books which he has composed which bear witness to his powers. So this man, when he realized to what extent corruption had overcome this treatise, assembled for it what he could of the copies found in his time. Hence it was possible for him, through the copies he

[9] The diagrams are those associated with propositions 11, 12, and 13 of the *Conics*, book 1.

had assembled and his prowess in the science of geometry, to restore the first four books of this treatise. But in doing so he followed the method of one who does not merely seek to report his restorations on the basis of what Apollonius wrote: rather he collected [manuscripts], preferred [readings], and employed his intelligence in what he could not correct in reporting the exact words of Apollonius on the topic, until he discovered proof for it.

Thus the investigators of the science of conics confined themselves, after Eutocius, to reading the four books which he had corrected; this is in accord with what Galen says in his censure of the geometers of his time in the work *Water, Air and Places*,[10] which is evidence for the small number of those geometers whose mind aspired to investigate the science of conics at that time, to say nothing of those who lived after Eutocius.

But as for the people of our time, there are few of the geometers among them who possess understanding of the treatise of Euclid on geometry, not to mention what is beyond that. Indeed some of them, in the feebleness of their understanding, have failed to comprehend the beginning of Euclid's work, to say nothing of what is after it, and have replaced Euclid's words there with words of the utmost stupidity and incorrectness. There were some among them who went so far as to compose geometrical propositions which they proved—in their own opinion—with proofs contradicting the proofs of Euclid: to the extent that some of them declared, in their proofs, that a cone is half the cylinder.[11] Some of this class of people whom we have described recognized their error after an interval of time had elapsed, and renounced it. But some of them continued in their error and persisted in it: their books are [still] to be found in our time, and hence we have refrained from reporting their error.

Now we had got hold of seven books of the eight books which Apollonius composed on conics in the form in which he had composed them. So we wanted to translate and understand them, but the task proved impossible for us because of the excessive number of errors which had accrued in that treatise for the reasons we have described. But we persisted in that for a long time. Then al-Ḥasan b. Mūsā,[12] through his prowess and superiority in the science of geometry, succeeded in the theory of the science of the section of the cylinder,[13] when it is cut by a plane not parallel to its base: the circumference of [that] section is a closed curve. He discovered the theory of it and the theory of the basic *symptomata* of the diameters, axes and chords which occur in it. Thus he discovered how to measure it and was able to make that an introduction to and a way of studying the science of conic sections, for he thought that that would be easier for him in his investigation,

[10] This is in Galen's commentary to Hippocrates's *On Airs, Waters, and Places* (of which only the Arabic translation is extant).

[11] It has long been known that the volume of a cone is one-third the volume of a cylinder of the same base and height.

[12] One of the three sons of Mūsā ibn Shākir that comprised the Banū Mūsā.

[13] This is essentially the opposite of what happened in the Greek tradition, wherein the section of the cone was studied first, by Apollonius and his predecessors, and the section of the cylinder was studied later, by Serenus. The nature of the Banū Mūsā's introduction here strongly suggests they were unaware of Serenus's text.

and more like the [correct] way of proceeding in order in this science. Then he investigated the science of the conic section of a cylinder[14] when the line bounding it is a closed curve, and found that shape of the section of the cylinder, the theory of which he had discovered, is identical with the shape of a section of the cone. And he discovered the proof of the fact that to every section falling within the cylinder in the way we described[15] there corresponds some cone in which falls an identical section, and that to every elliptical section of this type there corresponds some cylinder which contains the equivalent section. So al-Ḥasan at that point composed a treatise on what he had discovered of that science and died—may God have mercy on him.

Then Aḥmad b. Mūsā[16] managed to travel to Syria as the one in charge of its post; for he intended to search for manuscripts of that work [Apollonius's *Conics*] in the hope that he would collect thence material for it which would enable him to correct it. But that proved impossible for him. But he got hold of one manuscript of the four books of Apollonius's treatise which Eutocius had restored, although it too had accumulated errors after Eutocius, for the reasons we described. So when Aḥmad got this manuscript, he began to comment on the treatise, beginning with the first four books which Eutocius had restored, since he found that the errors in these were fewer than those in the original of Apollonius's treatise.[17] So he expended toil and hardship on understanding them, until he was done with them. Then his departure from Syria to Iraq took place, and when he returned to Iraq, he went back to commenting on the rest of the seven books which had come down to us from the original treatise of Apollonius. We have already described the state of corruption into which this treatise had fallen due to the multiplicity of errors. However Aḥmad, due to his understanding of the four books which Eutocius restored, had acquired control over the understanding of the rest of the treatise, experience with it, and understanding of the methods employed by Apollonius and the basic principles he set down. Thus, by these means, he was enabled to understand the three remaining books of the seven, so that he completely comprehended them. And he did something for this treatise which is of great use in facilitating the comprehension of it for anyone who wants to read it—something which Apollonius did not do when he composed it, nor Eutocius when he restored it. Namely, he examined every premise which one needs for the proof of each of the propositions, mentioned it explicitly in the place where one needs it, and described its position in the treatise.

In what follows, we include selected definitions and theorems from book 1 of the *Conics*, some of the theorems without proofs, interspersed with commentary of Eutocius (shown in a different typeface). As Apollonius noted in his opening letter, book 1 contains the basic results on the generation of the parabola, the ellipse, and the hyperbola.

[14] Presumably, this means the ellipse, that conic section that is also a section of the cylinder.

[15] That is, not parallel to the base of the cylinder.

[16] Another of the three sons of Mūsā ibn Shākir.

[17] Presumably, this means in the copy they had found earlier.

First Definitions

1. If from a point a straight line is joined to the circumference of a circle which is not in the same plane with the point, and the line is produced in both directions, and if, with the point remaining fixed, the straight line being rotated about the circumference of the circle returns to the same place from which it began, then the generated surface composed of the two surfaces lying vertically opposite one another, each of which increases indefinitely as the generating straight line is produced indefinitely, I call a *conic surface*, and I call the fixed point the *vertex*, and the straight line drawn from the vertex to the center of the circle I call the *axis*.
2. And the figure contained by the circle and by the conic surface between the vertex and the circumference of the circle I call a *cone*, and the point which is also the vertex of the surface I call the *vertex* of the cone, and the straight line drawn from the vertex to the center of the circle I call the *axis*, and the circle I call the *base* of the cone.
3. I call *right cones* those having axes perpendicular to their bases, and I call *oblique cones* those not having axes perpendicular to their bases.
4. Of any curved line which is in one plane, I call that straight line the *diameter* which, drawn from the curved line, bisects all straight lines drawn to this curved line parallel to some straight line; and I call the end of the diameter situated on the curved line the *vertex* of the curved line, and I say that each of these parallels is drawn *ordinatewise* to the diameter.
5. Likewise, of any two curved lines lying in one plane, I call that straight line the *transverse diameter* which cuts the two curved lines and bisects all the straight lines drawn to either of the curved lines parallel to some straight line; and I call the end of the transverse diameter situated on the curved lines the *vertices* of the curved line; and I call that straight line the *upright diameter* which, lying between the two curved lines, bisects all the straight lines intercepted between the curved lines and drawn parallel to some straight line; and I say that each of the parallels is drawn ordinatewise to the transverse or upright diameter.
6. The two straight lines, each of which, being a diameter, bisects the straight lines parallel to the other, I call the *conjugate diameters* of a curved line and of two curved lines.
7. And I call that straight line the *axis* of a curved line and of two curved lines which being a diameter of the curved line or lines cuts the parallel straight lines at right angles.
8. And I call these straight lines the *conjugate axes* of a curved line and of two curved lines which, being conjugate diameters, cut the straight lines parallel to each other at right angles.

Eutocius's Commentary on the Definitions

"If from a point a straight line is joined to the circumference of a circle," etc.[18] **For let the circle be AB, the center of which be C, and let there be some point D not**

[18] Eutocius is quoting Apollonius's definition of a conic surface above.

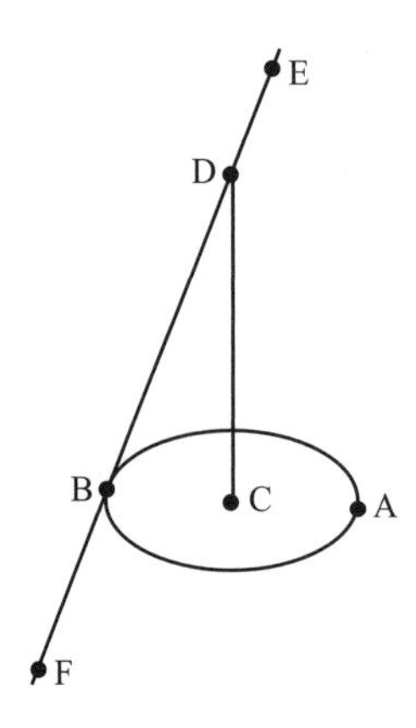

Figure 5.1.1.

in the plane of the circle, and let DB, having been joined, be projected to infinity in both directions as towards E, F [figure 5.1.1]. Indeed, if, with D being fixed, DB be carried around, until B, having been carried around upon the perimeter of the circle AB, is restored to its original position from which it began to be revolved, it will generate a certain surface, which is comprised of two surfaces being joined to one another at D, which he calls a *conic surface*. But he says, that also it extends to infinity, on account of the fact that the straight line drawing it—for instance, DB—is projected infinitely. And he calls D the *vertex* of the surface, and DC the *axis*.

And he calls a *cone* the shape which is bounded by both the circle AB and the surface, which the straight line DB alone draws, and calls D the *vertex* of the cone, and DC the *axis*, and the circle AB the *base*.

And if, on the one hand, DC is orthogonal to the circle AB,[19] he calls the cone *right*; but if, on the other hand, it is not orthogonal, he calls the cone *oblique*; but an oblique cone will be generated, whenever upon taking the circle, we set up a straight line from the center itself not orthogonal to the plane of the circle, but from the raised point of the set-up line we join a straight line to the circle, and rotate the joined straight line around the circle, with the point at the point on the set-up straight line remaining fixed: for the produced shape will be an oblique cone.

But it is manifest that the rotating straight line becomes greater and smaller during this rotation, and in certain locations also will be equal at one point and another point of the circle. It is proved thus: if straight lines be drawn from the vertex to the base of an oblique cone, of all the straight lines being drawn from the vertex to the base, one is least, and one is greatest; but two alone are equal, one on each side of the least and the greatest; but always the one nearer to the least is lesser than the one farther. Let there be an oblique cone, whose base is the circle ABC, and whose vertex is the point D. The line drawn perpendicular from the vertex of the oblique cone to the underlying plane[20] will fall either at the perimeter of the

[19] That is, to the plane of the circle.

[20] The plane of the base circle.

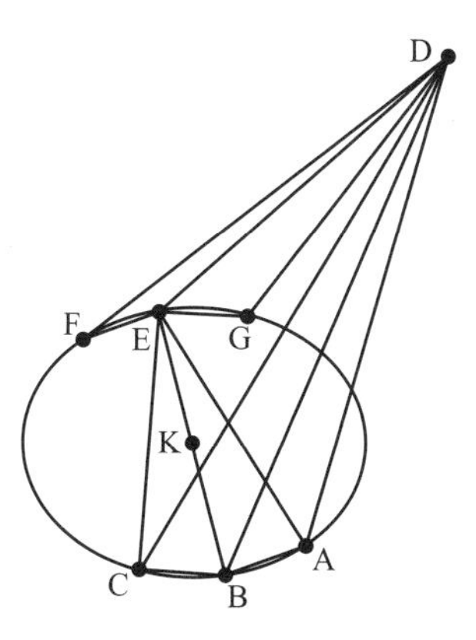

Figure 5.1.2.

circle $ABCFG$, or outside it, or inside it. Let DE first fall at the perimeter [figure 5.1.2], and let the center of the circle be taken, and let it be K, and let EK from E to K be joined, and projected to B, and let BD be joined, and let there be taken two equal arcs EF, EG, one on each side of E, and on each side of B as AB, BC, and let EF, EG, DF, DG, EA, EC, AB, BC, DA, and DC be joined. So since EF equals EG (for they are subtended by equal arcs), but also D is common and perpendicular, therefore base(DF) equals base(DG). Again, since arc(AB) equals arc(BC), and the diameter of the circle is BE, the remainder arc EFC equals the remainder arc EGA, so also AE equals EC. But also ED is common and perpendicular: therefore the base DA is equal to the base DC. And likewise, all those equally distant from either DE or DB will be shown to be equal. Again, since of the triangle DEF, the angle under DEF is a right angle, DF is greater than DE. And again, since the straight line EA is greater than the straight line EF, and since the arc EFA is greater than the arc EF, but also DE is common and orthogonal, DF is therefore less than DA. Also on account of the same things, DA is less than DB. So since DE was shown to be less than DF, DF less than DA, and DA less than DB; DE is the smallest, DB is largest, but always the one nearer to DE is less than the one farther.

But indeed let the perpendicular DE fall outside of the circle $ABCGF$ [figure 5.1.3], and let again the center of the circle K be taken, and let EK be joined and projected to B, and let DB, DH be joined, and let two equal arcs HF, HG be taken, one on each side of H, and let two equal arcs AB, BC be taken, one on each side of B, and let EF, EG, FK, GK, DF, DG, AB, BC, KA, KC, DK, DA, AC be joined. So since the arc HF is equal to the arc HG, the angle under HKF is therefore also equal to the angle under HKG. So since the straight line FK is equal to the straight line KG—for they are both radii—but KE is common, and the angle under FKE is common to the angle under GKE, and the base ZE is equal to the base GE. So since the straight line HE is equal to the straight line GE, and ED is common and orthogonal, the base DF is therefore equal to the base DG. Again, since the arc BA is equal to the arc BC, the angle under AKB is also therefore equal to the angle under CKB, so that also the remaining angle under AKE is equal to the remaining angle under

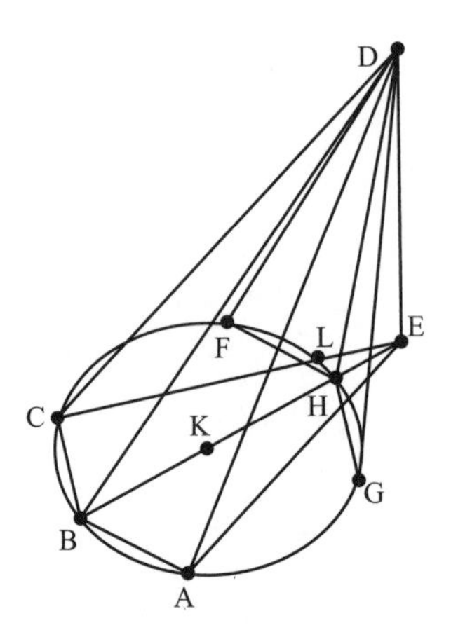

Figure 5.1.3.

CKE. So since the straight line AK is equal to the straight like CK–for they are both radii–but KE is common, the two are equal to the two, and the angle under AKE is equal to the angle under CKE: and the base AE is therefore equal to the base CE. So since the straight line AE is equal to the straight line CE, and the straight line ED is both common and orthogonal, the base DA is therefore equal to the base DC. Likewise also all the straight lines that are equidistant from DB or DH will be shown to be equal. Also, since EH is less than EF, but also ED is common and orthogonal, the base DH is therefore less than the base DF. Again, since the segment from E touching the circle is bigger than all those falling towards the arc, and it was shown in the third book of the *Elements*,[21] that the rectangle contained by AE, EL is equal to the square on EF, whenever EF is tangent, it is manifest that, as AE is to EF, so too is EF to EL. But EF is greater than EL: for always the one nearer to the least is less than the one farther: also AE is therefore greater than EF. So since EF is smaller than EA, and ED common and orthogonal, the base DF is therefore smaller than the base DA. Again, since AK is equal to KB, but KE is common, the two straight lines AK, KE are therefore equal to EK, KB, that is, to the whole EKB. But AK, KE are bigger than AE: and BE is therefore greater than AE. Again, since AE is smaller than EB, but ED is common and orthogonal, the base DA is therefore smaller than the base BD. So since DH is smaller than DZ, and DZ smaller than DA, and DA smaller than DB, DH is smallest and DB the biggest, and the nearer is always smaller than the farther.

The third case, where the perpendicular DE falls inside the circle $ABCGF$, is proved similarly.

"Of any curved line which is in one plane, I call that straight line the *diameter*," etc.[22] He said "in one plane" because of the helix of the cylinder and the sphere: for they are not in one plane. But what he means is as follows: let there be a curved

[21] *Elements* 3:36.

[22] Eutocius is quoting definition 4 above.

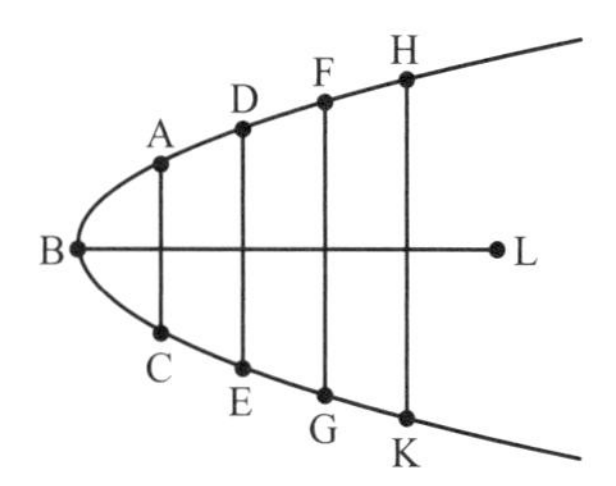

Figure 5.1.4.

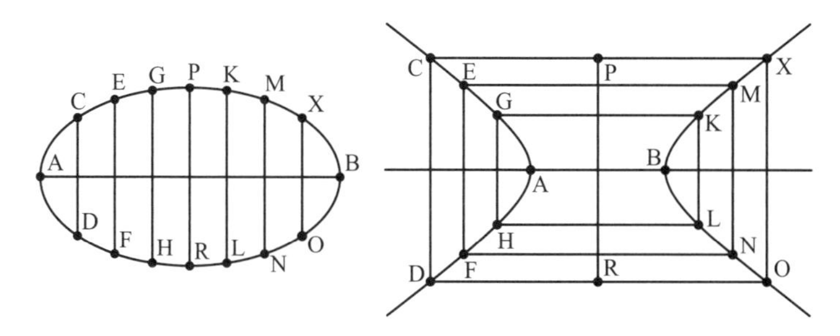

Figure 5.1.5.

figure *ABC* and in it some parallel straight lines *AC*, *DE*, *FG*, *HK*, and let from *B* the straight line *BL* be drawn, cutting them in two [figure 5.1.4]. So he says, that the *diameter* of the curve *ABC* is *BL*, the *vertex* *B*, and each of the lines *AC*, *DE*, *FG*, *HK* have been drawn ordinatewise to *BL*. But if *BL* bisects and cuts the parallels[23] at right angles, it is called the *axis*.

"Likewise of any two curved lines," etc.[24] For if we consider the curves *A*, *B* and in them the parallels *CD*, *EF*, *GH*, *KL*, *MN*, *XO* and the straight line *AB*, having been produced in each direction and cutting the parallels in two, I call *AB*, he says, the *transverse diameter*; the points *A* and *B* the *vertices* of the curves; and the lines having been drawn ordinatewise to *AB* the lines *CD*, *EF*, *GH*, *KL*, *MN*, *XO* [figure 5.1.5]. And if it (*AB*) also bisects them at right angles, it is called the *axis*. But if a certain straight line, such as *PR*, having been drawn through *CX*, *EM*, *GK*, bisects the parallels to *AB*, then *PR* is called the *upright diameter*, and each of the lines *CX*, *EM*, *GK*, [he says] have been drawn ordinatewise to the upright diameter *PR*. But if it bisects it at right angles, it is called *the upright axis*; and if *AB*, *PR* bisect the parallels of each other, they are called the *conjugate diameters*, but if they bisect at right angles, they are called the *conjugate axes*.

Propositions

Proposition 4. *If either one of the vertically opposite surfaces is cut by some plane parallel to the circle along which the straight line generating the surface is moved, the plane cut off within the surface will be a circle having its center on the axis, and the figure contained by*

[23] That is, these ordinates.

[24] Eutocius here explains definitions 5–8.

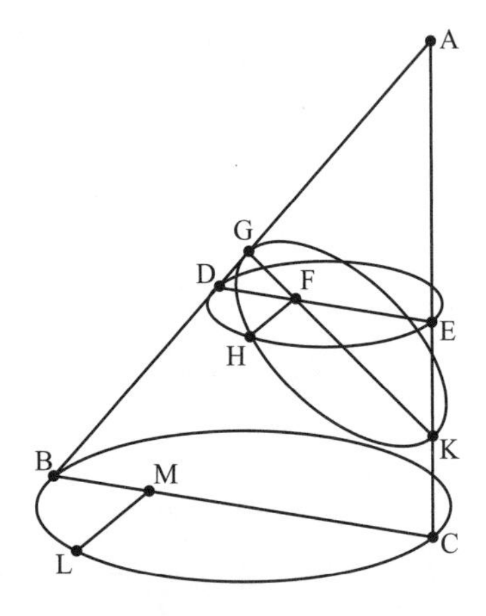

Figure 5.1.6.

the circle and the conic surface intercepted by the cutting plane on the side of the vertex will be a cone.

We skip the proof of this geometrically obvious proposition, but we do provide the proof of the next one.

Proposition 5. *If an oblique cone is cut by a plane through the axis at right angles to the base, and is also cut by another plane on the one hand at right angles to the axial triangle, and on the other cutting off on the side of the vertex a triangle similar to the axial triangle and lying subcontrariwise, then the section is a circle, and let such a section be called subcontrary.*

Let there be an oblique cone whose vertex is the point A and whose base is the circle BC, and let it be cut through the axis by a plane perpendicular to the circle BC, and let it make as a section the triangle ABC [figure 5.1.6]. Then let it also be cut by another plane perpendicular to the triangle ABC and cutting off on the side of the point A the triangle AKG similar to the triangle ABC and lying subcontrariwise, that is, so that the angle AKG is equal to the angle ABC. And let it make as a section on the surface, the line GHK. I say that the line GHK is a circle.

For let any points H and L be taken on the lines GHK and BC, and from the points H and L let perpendiculars be dropped to the plane through the triangle ABC. Then they will fall to the common sections of the planes. Let them fall as for example FH and LM. Therefore FH is parallel to LM. Then let the straight line DFE be drawn through F parallel to BC; and FH is also parallel to LM. Therefore the plane through FH and DE is parallel to the base of the cone. Therefore it is a circle whose diameter is the straight line DE [book 1, proposition 4].

Therefore rect(DF, FE) equals sq(FH). And since ED is parallel to BC, angle ADE is equal to angle ABC. And angle AKG is supposed equal to angle ABC. And therefore angle AKG is equal to angle ADE. And the vertical angles at the point F are also equal. Therefore triangle DFG is similar to the triangle KFE, and so EF is to FK as GF is toFD. Therefore, rect(EF, FD) equals rect(KF, FG). But it has been shown that sq(FH)

equals rect(EF, FD); and therefore rect(KF, FG) equals sq(FH). Likewise then all the perpendiculars drawn from the line GHK to the straight line GK could also be shown to be equal in square to the rectangle, in each case, contained by the segments of the straight line GK. Therefore the section is a circle whose diameter is the straight line GK.

In the next four propositions, Apollonius defines the parabola, ellipse, and hyperbola as sections of a cone produced by planes that cut the base of the cone in a straight line perpendicular to the base of the axial triangle. If the diameter of the section is parallel to one side of the axial triangle, then the section is a parabola; if it intersects both sides of the axial triangle, the section is an ellipse; and if the diameter produced meets one side of the axial triangle beyond the vertex, the section is a hyperbola. Apollonius distinguishes between a single branch of a hyperbola and the more familiar double-branched hyperbola, which he calls the opposite sections. Along with the definitions, Apollonius develops the "symptoms" of the curve, the relationships between the ordinate and abscissa of an arbitrary point on the curve that are equivalent to our equations of these curves.

Proposition 11 [Parabola]. *If a cone is cut by a plane through its axis, and also cut by another plane cutting the base of the cone in a straight line perpendicular to the base of the axial triangle, and if, further, the diameter of the section is parallel to one side of the axial triangle, and if any straight line is drawn from the section of the cone to its diameter such that this straight line is parallel to the common section of the cutting plane and of the cone's base, then this straight line to the diameter will be equal in square to the rectangle contained by (a) the straight line from the section's vertex to where the straight line to the diameter cuts it off and (b) another straight line which has the same ratio to the straight line between the angle of the cone and the vertex of the section as the square on the base of the axial triangle has to the rectangle contained by the remaining two sides of the triangle. And let such a section be called a parabola.*

Let there be a cone whose vertex is the point A, and whose base is the circle BC, and let it be cut by a plane through its axis, and let it make as a section the triangle ABC [figure 5.1.7]. And let it also be cut by another plane cutting the base of the cone in a straight line DE perpendicular to the straight line BC, and let it make as a section on the surface of the cone the line DFE, and let the diameter of the section FG be parallel to one side AC of the axial triangle. And let the straight line FH be drawn from the point F perpendicular to the straight line FG, and let it be contrived that sq(BC) is to rect(BA, AC) as FH is to FA. And let some point K be taken at random on the section, and through K let the straight line KL be drawn parallel to the straight line DE. I say that sq(KL) equals rect(HF, FL).[25]

For let the straight line MN be drawn through L parallel to the straight line BC. And the straight line DE is also parallel to the straight line KL. Therefore the plane through KL and MN is parallel to the plane through BC and DE, that is, to the base of the cone. Therefore the plane through KL and MN is a circle whose diameter is MN [book 1, proposition 4]. And KL is perpendicular to MN since DE is also perpendicular to BC. Therefore rect(ML, LN)

[25] If we set $KL = y$, $LF = x$, and $HF = p$, Apollonius's "symptom" of this curve is the equation $y^2 = px$.

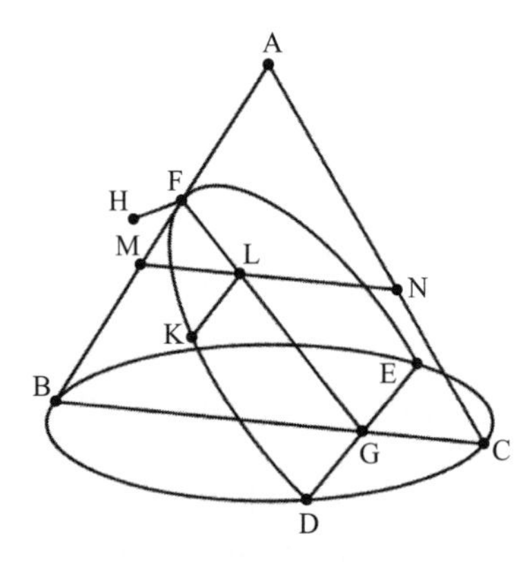

Figure 5.1.7.

equals sq(KL). And since sq(BC) is to rect(BA, AC) as HF is to FA, and sq(BC) is to rect(BA, AC) as the ratio of BC to CA compounded[26] with the ratio of BC to BA,[27] therefore HF is to FA as the ratio of BC to CA compounded with the ratio of BC to BA.

But BC is to CA as MN is to NA and as ML is to LF, and BC is to BA as MN is to MA or as LM is to MF or as NL is to FA. Therefore HF is to FA as the ratio of ML to LF compounded with the ratio of NL to FA. But rect(ML, LN) is to rect(LF, FA) as the ratio of ML to LF compounded with the ratio of LN to FA. Therefore HF is to FA as rect(ML, LN) is to rect(LF, FA). But with the straight line FL taken as a common height, HF is to FA as rect(HF, FL) is to rect(LF, FA), therefore rect(ML, LN) is to rect(LF, FA) as rect (HF, FL) is to rect(LF, FA). Therefore rect(ML, LN) equals rect(HF, FL). But rect(ML, LN) equals sq(KL), therefore also sq(KL) equals rect(HF, FL).

And let such a section be called a *parabola*,[28] and let HF be called the straight line to which the straight lines drawn ordinatewise to the diameter FG are applied in square, and let it also be called the upright side.[29]

Let it have been made, that sq(BC) is to rect(BA, AC) as HF is to FA: the thing being said is manifest, unless someone wants to comment on it. For let rect(OP, PR) be equal to rect(BA, AC), and let it, being projected along PR, make PS as width and be equal[30] to the square on BC; and let it have happened that OP is to PS as AF is to FH. The thing being sought has therefore happened. For since it is the case, that OP is to PS as AF is to FH; *invertendo*, SP is to PO as HF is to FA. But SP is to PO as SR is to PO as sq(BC) is to rect(BA, AC) [figure 5.1.8]. This is useful also in the following two theorems.

But sq(BC) has to rect(BA, AC) a ratio compounded from that which BC has to CA and BC to BA; it has been shown in the twenty-third theorem of the sixth

[26] See Eutocius's commentary below for a discussion of compound ratios.

[27] *Elements* 6:23.

[28] The name *parabola* comes from the Greek word *parabolē*, meaning "applied," because the square on the ordinate y is equal to the rectangle applied to the abscissa x.

[29] This line HF is today called the *parameter* or *latus rectum*.

[30] That is, the applied area is equal.

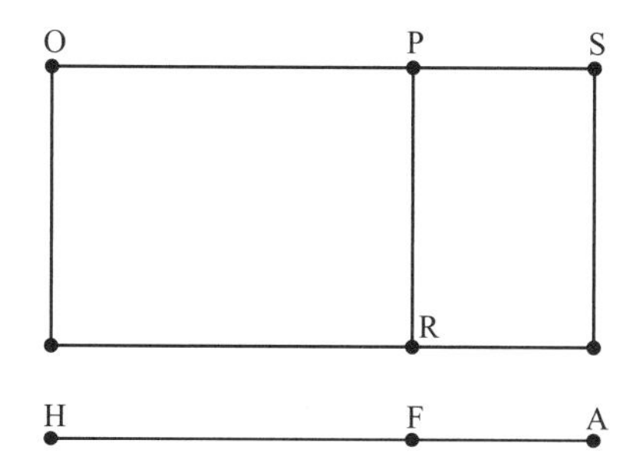

Figure 5.1.8.

book of the *Elements*, that equiangular parallelograms have to one another a ratio compounded out of that of the sides; but since it is discussed too inductively and not in the necessary manner by the commentators, we researched it; and it is written in our published work on the fourth theorem of the second book of Archimedes's *On the Sphere and the Cylinder*, and also in the scholia of the first book of Ptolemy's *Syntaxis*.[31] But it is a good idea that this be written down here also, because readers do not always read it even in those works, and also because nearly the entire treatise of the *Conics* makes use of it.

A ratio is said to be compounded from ratios, whenever the sizes of the ratios, being multiplied into themselves, make something, with "size" of course meaning the number after which the ratio is named. So it is possible in the case of multiples that the size be a whole number, but in the case of the remaining relations it is necessary that the size must be a number plus part or parts, unless perhaps one wishes the relation to be irrational, such as are those according to the incommensurable magnitudes. But in the case of all the relations, it is manifest that the product of the size itself with the consequent of the ratio makes the antecedent.

Accordingly, let there be a ratio of A to B, and let some mean of them be taken, as it chanced, as C, and let D equals size($A:C$), and E equals size($C:B$), and let D, multiplying E, make F. I say, that the size of the ratio $A:B$ is F, F equals size($A:B$), that is, that F, multiplying B, makes A. Indeed, let F, multiplying B, make G [figure 5.1.9]. So since D, multiplying E has made F, and multiplying C has made A, therefore it is, that E is to F as C is to G. *Alternando*, E is to C as F is to G. But E is to C as F is to A. Therefore H equals A, so that F, multiplying B, has made A.

But do not let this confuse those reading that this has been proved through arithmetic, for both the ancients made use of such proofs, being mathematical rather than arithmetical on account of the proportions, and that the thing being sought is arithmetical. For both ratios and sizes of ratios and multiplications, first begin by numbers, and through them by magnitudes. As someone once said[32] "for these mathematical studies appear to be related."

[31] The *Almagest*.

[32] A quote attributed to Archytas of Tarentum.

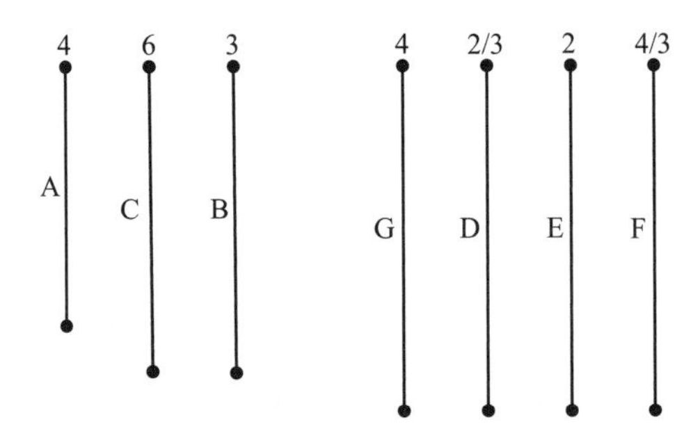

Figure 5.1.9. This diagram, unlike most in the manuscripts, has in addition to the labels, numerical values. The values assigned to *A*, *B*, and *C* are clear: 4, 3, and 6 respectively. However, the values for the remaining magnitudes were either difficult to read or incorrect. We have corrected them here so that they work with the assigned values of *A*, *B*, and *C*.

Proposition 12 [Hyperbola]. *Let a cone be cut by a plane through the axis, and let it be cut by another plane cutting the base of the cone in a straight line perpendicular to the base of the axial triangle, and let the diameter of the section, when produced, meet one side of the axial triangle beyond the vertex of the cone; then if any straight line be drawn from the section of the cone parallel to the common section of the cutting plane and the base of the cone as far as the diameter of the section, its square will be equal to the area applied to a certain straight line [the parameter]; this line is such that the straight line subtending the external angle of the triangle, lying in the same straight line with the diameter of the section, will bear to it the same ratio as the square on the line drawn from the vertex of the cone parallel to the diameter of the section as far as the base of the triangle bears to the rectangle bounded by the segments of the base made by the line so drawn; the breadth of the applied figure will be the intercept made by the ordinate on the diameter in the direction of the vertex of the section; and the applied figure will exceed by a figure similar and similarly situated to the rectangle bounded by the straight line subtending the external angle of the triangle and the parameter of the ordinates; and let such a section be called a hyperbola.*

Let there be a cone whose vertex is the point A and whose base is the circle BC, and let it be cut by a plane through its axis, and let it make as a section the triangle ABC [figure 5.1.10]. And let the cone also be cut by another plane cutting the base of the cone in the straight line DE perpendicular to BC, the base of the triangle ABC, and let this second cutting plane make as a section on the surface of the cone the line DFE, and let the diameter of the section FG when produced meet AC, one side of the triangle ABC, beyond the vertex of the cone at the point H. And let the straight line AK be drawn through A parallel to the diameter of the section FG, and let it cut BC. And let the straight line FL be drawn perpendicular to FG, and let it be contrived that sq(KA) is to rect(BK, KC) as FH is to FL. And let some point M be taken at random on the section, and through M let the straight line MN be drawn parallel to DE, and through N let the straight line NOX be drawn parallel to FL. And let the straight line HL be joined and produced to X, and let straight lines LO and XP be drawn through L and X parallel to FN. I say that

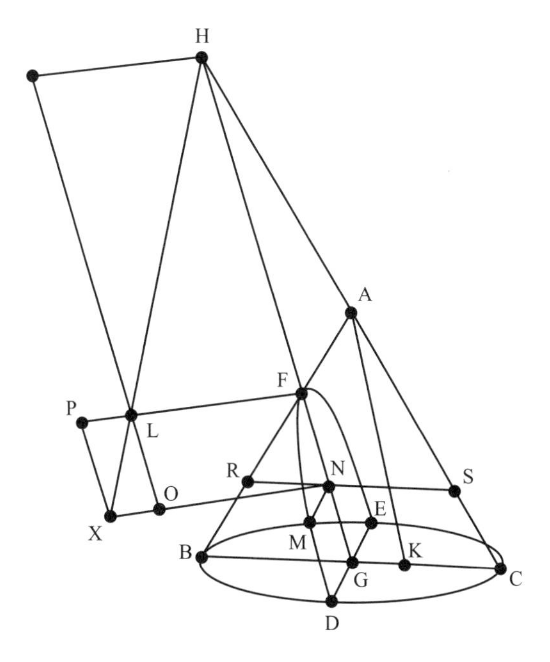

Figure 5.1.10.

MN is equal in square to the parallelogram FX which is applied to FL, having FN as breadth, and projecting beyond by a figure LX similar to the rectangle contained by HF and FL.[33]

For let the straight line RNS be drawn through N parallel to BC; and NM is also parallel to DE. Therefore the plane through MN and RS is parallel to the plane through BC and DE, that is, to the base of the cone. Therefore if the plane is produced through MN and RS, the section will be a circle whose diameter is the straight line RNS [book 1, proposition 4]. And MN is perpendicular to it. Therefore, rect(RN, NS) equals sq(MN). And since sq(AK) is to rect(BK, KC) as FH is to FL, and sq(AK) is to rect(BK, KC) as the ratio of AK to KC compounded with the ratio of AK to KB, therefore also FH is to FL as the ratio of AK to KC compounded with the ratio of AK to KB. But AK is to KC as HG is to GC as HN is to NS, and AK is to KB as FG is to GB as FN is to NR. Therefore HF is to FL as the ratio of HN to NS compounded with the ratio of FN to NR.

And rect(HN, NF) is to rect(SN, NR) as the ratio of HN to NS compounded with the ratio of FN to NR. Therefore also rect(HN, NF) is to rect(SN, NR) as HF is to FL and as HN is to NX. But, with the straight line FN taken as a common height, HN is to NX as rect(HN, NF) is to rect(FN, NX). Therefore also rect(HN, NF) is to rect(SN, NR) as rect(HN, NF) is to rect(XN, NF). Therefore rect(SN, NR) equals rect(XN, NF). But it was shown that sq(MN) equals rect(SN, NR); therefore also sq(MN) equals rect(XN, NF). But

[33] We set $FN = x$, $MN = y$, $FL = p$, and $FH = 2a$. Since the rectangle LX is similar to the rectangle contained by HF and FL, we have $LP : LO = FL : FH$, or $LP = \frac{FL}{FH} \times LO = \frac{FL}{FH} \times FN$. Since the area of the rectangle FX is $FN \times FP = FN(FL + LP)$, we can translate Apollonius's symptom $MN^2 = FX$, or $MN^2 = FN \times FP = FN(FL + LP)$, as $y^2 = x(p + \frac{p}{2a}x)$, our modern equation for a hyperbola with vertex at the origin.

the rectangle contained by XN and NF is the parallelogram XF. Therefore the straight line MN is equal in square to XF which is applied to the straight line FL, having FN as breadth, and projecting beyond by the parallelogram LX similar to the rectangle contained by HF and FL.

And let such a section be called a hyperbola,[34] and let LF be called the straight line to which the straight lines drawn ordinatewise to FG are applied in square;[35] and let the same straight line also be called the upright side, and the straight line FH the transverse side.[36]

Proposition 13 [Ellipse]. *Let a cone be cut by a plane through the axis, and let it be cut by another plane meeting each side of the axial triangle, being neither parallel to the base nor subcontrary, and let the plane containing the base of the cone meet the cutting plane in a straight line perpendicular either to the base of the axial triangle or to the base produced; then if a straight line be drawn from any point of the section of the cone parallel to the common section of the planes as far as the diameter of the section, its square will be equal to an area applied to a certain straight line [the parameter]; this line is such that the diameter of the section will bear to it the same ratio as the square on the line drawn from the vertex of the cone parallel to the diameter of the section as far as the base of the triangle bears to the rectangle contained by the intercepts made by it on the sides of the triangle; the breadth of the applied figure will be the intercept made by it on the diameter in the direction of the vertex of the section; and the applied figure will be deficient by a figure similar and similarly situated to the rectangle bounded by the diameter and the parameter; and let such a section be called an ellipse.*

Let there be a cone whose vertex is the point A and whose base is the circle BC, and let it be cut by a plane through its axis, and let it make as a section the triangle ABC [figure 5.1.11]. And let it also be cut by another plane on the one hand meeting both sides of the axial triangle and on the other extended neither parallel to the base of the cone nor subcontrariwise, and let it make as a section on the surface of the cone the line DE. And let the common section of the cutting plane and of the plane the base of the cone is in be the straight line FG, perpendicular to the straight line BC, and let the diameter of the section be the straight line ED. And let the straight line EH be drawn from E perpendicular to ED, and let the straight line AK be drawn through A parallel to ED, and let it be contrived that sq(AK) is to rect(BK, KC) as DE is to EH. And let some point L be taken on the section, and let the straight line LM be drawn through L parallel to FG. I say that the straight line LM is equal in square to some area which is applied to EH, having EM as breadth and deficient by a figure similar to the rectangle contained by DE and EH.[37]

[34] The name *hyperbola* comes from the Greek *huperbolē*, meaning "exceeding" because the square on the ordinate is equal to a rectangle applied to the line EH with breadth equal to the abscissa but exceeding by another rectangle.

[35] That is, as in the case of the parabola, this line segment is called the parameter.

[36] In modern terms, FH, the distance between the two branches of the hyperbola, is called the major axis.

[37] We set $EM = x$, $LM = y$, $EH = p$, and $DE = 2a$. Since the rectangle contained by OH and XO (the deficiency) is similar to the rectangle contained by DE and EH, we have $DE : EH = XO : OH = EM : OH$, or $OH = \frac{EH}{DE} \cdot EM$. Since the area of the applied figure is $EM \times EO = EM \cdot (EH - OH)$, we can translate Apollonius's symptom $LM^2 = EM \times EO = EM \cdot (EH - OH)$ as $y^2 = x(p - \frac{p}{2a}x)$, our modern equation for an ellipse with vertex at the origin.

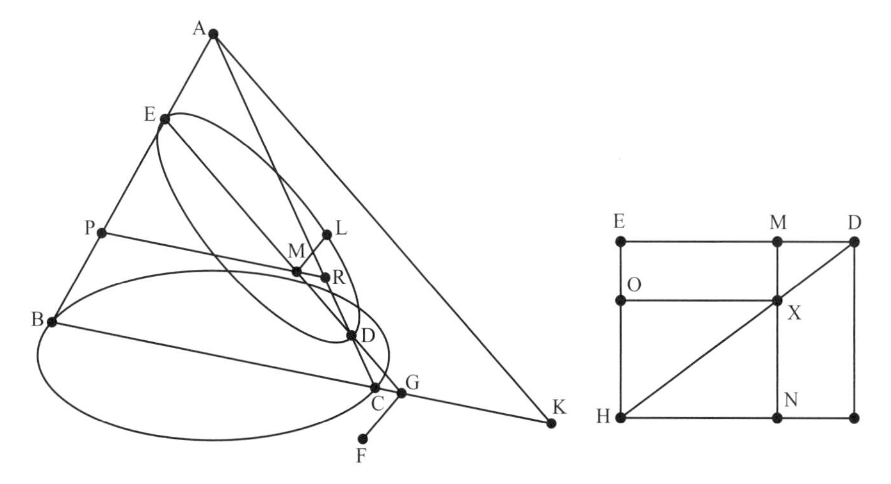

Figure 5.1.11. The rectangular figure on the right has E, D, and M as they are in the figure on the left and is in a plane perpendicular to the plane of the ellipse. Standard diagrams generally draw this plane as being coplanar with the axial triangle ABC, but in reality the figure could "spin" around ED. We have separated the figure from the cone diagram here to make it easier to view.

For let the straight line DH be joined, and on the one hand let the straight line MXN be drawn through M parallel to HE, and on the other let straight lines HN and XO be drawn through H and X parallel to EM, and let the straight line PMR be drawn through M parallel to BC. Since then PR is parallel to BC, and LM is also parallel to FG, therefore the plane through LM and PR is parallel to the plane through FG and BC, that is, to the base of the cone. If therefore a plane is extended through LM and PR, the section will be a circle whose diameter is PR [book 1, definition 4]. And LM is perpendicular to it. Therefore rect(PM, MR) equals sq(LM). And since sq(AK) is to rect(BK, KC) as ED is to EH, and sq(AK) is to rect(BK, KC) as the ratio of AK to KB compounded with the ratio of AK to KC, but AK is to KB as EG is to GB as EM is to MP, and AK is to CK as DG is to GC as DM is to MR, therefore DE is to EH as the ratio of EM to MP compounded with the ratio of DM to MR.

But rect(EM, MD) is to rect(PM, MR) as the ratio of EM to MP compounded with the ratio of DM to MR. Therefore rect(EM, MD) is to rect(PM, MR) as DE is to EH as DM is to MX. And, with the straight line ME taken as a common height, DM is to MX as rect(DM, ME) is to rect(XM, ME). Therefore also rect(EM, MD) is to rect(PM, MR) as rect(DM, ME) is to rect(XM, ME). Therefore rect(PM, MR) equals rect(XM, ME). But it was shown that rect(PM, MR) equals sq(LM); therefore also rect(XM, ME) equals sq(LM). Therefore the straight line LM is equal in square to the parallelogram MO which is applied to the straight line HE, having EM as breadth and deficient by the figure ON similar to the rectangle contained by DE and EH.

And let such a section be called an ellipse,[38] and let EH be called the straight line to which the straight lines drawn ordinatewise to DE are applied in square,[39] and

[38] The name *ellipse* comes from the Greek *elleipsis*, meaning "deficient" because the square on the ordinate is equal to a rectangle applied to the line *HE* with breadth equal to the abscissa but deficient by another rectangle.

[39] The parameter.

let the same straight line be called the upright side, and the straight line ED the transverse side.[40]

It is necessary to point out, that this theorem has three diagrams, as has been said often in the case of the ellipse: for D either falls on AC above C, or at C itself, or meets AC, having been projected, outside.

Proposition 14 [Opposite Sections]. *If the vertically opposite surfaces are cut by a plane not through the vertex, the section on each of the two surfaces will be that which is called the hyperbola; and the diameter of the two sections will be the same straight line; and the straight lines, to which the straight lines drawn to the diameter parallel to the straight line in the cone's base are applied in square, are equal; and the transverse side of the figure, that between the vertices of the sections, is common. And let such sections be called opposite.*

Let there be the vertically opposite surfaces whose vertex is the point A, and let them be cut by a plane not through the vertex, and let it make as sections on the surface the lines DEF and GHK. I say that each of the two sections DEF and GHK is the so-called hyperbola.

For let there be the circle $BDCF$ along which the line generating the surface moves, and let the plane $XGOK$ be extended parallel to it on the vertically opposite surface; and the straight lines FD and GK are common sections of the sections GHK and FED, and of the circles [book 1, proposition 4]. Then they will be parallel. And let the straight line LAU be the axis of the conic surface, and the points L and U be the centers of the circles, and let a straight line drawn from L perpendicular to the straight line FD be produced to the points B and C, and let a plane be produced through the straight line BC and the axis. Then it will make as sections in the circles the parallel straight lines XO and BC, and on the surface the straight lines BAO and CAX [figure 5.1.12].

Then the straight line XO will be perpendicular to the straight line GK, since the straight line BC is also perpendicular to the straight line FD, and each of the two is parallel to the other. And since the plane through the axis meets the sections in the points M and N within the [curved] lines [FD and GK], it is clear that the plane through the axis also cuts the [curved] lines. Let it cut them at H and E; therefore M, E, H, and N are points on the plane through the axis and in the plane the lines are in; therefore the line $MEHN$ is a straight line. It is also evident both that X, H, A, and C are in a straight line and B, E, A, and O also. For they are both on the conic surface and in the plane through the axis.

Let then the straight lines HR and EP be drawn from H and E perpendicular to HE, and let the straight line SAT be drawn through A parallel to $MEHN$, and let it be contrived that HE is to EP as sq(AS) is to rect(BS, SC), and EH is to HR as sq(AT) is to rect(OT, TX). Since then a cone, whose vertex is the point A and whose base is the circle BC, has been cut by a plane through its axis, and it has made as a section the triangle ABC; and it has also been cut by another plane cutting the base of the cone in the straight line DMF perpendicular to the straight line BC, and it has made as a section on the surface the line DEF; and the diameter ME produced has met one side of the axial

[40] Today, this line is generally called the major axis of the ellipse.

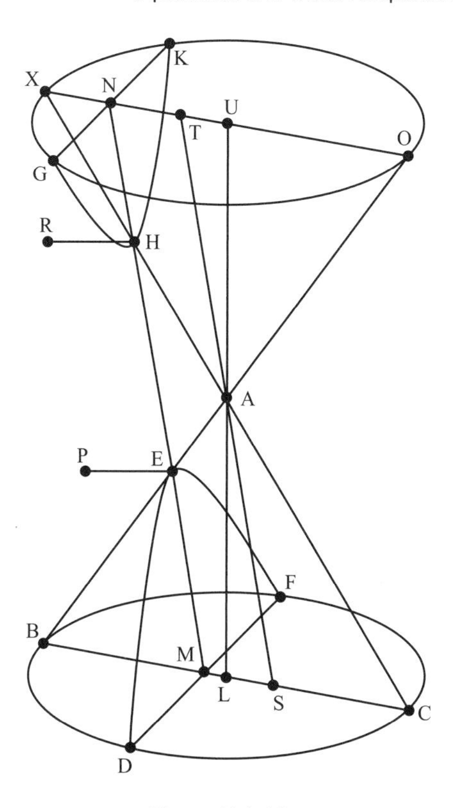

Figure 5.1.12.

triangle beyond the vertex of the cone, and through the point A the straight line AS has been drawn parallel to the diameter of the section EM, and from E the straight line EP has been drawn perpendicular to the straight line EM, and EH is to EP as sq(AS) is to rect(BS, SC), therefore the section DEF is a hyperbola [book 1, proposition 12], and EP is the straight line to which the straight lines drawn ordinatewise to EM are applied in square, and the straight line HE is the transverse side of the figure. And likewise GHK is also a hyperbola whose diameter is the straight line HN and whose straight line to which the straight lines drawn ordinatewise to HN are applied is HR, and the transverse side of whose figure is HE.

I say that the straight line HR is equal to the straight line EP. For since BC is parallel to XO, AS is to SC as AT is to TX, and AS is to SB as AT is to TO. But sq(AS) is to rect(BS, SC) as the ratio of AS to SC compounded with the ratio of AS to SB, and sq(AT) is to rect(XT, TO) as the ratio of AT to TX compounded with the ratio of AT to TO; therefore sq(AS) is to rect(BS, SC) as sq(AT) is to rect(XT, TO). Also sq(AS) is to rect(BS, SC) as HE is to EP, and sq(AT) is to rect(XT, TO) as HE is to HR. Therefore also HE is to EP as EH is to HR. Therefore EP equals HR.

And it was necessary to likewise show that, sq(AS) is to rect(BS, SC) as sq(AT) is to rect(XT, TO). For since BC is parallel to XO, CS is to SA as XT is to TA, and on account of this, AS is to SB as AT is to TO; therefore, through equality, CS is to SB as XT is to TO. Therefore also sq(CS) is to rect(CS, SB) as sq(XT) is to rect(XT, TO). But

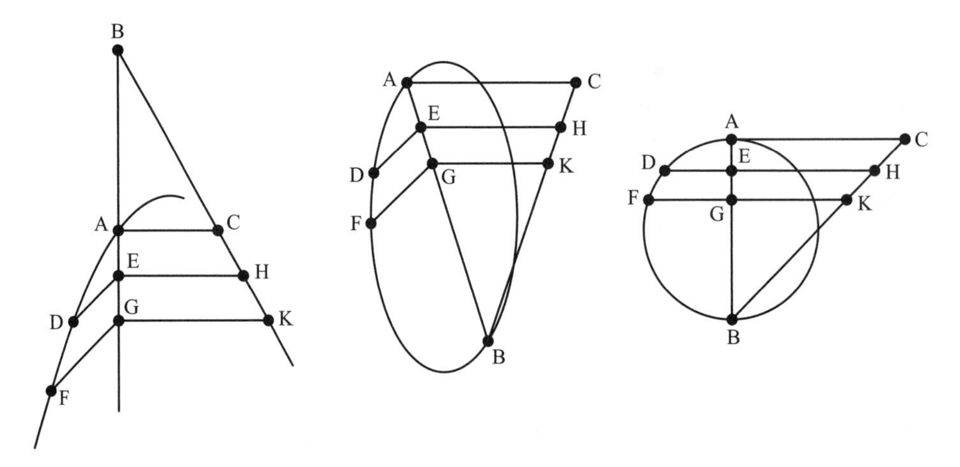

Figure 5.1.13.

on account of similar triangles, sq(AS) is to sq(SC) as sq(AT) is to sq(XT); therefore, through equality, sq(AS) is to rect(BS, SC) as sq(AT) is to rect(XT, TO).

And so, sq(AS) is to rect(BS, SC) as HE is to EP, but also, sq(AT) is to rect(XT, TO) as HE is to HR; and therefore, HE is to EP as EH is to HR. Therefore EP is equal to HR. But it[41] does not have cases, and the investigation is manifestly continuous with the three before it: for similarly, by means of them, he investigates that the diameter of the opposite sections is the principal one, and also investigates the parameters.

Second Definitions

9. Let the midpoint of the diameter of both the hyperbola and the ellipse be called the *center* of the section, and let the straight line drawn from the center to meet the section be called the *radius* of the section.
10. And likewise let the midpoint of the transverse side of the opposite sections be called the *center*.
11. And let the straight line drawn from the center parallel to an ordinate, being a mean proportional to the sides of the figure and bisected by the center, be called the *second diameter*.

Proposition 21. *If in a hyperbola or ellipse or in the circumference of a circle straight lines are dropped ordinatewise to the diameter, the squares on them will be to the areas contained by the straight lines cut off by them beginning from the ends of the transverse side of the figure, as the upright side of the figure is to the transverse, and to each other as the areas contained by the straight lines cut off (abscissas), as we have said.*

Let there be a hyperbola or ellipse or circumference of a circle whose diameter is AB and whose parameter is the straight line AC, and let the straight lines DE and FG be dropped ordinatewise to the diameter [figure 5.1.13]. I say that sq(FG) is to rect(AG, GB) as AC is to AB and [that] sq(FG) is to sq(DE) as rect(AG, GB) is to rect(AE, EB).

[41] That is, this theorem.

For let the straight line BC determining the figure be joined, and through E and G let the straight lines EH and GK be drawn parallel to the straight line AC. Therefore sq(FG) equals rect(KG, GA) and sq(DE) equals rect(HE, EA). And since KG is to GB as CA is to AB, and with AG taken as common height, KG is to GB as rect(KG, GA) is to rect(BG, GA), therefore CA is to AB as rect(KG, GA) is to rect(BG, GA), or CA is to AB as sq(FG) is to rect(BG, GA).[42] Then also for the same reasons, CA is to AB as sq(DE) is to rect(BE, EA). And therefore sq(GF) is to rect(BG, GA) as sq(DE) is to rect(BE, EA); alternately sq(FG) is to sq(DE) as rect(BG, GA) is to rect(BE, EA).

The next two propositions demonstrate how to find tangents to the parabola and to the central conics. For the parabola with equation $x^2 = py$, to find the tangent line at a point $C = (x, y)$, we find the point $A = (0, -y)$ on the axis and join CA. For the ellipse and hyperbola, in figure 5.1.15, we set $AD = x$ and $AE = t$. This implies that in the ellipse $BD = 2a - x$ and $BE = 2a + t$, while in the hyperbola $BD = 2a + x$ and $BE = 2a - t$. Therefore, in the ellipse, the ratio $BD : DA = BE : EA$ becomes $(2a - x)/x = (2a + t)/t$, while in the hyperbola that same ratio becomes $(2a + x)/x = (2a - t)/t$. Solving for t gives $t = ax/(a - x)$ for the ellipse and $t = ax/(a + x)$ for the hyperbola. In each case, once we know $t = AE$, we can draw the tangent line. In all three cases, it is straightforward to confirm these solutions using calculus.

Proposition 33. *If on a parabola some point is taken, and from it an ordinate is dropped to the diameter, and, to the straight line cut off by it on the diameter from the vertex, a straight line in the same straight line from its extremity is made equal, then the straight line joined from the point thus resulting to the point taken will touch the section.*[43]

Let there be a parabola whose diameter is the straight line AB, and let the straight line CD be dropped ordinatewise, and let the straight line AE be made equal to the straight line ED, and let the straight line AC be joined [figure 5.1.14]. I say that the straight line AC produced will fall outside the section.

For if possible, let it fall within, as the straight line CF, and let the straight line GB be dropped ordinatewise. And since the ratio of sq(BG) to sq(CD) is greater than that of sq(FB) to sq(CD), but sq(FB) is to sq(CD) as sq(BA) is to sq(AD), and sq(BG) is to sq(CD) as BE is to DE, therefore the ratio of BE to DE is greater than that of sq(BA) to sq(AD). But BE is to DE as 4 rect(BE, EA) is to 4 rect(DE, EA); therefore also the ratio of 4 rect(BE, EA) to 4 rect(DE, EA) is greater than that of sq(AB) to sq(AD). Therefore alternately the ratio of 4 rect(BE, EA) to sq(AB) is greater than that of 4 rect(DE, EA) to sq(AD); and this is absurd; for since AE equals DE, hence 4 rect(DE, EA) equals sq(AD). But 4 rect(BE, EA) is less than sq(AB), for E is not the midpoint of AB. Therefore the straight line AC does not fall within the section; therefore it touches it.

Proposition 34. *If on a hyperbola or ellipse or circumference of a circle some point is taken, and if from it a straight line is dropped ordinatewise to the diameter, and if the straight lines which the ordinatewise line cuts off from the other ends of the figure's transverse side have*

[42] Using the same modern notation as before, this result shows that the symptoms of the ellipse and hyperbola can be rewritten as $y^2 = \frac{p}{2a}x_1x_2$, where x_1 and x_2 are the distances of the point G from the two ends of the axis of the curve.

[43] We call this line a tangent line to the parabola.

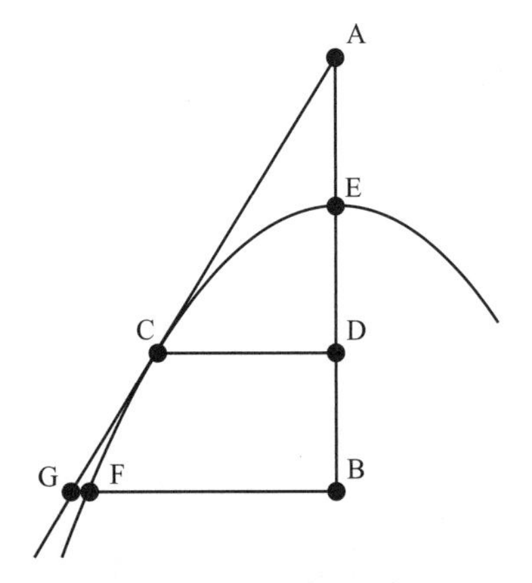

Figure 5.1.14.

to each other a ratio which other segments of the transverse side have to each other, so that the segments from the vertex are corresponding, then the straight line joining the point taken on the transverse side and that taken on the section will touch the section.

Let there be a hyperbola or ellipse or circumference of a circle whose diameter is the straight line AB, and let some point C be taken on the section, and from C let the straight line CD be drawn ordinatewise, and let it be contrived that BD is to DA as BE is to EA,[44] and let the straight line EC be joined [figure 5.1.15]. I say that the straight line CE touches the section.

For if possible, let it cut it, as the straight line ECF, and let some point F be taken on it, and let the straight line GFH be dropped ordinatewise, and let the straight lines AL and BK be drawn through A and B parallel to the straight line EC, and let the straight lines DC, BC, and GC be joined and produced to the points K, X, and M. And since BD is to DA as BE is to EA, but BD is to DA as BK is to AN, and BE is to AE as BC is to CX as BK is to XN, therefore BK is to AN as BK is to XN; therefore AN equals NX. Therefore rect(AN, NX) is greater than rect(AO, OX). Therefore the ratio of NX to XO is greater than that of OA to AN. But NX is to XO as KB is to BM; therefore the ratio of KB to BM is greater than that of OA to AN. Therefore rect(KB, AN) is greater than rect(BM, OA). And so the ratio of rect(KB, AN) to sq(CE) is greater than that of rect(BM, OA) to sq(CE).

But rect(KB, AN) is to sq(CE) as rect(BD, DA) is to sq(DE) through the similarity of triangles BKD, ECD, and NAD,[45] and rect(BM, OA) is to sq(CE) as rect(BG, GA) is to sq(GE); therefore the ratio of rect(BD, DA) to sq(DE) is greater than that of rect(BG, GA) to sq(GE). Therefore *alternando* the ratio of rect(BD, DA) to rect(BG, GA) is greater than that of sq(DE) to sq(GE). But rect(BD, DA) is to rect(AG, GB) as sq(CD) is to sq(GH) [book 1, proposition 21], and sq(DE) is to sq(EG) as sq(CD) is to sq(FG); therefore also the ratio of sq(CD) to sq(HG) is greater than that of sq(CD) to sq(FG). Therefore HG

[44] For the hyperbola, find E on BA such that $(BD+DA):DA=BA:EA$; then $BD:DA=(BA-EA):EA=BE:EA$. For the ellipse, find E on BA extended so that $(BD-DA):DA=BA:EA$; then $BD:DA=(BA+EA):EA=BE:EA$.

[45] See Eutocius's commentary below.

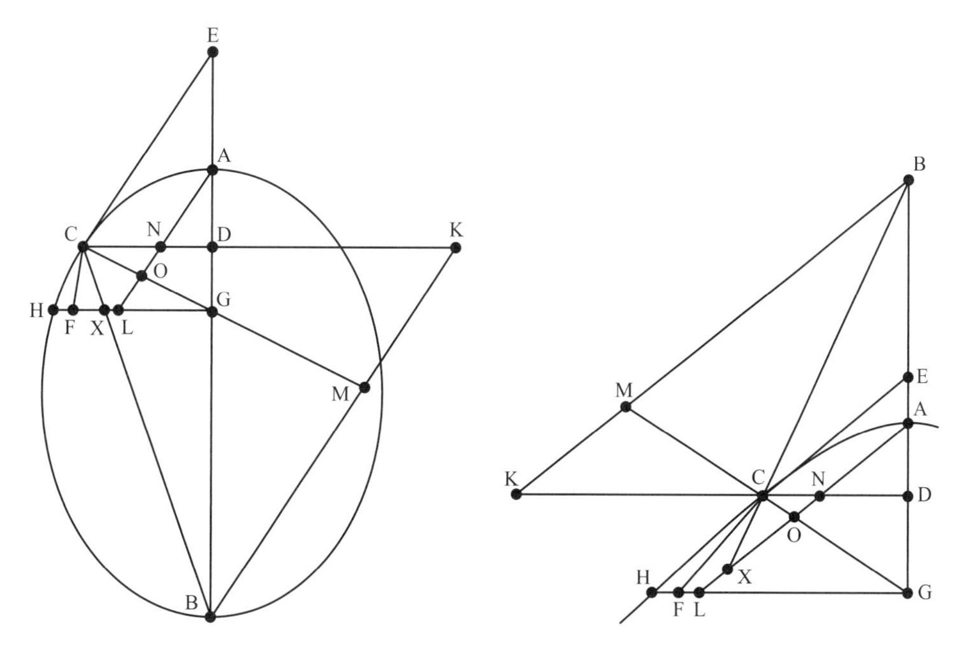

Figure 5.1.15.

is less than FG; and this is impossible. Therefore the straight line EC does not cut the section; therefore it touches it.

"But rect(KB, AN) is to sq(CE) as rect(BD, DA) is to sq(DE)" So since, on account of the lines AN, EC, KB being parallel, AN is to EC as AD is to DE; but EC is to KB and ED is to DB. Therefore *ex aequali*, AN is to KB as AD is to DB; and therefore sq(AN) is to rect(AN, KB) as sq(AD) is to rect(AD, DB). But sq(EC) is to sq(AN) as sq(ED) is to sq(DA); therefore *ex aequali*, sq(EC) is to rect(AN, KB) as sq(ED) is to rect(AD, DB); and *alternando*, rect(KB, AN) is to sq(EC) as rect(BD, DA) is to sq(ED).

In propositions 46 through 51, Apollonius shows that besides the principal diameters of the conics discussed earlier, other diameters and pairs of diameters have the same properties as originally shown. He summarizes these results in the porism to proposition 51.

Porism. And with these things shown [propositions 46–51], it is at once evident that in the parabola each of the straight lines drawn off parallel to the original diameter is a diameter, but in the hyperbola and ellipse and opposite sections, each of the straight lines drawn through the center is a diameter; and that in the parabola the straight lines dropped to each of the diameters parallel to the tangents will equal in square the rectangles applied to it, but in the hyperbola and opposite sections they will equal in square the areas applied to the diameter and projecting beyond by the same figure, but in the ellipse the areas applied to the diameter and defective by the same figure; and that all the things which

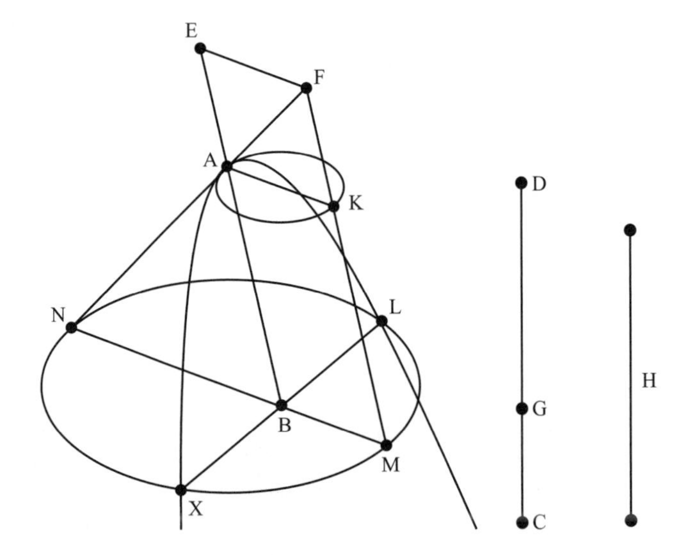

Figure 5.1.16.

have been already proved about the sections as following when the principal diameters are used, will also, those very same things, follow when the other diameters are taken.

Proposition 52. *Given a straight line in a plane bounded at one point, to find in the plane the section of a cone called parabola, whose diameter is the given straight line, and whose vertex is the end of the straight line, and where whatever straight line is dropped from the section to the diameter at a given angle, will equal in square the rectangle contained by the straight line cut off by it from the vertex of the section and by some other given straight line.*

Let there be a straight line AB given in position and bounded at the point A, and another straight line CD given in magnitude, and first let the given angle be a right angle [figure 5.1.16]; it is required then to find a parabola in the plane of reference whose diameter is the straight line AB and whose vertex is the point A, and whose upright side is the straight line CD, and where the straight lines dropped ordinatewise will be dropped at a right angle, that is, so that AB is the axis (definition 7).

Let AB be produced to E, and let CG be taken as the fourth part of CD, and let EA be greater than CG, and let CD to H be as H to EA. Therefore CD is to EA as sq(H) is to sq(EA), and CD is less than 4 EA; therefore also sq(H) is less than 4 sq(EA). Therefore H is less than 2 EA; and so the two straight lines EA are greater than H. It is therefore possible for a triangle to be constructed from H and two straight lines EA. Then let the triangle EAF be constructed on EA at right angles to the plane of reference so that EA equals AF, and H equals FE, and let the straight line AK be drawn parallel to FE, and FK to EA, and let a cone be conceived whose vertex is the point F and whose base is the circle about diameter KA, at right angles to the plane through AFK. Then the cone will be a right cone (definition 3); for AF equals FK.

And let the cone be cut by a plane parallel to the circle KA, and let it make as a section the circle MNX [book 1, proposition 4], at right angles clearly to the plane through MFN, and let the straight line MN be the common section of the circle MNX and of the triangle MFN; therefore it is the diameter of the circle. And let the straight line XL be the common section of the plane of reference and of the circle. Since then circle MNX is at right angles to triangle MFN, and the plane of reference is also at right angles to triangle MFN, therefore the straight line LX, their common section, is at right angles to triangle MFN, that is, to triangle KFA; and therefore it is perpendicular to all the straight lines touching it and in the triangle; and so it is perpendicular to both MN and AB.

Again since a cone, whose base is the circle MNX and whose vertex is the point F, has been cut by a plane at right angles to the triangle MFN and makes as a section circle MNX, and since it has also been cut by another plane, the plane of reference, cutting the base of the cone in a straight line XL at right angles to MN which is the common section of the circle MNX and the triangle MFN, and the common section of the plane of reference and of the triangle MFN, the straight line AB, is parallel to the side of the cone FKM, therefore the resulting section of the cone in the plane of reference is a parabola, and its diameter AB [book 1, proposition 11], and the straight lines dropped ordinatewise from the section to AB will be dropped at right angles; for they are parallel to XL which is perpendicular to AB. And since CD is to H as H is to EA, and EA, AF, FK are all equal, as are H, EF, AK, therefore CD is to AK as AK is to AF. And therefore CD is to AF as sq(AK) is to sq(AF) or rect(AF, FK).

Therefore CD is the upright side of the section; for this has been shown in the eleventh theorem [book 1, proposition 11].

This proposition shows that given a straight line, one endpoint, and another given straight line, one can construct a cone and a cutting plane so that the section formed by that plane in the cone is a parabola with the given line as diameter, the endpoint of the line as vertex, and the second given line as parameter, where the angle between the ordinate and abscissa is a right angle. That is, given those three items, one can now assert the construction of such a parabola, just as one could in the *Elements* assert the construction of a circle with given center and radius. In the next proposition, Apollonius solves the same problem when the given angle is not right. Then, in several further propositions, Apollonius shows that given a pair of vertices at the ends of a given line segment and a parameter, one can also determine a cone and cutting plane whose section is a hyperbola (propositions 54, 55), an ellipse (propositions 56, 57, 58), or opposite sections (proposition 59) with given vertices, axis, and parameter. Hence one can also assert the construction of an ellipse or hyperbola or opposite sections. Apollonius then concludes book 1 with proposition 60.

Proposition 60. *Given two straight lines bisecting each other, to describe about each of them opposite sections, so that the straight lines are their conjugate diameters and the diameter of one pair of the opposite sections is equal in square to the figure of the other pair, and likewise the diameter of the second pair of opposite sections is equal in square to the figure of the first pair.*

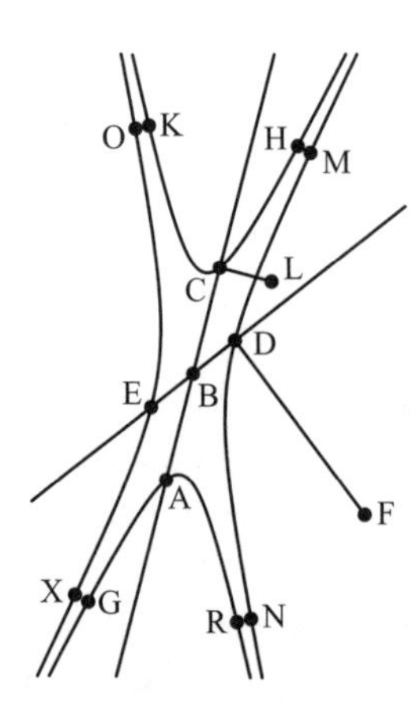

Figure 5.1.17. The "figures" referred to by Apollonius here are *ED* and *DF* for the hyperbolas *MDN* and *OEX*, and *AC* and *CL* for *KCH* and *GAR*. (Note that the segment *DF* is drawn half its correct length, so it would fit within the size of the rest of the diagram.) While Apollonius has yet to define the asymptotes for hyperbola (this comes in book 2), it is evident that the conjugate sections share asymptotes.

Let there be the two given straight lines AC and DE bisecting each other [figure 5.1.17]; then it is required to describe opposite sections about each of them as a diameter so that the straight lines AC and DE are conjugates in them, and DE is equal in square to the figure about AC, and AC is equal in square to the figure about DE.

Let rect(AC, CL) be equal to sq(DE), and let LC be perpendicular to CA. And given two straight lines AC and CL perpendicular to each other, let the opposite sections RAG and HCK be described whose transverse diameter will be CA and whose upright side will be CL, and where the ordinates from the sections to CA will be dropped at a given angle [book 1, proposition 59]. Then the straight line DE will be the second diameter of the opposite sections [definition 11]; for it is the mean proportional between the sides of the figure, and parallel to an ordinate; it has been bisected at B.

Then again let rect(DE, DF) be equal to sq(AC), and let DF be perpendicular to DE. And given two straight lines ED and DF lying perpendicular to each other, let the opposite sections MDN and OEX be described, whose transverse diameter will be DE and the upright side of whose figure will be DF, and where the ordinates from the sections will be dropped to DE at a given angle [book 1, proposition 59]; then the straight line AC will also be a second diameter of the sections MDN and XEO. And so AC bisects the parallels to DE between the sections RAG and HCK, and DE the parallels to AC; and this it was required to do. And let such sections be called *conjugate*.

5.2 Apollonius, *Conics*, 2–4

Book 2

Book 2 of the *Conics* deals with asymptotes to the hyperbola as well as methods of determining diameters and axes of the sections.

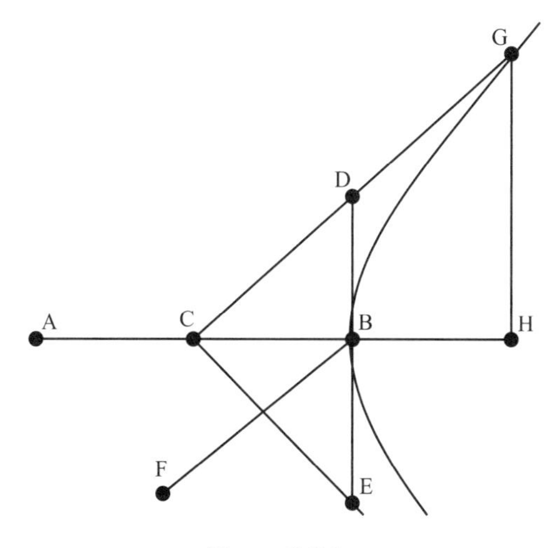

Figure 5.2.1.

Apollonius's Opening Letter

Apollonius to Eudemus. If you are well, well and good, and I, too, fare pretty well. I have sent you my son Apollonius bringing you the second book of the conics as arranged by us. Go through it then carefully and acquaint those with it worthy of sharing in such things. And Philonides,[46] the geometer, I introduced to you in Ephesus, if he ever happen about Pergamum, acquaint him with it too. And take care of yourself, to be well. Good-bye.

Proposition 1. *If a straight line touches a hyperbola at its vertex, and from it on both sides of the diameter a straight line is cut off equal in square to the fourth of the figure, then the straight lines drawn from the center of the section to the ends thus taken on the tangent will not meet the section.*

Let there be a hyperbola whose diameter is the straight line AB and center C, and upright side the straight line BF; and let the straight line DE touch the section at B, and let the squares on BD and BE each be equal to the fourth of the figure AB, BF, and the straight lines CD and CE be joined and produced [figure 5.2.1]. I say that they will not meet the section.

For if it is possible, let CD meet the section at G, and from G let the straight line GH be dropped ordinatewise; therefore it is parallel to DB. Since then AB is to BF as sq(AB) is to rect(AB, BF), but sq(CB) equals one-fourth of sq(AB), and sq(BD) equals one-fourth of rect(AB, BF), therefore AB is to BF as rect(AH, HB) is to sq(HG); therefore sq(CH) is to sq(HG) as rect(AH, HB) is to sq(HG). Therefore rect(AH, HB) equals sq(CH); and this is absurd.[47] Therefore the straight line CD will not meet the section. Then likewise we could show that neither does CE; therefore the straight lines CD and CE are asymptotes to the section.

[46] Philonides (200–130 BCE) of Laodicea, in Syria, was an Epicurean philosopher and mathematician who lived in the Seleucid court during the reigns of Antiochus IV, Epiphanes, and Demetrius I Soter.

[47] *Elements* 2:6.

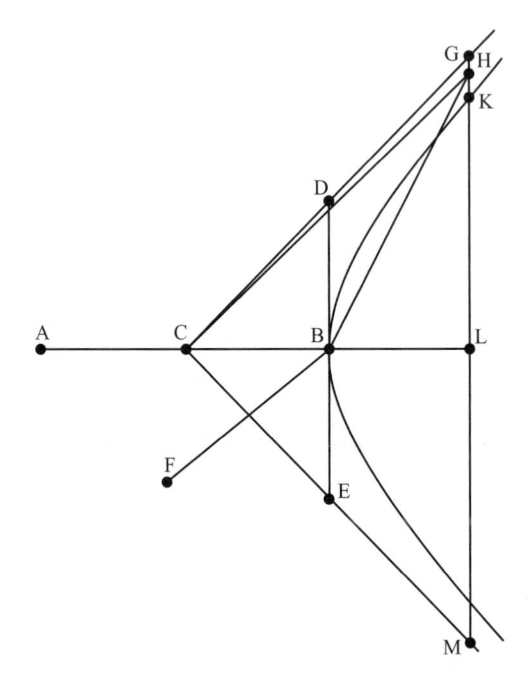

Figure 5.2.2.

In modern terms, since $AB = 2a$ and $BF = p$, the parameter, the squares on BD and BE are each equal to $\frac{1}{4}(2ap) = \frac{pa}{2}$. This value is generally named b^2; thus $BD = b$ and the slopes of the asymptotes are $\pm\frac{b}{a}$.

Proposition 2. *With the same things it is to be shown that a straight line cutting the angle contained by the straight lines DC and CE is not another asymptote.*

For if possible, let CH be it, and let the straight line BH be drawn through B parallel to CD, and let it meet CH at H, and let DG be made equal to BH, and let GH be joined and produced to the points K, L, and M [figure 5.2.2]. Since then BH and DG are equal and parallel, DB and HG are also equal and parallel. And since AB is bisected at C and a straight line BL is added to it, rect(AL, LB) together with sq(CB) equals sq(CL).[48] Likewise then, since GM is parallel to DE, and DB equals BE, therefore also GL equals KM. And since GH equals DB, therefore GK is greater than DB. And also KM is greater than BE, since also LM is greater than BE; therefore rect(MK, KG) is greater than rect(DB, BE), that is, greater than sq(DB).

Since then AB is to BF as sq(CB) is to sq(BD) [book 2, proposition 1], but AB is to BF as rect(AL, LB) is to sq(LK) [book 1, proposition 21], and sq(CB) is to sq(BD) as sq(CL) is to sq(LG), therefore also sq(CL) is to sq(LG) as rect(AL, LB) is to sq(LK). Since then the whole of sq(LC) is to the whole of sq(LG) as the part subtracted rect(AL, LB) is to the part subtracted sq(LK), therefore also, sq(LC) is to sq(LG) as the remainder sq(CB) is to

[48] *Elements* 2:6.

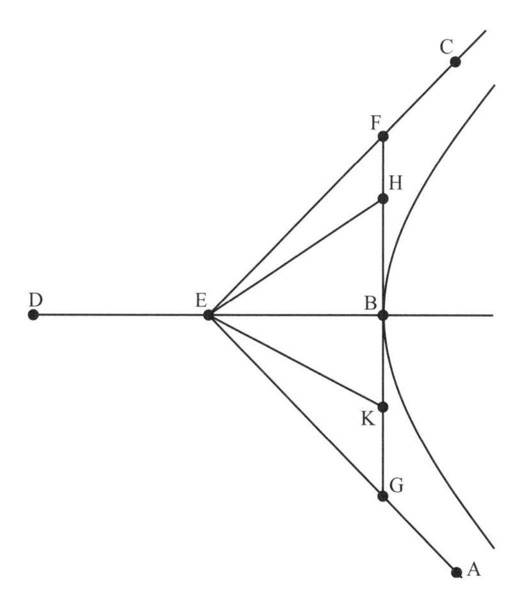

Figure 5.2.3.

the remainder rect(MK, KG), that is, sq(CB) is to rect(MK, KG) as sq(CB) is to sq(DB). Therefore sq(DB) equals rect(MK, KG); and this is absurd; for it has been shown to be greater than it. Therefore the line CH is not an asymptote to the section.

This theorem does not have cases. Nevertheless, BH always cuts the section at two points. For since it [BH] is parallel to CD, it will intersect CH: so that it will intersect the section first.[49]

Proposition 3. *If a straight line touches a hyperbola, it will meet both of the asymptotes and it will be bisected at the point of contact, and the square on each of its segments will be equal to the fourth of the figure resulting on the diameter*[50] *drawn through the point of contact.*

Let there be the hyperbola ABC, and its center E, and asymptotes FE and EG, and let some straight line HK touch it at B [figure 5.2.3]. I say that the straight line HK produced will meet the straight lines FE and EG.

For if possible, let it not meet them, and let EB be joined and produced, and let ED be made equal to EB; therefore the straight line BD is the diameter. Then let the squares on HB and BK each be made equal to the fourth of the figure on BD and the parameter, and let EH and EK be joined. Therefore they are asymptotes [book 2, proposition 1]; and this is absurd [book 2, proposition 2]; for FE and GE are supposed asymptotes. Therefore KH produced will meet the asymptotes EF and FG at F and G.

I say then also that the squares on BF and BG will each be equal to the fourth of the figure on BD. For let it not be, but if possible, let the squares on BH and BK each be equal

[49] That is, "before" it intersects CH at H.

[50] The "figure resulting on the diameter" is the rectangle on the diameter and the parameter.

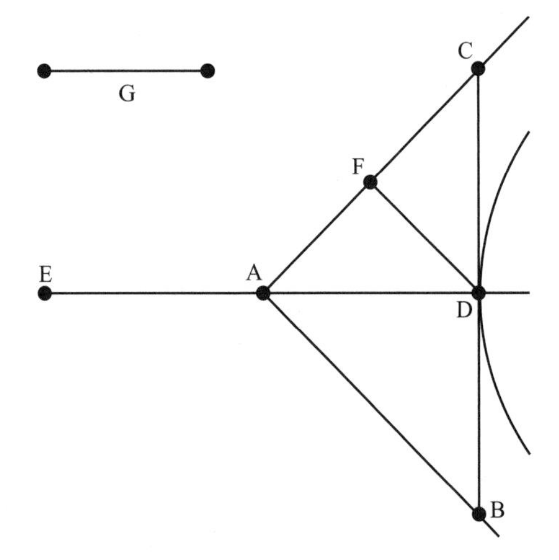

Figure 5.2.4.

to the fourth of the figure. Therefore HE and EK are asymptotes [book 2, proposition 1]; and this is absurd [book 2, proposition 2]. Therefore the squares on FB and BG will each be equal to the fourth of the figure on BD.

Now that asymptotes have been defined, Apollonius shows how to construct a hyperbola through a given point with given asymptotes, thus providing a new construction postulate.

Proposition 4.[51] *Given two straight lines containing an angle and a point within the angle, to describe through the point the section of a cone called hyperbola so that the given straight lines are its asymptotes.*

Let there be the two straight lines AC and AB containing a chance angle at A, and let some point D be given, and let it be required to describe through D a hyperbola to the asymptotes CA and AB [figure 5.2.4]. Let the straight line AD be joined and produced to E, and let AE be made equal to DA, and let the straight line DF be drawn through D parallel to AB, and let FC be made equal to AF, and let CD be joined and produced to B, and let it be contrived that rect(DE, G) equals sq(CB), and with AD extended let a hyperbola be described about it through D so that the ordinates are equal in square to the areas applied to G and exceeding by a figure similar to rectangle DE, G. Since then DF is parallel to BA, and CF equals FA, therefore CD equals DB; and so sq(CB) is equal to 4 sq(CD). And sq(CB) equals rect(DE, G); therefore the squares on CD and DB are each equal to the fourth part of the figure DE, G. Therefore the straight lines AB and AC are asymptotes to the hyperbola described.

Proposition 5. *If the diameter of a parabola or hyperbola bisects some straight line within the section, the tangent to the section at the end of the diameter will be parallel to the bisected straight line.*

[51] Some scholars believe that this proposition was an interpolation by Eutocius himself.

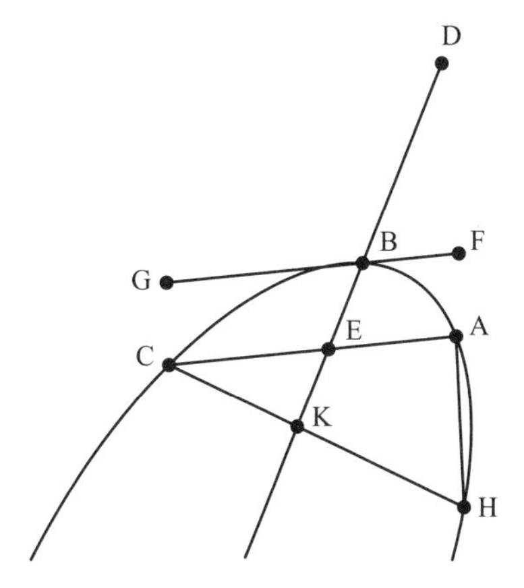

Figure 5.2.5.

Let there be the parabola or hyperbola ABC whose diameter is the straight line DBE, and let the straight line FBG touch the section, and let some straight line AEC be drawn in the section making AE equal to EC [figure 5.2.5]. I say that AC is parallel to FG.

For if not, let the straight line CH be drawn through C parallel to FG, and let HA be joined. Since then ABC is a parabola or hyperbola whose diameter is DE, and tangent FG, and CH is parallel to it, therefore CK equals KH [book 1, propositions 46, 47]. But also CE equals EA. Therefore AH is parallel to KE; and this is absurd; for produced it meets BD.

Proposition 8. *If a straight line meets a hyperbola in two points, produced both ways it will meet the asymptotes, and the straight lines cut off on it by the section from the asymptotes will be equal.*

Let there be the hyperbola ABC, and the asymptotes ED and DF, and let some straight line AC meet ABC [figure 5.2.6]. I say that produced both ways it will meet the asymptotes.

Let AC be bisected at G, and let DG be joined. Therefore it is a diameter of the section; therefore the tangent at B is parallel to AC. Then let HBK be the tangent; then it will meet ED and DF [book 2, proposition 3]. Since then AC is parallel to KH, and KH meets DK and DH, therefore also AC will meet DE and DF. Let it meet them at E and F; and HB is equal to HK [book 2, proposition 3]; therefore also FG equals GE. And so also CF equals AE.

Proposition 10. *If some straight line cutting the section meets both of the asymptotes, the rectangle contained by the straight lines cut off between the asymptotes and the section is equal to the fourth of the figure resulting on the diameter bisecting the straight lines drawn parallel to the drawn straight line.*

Proposition 12. *If two straight lines at chance angles are drawn to the asymptotes from some point of those on the section, and parallels are drawn to the two straight lines from some point of those on the section, then the rectangle contained by the parallels will be equal to that contained by those straight lines to which they were drawn parallel.*

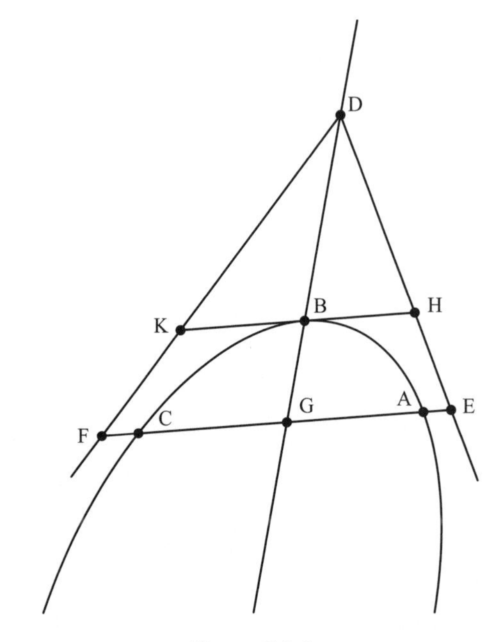

Figure 5.2.6.

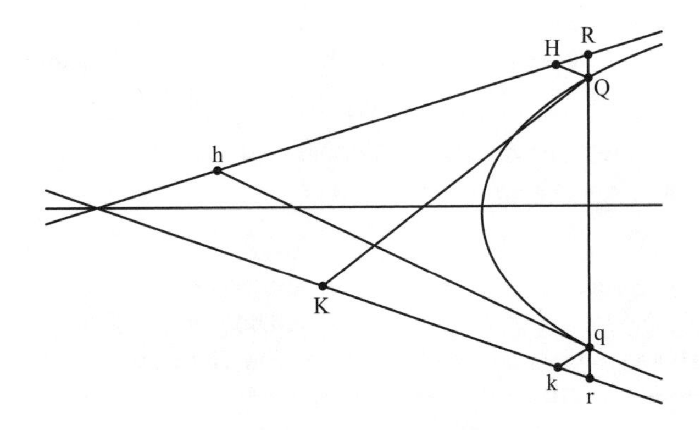

Figure 5.2.7.

Proposition 10 shows that the product $QR \cdot Qr$ and the product $qr \cdot qR$ are both equal to $BK^2 = b^2$, for any points Q, q on the hyperbola at opposite ends of an ordinate. So $QR : qR = qr : Qr$ [figure 5.2.7]. In proposition 12, we draw from Q, q a pair of parallel lines to each of the asymptotes, intersecting one at H, h, respectively, and the other at K, k. Then, since $RQ : Rq = HQ : hq$ and $qr : Qr = qk : QK$, so it follows that $HQ : hq = qk : QK$ or $HQ \cdot QK = hq \cdot qk$. In other words, the product of the lengths of the two lines drawn from any point on the hyperbola in given directions to the asymptotes is a constant. In modern notation, if we consider a rectangular hyperbola with asymptotes the x and y axes, this result shows that the equation of the hyperbola

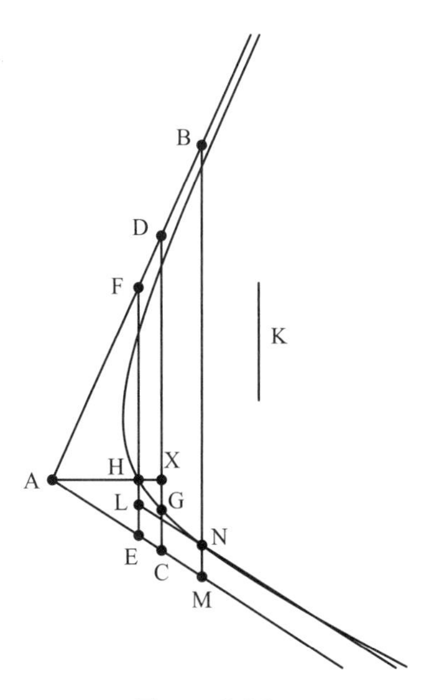

Figure 5.2.8.

can be written as $xy = c$. For Apollonius, these results give him a new way to define a hyperbola.

Proposition 14. *The asymptotes and the section, if produced indefinitely, draw nearer to each other; and they reach a distance less than any given distance.*

Let there be a hyperbola whose asymptotes are AB and AC, and a given distance K. I say that AB and AC and the section, if produced, draw nearer to each other and will reach a distance less than K.

For let EHF and CGD be drawn parallel to the tangent, and let AH be joined and produced to X [figure 5.2.8]. Since then rect(CG, GD) equals rect(FH, HE) [book 2, proposition 10], therefore DG is to FH as HE is to CG. But DG is greater than FH; therefore also HE is greater than CG. Then likewise we could show that the succeeding straight lines are less. Then let the distance EL be taken less than K, and through L let LN be drawn parallel to AC; therefore it will meet the section. Let it meet it at N, and through N let MNB be drawn parallel to EF. Therefore MN equals EL and so MN is less than K.

Proposition 44. *Given a section of a cone, to find a diameter.*

Let there be the given conic section on which are the points A, B, C, D, and E [figure 5.2.9]. Then it is required to find a diameter. Let it have been done, and let it be CH. Then with DF and EH drawn ordinatewise and produced, DF equals FB and EH equals HA [definition 4]. If then we fix the straight lines BD and EA in position to be parallel, the points H and F will be given. And so HFC will be given in position.

Then it will be constructed thus: Let there be the given conic section on which are the points A, B, C, D, and E, and let the straight lines BD and AE be drawn parallel and be

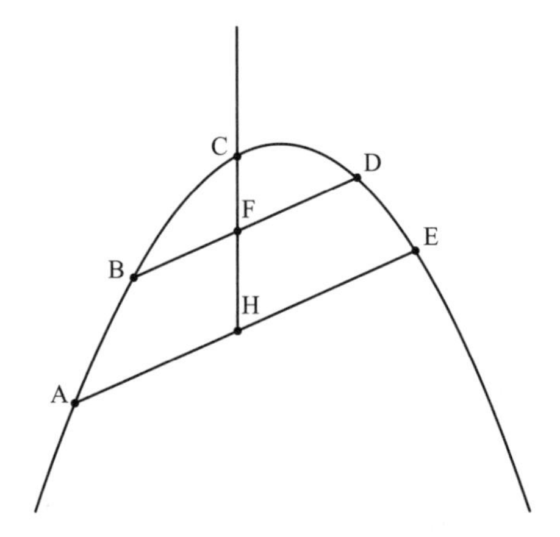

Figure 5.2.9.

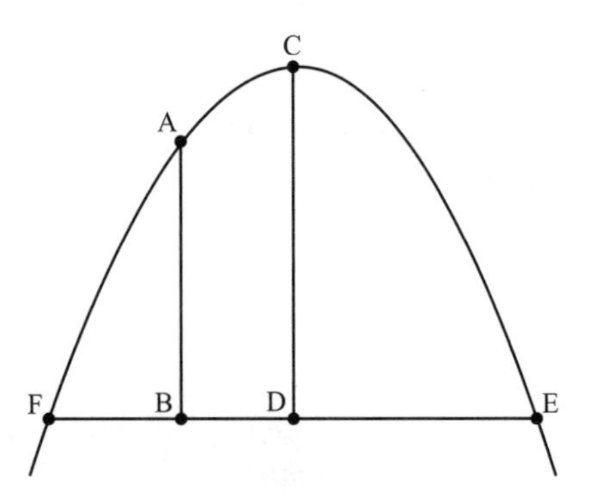

Figure 5.2.10.

bisected at F and H. And the straight line FH joined will be a diameter of the section. And in the same way we could also find an indefinite number of diameters.

Proposition 45. *Given an ellipse or hyperbola, to find the center.*

And this is evident; for if two diameters of the section, AB and CD, are drawn through [book 2, proposition 44], the point at which they cut each other will be the center of the section, as indicated.

Proposition 46. *Given a section of a cone, to find the axis.*

Let the given section of a cone first be a parabola, on which are the points F, C, and E. Then it is required to find its axis.

For let AB be drawn as a diameter of it [book 2, proposition 44]. If then AB is an axis, what was enjoined would have been done; but if not, let it have been done, and let CD be the axis; therefore the axis CD is parallel to AB [book 1, proposition 51 porism] and bisects the straight lines drawn perpendicular to it [definition 7]. And the perpendiculars to CD are also perpendiculars to AB; and so CD bisects the perpendiculars to AB [figure 5.2.10]. If then I fix EF, a perpendicular to AB, it will be given in position, and therefore ED equals

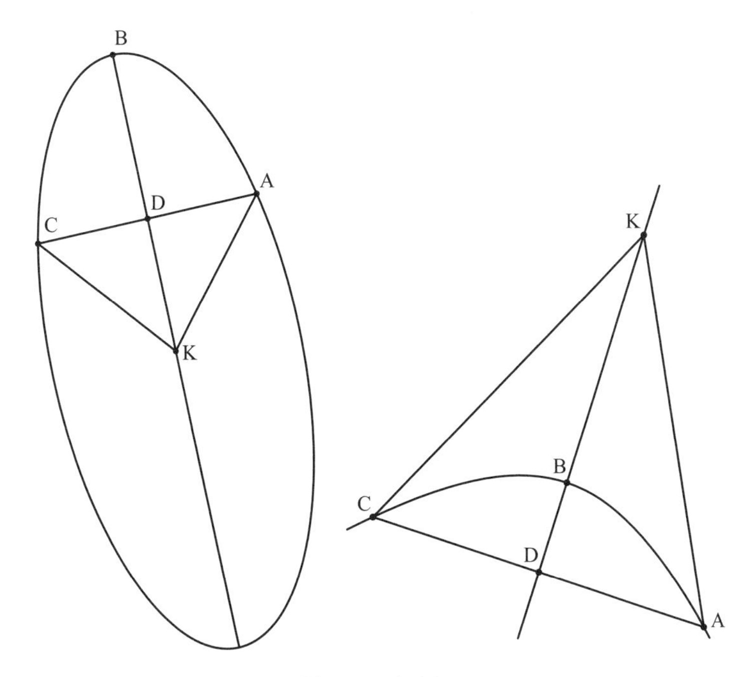

Figure 5.2.11.

DF; therefore the point D is given. Therefore through the given point D, CD has been drawn parallel to AB which is in given in position; therefore CD is given in position.

Then it will be constructed thus: Let there be the given parabola on which are the points F, E, and A, and let AB, diameter of it, be drawn [book 2, proposition 44], and let BE be drawn perpendicular to it, and let it be produced to F. If then EB equals BF, it is evident that AB is the axis [definition 7]; but if not, let EF be bisected by D, and let CD be drawn parallel to AB. Then it is evident that CD is the axis of the section; for being parallel to a diameter, that is, being a diameter [book 1, proposition 51 porism], it bisects EF at right angles. Therefore CD has been found as the axis of the given parabola [definition 7].

And it is evident that the parabola has only one axis. For if there is another, as AB, it will be parallel to CD [book 1, proposition 51 porism]. And it cuts EF, and so it also bisects it [definition 4]. Therefore BE equals BF; and this is absurd.

Proposition 47. *Given a hyperbola or ellipse, to find the axis.*

Let there be the hyperbola or ellipse ABC; then it is required to find its axis. Let it have been found and let it be KD, with K the center of the section; therefore KD bisects the ordinates to itself and at right angles [definition 7]. Let the perpendicular CDA be drawn, and let KA and KC be joined [figure 5.2.11]. Since then CD equals DA, therefore CK equals KA. If then we fix the given point C, CK will be given. And so the circle described with center K and radius KC will also pass through A and will be given in position. And the section ABC is also given in position; therefore the point A is given. But the point C is also given; therefore CA is given in position. Also, CD equals DA, therefore the point D is given. But K is also given; therefore DK is given in position.

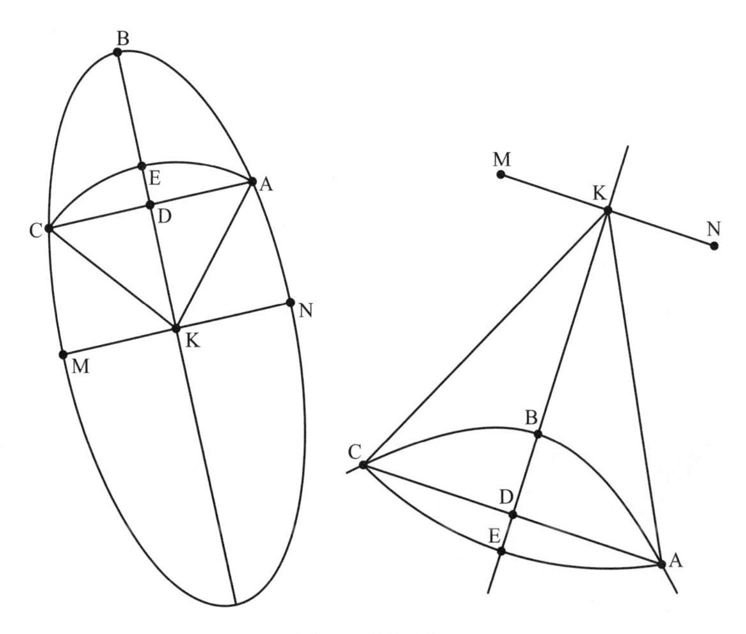

Figure 5.2.12.

Then it will be constructed thus: Let there be the given hyperbola or ellipse ABC, and let K be taken as its center; and let a point C be taken at random on the section, and let the circle CEA, with center K and radius KC, be described, and let CA be joined and bisected at D, and let KC, KD, and KA be joined, and let KD be drawn through to B [figure 5.2.12]. Since then AD equals DC, and DK is common, therefore the two straight lines CD and DK are equal to the two straight lines AD and DK, and base KA equals base KC. Therefore KBD bisects ADC at right angles. Therefore KD is an axis [definition 7]. Let MKN be drawn through K parallel to CA; therefore MN is the axis of the section conjugate to BK [definition 8].

Note that MN is not needed in this proposition but is needed in the next one.

Proposition 48. *Then with these things shown, let it be next in order to show that there are no other axes of the same sections.*

For if possible, let there also be another axis KG. Then in the same way as before, with AH drawn perpendicular, AH equals HL [definition 4]; and so also AK equals KL [figure 5.2.13]. But also AK equals KC; therefore KL equals KC; and this is absurd.

Now that the circle AEC does not hit the section also in another point between the points A, B, and C, is evident in the case of the hyperbola; and in the case of the ellipse let the perpendiculars CR and LS be drawn. Since then KC equals KL, for they are radii, also sq(KC) equals sq(KL). But sq(CR) together with sq(RK) equals sq(CK), and sq(KS) together with sq(SL) equals sq(LK); therefore sq(CR) with sq(RK) equals sq(KS) with sq(SL). Therefore the difference between sq(CR) and sq(SL) is equal to the difference between sq(KS) and sq(RK). Again, since rect(MR, RN) together with sq(RK) equals

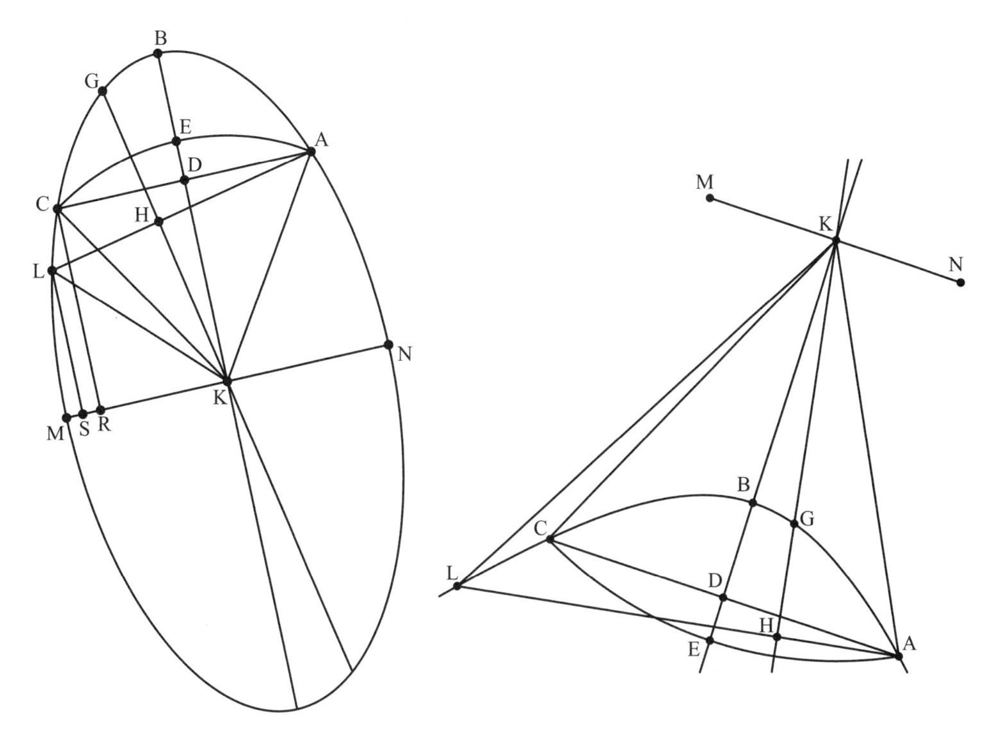

Figure 5.2.13.

sq(KM), and also rect(MS, SN) with sq(SK) equals sq(KM),[52] therefore rect(MR, RN) together with sq(RK) equals rect(MS, SN) together with sq(SK). Therefore the difference between sq(SK) and sq(RK) equals the difference between rect(MR, RN) and rect(MS, SN). And it was previously shown that the difference between sq(CR) and sq(SL) is equal to the difference between sq(KS) and sq(RK). Therefore the difference between sq(CR) and sq(SL) is equal to the difference between rect(MR, RN) and rect(MS, SN).

And since CR and LS are ordinates, sq(CR) is to rect(MR, RN) as sq(SL) is to rect (MS, SN) [book 1, proposition 21]. But the same difference was also shown for both, therefore sq(CR) equals rect(MR, RN) and sq(SL) equals rect(MS, SN). Therefore the line LCM is a circle; and this is absurd, for it is supposed an ellipse.

Apollonius's comment on the "obviousness" for the case of the hyperbola, in some ways, presages the line of investigation of book 4. Here, we have a more specific case where the hyperbola and the circle are concentric, whereas in book 4, this is not necessarily the case. The first contradictory result, that $KL = KC$, seems puzzling when first encountered. Just why it is absurd seems to depend on the rest of Apollonius's argument: that the circle AEC does not hit the section at any points between A, B, and C besides of course A and C themselves. To clarify what Apollonius means when he says there are no other axes: the conjugate axis (MN in figure 5.2.13) is linked by

[52] *Elements* 2:5.

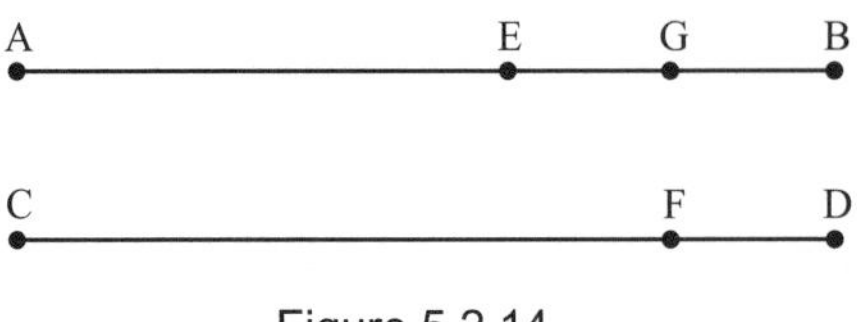

Figure 5.2.14.

virtue of this conjugacy to the other axis *BK*. Apollonius is showing that no other axis is like *BK*, and hence no other is like *MN*.

Let *AB* and *CD* be two equal magnitudes, and let them be cut into unequal sections at *E* and *F*. I say that the difference between *AE* and *FC* is the same as the difference between *EB* and *FD* [figure 5.2.14].

For let *AG* be made equal to *CF*: therefore *EG* is the excess of *AG*, *AE*, that is, of *CF*, *AE*: for the magnitude *AG* is equal to *CF*. But also *AB* is equal to *CD*, therefore also the remainder *GB* is equal to the remainder *FD*, so that *EG* is the remainder of *EB* and *BG*, or of *EB* and *FD*.[53]

But let there be four magnitudes *AE*, *EB*, *CF* and *FD*, and let *AE* differ from *CF* by the difference of *EB* and *FD*. I say that the sum of *AE*, *EB* is equal to the sum of *CF*, *FD*. For let *CF* be made equal to *AG*: therefore *EG* is the excess of *AE* and *CF*. But it was established that *EA*, *CF* and *EB*, *FD* differ from each other by the same [amount]: therefore *GB* is equal to *FD*. And also *AG* [is equal] to *CF*: therefore *AB* is equal to *CD*.

Indeed it is clear, that if a first exceeds a second by something, and if a third exceeds a fourth by the same, then the first and the fourth are equal to the second and the third, according to the so-called arithmetic mean. For if the same things are supposed, as the first is to the third, so too is the second to the fourth, the first will be equal to the third, and the second to the third. For it is possible to show this in the other cases through the proof in the twenty-fifth theorem of the fifth book of Euclid's *Elements*: If four magnitudes be proportional, the first and the fourth will be greater than the remaining two.

Book 3

The aim of book 3 of the *Conics* seems to be to present the results necessary to deal with the three- and four-line locus problems that, Apollonius claims, had been only partially solved by Euclid. However, Apollonius does not deal with these directly, so we conclude our excerpts from this book with a discussion of how the results presented enable the problem to be solved completely. Another development in book 3 is the appearance of the foci of the ellipse and hyperbola, although the points are not given any special name. It is also curious that in this book alone, there is no introductory

[53] Since *BG* is equal to *FD*.

letter. However, Eutocius does have an introductory letter to his commentary, so we begin with that.

The third book of the *Conics*, my dearest friend Anthemius, was regarded with great care by the ancients, as the many different editions of it make clear; but it neither has an introductory letter, as do the others, nor are worthy scholia to this book found written by our predecessors, although the things in it are worthy of consideration, as also Apollonius himself said in the introduction to all the books. But all the proofs are set down clearly for you by us, based on the previous books and the scholia to them.

Proposition 1. *If straight lines, touching a section of a cone or circumference of a circle, meet, and diameters are drawn through the points of contact meeting the tangents, the resulting vertically related triangles will be equal.*

Let there be the section of a cone or circumference of a circle AB, and let AC and BD, meeting at E, touch AB, and let the diameters of the section CB and DA be drawn through A and B, meeting the tangents at C and D [figure 5.2.15]. I say that tri(ADE) equals tri(EBC).

For let AF be drawn from A parallel to BD; therefore it has been dropped ordinatewise. Then in the case of the parabola, pllg($ADBF$) is equal to tri(ACF) [book 1, proposition 42], and, with the common area $AEBF$ subtracted, tri(ADE) equals tri(CBE).

And in the case of the others let the diameters meet at center G. Since then AF has been dropped ordinatewise, and AC touches, rect(FG, GC) equals sq(BG) [book 1, proposition 37]. Therefore FG is to GB as BG is to GC; therefore also FG is to GC as sq(FG) is to sq(GB). But sq(FG) is to sq(GB) as tri(AGF) is to tri(DGB), and FG is to GC as tri(AGF) is to tri(AGC); therefore also tri(AGF) is to tri(AGC) as tri(AGF) is to tri(DGB). Therefore tri(AGC) equals tri(DGB). Let the common area $DECG$ in the case of the hyperbola and $AGBE$ in the case of the ellipse be subtracted. Therefore as remainders, tri(AED) equals tri(CEB).

Proposition 17. *If two straight lines touching a section of a cone or circumference of a circle meet, and two points are taken at random on the section, and from them in the section are drawn parallel to the tangent straight lines cutting each other and the line of the section, then as the squares on the tangents are to each other, so will the rectangles contained by the straight lines taken similarly.*

Let there be the section of a cone or circumference of a circle AB; and tangents to AB, AC, and CB, meeting at C; and let points D and E be taken at random on the section, and through them let $EFIK$ and $DFGH$ be drawn parallel to AC and CB [figure 5.2.16]. I say that sq(CA) is to sq(CB) as rect(KF, FE) is to rect(HF, FD).

For let the diameters $ALMN$ and $BOXP$ be drawn through A and B, and let the tangents and parallels be produced to the diameters, and let DX and EM be drawn from D and E parallel to the tangents; then it is evident that KI is equal to IE and HG is equal to GD [book 1, propositions 46, 47]. Since then KE has been cut equally at I and unequally at F, rect(KF, FE) with sq(FI) equals sq(EI).[54] And since the triangles are similar because

[54] *Elements* 2:5.

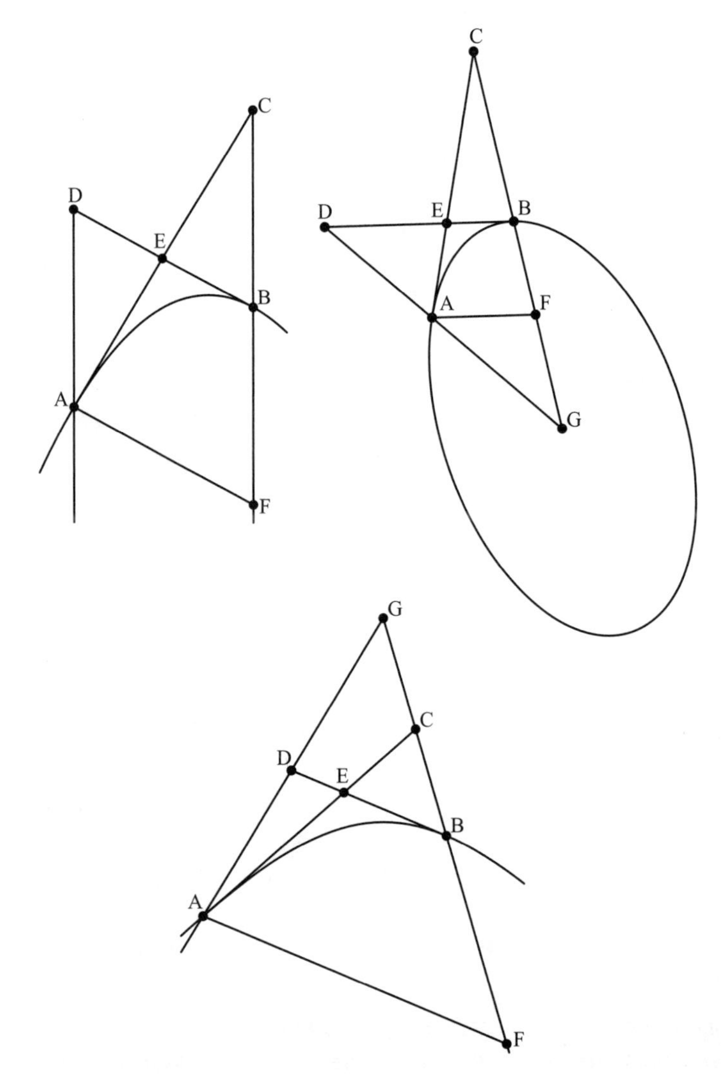

Figure 5.2.15.

of the parallels, whole sq(EI) is to whole tri(IME) as part subtracted sq(IF) is to part subtracted tri(FIL). Therefore also remainder rect(KF, FE) is to remainder qdrl(FM) as whole sq(EI) is to whole tri(IME).

But sq(EI) is to tri(IME) as sq(CA) is to tri(CAN). Therefore rect(KF, FE) is to qdrl(FM) as sq(CA) is to tri(CAN). But tri(CAN) equals tri(CPB) [book 3, proposition 1], and qdrl(FM) equals qdrl(FX) [book 3, proposition 3]; therefore rect(KF, FE) is to qdrl(FX) as sq(CA) is to tri(CPB). Then likewise it could be shown that rect(HF, FD) is to qdrl(FX) as sq(CB) is to tri(CPB). Since then rect(KF, FE) is to qdrl(FX) as sq(CA) is to tri(CPB), and inversely qdrl(FX) is to rect(HF, FD) as tri(CPB) is to sq(CB). Therefore *ex aequali*, sq(CA) is to sq(CB) as rect(KF, FE) is to rect(HF, FD).

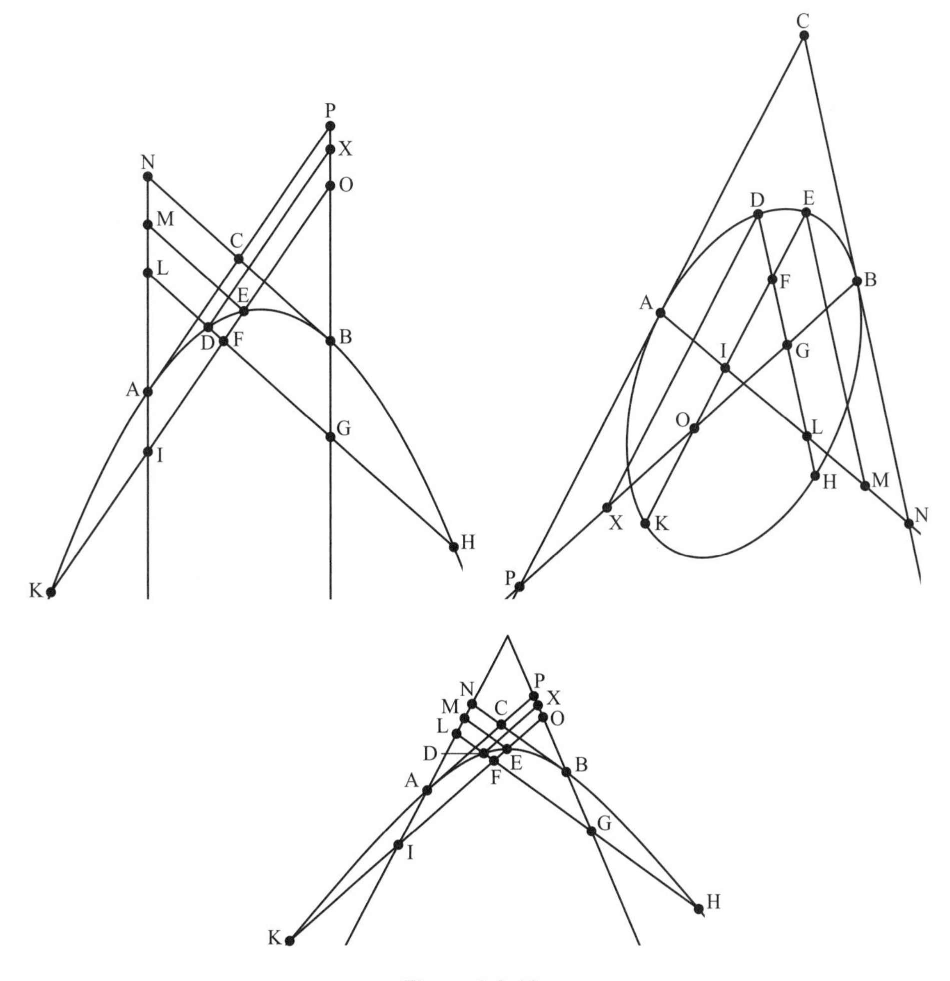

Figure 5.2.16.

This theorem also was set down similarly to the one before it, the very one which we, having set them aside there as cases, have written. If, in the case of the ellipse or the perimeter of the circle, the diameters being drawn through the points of section be parallel to the tangents BC, CA, and so it is, that as sq(CA) is to sq(CB), so rect(KF, FE) is to rect(DF, FH). For let ordinates DP and HM be drawn through D and H [figure 5.2.17]. So since it is, that as sq(AC) is to sq(CB), so too is sq(BN) to sq(NA), that is, rect(AN, NL); but as sq(BN) is to rect(AN, NL), so sq(DP), that is, sq(FO), is to rect(AP, PL), and sq(EO) to rect(AO, OL); and therefore the remainder is to the remainder, as the whole is to the whole. But if, from sq(EO), sq(DP) be subtracted, that is, sq(FO) [be subtracted], rect(KF, FE) remains: for KO is equal to OE: but if, from rect(AO, OL) be subtracted rect(AP, PL), then rect(MO, OP) remains, that is, rect(HF, FD) [remains]: for AP is equal to ML and PN is equal to NM. Therefore it is, that as sq(CA) is to sq(CB), the remainder, rect(KF, FE), is to rect(DF, FH).

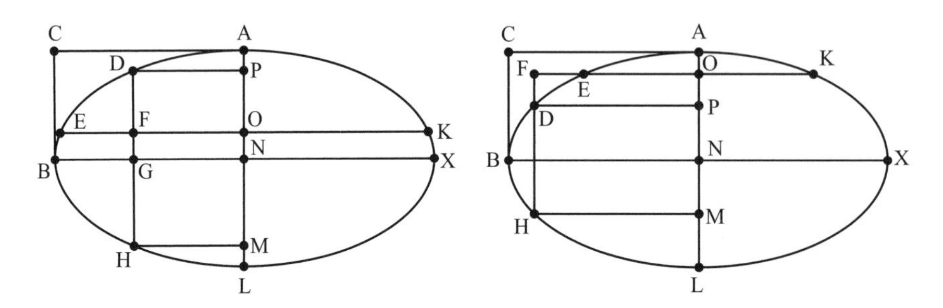

Figure 5.2.17.

But whenever the point F is outside the section, it is necessary to invert the additions and subtractions.

We can consider this result as the generalization of *Elements* 3:35: the rectangles contained by the segments of intersecting chords are shown to have the same ratio as the squares of two tangents drawn from a point to which the chords are parallel—the tangents are equal in the case of the circle, of course, and thus the rectangles.

We include the next proposition because it is used in the proof of proposition 1 of book 4. We omit its proof.

Proposition 37. *If two straight lines touching a section of a cone or circumference of a circle or opposite sections meet, and a straight line is joined to their points of contact, and from the point of meeting of the tangents some straight line is drawn across cutting the line of the section at two points, then as the whole straight line is to the straight line cut off outside, so will the segments produced by the straight line joining the points of contact be to each other.*

Proposition 45. *If in a hyperbola or ellipse or circumference of a circle or opposite sections straight lines are drawn from the vertex of the axis at right angles, and a rectangle equal to the fourth part of the figure is applied to the axis on each side and exceeding by a square figure in the case of the hyperbola and opposite sections, but deficient in the case of the ellipse, and some straight line is drawn tangent to the section and meeting the perpendicular straight lines, then the straight lines drawn from the points of meeting to the points produced by the application make right angles at the aforesaid points.*

Let there be one of the sections mentioned whose axis is AB, and AC and BD are at right angles, and CED tangent, and let the rectangle AF, FB and the rectangle AG, GB equal to the fourth part of the figure be applied on each side, as has been said, and let CF, CG, DF, and DG be joined [figure 5.2.18]. I say that angle CFD and angle CGD are each a right angle.

For since it has been shown that rect(AC, BD) equals a fourth of the figure on AB,[55] and since also rect(AF, FB) equals a fourth of the figure on AB, therefore rect(AC, BD)

[55] This was shown in *Elements* 3:42, but it is straightforward to demonstrate it analytically.

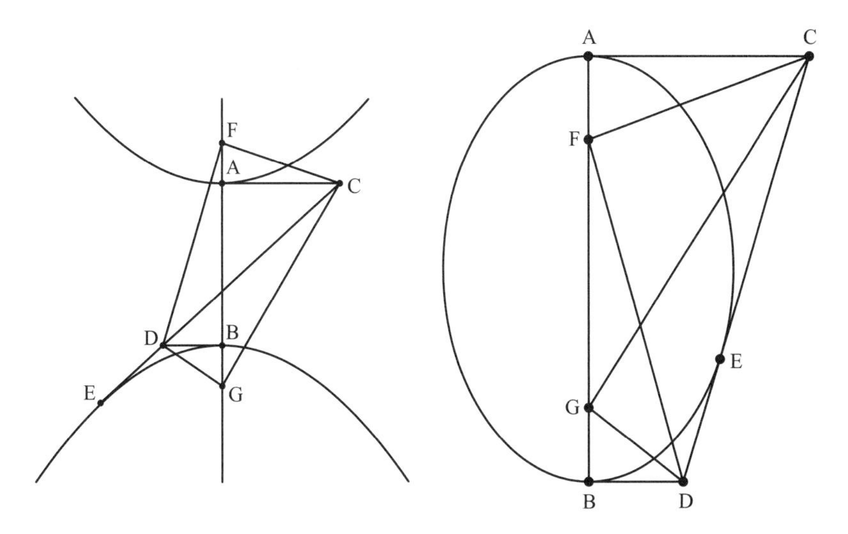

Figure 5.2.18.

equals rect(AF, FB). Therefore, AC is to AF as FB is to BD. And the angles at points A and B are right; therefore angle ACF equals angle BFD, and angle AFC equals angle FDB. And since angle CAF is right, therefore angles ACF and AFC together make a right angle. And it has also been shown that angle ACF equals angle DFB; therefore angles AFC and DFB together make a right angle. Therefore angle DFC is a right angle. Then likewise it could also be shown that angle CGD is a right angle.

The "points produced by the application" today we call the foci of the conic sections. If, in the ellipse, we take AB as the x axis with the origin at A, then the point B has coordinate $2a$, the center of the ellipse has coordinate a, and the two foci are the points satisfying $x(2a-x)=b^2$, where, recall, one-fourth of the "figure" is $\frac{1}{4}2ap=\frac{pa}{2}=b^2$. The two solutions of this quadratic equation are $x=a\pm\sqrt{a^2-b^2}=a\pm c$, if we set $c^2=a^2-b^2$. Thus the two foci of the ellipse have coordinates $a+c$ and $a-c$ and are both at distance c from the center. In the case of the hyperbola, with one vertex at the origin, if we set $c^2=a^2+b^2$, the two foci are also at distances c from the origin and therefore at distance $c-a$ from the vertices.

Proposition 46. *With the same things being so, the straight lines joined make equal angles with the tangents.* [Angle ACF equals angle DCG, and angle CDF equals angle BDG; see figure 5.2.18.]

Proposition 47. *With the same things being so, the straight line drawn from the point of meeting of the joined straight lines to the point of contact will be perpendicular to the tangent.* [CG and FD meet at H; CD and BA produced meet at K, and EH is joined; then EH is perpendicular to CD; see figure 5.2.19.]

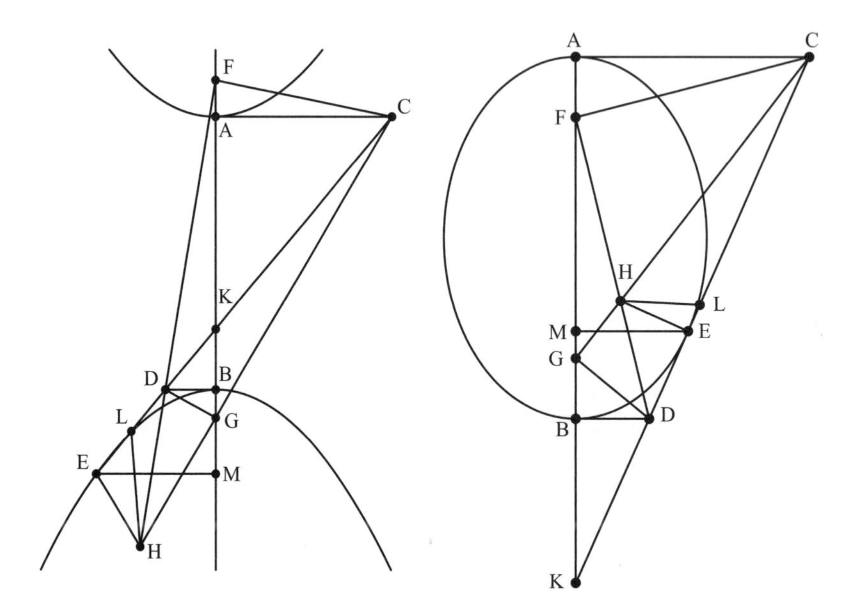

Figure 5.2.19.

Propositions 46 and 47 are easily proved analytically. Proposition 48 gives the familiar result that the lines from two foci to any point on the ellipse (or hyperbola) make equal angles with the tangent to the curve at that point.

Proposition 48. *With the same things being so, it must be shown that the straight lines drawn from the point of contact to the points produced by the application make equal angles with the tangent.*

For let the same things be supposed, and let EF and EG be joined [figure 5.2.20]. I say than angle CEF is equal to angle GED.

For since angles DGH and DEH are right angles [book 3, propositions 45, 47], the circle described about DH as a diameter will pass through the points E and G; and so angle DHG equals angle DEG, for they are in the same segment. Likewise then also angle CEF equals angle CHF. But angle CHF equals angle DHG, for they are vertical angles; therefore also angle CEF equals angle DEG.

Proposition 49. *With the same things being so, if from one of the points of application a perpendicular is drawn to the tangent, then the straight lines from that point to the ends of the axis make a right angle.* [If GH is drawn from G perpendicular to CD and AH and BH are joined, then angle AHB is a right angle; see figure 5.2.21.]

Proposition 50. *With the same things being so, if from the center of the section there falls to the tangent a straight line parallel to the straight line drawn through the point of contact and one of the points of application, then it will be equal to one half the axis.* [With H the center, let EF be joined and let DC and BA meet at K. Through H draw HL parallel to EF. Then HL equals HB; see figure 5.2.22.]

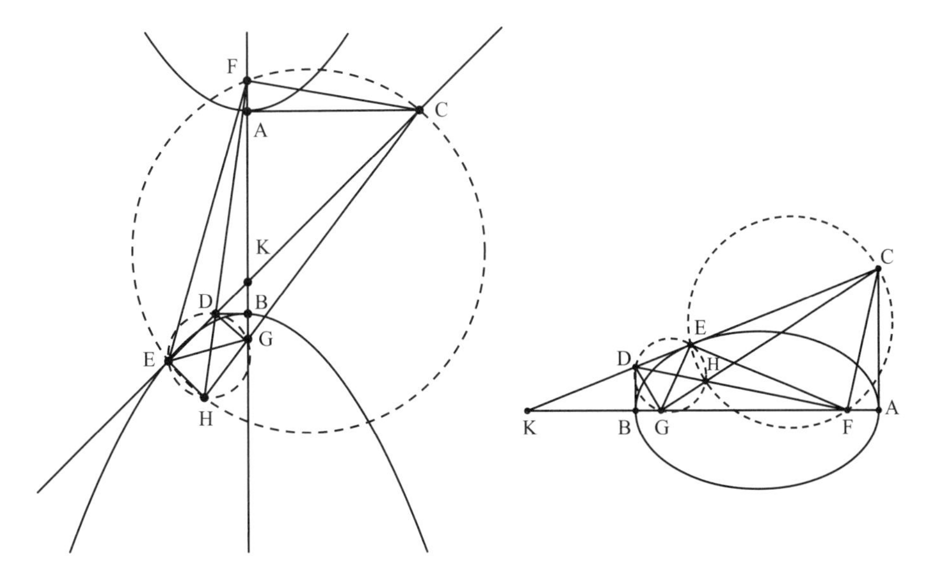

Figure 5.2.20.

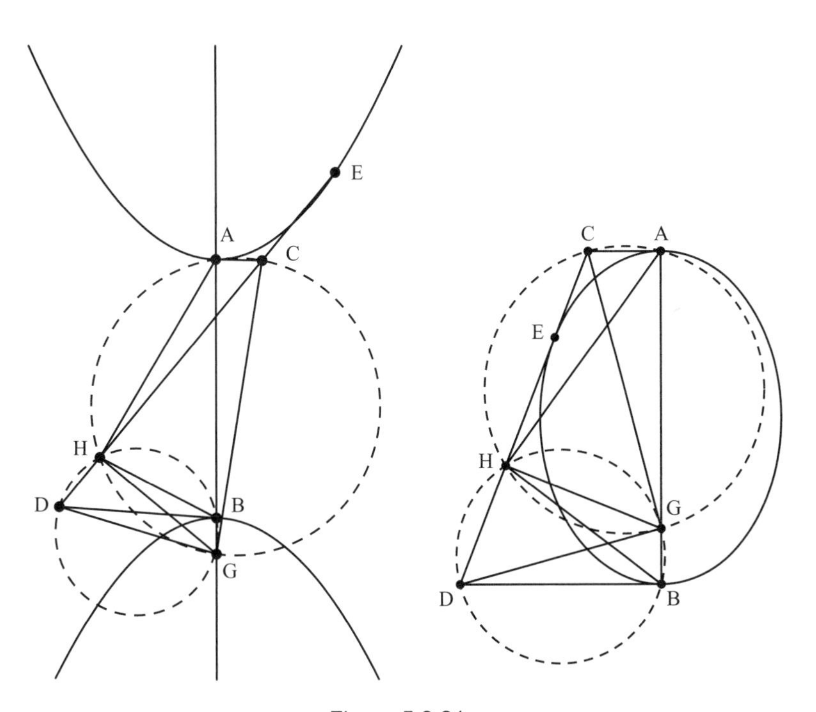

Figure 5.2.21.

Proposition 51. *If a rectangle equal to the fourth part of the figure is applied from both sides to the axis of a hyperbola or opposite sections and exceeding by a square figure, and straight lines are deflected from the resulting points of application to either one of the sections, then the greater of the two straight lines exceeds the less by exactly as much as the axis.*

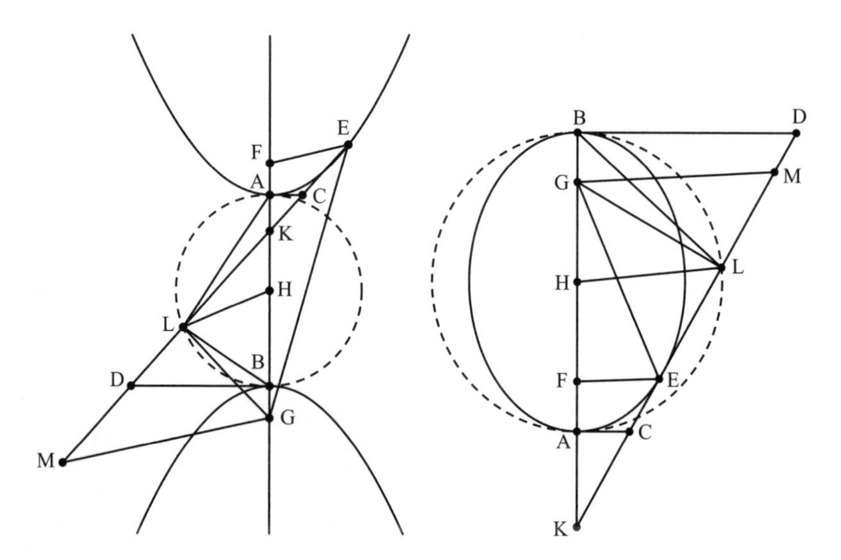

Figure 5.2.22.

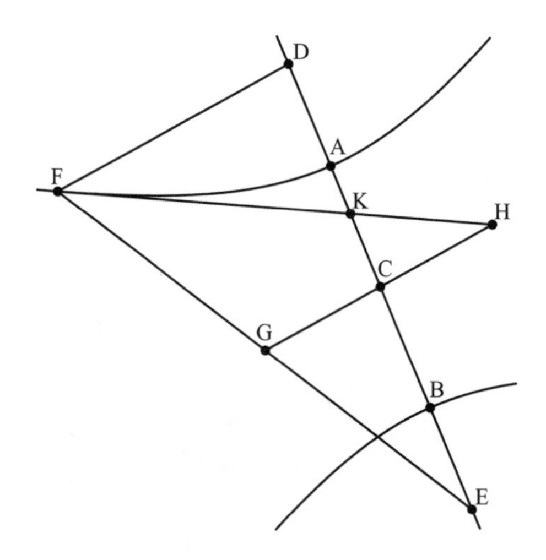

Figure 5.2.23.

For let there be a hyperbola or opposite sections whose axis is AB and center C, and let each of the rectangles AD, DB and AE, EB be equal to the fourth part of the figure, and from points E and D let the straight lines EF and FD be deflected to the line of the section [figure 5.2.23]. I say that EF equals FD, AB taken together.

Let FKH be drawn tangent through F, and GCH through C parallel to FD; therefore angle KHG equals angle KFD; for they are alternate. And angle KFD equals angle GFK [book 3, proposition 48]; therefore GF equals GH. But GF is equal to GE, since also AE is equal to BD and AC is equal to CB and EC is equal to CD; and therefore GH is equal

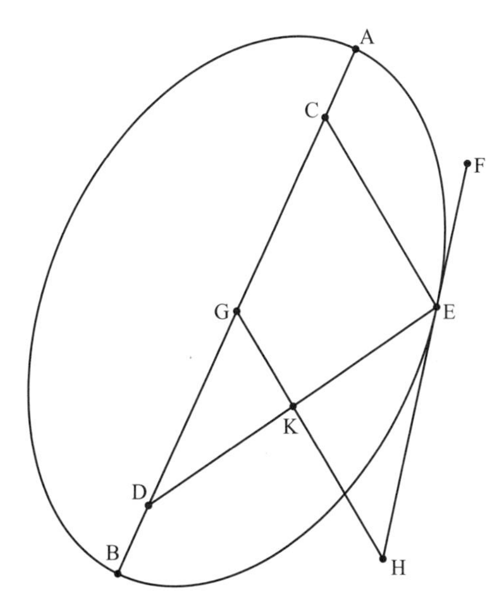

Figure 5.2.24.

to EG. And so FE is double GH. And since it has been shown [book 3, proposition 50] that CH equals CB, therefore FE is twice GC and CB taken together. But FD is twice GC, and AB is twice CB, therefore FE equals FD and AB taken together. And so EF is greater than FD by AB.

Proposition 52. *If in an ellipse a rectangle equal to the fourth part of the figure is applied from both sides to the major axis and deficient by a square figure, and from the points resulting from the application straight lines are deflected to the line of the section, then they will be equal to the axis.*

Let there be an ellipse, whose major axis is AB, and let each of the rectangles AC, CB and AD, DB be equal to the fourth of the figure, and from C and D let the straight lines CE and ED have been deflected to the line of the section [figure 5.2.24]. I say that CE and ED taken together are equal to AB.

Let FEH be drawn tangent, and G be the center, and through it let GKH be drawn parallel to CE. Since then angle CEF equals angle HEK [book 3, proposition 48], and angle CEF equals angle EHK, therefore also angle EHK equals angle HEK. Therefore also HK equals KE. And since AG equals GB, and AC equals DB, therefore also GC equals GD; and so also EK equals KD. And for this reason ED is double HK, and EC is double KG, and ED and EC together are double GH. But also AB is double GH [book 3, proposition 50]; therefore AB is equal to ED and EC taken together.

Readers will likely recognize the similarity between propositions 51 and 52 and a modern view of the hyperbola and ellipse as being a locus of points with differences or sums of distances to the focal points being constant. Choose two points F_1 and F_2, and choose some length $2a$ that exceeds the distance between F_1 and F_2. We consider the set of points X for which $d(F_1,X) \mp d(F_2,X) = 2a$. Then in the subtractive case the resulting curve(s) is the hyperbola or opposite sections, and in the additive case the resulting curve is the ellipse. It is critical to note, however, that Apollonius himself does not conceive of the hyperbola or ellipse in this fashion: they are sections of cones, having fundamental properties established in book 1. These two propositions reveal their further properties, not their genesis. The word "focus" for what Apollonius calls the "points produced by the application" was first used by Johannes Kepler in 1604.

Proposition 54. *If two tangents to a section of a cone or to a circumference of a circle meet, and through the points of contact parallels to the tangents are drawn, and from the points of contact, to the same point of the line of the section, straight lines are drawn across cutting the parallels, then the rectangle contained by the straight lines cut off to the square on the straight line joining the points of contact has a ratio compounded of the ratio which the inside segment line joining the point of meeting of the tangents and the midpoint of the straight line joining the points of contact has in square to the remainder, and of the ratio which the rectangle contained by the tangents has to the fourth part of the square on the straight line joining the points of contact.*

Let there be a section of a cone or circumference of a circle ABC and tangents AD and CD, and let AC be joined and bisected at E, and let DBE be joined, and let AF be drawn from A parallel to CD, and CG from C parallel to AD, and let some point H on the section be taken, and let the straight lines AH and CH be joined and produced to G and F [figure 5.2.25]. I say that rect(AF, CG) is to sq(AC) as the ratio of sq(EB) to sq(BD) compounded with the ratio of rect(AD, DC) to a fourth of sq(AC) or rect(AE, EC).

For let $KHOXL$ be drawn from H parallel to AC, and from B, MBN parallel to AC; then it is evident that MN is tangent. Since then AE equals EC, also MB equals BN. And KO equals OL, and HO equals OX, and KH equals XL. Since then MB and MA are tangents and KHL has been drawn parallel to MB, sq(AM) is to sq(MB) as sq(AK) is to rect(XK, KH) [book 3, proposition 16], or sq(AM) is to rect(MB, BN) as sq(AK) is to rect(LH, HK). And rect(NC, AM) is to sq(AM) as rect(LC, AK) is to sq(AK); therefore *ex aequali* rect(NC, AM) is to rect(MB, BN) as rect(LC, AK) is to rect(LH, HK).

But rect(LC, AK) is to rect(LH, HK) as the ratio of LC to LH compounded with the ratio of AK to HK, or rect(LC, AK) is to rect(LH, HK) as the ratio of FA to AC compounded with the ratio of GC to CA, which is the same as the ratio of rect(GC, FA) to sq(CA). Therefore rect(NC, AM) is to rect(MB, BN) as rect(GC, FA) is to sq(CA). But with rect(ND, DM) taken as mean, rect(NC, AM) is to rect(MB, BN) as the ratio of rect(NC, AM) to rect(ND, DM) compounded with the ratio of rect(ND, DM) to rect(MB, BN); therefore rect(GC, FA) is to sq(CA) as the ratio of rect(NC, AM) to rect(ND, DM) compounded with the ratio of rect(ND, DM) to rect(MB, BN). But rect(NC, AM) is to rect(ND, DM) as sq(EB) is to sq(BD), and rect(ND, DM) is to rect(NB, BM) as rect(CD, DA) is to rect(CE, EA); therefore rect(AF, CG) is to sq(AC) as the ratio of sq(EB) to sq(BD) compounded with the ratio of rect(AD, DC) to rect(CE, EA).

"And rect(NC, AM) is to sq(AM) as rect(LC, AK) is to sq(AK). . . ." For since it is, as AD is to DM, so too is CD to DN, *convertendo*, as DA is to AM, so too is DC to CN. Because of the same reasons and *invertendo*, as KA is to AD, so too is LC to CD: therefore *ex aequali*, as MA is to AK, so too is NC to CL: and *alternando*, as

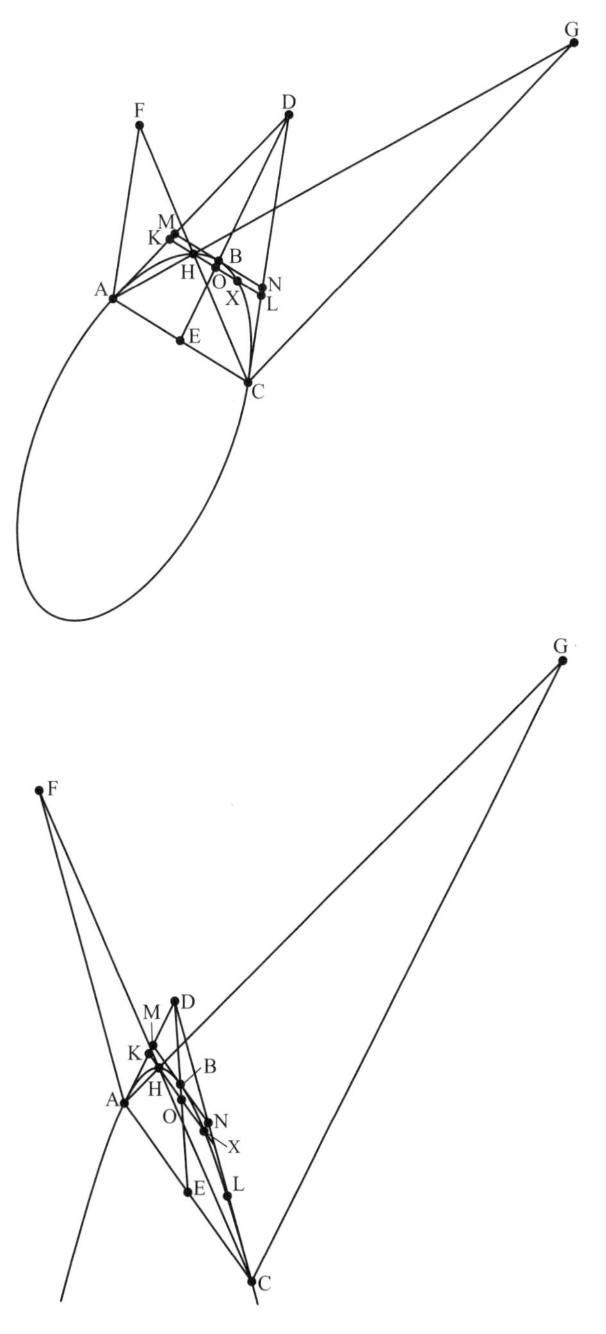

Figure 5.2.25.

MA is to NC, so too is KA to LC. Therefore also as rect(NC, AM) is to sq(AM), so too is rect(LC, KA) to the sq(KA).

"But rect(NC, AM) is to rect(ND, DM) as sq(EB) is to sq(BD)...." For since rect(AM, CN) has to rect(ND, DM) a ratio compounded from the ratio of AM to MD and the ratio of CN to ND; but as AM is to MD, so too is EB to BD; and as CN is to ND, so too is EB to BD; therefore rect(AM, CN) has to rect(ND, DM) a ratio double

that which EB has to BD. But sq(EB) has to sq(BD) a ratio double that of EB to BD: and therefore as rect(AM, CN) is to rect(ND, DM), so too is sq(EB) to sq(BD).

"And rect(ND, DM) is to rect(NB, BM) as rect(CD, DA) is to rect(CE, EA). . . ." For since rect(ND, DM) has to rect(NB, BM) a ratio compounded out of the ratios of DN to NB and DM to MB, but as DN is to NB, so too is DC to CE, and as DM is to MB, so too is DA to AE, therefore it will have a ratio compounded out of the ratio of DC to CE and DA to AE, which is the same as that which rect(CD, DA) has to rect(CE, EA). Therefore as rect(ND, DM) is to rect(NB, BM), so too is rect(CD, DA) to rect(CE, EA).

As noted earlier, although Apollonius mentioned the three- and four-line locus problems in his introduction to the *Conics* and said that the results of book 3 would enable the construction of these loci to be completed, it is curious that he does not mention these problems explicitly in this book. Nevertheless, the three-line locus property of conics is easily deduced for the ellipse, hyperbola, parabola, and circle from book 3, proposition 54, while for opposite sections it can be deduced from book 3, propositions 55 and 56. The three-line locus property of conics can be stated thus: Any conic section or circle or pair of opposite sections can be considered as the locus of points whose distances from three given fixed straight lines (the distances being either perpendicular or at a given constant angle to each of the given straight lines, although the constant angle may be different for each of the three straight lines) are such that the square of one of the distances is always in a constant ratio to the rectangle contained by the other two distances.[56]

It is shown in book 3, proposition 54, that in the case of conic sections and circles, rect(AF, CG) : sq(AC) = sq(EB) : sq(BD) compounded with rect(AD, DC) : one-fourth sq(AC). Now if we consider the straight lines AD, DC, and AC as fixed and given and therefore straight line DE as fixed and given as bisecting AC, then it is evident that the straight lines AC, EB, BD, AD, DC, and therefore the squares on them and the rectangles contained by them, are also fixed and given [figure 5.2.26]. Then although as the point H is taken at different points along the conic, the straight lines AF and CG change in magnitude, nevertheless the magnitude of rect(AF, CG), because of the above proportion, remains constant.

For let HX be drawn parallel to BE, and HY to AD, and HZ to DC. Then HX is the distance from H to AC at a given angle, and AY because of parallels represents the distance from H to AD at another given angle, and ZC represents the distance from H to DC at another given angle. Then by similar triangles, $CZ : ZH = AC : AF$ and $AY : YH = AC : CG$; therefore compounding, rect(CZ, AY) : rect(ZH, YH) = sq(AC): rect(AF, CG). Now we have seen that rect(AF, CG) is a constant magnitude as the point H changes, and that sq(AC) is constant; therefore their ratio is constant. Therefore rect(CZ, AY) : rect(ZH, YH) is a constant ratio (statement 1).

[56] The four-line locus problem can be deduced from the three-line problem by considering it as multiple instances of the latter and combining the appropriate results. For details, see Apollonius of Perga, *Conics*, *Books I–III*, trans. R. Catesby Taliaferro (Santa Fe, NM: Green Lion, 2000), 267–75.

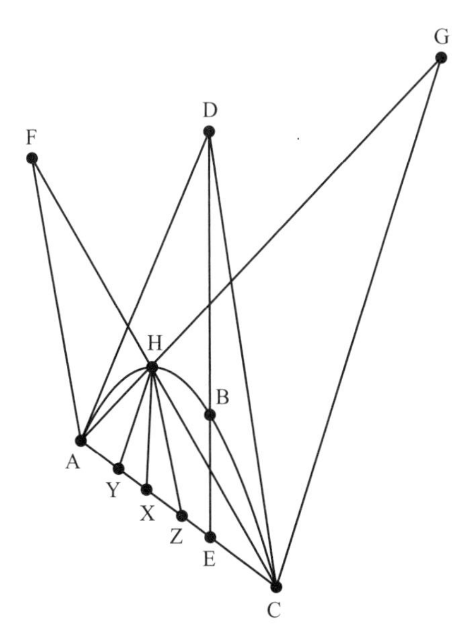

Figure 5.2.26.

Again by similar triangles, $ZH:HX=CD:DE$ and $YH:HX=AD:DE$; therefore compounding, rect(ZH, YH) : sq(HX) = rect(CD, AD) : sq(DE). But rect(CD, AD) and sq(DE) are constant magnitudes as the point H changes; therefore their ratio is constant. Therefore rect(ZH, YH) : sq(HX) is a constant ratio (statement 2).

Compounding statements (1) and (2), we get a constant ratio; that is, rect(CZ, AY): sq(HX) is a constant ratio. In other words, as the point H changes, the rectangle contained by the distances from H to two of the given straight lines (at given angles to those straight lines) has a constant ratio to the square on the distance to the third straight line (at a given angle to that straight line). And it can be also be proved by means of similar triangles that if any three angles are chosen for the distances other than those chosen here for demonstration, then the corresponding ratio will be constant, although not equal.

Book 4

This book deals mostly with the number of points in which two different conic sections can intersect or be tangent. In particular, proposition 25 shows that two conics can not intersect at more than four points, while the two following propositions deal with the question of tangents. More specialized results are found in propositions 28 through 54, while propositions 55 through 57 are analogous to propositions 25 through 27 but deal with the case of opposite sections. We present proofs only for propositions 1, 9, and 25, and give just the statements and diagrams of some other propositions. Eutocius gives an introduction to this book, which we present below (shown in alternate typeface), but comments on only a very few of the propositions, giving some alternate proofs.

Apollonius's Opening Letter

Apollonius to Attalus:[57] Greetings!

Earlier, I presented the first three books of my eight book treatise on conics to Eudemus of Pergamum, but with his having passed away I decided to write out the remaining books for you, because of your earnest desire to have them. To start, then, I am sending you the fourth book. This book treats of the greatest number of points at which sections of a cone can meet one another or meet a circumference of a circle, assuming that these do not completely coincide, and, moreover, the greatest number of points at which a section of a cone or circumference of a circle can meet the opposite sections. Besides these questions, there are more than a few others of similar character. Conon of Samos presented the first mentioned question to Thrasydaeus without giving a correct demonstration, for which he was rightly attacked by Nicoteles of Cyrene.[58] As for the second question, Nicoteles, in replying to Conon, only mentions that it can be demonstrated, but I have found no demonstration either by him or by anyone else. And all these things just spoken of, whose demonstrations I have not found anywhere, require many and various striking theorems, of which most happen to be presented in the first three books of my treatise on conics, and the rest in this book. The investigations of these theorems is also of considerable use in the synthesis of problems and limits of possibility. So, Nicoteles was not speaking truly when, for the sake of his argument with Conon, he said that none of the things discovered by Conon were of any use for limits of possibility; but even if the limits of possibility are able to be obtained completely without these things, yet, surely, some matters are more readily perceived by means of them, for example, whether a problem might be done in many ways, and in how many ways, or again, whether it might not be done at all. Moreover, this preliminary knowledge brings with it a solid starting point for investigations, and the theorems are useful for the analysis or limits of possibility. But apart from such usefulness, these things are also worthy of acceptance for the demonstrations themselves: indeed, we accept many things in mathematics for this and no other reason.

The fourth book, my dear friend Anthemius, has an examination of the various ways the sections of a cone intersect each other and the perimeter of a circle, either as tangent or as cutting; but it is also elegant and clear to those chancing upon it, and especially from the different editions, but are not wanting of scholia: for the marginalia fill the wanting. And all the things in it were shown by a reduction to absurdity, just as also Euclid showed the things concerning the sections of the circle and the points of contact. But the manner of proof[59] is useful and necessary; this [manner of proof] also was pleasing to Aristotle and the geometers and especially to Archimedes. So it will be possible for you, reading the four books, through the diligent study of the *Conics*, to investigate analytically, and to construct the proposed: on which account Apollonius himself in the beginning of the book says that the four books consist of a course in the elements, but that the remaining books are a fuller treatment. So read it carefully, and if you want me to give a presentation of the following on the same model, I will do it, God willing. Be well.

[57] Nothing is known about Attalus, the person to whom the opening letters of books 4–7 are addressed.

[58] Conon was mentioned favorably by Archimedes, but nothing is known of either Thrasydaeus or Nicoteles.

[59] That is, proof by contradiction.

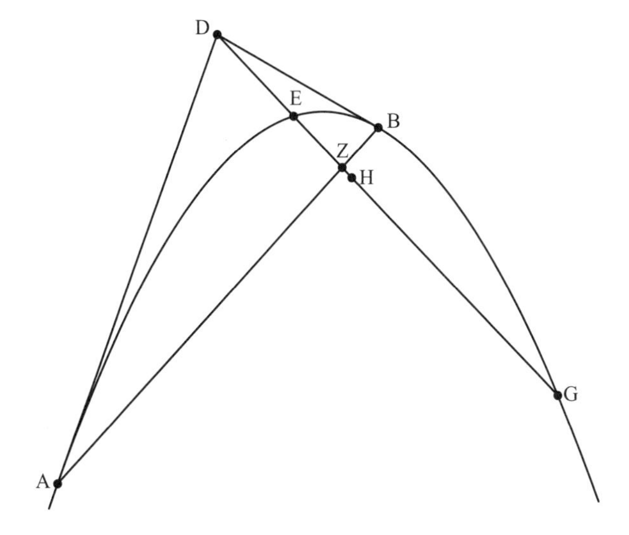

Figure 5.2.27.

Proposition 1. *If a point be taken outside a section of a cone or circumference of a circle, and from this same point two lines are extended towards the section, of which one touches the section and the other cuts the section at two points, and if the line cut off inside the curve be divided in that ratio which the whole line cut off has to the part outside bounded between the point and the curve, so that the homologous lines are at the same point, then the line drawn from the point of contact to the point of division [i.e., the point dividing the inside segment] will intersect the curve, and the line drawn from the point of intersection will touch the curve.*

For let ABG be a conic section or circumference of a circle, and let the point D be taken outside the section; from D let DB touch the section at point B and let DEG cut the section at points E, G, and let GZ have to ZE the ratio which GD has to DE [figure 5.2.27]. I say that the line drawn from B to Z will intersect the section, and the line drawn from the point of intersection D will touch the section.

For let DA be drawn from D touching the section, and let BA be joined cutting EG, if possible, not at Z, but at H. Now, since BD, DA touch the section, BA is drawn from the points of contact, and GD goes through AB, cutting the section at G, E, and intersecting AB at H: so as GD is to DE, so too is GH to HE [book 3, proposition 37]. But this is absurd, for it was assumed that as GD is to DE, so is GZ to ZE.

Therefore, BA does not cut GE in a different point from Z; therefore, it cuts GE at Z.

Proposition 9. *If from the same point two lines are drawn each cutting a section of a cone or circumference of a circle in two points, and if the segments cut off inside are divided in the same ratio as the wholes are to the segments cut off outside, so that the homologous lines are at the same point, then the line drawn through the points of division will intersect the section in two points, and lines drawn from the points of intersection to the point outside will touch the curve.*

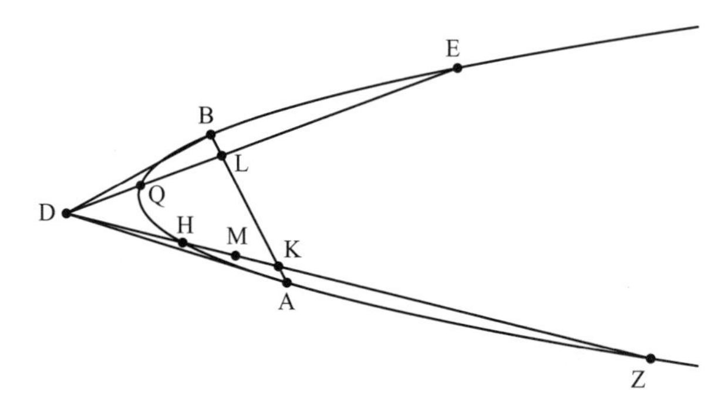

Figure 5.2.28.

Let AB be a curve such as we have described, and from a point D let DE and DZ be drawn cutting the curve at Q, E and at Z, H, respectively [figure 5.2.28]. Furthermore, let the ratio that EL has to LQ be the same ratio that DE has to QD, and that which ZK has to KH be the same that DZ has to DH. I say that the line joined from L to K will intersect the section at both ends, and the lines joined from the points of intersection will touch the section.

For since ED and ZD both cut the section at two points, it is possible to draw a diameter of the section through D, and, with that, also lines touching the section on either side. Let lines DB and DA be drawn touching the section, and join BA not passing through L, K, if possible, but through only one of the two, or through neither.

First, let it pass through L only and let it cut ZH at M. As ZD is to DH, therefore, so is ZM to MH [book 3, proposition 37]. But this is absurd, for it has been assumed that as ZD is to DH, so is ZK to KH.

If BA passes through neither L nor K, then the absurdity occurs with regards to each line DE, DZ.

Proposition 25. *A section of a cone does not cut a section of a cone or circumference of a circle at more than four points.*

For, if possible, let them cut at five points A, B, G, D, E, and let the points of intersection A, B, G, D, E be taken in succession so that no point of intersection between them is left out, and let AB, GD be joined and produced. So, these will intersect outside the sections in the cases of the parabola and hyperbola [book 2, propositions 24, 25].[60] Let them intersect at L, and let the ratio that AL has to LB be that which AO has to OB, [and] the ratio that DL has to LG be that which DP has to PG. Therefore, the line joined from P to O and produced will intersect the section on each side, and the lines joining the points of intersection and L will touch the section [book 4, proposition 9]. So let it intersect at Q, R, and let QL, LR

[60] Conics, book 2, propositions 24 and 25, are as follows: "If two straight lines meet a parabola/hyperbola each at two points, and if a point of meeting of neither one of them is contained by the points of meeting of the other, then the straight lines will meet each other outside the section (but [in the case of the hyperbola] within the angle containing the section)."

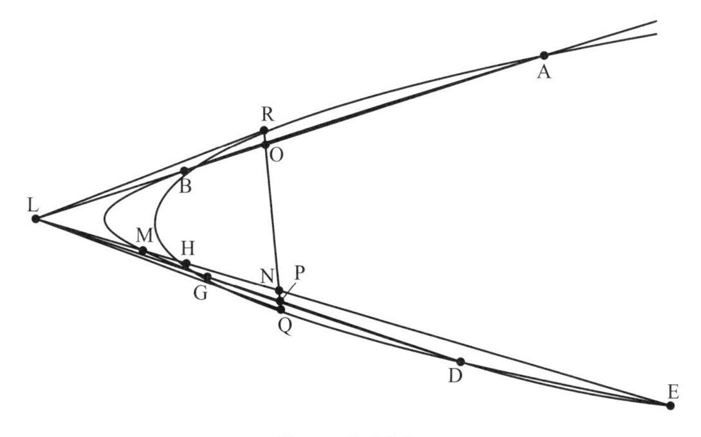

Figure 5.2.29.

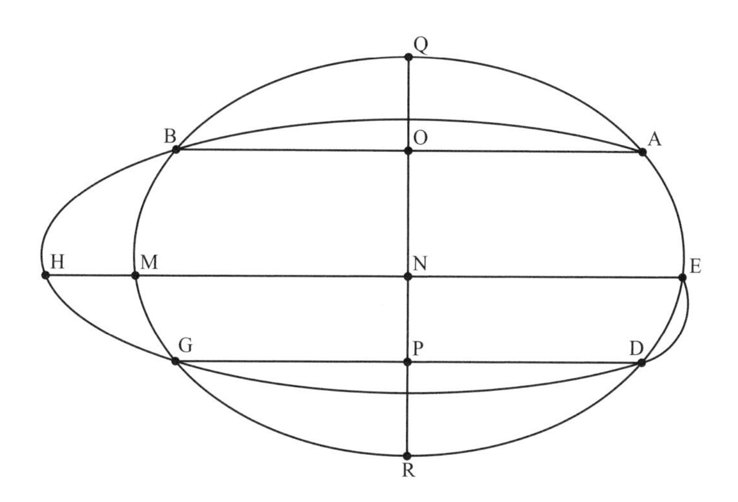

Figure 5.2.30.

be joined [figure 5.2.29]. Hence, they touch the section. Therefore, seeing that there is no point of intersection between *B*, *G*, the line *EL* cuts each of the sections. Let it cut at *M*, *H*. Therefore, in one of the sections *EN* will be to *NH* as *EL* is to *LH*, while in the other *EN* will be to *NM* as *EL* is to *LM*. But this is absurd,[61] so that what was assumed at the start is absurd.

If *AB*, *DG* be parallel, the sections will, of course, be ellipses or a circumference of a circle [figure 5.2.30]. Let *AB*, *GD* be cut in two equal parts at *O*, *P*, and let *OP* be joined and produced on each side. It will, then, intersect the sections. So, let it intersect at *Q*, *R*. *QR* will then be a diameter of the sections, and *AB*, *GD* set up ordinatewise [book 2,

[61] From the two proportions $EN:NH=EL:LH$ and $EN:NM=EL:LM$, it follows that $LH:NH=EL:EN=LM:NM$, or, $LH:HN=LM:MN$. Hence, two distinct points *H* and *M* must divide the segment *LM* in the same ratio. This is the absurdity to which Apollonius most likely refers.

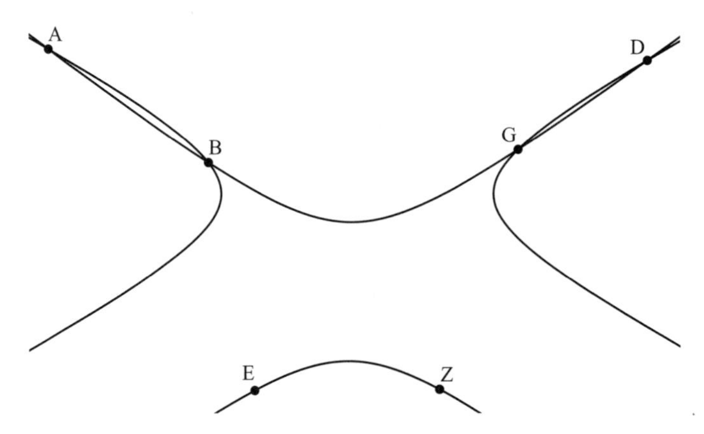

Figure 5.2.31. Book 4, proposition 55, statement 1

proposition 28]. Let $ENMH$ be drawn from E parallel to AB, GD. Therefore, EMH cuts QR and each of the curves, because there are no other intersections besides A, B, G, D. In one of the sections, then NM will equal EN, while in the other NE will equal NH [book 1, definition 4], so that NM equals NH, which is absurd.

Proposition 26. *If, of the curves mentioned above, some touch at one point, then they will not meet each other in more than two other points.*

Proposition 27. *If, of the curves mentioned above, some touch one another at two points, they will not meet one another at another point.*

Proposition 55. *Opposite sections will not intersect opposite sections at more points than four.*

[Statement 1] For let there be one pair of opposite sections AB, GD, and another pair of opposite sections $ABGD$, EZ, and, to start, let $ABGD$ cut each of AB, GD at four points A, B, G, D containing convexities turned oppositely [figure 5.2.31]. Therefore, the opposite section of $ABGD$, that is, EZ, will not intersect AB, GD [book 4, proposition 43].

[Statement 2] But let $ABGD$ cut AB at A, B, and G at one point G [figure 5.2.32]. Therefore EZ does not intersect section G [book 4, proposition 41]. If EZ meets AB, it will meet at one point only; for if it meets at two points, its opposite section, ABG, will not meet the other opposite section [book 4, proposition 43]. But it has been assumed that it meets it at one point G.

[Statement 3] If ABG cuts ABE at two points A, B, while EZ meets ABE at one point [figure 5.2.33]. EZ will not intersect section D [book 4, proposition 41], whereas, intersecting ABE, it will not intersect ABE at more points than two.[62]

[Statement 4] If $ABGD$ cuts each of two opposite sections at one point [figure 5.2.34], EZ will intersect neither at two points [book 4, proposition 42].

[62] Since *ABE* meets *ABG*, it cannot, by book 4, proposition 37, meet the opposite section of *ABG*, (i.e., *EZ*), at more points than two.

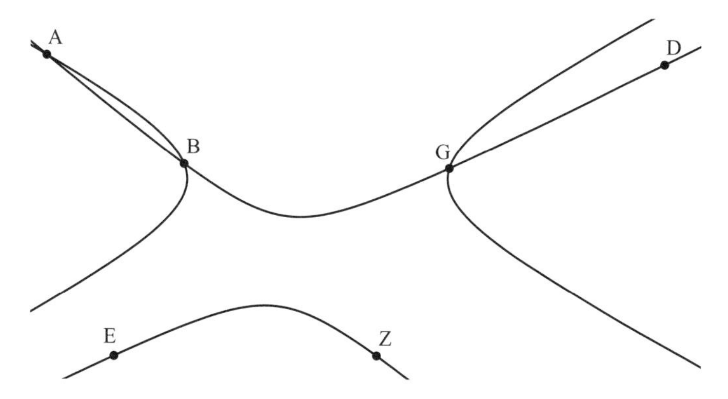

Figure 5.2.32. Book 4, proposition 55, statement 2

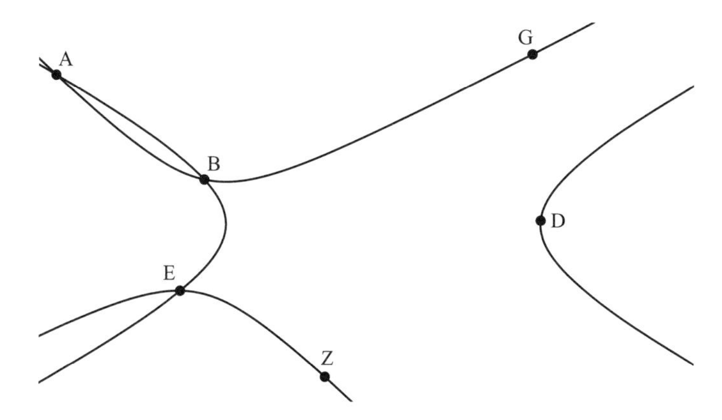

Figure 5.2.33. Book 4, proposition 55, statement 3

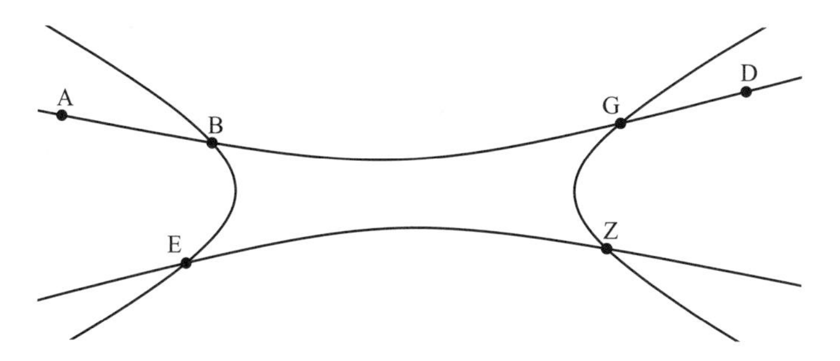

Figure 5.2.34. Book 4, proposition 55, statement 4

[Statement 5] If the sections have their concavities in the same direction, and one cuts the other at four points A, B, G, D [figure 5.2.35], EZ will not intersect the other opposite section [book 4, proposition 44]. Of course, EZ will not intersect AB; for, again, AB will not intersect the opposite sections $ABGD$, EZ at more points than four [book 4, proposition 38]; neither will GD intersect EZ.

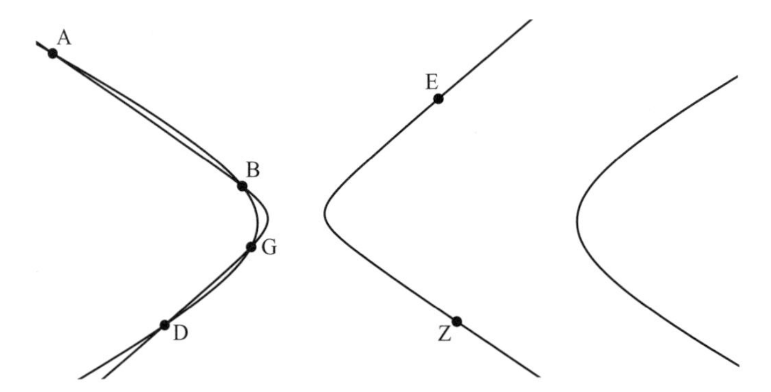

Figure 5.2.35. Book 4, proposition 55, statement 5

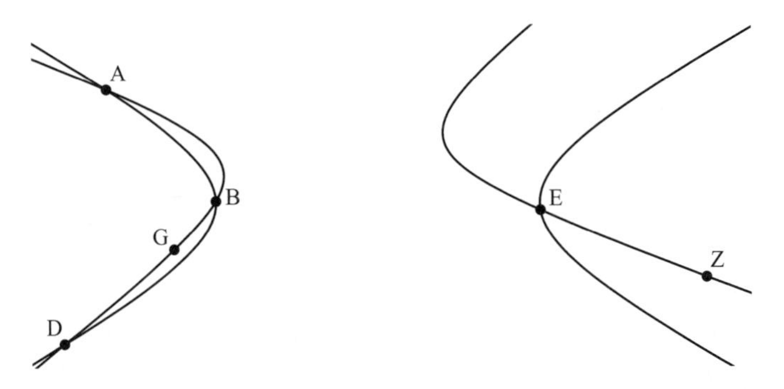

Figure 5.2.36. Book 4, proposition 55, statement 6

[Statement 6] If $ABGD$ meets the other section at three points [figure 5.2.36], EZ will intersect the other at one point only [book 4, proposition 46]. And we will say the same things as before for the remaining cases.

So, since what was proposed is clear in all possible configurations, opposite sections will not meet opposite sections at more points than four.

One way to look at these results from book 4 is to recall that five points uniquely determine a conic. Thus, if two conics intersected at five or more points, they would be the same conic. The five-point view provides a handy way to generate figures demonstrating each of the possibilities. Let one conic be chosen, and select up to four points on it to be where another conic intersects. Then select at least one point not on the conic, such that the total number of selected points is five, and describe the conic given by those five points. By adjusting the relative positions of the chosen points, any number of example diagrams may be easily produced; indeed, this is how we produced figures 5.2.31–5.2.36 for proposition 55.

The last two propositions in book 4 just clarify proposition 55 in the case where opposite sections are tangent to opposite sections.

Proposition 56. *If opposite sections touch opposite sections at one point, they will not intersect at more than two other points.*

Proposition 57. *If opposite sections touch opposite sections at two points, they will not intersect at another point.*

5.3 Apollonius, *Conics*, 5–7

Book 5

The major topic of book 5 is finding minimum lines to conic sections from points on the axes. A minimum line is the shortest line to the conic section of all lines that could be drawn meeting certain conditions.

Apollonius's Opening Letter

From Apollonius to Attalus, peace be on you. In this fifth book I have composed propositions on the maximum and minimum lines. You should realize that our predecessors and contemporaries paid (a little) attention only to the minimum lines: they proved thereby which straight lines are tangent to the sections and also the reverse, i.e., what properties are possessed by the tangents to the sections such that when those properties are possessed by lines they are tangents. But as for us, we have proven those things in book I without making use in our proof that of the topic of minimum lines; for we wanted to make the place where those things were put near to our discussion of the derivation of the three sections, in order to show in this way that in each of the sections there may occur an infinite number of properties and necessities of these things, as is the case with the original diameters. As for the propositions in which we speak of the minimum lines, we have separated them out and treated them individually, after much investigation, and have attached the discussion of them to the discussion of the maximum lines which we have mentioned above, because of our opinion that students of this science need them for the knowledge of analysis and determination of problems and their synthesis, not to speak of the fact that they are one of the subjects which deserve investigation in their own right. Farewell.

Proposition 4. *If a point is marked, on the axis of a parabola, the distance of which from the vertex of the section is equal to half the* latus rectum, *and lines are drawn from that point to the section, then the minimum of these lines is the line drawn to the vertex of the section, and those closer to this line will be smaller than those farther from it; and their squares will exceed the square on it (the line to the vertex) by the equivalent of the square of the amount cut off on the axis towards the vertex by the perpendiculars drawn to the axis from the end of each of them.*

Let the axis of the parabola be ΓE, and let ΓZ be equal to half the *latus rectum*, and let there be drawn from point Z to the section $AB\Gamma$ lines $ZH, Z\Theta, ZB, ZA$ [figure 5.3.1]. Then I say that the least of the lines drawn from point Z to section $AB\Gamma$ is line ΓZ, and that those lines nearer to ΓZ are smaller than those farther from it; and that the square

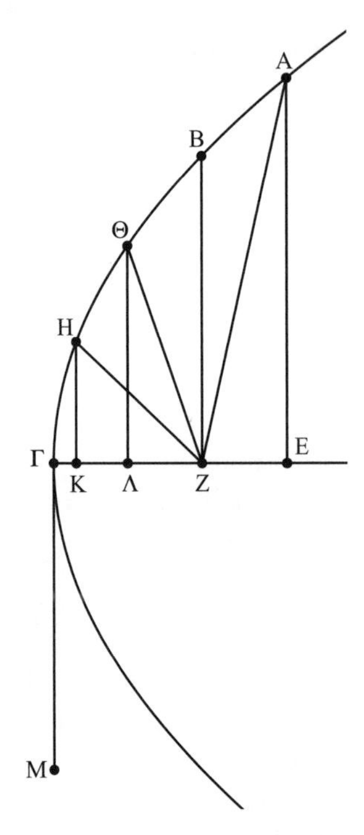

Figure 5.3.1.

on each of them is equal to the square on ΓZ plus the square on the line between point Γ and the foot of the perpendicular from it (the end of the line).

We draw perpendiculars $HK, \Theta\Lambda, AE$. Let half of the *latus rectum* be ΓM. Then $\Gamma Z = \Gamma M$. And 2 rect($\Gamma M, \Gamma K$) equals sq(KH), as is proven in proposition 11 of book I. But 2 rect($\Gamma M, \Gamma K$) equals 2 rect($Z\Gamma, \Gamma K$). Therefore sq(KH) equals 2 rect($Z\Gamma, \Gamma K$), and 2 rect($Z\Gamma, \Gamma K$) together with sq(KZ) equals sq(ZK) and sq(KH), taken together. But these two squares, sq(ZK) and sq(KH), equal sq(ZH). Therefore 2 rect($Z\Gamma, \Gamma K$) together with sq(ZK) equals sq(ZH).

Therefore the square on ZH exceeds the square on $Z\Gamma$ by the square on ΓK. And it will be proved from this that ΘZ is greater than ZH and ZH is greater than $Z\Gamma$.[63] So the line $Z\Gamma$ is shortest, and those lines closer to it are shorter than those farther. And it is proven that the excess of the square of each of them over the square on the shortest line is of the amount of the square on the line cut off from the axis towards the vertex of the section by the perpendiculars from the ends of the lines.

Propositions 5 and 6 give similar results for the hyperbola and the ellipse.

Proposition 7. *If a point is marked on the minimum line mentioned in one of the three sections, and lines are drawn from it to the section, then the shortest of them is the line*

[63] By a similar argument, sq($Z\Theta$) equals the sum of sq($Z\Gamma$) and sq($\Gamma\Lambda$).

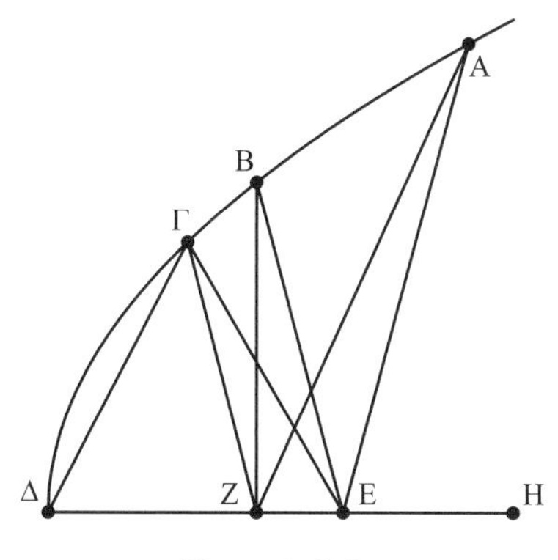

Figure 5.3.2.

between that point and the vertex of the section, and those of the other lines drawn in that half of the section closer to it (the line to the vertex) are shorter than those farther (from it).

Let there be a conic section $AB\Gamma\Delta$, with axis ΔH. Let the minimum line be ΔE. Let there be an arbitrary point on ΔE, namely, Z. We draw from it lines to the section, namely, $Z\Gamma$, ZB, ZA [figure 5.3.2]. Then I say that ΔZ is the shortest of them, and that those closer to it are smaller than those farther from it.

We join the line ΓE. Then ΓE is greater than $E\Delta$. Hence angle $\Gamma\Delta E$ is greater than angle $\Delta\Gamma E$. By how much the more is angle $Z\Delta\Gamma$ greater than $\Delta\Gamma Z$. Therefore ΓZ is greater than $Z\Delta$. Furthermore, BE is greater than $E\Gamma$, so angle $B\Gamma E$ is greater than angle ΓBE. So by how much the more is angle ΓBZ less than angle $B\Gamma Z$. Therefore BZ is greater than $Z\Gamma$. Similarly also it will be proven that AZ is greater than BZ. So line ΔZ is the shortest of the lines drawn from point Z to the section, and as for the other lines, those of them closer to line ΔZ are shorter than those farther [from it].

Proposition 8. *If a point is marked on the axis of a parabola, the distance of which from the vertex of the section is greater than half the* latus rectum, *and there is cut off on the axis, from that point which was marked on it, towards the vertex of the section, a line equal to half the* latus rectum, *and from the (other) end of that line which was cut off there is drawn a perpendicular to the axis, and that perpendicular is extended to meet the section, and there is drawn from the place where it meets the section a line to the point which had been marked, then that line is the shortest of the lines drawn from that point which was marked on the axis of the section, and of all the other lines on both sides (of it), those closer to it are shorter than those farther (from it).*

Let there be a parabola $AB\Gamma$, with axis $\Gamma\Delta$, and let the line ΓE be longer than half the *latus rectum*, and let half the *latus rectum* be ZE. We draw perpendicular ZH to ΓE, and join the line EH [figure 5.3.3]. Then I say that line EH is the shortest of the lines drawn from point E to the section. And as for the other lines drawn from $AB\Gamma$ to point E, those of them closer to line EH are shorter than those farther from it, on both sides of EH.

So we draw from point E on the section lines $EK, E\Lambda, E\Theta, EA$. Then I say also that the square on each of these exceeds the square on EH by an amount equal to the square

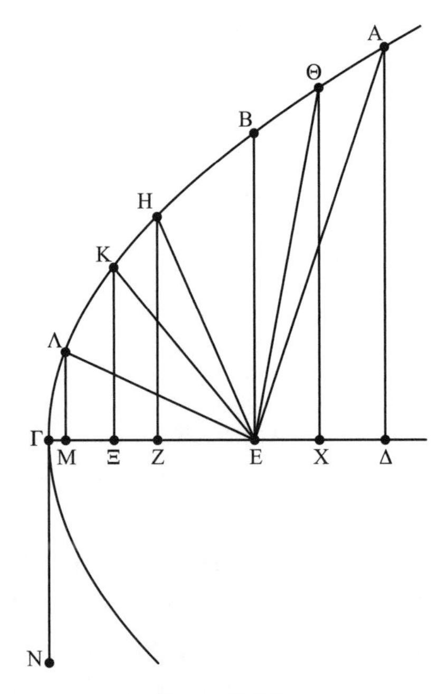

Figure 5.3.3.

on the line between the foot of the perpendicular from it (K, Λ, Θ, A) and point Z. So let us draw the perpendiculars $(K\Xi, \Lambda M, \Theta X, A\Delta)$, and let BE be perpendicular (to the axis), and let ΓN be half the *latus rectum*. Then 2 rect($\Gamma N, \Gamma\Xi$) equals sq($K\Xi$), as is proven in proposition 11 of book I, and 2 rect($\Gamma N, \Gamma\Xi$) equals 2 rect($EZ, \Gamma\Xi$). We make 2 rect($EZ, Z\Xi$), sq(EZ), and sq($Z\Xi$) common. Then 2 rect($EZ, \Gamma\Xi$), 2 rect($EZ, Z\Xi$), sq(EZ), and sq($Z\Xi$) together equal sq($K\Xi$) with sq(ΞE), which equals sq(KE).

But 2 rect($EZ, \Gamma\Xi$) with 2 rect($EZ, Z\Xi$) equals 2 rect($\Gamma Z, EZ$). Therefore sq(KE) equals 2 rect($\Gamma Z, EZ$) together with sq($Z\Xi$) and sq(EZ). But 2 rect($\Gamma Z, ZE$) equals sq(ZH) because ZE equals ΓN. Therefore sq(ZH), sq(ZE), and sq($Z\Xi$) taken together equal sq(EH). But sq(ZH) with sq(ZE) equals sq(EH). Therefore sq(KE) equals sq(EH) and sq($Z\Xi$) taken together. Therefore the amount by which sq(KE) exceeds sq(EH) equals sq($Z\Xi$). Similarly also it will be proven that the difference between sq($E\Lambda$) and sq(EH) equals sq(MZ).

And since 2 rect($\Gamma Z, ZE$) equals sq(ZH) (since ZE equals ΓN), therefore the difference between sq(ΓE) and sq(EH) equals sq(ΓZ). And $Z\Xi$ is less than ZM, which is less than $Z\Gamma$. Therefore line EH is the least of the lines drawn from point E to the section on the side of Γ. Furthermore, sq(BE) equals 2 rect($\Gamma N, \Gamma E$), which equals 2 rect($\Gamma E, EZ$). And 2 rect($\Gamma Z, ZE$) equals sq(ZH). Therefore sq(BE) equals sq(HE) with sq(EZ). Therefore the amount by which sq(BE) exceeds sq(BH) is equal to sq(ZE).

Furthermore, sq($X\Theta$) equals 2 rect($\Gamma X, ZE$), because ZE equals ΓN. We make sq(XE) common (to both sides). Then 2 rect($\Gamma Z, ZE$) with sq(ZE) and sq(ZX) equals sq($E\Theta$). But 2 rect($\Gamma Z, ZE$) with sq(ZE) equals sq(EH). Therefore sq(ZX) equals the difference between sq($E\Theta$) and sq(EH). Similarly it will also be proven that sq(ΔZ) is the difference between sq(AE) and sq(EH). But ΔZ is greater than ZX, which is greater

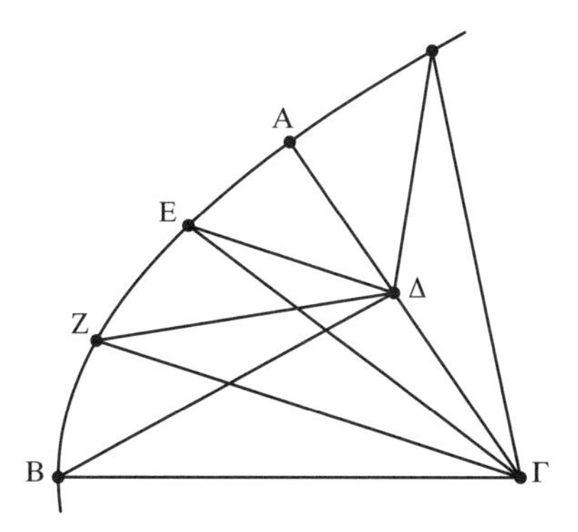

Figure 5.3.4.

than ZE. Therefore line EH is the least of the lines drawn from the point E to the section, and those closer to it are smaller than those farther; and the difference between the square on each of them and the square on it is equal to the square on the line between the foot of the perpendicular from it and point Z.

At first, it might seem surprising to have half the *latus rectum* laid down as ZE, being measured from E rather than from Γ. However, we can see that Apollonius's deduction works. We can also demonstrate this via calculus; if we let the parabola have *latus rectum* b and equation $bx = y^2$, suppose that $(a, 0)$ is a given point on the axis, and that $(x, \sqrt{bx})$ is an arbitrary point on the parabola. Hence $a \geq 0$ and $x \geq 0$. The hypothesis of this proposition is that $a > b/2$. We wish to minimize the square of the distance D: $D^2 = (x-a)^2 + bx$. This yields $2DD' = b + 2(x-a)$, which is minimized when $x = a - b/2$. We can then immediately see that if $a = b/2$, the case considered in proposition 4, the desired point is the vertex of the parabola. The case of $a < b/2$ is considered in proposition 7; we can see it via calculus as the endpoint solution when $x = 0$.

Proposition 12. *If a point is marked on one of the lines which has been proven to be a minimum of lines drawn from some point on the axis to one of the (three) sections, and lines are drawn from that point to the section on one side, then the shortest of them is the portion of the minimum line adjoining the section, and those closer to it are shorter than those farther from it.*

Let there be a conic section AB with axis $B\Gamma$, and the minimum line drawn from some point on it ΓA. We mark on it [ΓA] an arbitrary point Δ [figure 5.3.4]. Then I say that line ΔA is the shortest of the lines drawn from point Δ in that part of the section.

We draw lines $\Delta E, \Delta Z, \Delta B$, and join lines $Z\Gamma, \Gamma E$ and lines AE, EZ, ZB. Then $E\Gamma$ is greater than ΓA, so angle ΓAE is greater than angle ΓEA. But angle ΓEA is greater than angle $AE\Delta$, so angle $EA\Delta$ is much greater than angle $AE\Delta$. Therefore $E\Delta$ is greater than ΔA. Furthermore, $Z\Gamma$ is greater than ΓE, so angle $ZE\Gamma$ is greater than angle $EZ\Gamma$. Therefore angle $ZE\Delta$ is much greater than angle $EZ\Delta$. Therefore $Z\Delta$ is greater than ΔE. Similarly it will also be proven that $B\Delta$ is greater than ΔZ. Therefore line $A\Delta$ is the smallest of the lines drawn in this part of the section, and those [lines] closer to it are smaller than

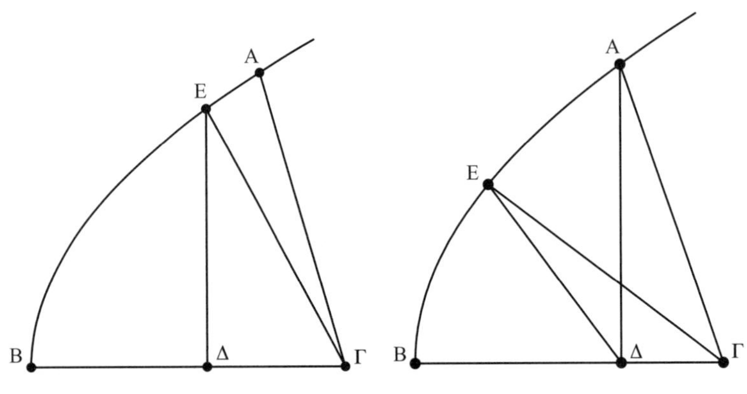

Figure 5.3.5.

those farther [from it]. Similarly it will also be proven concerning those lines when they are drawn in the other part of the section.

Note that this proposition does not state how to find the minimum line that the point is chosen on, merely that *if* it is chosen on the minimum line, the new minimum is a part of the existing minimum. Proposition 62 shows how to do this in the case of a parabola.

Proposition 13. *If there is drawn from a point on the axis of a parabola the minimum of the lines drawn from that point to the section so as to form an angle with the axis, then that angle which it forms with the axis will be acute; and if a perpendicular is dropped from its end to the axis, then that perpendicular cuts off a portion equal to half the* latus rectum.

Let there be a parabola AB with axis $B\Gamma$, and let $A\Gamma$ be a minimum line from the axis to the parabola [figure 5.3.5]. $A\Gamma$ is the minimum, so $B\Gamma$ is greater than half the *latus rectum*. For if it were not greater than it, it would either be equal to it or less than it. But if it were equal to it, $B\Gamma$ would be the minimum, as is proven in proposition 4 of this book. But that is not so, for the minimum is $A\Gamma$. And if $B\Gamma$ were less than half the *latus rectum*, then, when a line equal to half the *latus rectum* was cut off from the axis, the point at which the cut was made would be beyond point Γ. So it could be proven from proposition 4 of this book that the line $B\Gamma$ is smaller than line ΓA.

So line $B\Gamma$ is not smaller than half the *latus rectum*. And we have proven that it is not equal to it. Therefore it is greater than it. So let the line equal to half the *latus rectum* be line $\Gamma\Delta$. Then I say that the perpendicular drawn from point Δ meets point A. For if that is not so, let the perpendicular be ΔE. Then $E\Gamma$ is the shortest of the lines drawn from point Γ to the section, as is proven in proposition 8 of this book. But $A\Gamma$ was the minimum: that is absurd. So the perpendicular drawn from point Δ meets A, and line $\Delta\Gamma$ is equal to half the *latus rectum*, and angle $A\Gamma B$ is acute.

Proposition 24. *If a point is marked on a parabola, then only one of the minimum lines drawn from the axis meets it.*

Let the section be the parabola AB, with axis BG. We mark on the section point A [figure 5.3.6]. Then I say that only one of the minimum lines can be drawn from the axis to point A.

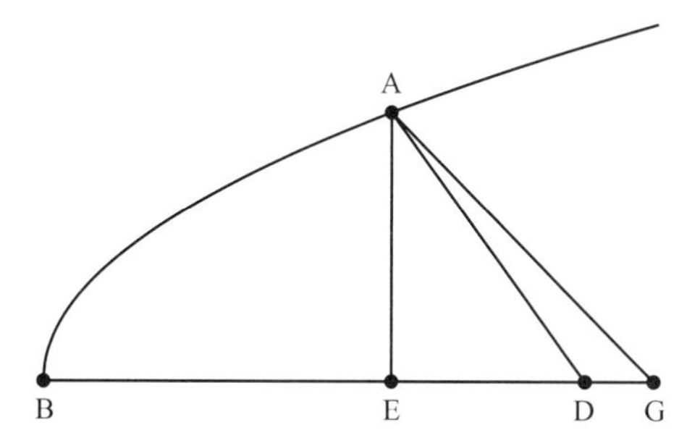

Figure 5.3.6.

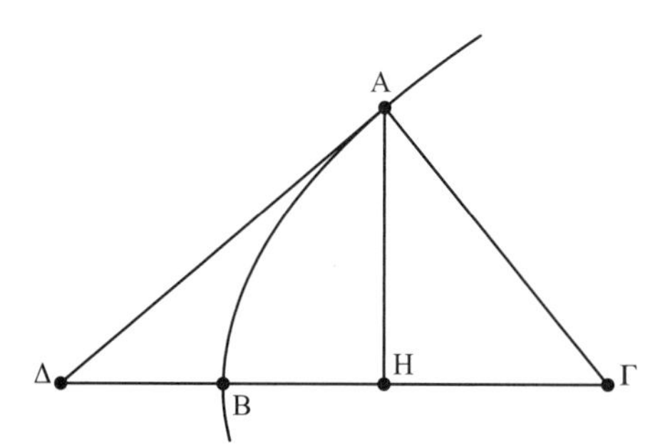

Figure 5.3.7.

For, if possible, let us draw two minimum lines, AG, AD. We draw from point A a perpendicular to BG, namely, AE. Then line ED equals half the *latus rectum*, as is proved in proposition 13 of this book. And similarly also EG equals half the *latus rectum*; that is absurd. So only one of the minima can be drawn from the axis to point A.

Propositions 25 and 26 present the same result for a hyperbola and an ellipse respectively.

Proposition 27. *The line drawn from the end of one of the minimum lines we mentioned tangent to the section is perpendicular to the minimum line.*

Let the section be, first a parabola, AB, with axis $B\Gamma$ [figure 5.3.7]. Then I say that the line drawn from the end of a minimum line tangent to section AB is perpendicular to the minimum line. Now if the minimum line is part of line $B\Gamma$, then what we said is obvious. But if the minimum is $A\Gamma$, we draw from point A a line tangent to section AB, namely, $A\Delta$: then I say that angle $\Delta A\Gamma$ is right.

We draw perpendicular AH. Then line ΓH equals half the *latus rectum*, as is proven in proposition 13 of this book. Furthermore, $A\Delta$ is tangent to a parabola, and perpendicular AH has been drawn from point A to the axis. Therefore ΔB equals BH, as is proven in proposition 35 of book I. Therefore the ratio of ΓH to the *latus rectum* equals that of BH to $H\Delta$. Therefore rect($\Gamma H, H\Delta$) equals the product of BH and the *latus rectum*. But the product of BH and the *latus rectum* equals sq(AH). Therefore sq(AH) equals rect($\Gamma H, H\Delta$). And angle $AH\Delta$ is right. So angle $\Delta A\Gamma$ is right.

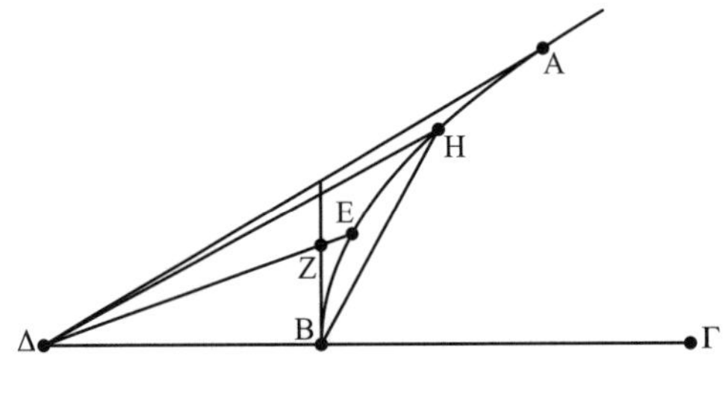

Figure 5.3.8.

This proposition shows that a tangent to a parabola drawn at the end of a minimum line meets the minimum line at a right angle, and hence we can say that the minimum line is a normal. Apollonius proves analogous results for hyperbolas and ellipses in the next three propositions. He then continues with similar results in propositions 31 and 32:

Proposition 31. *If a perpendicular is drawn at the point where a minimum line meets the section, it is tangent to the section.*

Proposition 32. *If a tangent is drawn at a point on the section, and a perpendicular is drawn from the point of tangency to the axis, this perpendicular is a minimum.*

Proposition 34. *If a point is marked outside a conic section on a maximum or minimum line extended, then the smallest length intercepted between that point and the section (on lines drawn from that point on either side of the section but not extended to cut the section in more than one point) is the line which is the extension of the maximum or minimum line; and of the other lines, those closer to it are smaller than those farther from it.*

Let there be a conic section, AB, with a maximum or minimum line in it, $B\Gamma$. Let it be extended in a straight line, and let us mark on it, after it is extended outside the section, an arbitrary point Δ. We draw from point Δ to the section lines ΔA, ΔH, ΔE: let each of them cut the section in one point only [figure 5.3.8]. Then I say that line ΔB is the smallest of the lines drawn from point Δ to the section, and that, of the other lines, those of them closer to it are smaller than those farther from it.

We draw BZ tangent to the section. Then angle $ZB\Delta$ is right, because of what was proven in propositions 27, 28, 29, and 30 of this book. Therefore ΔZ is greater than ΔB. Therefore line ΔE is much greater than line ΔB. So we join line HB and line HE. Then angle ΔEH will be obtuse, so ΔH is greater than ΔE. Similarly also it will be proven that ΔA is greater than ΔH. And similarly it is possible for us to prove the same concerning the lines drawn to the other side of ΔB.

In the next several propositions, Apollonius proves separately results about minima for parabolas, ellipses, and hyperbolas. We present only results for parabolas.

Proposition 58. *For every point marked outside one of the conic sections, provided that it is not on the axis wherever the axis is extended in a straight line, it is possible for us to*

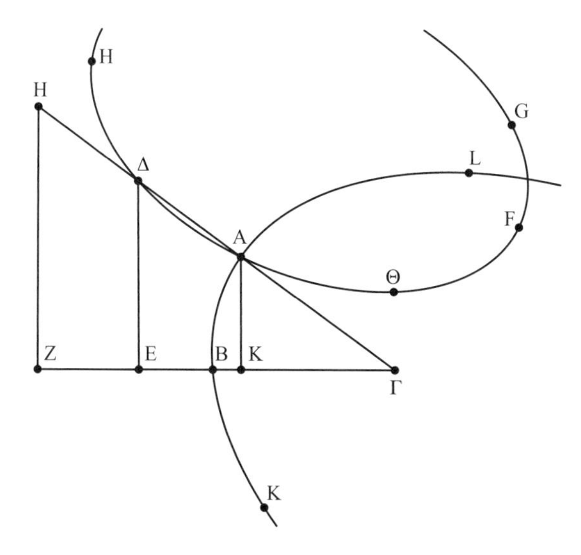

Figure 5.3.9.

draw from it (that point) some line such that the part of it which falls between the section and its axis is one of the minima.

Let the section be first a parabola, AB, with its axis, extended, ΓZ. We mark outside of the section point Δ, not on the axis [figure 5.3.9]. Then I say that there can be drawn from point Δ a line such that the part of which falls between AB and $B\Gamma$ is one of the minima.

For we draw perpendicular ΔE to line ΓZ, wherever it falls on it. Let EZ be equal to half the *latus rectum*, and let ZH be a perpendicular to $Z\Gamma$. We construct a hyperbola, $\Delta A\Theta$, passing through point Δ, with asymptotes HZ, $Z\Gamma$, as is shown in proposition 4 of book 2. Then it will cut the parabola: let it cut it in point A. We join line ΔA and extend it to points H, Γ, and drop a perpendicular from point A on to ΓZ, namely, AK. Then ΔH equals $A\Gamma$, as is proven in proposition 8 of book 2. Therefore ZE equals $K\Gamma$. But ZE equals half the *latus rectum.* Therefore $K\Gamma$ equals half the *latus rectum*. And KA is a perpendicular. Therefore $A\Gamma$ is one of the minima, as is proven in proposition 8 of this book.

Proposition 62. *It is possible for us to pass one of the minima through any point which is between one of the conic sections and its axis.*

Let the section be, first, a parabola, AB, with axis BH. We mark in the place mentioned point Γ. Then I say that it is possible for us to pass through point Γ one of the minima.

We draw from point Γ perpendicular $\Gamma\Delta$. Let half the *latus rectum* be ΔE. We draw from point E a perpendicular to ΔH, namely, $E\Theta$, and construct a hyperbola passing through point Γ with asymptotes ΘE, EH: then this hyperbola will cut the parabola. So let it cut it in point A, and let the hyperbola be $A\Gamma$. We join line $A\Gamma$ and extend it to meet $E\Delta$ in H and to meet $E\Theta$ in Θ [figure 5.3.10]. Then I say that AH is one of the minima.

We draw perpendicular AZ. Then ΓH equals ΘA, as is proven in proposition 8 of book 2. Therefore ΔH equals EZ. But $E\Delta$ is half the *latus rectum*. Therefore ZH is half

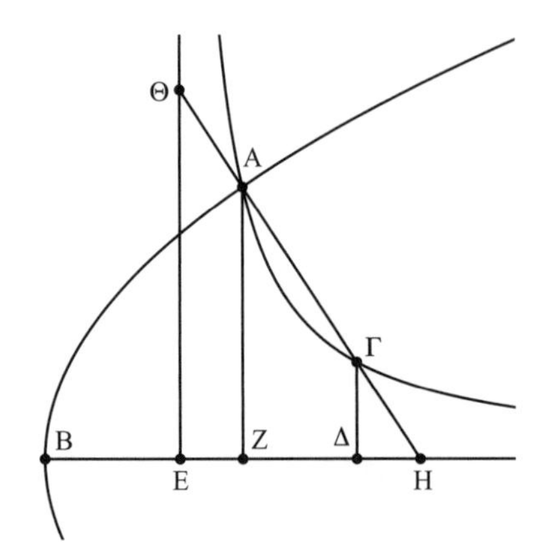

Figure 5.3.10.

the *latus rectum*. Therefore line AH is one of the minima, as is proven in proposition 8 of this book.

Book 6

This book deals mainly with the questions of equality and similarity of conic sections. We have left out results dealing with segments of conic sections.

Apollonius's Opening Letter

From Apollonius to Attalus: peace be on you. I have sent you the sixth book of the Conics. My aim in it is to report on conic sections which are equal to each other and those unequal to each other, and on those similar to each other and dissimilar to each other, and on segments of conic sections. In this we have enunciated more than what was composed by others among our predecessors. In this book there is also how to find a section in a given right cone equal to a given section, and how to find a right cone, containing a given conic section, similar to a given cone. What we have stated on this subject is fuller and clearer than the statements of our predecessors. Farewell.

Definitions

1. Conic sections which are called *equal* are those which can be fitted, one on another, so that one does not exceed the other. Those which are said to be *unequal* are those for which that is not so.
2. And *similar* conic sections are such that, when ordinates are drawn in them to fall on the axes, the ratios of the ordinates to the lengths they cut off from the axes from the vertex of the section are equal to one another, while the ratio to each other of the portions which the ordinates cut off from the axes are equal ratios. Sections which are *dissimilar* are those in which what we stated does not occur.
9. A conic section is said to be *placed in* a cone, or a cone is said to *contain* a conic section, when the whole of the section is in the surface bounding the cone between

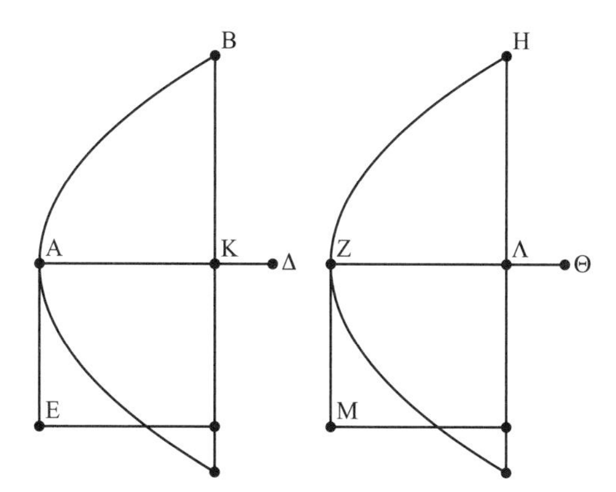

Figure 5.3.11.

its vertex and its base; or in that surface after it has been produced beyond the base, so that the whole of the section is in the surface below the base, or else some of the section is in this surface and some in the other surface.

10. Right cones which are said to be *similar* are those for which the ratios of their axes to the diameters of their bases are equal.
11. The figure which I call the *figure of the section constructed on the axis or on the diameter* is that contained by the axis or diameter together with the *latus rectum.*

Proposition 1. *Parabolas in which the* latera recta *which are the parameters of the perpendiculars (ordinates) to the axes are equal, are themselves equal; and if parabolas are equal, their* latera recta *are equal.*

Let there be two parabolas, with axes $A\Delta, Z\Theta$, and equal *latera recta*, AE, ZM. Then I say that these sections are equal.

When we apply axis $A\Delta$ to axis $Z\Theta$, then the section will coincide with the section so as to fit on it. For if it does not fit on it, let there be a part of section AB which does not fit on section ZH. We mark point B on the part of it which does not coincide with ZH, and draw from it [to the axis] perpendicular BK, and complete rectangle KE [figure 5.3.11]. We make $Z\Lambda$ equal to AK, and draw from point Λ a perpendicular to the axis, ΛH, and complete rectangle ΛM.

Then lines KA, AE, are equal to lines $\Lambda Z, ZM$, each to its correspondent. Therefore rectangle KE equals rectangle ΛM. And line KB is equal in square to rectangle EK, as is proven in proposition 11 of book 1. And similarly too line ΛH is equal in square to rectangle ΛM. Therefore KB equals ΛH. So when the axis (of one section) is applied to the axis (of the other), line AK will coincide with line $Z\Lambda$, and line KB will coincide with line ΛH, and point B will coincide with point H. But it (point B) was supposed not to fall on the section ZH: that is absurd. So it is impossible for the section AB to not equal the section ZH.

Furthermore, we make the section AB equal to the section ZH, and make AK equal to line $Z\Lambda$, and draw the perpendiculars to the axis from points K, Λ, and complete rectangles $EK, M\Lambda$; then section AB will coincide with section ZH, and therefore axis

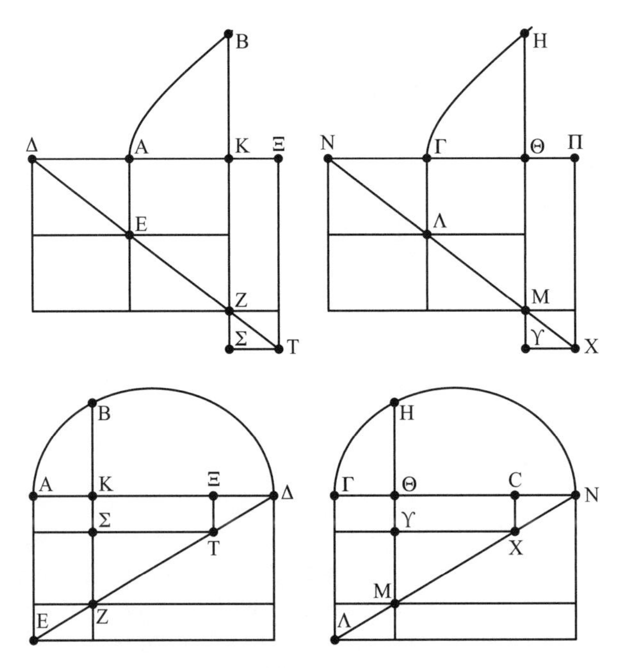

Figure 5.3.12.

ΛK will coincide with axis $Z\Lambda$. For if it does not coincide with it, parabola ZH has two axes, which is impossible. So let AK coincide with $Z\Lambda$. Then point K will coincide with point Λ, because AK equals $Z\Lambda$. Therefore BK equals ΛH. Therefore rectangle EK equals rectangle ΛM. And AK equals $Z\Lambda$. Therefore AE equals ZM.

Proposition 2. *If the figures that are constructed on the transverse axes of hyperbolas or ellipses are equal and similar, then the sections [themselves] will be equal; and if the sections [either hyperbolas or ellipses] are equal, then the figures constructed on their transverse axes are equal and similar, and their situation is similar.*

Let there be two hyperbolas or ellipses, $AB, \Gamma H$, with axes $AK, \Gamma\Theta$. Let the figures which are constructed on their transverse axes be equal and similar: these are $\Delta E, N\Lambda$. Then I say that sections $AB, \Gamma H$ are equal.

We apply axis AK to axis $\Gamma\Theta$; then section AB will coincide with section ΓH. For if that is not so, let a part of section AB not coincide with section ΓH. We mark point B on that part, and draw from it a perpendicular to the axis, BK, and complete the delineation of [the corresponding] rectangle ΔZ [figure 5.3.12]. We make $\Gamma\Theta$ equal to line AK and draw from point Θ a perpendicular to $\Gamma\Theta$, namely, ΘH, and complete the delineation of [the corresponding] rectangle NM.

Then lines AE, AK are [respectively] equal to lines $\Lambda\Gamma, \Gamma\Theta$. Therefore rectangle EK equals rectangle $\Lambda\Theta$. Furthermore, rectangles $\Lambda M, EZ$ are similar and similarly situated, because they are similar to rectangles $\Delta E, N\Lambda$ [respectively]; and AK equals $\Gamma\Theta$. Therefore rectangle EZ equals rectangle ΛM. And rectangles $KE, \Theta\Lambda$ were [already proved] equal. Therefore rectangle AZ equals rectangle ΓM. And the lines which are equal to them in square are [respectively] $BK, \Theta H$, as is proven in propositions 12 and 13 of book 1. So when the axis is applied to the axis, line BK will coincide with line ΘH, and point B will coincide with point H. But it [B] was supposed not to fall on section ΓH; that is absurd. So the whole of section AB will fit on section ΓH.

Furthermore, we make the two sections equal, and make lines $AK, \Gamma\Theta$ equal, and draw from them perpendiculars $KB, \Theta H$, and complete the delineation of [the rectangles] $\Delta E, \Delta Z, N\Lambda$, and NM: then section AB will fit on section ΓH, and axis AK will coincide with axis $\Gamma\Theta$. For if it did not coincide with it, then the hyperbola would have two axes, and the ellipse three axes, which is impossible. So line AK coincides with line $\Gamma\Theta$; and it is equal to it. So point K will coincide with point Θ, and line KB will coincide with line ΘH, and [hence] point B will coincide with point H, and line KB will fit on line $H\Theta$. Therefore KB equals $H\Theta$. For that reason rectangle AZ equals rectangle ΓM. But AK equals $\Gamma\Theta$. Therefore KZ equals ΘM.

Furthermore, we make $A\Xi$ equal to $\Gamma\Pi$; then it will be proven, as we proved above, that ΞT equals ΠX. Therefore ΣZ equals $M\Upsilon$. And ΣT equals ΥX. Therefore rectangles ZT, MX are equal and similar. Therefore rectangle ΔE is equal and similar to rectangle $N\Lambda$. And also rectangle ΔZ is equal and similar to rectangle NM. But KZ equals ΘM. Therefore ΔK equals $N\Theta$. But it was [assumed] that AK equals $\Gamma\Theta$. Therefore ΔA equals $N\Gamma$. And rectangle ΔE is equal and similar to rectangle $N\Lambda$. Therefore AE equals $\Gamma\Lambda$. Therefore rectangle ΔE equals rectangle $N\Lambda$. And these are the figures [constructed] on the axes.

Proposition 11. *Every parabola is similar to every parabola.*

Let there be two parabolas, $AB, \Gamma\Delta$, with axes $AK, \Gamma O$. Then I say that the sections are similar.

We make their *latera recta* $A\Pi, \Gamma P$, and make AK to $A\Pi$ equal to ΓO to ΓP. We cut AK at two points, Z, Θ, arbitrarily, and cut ΓO into the same number of segments, with the same ratio [as AK and its segments], at points M, Ξ [figure 5.3.13]. We draw from axes $AK, \Gamma O$, perpendiculars $ZE, \Theta H, KB, M\Lambda, N\Xi, \Delta O$ [and extend them to meet the sections again in $I, \Sigma, T.\Upsilon, \Phi, X$]. Then ΠA is to AK as ΓP is to ΓO. And line KB is the mean proportional between lines $A\Pi, AK$, while line $O\Delta$ is the mean proportional between lines $\Gamma P, \Gamma O$, because of what is proven in proposition 11 of book 1. Therefore KB is to KA as ΔO is to $O\Gamma$. And BT equals $2BK$, while ΔX equals $2\Delta O$. Therefore BT is to AK as ΔX is to ΓO.

Furthermore, ΠA is to AK as ΓP is to ΓO. And AK is to $A\Theta$ as $O\Gamma$ is to $\Gamma\Xi$. Therefore $A\Pi$ is to $A\Theta$ as ΓP is to $\Gamma\Xi$. Hence it will be proven, as we proved above, that $H\Sigma$ is to $A\Theta$ as $N\Phi$ is to $\Gamma\Xi$. And similarly too it will be proven that EI is to ZA as $\Lambda\Upsilon$ is to $M\Gamma$.

Therefore the ratio of [each of] the lines $BT, H\Sigma, EI$, which are perpendiculars to the axis, to the amounts which they cut off from the axis, namely, $AK, A\Theta, AZ$, is [respectively] equal to the ratio of lines $\Delta X, N\Phi, \Lambda Y$, which are perpendiculars to the axis, to the amounts which they cut off from the axis, namely, $O\Gamma, \Xi\Gamma, M\Gamma$. And the ratios of the segments cut off from one of the axes to the segments cut off from the other are equal. So section AB is similar to section $\Gamma\Delta$.

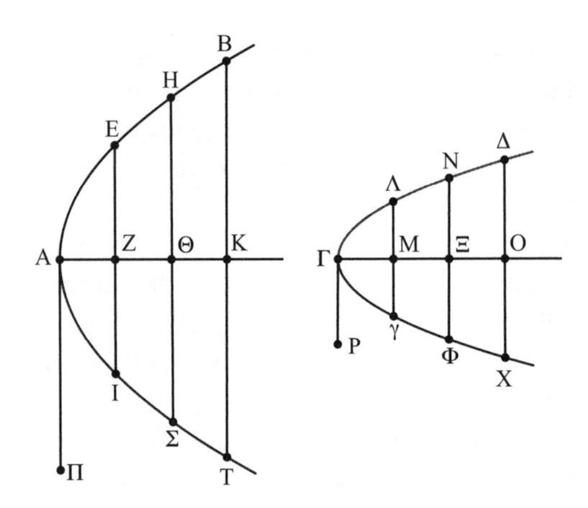

Figure 5.3.13.

In modern terms, two conics are similar if there is a similarity transformation between them, a function taking (x,y) to (kx,ky) for some k. Thus, given the parabolas $y^2=px$ and $y^2=qx$, consider the transformation $x \to \frac{p}{q}x$, $y \to \frac{p}{q}y$. The transformed parabola is then $\frac{p^2}{q^2}y^2 = \frac{p^2}{q}x$, or $y^2=qx$. Thus the similarity transformation has taken the first parabola to the second, and similarity is proved.

Proposition 12. *Hyperbolas and ellipses in which the figures constructed on their axes are similar are also [themselves] similar; and if the sections are similar, then the figures constructed on their axes are similar.*

Let there be two hyperbolas or ellipses, with the figures constructed on their axes being similar: these are sections $AB, \Gamma\Delta$ with axes $AK, \Gamma O$, and transverse diameters $A\Pi, P\Gamma$.

We cut off from the axes segments $AK, \Gamma O$, and let AK to $A\Pi$ equal ΓO to ΓP. We cut AK arbitrarily at points Z, Θ, and cut ΓO into the same number of segments as AK and in the same ratios, at points M, Ξ [figure 5.3.14]. We draw from points Z, Θ, K, M, Ξ, O perpendiculars to the two axes, BK, ΘH, ZE, $O\Delta$, ΞN, $M\Lambda$, [and extend them to meet the sections again in T, Σ, I, X, Φ, Y]. Then, because the figures of the two sections are similar, sq(BK) is to rect($\Pi K, KA$) as sq(ΔO) is to rect($OP, O\Gamma$), as may be proven from proposition 21 of book 1. But rect($\Pi K, KA$) is to sq(KA) as rect($OP, O\Gamma$) is to sq($O\Gamma$). Therefore sq(BK) is to sq(KA) as sq(ΔO) is to sq($O\Gamma$). Therefore BK is to KA as ΔO is to $O\Gamma$ and BT is to KA as ΔX is to $O\Gamma$.

Furthermore, ΠA is to AK as $P\Gamma$ is to ΓO and KA is to $A\Theta$ as $O\Gamma$ is to $\Gamma\Xi$. Therefore $A\Pi$ is to $A\Theta$ as $P\Gamma$ is to $\Gamma\Xi$. Hence it will be proven, as we proved above, that $H\Sigma$ is to ΘA as $N\Phi$ is to $\Xi\Gamma$, and that EI is to ZA as ΛY is to $M\Gamma$. Therefore the ratios of the perpendiculars $BT, H\Sigma, EI$, to the amounts they cut off from the axis, namely, $AK, A\Theta, AZ$, are [respectively] equal to the ratios of the perpendiculars $\Delta X, N\Phi, \Lambda Y$, to the amounts they cut off from the axis, namely, $O\Gamma, \Xi\Gamma, M\Gamma$. And the ratios of the parts of AK (which is the axis of AB) which the perpendiculars cut off, to the parts of ΓO (which is the axis of $\Gamma\Delta$) which the perpendiculars cut off are equal. So section AB is similar to section $\Gamma\Delta$.

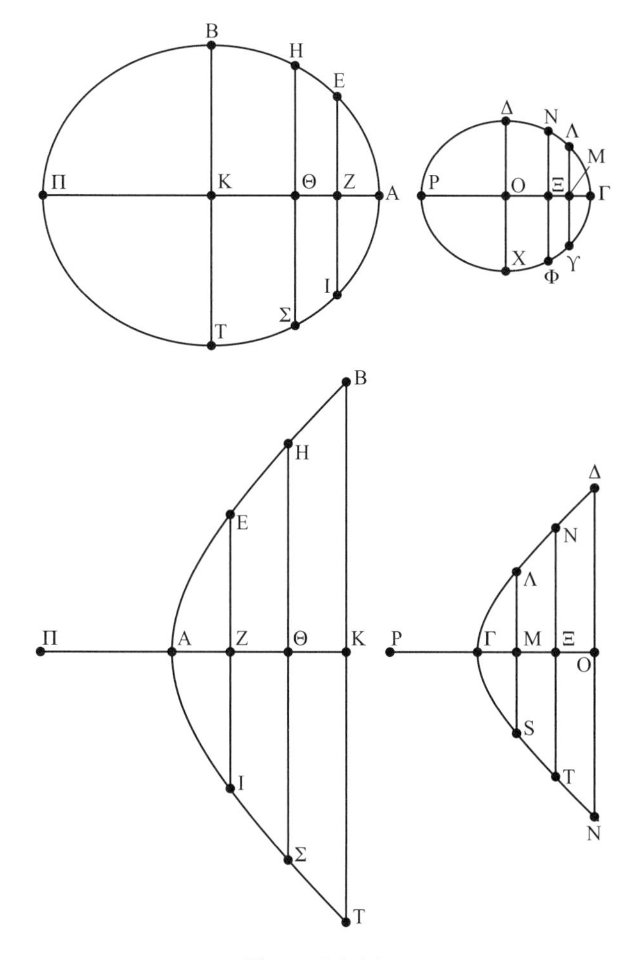

Figure 5.3.14.

Furthermore, we make section AB similar to section $\Gamma\Delta$. Then, since the two sections are similar, we draw in section AB some perpendiculars to the axis, $BT, H\Sigma, EI$, and in section $\Gamma\Delta$ perpendiculars $\Delta X, N\Phi, \Lambda Y$, and let the ratios of these perpendiculars to the amounts they cut off from the two axes be equal [respectively], and likewise the ratios of the parts they [the perpendiculars] cut off from one of the two axes to the parts they [the corresponding perpendiculars] cut off from the other axis; then BK is to AK as ΔO is to $O\Gamma$, and KA is to $A\Theta$ as $O\Gamma$ is to $\Gamma\Xi$, and $A\Theta$ is to ΘH as $\Gamma\Xi$ is to $N\Xi$. Therefore BK is to ΘH as ΔO is to $N\Xi$. And sq(BK) is to sq($H\Theta$) as sq(ΔO) is to sq($N\Xi$).

Therefore rect($\Pi K, KA$) is to rect($\Pi\Theta, \Theta A$) as rect($PO, O\Gamma$) is to rect($P\Xi, \Xi\Gamma$), because of what was proven in proposition 21 of book 1. And because KA is to $A\Theta$ as $O\Gamma$ is to $\Gamma\Xi$, [and KA is to $A\Pi$ as $O\Gamma$ is to ΓP], $K\Pi$ is to $\Pi\Theta$ as PO is to $P\Xi$, and [hence] $\Pi\Theta$ is to ΘK as $P\Xi$ is to $O\Xi$. But $K\Theta$ is to $A\Theta$ as $O\Xi$ is to $\Xi\Gamma$. Therefore $\Pi\Xi$ is to ΞA as $P\Xi$ is to $\Xi\Gamma$. And [hence] rect($\Pi\Theta, \Theta A$) is to sq(ΘA) as rect($P\Xi, \Xi\Gamma$) is to sq($\Xi\Gamma$). But sq($A\Theta$) is to sq(ΘH) as sq($\Gamma\Xi$) is to sq($N\Xi$). Therefore rect($\Pi\Theta, \Theta A$) is to sq(ΘH) as

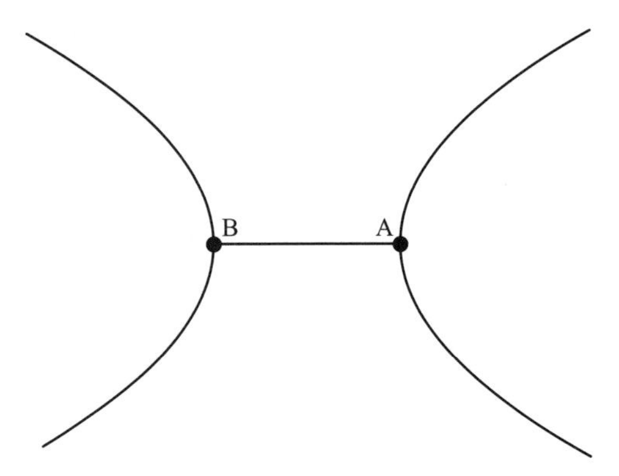

Figure 5.3.15.

rect($P\Xi, \Xi\Gamma$) is to sq(ΞN). But the ratio of rect($\Pi\Theta, \Theta A$) to sq(ΘH) equals the ratio of ΠA to the *latus rectum* [of AB], as is proven in proposition 21 of book 1. And the ratio of rect($P\Xi, \Xi\Gamma$) to sq(ΞN) equals the ratio of $P\Gamma$ to the *latus rectum* [of $\Gamma\Delta$], because, likewise, of what is proven in proposition 21 of book 1. So the figures constructed on ΠA and $P\Gamma$ are similar.

If two ellipses with equations $y^2 = x(p - \frac{p}{2a}x)$ and $y'^2 = x'(p' - \frac{p'}{2a'}x')$ are similar, then, according to this result, $\frac{p}{a} = \frac{p'}{a'}$. But this means that the similarity transformation $x \to x' = \frac{a'}{a}x$, $y \to y' = \frac{a'}{a}y$ transforms one of these ellipses to the other. The result for hyperbolas is identical.

Furthermore, although it seems obvious, in propositions 14 and 15, Apollonius proves that a parabola is not similar to an ellipse or a hyperbola and that a hyperbola is not similar to an ellipse. In proposition 25, he shows additionally that it is impossible for any part of the three conic sections to be an arc of a circle. On the other hand, proposition 16 shows that opposite sections are similar and equal, that is, that Apollonius considers opposite sections as two separate curves, not a single curve with two branches.

Proposition 16. *Opposite sections are similar and equal.*

Let there be two opposite sections A, B with axis AB. Then I say that sections A and B are similar and equal. For the *latera recta* of sections A and B are equal, as is proven in the proof of proposition 14 of book 1. And line AB is a side common to their two figures [figure 5.3.15]. So the figures constructed on the axis of sections A and B are similar and equal. Therefore section A is similar to section B and is equal to it, as is proven in proposition 12 of this book.

Proposition 28. *We want to show how to find, in a given right cone, a parabola equal to a given parabola.*

Let the given right cone be a cone, through the axis of which passes triangle $AB\Gamma$. Let the given parabola be section ΔE, with axis $\Delta\Lambda$ and *latus rectum* ΔZ, and let ΔZ be to AH as sq(ΓB) is to rect($AB, A\Gamma$). We draw line $H\Theta$ parallel to line $A\Gamma$. We cut the cone with

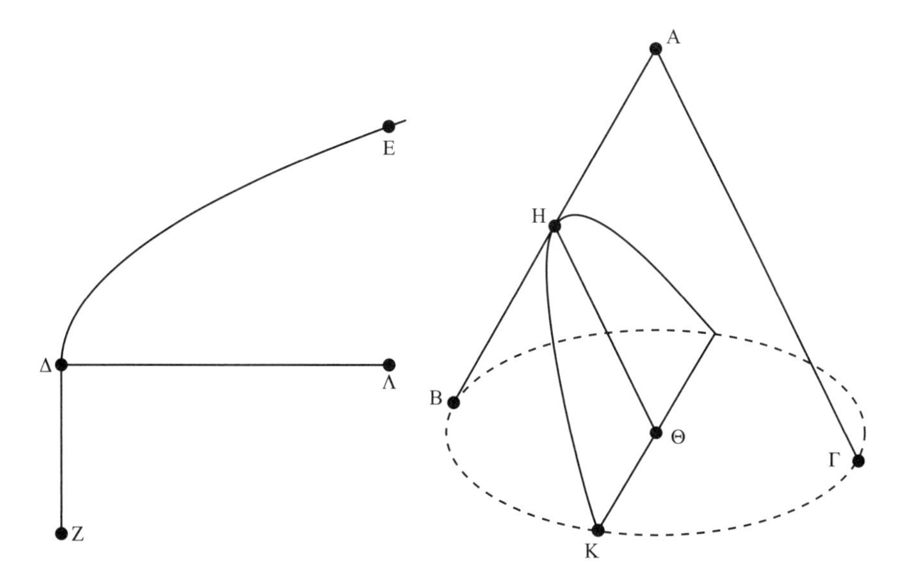

Figure 5.3.16.

a plane passing through line $H\Theta$ and erected at right angles to plane $AB\Gamma$; let [this plane] generate section KH, the axis of which is $H\Theta$ [figure 5.3.16]. Then I say that section KH equals section ΔE.

The perpendiculars drawn in section KH to line $H\Theta$ are equal in square to the rectangle which is applied to a line whose ratio to AH is equal to the ratio of sq($B\Gamma$) to rect($AB, A\Gamma$), as is proven in proposition 11 of book 1. But the ratio of ΔZ to AH also equals that of sq($B\Gamma$) to rect($AB, A\Gamma$). Therefore ΔZ equals the *latus rectum* of section KH. And it was proven in proposition 1 of this book that, when that is the case, the two sections are equal. Therefore section ΔE equals section KH.

Then I say that no other section, apart from this one, can be found in [this] cone such that the point of its vertex (which is the end of the axis) lies on line AB[64] [and such that it is equal to section ΔE]. For if it is possible to find another parabola equal to section ΔE, then its plane cuts the plane of the triangle passing through the axis of the cone at right angles, and the axis of the section lies in the plane of triangle $AB\Gamma$, because the cone is a right cone (and similarly for the axis of every section in a right cone).

So if it is possible for another section whose vertex lies on line AB to be equal to section ΔE, then its axis is parallel to line $A\Gamma$, and the point of its vertex is different from point H. And the ratio of its *latus rectum* to the line cut off by that section from line AB adjacent to point A equals that of sq($B\Gamma$) to rect($AB, A\Gamma$). But this latter ratio equals that of ΔZ to AH. Therefore the line ΔZ is not equal to the *latus rectum* of the other section. But the two sections are [supposed to be] equal; that is absurd, because of what was proven in

[64] The restriction to line *AB* may seem trivial (for since all axial triangles in a given right cone are congruent, it is obviously possible to generate an identical section from any one). However, if one considers the problem for the oblique cone, the restriction is necessary.

proposition 1 of this book. So there cannot be found on line AB the vertex of the axis of another section equal to section ΔE.

Contrast this proposition with book 1, proposition 52, which gives two things: the diameter of a parabola, and the requirement that the ordinates meet the diameter at right angles (thus making the diameter actually the axis; proposition 53 then removes this requirement). The cone, and hence a section of it, is constructed so that this section is a parabola and has the required diameter. And in doing so, the construction makes evident that the cone constructed is a right cone. In this proposition, two things are given: a parabola and a right cone; it is required to cut the cone such that the resulting section is a parabola equal to the given parabola. Propositions 29 and 30 of book 6 do similarly for the case of the hyperbola and ellipse; as with book 6, proposition 28, and book 1, propositions 52 and 53, as noted above, we can contrast book 6, propositions 29 and 30, with book 1, propositions 54–55, and 56–57, respectively.

Book 7

This book treats various properties of diameters, particularly those dealing with conjugate diameters. It evidently is designed to help prepare the reader for the lost book 8.

Apollonius's Opening Letter

From Apollonius to Attalus. Peace be on you. I have sent to you with this letter of mine the seventh book of the treatise on Conics. In this book are many wonderful and beautiful things on the topics of diameters and the figures constructed on them, set out in detail. All of this is of great use in many types of problems, and there is much need for it in the kind of problems which occur in conic sections which we mentioned, among those which will be discussed and proven in the eighth book of this treatise (which is the last book in it). I shall strive to send that to you speedily. Farewell.

Proposition 1. *If the axis of a parabola is extended in a straight line outside of the section to a point such that the part of it which falls outside of the section is equal to the* latus rectum, *and furthermore a line is drawn from the vertex of the section to any point on the section, and a perpendicular is drawn to the axis drawn from where it meets the section, then the line which was drawn from the vertex is equal in square to the rectangle contained by (1) the line between the foot of the perpendicular and the vertex of the section, and (2) the line between the foot of the perpendicular and the point to which the axis was extended.*

Let there be a parabola, AB, with axis $A\Gamma$. We extend line ΓA to Δ; I let $A\Delta$ be equal to the *latus rectum*. We draw from point A line AB in any position (so as to cut the curve again in B), and draw line $B\Gamma$ as perpendicular to $A\Gamma$ [figure 5.3.17]. Then I say that sq(AB)= rect($\Delta\Gamma, A\Gamma$).

Line $A\Gamma$ is the axis of the section, and line $B\Gamma$ is perpendicular to it, and line $A\Delta$ is equal to the *latus rectum*. Therefore sq($B\Gamma$) equals rect($\Delta A, A\Gamma$), as is proven in proposition 11 of book 1. So we make sq($A\Gamma$) common. Then sq($A\Gamma$) together with sq(ΓB) equals

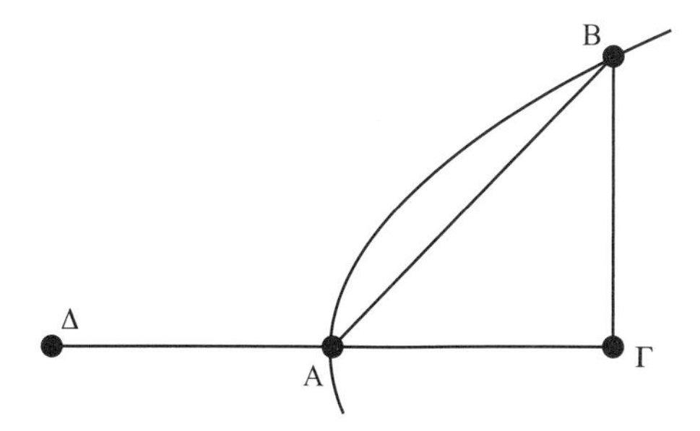

Figure 5.3.17.

rect($\Delta A, A\Gamma$) together with sq($A\Gamma$). But the squares on $A\Gamma$ and ΓB equal the square on AB, and rect($\Delta A, A\Gamma$) together with sq($A\Gamma$) equals rect($\Delta\Gamma, \Gamma A$). Therefore sq(AB) equals rect($\Delta\Gamma, \Gamma A$).

Propositions 2 and 3 are similar, with proposition 2 dealing with the hyperbola and proposition 3 with the ellipse. We present only proposition 3.

Proposition 3. *If a line is constructed on the extension of one of the axes of an ellipse, whichever axis it may be,*[65] *and one of its ends is one of the ends of the transverse diameter, while the other end is outside of the section, and the ratio of it (the line) to the line between its other end and the remaining end of the (transverse) diameter is equal to the ratio of the transverse diameter to the* latus rectum, *and a line is drawn, from the end common to the (transverse) diameter and the line constructed on the axis, to any point on the section, and from the place where it meets (the section) a perpendicular is drawn to the axis; then the ratio of the square on the line which was drawn (to the section) to the rectangle contained by the two lines between the foot of the perpendicular and the two ends of the (first) line which was constructed on the axis is equal to the ratio of the transverse diameter to the line between those two ends of the transverse diameter and the line which was constructed that are different from each other.*[66] *Let the line which was constructed be called the "homologue."*

Let there be an ellipse, with axis $A\Gamma$ and figure $\Gamma\Delta$. Let the line constructed on the extension of the axis be $A\Theta$, and let $\Gamma\Theta$ be to ΘA as ΓA is to $A\Delta$. From point A let a line, AB, be drawn to the section, and let us draw BE perpendicular to the axis [figures 5.3.18, 5.3.19]. Then I say that sq(AB) is to rect($\Theta E, EA$) as $A\Gamma$ is to $\Gamma\Theta$.

We make rect(AE, EZ) equal to sq(BE). Then rect(AE, EZ) is to rect($AE, E\Gamma$) as sq(BE) is to rect($AE, E\Gamma$). But sq(BE) is to rect($AE, E\Gamma$) as the ratio of the *latus rectum* (which is $A\Delta$) to the transverse diameter (which is $A\Gamma$), as is proven in proposition 21 of book 1. Therefore rect(AE, EZ) is to rect($AE, E\Gamma$) as ΔA is to $A\Gamma$ which is as ZE is to $E\Gamma$, and ΔA is to $A\Gamma$ as $A\Theta$ is to $\Theta\Gamma$. Therefore ZE is to $E\Gamma$ as $A\Theta$ is to $\Theta\Gamma$. Therefore *dividendo*, $Z\Gamma$ is to ΓE as $A\Gamma$ is to $\Gamma\Theta$. Therefore *dividendo* [figure 5.3.18] and *componendo* [figure 5.3.19], ZA is to ΘE as $A\Gamma$ is to $\Gamma\Theta$.

[65] Either the major axis [figure 5.3.18] or the minor axis [figure 5.3.19].
[66] That is, the complete extended line.

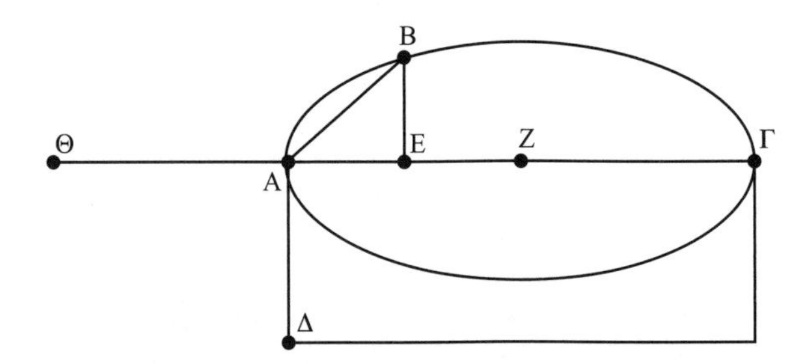

Figure 5.3.18.

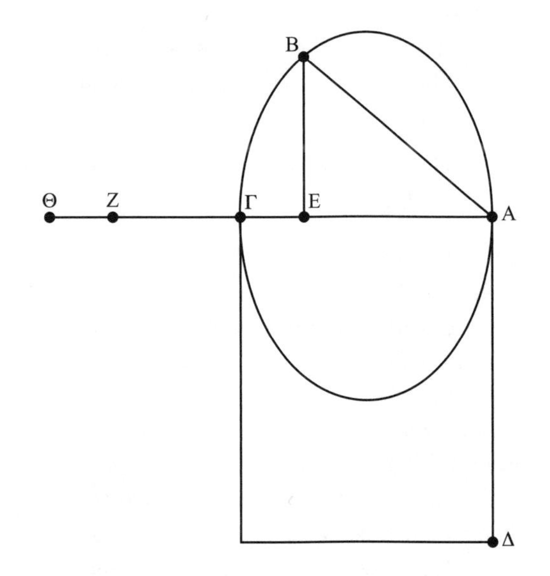

Figure 5.3.19.

But, when we make AE a common height, ZA is to ΘE as rect(ZA, AE) is to rect($E\Theta, EA$). Therefore $A\Gamma$ is to $\Gamma\Theta$ as rect(ZA, AE) is to rect($AE, E\Theta$). But rect(ZA, AE) equals sq(AB). Therefore sq(AB) is to rect($AE, E\Theta$) as $A\Gamma$ is to $\Gamma\Theta$.

As before, since propositions 6 and 7, dealing with the hyperbola and the ellipse, are similar, we present only proposition 7. It uses the homologue defined in proposition 3.

Proposition 7. *If there are constructed on the extension of the axis of an ellipse two lines at the two ends of it (the axis), each of them being equal to the homologue, and two conjugate diameters are drawn in the section, and, from the vertex of the section, a line is drawn parallel to one of the conjugate diameters so as to meet the section (again), and, from the place where it meets (the section), a perpendicular is drawn to the axis, then the ratio of the diameter which is not parallel to the line which was drawn, to the other diameter, is equal in square to the ratio to each other of the two parts (of the line between the ends of the two homologues which are not the ends of the diameter) into which it is cut by the perpendicular: according to how the homologues are placed. If they are found*

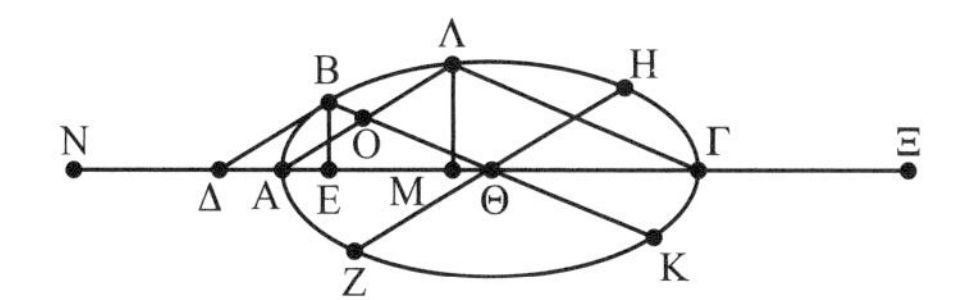

Figure 5.3.20.

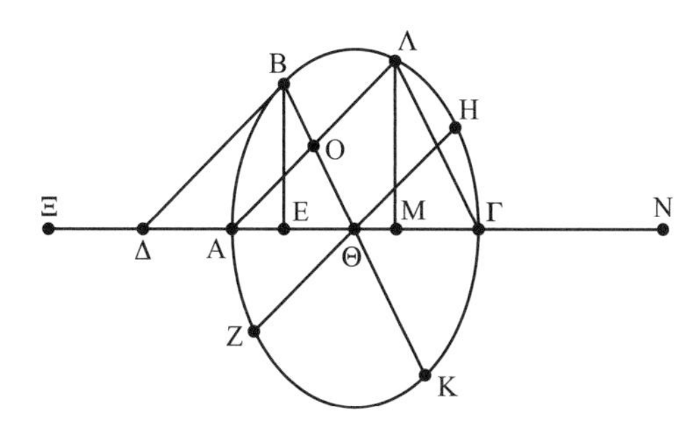

Figure 5.3.21.

on the major axis [figure 5.3.20], *they are outside the section, and if on the minor axis* [figure 5.3.21], *then they are on the axis itself. And the ratio of the diameter mentioned to the parameter of the ordinates falling on it (which are parallel to the other [conjugate] diameter) is (also) equal to the ratio mentioned.*

Let there be an ellipse, with axis $A\Gamma$. Let the two homologues be AN, $\Gamma\Xi$. Let diameters ZH, BK be conjugate, in any position. We draw line $A\Lambda$ parallel to the diameter ZH, and draw from point Λ a perpendicular to the axis, namely, ΛM [figures 5.3.20, 5.3.21]. Then I say that sq(BK) is to sq(ZH) as $M\Xi$ is to MN, and that the ratio of KB to the parameter of the lines drawn to it in the section parallel to ZH—which parameter is the *latus rectum*—also equals the ratio $M\Xi$ to MN.

We join $\Gamma\Lambda$, and draw perpendicular BE from point B, and draw from it, too, a line parallel to line ZH, namely, $B\Delta$. Then that line ($B\Delta$) is tangent to the section. And since $\Gamma\Theta$ equals ΘA and ΛO equals OA, $\Gamma\Lambda$ is parallel to $B\Theta$. Therefore ΔE is to $E\Theta$ as AM is to $M\Gamma$, because of the similarity of triangles. But ΔE is to $E\Theta$ as sq(ΔB) is to sq(ΘH), because of what is proven in proposition 4 of this book. Therefore AM is to $M\Gamma$ as sq(ΔB) is to sq(ΘH). And since sq($B\Theta$) is to sq($B\Delta$) as sq($\Gamma\Lambda$) is to sq($A\Lambda$), because of the similarity of the two triangles, and sq($B\Delta$) is to sq(ΘH) as AM is to $M\Gamma$, therefore the ratio of sq(ΘB) to sq(ΘH) is compounded of the ratios of sq($\Gamma\Lambda$) to sq($A\Lambda$) and AM to $M\Gamma$.

But the ratio sq($\Gamma\Lambda$) to sq($A\Lambda$) is compounded of the ratios of sq($\Gamma\Lambda$) to rect($\Gamma M, M\Xi$) with rect($\Gamma M, M\Xi$) to rect($AM.MN$) and with rect(AM, MN) to sq($A\Lambda$). Therefore the ratio of sq(ΘB) to sq(ΘH) is compounded of the ratios sq($\Gamma\Lambda$) to rect($\Gamma M, M\Xi$), rect($\Gamma M, M\Xi$) to rect(AM, MN), rect(AM, MN) to sq($A\Lambda$), and AM to $M\Gamma$. But sq($\Gamma\Lambda$) is to rect($\Gamma M, M\Xi$) as $A\Gamma$ is to $A\Xi$, as is proven in proposition 3 of this book; and rect(AM, MN) is to sq($A\Lambda$) as ΓN is to $A\Gamma$, as is also proven in proposition 3 of this book;

and the ratio of rect($\Gamma M, M\Xi$) to rect(AM, MN) is compounded of the ratio of ΓM to AM with that of $M\Xi$ to MN. Therefore the ratio of sq(ΘB) to sq(ΘH) is compounded of the ratios $A\Gamma$ to $A\Xi$, ΓN to $A\Gamma$, ΓM to AM, $M\Xi$ to MN, and AM to $M\Gamma$.

And the ratio which is composed of these latter ratios which we mentioned is equal to that of $M\Xi$ to MN, because the part of it, the ratio of ΓN to $A\Gamma$, when compounded with the ratio of $A\Gamma$ to $A\Xi$ equals the ratio of ΓN to $A\Xi$, and ΓN equals $A\Xi$; and as for the part of it, the ratio of ΓM to AM, when compounded with the ratio of AM to ΓM, equals the ratio of ΓM to itself. Therefore the ratio composed of these ratios equals the remaining ratio, which is that of $M\Xi$ to MN. Therefore sq(BK) is to sq(ZH) as $M\Xi$ is to MN and sq($B\Theta$) is to sq(ΘH) as ΞM is to MN. And, furthermore, sq(BK) is to sq(ZH) as the ratio of KB to the line by which the lines drawn from the section to KB parallel to line ZH are equal in square (to the intercepts).

Therefore the ratio of line BK to the parameter of the ordinates falling on it equals the ratio of $M\Xi$ to MN. Hence it will be proven that, if the perpendicular falling from point Λ on the axis passes through center Θ, then diameter KB will be equal to diameter ZH, because $M\Xi$ equals MN.

Proposition 12. *In any ellipse, the sum of the squares on any two of its conjugate diameters whatever is equal to the sum of the squares on its two axes.*

Let the figure for the ellipse [figures 5.3.20, 5.3.21] be as it was in proposition 7 of this book. Then the axis is $A\Gamma$, the two conjugate diameters BK, ZH, and the two homologues AN, $\Gamma\Xi$. And the ratio of sq($A\Gamma$) to the square on the other of the two axes of the section equals the ratio of $A\Gamma$, which is the transverse diameter, to the *latus rectum* (corresponding to it), as is proven in proposition 15 of book 1. But the ratio of $A\Gamma$ to its *latus rectum* equals the ratio of ΓN to AN, because AN is the homologue. And AN equals $\Gamma\Xi$. Therefore the ratio of sq($A\Gamma$) to the square on the other of the two axes of the section equals the ratio of $N\Gamma$ to $\Gamma\Xi$. And for that reason the ratio of sq($A\Gamma$) to the sum of sq($A\Gamma$) and the square on the other of the two axes of the section equals the ratio of $N\Gamma$ to $N\Xi$.

Furthermore, sq($A\Gamma$) is to sq(BK) as $N\Gamma$ is to $M\Xi$, as is proven in the proof of proposition 8 of this book. And sq(BK) is to sq(BK) and sq(ZH) together as $M\Xi$ is to $M\Xi$ and NM, taken together, because it was proven in proposition 7 of this book that sq(BK) is to sq(ZH) as $M\Xi$ is to MN. But $M\Xi$ and NM together equal ΞN. Therefore, *ex aequali*, sq($A\Gamma$) is to sq(BK) and sq(ZH) taken together as $N\Gamma$ is to $N\Xi$. And we had (already) proven that the ratio of $N\Gamma$ to $N\Xi$ equals the ratio of sq($A\Gamma$) to the sum of the squares on the two axes. Therefore the sum of the squares on the two axes equals the sum of sq(BK) and sq(ZH).

Proposition 13 is analogous to proposition 12 but for hyperbolas. It asserts that the difference between the squares on any pair of conjugate diameters is equal to the difference between the squares on its axes. Propositions 21 through 24 consider the ratios of conjugate diameters in hyperbolas and ellipses. We state just the first three, without proofs, and give a proof of the first part of proposition 24.

Proposition 21. *If there is a hyperbola, and its transverse axis is greater than its erect axis, then*

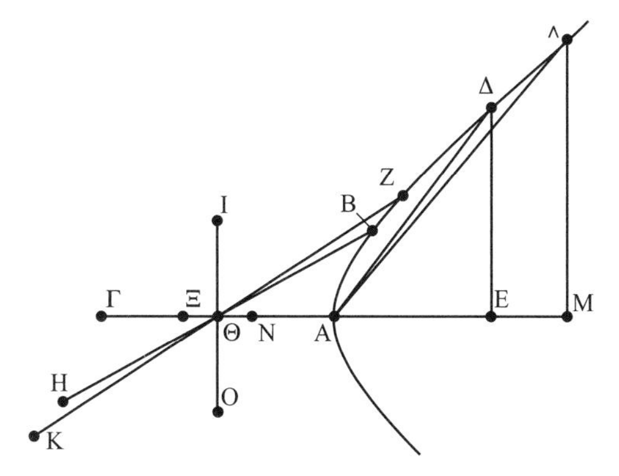

Figure 5.3.22.

- *the transverse diameter of each pair of conjugate diameters among its other diameters is greater than the erect diameter of that pair;*
- *the ratio of the greater axis to the lesser axis is greater than the ratio of the transverse diameter to the erect diameter among the other conjugate diameters;*
- *and the ratio of a transverse diameter nearer to the greater axis to the erect diameter conjugate with it is greater than the ratio of a transverse diameter farther from that axis to the erect diameter conjugate with it.*

Let there be a hyperbola with axes $A\Gamma$, IO, and let there be two other transverse diameters, BK, BH, and let $A\Gamma$ be greater than IO [figure 5.3.22]. Then I say that

- BK is greater than the erect diameter conjugate with it;
- diameter ZH is also greater than the erect diameter conjugate with it;
- the ratio of $A\Gamma$ to OI is greater than the ratio of BK to the erect diameter conjugate with it, and also greater than the ratio of ZH to the erect diameter conjugate with it;
- and the ratio of BK to the erect diameter conjugate with it is greater than the ratio of ZH to the erect diameter conjugate with it.

Proposition 22. *If there is a hyperbola, and its transverse axis is shorter than its erect axis, then*

- *the transverse diameter of each pair of diameters among the other conjugate diameters is shorter than the erect diameter of that pair;*
- *the ratio of the shorter axis to the longer axis is less than the ratio of any of the other transverse diameters to the erect diameter conjugate with it;*
- *and the ratio of the transverse diameter nearer to the shorter axis to the erect diameter with it is less than the ratio of a transverse diameter farther from the axis to the diameter conjugate with it.*

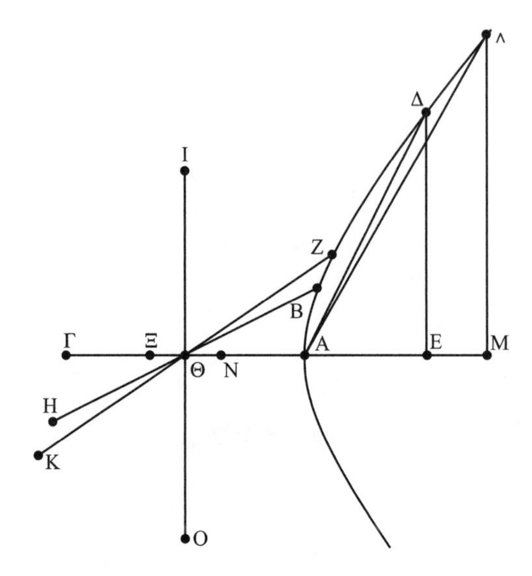

Figure 5.3.23.

Let there be a hyperbola with axes $A\Gamma$, OI, and center Θ, and with two diameters BK, ZH, and let the transverse axis $A\Gamma$ be shorter than the erect axis OI [figure 5.3.23]. Then I say that

- each of BK and ZH is shorter than the erect diameter conjugate with it;
- the ratio of $A\Gamma$ to OI is less than the ratio of BK to the erect diameter conjugate with it; and is less than the ratio of ZH to the erect diameter conjugate with it;
- and the ratio of BK to the erect diameter conjugate with it is less than the ratio of ZH to the erect diameter conjugate with it.

Proposition 23. *If the axes of a hyperbola are equal, then every conjugate pair of its diameters is equal.*

Here, the conditions of the previous two propositions are both fulfilled, provided that we replace "less than" with "less than or equal to" and the same for "greater than." Then the conclusion is obvious.

Proposition 24. *If there is an ellipse, and conjugate diameters are drawn in it, then*

- *the ratio of the greater of each pair of conjugate diameters to the lesser is less than the ratio of the major axis to the minor axis;*
- *and for any two pairs of conjugate diameters, the ratio of the greater diameter which is nearer to the major axis than the other greater diameter to the lesser diameter conjugate with it is greater than the ratio of the greater diameter which is farther from the major axis to the lesser diameter conjugate with it.*

Let the major of the two axes of the ellipse be AB, and its minor axis $\Gamma\Delta$, and (two pairs of) its conjugate diameters EZ, HK and $N\Xi$, $O\Pi$. Let EZ be greater than HK,

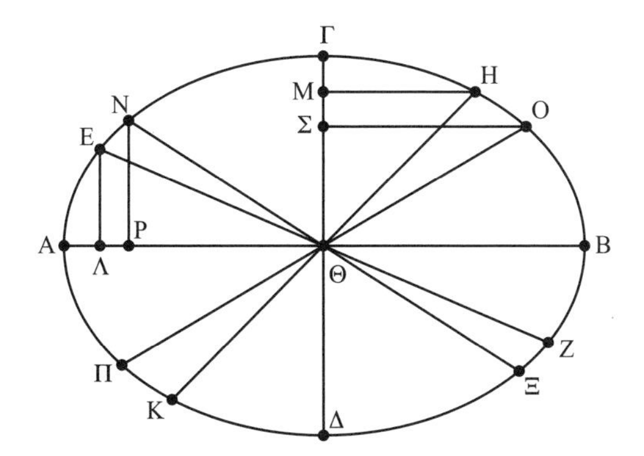

Figure 5.3.24.

its conjugate, and $N\Xi$ be greater than $O\Pi$, its conjugate, (and let EZ be closer to the major axis than $N\Xi$) [figure 5.3.24]. We draw from points E, N perpendiculars to axis AB, namely, $E\Lambda$, NP, and draw from points H, O perpendiculars to $\Gamma\Delta$, namely, HM, $O\Sigma$. Then rect($A\Theta, \Theta B$) is to sq($\Theta\Gamma$) as rect($A\Lambda, \Lambda B$) is to sq(ΛE), as is proved in proposition 21 of book 1. But rect($A\Theta, \Theta B$) is greater than sq($\Theta\Gamma$); therefore rect($A\Lambda, \Lambda B$) is greater than sq(ΛE). Therefore $A\Theta$ is greater than ΘE, and hence AB is greater than EZ.

Furthermore, rect($\Gamma\Theta, \Theta\Lambda$) is to sq($\Theta B$) as rect($\Gamma M, M\Delta$) is to sq($MH$). But rect($\Gamma\Theta, \Theta\Delta$) is less than sq($\Theta B$). So rect($\Gamma M, M\Delta$) is less than sq($MH$). Therefore $\Theta\Delta$ is less than ΘH, and hence $\Gamma\Delta$ is less than KH. But it was proved that AB is greater than EZ. Therefore the ratio of AB to $\Gamma\Delta$ is greater than the ratio of EZ to KH. And diameter EZ is conjugate with diameter KH.

The second part of the proof shows that the ratio of EZ to its conjugate HK is greater than the ratio of ΞN to its conjugate $O\Pi$. Note that, in analogy to proposition 23, if the two axes of the ellipse are equal, then the curve is a circle and the result is trivially true.

Proposition 25. *In every hyperbola, the line equal to the sum of its two axes is less than the line equal to the sum of any other pair whatever of its conjugate diameters; and the line equal to the sum of a transverse diameter closer to the greater axis plus its conjugate diameter is less than the line equal to the sum of a transverse diameter farther from the greater axis plus its conjugate diameter.*

We skip the proof of proposition 25 but give the proof of proposition 26, its analog for the case of an ellipse.

Proposition 26. *In every ellipse, the sum of its two axes is less than the sum of any conjugate pair of its diameters; and the sum of any conjugate pair of its diameters which is closer to the two axes is less than the sum of any conjugate pair of its diameters farther from the two axes; and the sum of the conjugate pair of its diameters each of which is equal to the other is greater than that of any other conjugate pair of its diameters.*

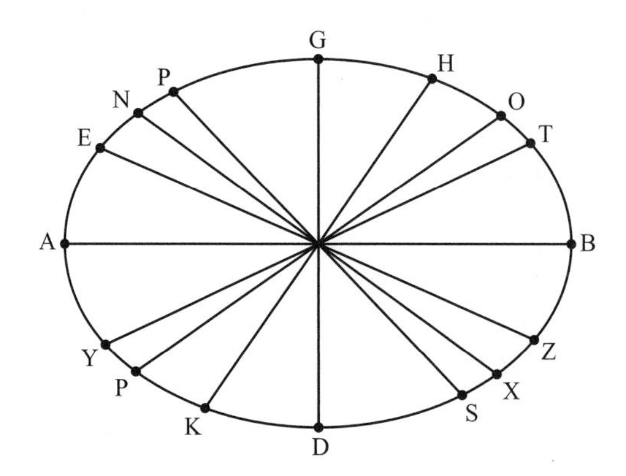

Figure 5.3.25.

Let there be an ellipse with major axis AB and minor axis GD, and conjugate diameters EZ, KH, and NX, OP, and YT, PS, and let EZ be greater than its conjugate KH, let XN be greater than its conjugate OP, and let PS be equal to its conjugate YT [figure 5.3.25]. Then I say that the line equal to the sum of the two axes AB, GD is less than the line equal to the sum of the two diameters EZ, HK, and that equal to the sum of the two diameters NX, OP; and that the greater of the sums of the pairs of conjugate diameters is the line equal to the sum of the two diameters PS, YT.

The ratio of AB to GD is greater than the ratio of EZ to KH, as is proven in proposition 24 of this book. Therefore the ratio of the square on the sum of the lines AB, GD to the sum of sq(AB) and sq(GD) is less than the ratio of the square of the sum of the lines EZ, KH to the sum of sq(EZ) and sq(KH). But the sum of sq(EZ) and sq(KH) equals the sum of sq(AB) and sq(GD), as is proven in proposition 12 of this book. Hence the square on the sum of AB, GD is less than the square on the sum of the lines EZ, KH. Therefore the line equal to the sum of the two axes AB and GD is less than the line equal to the sum of the two diameters EZ and KH. Similarly too it will be proven that the line equal to the sum of the lines EZ, HK is less than the line equal to the sum of the two diameters PS, YT.

Proposition 31. *When a pair of conjugate diameters is drawn in an ellipse or between conjugate opposite sections (i.e., hyperbolas), then the parallelogram bounded by that pair of diameters with angles equal to the angles formed by the diameters at the center is equal to the rectangle bounded by the two axes.*

Let there be an ellipse [figure 5.3.26] or conjugate opposite sections [figure 5.3.27], with center Θ, and axes AB, $\Gamma\Delta$, and with one pair of its conjugate diameters $Z\Lambda$, ΞN. Let tangents (to these sections) pass through points Z, Λ, Ξ, N, namely, HP, KM, HK, PM. Then lines HP and KM are parallel to diameter ΞN, and lines HK and PM are parallel to diameter $Z\Lambda$, as is proven in propositions 5 and 20 of book 2. So quadrilateral HM is a parallelogram, and its angles are equal to the angles formed by diameters $Z\Lambda, \Xi N$ at center Θ. Then I say that quadrilateral MH is equal to the rectangle bounded by the two axes $AB, \Gamma\Delta$.

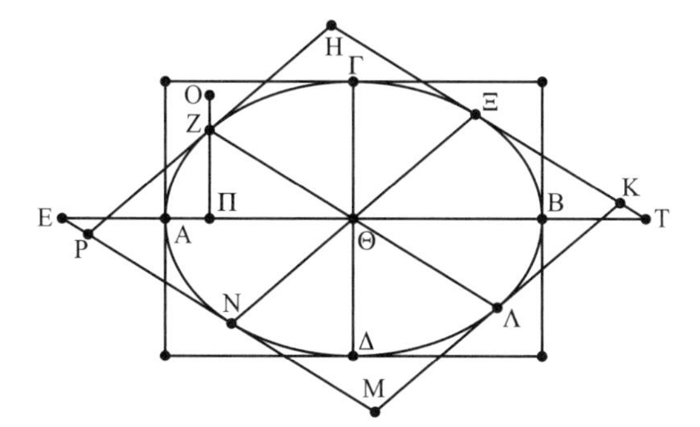

Figure 5.3.26.

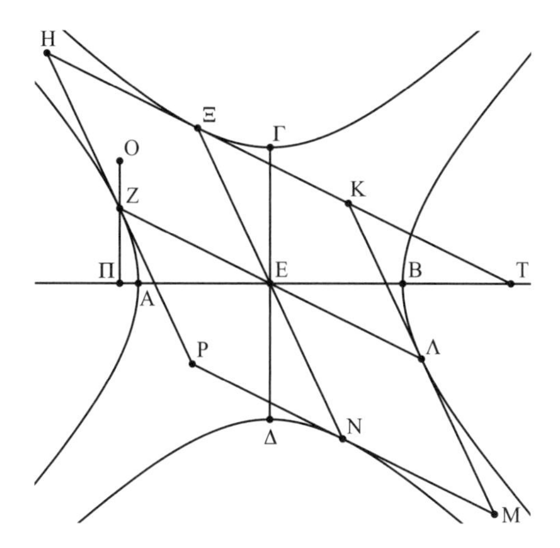

Figure 5.3.27.

We draw from point Z a perpendicular to $B\Theta A$, namely, $Z\Pi$, and make line ΠO a mean proportional between lines $E\Pi$ and $\Pi\Theta$. Then sq($A\Theta$) is to sq($\Theta\Gamma$) as rect($\Theta\Pi, \Pi E$) is to sq($Z\Pi$), as is proven in proposition 37 of book I. But rect($\Theta\Pi, \Pi E$) equals sq(ΠO). Therefore sq($A\Theta$) is to sq($\Theta\Gamma$) as sq(ΠO) is to sq($Z\Pi$), and sq($A\Theta$) is to rect($A\Theta, \Theta\Gamma$) as rect($O\Pi, \Theta E$) is to rect($Z\Pi, \Theta E$). And *permutando*, sq($A\Theta$) is to rect($O\Pi, \Theta E$) as rect($A\Theta, \Theta\Gamma$) is to rect($Z\Pi, \Theta E$).

But sq($A\Theta$) equals rect($E\Theta, \Theta\Pi$), as is proven in proposition 37 of book 1. Therefore rect($E\Theta, \Theta\Pi$) is to rect($O\Pi, \Theta E$) as rect($A\Theta, \Theta\Gamma$) is to rect($Z\Pi, \Theta E$). And line $\Theta\Xi$ is parallel to line ZE. Therefore sq(ZE) is to sq($\Theta\Xi$) as $E\Pi$ is to $\Pi\Theta$, as is proven in proposition 4 of this book. And tri(ΘZE) is to tri($\Xi\Theta T$) as sq(ZE) is to sq($\Theta\Xi$), because the two triangles are similar. Therefore tri(ΘZE) is to tri($\Xi\Theta T$) as $E\Pi$ is to $\Pi\Theta$, and 2 tri(ΘZE) is to 2 tri($\Xi\Theta T$) as $E\Pi$ is to $\Pi\Theta$.

But quadrilateral $\Xi\Theta ZH$ is a mean proportional between 2 tri(ΘZE) and 2 tri($\Xi\Theta T$). And similarly $O\Pi$ is a mean proportional between lines $E\Pi$ and $\Pi\Theta$. Therefore 2 tri(ΘZE) is to pllg(ΘH) as $O\Pi$ is to $\Pi\Theta$. But $O\Pi$ is to $\Pi\Theta$ as rect($O\Pi, \Theta E$) is to rect($\Pi\Theta, \Theta E$). Therefore 2 tri(ΘZE) is to qdrl(ΘH) as rect($O\Pi, \Theta E$) is to rect($\Pi\Theta, \Theta E$). And we had already proven that rect($O\Pi, \Theta E$) is to rect($\Pi\Theta, \Theta E$) as rect($Z\Pi, \Theta E$) is to rect($A\Theta, \Theta\Gamma$). Therefore 2 tri(ΘZE) is to qdrl(ΘH) as rect($Z\Pi, \Theta E$) is to rect($A\Theta, \Theta\Gamma$). But 2 tri(ΘZE) equals rect($Z\Pi, \Theta E$). Therefore quadrilateral ΘH equals rect($A\Theta, \Theta\Gamma$), and hence 4 quadrilateral ΘH, which is quadrilateral HM, equals 4 rect($A\Theta, \Theta\Gamma$), which equals the rectangle bounded by the two axes AB, $\Gamma\Delta$. So quadrilateral MH equals the rectangle bounded by the two axes $AB, \Gamma\Delta$.

5.4 Apollonius, *On Cutting Off a Ratio*

This is the only other of Apollonius's works to survive in its entirety but, like books 5–7 of the *Conics*, only in an Arabic translation. This was one of the works included by Pappus in his *Treasury of Analysis*, discussed in book 7 of the *Collection*. In *Cutting Off a Ratio*, a work in two books, Apollonius deals with a general problem: given two straight lines and a fixed point on each line, to draw through a given point a line that cuts off segments from each line (measured from the fixed points) bearing a given ratio to one another. To accomplish this, Apollonius considers numerous cases and subcases, beginning with the situation where the two given lines are parallel and then moving to the situation where they intersect. The various cases deal with the differing positions of the given straight lines relative to the given point and the different ways the line is drawn from the given point to the given lines. For each case, Apollonius begins with an analysis, which sometimes leads to a necessary condition for the problem to be possible. He then concludes with a synthesis, generally showing that the given solution is unique. So this book was an ideal candidate for Pappus to include in his *Treasury*, since it displays in overwhelming detail the basic method of analysis and synthesis. We present here just two cases from book 1. In the first case, the two given lines are parallel, while in the second they intersect. Note that Apollonius uses the basic language of Euclid's *Data*; that is, he determines line by line and point by point which elements of the diagram are "given," that is, are determined by what has already been shown.

Two unbounded straight lines being positioned in a plane, either parallel or intersecting, on each of which a point is given, and a ratio being given, and a point that is not on either of the lines being given: to draw a straight line from the given point across the positioned lines, cutting them off so that the ratio to one another of the two segments adjacent to the two given points on the two lines is the same as the given ratio.

Book 1

First, let the two lines be parallel, and let them be the lines AB, CD. Let the given point on line AB be point E and on CD point G. And let the line that passes through these terminal

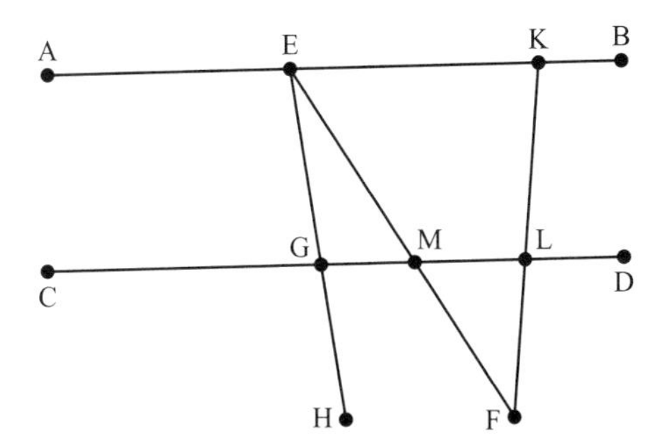

Figure 5.4.1.

points on the lines be EH [figure 5.4.1]. Then the given point is either within angle DGH or within angle BEG and angle DGE, or within angles in adjacent positions.

Locus One. And first let it be within the angle DGH at point F. Then the lines that are drawn from point F, which cut off from the two lines adjacent to the two points E and G segments whose ratio to one another is the same as a given ratio, occur in three ways. Either they cut off segments from EB, GD, or from EA, GD, or from EA, GC.

Case 1. So let the line be drawn and fall in the first way as line FK, cutting off from the two lines EB, GD a ratio of EK to GL the same as a given ratio. And let the line EF be joined. So EF has been given in position. But CD is also given in position. So point M is given. But each of the two points E, F is given. So the ratio of EM to MF is given. And *componendo*, the ratio of EF to FM is given. And EF is to FM as EK is to ML. So the ratio of EK to ML is given. But the ratio of EK to GL is given. So the ratio of GL to ML is given. And *separando*, the ratio of GM to ML is given. And the line GM is given in magnitude, since points G and M are both given. So the line ML is given in magnitude and position. But point M is given. So L too is given. But point F is also given. So line FK is given. And since ML is less than GL, the ratio of EK to ML is greater than the ratio of EK to GL. But EK is to ML as EF is to FM. So the ratio of EF to FM is greater than the ratio of EK to GL, which is equal to the given ratio. So for that reason it is necessary that the ratio being given is less than the ratio of EF to FM.

And the synthesis of the problem is as follows [figure 5.4.2]: let EF be joined. It has become clear that the ratio being given must be less than the ratio of EF to FM. So let it be the ratio of N to JO, and let it be made that, as EF is to FM, so is N to JP, and as OP is to PJ, so is GM to ML. And let FL be joined and produced in a straight line. I say that the line FK alone does what the problem required. This will be shown as follows. Since GM is to ML as OP is to PJ, *componendo*, GL is to LM as JO is to JP. Inversely, LM is to GL as PJ is to JO. But also, since EF is to FM as EK is to ML, we know that EK is to ML as N is to JP. But ML is to LG as PJ is to JO. So *ex aequali*, EK is to GL as N is to JO. Thus, from point F a straight line FK has been drawn that cuts off a ratio EK to GL the same as the given ratio. Thus line FK does what the problem requires.

And I say that it alone does this. For, if it be possible for another to do it, then let line FX be drawn cutting off a ratio EX to GR which is the same as the given ratio [figure 5.4.2].

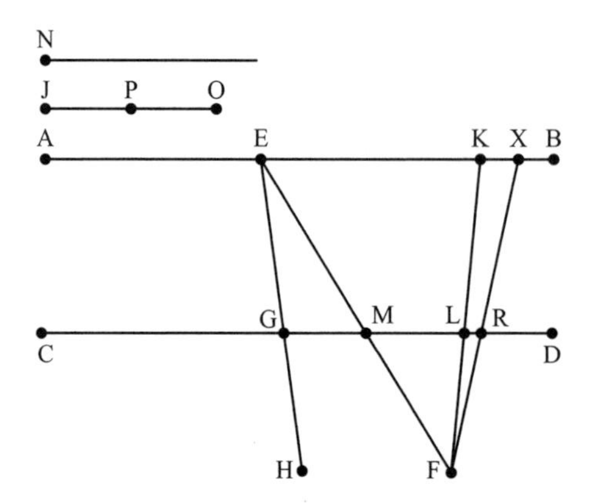

Figure 5.4.2.

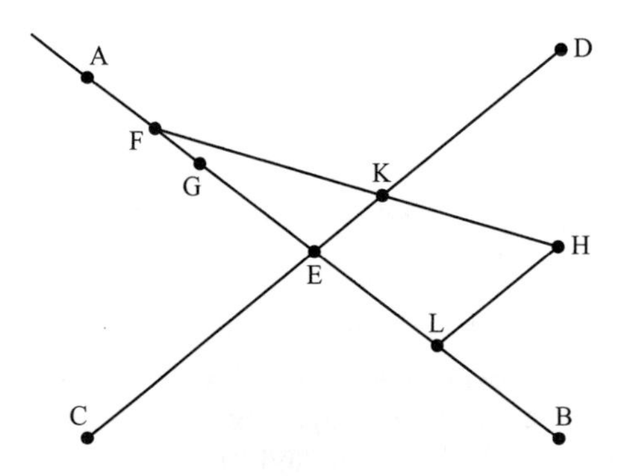

Figure 5.4.3.

So, since LM is less than LG, the ratio of RL to LM is greater than the ratio of RL to LG. And *componendo*, the ratio of RM to ML is greater than the ratio of RG to LG. And RM is to ML as XE is to EK. So it is greater than the ratio of RG to LG. And *alternando*, the ratio of XE to RG is greater than the ratio of EK to GL. So line FX does not do what the problem requires. Likewise it is clear that no line other than FK does. Therefore, FK alone does what the problem requires. And it is clear that the lines nearer point G cut off ratios less than those farther from it.

In this second case, Apollonius makes use of the procedure of application of areas, discussed by Euclid in *Elements* 6, propositions 28–29 and in *Data*, propositions 57–59.

Let the lines AB, CD cut one another at point E. Let the terminal point on line AB be point G, and on line CD point E [figure 5.4.3]. Now the given point is either within angle DEB or within angle AEC or within adjacent angles.

Locus Four. Then first let it be within angle DEB as point H. The straight lines that are drawn from point H, cutting off lines next to the points E and G in a given ratio, occur in four ways. Either they will cut off from the two lines ED, GA, or from ED, GE, or from EC, GB, or from ED, GB.

Case 1. So let a line HF be drawn in the first way, cutting off from the two lines ED, GA a ratio of EK to GF the same as a given ratio. And let a line HL be drawn parallel to line DE. So point L is given. And let it be made that, as EK is to GF, so is LH to GA. Since the ratio of LH to AG is given, and line LH is given, GA also is given in magnitude and position. But point G is given. So point A is given. But point L is also given. So line AL is given. And because LH is to GA as EK is to GF, then, *alternando*, LH is to EK as AG is to GF. However, LH is to EK as LF is to FE. So LF is to FE as AG is to GF. And, *convertando*, FL is to LE as GA is to AF. So rect(AG, EL) is equal to rect(LF, FA). However, rect(AG, EL) is given. So rect(LF, FA) is given. And to a given line, namely AL, there has been applied a rectangle equal to a given rectangle and deficient by a square. So each of AF, FL is given. So point F is given. But point H is also given. So line FH is given.

Now let it be shown that the application is a possibility. It is necessary in the synthesis to apply to the line LA a rectangle equal to rect(LE, GA) and deficient from the complete line by a square. That is possible since rect(LG, GA) is greater than rect(LE, GA).

And this problem will be synthesized thus: Let HL remain parallel as before. And let the given ratio be the ratio of M to N. Let the ratio of LH to GA be made equal to the ratio of M to N. And let there be applied to line AL the rect(LF, FA) equal to rect(AG, EL) and deficient by a square. And let FH be joined. Then I say that line FH alone does what the problem requires. That is, KE is to GF as M is to N.

For since rect(LF, FA) is equal to rect(AG, EL), we have FL is to LE as GA is to AF. And, *convertando*, LF is to FE as AG is to GF. However, LF is to FE as LH is to EK. So LH is to EK as AG is to GF. And, *alternando*, LH is to GA as EK is to GF. However, it was made that LH is to GA as M is to N. So EK is to GF as M is to N. So line FH does what the problem requires.

I say that it alone does so. For if it be possible for another to do so, let another line be drawn, such as HJ [figure 5.4.4]. And because rect(LG, GA) is greater than rect(LF, FA), also rect(LF, FA) is greater than rect(LJ, JA). And rect(LF, FA) is equal to rect(GA, LE). So rect(GA, LE) is greater than rect(LJ, JA). So the ratio of JL to LE is less than the ratio of GA to AJ. And the ratio of LJ to JE, *convertando*, is greater than the ratio of AG to GJ. However LJ is to JE as LH is to ES. So the ratio of LH to ES is greater than the ratio of AG to GJ. And, *alternando*, the ratio of LH to GA is greater than the ratio of ES to GJ. However, LH is to GA as M is to N. So the ratio of M to N is greater than the ratio of ES to GJ. And it is clear that the lines near to point E cut off ratios greater than the ratios that the lines distant from it cut off.

5.5 Nicomedes, *On the Conchoid* (from Eutocius)

Nothing is known about the life of Nicomedes, except that he was a contemporary of both Eratosthenes and Archimedes and therefore lived from about 280 BCE to 200 BCE. He is best known for his treatise on the properties of the curve called the conchoid. Although the actual treatise is lost, some of it is preserved in the

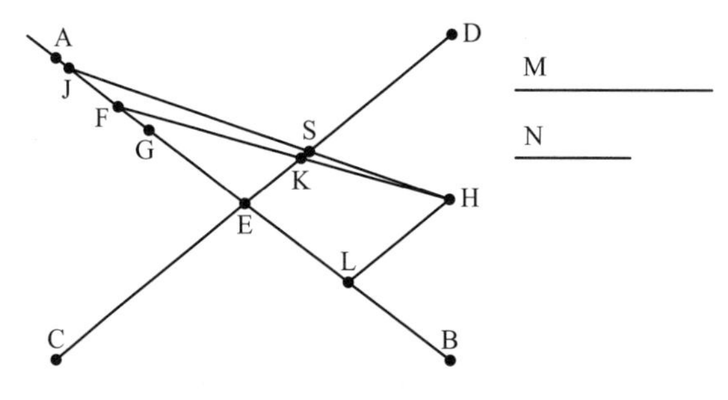

Figure 5.4.4.

commentary of Eutocius on Archimedes's *Sphere and Cylinder II*, as well as in book 4 of Pappus's *Collection*. In particular, we learn how Nicomedes used the conchoid to solve the problems of trisecting an angle and doubling a cube. Until recently, there was no evidence of anyone studying this curve between late antiquity and the seventeenth century, at which time it was discussed by Viète, Descartes, and Newton, among others. Recently, however, a treatment of the conchoid and its use in solving the two classical problems was discovered in *The Rectifying of the Curved*, a book in Hebrew written by the fourteenth-century Spanish mathematician and philosopher Alfonso of Valladolid, the Christian name of the converted Jew Abner of Burgos.[67]

In what follows, we quote from Eutocius's *Commentary* on the use of the conchoid in solving the cube duplication problem, or, equivalently, the finding of two mean proportionals between two given lines. But we begin with a description of the mechanical device Nicomedes invented to draw this curve, as well as a modern definition. AB is a ruler with a slot in it parallel to its length, and FE is a second ruler fixed perpendicular to AB with a fixed peg P in it. A third ruler PC pointed at C has a slot in it parallel to its length that fits the peg P. D is a fixed peg on PC in line with the slot that can move freely along the slot in AB. If the ruler PC moves so that the peg D moves along the slot in AB, the extremity C of the ruler describes the conchoid curve [figure 5.5.1]. The straight line AB is called the *ruler*, the fixed point P the *pole*, and the constant CD the *distance*.

In modern terms, we define the conchoid by beginning with a straight line AB, called the *ruler*, a point P outside of it, called the *pole*, and distance d. Then the conchoid is the locus of all points lying at the given distance d from the ruler AB along the segment that connects them to the pole P [figure 5.5.2]. Note that in the modern definition, the curve has two branches, one on each side of the ruler, to which both are asymptotic. If P is the origin and AB is the line $y=a$, then the polar equation of the curve is $r=a\sec\theta+d$. Nicomedes himself perhaps referred to the second branch in his lost treatise, the one on the same side of the ruler as the pole, but in his solutions of the two problems, he used only the first branch. Note that the second branch has three different possible forms, depending on the relationship between a and d: if $a<d$, it

[67] For some excerpts from this work, see Victor J. Katz et al., *Sourcebook in the Mathematics of Medieval Europe and North Africa* (Princeton, NJ: Princeton University Press, 2016), pp. 345–53.

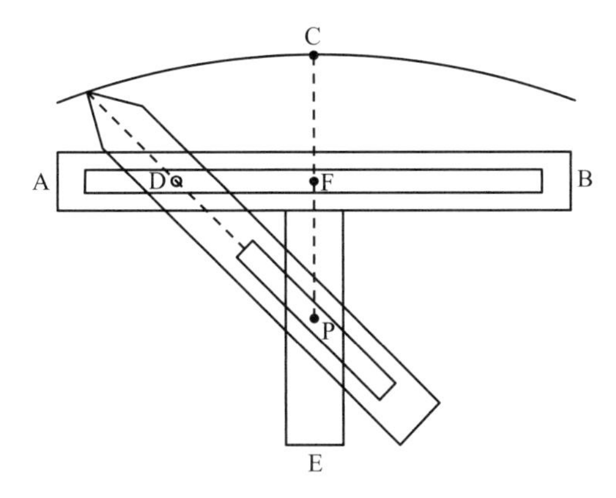

Figure 5.5.1.

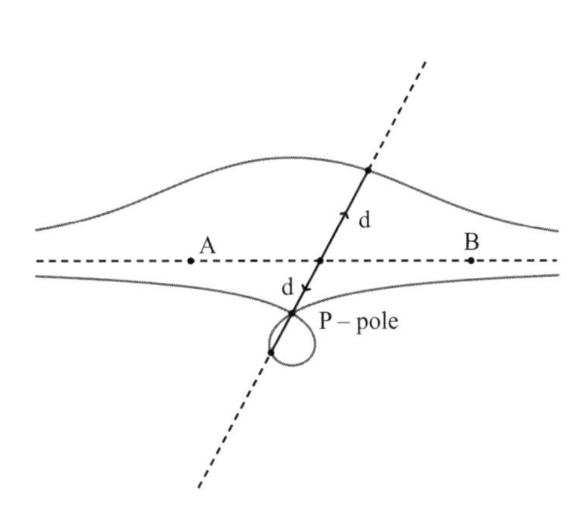

Figure 5.5.2.

has a loop; if $a = d$, then P is a cusp; and if $a > d$, the curve is smooth. The first branch does not change topologically.

The use of the conchoid in constructions is usually as follows: Given two straight lines ℓ, m, meeting at a given angle, and a point P outside the angle, one can draw a line through P cutting ℓ and m so that a segment of given length d is intercepted between ℓ and m. This is achieved by constructing a conchoid with ruler ℓ, pole P, and distance d. The line through P and through the intersection point of this conchoid and m satisfies the conditions. These two diagrams show two different possibilities for the location of the lines and the pole. To help with visualizing the situation that Eutocius describes below, we label the second diagram with the same letters as in Eutocius's diagram [figures 5.5.3, 5.5.4].

We now continue with Eutocius's report on Nicomedes's own work on the use of the conchoid to find two mean proportionals between two given lines. This is found in Eutocius's *Commentary* to Archimedes's *Sphere and Cylinder II*, proposition 1.

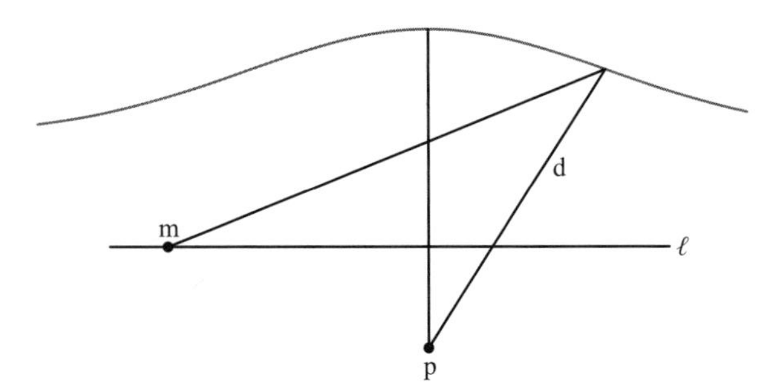

Figure 5.5.3.

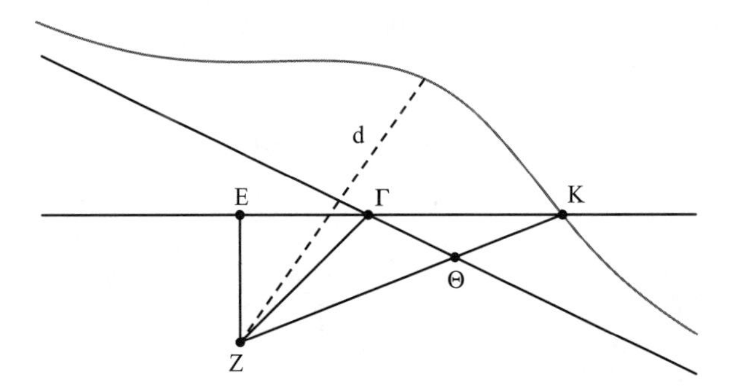

Figure 5.5.4.

Let two lines be given at right angles to each other, $\Gamma\Lambda$, ΛA, between whom it is required to find continuous two mean proportionals, and let the parallelogram $AB\Gamma\Lambda$ be completed, and let each of AB, $B\Gamma$ be bisected by the points Δ, E [figure 5.5.5]. Having joined $\Delta\Lambda$, let it be produced and let it meet ΓB, produced as well at H. Let EZ be drawn at right angles to $B\Gamma$, and let ΓZ be produced, being equal to $A\Delta$. Let ZH be joined, and let $\Gamma\Theta$ be drawn parallel to it. There being an angle, the one contained by $K\Gamma$, $\Gamma\Theta$, let $Z\Theta K$ be drawn through, from Z, a given point, making ΘK equal to $A\Delta$ or to ΓZ; for it was proved through the conchoid that this is possible. Having joined $K\Lambda$, let it be produced and let it meet AB produced at M. I say that it is: as $\Gamma\Lambda$ is to $K\Gamma$, so is $K\Gamma$ to MA and MA to $A\Lambda$.

Since $B\Gamma$ has been bisected by E, and $K\Gamma$ is added to it, therefore rect$(BK, K\Gamma)$ with sq(ΓE) is equal to sq(EK).[68] Let sq(EZ) be added as common. Therefore rect$(BK, K\Gamma)$ with sq(ΓE) and sq(EZ), that is, with sq(ΓZ), is equal to sq(KE) and sq(EZ), that is, to sq(KZ). And since as MA is to AB, so is $M\Lambda$ to ΛK, but as $M\Lambda$ is to ΛK, so is $B\Gamma$ to ΓK. Therefore also, as MA is to AB, so is $B\Gamma$ to ΓK. And $A\Delta$ is half AB, while ΓH is twice $B\Gamma$ (since $\Gamma\Lambda$ is twice ΔB). Therefore it shall also be: as MA is to $A\Delta$, so is $H\Gamma$ to $K\Gamma$. But as $H\Gamma$ is to ΓK, so is $Z\Theta$ to ΘK (through the parallels HZ, $\Gamma\Theta$). Therefore *componendo*, too: as $M\Delta$ is to ΔA, so is ZK to $K\Theta$. But $A\Delta$, in turn, is assumed equal to ΘK, since $A\Delta$ is equal to ΓZ, as well. Therefore $M\Delta$, as well, is equal to ZK; therefore sq$(M\Delta)$, too, is equal to sq(ZK). And rect(BM, MA) with sq(ΔA) is equal to sq$(M\Delta)$, while rect$(BK, K\Gamma)$

68 *Elements* 2:6.

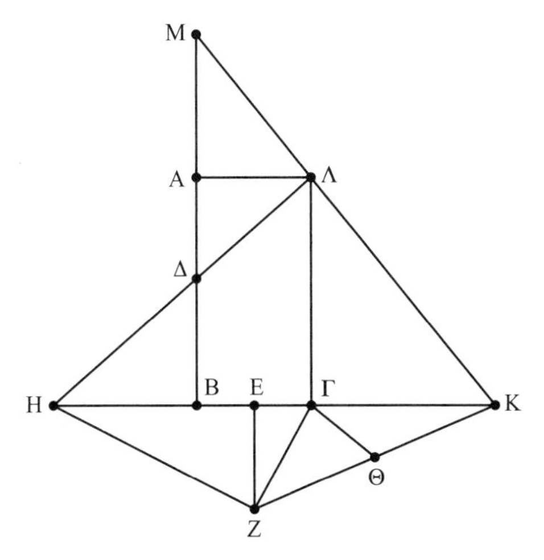

Figure 5.5.5.

with sq(ΓZ), was proved equal to sq(ZK), of which sq($A\Delta$) is equal to sq(ΓZ), for $A\Delta$ was assumed equal to ΓZ. Therefore rect(BM, MA), too, is equal to rect($BK, K\Gamma$). Therefore as MB is to BK, so is $K\Gamma$ to AM. But as BM is to BK, so is $\Gamma\Lambda$ to ΓK. Therefore also, as $\Lambda\Gamma$ is to ΓK, so is ΓK to AM. And it is also: as $\Lambda\Gamma$ is to ΓK, so is MA to $A\Lambda$. Therefore also: as $\Lambda\Gamma$ is to ΓK, so is ΓK to AM and AM to $A\Lambda$.

5.6 Pappus, *Collection*, 4, Quadratrix

Nicomedes is also associated with the quadratrix, but Proclus claims that it was invented by Hippias, while Pappus notes in his discussion of this curve that Dinostratus was also involved in its discovery. We present here parts of book 4 of Pappus's *Collection* that deal with this curve, which earlier mathematicians used to rectify an arc of the quadrant and to square the circle. We include his report from Sporus, perhaps Pappus's teacher, detailing his objections to the use of this curve to solve the circle quadrature problem.

For the squaring of the circle a certain line has been taken up by Dinostratus,[69] and Nicomedes and some other more recent (mathematicians). It takes its name from the *symptoma* concerning it. For it is called "quadratrix" by them, and it has a genesis of the following sort.

Set out a square $ABCD$ and describe the arc BED of a circle with center A, and assume that AB moves in such a way that while the point A remains in place, the point B travels along the arc BED, whereas BC follows along with the traveling point B down the straight line BA, remaining parallel to AD throughout, and that in the same time both AB, moving uniformly, completes the angle BAD, i.e.: the point B completes the arc BED, and BC

[69] Dinostratus (390–320 BCE) is thought to be one of the inventors of the quadratrix.

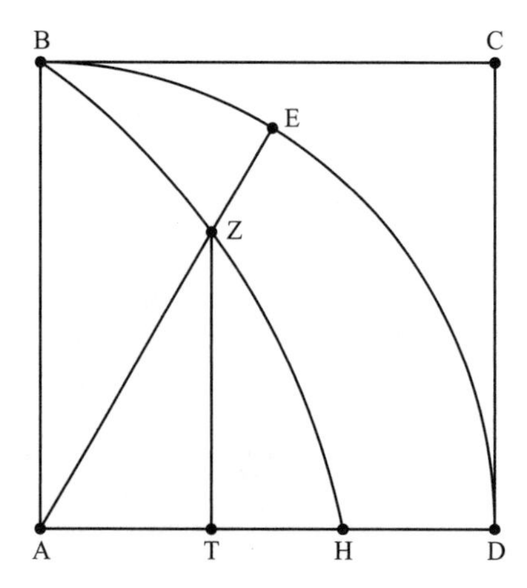

Figure 5.6.1.

passes through the straight line BA, i.e.: the point B travels down BA. Clearly it will come to pass that both AB and BC reach the straight line AD at the same time. Now, while a motion of this kind is taking place, the straight lines BC and BA will intersect each other during their traveling in some point that is always changing its position together with them. By this point a certain line such as BZH is described in the space between the straight lines BA and AD and the arc BED, concave in the same direction as BED, which appears to be useful, among other things, for finding a square equal to a given circle.

And its principal *symptoma* is of the following sort. Whichever arbitrary straight line is drawn through in the interior toward the arc, such as AZE, the straight line BA will be to the straight line ZT as the whole arc BED is to the arc ED [figure 5.6.1]. For this is obvious from the genesis of the line.

Sporus,[70] however, is with good reason displeased with it, on account of the following observations. For, first of all, he takes into the assumption the very thing for which it [the quadratrix] seems to be useful. For how is it possible when two points start from B, that they move, the one along the straight line to A, the other along the arc to D, and come to a halt at their respective endpoints at the same time, unless the ratio of the straight line AB to the arc BED is known beforehand? For the velocities of the motions must be in this ratio, also. Also, how do they think that they come to a halt simultaneously, when they use indeterminate velocities, except that it might happen sometime by chance; and how is that not absurd?

Furthermore, however, its endpoint, which they use for the squaring of the circle, that is, the point in which it intersects the straight line AD, is not found by the above generation of the line. Consider what is being said, however, with reference to the diagram set forth. For when the straight lines CB and BA, traveling, come to a halt simultaneously, they will both reach AD, and they will no longer produce an intersection in each other.

[70] Sporus (240–300 CE) was probably born in Bithynia, now part of Turkey. He was perhaps a teacher of Pappus. He worked on the classical problems of squaring the circle and duplicating the cube.

For the intersecting stops when AD is reached, and this last intersection would have taken place as the endpoint of the line, the point where it meets the straight line AD. Except if someone were to say that he considers the line to be produced, as we assume straight lines to be produced, up to AD. This, however, does not follow from the underlying principles, but one proceeds just as if the point H were taken after the ratio of the arc to the straight line had been taken beforehand. Without this ratio being given, however, one must not, trusting in the opinion of the men who invented the line, accept it, since it is rather mechanical. Much rather, however, one should accept the problem that is shown by means of it.

Given the quadratrix, Pappus proved by a *reductio* argument in proposition 26 that as the arc DEB is to the line BA, so is BA to AH. He then continued with proposition 27:

It is obvious, also, however, that when a straight line is taken as the third proportional to the straight lines AH and BA, it will be equal to the arc DEB, and is fourfold to the circumference of the whole circle. When, however, a straight line equal to the circumference of the circle has been found, it is very clear that it is rather easy indeed to put together a square equal to the circle itself. For the rectangle between the circumference of the circle and the radius is two times the circle, as Archimedes has shown (*Measurement of the Circle*, proposition 1).

In two further arguments, given in propositions 28 and 29, Pappus shows that the quadratrix can be determined, through a geometrical analysis, from either an Apollonian helix or an Archimedean spiral.

5.7 Diocles, *On Burning Mirrors*

Little is known about Diocles's life, except that he was probably born around 240 BCE and that he worked for part of his life in Arcadia, in the central Peloponnese. He is best known for his book *On Burning Mirrors*, which for many years was thought lost and was known only through mention of it by Eutocius in his commentary on Archimedes's *On the Sphere and Cylinder*. Recently, however, an Arabic translation was discovered, and we present excerpts from that work below. The question of burning mirrors had been raised earlier, that is, the question of whether one can construct a mirror surface that reflects all of the rays of the sun to a given point and thereby cause burning. In fact, there are stories that Archimedes constructed such a mirror and was able to use it to burn Roman ships during the siege of Syracuse, although this is a myth. In any case, Diocles demonstrates in his treatise the focal property of the parabola, a property that was not mentioned by Apollonius in his *Conics*, the property that allows a burning mirror to be constructed. He also shows how to construct a parabola with given focal length, using in effect the focus-directrix property of a parabola, that the points of the parabola are equally distant from the focus and a given straight line called the directrix. He then continues by describing how to use conic sections in doubling a cube, as well as using a curve called the cissoid to find two mean proportionals between two given lines.

In the name of God, the merciful, the compassionate, O God, grant long life.

The book of Diocles on burning mirrors.

He said: Pythion the Thasian geometer wrote a letter to Conon in which he asked him how to find a mirror surface such that when it is placed facing the sun the rays reflected from it meet the circumference of a circle. And when Zenodorus (2nd c. BCE) the astronomer came down to Arcadia and was introduced to us, he asked us how to find a mirror surface such that when it is placed facing the sun the rays reflected from it meet at a point and thus cause burning. So we want to explain the answer to the problem posed by Pythion and to that posed by Zenodorus; in the course of this we shall make use of the premises established by our predecessors. One of these two problems, namely, the one requiring the construction of a mirror which makes all the rays meet in one point, is the one which was solved practically by Dositheus.[71] The other problem, since it was only theoretical, and there was no argument worthy to serve as proof in its case, was not solved practically. We have set out a compilation of the proofs of both these problems and elucidated them.

The burning-mirror surface submitted to you is the surface bounding the figure produced by a section of a right-angled cone (parabola) being revolved about the line bisecting it (its axis). It is a property of that surface that all the rays are reflected to a single point, namely, the point (on the axis) whose distance from the surface is equal to a quarter of the line which is the parameter of the squares on the perpendiculars drawn to the axis (the ordinates). Whenever one increases that surface by a given amount, there will be a (corresponding) increase in the above-mentioned conic section. So the rays reflected from that additional (surface) will also be reflected to exactly the same point, and thus they will increase the intensity of the heat around that point. The intensity of the burning in this case is greater than that generated from a spherical surface, for from a spherical surface the rays are reflected to a straight line, not to a point, although people used to guess that they are reflected to the center; the rays which meet at one place in that (spherical) surface are reflected from the surface (consisting) of a spherical segment less than half the sphere, and (even) if the mirror consists of half the sphere or more than half, only those rays reflected from less than half the sphere are reflected to that place.

. . .

Perhaps you would like to make two examples of a burning-mirror, each having a diameter of two cubits, one constructed on the circumference of a circle (i.e., spherical), the other on a section of a right-angled cone (i.e., parabolic), so that it may be possible for you to measure the burning-power of each of them by the degree of its efficiency. So one knows the base of their burning-powers, and (then) measuring the (relation between) the burning (power) of one and that of the other is a matter requiring observation; that is to say, if the mirror-surface with a diameter of the amount of one foot burns the whole of the burning-area which heats up in (a piece of) wood, then it is more likely to burn (it) easily when its diameter is seven times that amount. For when the burning-power is multiplied

[71] Archimedes addressed several of his treatises to Dositheus.

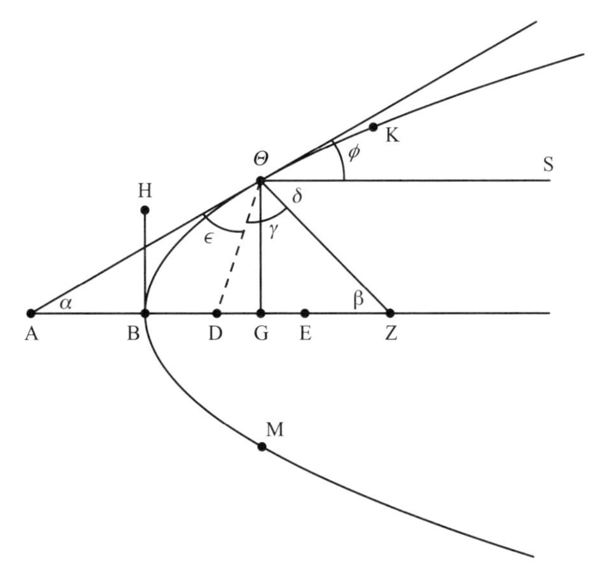

Figure 5.7.1.

by seven, the difference between it (such a mirror) and the original mirror must be very great.

We believe that it is possible to make a burning-instrument of glass such that it has a special property, namely, that one can make lamps from it which produce fire in temples and at sacrifices and immolations, so that the fire is clearly seen to burn the sacrificial victims; this occurs, as we are informed, in certain remote cities, especially on the days of great celebration; this causes the people of those cities to marvel. That is something which we too shall do.

Let there be a parabola KBM, with axis AZ, and let half the parameter of the squares on the ordinates be line BH [figure 5.7.1]. Let BE on the axis be equal to BH, and let BE be bisected at point D.[72] Let us draw a line tangent to the section at an arbitrary point, namely, line ΘA, and draw line ΘG as ordinate to AZ. Then we know that AB equals BG and that the line drawn from Θ perpendicular to ΘA meets AZ beyond E. So let us draw $Z\Theta$ perpendicular to ΘA, and join ΘD. Then GZ equals BH;[73] since HB equals BE, also GZ equals BE. We subtract GE, common to GZ and BE, then the remainders GB, EZ are equal. But GB equals BA, so AB equals EZ. And BD equals DE, because BE is bisected at D, so the sums AD, DZ are equal. And because triangle $A\Theta Z$ is right-angled and its base AZ is bisected at D, AD equals $D\Theta$ and also DZ. So angle β equals angle γ and angle α equals angle ϵ. So let a line parallel to AZ pass through Θ, namely, line ΘS. Then angle β is equal to angle δ, which is alternate to it, and angle β equals angle γ, so angle

[72] If the equation of the parabola is $y^2 = px$, then $BH = BE = \frac{p}{2}$ and $BD = \frac{p}{4}$. D is today known as the focus of the parabola.

[73] $\Theta G^2 = AG \cdot GZ$, because $A\Theta Z$ is a right triangle. Also, $\Theta G^2 = 2BH \cdot BG$, because Θ is on the parabola. So $2BH \cdot BG = AG \cdot GZ = 2BG \cdot GZ$, and $GZ = BH$.

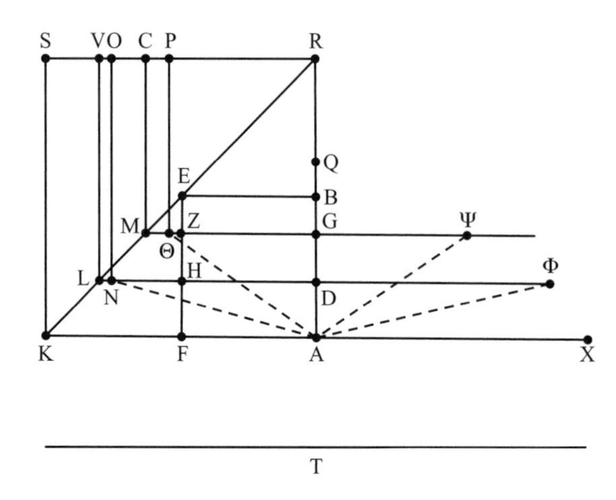

Figure 5.7.2.

γ equals angle δ also. And since angle $Z\Theta C$ is equal to angle $Z\Theta A$, being right angles, so the remainder, angle ϕ and angle ϵ, are equal.

So when line $S\Theta$ meets line $A\Theta$, it is reflected to point D, forming equal angles, ϕ and ϵ, between itself and the tangent $A\Theta$. Hence it has been shown that if one draws from any point on KBM a line tangent to the section, and draws the line connecting the point of tangency with point D, e.g., line ΘD, and draws line $S\Theta$ parallel to AZ, then in that case line $S\Theta$ is reflected to point D, i.e., the line passing through point Θ is reflected at equal angles from the tangent to the section. And all parallel lines from all points on KBM have the same property, so, since they make equal angles with the tangents, they go to point D.

Hence, if AZ is kept stationary, and KBM revolved about it until it returns to its original position, and a concave surface of brass is constructed on the surface described by KBM, and placed facing the sun, so that the sun's rays meet the concave surface, they will be reflected to point D, since they are parallel to each other. And the more the (reflecting) surface is increased, the greater will be the number of rays reflected to point D.

We skip Diocles's construction of a burning mirror based on the circumference of a circle and proceed to show how he constructs a parabola with a given focus.

How do we shape the curvature of the burning-mirror when we want the point at which the burning occurs to be at a given distance from the center of the surface of the mirror? We draw with a ruler on a given board a line equal to the distance we want; that is line AB [figure 5.7.2]. We make AK twice BA and erect BE perpendicular and equal to AB; we join EK. We make AF equal to BE, and join EF; then $ABEF$ is a square, and also EF is equal to FK. We mark on BA two points, G, D, and make EZ equal to BG and HE equal to BD. We join ZG, HD, and produce them on both sides; let them meet EK in M, L. Then if, with A as center and GM as radius, we draw a circle, it cuts GM; let it cut it in Θ. Then we continue to draw it in the same way until it cuts it in Ψ. Again, if, with center A and radius DL, we draw a circle, it cuts DL; let it cut in N. Then we continue to draw it about center

A until it cuts it [DL] again in Φ. Then we draw AX as an extension in a straight line of KA and make it [AX] equal to it [KA]. Then points K, N, Θ, B, Ψ, Φ, X lie on a parabola.

For we produce AB to R, letting BR equal AB; let us draw RS perpendicular to AB and equal to KA, and join SK, and draw from points L, M, Θ, N to line RS perpendiculars LV, MC, NO, ΘP. Then when KE is produced in a straight line, it passes through R. So VL equals LD and MC equals MG, because KER is a diagonal of square AS. But LD equals NA and MG equals ΘA, and also LV equals LD and MC equals MG. So AN, LV, NO are all equal, as are $A\Theta$, MC, and ΘP. And AK equals KS and AR is bisected at B. Since that is so, points B, Θ, N, K lie on a parabola, as we shall prove subsequently. Similarly points Ψ, Φ, X also (lie on the parabola).

The argument that follows shows that the points Diocles has constructed, being equidistant from the focus A and the directrix, line RS, lie on a parabola whose vertex is B and whose parameter is four times the length of AB.

So if we mark numerous points on AB, and draw through the lines parallel to AK, and mark on the lines points corresponding to the other points (i.e., Θ, N, etc.), and bend along the resultant points a ruler made of horn, fastening it so that it cannot move, then draw a line along it and cut the board along that line, then shape the curvature of the figure we wish to make to fit that template, the burning from that surface will occur a point A, as was proved in the first proposition.

Let the diagonal of square AS be line RK, and let us bisect AR at B and mark some point, G, between A and B. Let two lines parallel to AK, BE and GM, pass through points B and G, and let a line parallel to AB, EF pass through point E. Then $MG-MZ$ equals BE. And because of what was stated in the proposition preceding this one, AE is a square. So BE equals AB, and EZ equals both ZM and BG; therefore MZ is equal to BG. So $MG-BG$ [$=MG-MZ=ZG$] equals AB. Let BQ be equal to GB, Then QA [$=AB+BQ=ZG+MZ$] equals MG.

So the circle constructed on center A with radius equal to MG passes through Q. Then I say that it (the circle) cuts line MG between points M and G. For if it were to pass through M or fall beyond M, its radius would be longer than GM, since angle G is right, and that is impossible, since we have made it (the radius) equal to it (GM). So the circumference of the above-mentioned circle cuts MG between points M and G. Then let it cut it in Θ, and let us draw a line, ΘA, joining points Θ and A. Then $A\Theta$ equals MG. And we have shown that AQ also equals MG. So ΘA equals AQ.

And since QB equals BG, we have the sum of 4 rect(AB, BG) and sq(GA) equals sq(AQ).[74] But sq(AQ) equals sq($A\Theta$), and the sum of sq(ΘG) and sq(GA) equals sq($A\Theta$), because angle G is right. So the sum of 4 rect(AB, BG) and sq(GA) equals the sum of sq(ΘG) and sq(GA). And when we eliminate sq(GA), common, 4 rect(AB, BG) equals sq(ΘG), remainders. And 4 rect(AB, BG) equals rect($BG, 4AB$). Let $4AB$ equal T. Then sq(ΘG) equals rect(BG, T), and also sq(AK) [$=(2AB)^2=AB\cdot 4AB$] equals rect(AB, T), because SA is a square and AR is bisected at B. So the parabola that passes through

[74] Since $AQ=AG+2BG$, it follows that $AQ^2=AG^2+4AG\cdot BG+4BG^2=AG^2+4(BG(AG+BG))=AG^2+4BG\cdot AB$.

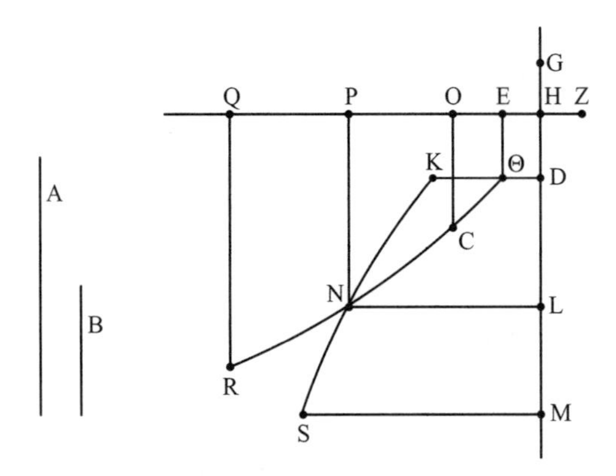

Figure 5.7.3.

points B and K will also pass through Θ, and the parameter of the squares on the ordinates is line T, which is what we wanted to prove.

After giving a new argument solving Archimedes's construction problem of cutting a given sphere by a plane so that the two segments of the sphere bear a given ratio to each other, Diocles next shows how to double a cube.

We make the given line A. Then we must seek another line such that the cube of A is twice that line's cube. We construct two lines, GD, EZ, intersecting at right angles; let GH, HD each equal one-fourth of A. We make B equal to one-half of A, and EH, ZH each equal to one-fourth of B. Draw ΘD, $E\Theta$ perpendicular to GD, EZ, and make DK equal to GD [figure 5.7.3]. Then it is obvious that ZE is equal to $E\Theta$, since A is double B. Produce each of GD, ZE in a straight line to points M, Q and mark on lines DM, EQ many points close to each other, as L, M, O, P, Q. Draw from points L, M on line HM perpendiculars LN, MS, and from points O, P, Q perpendiculars to line HQ, namely, OC, PN, QR.

Now we made the line drawn from D to K equal to GD. So we set D as center, and with radius equal to GL draw a circle, and mark N at the place where it cuts LN. Similarly we draw a circle with center D and radius GM, and mark S at the place where it cuts MS. Let us draw with the curved ruler a line passing through points K, N, S and the other points marked in this way; that is, line KNS.[75] We operate similarly on the other line; we draw a circle with center E and radius ZO; let it cut OC at C. Similarly we adopt the same center E, and radius ZP, ZQ (in turn), and draw circles; let these cut PN, QR in N, R, respectively. Likewise we draw with the curved ruler a line passing through points Θ, C, N, R, namely, line ΘCNR.[76] Then lines KNS, ΘCNR cut one another at some point; let them cut at point N. Draw perpendicular NL to line HM and perpendicular NP to line HQ.

Then I say that the cube on A is twice the cube on NL. For GL is equal to DN, so 4 rect(LH, HD) added to sq(DL) is equal to the sum of the squares on DL and KN. We subtract sq(DL), which is common, and the remainders are equal: 4 rect(LH, HD), which

[75] As above, this is the method of constructing a parabola with given focus and directrix. Here the focus is the point D and the directrix is a line through G parallel to EZ.

[76] Here the parabola drawn has focus E with directrix a line through Z parallel to GD.

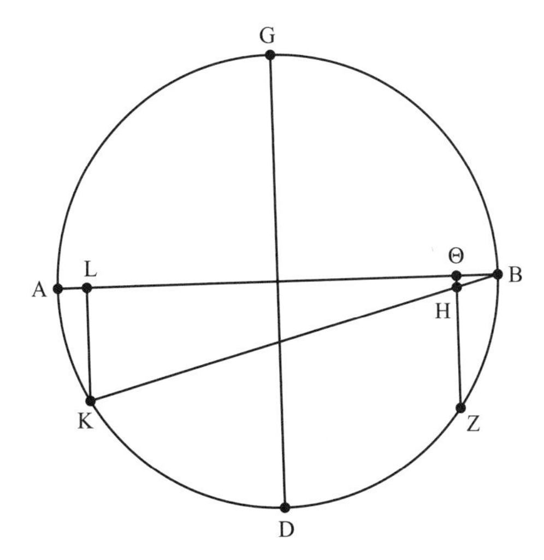

Figure 5.7.4.

equals rect(A, HL), is equal to sq(LN). So as A is to LN so is NL to LH. Similarly, we prove that rect(B, HP), or rect(B, LN), equals sq(PN), which is in turn equal to sq(HL). Thus as NL is to LH, so is LH to B. And it has been shown that as NL is to LH, so is A to NL. So between A and B are the two mean proportionals LN, LH. So the cube of A is to the cube of NL as A is to B. But A is twice B. So the cube of A is twice the cube of NL. So NL is the required line. . . . And furthermore it is obvious that lines ΘCNR, KNS are parabolas, and that each of them passes through point N. Hence the conic sections necessarily intersect each other.

Diocles next presents a new method to determine two mean proportionals between two given lines. First, he proves a preliminary result.

One may also find two intermediate lines by the following method: We draw circle $ABGD$; let two diameters in the circle, AB and GD, cut one another at right angles. We cut off two equal arcs, KD and ZD, from the circle, and draw perpendicular $Z\Theta$ from Z to AB. Join BK [figure 5.7.4]. Then I say that $Z\Theta$ and ΘB are continuous proportionals between $A\Theta$ and ΘH. The reason that as $A\Theta$ is to ΘZ, so is ΘZ to ΘB is clear. But I say also that as $Z\Theta$ is to ΘB, so is ΘB to ΘH. For let perpendicular KL be drawn from K to AB. Then as $B\Theta$ is to ΘH, so is BL to LK. But BL equals $A\Theta$ and LK equals $Z\Theta$, since arc ZD equals arc DK and $Z\Theta$, KL are perpendiculars. So as $B\Theta$ is to ΘH, so is $A\Theta$ to ΘZ. And, as we said, $A\Theta$ is to ΘZ as $Z\Theta$ is to $B\Theta$. So the four lines $A\Theta$, ΘZ, ΘB, ΘH are in continuous proportion.

Diocles then constructs a curve called the cissoid that will enable him to determine the two mean proportionals. He constructs the curve point by point, connecting them via his curved ruler. We can determine the equation of the curve as follows: Put the origin at the point B of the circle, and assume the radius of the circle is a [figure 5.7.5]. Then if point P on the curve has coordinates (x, y), we have $BK = x$, $KP = y$, $AK = 2a - x$, and $ZK = \sqrt{a^2 - (a-x)^2} = \sqrt{2ax - x^2}$. Since ZK and KB are continuous proportionals between AK and KP, we have $AK : ZK = KB : KP$, or $AK \cdot KP = ZK \cdot KB$. This translates

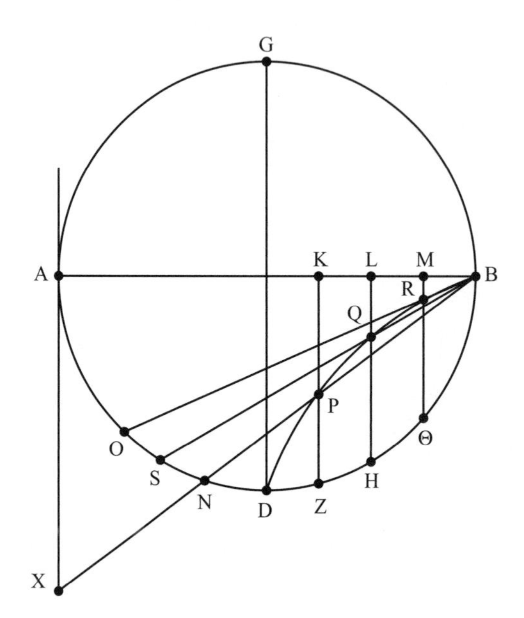

Figure 5.7.5.

into $(2a-x)y=\sqrt{2ax-x^2}x$, or $(2a-x)^2y^2=(2ax-x^2)x^2$, or, finally, $2ay^2=(x^2+y^2)x$. We can also write the equation of the cissoid as

$$y^2=\frac{x^3}{2a-x}.$$

If we draw the tangent to the circle at A, the cissoid is today defined as the locus of points P such that the distance $\overline{BP}$ is equal to the distance $\overline{NX}$, where N is the intersection of the line through B and P with the circle and X is the intersection of that line with the tangent line. Note that if P has coordinates (x_1, y_1), then N has coordinates $(2a-x_1, \sqrt{2ax_1-x_1^2})$, while X has coordinates $(2a, \frac{\sqrt{2ax_1-x_1^2}}{2a-x_1}2a)$. One can then show that the cissoid with the equation given above has this locus property.

Let there again be a circle $ABGD$ with two diameters AB, GD cutting one another at right angles. Let us cut off from the circle successive equal arcs DZ, ZH, $H\Theta$, and draw perpendiculars ZK, HL, ΘM to line AB [figure 5.7.5]. Cut off from the other quadrant of the circle (i.e., AD), beginning from point D, arcs equal in size and also in number to arcs DZ, ZH, ΘH, namely, arcs DN, NS, SO. Let the line joining B to N cut ZK at P, and the line joining B to S cut HL at Q, and the line joining B to O cut ΘM at R. Then it has been shown above that ZK and KB are continuous proportionals between AK and KP. Similarly HL and LB are continuous proportionals between AL and LQ, and ΘM and MB are continuous proportionals between AM and MR. So if we construct the perpendiculars closer than those we mentioned, and mark points on them as we marked P, Q, R, and draw through all these points by means of the curved ruler line $BRQPD$, then it is obvious

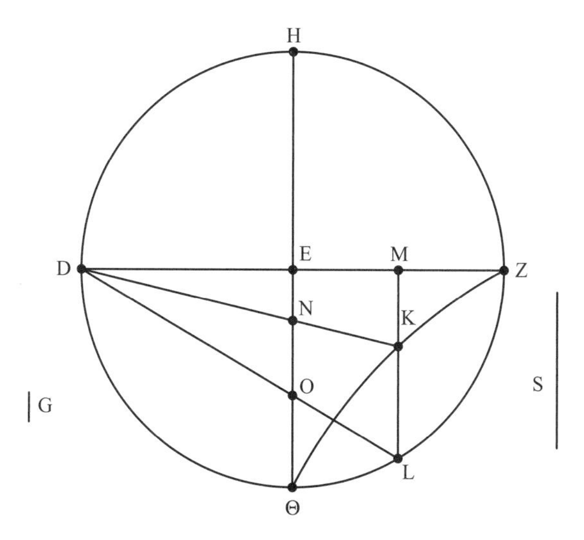

Figure 5.7.6.

that if we mark on it a point, e.g., P, and draw perpendicular PK from P to AB, the result is that ZK and KB are continuous proportionals between AK and KP.

The next paragraph is taken directly from Eutocius, who makes clearer than Diocles himself that what has been done will produce two mean proportionals between two given lines.

Having made these preliminary constructions, let the two given lines (whose mean proportionals it is required to find) be S, G, and let there be a circle, in which two diameters at right angles to each other are DZ and $H\Theta$. Let the line produced through the continuous points be drawn, as has been said, namely, $ZK\Theta$, and let it come to be: as S is to G so is DE (the radius a of the circle) to EN [figure 5.7.6]. Joining DN and producing it, let it cut the line at K, and let ML be drawn through K parallel to $H\Theta$. Therefore, through what has been proved above, ML, MZ are mean proportionals between DM and MK. And it is: as DM is to MK, so is DE to EN, and as DE is to EN, so is S to G. If we insert means between S, G in the same ratio as DM, ML, MZ, MK, e.g., C, T, we shall have taken C, T as mean proportionals between S and G, which it was required to find.[77]

5.8 Pappus, *Collection*, 5, Hexagonal Form of Honeycombs

Book 5 of Pappus's *Collection* is devoted to isoperimetry, determining shapes with the largest area bounded by a given perimeter. The introduction is a wonderful essay on the practical intelligence of bees.

[77] We begin by finding C such that $S:C=DM:ML$ and then find T by solving $C:T=ML:MZ$. Thus, we are just rescaling the original continuous proportion.

Though God has given to men, most excellent Megethion,[78] the best and most perfect understanding of wisdom and mathematics. He has allotted a partial share to some of the unreasoning creatures as well. To men, as being endowed with reason, He granted that they should do everything in the light of reason and demonstration, but to the other unreasoning creatures He gave only this gift, that each of them should, in accordance with a certain natural forethought, obtain so much as is needful for supporting life. This instinct may be observed to exist in many other species of creatures, but it is specially marked among bees. Their good order and their obedience to the queens who rule in their commonwealths are truly admirable, but much more admirable still is their emulation, their cleanliness in the gathering of honey, and the forethought and domestic care they give to its protection. Believing themselves, no doubt, to be entrusted with the task of bringing from the gods to the more cultured part of mankind a share of ambrosia in this form, they do not think it proper to pour it carelessly into earth or wood or any other unseemly and irregular material, but, collecting the fairest parts of the sweetest flowers growing on the earth, from them they prepare for the reception of the honey the vessels called honeycombs, [with cells] all equal, similar and adjacent, and hexagonal in form. That they have contrived this in accordance with a certain geometrical forethought we may thus infer. They would necessarily think that the figures must all be adjacent one to another and have their sides common, in order that nothing else might fall into the interstices and so defile their work.

Now there are only three rectilineal figures which would satisfy the condition, I mean regular figures which are equilateral and equiangular, inasmuch as irregular figures would be displeasing to the bees. For equilateral triangles and squares and hexagons can lie adjacent to one another and have their sides in common without irregular interstices. For the space about the same point can be filled by six equilateral triangles and six angles, of which each is $\frac{2}{3}$ right angle, or by four squares and four right angles, or by three hexagons and three angles of a hexagon, of which each is $1\frac{1}{3}$ right angle. But three pentagons would not suffice to fill the space about the same point, and four would be more than sufficient; for three angles of the pentagon are less than four right angles (inasmuch as each angle is $1\frac{1}{5}$ right angle), and four angles are greater than four right angles. Nor can three heptagons be placed about the same point so as to have their sides adjacent to each other; for three angles of a heptagon are greater than four right angles (inasmuch as each is $1\frac{3}{7}$ right angle). And the same argument can be applied even more to polygons with a greater number of angles. There being, then, three figures capable by themselves of filling up the space around the same point, the triangle, the square, and the hexagon, the bees in their wisdom chose for their work that which has the most angles, perceiving that it would hold more honey than either of the two others.

Bees, then, know just this fact which is useful to them, that the hexagon is greater than the square and the triangle and will hold more honey for the same expenditure of material in constructing each. But we, claiming a greater share in wisdom than the bees, will investigate a somewhat wider problem, namely, that of all equilateral and equiangular

[78] Nothing is known of Megethion.

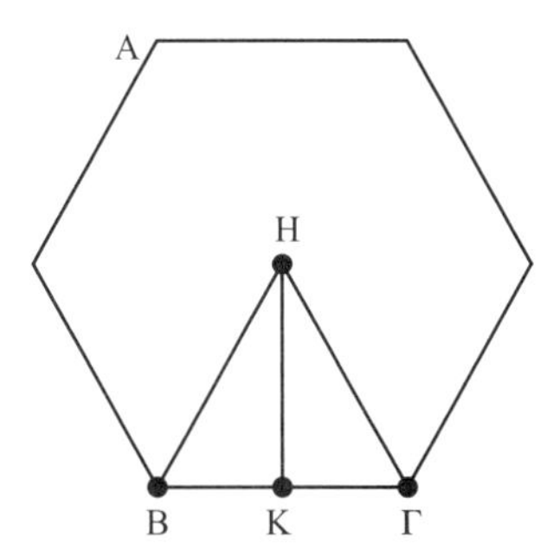

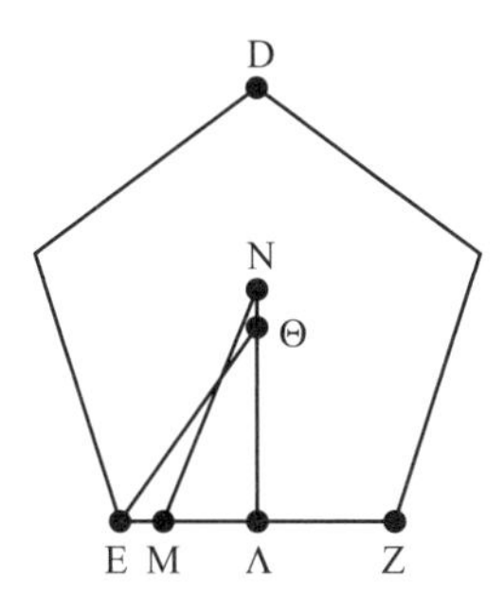

Figure 5.9.1.

plane figures having an equal perimeter, that which has the greater number of angles is always greater, and the greatest of them all is the circle having its perimeter equal to them.

Much of the remainder of book 5 is a description of the work of Zenodorus on isoperimetry, as described by Theon in his commentary to Ptolemy's *Almagest*. We present this next.

5.9 Zenodorus, *On Isoperimetric Figures*

Zenodorus (ca. 200–140 BCE) wrote a work titled *On Isoperimetric Figures*. It is preserved only in the works of Theon (4th c. CE), in his *Commentary on Ptolemy's Syntaxis*, and Pappus's *Collection*. The central goal of this work is to discover which of all plane figures with the same perimeter has the largest area. The extract here is from Theon's work.

We shall give the proof of these propositions in a summary taken from the proofs by Zenodorus in his book *On Isoperimetric Figures*.

Of all rectilinear figures having an equal perimeter—I mean equilateral and equiangular figures–the greatest is that which has most angles.

For let $AB\Gamma$, DEZ be equilateral and equiangular figures having equal perimeters, and let $AB\Gamma$ have the more angles [figure 5.9.1]. I say that $AB\Gamma$ is the greater. For let H, Θ be the centers of the circles circumscribed about the polygons $AB\Gamma, DEZ$, and let $HB, H\Gamma, \Theta E, \Theta Z$ be joined. And from H, Θ let $HK, \Theta\Lambda$ be drawn perpendicular to $B\Gamma, EZ$. Then since $AB\Gamma$ has more angles than DEZ, $B\Gamma$ is contained more often in the perimeter of $AB\Gamma$ than EZ is contained in the perimeter of DEZ. And the perimeters are equal. Therefore EZ is greater than $B\Gamma$; and therefore $E\Lambda$ is greater than BK. Let ΛM be placed equal to BK, and let ΘM be joined. Then since the straight line EZ bears to the perimeter of the polygon DEZ the same ratio as the angle $E\Theta Z$ bears to four right angles—owing to the fact that the polygon is equilateral and the sides cut off equal arcs from the circumscribing circle, while the angles at the center are in the same ratio as the arcs on which they stand[79]—and the perimeter of DEZ, that is, the

[79] *Elements* 3:26.

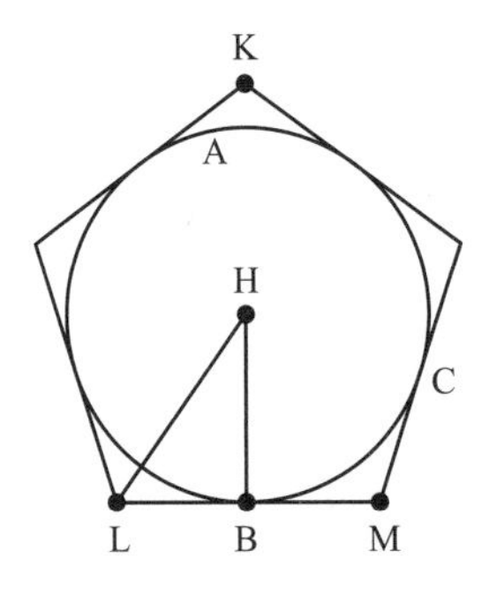

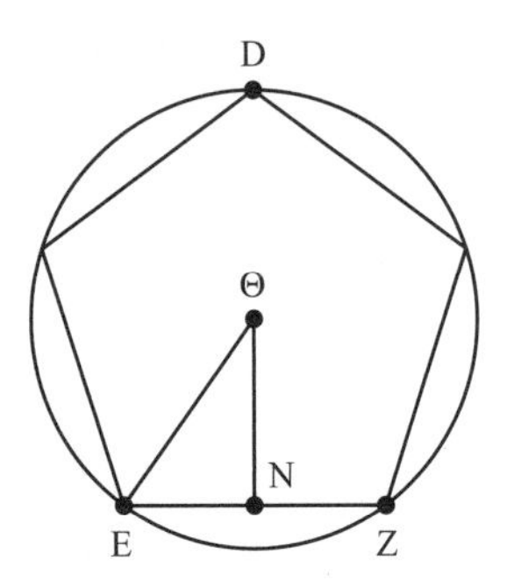

Figure 5.9.2.

perimeter of $AB\Gamma$, bears to $B\Gamma$ the same ratio as four right angles bear to the angle $BH\Gamma$, therefore *ex aequali*,[80] EZ is to $B\Gamma$ as angle $E\Theta Z$ is to angle $BH\Gamma$, that is, $E\Lambda$ is to ΛM as angle $E\Theta Z$ is to angle $BH\Gamma$, or $E\Lambda$ is to ΛM as angle $E\Theta\Lambda$ is to angle BHK. And since $E\Lambda$ has a greater ratio to ΛM than angle $E\Theta\Lambda$ has to angle $M\Theta\Lambda$, as we shall prove in due course, and $E\Lambda$ is to ΛM as angle $E\Theta\Lambda$ is to angle BHK, therefore, angle $E\Theta\Lambda$ has a greater ratio to angle BHK than angle $E\Theta\Lambda$ has to $M\Theta\Lambda$. Therefore angle $M\Theta\Lambda$ is greater than angle BHK.

Now the right angle at Λ is equal to the right angle at K. Therefore the remaining angle HBK is greater than the angle $\Theta M\Lambda$.[81] Let the angle ΛMN be placed equal to the angle HBK, and let $\Lambda\Theta$ be produced to N. Then since the angle HBK is equal to the angle $NM\Lambda$, and the angle at Λ is equal to the angle at K, while BK is equal to the side $M\Lambda$, therefore HK is equal to $N\Lambda$.[82] Therefore HK is greater than $\Theta\Lambda$. Therefore the rectangle contained by the perimeter of $AB\Gamma$ and HK is greater than the rectangle contained by the perimeter of DEZ and $\Theta\Lambda$. But the rectangle contained by the perimeter of $AB\Gamma$ and HK is double of the polygon $AB\Gamma$, since the rectangle contained by $B\Gamma$ and HK is double of the triangle $HB\Gamma$;[83] and the rectangle contained by the perimeter of DEZ and $\Theta\Lambda$ is double of the polygon DEZ. Therefore the polygon $AB\Gamma$ is greater than DEZ.

This having been proved, I say that if a circle has an equal perimeter with an equilateral and equiangular rectilineal figure, the circle shall be the greater.

For let ABC be a circle having an equal perimeter with the equilateral and equiangular rectilineal figure DEZ. I say that the circle is the greater [figure 5.9.2]. Let H be the center of the circle ABC, Θ the center of the circle circumscribing the polygon DEZ; and let there be circumscribed about the circle ABC the polygon KLM similar to DEZ, and let HB be joined, and from Θ let ΘN be drawn perpendicular to EZ, and let HL, ΘE be joined. Then since the perimeter of the polygon KLM is greater than the perimeter of the circle ABC, as Archimedes proves in his work *On the Sphere and Cylinder* [proposition 1], while the perimeter of the circle ABC is equal to the perimeter of the polygon DEZ, therefore

[80] *Elements* 5:17.
[81] *Elements* 1:32.
[82] *Elements* 1:26.
[83] *Elements* 1:41.

the perimeter of the polygon *KLM* is greater than the perimeter of the polygon *DEZ*. And the polygons are similar; therefore *BL* is greater than *NE*. And the triangle *HLB* is similar to the triangle *ΘEN*, since the whole polygons are similar; therefore *HB* is greater than *ΘN*. And the perimeter of the circle *ABC* is equal to the perimeter of the polygon *DEZ*. Therefore the rectangle contained by the perimeter of the circle *ABC* and *HB* is greater than the rectangle contained by the perimeter of the polygon *DEZ* and *ΘN*. But the rectangle contained by the perimeter of the circle *ABC* and *HB* is double of the circle *ABC* as was proved by Archimedes, whose proof we shall set out next;[84] and the rectangle contained by the perimeter of the polygon *DEZ* and *ΘN* is double of the polygon *DEZ*.[85] Therefore the circle *ABC* is greater than the polygon *DEZ*, which was to be proved.

We give only the statements of the next two theorems.

Now I say that, of all rectilineal figures having an equal number of sides and equal perimeter, the greatest is that which is equilateral and equiangular.

Now I say that, of all solid figures having an equal surface, the sphere is the greatest; and I shall use the theorems proved by Archimedes in his work On the Sphere and Cylinder.

5.10 Pappus, *Collection*, 4, Kinds of Geometrical Research

In this very famous passage, Pappus distinguished between three types of geometrical research according to the types of curves needed for construction and problem solving. These are plane (circle and straight line), solid (using, in addition, conics), and linear (using "higher" curves).

When the ancient geometers wished to trisect a given rectilinear angle, they got into difficulties for a reason such as the following. We say that there are three kinds of problems in geometry, and that some of the problems are called "plane," others "solid," and yet others "linear." Now, those that can be solved by means of straight lines and circle, one might fittingly call "plane." For the lines by means of which problems of this sort are found have their genesis in the plane as well. All those problems, however, that are solved when one employs for their invention either a single one or even several of the conic sections, have been called "solid." For it is necessary to use the surfaces of solid figures—I mean, however, (surfaces) of cones—in their construction. Finally, as a certain third kind of problems the so-called "linear" kind is left over. For different lines, besides the ones mentioned, are taken for their construction, which have a more varied and forced genesis, because they are generated out of less structured surfaces, and out of twisted motions. Of such a sort, however, are both the lines found on the so-called loci on surfaces and also others, more varied than those and many in number, which were found by Demetrius of Alexandria[86] in the "linear constitutions," and by Philo of Tyana,[87] from the twisting of

[84] *Measurement of a Circle*, proposition 1.

[85] *Elements* 1:41.

[86] Nothing is known about Demetrius, except that he lived sometime between 200 BCE and 100 CE.

[87] Nothing is known of Philo of Tyana, except that he lived sometime between 200 BCE and 100 CE.

both plectoids and all sorts of other surfaces on solids and which have many astonishing *symptomata* about them. And some of them were deemed, by the more recent geometers, worthy of rather extensive discussion, and a certain one of them is the line that was also called "the paradox" by Menelaus. And of this same kind, i.e., the linear kind, are also the other spiral lines, the quadratrices, and the conchoids and the cissoids.[88]

Somehow, however, an error of the following sort seems to be not a small one for geometers, namely, when a plane problem is found by means of conics or of linear devices by someone, and summarily, whenever it is solved from a non-kindred kind, such as is the problem on the parabola in the fifth book of Apollonius's *Conics*[89] and the *neusis* of a solid on a circle, which was taken by Archimedes in the book about the spiral. For it is possible to find the theorem written down by him without using a solid. I mean in fact it is possible to show that the circumference of the circle in the first rotation of the spiral is equal to the straight line drawn at right angles to the generator of the spiral up to the point of intersection with the tangent of the spiral.[90]

Now, since a difference of such a sort belongs to problems, the earlier geometers were not able to find the above-mentioned problem on the angle, given that it is by nature solid, and they sought it by means of plane devices. For the conic sections were not yet common knowledge for them, and on account of this they got into difficulties. Later, however, they trisected the angle by means of conic sections, using for the invention the *neusis* described in what follows.

5.11 Pappus, *Collection*, 4, Angle Trisection

In paragraphs 38–40 of book 4 of the *Collection*, Pappus presents a method of trisecting an angle, assuming one can place a straight line of given length between two perpendicular straight lines in such a way that the line when extended passes through a given point. This is called *neusis*, or *verging*.

38. Now, when this has been shown, a given rectilinear angle is trisected in the following way. Let the angle ABC, first, be acute, and from a certain point A draw the perpendicular AC, and when the rectangle CZ is completed, produce ZA toward E, and since CZ is a rectangle, place the straight line ED between the straight lines EA, AC, verging toward B and equal to two times AB (for that this can come about has been written down above) [figure 5.11.1]. I claim in fact that the angle EBC is the third part of the given angle ABC.

For bisect ED in H, and join AH. Then the three straight lines DH, HA, and HE are equal. Therefore, DE is twice AH. But it is twice AB, also. Therefore, BA is equal to AH, and the angle ABD is equal to the angle AHD. However, the angle AHD is two times the angle AED, i.e., two times the angle DBC. Therefore, the angle ABD is two times the angle DBC, also. And when we bisect the angle ABD, the angle ABC will be trisected.

[88] These curves are discussed in sections 5.5, 5.6, and 5.7.

[89] The reference is perhaps to *Conics*, book 5, proposition 62, where it is possible to make a plane construction, although Apollonius does not.

[90] The reference is to Archimedes, *On Spirals*, proposition 18. But exactly what Pappus means by his statement is unclear, since Archimedes's proof is certainly "solid" in Pappus's sense.

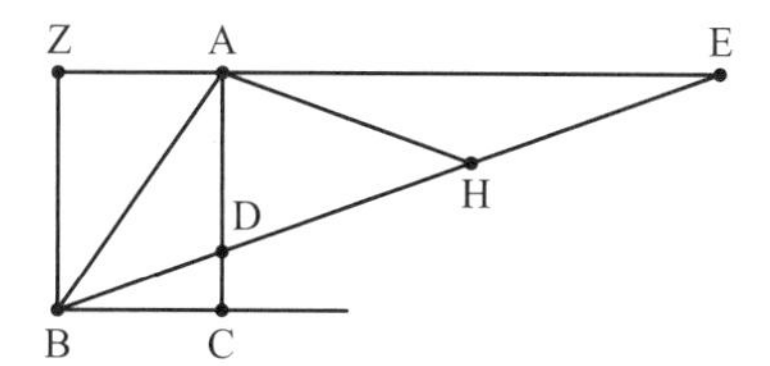

Figure 5.11.1.

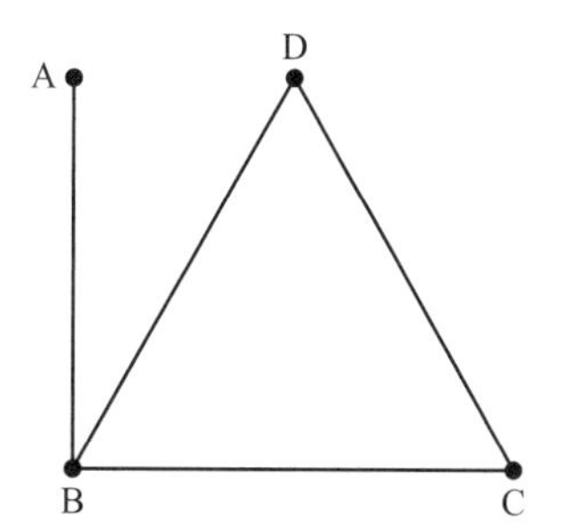

Figure 5.11.2.

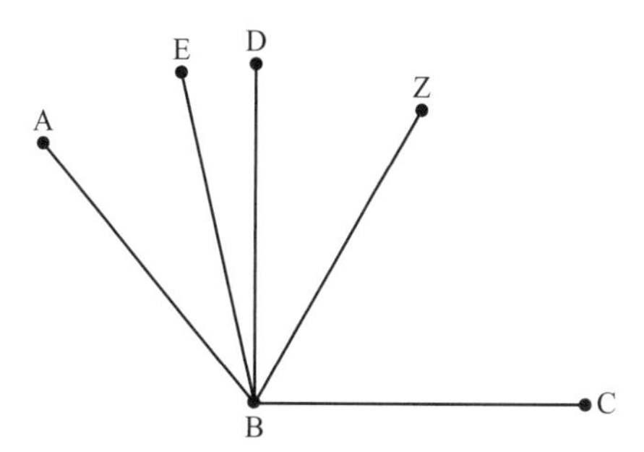

Figure 5.11.3.

39. When, however, the given angle happens to be a right angle, we will cut off a certain segment BC and describe over it the equilateral triangle BDC. And when we bisect the angle DBC, we will have trisected the angle ABC [figure 5.11.2].

40. Finally, let the angle be obtuse, and draw BD at right angles to CB, and, on the one hand, cut off the angle DBZ as a third part of the angle DBC, and on the other hand, the angle EBD as the third part of the angle ABD (for I have shown these two constructions above) [figure 5.11.3]. Then the angle EBZ is the third part of the whole angle ABC as well. When, however, we erect an angle equal to the angle EBZ along both AB and BC, we will trisect the given angle.

This *neusis* construction in paragraph 38 can be accomplished through the use of the conchoid of Nicomedes, although there is no direct evidence that Nicomedes did

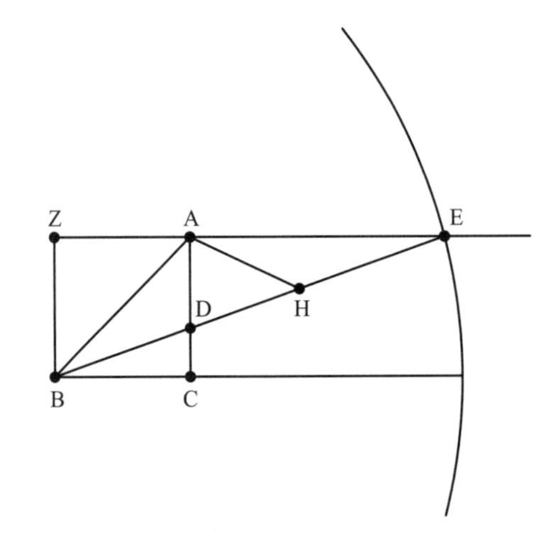

Figure 5.11.4.

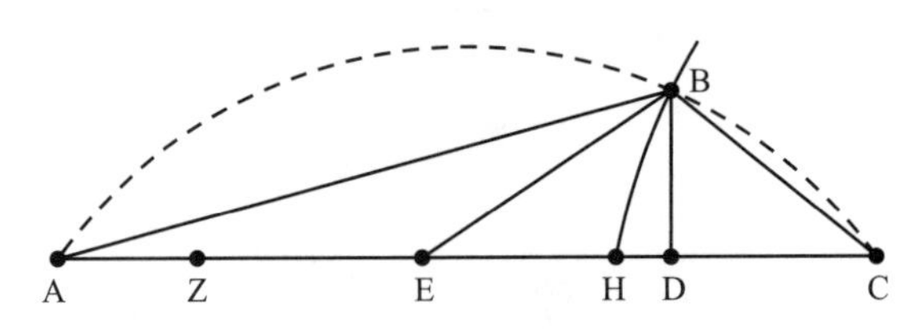

Figure 5.11.5.

this. Given figure 5.11.4, we take B as the pole of the conchoid, AC the ruler, and $2AB$ the distance. Then we draw the conchoid with those parameters and choose E to be the point where the conchoid intersects the line ZA extended. The proof then follows as before. Pappus himself shows than one can also trisect an angle through the use of solid loci, that is, conic sections.

43. The third part of a given arc is cut off in a different way, also, without the *neusis*, by means of a solid locus of the following sort.

Assume that the straight line through A and C is given in position, and that the angle ABC has been bent over the points A and C given on it, making an angle ACB that is two times the angle CAB [figure 5.11.5]. I claim that B lies on a uniquely determined hyperbola.

Draw the perpendicular BD, and cut off DE, equal to CD. Then BE, when it has been joined, will be equal to AE.[91] Position EZ as equal to DE, also. Then CZ is three times CD. Let AC be three times CH, also.[92]

[91] Since triangle EBD is congruent to triangle BDC, angle BEC = angle BCA, which equals twice angle BAE. Since angle BEC equals the sum of angles BAE and ABE, we have angle BAE = angle ABE and triangle ABE is isosceles.

[92] That is, choose H so that $AC = 3CH$.

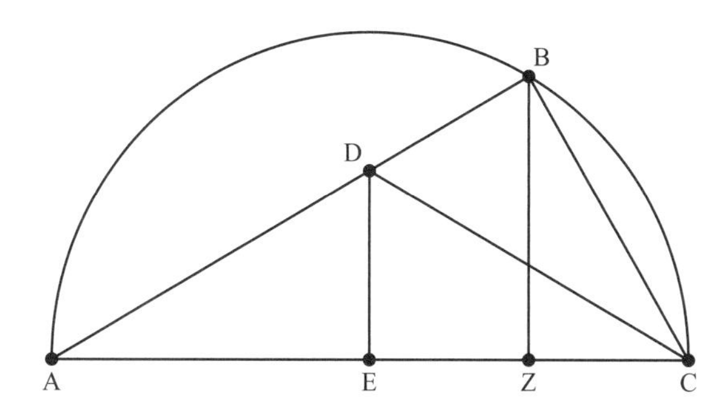

Figure 5.11.6.

Now H will be given, and the remaining AZ will be three times HD.[93] And since sq(BD) is the difference between sq(BE) and sq(EZ), whereas rect(DA, AZ) is the difference of these also,[94] rect(DA, AZ), that is, three times rect(AD, DH), will therefore be equal to sq(BD). Therefore, B lies on a hyperbola, the *latus transversum* [diameter] of which is AH, and the *latus rectum* three times AH.[95] And it is obvious that the point C cuts off half the *latus transversum* AH on the straight line CH drawn to the vertex H of the conic section. And the synthesis is obvious. For one will have to divide AC so that AH is two times HC, and describe the hyperbola through H with axis AH, the *latus transversum* of which is three times AH, and to show that it creates the above-mentioned twofold ratio of the angles. And that the hyperbola described in this way cuts off the third part of the given circular arc is rather easy to understand when the points A and C are posited as the endpoints of the arc.[96]

44. Some have set out the analysis of trisecting an angle or an arc in yet another way without a *neusis*. Let the argument be about an arc, however. For it makes no difference whether one divides an angle or an arc.

Assume that it has turned out that way in fact, and that BC has been cut off as the third part of the arc ABC, and join AB, BC, and CA [figure 5.11.6]. Then the angle ACB is two times the angle BAC. Bisect the angle ACB by CD, and draw the perpendiculars DE and ZB. Then AD is equal to DC, so that AE is equal to EC, also. Therefore, E is given. Now, since AD is to DB, that is, AE to EZ, as AC is to CB,[97] *alternando*, BC is therefore to EZ as CA is to AE, also. CA is twice AE, however. Therefore, BC is twice EZ as well.

[93] $AZ = AC - CZ = 3(CH - CD) = 3HD$.

[94] By *Elements* 2:6, with DZ the bisected line (at E) and ZA the added line, $DA \cdot AZ + ZE^2 = AE^2$, or $DA \cdot AZ = AE^2 - ZE^2 = BE^2 - EZ^2$.

[95] By *Conics*, book 1, proposition 21, and its converse, if $B = (x, y)$ lies on the hyperbola, then, in modern terms, $y^2 = \frac{p}{2a} x_1 x_2$, where x_1 and x_2 are the distances from point D to the two ends H and A of the diameter, which has a length equal to $2a$, and where the parameter is p (equal to the *latus rectum*). In this case, since $BD^2 = 3AD \cdot DH$, the equation is $y^2 = 3x_1 x_2$, so $3 = \frac{p}{2a}$ and the parameter p is equal to $3AH$, as stated.

[96] If we begin with the arc AC, then going backward through the analysis gives us the hyperbola with vertex H and the intersection point B of this hyperbola with the arc so that arc BC is one-third of arc AC.

[97] Since DC bisects angle ACB, by *Elements* 6:3 $AC : BC = AD : DB$.

Therefore, the square over BC, that is, the sum of the squares over BZ and ZC is four times the square over EZ.

Now, since the two points E and C are given, and BZ is at right angles to AC, and the ratio of the square over EZ to the sum of the squares over BZ and ZC is given, B lies therefore on a hyperbola.[98] But it also lies on an arc that is given in position. Therefore, B is given. And the synthesis is obvious.[99]

5.12 Pappus, *Collection*, 7, Analysis and Synthesis

Book 7 of the *Collection* deals with what Pappus calls the *Domain of Analysis*, a collection of thirty-three books written by three authors, Euclid, Apollonius, and Aristaeus, most of which are no longer extant. But Pappus used his book 7 to comment on these works, giving a large collection of lemmas to the theorems there, presumably to help students understand the material. These lemmas are today often the only remaining account of the contents of these books. Book 7 begins, however, with a definition of analysis and synthesis, procedures that were used by the Greek mathematicians in their theoretical work.

That which is called the *Domain of Analysis*, my son Hermodorus,[100] is, taken as a whole, a special resource that was prepared, after the composition of the *Common Elements*, for those who want to acquire a power in geometry that is capable of solving problems set to them; and it is useful for this alone. It was written by three men: Euclid the Elementarist, Apollonius of Perge, and Aristaeus the elder, and its approach is by analysis and synthesis.

Now, analysis is the path from what one is seeking, as if it were established, by way of its consequences, to something that is established by synthesis. That is to say, in analysis we assume what is sought as if it has been achieved, and look for the thing from which it follows, and again what comes before that, until by regressing in this way we come upon some one of the things that are already known, or that occupy the rank of a first principle. We call this kind of method "analysis," as if to say *anapalin lysis* (reduction backward). In synthesis, by reversal, we assume what was obtained last in the analysis to have been achieved already, and, setting now in natural order, as precedents, what before were following, and fitting them to each other, we attain the end of the construction of what was sought. This is what we call "synthesis."

There are two kinds of analysis: one of them seeks after truth, and is called "theorematic"; while the other tries to find what was demanded, and is called "problematic." In the case of the theorematic kind, we assume what is sought as a fact and true, then, advancing through its consequences, as if they are true facts according to the hypothesis, to something established; if this thing that has been established is a truth, then that which was sought will also be true, and its proof the reverse of the analysis; but if we should meet

[98] Since BC is twice EZ, this is the hyperbola with focus C, directrix DE, and eccentricity 2.

[99] If we begin with the arc AC, we bisect it at E and draw the perpendicular ED. Then we describe the hyperbola with directrix ED, focus C, and eccentricity 2. This hyperbola intersects the given arc AC in B, which divides the arc in the ratio 2:1. The proof of this is simply the reverse of the analysis.

[100] Nothing more is known of Hermodorus.

with something established to be false, then the thing that was sought too will be false. In the case of the problematic kind, we assume the proposition as something we know, then, proceed through its consequences, as if true, to something established; if the established thing is possible and obtainable, which is what mathematicians call "given," the required thing will also be possible, and again the proof will be the reverse of the analysis; but should we meet with something established to be impossible, then the problem too will be impossible. Diorism is the preliminary distinction of when, how, and in how many ways the problem will be possible. So much, then, concerning analysis and synthesis.

The order of the books of the *Domain of Analysis* alluded to above is this: Euclid, *Data*, one book; Apollonius, *Cutting Off of a Ratio*, two; *Cutting Off of an Area*, two; *Determinate Section*, two; *Tangencies*, two; Euclid, *Porisms*, three; Apollonius, *Neuses*, two; by the same, *Plane Loci*, two, *Conics*, eight; Aristaeus, *Solid Loci*, five; Euclid, *Loci on Surfaces*, two; Eratosthenes,[101] *On Means*, one. These make up 32 books. I have set out epitomes of them, as far as the *Conics* of Apollonius, for you to study, with the number of the dispositions and diorisms and cases in each book, as well as the lemmas that are wanted in them, and there is nothing wanting for the working through of the books, I believe, that I have left out.

5.13 Pappus, *Collection*, 7, Three- and Four-Line Locus and Guldin's Theorem

As part of his discussion of Apollonius's *Conics*, in book 7 of the *Collection*, Pappus discusses in some detail the problem of the locus on three and four lines and its generalization to more lines. As noted at the end of *Conics*, book 3 (section 5.2), it is possible to give a complete solution of that problem for three and four lines using some of the results of that book. But Pappus complains that no one has yet described the locus when there are more than four lines. He then follows this with a statement of a theorem he claims is "much more valuable," the theorem today known as Guldin's theorem, after Paul Guldin (1577–1643): If any plane figure revolves about an external axis in its plane, the volume of the solid figure so generated is equal to the product of the area of the figure and the distance traveled by the center of gravity of the figure. Pappus, however, presents no proof of this result.

Thus Apollonius. The locus on three and four lines that he says, in his account of the third book [of the *Conics*], was not completed by Euclid, neither he nor anyone else would have been capable of; no, he could not have added the slightest thing to what was written by Euclid, using only the conics that had been proved up to Euclid's time, as he himself confesses when he says that it is impossible to complete it without what he was forced to write first. But either Euclid, out of respect for Aristaeus as meritorious for the conics he had published already, did not anticipate him, or, because he did not desire to commit to writing the same matter as he (Aristaeus)—for he was the fairest of men, and kindly to

[101] It is a bit strange that Pappus includes Eratosthenes here, even though he is not mentioned in the opening paragraph.

everyone who was the slightest bit able to augment knowledge, as one should be, and he was not at all belligerent, and though exacting, not boastful, the way this man (Apollonius) was—he wrote only as far as it was possible to demonstrate the locus by means of the other's *Conics*, without saying that the demonstration was complete. For had he done so, one would have had to convict him, but as things stand, not at all. And in any case, Apollonius himself is not castigated for leaving most things incomplete in his *Conics*. He was able to add the missing part to the locus because he had Euclid's writings on the locus already before him in his mind, and had studied for a long time in Alexandria under the people who had been taught by Euclid, where he also acquired this so great condition of mind, which was not without defect.

This locus on three and four lines that he boasts of having augmented instead of acknowledging his indebtedness to the first to have written on it is like this: If three straight lines are given in position, and from some single point straight lines are drawn onto the three at given angles, and the ratio of the rectangle contained by two of the lines drawn onto them to the square of the remaining one is given, the point will touch a solid locus given in position, that is, one of the three conic curves. And if straight lines are drawn at given angles onto four straight lines given in position, and the ratio of the rectangle contained by two of the lines that were drawn to the rectangle contained by the other two that were drawn is given, likewise the point will touch a section of a cone given in position.

Now if they are drawn onto only two lines, the locus has been proved to be plane, but if onto more than four, the point will touch loci that are as yet unknown, but just called "curves," and whose origins and properties are not yet known. They have given a synthesis of not one, not even the first and seemingly the most obvious of them, or shown it to be useful. The propositions of these loci are: If straight lines are drawn from some point at given angles onto five straight lines given in position, and the ratio is given of the rectangular parallelepiped solid contained by three of the lines that were drawn to the rectangular parallelepiped solid contained by the remaining two lines that were drawn and some given one, the point will touch a curve given in position. And if onto six, and the ratio of the aforesaid solid contained by the three to that by the remaining three is given, again the point will touch a curve given in position. If onto more than six, one can no longer say "the ratio is given of the something contained by four to that by the rest," since there is nothing contained by more than three dimensions.

Our immediate predecessors have allowed themselves to admit meaning to such things, though they express nothing at all coherent when they say "the thing contained by these," referring to the square of this line or the rectangle contained by these. But it was possible to enunciate and generally to prove these things by means of compound ratios, both for the propositions given above, and for the present ones, in this way:

If straight lines are drawn from some point at given angles onto straight lines given in position, and there is given the ratio compounded of that which one drawn line has to one, and another to another, and a different one to a different one, and the remaining one to a given, if there are seven, but if eight, the remaining to the remaining one, the point will touch a curve given in position. And similarly for however many, even or odd in number. As I said, of not one of these that come after the locus on four lines have they made a synthesis so that they know the curve.

They who look at these things are hardly exalted, as were the ancients and all who wrote the finer things. When I see everyone occupied with the rudiments of mathematics and of the material for inquiries that nature sets before us, I am ashamed; I for one have proved things that are much more valuable and offer much application. In order not to end my discourse declaiming this with empty hands, I will give this for the benefit of the readers:

The ratio of solids of complete revolution is compounded of that of the revolved figures and that of the straight lines similarly drawn to the axes from the centers of gravity in them; that of solids of incomplete revolution from that of the revolved figures and that of the arcs that the centers of gravity in them describe, where the ratio of these arcs is, of course, compounded of that of the lines drawn and that of the angles of revolution that their extremities contain, if these lines are also at right angles to the axes. These propositions, which are practically a single one, contain many theorems of all kinds, for curves and surfaces and solids, all at once and by one proof, things not yet and things already demonstrated, such as those in the twelfth book of the *First Elements.*

The eight books of Apollonius's *Conics* contain 487 theorems or diagrams, and there are 70 lemmas, or things assumed in it.

5.14 Pappus, *Collection*, 7, Constructing Conics using Focus and Directrix

In the final section of book 7, Pappus shows how to construct a conic section from its directrix and focus. In the extant manuscripts, where the ratio of distance to focus to distance to directrix is discussed, there is some confusion as to whether a ratio greater than one gives an ellipse or a hyperbola, and similarly for a ratio smaller than one. We have corrected that in what follows.

Given two points A, B, and GD at right angles, let the ratio of the square of AD to the squares of GD, DB together be given. I say that G touches a section of a cone, whether the ratio is equal to equal or greater to less or less to greater.

Proposition 236a. *For first let the ratio be equal to equal.*

And since the square of AD equals the squares of GD, DB, let DE be made equal to BD [figure 5.14.1]. Then the rectangle contained by BA, AE equals the square of DG. Let AB be bisected by Z. Then Z is given. And AE is twice ZD. Hence the rectangle contained by BA, AE is twice the rectangle contained by AB, ZD. And twice AB is given. Therefore the rectangle contained by a given line and DZ equals the square of DG. Thus G touches a parabola given in position and passing through Z.

We clarify Pappus's argument that the locus of points equidistant from a given point and a given line ($GB = DA$) is a parabola and put it into a modern context. The point B will be the focus of the parabola, so we set $B = (f, 0)$. The directrix is the line through A perpendicular to line AB, and since the parabola will pass through the point Z, which is equidistant from A and B, we can take Z as the origin and therefore set $A = (-f, 0)$. Then $AB = 2f$. We let $G = (x, y)$. Therefore, $D = (x, 0)$ and

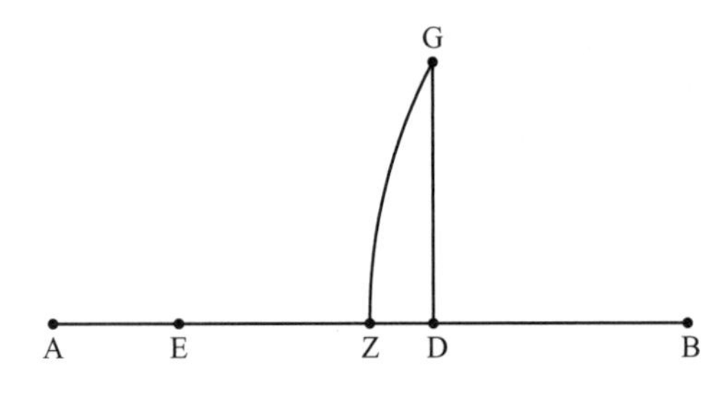

Figure 5.14.1.

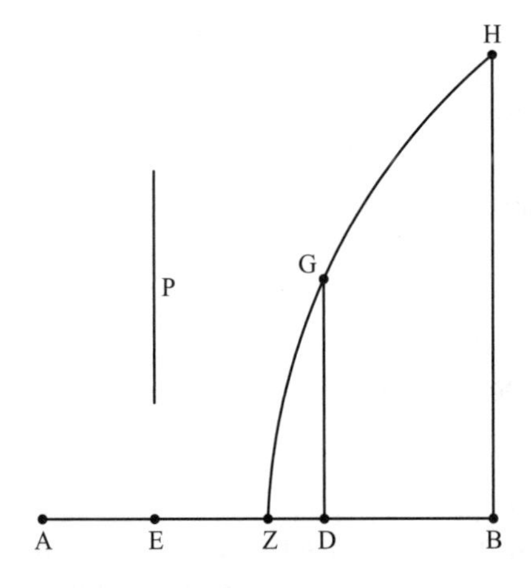

Figure 5.14.2.

$E=(2x-f, 0)$. Since $AD^2=BG^2=GD^2+DB^2$, we have $DG^2=BG^2-DB^2=AD^2-DB^2=(AD+DB)(AD-BD)=AB(AD-DE)=AB\cdot AE$. But $AE=(2x-f)-(-f)=2x=2\,ZD$. So $DG^2=AB\cdot AE=2\,AB\cdot ZD$, or $y^2=4f\cdot x$. Thus, any point satisfying the initial condition that its distances from the focus and the directrix are equal lies on the locus given by the equation $y^2=4fx$. From *Conics*, book 1, proposition 11, we know that the equation of a parabola is $y^2=px$, where p is the parameter. Thus, in this case, $p=4f$ and the focus B of the parabola lies at a distance $\frac{p}{4}$ from the vertex Z. That is, the parabola $y^2=4fx$ is the curve whose points are equidistant from the focus $(f, 0)$ and the directrix $x=-f$.

Proposition 236b. *The synthesis of the locus will be made as follows.*

Let the given points be A, B, and let the ratio be equal to equal, and let AB be bisected by Z [figure 5.14.2]. Let P be twice AB, and with ZB being a straight line given in position terminated at Z, and with P given in magnitude, let parabola HZ be drawn about axis ZB so that if a point such as G is taken on it, and perpendicular GD is drawn, the rectangle contained by P, ZD equals the square of DG. And let BH be drawn at right angles. I say that part GH of the parabola solves the locus.

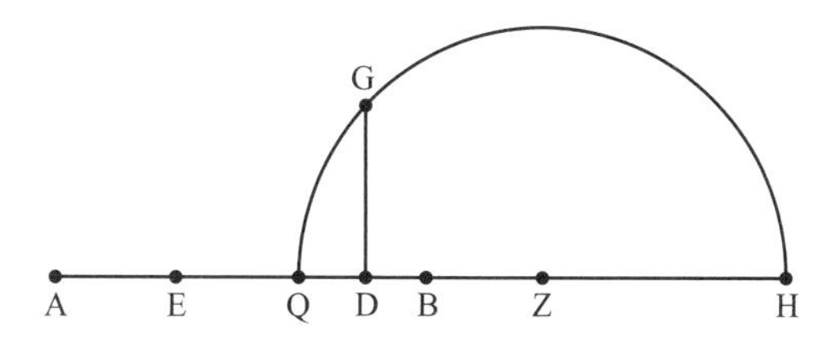

Figure 5.14.3.

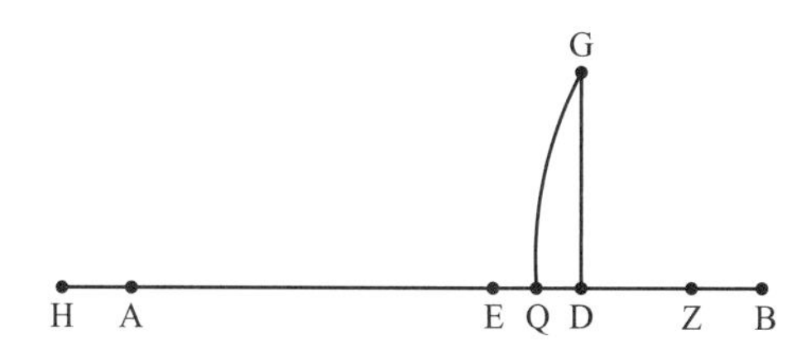

Figure 5.14.4.

For let the perpendicular GD be drawn, and let DE be made equal to BD. Then since AB is twice BZ, and EB twice BD, therefore AE too is twice ZD. Hence the rectangle contained by BA, AE equals twice the rectangle contained by AB, ZD, that is, the square of DG. Let the square of ED be added in common, which equals the square of DB. Therefore the sum, the square of AD, equals the squares of GD, DB. Thus curve ZGH solves the locus.

Proposition 237a, b. *Again let the two given points be A, B, and let DG be at right angles, and let the ratio of the square of AD to the squares of BD, DG be, in the first case, greater to less, in the second less to greater. I say that G touches a section of a cone, in the first case an ellipse* [figure 5.14.3], *in the second a hyperbola* [figure 5.14.4].

For since the ratio of the square of AD to the squares of BD, DG is given, let the ratio of the square of DE to the square of BD be the same as it. Now in the first case BD is less than DE, in the second BD is greater than DE. Then let DZ be made equal to ED. Since the ratio of the square of AD to the squares of GD, DB is given, and the ratio of the square of ED to the square of DB is the same as it, therefore the remainder, the ratio of the rectangle contained by ZA, AE to the square of DG is given. But since the ratio of ED to DB and of ZD to DB and so that of ZB to BD is given, let the ratio of AB to BH be the same as it. Hence the sum, the ratio of AZ to DH, is given. Again, since the ratio of ED to DB is given, therefore the ratio of EB to BD too is given. Let the ratio of AQ to BQ be the same as that of ED to DB. Then the ratio of AB to BQ too is given. Hence Q is given. And the remainder, the ratio of AE to QD, is given. Therefore also the ratio of the rectangle contained by ZA, AE to the rectangle contained by QD, DH is given. But the ratio of the rectangle contained by ZA, AE to the square of GD is given. Therefore the ratio of the rectangle contained by HD, DQ to the square of DG too is given. And Q, H are two given points. Hence in the first case G touches an ellipse, in the second a hyperbola.

We clarify Pappus's argument for the case of an ellipse. That is, we show why the locus of points, the ratio of whose distances from a given point (the focus) and a given line (the directrix) is a value smaller than 1, is an ellipse. As in the previous case, the point B is the focus and a line through A perpendicular to line AB is the directrix. We begin by assuming that the ratio of $\sqrt{BD^2+DG^2}=GB$, the distance of G from the focus, to AD, the distance of the arbitrary point G from the directrix, is e, where $e<1$. Therefore, the ratio of AD^2 to BD^2+DG^2 will be $\frac{1}{e^2}$. Then E is chosen so that $DE^2:BD^2=\frac{1}{e^2}$, or $DE:BD=\frac{1}{e}$. So if $G=(x,y)$ and $B=(r,0)$, then $D=(x,0)$, $BD=r-x$, and $DE=\frac{BD}{e}=\frac{r-x}{e}$. We choose Z so that $DZ=ED=\frac{r-x}{e}$. Then, since

$$\frac{AD^2}{GD^2+DB^2}=\frac{ED^2}{DB^2}=\frac{1}{e^2},$$

it follows that the ratio of the differences is also $\frac{1}{e^2}$. That is,

$$\frac{AD^2-ED^2}{GD^2+DB^2-DB^2}=\frac{(AD+DE)(AD-DE)}{GD^2}=\frac{(AD+DZ)(AE)}{GD^2}=\frac{AZ\cdot AE}{GD^2}=\frac{1}{e^2}.$$

Furthermore, since $ED:DB=1:e=ZD:DB$, we have $ZB:DB=(DZ-DB):DB=DZ:DB-DB:DB=\frac{1}{e}-1=\frac{1-e}{e}$. We next choose H so that $AB:BH=(1-e):e$. Then $AZ:DH=(AB+BZ):(BH+DB)=(1-e):e$. Also, since $ED:DB=1:e$, we have $EB:BD=(ED+BD):BD=ED:BD+BD:BD=\frac{1}{e}+1=\frac{1+e}{e}$. We next choose Q so that $AQ:BQ=1:e$. Therefore $AB:BQ=(AQ+BQ):BQ=AQ:BQ+BQ:BQ=\frac{1}{e}+1=\frac{1+e}{e}$. It then follows that $AE:QD=(AB-BE):(BQ-BD)=\frac{1+e}{e}$. Therefore,

$$\frac{AZ\cdot AE}{QD\cdot DH}=\frac{AZ}{DH}\cdot\frac{AE}{QD}=\frac{1-e}{e}\cdot\frac{1+e}{e}=\frac{1-e^2}{e^2}.$$

But we know that $AZ\cdot AE:GD^2=\frac{1}{e^2}$. Therefore

$$\frac{HD\cdot QD}{DG^2}=\frac{HD\cdot QD}{AZ\cdot AE}\cdot\frac{AZ\cdot AE}{GD^2}=\frac{e^2}{1-e^2}\cdot\frac{1}{e^2}=\frac{1}{1-e^2}.$$

In other words, $\frac{x_1x_2}{y^2}=\frac{1}{1-e^2}$, or $y^2=(1-e^2)x_1x_2$, where x_1, x_2 are the distances of the point D from the two ends of the axis Q, H of the curve. This relationship, as shown in *Conics*, book 1, proposition 21, is one of the symptoms of an ellipse. In particular, that proposition shows that $1-e^2=\frac{p}{2a}$, where p is the parameter of the ellipse and $2a$ is the length of the major axis, here the length of QH. It is straightforward to convert that symptom to the original symptom given in *Conics*, book 1, proposition 13 that can be expressed in the form $y^2=x(p-\frac{p}{2a}x)$. It is also straightforward to show from the above that $a=\frac{r}{1-e}$, where r is the coordinate of the focus B. If c is the distance of the focus B from the center Z of the ellipse, then $c=a-r=a-a(1-e)=ae$, showing that the eccentricity e of an ellipse is $e=\frac{c}{a}$. Also, we know that b, the length of half the minor axis of the ellipse, is given by $b^2=a^2-c^2$. Furthermore, if the point

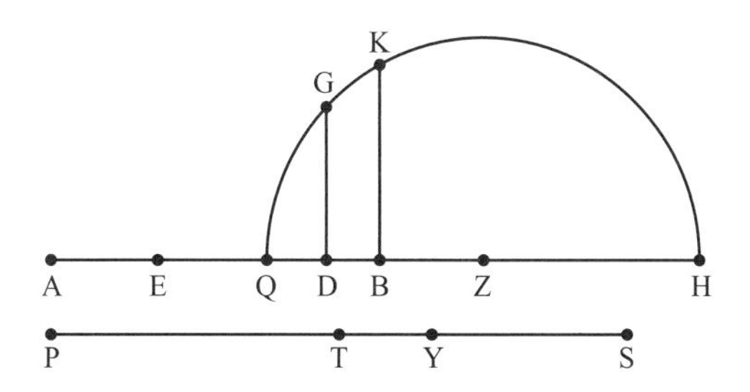

Figure 5.14.5.

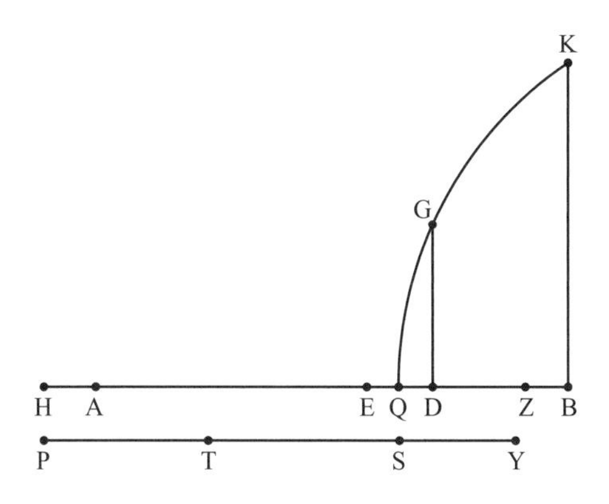

Figure 5.14.6.

(x, y) is the endpoint of the minor axis, then $x_1 = x_2 = a$, so $b^2 = \frac{p}{2a}a^2$, or $b^2 = \frac{pa}{2}$. Thus $1 - e^2 = 1 - \frac{c^2}{a^2} = \frac{a^2 - c^2}{a^2} = \frac{b^2}{a^2} = \frac{pa}{2a^2} = \frac{p}{2a}$ as claimed.

Proposition 237c, d. *The synthesis of the locus will be made as follows.*

Let the two given points be A, B, the given ratio that of the square of PT to the square of TS, in the first case greater to less, in the second less to greater. And let TY be made equal to PT, and let AB be made to BH as YS is to ST. And let AQ be made to QB as is PT to TS. And let there be drawn about axis QH, in the first case an ellipse [figure 5.14.5], in the second a hyperbola [figure 5.14.6], so that if a point such as G is taken on it, and perpendicular GD is drawn, the ratio of the rectangle contained by QD, DH to the square of DG is compounded out of that which TS has to SY and that which TS has to SP and that which the given ratio has, which is that of the square of PT to the square of TS. Let BK be drawn at right angles. I say that QK solves the assignment.

For let perpendicular GD be drawn, and let ZB be made to BD as AB is to BH, and ED to DB as AQ is to QB. Hence the ratio of DH to AZ is the same as that of HB to BA, that is, that of TS to SY, whereas the ratio of QD to AE is the same as that of TS to SP, for this was proved in the analysis. Hence the ratio of the rectangle contained by QD,

DH to the rectangle contained by *ZA*, *AE* is compounded out of that which *TS* has to *SY* and *TS* to *SP*. But since the rectangle contained by *QD*, *DH* has to the square of *DG* the ratio compounded out of that which *TS* has to *SY* and *TS* to *SP* and the given ratio, that of the square of *PT* to the square of *TS*, while the ratio of the rectangle contained by *QD*, *DH* to the square of *DG* is compounded out of that which the rectangle contained by *QD*, *DH* has to the rectangle contained by *ZA*, *AE* and the rectangle contained by *ZA*, *AE* to the square of *DG*, and the ratio of the rectangle contained by *QD*, *DH* to the rectangle contained by *ZA*, *AE* is the same as that compounded out of that which *TS* has to *SY* and *TS* to *SP*, therefore the remaining ratio of the rectangle contained by *EA*, *AZ* to the square of *DG* is the same as that of the square of *PT* to the square of *TS*, that is, that of the square of *ED* to the square of *DB*. And all to all, therefore, as is the square of *AD* to the squares of *BD*, *DG*, so is the square of *PT* to the square of *TS*, that is, the given ratio. Thus part *QK* of the section solves the locus.

5.15 Serenus, *On the Section of a Cylinder*

Serenus (ca. 300–360 CE) was a mathematician from Antinoöpolis, a city in Egypt. He wrote a commentary on Apollonius's *Conics*, which is no longer extant, but also wrote two works on related questions, including *On the Section of a Cylinder*, excerpted here. In the preface to that work, he noted that many students of geometry were under the erroneous impression that the oblique section of a cylinder was different from the oblique section of a cone, namely, the ellipse, whereas in fact it is the same curve. So the aim of this work was to prove that assertion.

Serenus begins by noting that a section of a cylinder formed by a cutting plane parallel to the base is, naturally, a circle. Like Apollonius, he defined a cylinder using a generating circle and lines from it, which sweep out a cylindrical surface; also like Apollonius, these cylinders can be right (occurring when the generating lines are perpendicular to the plane of the circle) or oblique (occurring when the generating lines are not perpendicular to the plane of the circle). Instead of Apollonius's axial triangle, we instead have an axial parallelogram: formed by a plane that passes through the axis of the cylinder, and hence cutting each base circle as its diameter.

In the case of the oblique cylinder, one particular cutting plane will give us the situation similar to *Conics*, book 1, proposition 5, namely, the subcontrary circle. But what about the section formed from a plane cutting the cylinder neither parallel to the base nor situated subcontrariwise? We include here propositions 14, 16, and 20, which collectively show that the ellipse is the non-subcontrary section of a cylinder.

Proposition 14. *If a cylinder is cut by a plane through the axis, and also cut by another plane cutting the plane of the base, and the common section of the base and the plane cutting the base is at right angles to the axial parallelogram or to it extended, and from the section some straight line be drawn to the diameter, parallel to the aforementioned common section of the planes, then the straight line drawn will be equal in square to an*

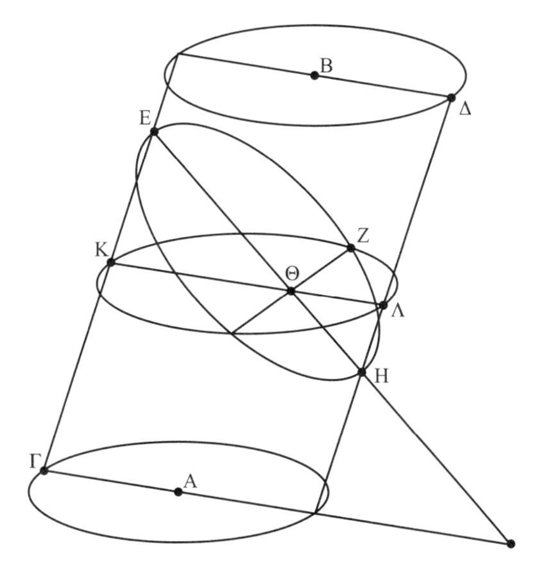

Figure 5.15.1.

area, which has a ratio to the rectangle contained by the sections of the diameter equal to the ratio which the square on the diameter of the section has to the square on the diameter of the base.

Let there be a cylinder, whose bases are the circles A and B, the parallelogram through the axis $\Gamma\Delta$, and let the cylinder be cut by a plane which cuts the base plane at right angles to ΓA produced, and let the resulting section be EZH, and the common section of the parallelogram and of the cutting plane be EH, being the diameter of the section, as has been shown [figure 5.15.1]. Taking a point Z at random on the section, let $Z\Theta$ be drawn from it, parallel to the common section of the cutting plane, at right angles to the diameter: therefore $Z\Theta$ intersects EH, as has been shown [figure 5.15.1]. I say that rect($E\Theta$, ΘH) has to sq($Z\Theta$) a ratio, that of the square on the diameter EH to the square on the diameter of the base.

Let $K\Theta\Lambda$ be drawn through Θ parallel to ΓA, and through $Z\Theta$, $K\Lambda$ let a cutting plane be drawn, making the section $KZ\Lambda$. So since $K\Lambda$ is parallel to ΓA, and $Z\Theta$ is parallel to the common section of the cutting plane and of the base plane, therefore also the planes through them are parallel: therefore the section $KZ\Lambda$ is a circle. Again, since $K\Lambda$ is parallel to ΓA, and $Z\Theta$ is parallel to the common section of the planes which are perpendicular to ΓA, therefore $Z\Theta$ is also perpendicular to $K\Lambda$. And $KZ\Lambda$ is a circle: therefore sq($Z\Theta$) is equal to rect($K\Theta$, $\Theta\Lambda$). Since KE is parallel to ΛH, therefore as $K\Theta$ is to $\Theta\Lambda$, so too is $E\Theta$ to ΘH: therefore rect($E\Theta$, ΘH) is similar to rect($K\Theta$, $\Theta\Lambda$). Therefore as rect($E\Theta$, ΘH) is to rect($K\Theta$, $\Theta\Lambda$), that is, to sq($Z\Theta$), so too is sq(EH) to sq($K\Lambda$), that is, to the square on the diameter of the base.

Proposition 16. *If a cylinder is cut by a plane which cuts the plane of the base, and the common section of the base and the cutting plane is at right angles to the axial*

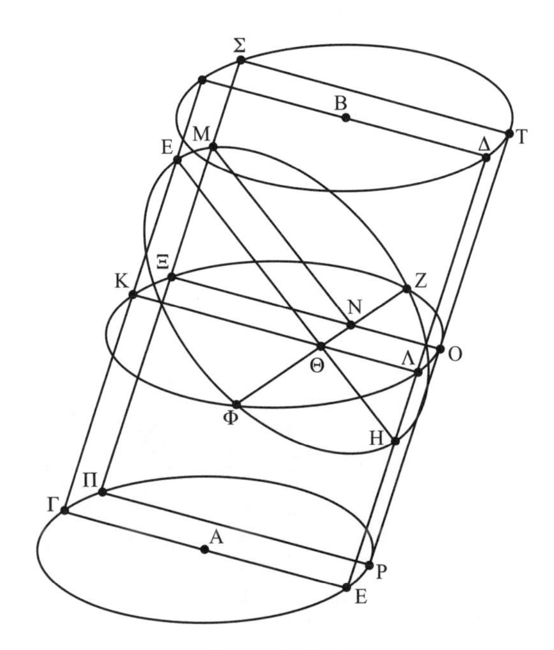

Figure 5.15.2.

parallelogram or to it extended, then the square on the segment drawn from the section to the diameter, parallel to the aforementioned common section of the planes, will be equal in square to an area, to which the rectangle contained by the sections of the diameter has a ratio, the ratio which the square on the diameter of the section has to the second diameter. And the segment drawn from the section to the second diameter, parallel to the diameter, will equal in square an area, to which the rectangle contained by the sections of the second diameter has a ratio, the ratio which the square on the second diameter has to the square on the diameter.

Let there be a cylinder, and let the construction be as in the fourteenth theorem [figure 5.15.2]. So since it was shown that rect($E\Theta$, ΘH) is to sq($Z\Theta$), as sq(EH) is to the square on the diameter of the base, which bisects EH and does so ordinatewise, as is shown in the ninth theorem; and the line bisecting the diameter and drawn ordinatewise is the second diameter, as was shown in the previous theorem; thus, as sq(EH) is to the square on the second diameter, so too would rect($E\Theta$, ΘH) be to sq($Z\Theta$): the very thing which it was necessary to show [figure 5.15.2].

But let it be hypothesized that Θ bisects the diameter EH, and that $Z\Theta\Phi$ is an ordinate: therefore $Z\Phi$ is the second diameter. Let MN be drawn from it on the section parallel to EH: I say that rect(ΦN, NZ) has to sq(MN) a ratio which the square on the second diameter ΦZ has to the square on the diameter of the section EH.

Through MN, let a plane be drawn cutting the cylinder and parallel to the parallelogram $\Gamma\Delta$: it will make a parallelogrammic section. Let it make $P\Sigma$, and let the common sections of it and the parallel circles be ΣT, ΞO, ΠP, and let MN be the common section of the parallelogram and the section EZH.

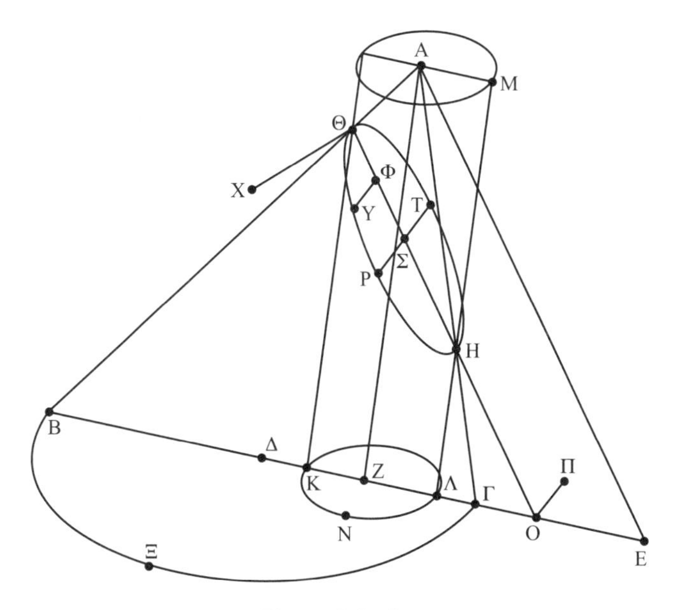

Figure 5.15.3.

So since the parallel planes $\Gamma\Delta$, $P\Sigma$ cut the plane $KZ\Lambda$, the common sections of them are parallel: therefore $K\Theta$ is parallel to $N\Xi$. But ΘE is also parallel to ΘE: therefore the angle $K\Theta E$ is equal to the angle ΞNM. And since the parallelogram $P\Sigma$ is equiangular to the parallelogram $\Gamma\Delta$, as was shown in the ninth theorem, therefore the angle $\Sigma\Pi P$ is equal to the angle $E\Gamma A$, that is, the angle $\Sigma\Xi N$ to the angle $EK\Theta$: therefore the triangles $EK\Theta$, $M\Xi N$ are similar to one another. Therefore, as $K\Theta$ is to ΘE, so too is ΞN to NM: and as sq($K\Theta$) is to sq(ΘE), that is, to the square on the second diameter ΦZ, that is, as the square on the second diameter ΦZ is to the square on the diameter EH, so too is sq(ΞN) to sq(NM). But sq($N\Xi$) is equal to rect(ΦN, NZ); for $KZ\Lambda$ is a circle, and ΘZ is at right angles to $K\Theta$ and ΞN. Therefore, as the square on the second diameter ΦZ is to the square on the diameter EH, so too is rect(ΦN, NZ) to sq(MN), which was to be shown.

Proposition 20. *I say next that it is possible to show a cone and a cylinder being cut together by one and the same ellipse.*

For let a scalene triangle $AB\Gamma$ be constructed on the base $B\Gamma$, which is bisected at Δ, let AB exceed $A\Gamma$, and on ΓA, at the point A, let the angle contained by ΓA, AE be constructed, being either greater or less than the angle $AB\Gamma$ [figure 5.15.3]. Let AE intersect $B\Gamma E$ at E, and let EZ be the mean proportional between BE, $E\Gamma$. Let AZ be joined, and in the triangle, let ΘH be drawn parallel to AE, and through the points Θ, H, draw the lines ΘK, ΛHM parallel to AZ. Let the parallelogram KM be completed, and through the plane BE, having been drawn at right angles to the plane BEA, let the circle $KN\Lambda$ have been drawn around the diameter $K\Lambda$; this circle will be the base of a cylinder, whose axial parallelogram is KM. In this same plane, let the circle $B\Xi\Gamma$ have been drawn around the diameter $B\Gamma$; it will be the base of a cone, whose axial triangle is $AB\Gamma$. Having extended ΘH to O, let $O\Pi$, being in the plane of the circle, be drawn at right angles to BE, and let a plane through the straight lines $O\Pi$, OH be drawn: it will make a section in the cone on the base $B\Xi\Gamma$. Let it make ΘPH: therefore the straight line ΘH is a diameter of the section. So with ΘH having been bisected at Σ, let the second diameter $P\Sigma T$, along with a randomly chosen line $Y\Phi$, be drawn ordinatewise to it.[102] And let it be contrived, that as the square on the diameter ΘH (of the section ΘPH) is to the square on the second diameter PT (of the same section), so too is the transverse side of the figure ΘH to the upright side ΘX.

So since ΘK is parallel to AZ, and ΘO is parallel to AE, therefore, as sq(AE) is to sq(EZ), so too is sq(ΘO) to sq(KO). But as sq(AE) is to rect(BE, $E\Gamma$), so too is sq(ΘH) (the diameter of the section of the cone) to sq(PT) (the second diameter of the same section). And as sq(ΘO) is to sq(OK), so too is sq(ΘH) to sq($K\Lambda$), that is, to the square on the diameter $H\Theta$ (of the section of the cylinder) to the second diameter (of the section of the cylinder), as was shown previously: therefore, the second diameter of the section of the cylinder is equal to PT, the second diameter of the section of the cone. And ΘH is bisected at Σ, and the second diameter of the section of the cylinder was drawn at right angles to ΘH, as also was PT: therefore PT is the second diameter both of the section of the cone and of the section of the cylinder. Therefore the point P is on the conic surface and the cylindrical surface. Again, since in the sections of both the cone and the cylinder, the diameters, ΘH and PT, are the same, therefore also the third proportional[103] is the same, that is, ΘX, the upright side of the figure. Therefore ΘX is, with respect to the section of the cylinder, the upright side of the figure. So since ΘH is to ΘX as rect($H\Phi$, $\Phi\Theta$) is to sq(ΦY), as it was shown also for the section of the cylinder, so too is the rectangle contained by the sections of the diameter to the square on the line drawn ordinatewise that makes the same sections of the diameter. And since, for the section of the cylinder, as the transverse side of the figure ΘH is to the upright side ΘX, so too is rect($H\Phi$, $\Theta\Phi$) to the square on the line equal to $Y\Phi$ which is drawn at equal angles to ΘH. But the segment equal to $Y\Phi$ and drawn at equal angles to it at Φ is not different than $Y\Phi$. Therefore ΦY is also in the section of the cylinder: therefore the point Y, being on the conic surface, is also on the cylindrical surface. Similarly it is shown for any other lines which we similarly draw ordinatewise. Therefore the curve ΘPH is in the surfaces of both figures: therefore the section ΘPH is one and the same in both figures. And since the angle ΓA, AE, that is, the angle AH, $H\Theta$, was constructed as being either greater or less than the angle at B, therefore the section is not subcontrary: therefore the section ΘPH is not a circle. Therefore ΘPH is an ellipse. And therefore, the section of the cone and cylinder so constructed is the ellipse itself: the very thing which was necessary to show.

[102] That is, drawn to the diameter ΘH. This was difficult to render: the essential points are that $P\Sigma T$ and $Y\Phi$ are drawn ordinatewise to the diameter, and that since Σ is where the diameter is bisected, the line $P\Sigma T$ is therefore the second diameter.

[103] The "third" proportional here, ΘX, satisfies the ratio $PT : \Theta X = \Theta X : \Theta H$. Previously in this proposition, Serenus uses "mean proportional" to refer to ΘX.

SOURCES, CHAPTER 5

5.1 The translation of the *Conics*, book 1, is by R. Catesby Taliaferro and is most easily found in Apollonius of Perga, *Conics, Books I–III*, ed. Dana Densmore (Santa Fe, NM: Green Lion Press, 2000). The translation of the Eutocius commentary is by Colin McKinney from the Greek original in Johan Ludvig Heiberg, ed., *Apollonii pergaei quae graece exstant cum commentariis antiquis* (Stuttgart: Teubner, 1974). The translation of the preface to the *Conics* by the Banū Mūsā is by Apollonius of Perga, *Conics, Books V to VII: The Arabic Translation of the Lost Greek Original in the Version of the Banū Mūsā*, vol. 2, ed. and trans Gerald J. Toomer (New York: Springer, 1990), 620–28.

5.2 The translation of Apollonius's *Conics*, books 2 and 3, is by R. Catesby Taliaferro from Apollonius of Perga, *Conics, Books I–III*. The translation of *Conics*, book 4, is by Michael Fried from Apollonius of Perga, *Conics, Book IV* (Santa Fe: Green Lion Press, 2002). The translation of the Eutocius commentary is by Colin McKinney from the Greek original in Heiberg, *Apollonii Pergaei*.

5.3 The translation of books 5–7 of the *Conics* is by Gerald J. Toomer from Apollonius of Perga, *Conics, Books V to VII*.

5.4 This translation is adapted from E. M. Macierowski, trans., *Apollonius of Perga, On Cutting Off a Ratio: An Attempt to Recover the Original Argumentation through a Critical Translation of the Two Extant Medieval Arabic Manuscripts* (Fairfield, CT: Golden Hind Press, 1987).

5.5 This translation is from Netz, *The Works of Archimedes*, Vol. 1, *The Two Books On the Sphere and the Cylinder*, ed. and trans. Reviel Netz (Cambridge: Cambridge University Press, 2004), 304–6.

5.6 This translation is from Pappus of Alexandria, *Book 4 of the Collection*, ed. and trans. Heike Sefrin-Weis (London: Springer, 2010), 131–36.

5.7 This translation is from Diocles, *On Burning Mirrors: The Arabic Translation of the Lost Greek Original*, ed. and trans. G. J. Toomer (New York: Springer, 1976). The final section is adapted from Archimedes, *The Works of Archimedes*, vol. 1, 280–81.

5.8 This translation from book 5 of Pappus's *Collection* is from Ivor Thomas, *Greek Mathematical Works, Vol. 2, Aristarchus to Pappus of Alexandria* (Cambridge, MA: Harvard University Press, 1980), 589–93.

5.9 This translation from Theon of Alexandria's *Commentary on Ptolemy's Syntaxis* is from Thomas, *Greek Mathematical Works* Vol. 2, 387–95.

5.10 This translation is from Pappus of Alexandria, *Book 4 of the Collection*, 144–46.

5.11 This translation is from Pappus of Alexandria, *Book 4 of the Collection*, 146–49, 152–55.

5.12 This translation is from Pappus of Alexandria, *Book 7 of the Collection: Part 1. Introduction, Text, and Translation*, ed. and trans. Alexander Jones (New York: Springer, 1986), 82–84.

5.13 This translation is from Pappus of Alexandria, *Book 7 of the Collection*, 118–24.

5.14 This translation is from Pappus of Alexandria, *Book 7 of the Collection*, 362–68.

5.15 This translation is by Colin McKinney from the Greek original in Johan Ludvig Heiberg, *Sereni Antinoensis opuscula* (Leipzig: Teubner, 1896), and from the French in Paul Ver Eecke, *Serenus d'Antinoë: Le Livre de la Section du Cylindre et le Livre de la Section du Cône* (Paris: Desclée de Brouwer, 1929).

6

Applied Geometry

The excerpts collected in this chapter embody key ideas in geometry, mostly directed at solving area and volume problems. Many of the passages we have included here are from works whose authors are unknown, but their ideas have been found in papyri or clay tablets discovered in Egypt or Mesopotamia and date from the time of Greek cultural influence in these areas. First, however, we look at two works that investigate how to increase the power of an artillery piece by determining how to double a cube. Then we look at Heron of Alexandria's *Metrica*, a book giving numerous methods for finding areas and volumes. Interspersed among the problems of the *Metrica* are problems with solutions from two other works that are attributed to Heron, but most likely were written by later authors: the *Geometrica* and the *Stereometrica*. The Babylonian tablet BM 34568, dating from the third century BCE, gives methods for determining the length and width of a rectangle under certain conditions, methods that appear again in various papyri from Egypt, written in Greek or an Egyptian script, over the next several centuries. In addition, we present problems from other papyri giving calculations of areas and volumes, often using techniques similar to those of Heron, followed by excerpts from some Latin works on surveying that give similar methods of measurement. It would appear that these papyri and the tablet were designed to teach methods of problem solving, but unfortunately too few remain for us to come to any definite conclusions about transmission from one part of the Greek world to another. We conclude with Polybius's description of the geometrical methods by which the Romans laid out their military camps, and some excerpts from Vitruvius's *On Architecture* discussing the mathematical knowledge that a Roman architect should have.

6.1 Philo of Byzantium, *Belopoeika*

Philo of Byzantium (ca. 240–200 BCE) was a Greek engineer who spent much of his life in Alexandria. He wrote a major work, *Compendium of Mechanics*, of which only a few parts are still extant. But among these is the *Belopoeika*, a work on artillery. One of the problems he dealt with was how to modify the dimensions of a catapult so that it would send a heavier missile to its target. After a long discovery and experimentation process by earlier engineers, he realized that a key element for attacking this problem was the size of the hole holding the torsion spring.

The object of artillery construction is to dispatch the missile at long range, to strike with powerful impact. To this end experimentation and most investigations have been directed. We shall recount to you exactly what we discovered at Alexandria through much association with the craftsmen engaged in such matters and through intercourse with many master craftsmen in Rhodes, from whom we understood that the most efficient engines more or less conformed to the method we are about to describe.

Reduce to units [*drachmae*] the weight of the stone for which the engine must be constructed. Make the diameter of the hole of as many finger breadths [*dactyls*] as there are units in the cube root of the number obtained, adding, furthermore, the tenth part of the root found. If the weight's root is not a whole number, take the nearest one; but, if it is the nearest above, endeavor to diminish proportionately the added tenth and, if it is the nearest below, to increase the added tenth.[1] Diameters of holes calculated by this method are as follows:

10 *minae*	–	11 *dactyls*
15 *minae*	–	$12\frac{3}{4}$ *dactyls*
20 *minae*	–	$14\frac{3}{4}$ *dactyls*
30 *minae*	–	$15\frac{3}{4}$ *dactyls*
50 *minae*	–	$19\frac{3}{4}$ *dactyls*
1 *talent*	–	21 *dactyls*
$2\frac{1}{2}$ *talents*	–	25 *dactyls*
3 *talents*	–	27 *dactyls*

The diameter of the circle to receive the spring is calculated by this method.[2]

It is possible, also, from one number, the smallest of those mentioned, I mean from the 10 *minae*, to determine the remaining diameters by geometrical construction by the doubling of the cube, as we showed in our first book. We shall not shrink from including it now. At ten the diameter of the 10-*minae* engine is suited to calculations involving the cube root (for $10 \times 100 = 1{,}000$, the cube root of which is 10 *dactyls*, or, rather, 11 *dactyls* when the tenth is added). Let there be a straight line, A, given, of this diameter, of which, for the sake of argument, we must find the double to the power 3. I put a line, B, double A and at right angles to it; from the end of B I drew at right angles another line, Γ, of unknown length [figure 6.1.1]. From the corner Θ I drew a straight line, K, and bisected it; let the point of bisection be K. With center K and radius $K\Theta$, I described a semicircle,

[1] A *mina* is a Greek unit of weight equal to approximately 1.25 lbs or 0.57 kg. A *drachma* is $\frac{1}{100}$ of a *mina*, which in turn is $\frac{1}{60}$ of a *talent*. A *dactyl* is a measure of length equal to about 19 mm or $\frac{3}{4}$ inch.

[2] Many of the numbers in the table are incorrect, if one calculates according to Philo's procedure. Probably errors have crept in during the transmission. In particular, given that the sizes of the shots are represented in archaeological findings from ancient sites, it is probable that the weights for the last two lines should be 2 and $2\frac{1}{2}$ *talents*, respectively. In that case the diameters are approximately correct.

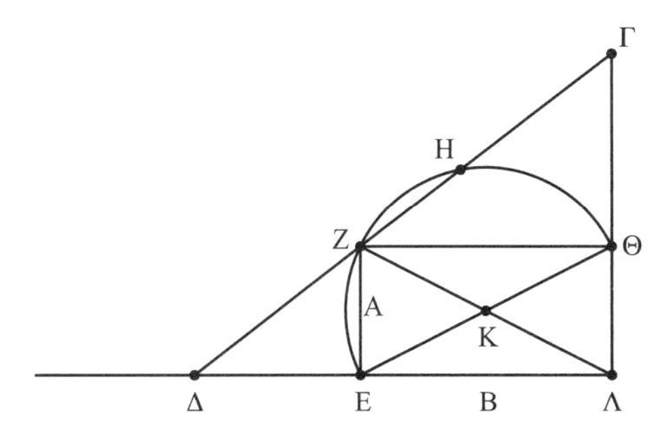

Figure 6.1.1.

cutting the corner Z. Taking a ruler, carefully straightened, I try to draw a connecting line, cutting both extensions [Γ and Δ] and keeping one part of the ruler at the corner Z. Let the ruler, then, be at Z. I juggle it about, always keeping one part of it touching the corner, until the section of the ruler from Γ to where it meets the edge of the semicircle, at H, is equal to the section extending from the junction at Δ to the corner Z.

Then the cube on ΔE is double the cube on EZ; the cube on $\Theta\Gamma$ is double the cube on $E\Delta$; and the cube on ΘZ is double the cube on $\Theta\Gamma$. The diameter of the circle to receive the spring is discovered by this method.[3]

To prove that the cube on ΔE is double the cube on $EZ = A$, we note first that since $\Gamma H = Z\Delta$, $H\Delta \cdot Z\Delta = Z\Gamma \cdot \Gamma H$. Since the constructed circle passes through Λ, we know from *Elements* 3:36 that $H\Delta \cdot Z\Delta = \Lambda\Delta \cdot E\Delta$ and $Z\Gamma \cdot H\Gamma = \Lambda\Gamma \cdot \Theta\Gamma$. It follows that $\Lambda\Delta \cdot E\Delta = \Lambda\Gamma \cdot \Theta\Gamma$ or that $\Lambda\Delta : \Lambda\Gamma = \Theta\Gamma : E\Delta$. By similar triangles, $\Lambda\Delta : \Lambda\Gamma = \Theta Z : \Theta\Gamma = E\Delta : EZ$. Therefore $\Theta Z : \Theta\Gamma = \Theta\Gamma : E\Delta = E\Delta : EZ$, or $\Lambda E : \Theta\Gamma = \Theta\Gamma : E\Delta = E\Delta : EZ$. It follows that $(E\Delta)^3 : (EZ)^3 = \Lambda E : EZ = 2$, as desired. The other relationships follow immediately. In modern terms, we can think of Philo's construction as finding the second intersection H of the circle centered at K with the rectangular hyperbola whose asymptotes are $\Lambda\Delta$ and $\Lambda\Gamma$ and that passes through the point Z. Note also that, as in other solutions to the problem of doubling the cube, the solution also amounts to finding two mean proportionals between a length and its double.

6.2 Heron, *Belopoeika*

Although many works are attributed to Heron of Alexandria, including the *Belopoeika* and the *Metrica*, we have essentially no biographical information for him other than that he lived in Alexandria. Otto Neugebauer asserted that because Heron in one of his works referred to a particular eclipse of the moon, and since his numerical values agreed with only one eclipse over a long period of time, he must have been alive in 62 CE when the eclipse took place.[4] Recently, however, several scholars have argued

[3] For example, if the diameter (without the added tenth) for 10 *minae* is given by $A = EZ$, then the diameter for 20 *minae* is the length of the first of the mean proportionals, namely, $E\Delta$.

[4] See Otto Neugebauer, *A History of Ancient Mathematical Astronomy* (New York: Springer, 1975), p. 846 and references cited there.

that there is no proof that Heron actually observed the eclipse himself and that he merely noted its occurrence from records. Thus all the eclipse proves is that Heron lived at some time after that date. We also know that Pappus refers to him, and Pappus was writing in the middle of the fourth century CE. Therefore, all we can conclude is that Heron lived and wrote sometime between the last third of the first century and the beginning of the fourth century, say, between 50 and 300 CE. However, since in problem 22 of book 1 of the *Metrica* Heron quotes the work *On the Chords in a Circle*, he was probably not familiar with Ptolemy's table of chords in the *Almagest*. Since the latter was written in the middle of the second century, it is reasonable to think that Heron lived before ca. 150 CE. Perhaps new research will enable a better determination of his dates to be made.

Heron is probably most famous for his rule for calculating the area of a triangle from the lengths of the three sides, but he produced many works in what may be called "applied mathematics." In what follows, Heron uses mathematics in the service of military technology. His use of mathematics in the construction of a catapult, described in his *Belopoeika*, is similar to that of Philo of Byzantium. However, he is much clearer on why one needs to find two mean proportionals in order to construct a new spring to project a missile of triple the size of a given one.

As we have spoken sufficiently and to the point about the construction and use of straight- and V-springs, we shall next describe their measurements. You must understand that the record of measurements was obtained from actual experiment. Older technicians, who only considered shape and design, did not have very satisfactory results in the projection of the missile, because they did not use harmonious measurements. Their successors, subtracting here and adding there, made their engines concordant and efficient. The above-mentioned engines (i.e., all their particular parts) are constructed from the diameter of the hole that receives the spring. The spring, in fact, is the initial guiding factor.

One must calculate the hole of the catapult thus. Multiply by one hundred the weight in *minae* of the stone to be discharged; find the cube root of the product; add to the result its tenth part, and make the diameter of the hole that number of finger breadths (*dactyls*). For instance, let the stone be of eighty *minae*; one hundred times this is eight thousand, of which the cube root is twenty; a tenth of this is two, which makes the answer twenty-two *dactyls*. Of this size the diameter of the hole is to be. If the product has no cube root, you must take the nearest figure and add the tenth part.

The hole of the straight-spring is calculated thus. Whatever the length of the arrow to be discharged, a ninth of it will be the diameter of the hole. For example, take a missile of three cubits, of which the ninth is eight *dactyls*. That will be the diameter of the hole.

It is possible, also, from one given diameter, to calculate the remaining diameters of stone-throwing engines by the doubling of the cube. When one efficient engine has been completed, it is possible to calculate others from it thus. Let the diameter of the engine be AB, and let it be required that we construct from it another engine throwing, let us suppose, a missile triple the size of the one mentioned. Now, since the spring is the cause of the discharge of the stone, the engine to be calculated will need a spring triple the size

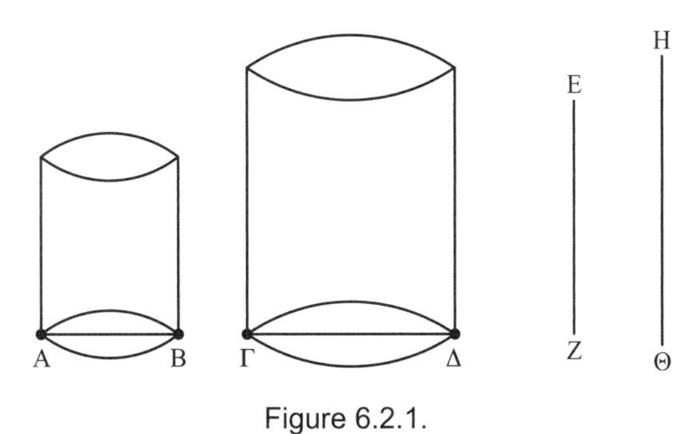

Figure 6.2.1.

of the one whose diameter is AB, and not with just any sort of hole, but with the spring's height proportionate to the hole, so that the cylinders formed by the springs are similar. Since the cylinders are similar to each other in the proportion of the cubes of their base diameters, let us imagine the diameter of the hole is found to be $\Gamma\Delta$. Of course, the cylinder on AB is to the cylinder on $\Gamma\Delta$ as the cube of AB is to the cube of $\Gamma\Delta$. Let $\Gamma\Delta$ be to EZ, and EZ to $H\Theta$ as AB is to $\Gamma\Delta$ [figure 6.2.1]. Then AB will be to $H\Theta$ as the cube of AB to the cube of $\Gamma\Delta$, and, therefore, as the cylinder on AB to the cylinder on $\Gamma\Delta$, so is AB to $H\Theta$. But the cylinder on AB is one-third of the cylinder on $\Gamma\Delta$, and AB is one-third of $H\Theta$; AB is given, so is $H\Theta$; and there are between AB and $H\Theta$ two mean proportionals $\Gamma\Delta$ and EZ.

Now, therefore, $\Gamma\Delta$ is given. For the instrumental construction, you must make $H\Theta$ three times AB, since the missile is to be three times as large, and find between AB and $H\Theta$ two mean proportionals $\Gamma\Delta$ and EA; then $\Gamma\Delta$ will be the diameter of the hole that is sought.[5]

We shall now explain how you must find the two mean proportionals between two straight lines. Set two given straight lines AB and $B\Gamma$ at right angles. It is required to find the two mean proportionals between these. Complete the rectangle $AB\Gamma\Delta$. Join $A\Gamma$, $B\Delta$; extend $\Delta\Gamma$, ΔA [figure 6.2.2]. Lay a ruler through point B, crossing these extensions, and move the ruler around point B until lines joining E to the points of intersection are equal to each other.

Suppose the ruler has assumed the position represented by the straight line ZBH. The other straight lines are EZ, EH. I affirm that the two mean proportionals (of AB, $B\Gamma$) are AZ, ΓH. If AB is first, second will be AZ, third ΓH, fourth $B\Gamma$. Since AE equals $E\Delta$ and EZ has been drawn, the product of ΔZ times ZA plus the square on AE equals the square on EZ.[6] Similarly, the product of ΔH times $H\Gamma$ with the square on ΓE equals the square on EH. And AE is equal to $E\Gamma$, EZ to EH. Therefore ΔZ times ZA will equal ΔH times $H\Gamma$. As $H\Delta$ is to ΔZ, so is AZ to ΓH. But $H\Delta$ is to ΔZ as AB to AZ, ZA to ΓH, and

[5] If we suppose AB is 11 *dactyls* long, then $H\Theta$ will be 33 *dactyls* long. Since $AB:\Gamma\Delta=\Gamma\Delta:EZ=EZ:H\Theta$, the diameter of $\Gamma\Delta$ is equal to $11\sqrt[3]{3}=15.86$. Presumably, one can get this value from the construction, at least approximately.

[6] By *Elements* 2:6, $\Delta Z\cdot ZA+AK^2=ZK^2$. If we add EK^2 to both sides, then by the Pythagorean theorem, $\Delta Z\cdot ZA+AE^2=EZ^2$, as claimed.

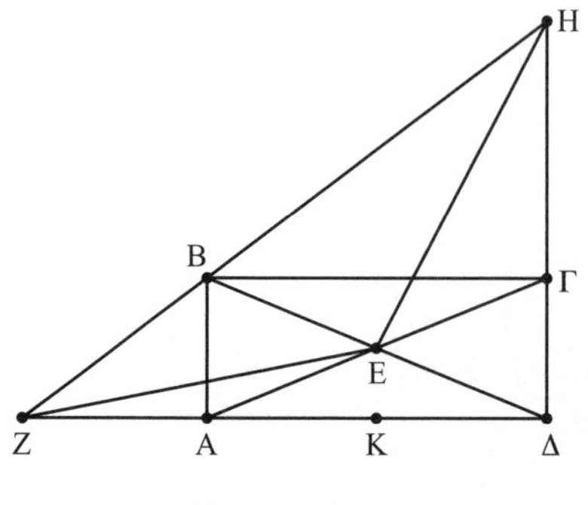

Figure 6.2.2.

$H\Gamma$ to ΓB; therefore, BA will be to AZ as $H\Gamma$ to ΓB; therefore, the two mean proportionals of AB and $B\Gamma$ are AZ and ΓH.

6.3 Heron, *Metrica*

The *Metrica* is Heron's major work on mensuration. Book 1 shows how to measure various plane figures, while book 2 gives rules for finding volumes of certain solid figures. Book 3 then deals with the division of both plane and solid figures into parts having given ratios to one another, thus reminding us of Euclid's own work on this subject. Heron does not generally give proofs, except for his proof of what is known as Heron's formula for determining the area of a triangle, given the lengths of the sides. But he does give arguments justifying his rules and also provides a few examples. He gives more examples for area calculations in the *Geometrica*, a work that is probably based on some of Heron's material but that was put together long after his death. Similarly, for volume calculations, there are further examples in the *Stereometrica*, but it has been speculated that this work too was written after Heron's lifetime. In what follows, the excerpts from the *Geometrica* and the *Stereometrica* are presented in a different type face. Note that in many of his discussions Heron uses the phrase "something is given," which reminds us of Euclid's *Data*. That is, he often first shows that with certain measurements known, other measurements are "given," that is, can be determined. He often follows this with a numerical example that shows exactly how to determine the desired measurement. It seems clear that Heron was writing this work for students to understand how to calculate areas and volumes—students who, in general, were not too concerned with the theory behind the calculations. Various similar measurement calculations are extant in papyrus fragments found in Egypt, dating from both before and after the probable dates of Heron (we include excerpts from some of these below). So Heron was part of a long tradition of geometrical problem solving.

Book 1

Problem 3. *Let the isosceles triangle $AB\Gamma$ have AB equal to $A\Gamma$ and, first, each of the two equal sides equal to 10 units, and second, $B\Gamma$ equal to 12 units* [figure 6.3.1]. *To find the area.*

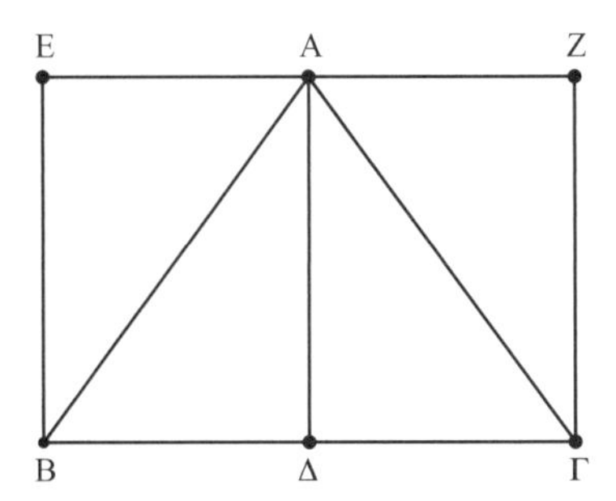

Figure 6.3.1.

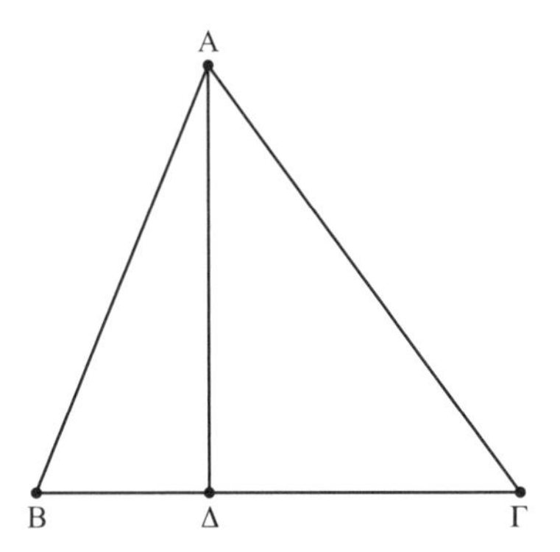

Figure 6.3.2.

Let line $A\Delta$ be drawn perpendicular to $B\Gamma$, and let a line EZ parallel to $B\Gamma$ be drawn through A, and also let two lines BE, ΓZ be drawn through B, Γ parallel to $A\Delta$. The parallelogram $B\Gamma EZ$ is then double the triangle $AB\Gamma$, since it has the same base and is in the same parallels. And since the triangle is isosceles and $A\Delta$ has been drawn perpendicular to the base, $B\Delta$ is equal to $\Delta\Gamma$. But $B\Gamma$ is 12 units, so $B\Delta$ is therefore 6 units. Since AB is 10 units, $A\Delta$ is therefore 8 units, since the square on AB is equal to the squares on $B\Delta$, ΔA. Also, BE will be 8 units, and $B\Gamma$ is 12 units. So the area of parallelogram $B\Gamma EZ$ is therefore 96 units. So the area of triangle $AB\Gamma$ is 48 units.

The method is as follows. Take half of 12; the result is 6. Multiply the 10 by itself, giving 100. Multiply the 6 by itself, which gives 36. Then 64 remains; take its root; the result is 8. And that is the length of the perpendicular $A\Delta$. Multiply 12 by 8; the result is 96. Then take half; the result is 48. And that will be the area of the triangle.

In problems 5 and 6, Heron calculates the area of triangles by use of *Elements* 2:12 and 2:13, the underlying technique we know today as the law of cosines.

Problem 5. *Let the acute angled triangle $AB\Gamma$ have one side AB of 13 units, another side $B\Gamma$ of 14 units, and $A\Gamma$ of 15 units* [figure 6.3.2]. *Find the area.*

It is manifest that the angle at B is acute, since the square on $A\Gamma$ is smaller than the sum of the squares on AB, $B\Gamma$. Let line $A\Delta$ be drawn perpendicular to $B\Gamma$. The square

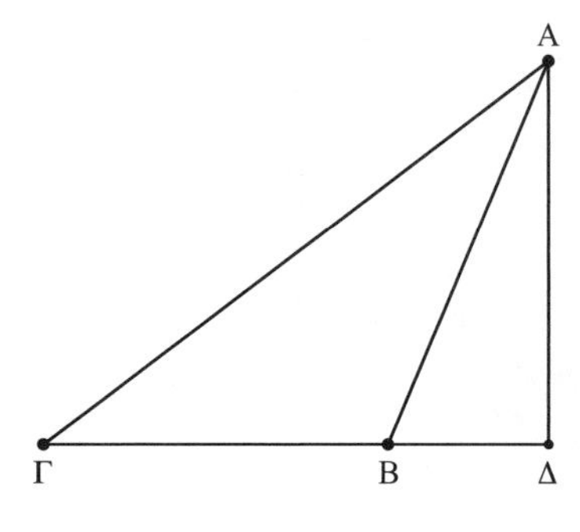

Figure 6.3.3.

on $A\Gamma$ is thus less than those on AB, $B\Gamma$ by twice the rectangle contained by ΓB, $B\Delta$, since this was demonstrated [*Elements* 2:13]. The sum of the squares on AB, $B\Gamma$ is 365 units, and the square on $A\Gamma$ is 225 units. So twice the rectangle contained by ΓB, $B\Delta$ is 140 units. The rectangle contained by ΓB, $B\Delta$ is thus 70 units, and $B\Gamma$ is 14 units. $B\Delta$ will therefore be 5 units. And since the square on AB is equal to those on $A\Delta$, ΔB, and since the square on AB is 169 units and that on $B\Delta$ is 25 units, so the square on $A\Delta$ is thus 144 units. So $A\Delta$ itself will be 12 units. But also $B\Gamma$ is 14 units. So the rectangle contained by $B\Gamma$, $A\Delta$ will thus be 168 units. This is double the triangle $AB\Gamma$. So triangle $AB\Gamma$ will be 84 units.

And this is the method. Multiply 13 by itself; there results 169. And 14 by itself; there results 196. And 15 by itself; there results 225. Add the 169 and the 196; the sum is 365. From that, subtract the 225; the remainder is 140. From that take its half; the result is 70. Apply to the 14 and there results 5. And multiply the 13 by itself, there results 169. From this subtract the 5 by itself; 144 remains. From this, take the side and there results 12. This will be the perpendicular. Multiply this by 14; the result is 168. From this, take half: 84. And this is the area.

Problem 6. *Let the obtuse-angled triangle $AB\Gamma$ have one side AB of 13 units, a second side $B\Gamma$ of 11 units, and $A\Gamma$ of 20 units* [figure 6.3.3]. *Find the perpendicular and the area.*

Prolong $B\Gamma$ until it meets the line $A\Delta$ which is perpendicular to it. The square on $A\Gamma$ is then greater than the squares on AB, $B\Gamma$ by twice the rectangle contained by ΓB, $B\Delta$ [*Elements* 2:12]. The square on $A\Gamma$ is 400 units, that on $B\Gamma$ is 121 units, and that on BA is 169. Twice the rectangle contained by ΓB, $B\Delta$ is therefore 110 units. The rectangle contained by ΓB, $B\Delta$ is therefore 55 units, and $B\Gamma$ is 11 units, so $B\Delta$ will therefore be 5 units. But also AB is 13 units. So $A\Delta$ will be 12 units. But also, $B\Gamma$ is 11 units. So the rectangle contained by $A\Delta$, $B\Gamma$ will be 132 units. This is double the triangle $AB\Gamma$. So triangle $AB\Gamma$ will therefore be 66 units.

And the method is as follows. Multiply 13 by itself; there results 169. And the 11 by itself; there results 121. And the 20 by itself; there results 400. Add the 169 and the 121; the sum is 290. Subtract this from the 400. 110 remains. Take half of this. The result is 55. Apply the 11; there results 5. And the 13 by itself is 169. Subtract the 5 by itself; 144 remains. From this, find its side; the result is 12. The perpendicular will be 12 units. Multiply this by the 11. The result is 132. Of this, take the half: 66. And this will be the area of the rectangle.

Heron's proof of his rule for finding the area of a triangle, given the lengths of the three sides, is unusual in Greek mathematics in that it uses the fourth power of a line segment, a seemingly "ungeometrical" concept. Nevertheless, the proof is clear. There is some indication that this proof is due ultimately to Archimedes, but it is not contained in any of his extant works. Note also that, here, as in subsequent results, Heron claims at the end of the proof that the desired area is "given." This is used in the sense of the same word in Euclid's *Data* and simply means that the area is determined. As part of his discussion, Heron presents a method for determining the square root of a nonsquare integer.

Problem 8. *There is a general method for finding the area of any triangle, the three sides being given, without utilizing the height.*

For example, if the sides of a triangle are 7, 8, and 9 units. Add the 7, 8, and 9; there results 24. Take half of this; there results 12. Subtract the 7 units, and 5 remains. Again, from the 12, subtract the 8, and 4 remains. Again, the 9, and 3 remains. Multiply the 12 by the 5, giving 60; multiply this by the 4, giving 240; and multiply again by the 3, giving 720. From this, take the square root, and it will be the area of the triangle. Since 720 does not have a rational square root, we shall approximate the root as follows. Since the nearest square to 720 is 729, and this has 27 as its root, divide the 720 by 27. The result is $26\frac{2}{3}$. Add 27; the result is $53\frac{2}{3}$. Take half of this; there results $26\frac{1}{2}\frac{1}{3}[=26\frac{5}{6}]$.[7] Thus the square root of 720 will be very nearly $26\frac{1}{2}\frac{1}{3}$. In fact, if we multiply $26\frac{1}{2}\frac{1}{3}$ by itself, there results $720\frac{1}{36}$, with the difference being $\frac{1}{36}$. If we wish to make the difference less than $\frac{1}{36}$, instead of 729 we take the value $720\frac{1}{36}$, and by the same method, we will find an approximation differing by much less than $\frac{1}{36}$.

The geometrical proof of this is as follows: The sides of a triangle being given, to find the area. Since if a perpendicular is drawn, its length is determined, so it is certainly possible to find the area of the triangle, but we want to determine the area without using the perpendicular. So let $AB\Gamma$ be the triangle, and let each of AB, $B\Gamma$, ΓA be given; to find the area. Let the circle ΔEZ be inscribed in the triangle with center H [*Elements* 4:4], and let AH, BH, ΓH, ΔH, EH, ZH be joined [figure 6.3.4].

Then the rectangle contained by $B\Gamma$, EH is double the triangle $BH\Gamma$; the rectangle contained by ΓA, ZH is double the triangle $A\Gamma H$, and that contained by AB, ΔH is double the triangle ABH. Therefore the rectangle contained by the perimeter of triangle $AB\Gamma$ and EH—that is, the radius of the circle ΔEZ—is double the triangle $AB\Gamma$. Let ΓB be produced, and let $B\Theta$ be made equal to $A\Delta$. Then $\Gamma B\Theta$ is half of the perimeter of triangle $AB\Gamma$—because $A\Delta$ is equal to AZ, ΔB is equal to BE, and $Z\Gamma$ is equal to ΓE—so the rectangle contained by $\Gamma\Theta$, EH is equal to triangle $AB\Gamma$. But the rectangle contained by $\Gamma\Theta$, EH is the square root of the square on $\Gamma\Theta$ by that on EH. It follows that the area of triangle $AB\Gamma$ multiplied by itself is equal to the square on $\Gamma\Theta$ by the square on EH.

[7] We use the notation $26\frac{1}{2}\frac{1}{3}$, as in the original, to denote $26+\frac{1}{2}+\frac{1}{3}$. This use of unit fractions was common through the Renaissance.

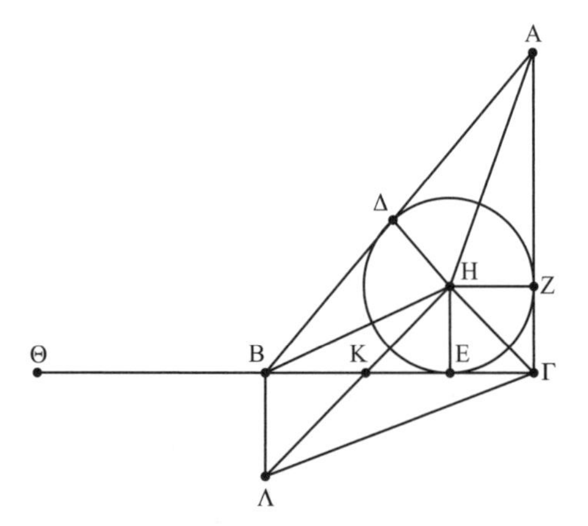

Figure 6.3.4.

Next, let line $H\Lambda$ be drawn at right angles to ΓH, and $B\Lambda$ at right angles to ΓB, and let $\Gamma\Lambda$ be joined. Then, since each of the two angles $\Gamma H\Lambda$, $\Gamma B\Lambda$ is right, a circle can be circumscribed about the quadrilateral $\Gamma HB\Lambda$. Therefore the angles ΓHB, $\Gamma\Lambda B$ are equal to two right angles. But the angles ΓHB, $AH\Delta$ are also equal to two right angles—because the angles at H are bisected by the lines AH, BH, ΓH, and the angles ΓHB, $AH\Delta$ are equal to $AH\Gamma$, ΔHB and so they are all in total equal to four right angles; therefore angle $AH\Delta$ is equal to angle $\Gamma\Lambda B$. Also, the right angle $A\Delta H$ is equal to the right angle $\Gamma B\Lambda$; so the triangle $AH\Delta$ is similar to the triangle $\Gamma B\Lambda$. Therefore, as $B\Gamma$ is to $B\Lambda$, so is $A\Delta$ to ΔH, that is, as $B\Theta$ is to EH. Therefore, *alternando*, as ΓB is to $B\Theta$, so is $B\Lambda$ to EH, that is, as BK is to KE, because $B\Lambda$ is parallel to EH. Then *componendo* as $\Gamma\Theta$ is to $B\Theta$, so is BE to EK. Therefore as the square on $\Gamma\Theta$ is to the rectangle contained by $\Gamma\Theta$, ΘB, so is the rectangle contained by BE, $E\Gamma$ to the rectangle contained by ΓE, EK, or to the square on EH. For in a right triangle, the line EH has been drawn from the right angle perpendicular to the base. Therefore the square on $\Gamma\Theta$ by the square on EH, whose square root is the area of triangle $AB\Gamma$, will be equal to the rectangle contained by $\Gamma\Theta$, ΘB by the rectangle contained by ΓE, EB. And each of $\Gamma\Theta$, ΘB, BE, ΓE is given; for $\Gamma\Theta$ is half of the perimeter of triangle $AB\Gamma$, while $B\Theta$ is the excess of half of the perimeter over ΓB, BE is the excess of half of the perimeter over $A\Gamma$, and $E\Gamma$ is the excess of half of the perimeter over AB, since $E\Gamma$ is equal to ΓZ, and $B\Theta$ to AZ, which is equal to $A\Delta$. Therefore the area of the triangle $AB\Gamma$ is given.

Then this will be synthesized as follows: Let the side AB be 13 units, the side $B\Gamma$ be 14 units, and $A\Gamma$ be 15 units. Add the 13, 14, and 15, and there results 42. Take half of that, which is 21. Then subtract the 13, so 8 remains; subtract the 14, so 7 remains; and subtract the 15, to 6 remains. Then multiply the 21 by the 8, and the result by 7, and that result by 6. The product is 7056. Of that, take the square root; the result is 84. So that is the area of the triangle.

Problem 11. *Given an isosceles trapezoid $AB\Gamma\Delta$ having AB equal to $\Gamma\Delta$, and such that each of those two lines are 13 units, with $A\Delta$ being 6 units and $B\Gamma$ being 16 units, to find the area and the perpendicular* [figure 6.3.5].

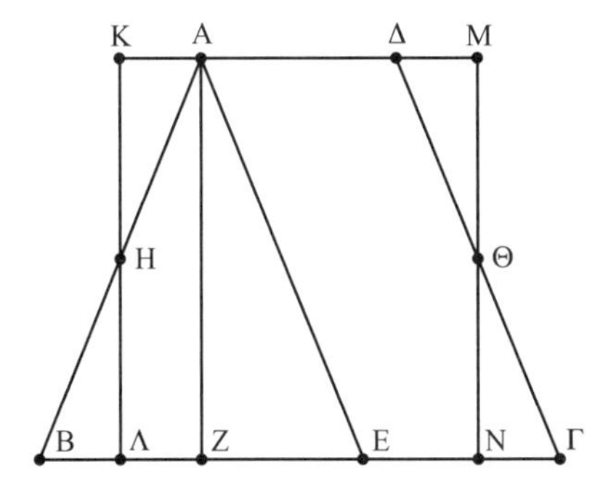

Figure 6.3.5.

Let the line AE be drawn parallel to $\Gamma\Delta$ and the line AZ be drawn perpendicular to $B\Gamma$. Then $AE\Gamma\Delta$ is a parallelogram. Then $A\Delta$ is equal to $E\Gamma$, and $\Gamma\Delta$ to AE, with AE being 13 units and $E\Gamma$ being 6 units. The remainder BE will therefore be 10 units. Also, since triangle ABE is isosceles having each side given, the perpendicular AZ will also be given, and this will be 12 units as has been demonstrated earlier. Let the sides AB, $\Gamma\Delta$ be cut into two equal parts at the points H, Θ, and let the lines $KH\Lambda$, $M\Theta N$ be drawn perpendicular to $B\Gamma$. Then triangle AKH is equal to triangle $BH\Lambda$ and $\Delta M\Theta$ is equal to $\Gamma N\Theta$. Thus if we add the rectilinear six-sided figure $AH\Lambda N\Theta\Delta$ to one of the equal parts and then to the other, we see that the parallelogram $K\Lambda MN$ is equal to the trapezoid $AB\Gamma\Delta$. And since AK is equal to $B\Lambda$ and ΔM to ΓN, the lines AK, ΔM are equal to $B\Lambda$, $N\Gamma$. Since $A\Delta$ added to $AK, \Delta M$ is KM and $B\Gamma$ less $B\Lambda, N\Gamma$ is KM, therefore $A\Delta$, $B\Gamma$ together equal twice KM. But each of $A\Delta$, $B\Gamma$ is given—they are 22 units—so twice KM is also equal to 22 units and KM itself is equal to 11 units. But we know that $K\Lambda$ is 12 units, since it is equal to AZ. Therefore the parallelogram $K\Lambda MN$ will be 132 units, and it is equal to the trapezoid $AB\Gamma\Delta$. So the trapezoid $AB\Gamma\Delta$ will also be 132 units.

As a consequence of this analysis, we have the synthesis. Subtract the 6 from the 16, giving 10; take half, which is 5, and multiply that by itself, resulting in 25. Multiply the 13 by itself, giving 169, and subtract the 25, giving 144. Take its square root, which is 12. The perpendicular will be 12 units. For the area, add the 16 and 6, giving 22. Take half, which is 11, and multiply by the perpendicular, which gives 132. This will be the area.

Problem 13. *Given an obtuse-angled trapezoid $AB\Gamma\Delta$ having the angle at B obtuse, with line AB being 13 units, line $\Gamma\Delta$ beings 20, $A\Gamma$ being 6, and $B\Delta$ being 17, to find the perpendicular and the area* [figure 6.3.6].

Let the line AE be drawn perpendicular to $B\Delta$ (extended) and the line AZ parallel to $\Gamma\Delta$. Then AZ will be 20 units, and $Z\Delta$ 6 units. The remainder BZ will be 11 units. Because the triangle ABZ is obtuse, we have AE equal to 12 units. And, similarly to what precedes, the rectangle contained by $B\Delta$, $A\Gamma$ combined and AE will be double the trapezoid $AB\Gamma\Delta$. So the area of the trapezoid will be 138 units.

And here is the synthesis. Subtract the 6 from the 17, giving 11. And the sides of the obtuse triangle being given, 13, 11, and 20, the perpendicular can be found. It is 12. Then

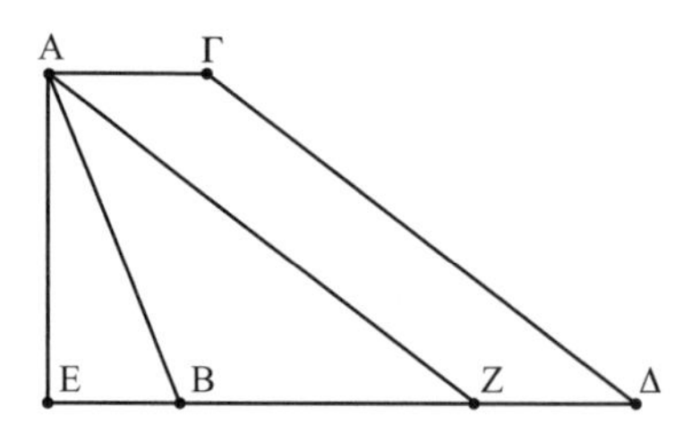

Figure 6.3.6.

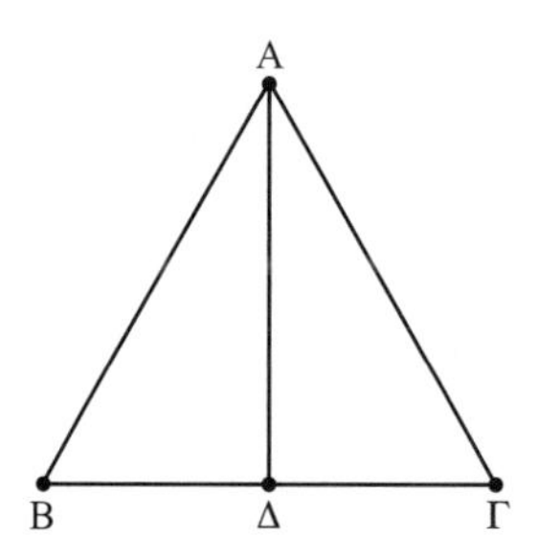

Figure 6.3.7.

add the 17 and 6 to give 23. Take its half, $11\frac{1}{2}$, and multiply by 12. The result is 138. This will be the area of the trapezoid.

The next example gives an unusual algorithm using fourth powers to determine the area A of an equilateral triangle of side a, namely, $A=\sqrt{\frac{3}{16}a^4}$. Of course, this is equivalent to the modern formula of $A=\frac{\sqrt{3}}{4}a^2$, but this algorithm appears in no other ancient source. It is also a bit curious that, in taking the square root of 1875, Heron does not appear to use his method described above. Perhaps that is because that method would give the approximation $43\frac{13}{43}$. Thus Heron evidently chose $\frac{1}{3}$ as the closest simple fraction to $\frac{13}{43}$. But in the example from the *Geometrica*, in boldface, Heron uses a different algorithm, namely, $A=(\frac{1}{3}+\frac{1}{10})a^2$ $[=\frac{13}{30}a^2]$, although this method gives the same value as the first algorithm, at least when Heron uses $\frac{1}{3}$ to approximate $\frac{13}{43}$. And he also calculates the altitude as $h=a-(\frac{1}{10}-\frac{1}{30})a$. In any case, Heron's rational approximation of $\frac{13}{30}=0.4333$ is quite close to $\frac{\sqrt{3}}{4}=0.4330$.

Problem 17. *Given an equilateral triangle with each side equal to 10 units, to find the area.*

Let this be $AB\Gamma$ [figure 6.3.7]. Let line $A\Delta$ be drawn perpendicular to ΓB. Since $B\Gamma$—that is, AB—is double $B\Delta$, the square on AB is quadruple that on $B\Delta$. So the square on $A\Delta$ is triple that on ΔB. But the quadruple of the square on ΔB is that on $B\Gamma$. Thus the square on $B\Gamma$ is $\frac{4}{3}$ of the square on $A\Delta$. Therefore, the square on $B\Gamma$ has a ratio to that on $A\Delta$ equal to the ratio of 4 to 3, and they are all multiplied by that on $B\Gamma$. So the square on $B\Gamma$ multiplied by itself has the same ratio to the square on $A\Delta$ multiplied by the square on $B\Gamma$. Thus the fourth power of $B\Gamma$ has the ratio to the square on $B\Gamma$ multiplied by the square on $A\Delta$ as 4 to 3, that is, as 16 to 12. But the square on $B\Gamma$ multiplied by the square on $A\Delta$ is the rectangle contained by $A\Delta$, $B\Gamma$ multiplied by itself, that is, two triangles multiplied by themselves. So the fourth power of $B\Gamma$ has a ratio to the two triangles multiplied by themselves as the ratio of 16 to 12. But two triangles multiplied by themselves are quadruple one triangle multiplied by itself. So the fourth power of $B\Gamma$ has the ratio to one triangle multiplied by itself as the ratio of 16 to 3. And the fourth power of $B\Gamma$ is given, since $B\Gamma$ is given. So the area of the triangle multiplied by itself is given, and thus the area of the triangle itself is given.

As a consequence of the analysis, the synthesis is as follow. Multiply the 10 by itself, giving 100. Multiply that by itself, giving 10,000. Of that, take $\frac{3}{16}$. The result is 1875. Of that, take the square root, and since it is not rational, it must be approximated as closely as possible, as we have seen. The area is then $43\frac{1}{3}$.

To find the area of a given equilateral triangle. Do this: multiply one of the sides with itself and multiply this result by $\frac{1}{3}+\frac{1}{10}$; this is the area of the equilateral triangle. For example, if each side is 10 units, to find the area do this. Multiply 10 by 10 giving 100; $\frac{1}{3}$ by 100, giving $33\frac{1}{3}$, and then $\frac{1}{10}$ by 100, giving 10. Add $33\frac{1}{3}$ to 10, giving $43\frac{1}{3}$. This many square units is the area of this equilateral triangle. To find the altitude of an equilateral triangle, subtract $\frac{1}{10}+\frac{1}{30}$ of the side, and know that the remainder is the altitude. Multiply then $\frac{1}{2}$ of the base by the altitude, and the product is the area. For example, in an equilateral triangle where each side is 10 units, $\frac{1}{10}$ of a side is 1; $\frac{1}{30}$ of 10 is $\frac{1}{3}$; subtract $1\frac{1}{3}$ from 10, giving $8\frac{2}{3}$; and this is the altitude. To find the area, do it this way: Take $\frac{1}{2}$ of the base, or 5, and multiply by $8\frac{2}{3}$, the altitude, which gives $43\frac{1}{3}$, so that is the area.

We next show Heron's calculation of the area of a regular pentagon in the *Metrica*. This is preceded by a lemma about an angle of $36°$ and followed by another problem from the *Geometrica* using a somewhat simpler algorithm.

Lemma. *Let the right triangle $AB\Gamma$ have the angle at Γ be right and that at A be two-fifths of a right angle* [figure 6.3.8]. *To show that the square on BA, $A\Gamma$ as one line is five times that on $A\Gamma$.*

Let $A\Gamma$ be prolonged to Δ, such that $\Gamma\Delta$ is equal to $A\Gamma$, and let $B\Delta$ be joined. First, AB is equal to $B\Delta$ and angle $AB\Gamma$ is equal to angle $\Gamma B\Delta$. But angle ΓBA is three-fifths of a right angle, since angle $BA\Gamma$ is two-fifths of a right angle, so angle $AB\Delta$ is six-fifths of a right angle. So $AB\Delta$ is an angle of the pentagon, and AB is equal to $B\Delta$. Therefore $A\Delta$ is cut in extreme and mean ratio, with the greater part being [equal to] AB, and $A\Gamma$ is half of $A\Delta$ [*Elements* 13:8]. So the square on BA, $A\Gamma$ taken together is five times the square on $A\Gamma$ [*Elements* 13:1].

Problem 18. *Given the equilateral and equiangular pentagon $AB\Gamma\Delta E$ with each of its sides equal to ten units, to find its area* [figure 6.3.9].

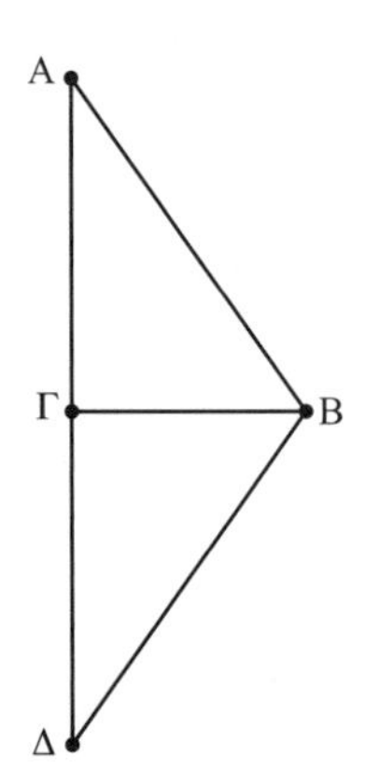

Figure 6.3.8.

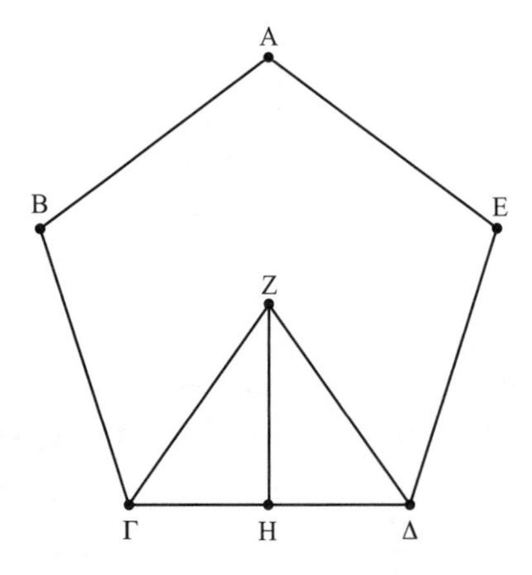

Figure 6.3.9.

Let Z be the center of the circle circumscribing the pentagon, and join ΓZ and $Z\Delta$. Let ZH be drawn perpendicular to $\Gamma\Delta$. Then the angle $\Gamma Z\Delta$ will be four-fifths of a right angle. So angle ΓZH is two-fifths of a right angle, and angle ΓHZ is right. Therefore the square on ΓZ, ZH taken together is five times the square on ZH [*lemma*]. But since it is not the case that one can find a number whose square is five times that of another number, it will be necessary to approximate. The [ratio of the] square 81 to the square 16 [is nearly 5]. So the number of ΓZ, ZH taken together will have a ratio to the number of ZH close to the ratio of 9 to 4; and *separando*, the ratio of the number of ΓZ to that of ZH will be close to that of 5 to 4. Therefore the ratio of the square on ΓZ to that on ZH will be close to 25 to 16; and *separando*, the ratio of the square on ΓH to that on ZH will be nearly 9 to 16. Thus the ratio of ΓH to HZ is about 3 to 4 and therefore the ratio of $\Gamma\Delta$ to ZH will be 6 to 4—that is, 3 to 2. So the ratio of the square on $\Gamma\Delta$ to the rectangle contained by $\Gamma\Delta$, ZH is also 3 to 2. But the square on $\Gamma\Delta$ is given; so the rectangle contained by $\Gamma\Delta$, ZH is also given, and it is double the triangle $\Gamma Z\Delta$. So the triangle $\Gamma Z\Delta$

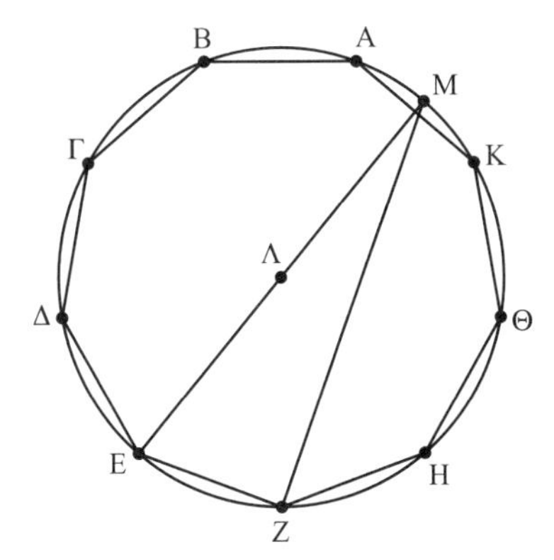

Figure 6.3.10.

is also given, and it is the fifth part of the pentagon $AB\Gamma\Delta E$. Thus the pentagon is also given.

The synthesis is as follows. Multiply the 10 by itself, giving 100. Of that, take a third, $33\frac{1}{3}$. Take that five times; the result if $166\frac{2}{3}$. And this will be the area of the pentagon very nearly.[8] And if one begins with another square that is nearly the quintuple of a square, we can find the area even more precisely.

To find the area of an equilateral pentagon, in which each side is 35 feet. I do it this way: 35 times 35 is 1225; 1225 times 12 is 14700, $\frac{1}{7}$ of 14,700 is 2100.[9] This is the area.

In the calculation of the area of a regular 9-gon, Heron quotes a work on chords, presumably a chord table written before Ptolemy's table in the *Almagest*, perhaps the work of Hipparchus (190–120 BCE). The line ZE in Heron's description would then be the chord of a 40° angle, which Heron here approximates as one-third of the diameter of the circle. Now, it is believed that Hipparchus's chord table calculated chords only at 7.5° intervals. Thus Heron would have needed to interpolate to find the chord of 40°. In any case, the correct ratio of that chord to the diameter of the circle is 0.3420, rather than the 0.3333 that Heron uses.

Problem 22. *Given the equilateral and equiangular 9-gon $AB\Gamma\Delta EZH\Theta K$, with each of its sides equal to ten units, to find its area* [figure 6.3.10].

Let a circle be circumscribed around the 9-gon, with center Λ. Join $E\Lambda$ and prolong it to M [on the circumference] and join MZ. The triangle EZM of the 9-gon is thus given, since it has been demonstrated in the work *On the Chords in a Circle* that ZE is the third

[8] The correct area of the pentagon of side a is $\frac{5\tan 54}{4}a^2 = 1.7205a^2$, since $ZH = \frac{1}{2}\Gamma\Delta\tan 54$. But Heron approximates ZH as $\frac{2}{3}a$, so he finds the area of the pentagon to be $\frac{5}{3}a^2$. Note that the tangent of 54° can be constructed. We find that $\tan 54 = \frac{3+\sqrt{5}}{\sqrt{10+2\sqrt{5}}}$.

[9] Heron here uses the approximation $\frac{12}{7} = 1.7143$ to the correct 1.7205, rather than the simpler fraction $\frac{5}{3} = 1.6667$.

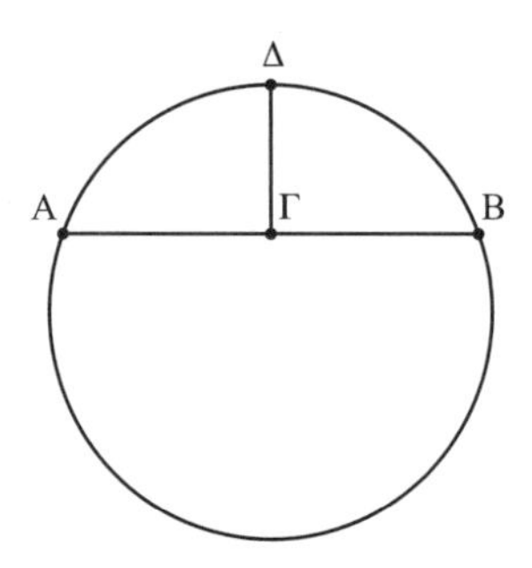

Figure 6.3.11.

part of EM, very nearly. The square on ME is therefore nine times that on EZ, and thus the square on MZ is eight times that on ZE, since the angle at Z, being inscribed in a semicircle, is right. The square on MZ thus has the ratio to that on ZE as 289 is to 36, very nearly. The ratio of MZ to ZE is as 17 is to 6, very nearly; thus the square on EZ has to the triangle EMZ a ratio of 36 to 51, that is, as 12 is to 17. Its ratio to the 9-gon is therefore as 12 is to $76\frac{1}{2}$ [since triangle $E\Lambda Z$ is half of triangle EMZ], that is, as 24 is to 153, or as 8 to 51, and the square on EZ is given. The 9-gon is therefore also given.

The synthesis is as follows. Multiply 10 by itself, giving 100. Multiply by 51, giving 5100. Of that, take the eighth, which is $637\frac{1}{2}$.[10] That will be the area of the 9-gon.

In problems 30 and 31, Heron describes an "ancient" and then a more precise method for finding the area of a segment of a circle. The first method is based on considering the circumference of a circle as three times the diameter, while the second one uses the Archimedean value of $\frac{22}{7}$. However, in neither case does Heron give an example other than the semicircle. We have therefore supplemented these with an example from the *Geometrica* (in boldface) that calculates the area of a different segment, followed by another example in which the perimeter of the segment is calculated.

Problem 30. *As for the segment of a circle smaller than a semicircle* [figure 6.3.11], *the Ancients measured it without too much care; for, adding the base and its perpendicular and taking half of this, they multiplied it by the perpendicular and they declared that the result was the area of the segment. These people seem to be the same as those who declare that the perimeter of the circle is three times the diameter.*

If we measure the semicircle according to such an assumption, it follows that the area of the semicircle is in agreement with that method. For example, let AB be the diameter of a semicircle, and $\Gamma\Delta$ the perpendicular, and let the diameter be 12 units. $\Gamma\Delta$ is then 6 units, and the circumference of the circle will be 36 units. Thus the circumference of the semicircle will be 18 units. But, since it has been demonstrated that the rectangle contained by the circumference and the radius is double the area, it follows that if we multiply the

[10] The actual area of the 9-gon of side length a is $\frac{9a^2}{4}$ tan 70. In this case, the area is 618.18, somewhat less than Heron's value.

18 by 6, and take its half, giving 54 units, then the area of the semicircle, under the given hypothesis, will be 54 units. And it will be the same thing if we add the 12 and the 6, giving 18, then take half, and multiply that by the perpendicular; the result is again 54.

Problem 31. *But those who are looking for greater precision add to the above area of the segment the fourteenth part of the square on half the base. So these people seem to have followed the other method, according to which the circumference of the circle is three times the diameter of the circle and greater by its seventh part.*

So if we suppose similarly, first that the diameter AB is 14 units and that the perpendicular $\Delta\Gamma$ is 7, the circumference of the semicircle will be 22 units; then multiply by the 7 and there results 154. Of this, take half, there results 77. And this is the area of the semicircle. But it will be the same thing if we proceed as follows: Add the 14 and the 7; then take half. The result is $10\frac{1}{2}$. Multiply by the 7; the result is $73\frac{1}{2}$. And the square on half of the base is 49; taking the 14th part, there results $3\frac{1}{2}$. If we add this to $73\frac{1}{2}$, there results 77. It is thus this method that it is necessary to use for segments that are less than a semicircle. On the other hand, this method does not apply for every segment, but only when the base of the segment is not more than triple the perpendicular, since certainly, if the base is 60 units and the perpendicular is 1, the rectangle contained by them will be 60 units, which is without doubt greater than the segment; and the fourteenth part of the square on half the base is also greater than the segment since this is $64\frac{4}{14}$ units. Thus the above method does not apply to every segment, but as has been said, only when the base is not greater than triple the perpendicular.

Given a circular segment smaller than a semicircle, whose base is 16 and height is 6, to find the area. Do it this way: Add the height and the base, which gives 22. Half of 22 is 11, by which you multiply by the height to have 66. Then half of the base is 8, which you square, giving 64, and take the fourteenth part, $4\frac{1}{2}\frac{1}{14}$. Add this to 66, giving $70\frac{1}{2}\frac{1}{14}$. That is the area of the segment.[11]

If you wish to find the perimeter [that is, the length of the arc] of this segment, do it this way: Multiply the base 16 by itself, giving 256. Multiply the height 6 by itself giving 36, which you then multiply by 4, giving 144. Add 256 and 144 to get 400 and take the square root, giving 20. Then take the fourth part of the height 6, giving $1\frac{1}{2}$, and add this to the 20, giving $21\frac{1}{2}$. This is the length of the perimeter.[12]

Two further examples from the *Geometrica* show that Heron could solve a quadratic equation involving lengths and areas.

Given the sum of the diameter, circumference, and area of a circle, to find each of them separately. It is done thus: Let the given sum be 212. Multiply this by 154; the result is 32648. To this add 841, making 33489, whose square root is 183. From this take away 29, leaving 154, whose eleventh part is 14; this will be the diameter of the circle. If you wish to find the circumference, take 29 from 183, leaving 154; double this, making 308, and take the seventh part, which is 44; this will be the

[11] The correct area of the segment is 70.7080, while Heron's value is 70.5714, quite a good approximation.

[12] The correct arc length is 21.45, while Heron's value is 21.5; again, quite a good approximation.

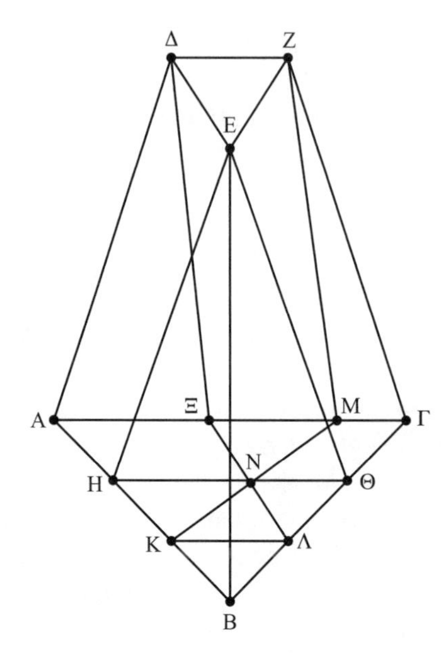

Figure 6.3.12.

circumference. To find the area, it is done thus: Multiply the diameter, 14, by the circumference, 44, making 616; take the fourth part of this, which is 154; this will be the area of the circle. The sum of the three numbers is 212.

Given a square with the sum of its area and perimeter equal to 896 feet, to find the area and perimeter separately. I do it thus: Take half of 4, giving 2; then 2 times 2 is 4, to which you add 896, producing 900. The square root of this is 30 feet. Then take half of 4, giving 2, subtract this from 4, giving 2, and subtract this from 30 giving 28. Then the area is the square of 28, namely, 784, while the perimeter is 112. And 784 + 112 = 896.

Book 2

Problem 6. *To measure a truncated pyramid with a triangular base, in which the upper base is also triangular and similar to the lower base.*

Let the lower base be the triangle $AB\Gamma$, and the upper base the triangle ΔEZ, which is similar to $AB\Gamma$. Let AB equal 18, $B\Gamma$ equal 24, $A\Gamma$ equal 36, ΔE equal 12. Therefore EZ is 16 and ΔZ is 24. Let the perpendicular from triangle ΔEZ to the base be 10 [figure 6.3.12].

Let AH equal ΔE and $\Gamma\Theta$ equal EZ; join line $H\Theta$ and bisect the lines $B\Theta$, BH at the points Λ, K, respectively. Through K draw KM parallel to $B\Gamma$; draw ΛN and extend it to Ξ and connect $K\Delta$. Now since triangles $AB\Gamma$ and ΔEZ are similar, AB is to ΔE as AB is to AH and $B\Gamma$ is to EZ as $B\Gamma$ is to $\Gamma\Theta$. But $A\Gamma$ is parallel to $H\Theta$. And since HK equals KB, and KNM is parallel to $B\Theta$, thus NH equals $N\Theta$. But also $B\Lambda$ equals $\Lambda\Theta$. And $AN\Xi$ is parallel to AB, but also $K\Lambda$ is parallel to $H\Theta$, thus to $A\Gamma$. So $AK\Lambda\Xi$ and $K\Delta\Gamma M$ are parallelograms. They have equal area since they are on the same base and between the same parallels. For the same reason, $HK\Lambda N$ equals $NK\Lambda\Theta$. Therefore the remainder parallelograms $AHN\Xi$ and $N\Theta M\Gamma$ are equal. And since AH equals $N\Xi$ equals ΔE, and $\Gamma\Theta$ equals MN equals EZ and they subtend equal angles, so is EM equal to ΔZ. And since $K\Lambda$ equals $A\Xi$ and also $M\Gamma$, then $A\Xi$ equals $M\Gamma$. Therefore $A\Gamma$, $M\Xi$ together equal $A\Gamma$, ΔZ together, and each sum is twice $\Gamma\Xi$. On the other side, since KB equals KH, so BA, HA taken together equals AB, ΔE together and each sum is twice AK or twice $\Xi\Lambda$. For the same reason, $B\Gamma$, EZ together are twice $\Lambda\Gamma$.

Now the truncated pyramid consists of the wedge, whose base is the parallelogram $AHN\Xi$ and whose top is the line segment ΔE, together with the wedge whose base is the parallelogram $MN\Theta\Gamma$ and whose top is the line EZ, and the prism whose base is the triangle $MN\Xi$ and whose top is triangle ΔEZ, and further the pyramid whose base is triangle $BH\Theta$ and vertex the point E. Now the volumes of the wedges whose bases are the parallelograms $AHN\Xi$ and $N\Theta\Gamma M$ and with the same height as the pyramid are together equal to the product of the area of parallelogram $NM\Gamma\Theta$ with that height. The volume of the prism whose base is triangle $MN\Gamma$ and upper base triangle ΔEZ is equal to the area of triangle $MN\Gamma$ multiplied by the height. Finally, the volume of the pyramid, whose base is triangle $H\Theta B$ and vertex the point E, equals one-third of the product of the area of triangle $BH\Theta$ and the height. But one-third of triangle $BH\Theta$ equals one and one-third of triangle $\Lambda N\Theta$. [So we add triangle $\Lambda N\Theta$ to quadrilateral $\Xi\Gamma\Theta N$, giving triangle $\Xi\Lambda\Gamma$.] The remaining one-third of triangle $\Lambda N\Theta$ equals one-twelfth of triangle $BH\Theta$. Therefore, the volume of the truncated pyramid is equal to the area of triangle $\Xi\Lambda\Gamma$ joined with one-twelfth of the triangle $BH\Theta$ and multiplied by the height.

Now the height is given. We also must show that the triangle $\Xi\Lambda\Gamma$ is given and the twelfth part of the triangle $BH\Theta$. But since the sum of AB and ΔE is given and it has been proved that $\Xi\Lambda$ is half of this, then $\Xi\Lambda$ is given. For the same reason, both $\Lambda\Gamma$ and $\Gamma\Xi$ are given. Therefore, the triangle $\Xi\Lambda\Gamma$ is given. On the other side, since BA and AH are given, also BH is given. For the same reason, $B\Theta$ is given. Furthermore, since $A\Gamma$ and $M\Xi$ are given, so is the sum of $A\Xi$ and $M\Gamma$ given, therefore $H\Theta$. So the triangle $H\Theta B$ is given, and therefore its twelfth.

We now calculate as follows: $(18+12)/2=15$; $(24+16)/2=20$; $(36+24)/2=30$. [These are the sides $\Xi\Lambda$, $\Lambda\Gamma$, $\Gamma\Xi$, respectively.] Now the area of the triangle with sides 15, 20, and 30 must be calculated. This is, as we learned earlier, approximately $131\frac{1}{4}$.[13] Furthermore, $18-12=6$, $24-16=8$, $36-24=12$. [These are the sides BH, $B\Theta$, $H\Theta$, respectively.] So we need the area of a triangle with sides 6, 8, and 12. As we have learned earlier, this is approximately 21.[14] And $\frac{1}{12}$ of this is $1\frac{1}{2}+\frac{1}{4}$. Adding this to $131\frac{1}{4}$ gives 133. We multiply this with the height, and that is the volume of the truncated pyramid $AB\Gamma\Delta EZ$.

We can easily derive from Heron's discussion here an algorithm to determine the volume of the frustum of a pyramid based on equilateral triangles. In fact, if the side of the lower triangle base is a_1, and the side of the upper triangle base is a_2, then the triangle equivalent to $\Xi\Lambda\Gamma$ has side $\frac{a_1+a_2}{2}$, while the side of the triangle equivalent to

[13] Using Heron's formula, the area of this triangle is 133.317, a bit more than Heron claims.

[14] Again using Heron's formula, the area is 21.33.

$BH\Theta$ has side $a_1 - a_2$. We also know that the area of an equilateral triangle of side s is $\frac{\sqrt{3}}{4}s^2$. It follows that the volume of the frustum, assuming its height is h, is

$$\left[\frac{\sqrt{3}}{4}\left(\frac{a_1+a_2}{2}\right)^2 + \frac{1}{12}\frac{\sqrt{3}}{4}(a_1-a_2)^2\right]h = \left[\frac{\sqrt{3}}{4}\left(\frac{a_1+a_2}{2}\right)^2 + \frac{1}{3}\frac{\sqrt{3}}{4}\left(\frac{a_1-a_2}{2}\right)^2\right]h.$$

And by a similar, but simpler, argument, we can see that a frustum of a pyramid whose top base and lower base are squares can be broken up into four wedges, one prism, and four pyramids, thus producing for the volume the algorithm expressed as

$$2a_2\frac{a_1-a_2}{2}h + a_2^2h + \frac{4}{3}\left(\frac{a_1-a_2}{2}\right)^2 h = \left[\left(\frac{a_1+a_2}{2}\right)^2 + \frac{1}{3}\left(\frac{a_1-a_2}{2}\right)^2\right]h,$$

with a_1, a_2 being the sides of the lower and upper square bases, respectively, and h the height. This result, in the simpler form $V = \frac{h}{3}(a_1^2 + a_1a_2 + a_2^2)$, was used in the Egyptian *Moscow Mathematical Papyrus* dating back to about 1850 BCE to calculate the volume of such a frustum. An analogous formula works for the volume of the frustum of a cone. In either case, if one is given the slant height ℓ of the frustum rather than the perpendicular height, we can find the latter via the Pythagorean theorem since $\ell^2 = h^2 + (\frac{a_1-a_2}{2})^2$, where in the case of a cone, a_1, a_2 are the diameters of the bottom and top bases, respectively. We next consider three problems from the *Stereometrica* illustrating the calculations of the volume of a frustum of a square pyramid, as well as the frustum of a cone.

Given the frustum of a pyramid with a square base, with each side of the base being 10 feet, each side of the upper base being 2 feet, and the slant height being 9 feet; to find the perpendicular height and the volume. I do it this way: 10 feet of the base less 2 of the upper base is 8 feet. Then the square on 8 is 64, of which half is 32. The square on the slant height of 9 is 81. Subtract 32 from 81 giving 49, the side of which is 7 feet. That is the perpendicular height. Since the perpendicular height is 7 feet, we find the volume in the following way: The sum of the upper base of 2 feet with the lower base of 10 feet is 12 feet. Half of 12 is 6, and the square on 6 is 36. Furthermore, 10 less 2 feet of the top base is 8 feet; half of 8 is 4, whose square is 16. One-third of 16 is $5\frac{1}{3}$ feet. Add this to the 36, giving $41\frac{1}{3}$ feet. Multiplying $41\frac{1}{3}$ by the 7 feet of the perpendicular gives $289\frac{1}{3}$ feet. That much is the volume of the frustum of the pyramid.

In the case of a cone, one needs the area of the circular bases. Heron uses the Archimedean value $\frac{22}{7}$ for π, which implies that the area of a circle of diameter d is $\frac{11}{14}d^2$. The two examples below show two different but equivalent versions of the algorithm to calculate the volume of the frustum of a cone.

To find the volume of the frustum of a cone from the diameters and the perpendicular. Suppose the diameter of the greater circle is 6 feet, of the smaller circle 2 feet, and suppose the perpendicular distance is 4 feet. I do it so: Find the square of 6, which is 36; the square of 2, which is 4; the sum of 36 and 4, which is 40; the product

of 6 and 2, which is 12; and the sum of 12 and 40, which is 52. Then the product of 52 by the perpendicular 4 gives 208. Then 11 times 208 is 2288, and $\frac{1}{42}$ of 2288 is $54\frac{20}{42} = 54\ \frac{1}{3}\ \frac{1}{7}$. So much is the volume.

We measure another frustum of a cone using the diameters and the slant height. Assume the diameter of the upper base is 4 feet, the slant height 15 feet, and the diameter of the lower base 28 feet. To find the volume, we proceed as follows: I subtract the upper diameter from the lower diameter, giving 24. Then half of 24 is 12 and the square of 12 is 144. I multiply the length of the slant height by itself, giving 225. Then 225 less 144 is 81, whose side is 9. This will be the perpendicular. The volume we will then find as follows: Add the diameters of the two bases, giving 32; half of 32 is 16. Then find the area of a circle whose diameter is 16 feet, by squaring the diameter, multiplying by 11, and dividing by 14. Once I have found that area, I subtract the upper diameter from the lower one, giving 24; take half of 24 to get 12. Then I find the area of this smaller circle, and when I have found that area, I take one-third of it. This I add to the area of the larger circle and multiply the result by the perpendicular. The volume of the frustum of the cone will result.

Problem 14. *To measure a segment of a cylinder formed by cutting it by a plane through the center of one of the bases.*

Suppose that the diameter of the base is 7 units, and the height of the segment is 20 units. Archimedes has demonstrated in the *Method* that such a segment is the sixth part of the solid parallelepiped whose square base is circumscribed around the base of the cylinder and has the same height as the segment. But the volume of the parallelepiped is given, so the volume of the segment of the cylinder is also given. Therefore it is necessary to multiply 7 by itself, and then multiply by the height. There results 980. One sixth of this is $163\frac{1}{3}$. This will be the volume of the segment of the cylinder.

Problems 16–19 calculate the volume of the regular solids, other than the cube. We just consider the tetrahedron and the dodecahedron.

Problem 16. *To measure the volume of a [regular] pyramid whose base is the triangle $AB\Gamma$ and whose vertex is the point Δ, and in which each side is 12 units* [figure 6.3.13].

Let the center of the circle circumscribed about triangle $AB\Gamma$ be E, and let ΔE, $E\Gamma$ by joined. The square on $B\Gamma$, that is, on $\Gamma\Delta$, is triple of that of ΓE (*Elements* 13:12). That on ΔE is therefore two-thirds of that on $\Gamma\Delta$. Since the square on $\Gamma\Delta$ is 144 units, that on ΔE is 96 units. And ΔE itself is a bit less than $9\frac{1}{2}\frac{1}{3}$.[15] Thus, since each of AB, $B\Gamma$, $\Gamma\Delta$ is given, and also the perpendicular ΔE is given, the volume of the pyramid is also given. To complete this, as we have seen earlier, multiply the surface of the equilateral triangle by $9\frac{1}{2}\frac{1}{3}$ and take one-third of the result. This will be the volume of the pyramid.

Problem 19. *To measure [the volume of] a [regular] dodecahedron in which each side is 10 units* [figure 6.3.14].

[15] The actual value is $4\sqrt{6} = 9.7980$.

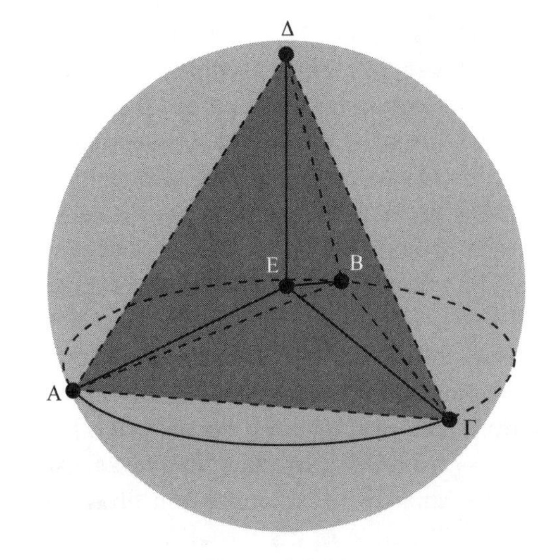

Figure 6.3.13.

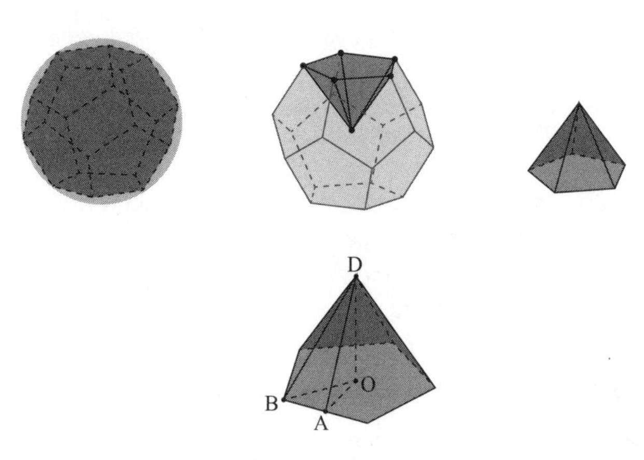

Figure 6.3.14.

We conceive of lines joining the center of the sphere [that circumscribes the dodecahedron] to every angle of one of the pentagons. There will be 12 pyramids having pentagonal bases with their vertex at the center of the sphere. But the side of the pentagon, in relation to the line drawn from the center of the sphere perpendicular to the pentagon, has a ratio of 8 to 9.[16] But the side of the pentagon is 10 units. Thus the perpendicular will be $11\frac{1}{4}$

[16] If p [$=DO$] is the line drawn from the center of the sphere perpendicular to the pentagon, then $p^2=R^2-r^2$, where $R=DB$ is the radius of the sphere and $r=BO$ is the radius of the circle circumscribing the pentagon. According to the porism to *Elements* 13:17, the side of the dodecahedron is equal to the greater segment of the side of the cube inscribed in the same sphere when that side is cut in extreme and mean ratio. Given that the square on the edge of the cube is $\frac{1}{3}$ the square on the diameter of the sphere, one can calculate that if a is the edge of the dodecahedron, then $R=\frac{a}{4}\sqrt{3}(1+\sqrt{5})$. Also, from book 1, problem 18 above, we can calculate that $r=a\frac{\sqrt{50+10\sqrt{5}}}{10}$. It follows that $p=\frac{1}{2}a\sqrt{\frac{25+11\sqrt{5}}{10}}$. Thus the ratio of p to a is $\frac{1}{2}\sqrt{\frac{25+11\sqrt{5}}{10}}=1.1135$ instead of Heron's value of $\frac{9}{8}=1.125$.

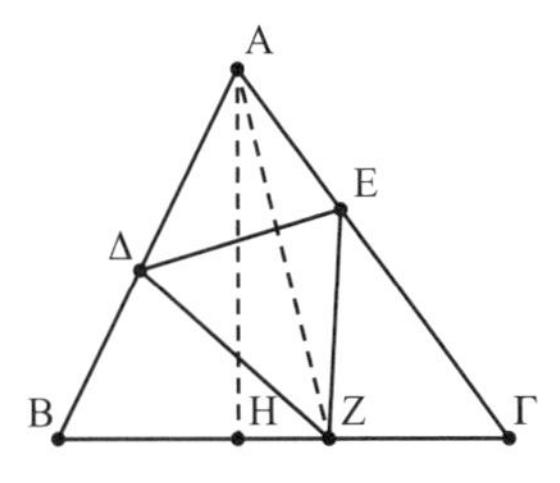

Figure 6.3.15.

units.[17] We then take the surface of the pentagon [book 1, problem 18] and multiply it by the perpendicular and then take one-third of the result. This will be the volume of one of the pyramids. If we multiply this by 12, we find the volume of the dodecahedron.[18]

Book 3

As in his proof of the triangle formula, in the following problem Heron also uses products of areas by areas.

Problem 4. *A triangle $AB\Gamma$ being given, to cut out a triangle ΔEZ [where Δ, E, Z are points on the sides] given in magnitude, such that the triangles that remain, $A\Delta E$, $B\Delta Z$, ΓEZ, are equal to each other* [figure 6.3.15].

If the sides AB, $B\Gamma$, ΓA are cut at the points Δ, Z, E such that the ratio of $A\Delta$ to $B\Delta$ equals that of BZ to $Z\Gamma$ and of ΓE to EA, then the triangles $A\Delta E$, $B\Delta Z$, and $Z\Gamma E$ will be equal to each other. Let AZ be joined. Since as BZ is to $Z\Gamma$, so is ΓE to EA, *componendo*, as $B\Gamma$ is to $Z\Gamma$, so is ΓA to AE. As the triangle $AB\Gamma$ is to $AZ\Gamma$, so is $AZ\Gamma$ to AZE; and *convertendo*, as triangle $AB\Gamma$ is to ABZ, so is $AZ\Gamma$ to $EZ\Gamma$, which is given. But triangle $AB\Gamma$ is also given; thus so is the area of $AB\Gamma$ multiplied by the area of $ZE\Gamma$; and it is equal to the area of triangle ABZ by the area of $AZ\Gamma$. The area of ABZ by the area of $AZ\Gamma$ is thus also given. But, with a line AH having been drawn perpendicular, the rectangle contained by ZB, AH is double the area of triangle ABZ and also the area of triangle $AZ\Gamma$ is double the rectangle contained by $Z\Gamma$, AH. Thus the rectangle contained by ZB, AH by the rectangle contained by AH, $Z\Gamma$ is given, that is, the square on AH by the rectangle contained by BZ, $Z\Gamma$; and that on AH is given, so that the rectangle on BZ, $Z\Gamma$ is also given; and $B\Gamma$ is given, so Z is given. The ratio of $B\Gamma$ to ΓZ is given, so also is that of ΓA to AE. And ΓA is given, so E is also given. For the same reasons also Δ is given. So ΔE, EZ, $Z\Delta$ are given in position.

As a consequence of the analysis, here is the synthesis. Suppose AB is 13 units, $B\Gamma$ 14 units, and ΓA 15 units. The triangle $AB\Gamma$ is therefore 84 units [in area]. But the triangle ΔEZ will be 24 units. So the triangles $A\Delta E$, ΔBZ, $EZ\Gamma$ remaining will each be 20 units. Multiply the 84 by the 20, which gives 1680, then multiply by 4, giving 6720. And since the

[17] The correct value is 11.135.

[18] Since the area of a pentagon of edge length 10 is 172.05, the volume of the dodecahedron is $4 \cdot 11.135 \cdot 172.05 = 7663$. Heron's calculation gives a volume of $4 \cdot 11.25 \cdot 166.67 = 7500$, in error by about 2.2%.

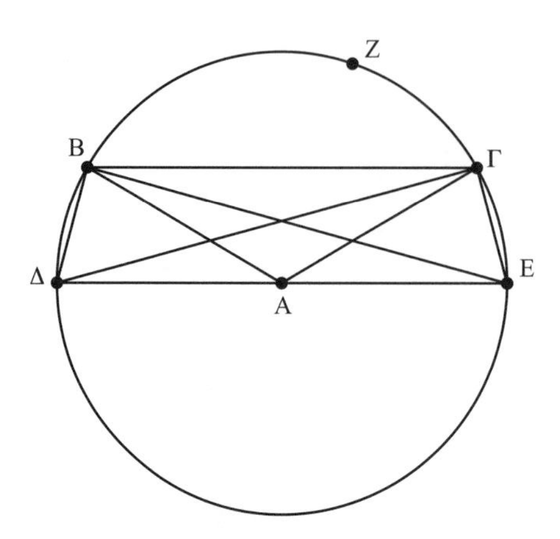

Figure 6.3.16.

perpendicular AH is 12 units, multiply it by itself, giving 144. Divide the 6720 by 144; the result is $46\frac{2}{3}$. But $B\Gamma$ is 14 units. So BZ will be, approximately, $8\frac{1}{2}$ units and $Z\Gamma$ will be $5\frac{1}{2}$ units. But as 14 is to $5\frac{1}{2}$, so is 15 to a certain other number. This will be $5\frac{25}{28}$. Again, as 14 is to $5\frac{1}{2}$, so is 13 to a certain other number. This will be $5\frac{3}{28}$. The result is that $B\Delta$ is $5\frac{3}{28}$.

Problem 18. *To divide a circle into three equal parts by two straight lines.*

Since it is evident that the problem is not rational, for practical convenience, we will solve it as exactly as possible, as follows. Let the circle be given, with center A, and let there be inscribed in it an equilateral triangle with side $B\Gamma$, and let the line ΔAE be drawn parallel to that side, and let $B\Delta$, $\Delta\Gamma$ be joined [figure 6.3.16].

I say that the segment $\Delta B\Gamma$ is, very nearly, a third part of the entire circle. Let BA, $A\Gamma$ be joined. The sector $ABZ\Gamma A$ is a third part of the whole circle. But triangle $AB\Gamma$ is equal to triangle $B\Gamma\Delta$. So the figure $\Delta BZ\Gamma$ is also a third part of the entire circle. And the excess of segment $\Delta B\Gamma$ over it is negligible relative to the entire circle.[19] Similarly, in inscribing another side of the equilateral triangle, we can cut out another third part. Thus that which is left will also be a third part of the entire circle.

6.4 Babylonian Tablet BM 34568

This Late Babylonian tablet probably dates from the second century BCE. Although its problems dealing with rectangles are mostly similar to those from Old Babylonia, the methods of solution are different. In particular, they no longer rely on "cut and paste" techniques so common in tablets from the second millennium BCE. Instead, they generally seem to rely on looking at appropriate diagrams, and just seeing the procedure that gives the result, or else using known geometric relationships, even without a diagram. As we note regarding other sources from Egypt and Rome, the methods and even the problems themselves are repeated in works dating from the

[19] The excess of segment $\Delta B\Gamma$ over $\frac{1}{3}$ of the circle is the area of the segment with base $B\Delta$. That area is $\frac{\pi-3}{12}r^2 = 0.0118r^2$, which is indeed negligible with respect to the entire circle of area πr^2.

next several centuries. For each of the following problems, we present a proposed explanation of the scribe's procedure. To help with understanding the fairly literal translation from the tablets, note that *accumulate* means "add," as does *join*; *steps of* means "times." and *lift* means "subtract."

Problem 3. The diagonal and the length I have accumulated: 9. 3 the width. What is the length and the diagonal? Since you do not know, 9 steps of 9, 81, and 3 steps of 3, 9. 9 from 81 you lift, remaining 72. 72 steps of $\frac{1}{2}$ you go: 36. 9 steps of what may I go so that 36? 9 steps of 4 you go; 36. 4 the length. 4 from 9 you lift, remaining 5. 5 the diagonal.

In this case, the scribe apparently understands, first, that the square on the sum of the diagonal and the length equals the square on the diagonal plus the square on the length plus twice the rectangle spanned by the diagonal and length, and second, that the square on the width is the difference between the square on the diagonal and the square on the length. It follows that subtracting the second square from the first gives twice the rectangle on the diagonal and the length plus twice the square on the length. Therefore, half of this total is the sum of the rectangle on the diagonal and the length plus the square on the length. Since this latter sum can be thought of as the rectangle on the length and the sum of the diagonal and the length, it follows that dividing by the sum of the diagonal and the length produces the length.

Problem 9. The length and the width I have accumulated: 14, and 48 the surface. Since you do not know, 14 steps of 14, 196. 48 steps of 4, 192. 192 from 196 you lift, remaining 4. What steps of what may I go so that 4? 2 steps of 2, 4. 2 from 14 you lift, remaining 12. 12 steps of $\frac{1}{2}$, 6. 6 the width. 2 to 6 you join: 8. 8 the length.

The procedure for this problem is easily shown in figure 6.4.1, where the square on the sum of the length and width less four times the surface is the square on the difference of the length and the width.

Problem 10. The length and the width I have accumulated: 23, and 17 the diagonal. Since you do not know, 23 steps of 23, 529. 17 steps of 17, 289. 289 from 529 you lift, remaining 240. 240 steps of 2, 480; 480 from 529 you lift, remaining 49. What steps of what may I go so that 49? 7 steps of 7, 49. 7 from 23 you lift, remaining 16. 16 steps of $\frac{1}{2}$ go, 8. 8 the width. 7 to 8 you join: 15. 15 the length.

In this case, the procedure follows from figure 6.4.2, where first, the square on the sum of the length and the width less the square on the diagonal is twice the area of the rectangle, and second, when we subtract four times the rectangle from the square on the sum of the length and the width, we are left with the square on the difference of the length and the width. Subtracting that difference from the sum gives twice the width. Note that with a slight change in figure 6.4.1, the combining of the two figures makes the Pythagorean theorem relationship among the diagonal, length, and width of a rectangle virtually obvious. This relationship was understood in the solution to problem 3.

Problem 13. The diagonal and length I have accumulated: 9. The diagonal and the width I have accumulated: 8. What is the width? Since you do not know, 9 steps of 9, 81. 8 steps of 8, 64. 81 and 64 you accumulate: 145. 1 from 145 you lift: remaining 144. What

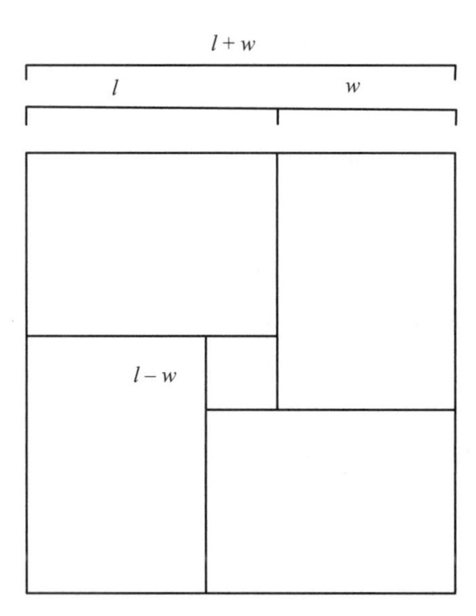

Figure 6.4.1.

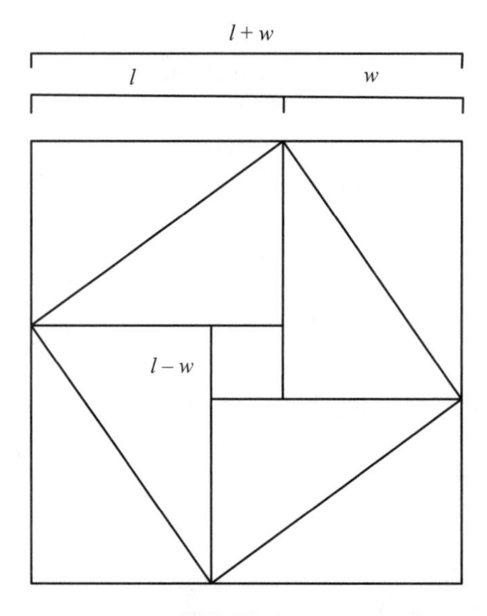

Figure 6.4.2.

steps of what may I go so that 144? 12 steps of 12, 144. 9 from 12 you lift: remaining 3, the width. 3 from 8 you lift: remaining 5, the diagonal. 5 from 9 you lift, remaining 4 the length.

Problem 14. The length, the width, and the diagonal I have accumulated: 70, and 420 the surface. What as much is the length, the width, and the diagonal? Since you do not know, 70 steps of 70, 4900. 420 steps of 2 you go, 840. 840 from 4900 lift: remaining 4060. 4060 steps of $\frac{1}{2}$ you go: 2030. 70 steps of what may I go so that 2030? 70 steps of 29, 2030. 29 the diagonal.

Problem 13 depends on the relationship expressed algebraically as

$$(\ell+w+d)^2=(\ell+d)^2+(w+d)^2-(\ell-w)^2,$$

while problem 14 depends on

$$(\ell+w+d)^2=2(\ell w+\ell d+dw+d^2).$$

How these relationships were developed is not clear, since there is no obvious geometric way to display them. In particular, note that the "1" in problem 13 ("1 from 145 you lift: remaining 144") represents the square of $\ell-w$.

6.5 *Papyrus Cairo*

The following geometric problems are from two Egyptian papyri written in Demotic script that are also discussed in chapter 2. All but the last problem are from *Papyrus Cairo J. E. 89127-30, 89137-43*, dated to the time of Ptolemy II Philadelphus, who ruled in Egypt from 286 to 243 BCE. These problems are similar to problems that appear on Babylonian tablets, both from two millennia earlier and from the time period of these papyri. However, the solutions here seem to be somewhat more detailed. The scribes also use approximate formulas that are found in the work of Heron, for example. And, in general, they give checks of the solution. The numbering of the problems is by Richard Parker, who was the first to translate and publish these.

Problem 30. A pole, the number of whose foot moved outward is 6 cubits and the number of whose peak will not be stated to you. If the number of lowering its top is 2 cubits, what was its peak? You shall reckon 6, 6 times: result 36. You shall reckon 2, 2 times: result 4. You shall add 36 to 4: result 40. You shall add 2 to 2: result 4. Take it to 40: result 10. The peak of the pole was 10.

The length h of the pole is the solution to the equation $h^2-(h-2)^2=6^2$. This reduces to $4h-4=36$, and the scribe proceeds to solve this equation in the usual way.

Problem 32. As for a piece of land that amounts to 100 square cubits that is square, if it is said to you: "Cause that it make a piece of land that amounts to 100 square cubits that is round," what is the diameter? Here is its plan. You shall add the $\frac{1}{3}$ of 100 to it: result $133\frac{1}{3}$. Cause that it reduce to its square root: result $11\frac{1}{2}\ \frac{1}{20}$. You shall say: "$11\frac{1}{2}\ \frac{1}{20}$ is the diameter of the piece of land that amounts to 100 square cubits." Here is its plan. You shall reckon $11\frac{1}{2}\ \frac{1}{20}$ 3 times: result $34\frac{1}{2}\ \frac{1}{10}\ \frac{1}{20}$. It is its circumference. Its $\frac{1}{3}$ is $11\frac{1}{2}\ \frac{1}{20}$. Its $\frac{1}{4}$ is $8\frac{1}{2}\ \frac{1}{10}\ \frac{1}{20}\ \frac{1}{120}$. You shall reckon $11\frac{1}{2}\ \frac{1}{20}$ $8\frac{1}{2}\ \frac{1}{10}\ \frac{1}{20}\ \frac{1}{120}$ times: result 100 square cubits again.

The idea is determine the diameter of a circular piece of land with an area of 100 square cubits. For the scribe, the diameter is obtained by taking the square root of the area plus its one-third: $d=\sqrt{A+\frac{A}{3}}$. He proves the result by the formula $A=\frac{C}{4}\cdot\frac{C}{3}$, where the circumference C is equal to $3d$—the relationship between circumference and diameter used extensively in ancient Mesopotamia. Note also the extensive use of unit fractions, the classical Egyptian method of expressing fractions.

Problem 34. A plot of land that amounts to 60 square cubits that is rectangular, the diagonal being 13 cubits. Now how many cubits does it make to a side? You shall reckon 13, 13 times: result 169. You shall reckon 60, 2 times: result 120. You shall add it to 169: result 289. Cause that it reduce to its square root: result 17. You shall take the excess of 169 against 120: result 49. Cause that it reduce to its square root: result 7. Subtract it from 17: remainder 10. You shall take to it $\frac{1}{2}$: result 5. It is the width. Subtract 5 from 17: remainder 12. It is the height. You shall say: "Now the plot of land is 12 cubits by 5 cubits." To cause that you know it. You shall reckon 12, 12 times: result 144. You shall reckon 5, 5 times: result 25. Result 169. Cause that it reduce to its square root: result 13. It is its diagonal of the plot.

The method of solution for problem 34 stems from Mesopotamia. The scribe here, as in Babylonian Tablet BM 34568, is using the geometric argument that, first, the square on the diagonal added to twice the rectangle is the square on the sum of the height and the width, and second, the square on the diagonal less twice the rectangle is the square on the difference of the height and the width. The next problem uses the same procedure, but the solution no longer consists of integers.

Problem 35. Another plot of land that amounts to 60 square cubits, whose diagonal is 15. What is its height and its width? You shall reckon 15, 15 times: result 225. You shall reckon 60, 2 times: result 120. Add the result to its companion number in them: result 345. Cause that they reduce to their square root: result $18\frac{1}{2}\ \frac{1}{12}$. You shall take the excess of 225 against 120: result 105. Cause that it reduce to is square root: result $10\frac{1}{4}$. You shall subtract it from $18\frac{1}{2}\ \frac{1}{12}$: remainder $8\frac{1}{3}$. Their half is $4\frac{1}{6}$. It is the width. Subtract $4\frac{1}{6}$ from $18\frac{1}{2}\ \frac{1}{12}$: remainder $14\frac{1}{3}\ \frac{1}{12}$. It is the height. To cause that you know it: You shall reckon $14\frac{1}{3}\ \frac{1}{12}$, $14\frac{1}{3}\ \frac{1}{12}$ times: result $207\frac{5}{6}$. You shall reckon $4\frac{1}{6}$, $4\frac{1}{6}$ times: result $17\frac{1}{6}$. Result 225. Cause that it reduce to its square root: result 15 again.

The scribe has cleverly used approximations in his final check on the result. In fact, although the square of $14\frac{1}{3}\ \frac{1}{12}$ is actually $207\frac{5}{6}\ \frac{1}{144}$, while the square of $4\frac{1}{6}$ is $17\frac{1}{3}\ \frac{1}{36}$, the scribe rounded off to make sure that the sum of the two squares is exactly 225.

Problem 36. A circular plot of land—if a large triangle is its middle, which is 12 divine cubits on each side, what is the area of the plot of land? Here is its plan. You shall say: "Now the plot makes 4 pieces of land. They are the 1 bare triangle and the 3 triangle segments also [figure 6.5.1]. Here is the area of the bare triangle."

You shall reckon 12, 12 times: result 144. You shall reckon 6, 6 times: result 36. Subtract it from 144: remainder 108. Cause that it reduce to its square root: result $10\frac{1}{3}\ \frac{1}{20}\ \frac{1}{120}$. It is the middle height of the bare triangle. The number of the base is 12 cubits. Their half is 6. Its middle height is $10\frac{1}{3}\ \frac{1}{20}\ \frac{1}{120}$. You shall reckon it 6 times: result $62\frac{1}{3}\ \frac{1}{60}$ cubits. It is the area of the triangle. You shall take $\frac{1}{3}$ to $10\frac{1}{3}\ \frac{1}{20}\ \frac{1}{120}$: result $3\frac{1}{3}\ \frac{1}{10}\ \frac{1}{60}\ \frac{1}{120}\ \frac{1}{180}$. It is the diameter of the triangle segment. Their measurements. Here is its plan. You shall add $3\frac{1}{3}\ \frac{1}{10}\ \frac{1}{60}\ \frac{1}{120}\ \frac{1}{180}$ to 12: result $15\frac{1}{3}\ \frac{1}{10}\ \frac{1}{60}\ \frac{1}{120}\ \frac{1}{180}$. You shall take to it one-half: result

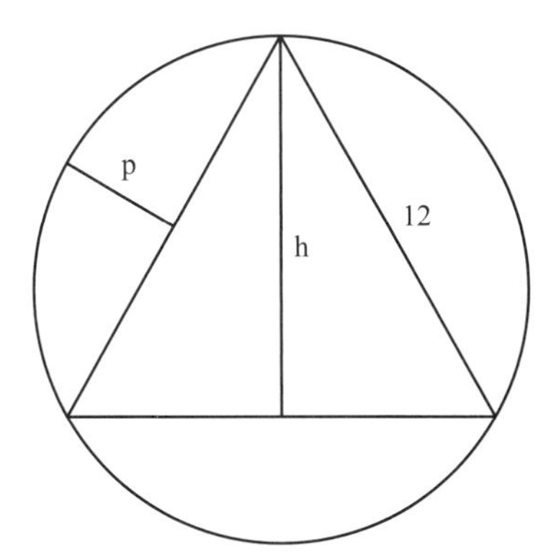

Figure 6.5.1.

$7\frac{2}{3}\ \frac{1}{20}\ \frac{1}{120}\ \frac{1}{240}\ \frac{1}{360}$. You shall reckon it $3\frac{1}{3}\ \frac{1}{10}\ \frac{1}{60}\ \frac{1}{120}\ \frac{1}{180}$ times: result $26\frac{5}{6}\ \frac{1}{10}$. You shall reckon it 3 times: result $80\frac{2}{3}\ \frac{1}{10}\ \frac{1}{30}$. It is the area of the three triangle segments. You shall add $80\frac{2}{3}\ \frac{1}{10}\ \frac{1}{30}$ to $62\frac{1}{3}\ \frac{1}{60}$: result $143\frac{1}{10}\ \frac{1}{20}$ square cubits. It is the area of the entire plot of land. To cause that you know it: the diameter of the bare triangle is $10\frac{1}{3}\ \frac{1}{20}\ \frac{1}{120}$ divine cubits. You shall add the diameter of the triangle segment, which is $3\frac{1}{3}\ \frac{1}{10}\ \frac{1}{60}\ \frac{1}{120}\ \frac{1}{180}$ to it: result $13\frac{5}{6}\ \frac{1}{45}$. It is the diameter of the circle rounded. You shall reckon it 3 times in order to know its circumference: result $41\frac{1}{2}\ \frac{1}{15}$. Its $\frac{1}{3}$ is $13\frac{5}{6}\ \frac{1}{45}$. Its $\frac{1}{4}$ is $10\frac{1}{3}\ \frac{1}{20}\ \frac{1}{120}$. You shall reckon $13\frac{5}{6}\ \frac{1}{45}$, $10\frac{1}{3}\ \frac{1}{20}\ \frac{1}{120}$ times: result $143\frac{5}{6}\ \frac{1}{10}\ \frac{1}{30}$. Excess $\frac{2}{3}\ \frac{1}{10}\ \frac{1}{20}$. It is the error in them of squaring.

The scribe determined the height (h) of the triangle and multiplied that by half the base ($\frac{b}{2}$) to get the area of the triangle. He next noted that the height (p) of a segment was $\frac{1}{3}$ of the height of the triangle and then found the area A of the segment by using the formula also used by Heron: $A = \frac{p+b}{2} \cdot p$. He finally multiplied that value by 3 and added the result to the area of the triangle to get the area of the circle. He checked this by calculating the area again, using the Mesopotamian result that the area of a circle is $\frac{C^2}{12}$, where the circumference C is equal to $3d$, the diameter d in this case being equal to $h+p$. The results differ by only a small amount. Surprisingly, perhaps, the scribe did not realize that, assuming $C=3d$, the area of the circle is exactly $b^2=144$.

Problem 40. Another pyramid that is 10 cubits by 10 cubits. Now it measures 10 at all its pieces [i.e., its base is a square of 10 to a side and its height is also 10]. Here is its plan. You shall reckon 10, 10 times: result 100. You shall reckon 100, 10 times: result 1,000. Its $\frac{1}{3}$ is the number $333\frac{1}{3}$ cubits. It is its number of cubic cubits.

The scribe used the well-known formula that the volume of a pyramid is $\frac{1}{3}$ times the area of the base times the height.

Problem 42 is from *Papyrus BM 10399*, which is Ptolemaic, probably somewhat later than *Papyrus Cairo*.

42. A mast . . . which is 100 divine cubits in length, which is 3 divine cubits in diameter at its foot, which is 1 divine cubit in diameter at its top, is encased in copper. If it happens that the mast is brought out of the casing and water is measured into the casing, how much water will it take? Also, what is the number of cubic cubits of the mast? The way of doing it: You shall add one for the diameter of its top to 3 for the diameter of its foot: result 4. You shall take its half: result 2. You shall reckon 2 two times: result 4. You shall take its $\frac{1}{4}$, whatever happened about the diameter. Then subtract their $\frac{1}{4}$ from them: remainder 3. You shall reckon 100 to 3 times because of the length of the mast, which is 100 divine cubits: result 300. It is the number of cubic cubits of the mast. The way of finding the number of *hins* that will go into the cubic cubit, 1 by 1, its depth 1. You know the number of palms that will go into the cubit of Thoth, amounting to 7. You shall reckon 7 to 7 times: result 49. You shall reckon 49 to 7 times: result 343. You shall say: "The *hin* is 1 palm by 1, its depth 1." You shall say: "Customarily 343 *hins* go into a cubic cubit 1 by 1, its depth 1." You shall reckon 343 to 300 times, which is the number of cubic cubits of the mast: result 102,900. The number of *hins* of water that will go into the casing of copper and that will go into the mast. Here is its plan amounting to *hins* of water 102,900.

The mast and its copper casing are both truncated cones. The scribe assumed that the volume of a truncated cone was equal to that of a cylinder whose diameter was the average of the upper and lower diameters of the cone. The calculation of the amount of water this would hold was then straightforward.

6.6 Columella, *Of Husbandry*

Born in what is now Spain, Lucius Columella (4–70 CE) served in the Roman army and then spent the rest of his life farming his estates in Latium. He is best known for his twelve-volume work on agriculture in the Roman Empire. It deals with many ideas important for farming, including soils, olive trees, cattle, sheep, chickens, and personnel management. For our purposes, a chapter in book 5 deals with the measurement of land in various shapes, and for each Columella produces an example of the procedure for calculating the area. In particular, he uses the method found in Heron's *Metrica* to calculate the area of a circular segment. At the beginning of this chapter, Columella records the various units of measurement common in the empire, and three in particular: the *jugerum* of 28,800 square feet, the *scrupulum* of 100 square feet, or $\frac{1}{288}$ of a *jugerum*, and the *uncia* of 2400 square feet, which equals 24 *scrupuli*, or $\frac{1}{12}$ of a *jugerum*.

Book 5, Chapter 2

On the Several Forms of Lands, and of Their Dimensions

All land is either square, or long, or shaped like a wedge, or triangular, or round, or exhibits the form of a semicircle, or of an arch of a circle, and sometimes also of several angles. The measuring of a square is very easy; for, seeing it is of the like number of feet on all sides, two sides are multiplied into one another; and what sum arises from the multiplication, that we call the number of square feet contained in it. As, if a place were a hundred feet every way, we multiply one hundred into a hundred, and they make ten thousand; therefore we will say, that that place has ten thousand square feet, which make a third part, and a seventy-second part of a *jugerum*; according to which proportion, we must make the computation and payment of any work done.

But if it be longer than it is broad, as for example, let the form of the *jugerum* have 240 feet in length, and 120 in breadth, as I said a little before; you shall multiply the feet in breadth with the feet in length thus: 120 times 240 amount to 28,800. We shall say, that a *jugerum* of land contains so many feet square. You shall proceed in like manner with all lands whose length is greater than their breadth.

But if the land be in the form of a wedge, suppose it be 100 feet long, and 20 feet broad on one part, and on the other, 10 feet; then we will add together the two breadths, and both sums will make 30 feet. The half of this sum is 15, by multiplying which with the length, we will make 1500 feet. We shall therefore say, that in this wedge, these are the square feet which will make one half uncia and three *scrupuli*, that is, $\frac{1}{24}$ part, and $\frac{3}{288}$ parts of a *jugerum*.

The standard ancient rule for calculating the area of a quadrilateral is to find the average of each pair of opposite sides and then multiply them together. In this case, since both of the longer sides of the wedge are the same 100 feet, their average is also 100. The average of the two short sides is 15, and so the area is calculated to be 1500 square feet. If we think of the wedge as an isosceles trapezoid, we would multiply the average, 15, of the two short sides by the height. In this case, the height is only a bit short of 100, namely, 99.875 feet, so Columella's answer is essentially correct.

But if you are to measure a triangle with three equal sides, you shall follow this method. Let the land be triangular, of 300 feet every way; multiply this number into itself, it will make 90,000 feet; take the third part of this sum, 30,000. Take also the tenth part, 9,000. Add both sums together; they will make 39,000. We will say, that this is the sum of the square feet in this triangle; which measure makes 1 *jugerum* and one-third of a *jugerum*, and the forty-eighth part of a *jugerum*.

The area of an equilateral triangle of side s is $\frac{\sqrt{3}}{4}s^2$. Columella calculates this area as $(\frac{1}{3}+\frac{1}{10})s^2$. Since $\frac{1}{3}+\frac{1}{10}=0.4333$, while $\frac{\sqrt{3}}{4}=0.4330$, Columella's method is quite accurate. We have seen this same approximation in other extracts as well.

But if the triangular land has unequal sides, as in the figure subjoined, which has a right angle, the computation shall be ordered and made in a different manner. Let the line of the one side, which makes the right angle, be 50 feet; and of the other, 100 feet. Multiply these two sums into one another. 50 times 100 make 5,000; the half of these makes 2,500, which part makes an *uncia* and a *scrupulum*, or $\frac{1}{12}$ part and 1/288 part of a *jugerum*.

The "figure subjoined" is just a right triangle, and the area is found correctly by taking half of the product of the two legs.

If the land shall be round, so as to have the appearance of a circle, take the feet thus: Let there be a round area, whose diameter contains 70 feet. Multiply this into itself; 70 times 70 make 4,900; multiply this sum by 11, they make 53,900 feet. I subtract [take] the fourteenth part of this sum, 3,850 feet. These I say are the square feet in this circle, which make an *uncia* and a half and $2\frac{1}{2}$ *scrupuli* of a *jugerum*.

Columella uses the standard Archimedean approximation for π, namely, $\frac{22}{7}$. Thus the area of circle of diameter d is calculated as $\pi\frac{d^2}{4}=\frac{11}{14}d^2$. In the next problem, he calculates the area of a semicircle as $\frac{11}{14}\frac{d^2}{2}$.

If the land be semicircular, whose base has 140 feet, and the breadth of the curvature 70 feet, you must multiply the breadth with the base: 70 times 140 makes 9,800. These multiplied by 11 make 107,800. The fourteenth part of this sum makes 7,700. We shall say, that these are the number of feet in this semicircle, which make three *unciae* and five *scrupuli*, i.e., $\frac{1}{4}$ part and $\frac{5}{288}$ parts of a *jugerum*.

But if it be less than a semicircle, we will measure the arch after this manner: Let there be an arch, whose base contains 16 feet, and the breadth 4 feet. I add the breadth to the base; both make 20 feet. These I multiply by four,[20] and they make 80 feet; the half of these is 40. Also the half of 16 feet, which make the base, is 8; these 8, multiplied into themselves, make 64; from these I take a fourteenth part, which makes 4 feet, and a little more. This you shall add to 40; both sums make 44. I say, that these are the square feet in that arch, which make half a *scrupulum*, i.e., $\frac{1}{576}$ part of a *jugerum*, less a twenty-fifth part of a *scrupulum*.

Columella's calculation of the area of a circular segment follows Heron's rule for this in the case where the base is not greater than triple the breadth, even though in this problem that condition is not satisfied. Heron's formula for the area A of a circular segment of base c and breadth h is $A=\frac{1}{2}(c+h)h+\frac{1}{14}(\frac{c}{2})^2$ [figure 6.6.1]. A modern calculation of the area would first find the radius of the circle as $R=\frac{1}{2h}[(\frac{c}{2})^2+h^2]=\frac{8^2+4^2}{8}=10$. Next, the angle θ of the segment is $2\cos^{-1}(\frac{R-h}{R})=2\cos^{-1}(.6)=1.8546$. Finally, the length of the corresponding arc is $s=R\theta=18.546$. The area A of the segment is then the area of the sector less the isosceles triangle formed by two radii and the base of the segment, or

$$A=\frac{1}{2}(Rs-c(R-h))=\frac{1}{2}(10\cdot 18.546-16\cdot 6)=44.73.$$

So Heron's formula is a rather good approximation, especially if one notes that $\frac{1}{14}\cdot 64=4.57$ is certainly more than a "little more" than 4. This would make the actual total from the formula 44.57, or about one-third of 1 percent too small.

If the land has six angles, it is reduced into square feet thus: Let there be a hexagon, with lines of 30 feet every way. I multiply one side into itself; 30 times 30 make 900. Of this sum I take a third part, 300, and a tenth part of the same, 90, which make 390. This must be multiplied by 6, because there are 6 sides, which, being reduced to one sum, make 2,340. Therefore we shall say, that there are so many square feet therein. Therefore, it will be an *uncia* of a *jugerum*, less half a *scrupulum*, and the tenth part of a *scrupulum*.

[20] It is not clear in this description where the "four" comes from. From Heron's formula for a circular segment, this "four" must be the breadth, rather than just a "four."

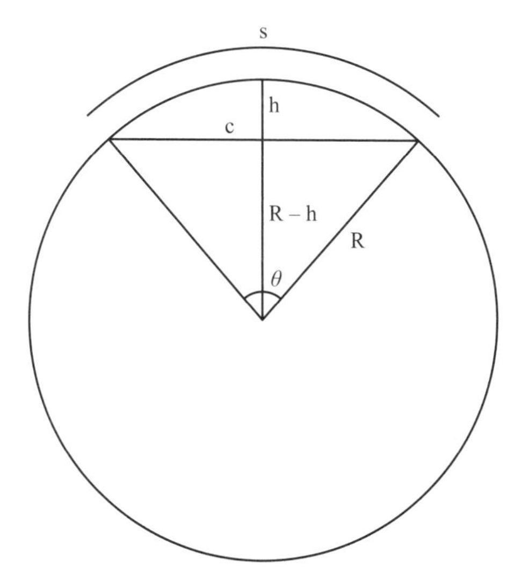

Figure 6.6.1.

Columella here calculates the area of a regular hexagon by calculating the area of one of the six equilateral triangles making it up as before and then multiplying by six.

6.7 *Ayer Papyrus*

The *Ayer Papyrus* is a Greek papyrus that was brought from Egypt to the Field Museum in Chicago by Mr. Ed. E. Ayer in 1895. He bought it from an Egyptian antiquities dealer, who said that it had been found at the Hawara archaeological site near the Fayum oasis, about 100 km southwest of Cairo. The papyrus is only a fragment, originally part of a longer papyrus roll. There are parts of seven problems on the papyrus. However, only a few words of the first, second, and fifth problems are readable, so these are left out in what follows. The third problem has only the final two lines, but since it has a diagram, it has a reasonable reconstruction (shown below in brackets). The fourth problem has text but no diagram, but again, it is not difficult to reconstruct what that would have been, especially since the sixth and seventh problems have both text and diagrams. The diagrams included below are, however, not the original ones, as they were rather rough and not drawn to scale. In particular, they included numbers for all parts of the diagrams, both what is assumed and what is to be found. To make the diagrams easier to follow, we have redrawn them but included only the given numerical values.

Although there is no positive information on the papyrus that would indicate a date, the forms of the letters and the seemingly close relationship of the problems to ones found in Heron's work have led scholars to believe that it was written in the first or

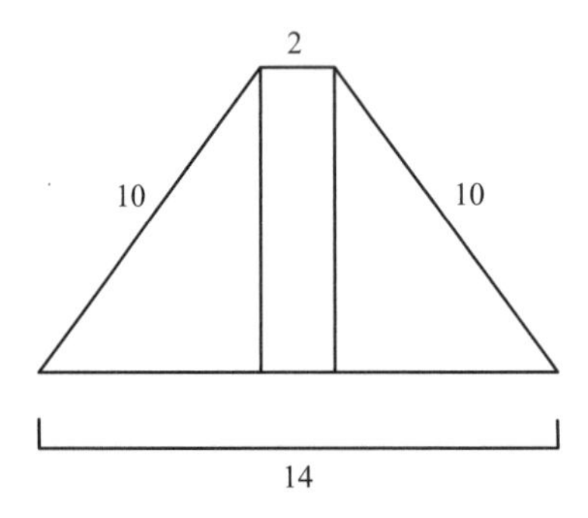

Figure 6.7.1.

early second century CE. The work was apparently a practical treatise on mensuration, giving methods for surveying farmlands in Egypt after the yearly Nile floods.

Problem 3. [If there be given an isosceles trapezoid such as the one drawn below, according to the conditions of the problem, the 10 squared is 100, and the 2 of the upper side from the 14 of the base leaves 12, $\frac{1}{2}$ of which is 6. This squared equals 36. Subtract this from 100; the remainder is 64, of which the square root is 8, which is the length of the perpendicular; $\frac{1}{2}$ of this is 4, and this by the 6 of the base equals 24; of so many acres is each of the right-angled triangles. And the 8 of the perpendicular by the 2 of the base is 16; of so many] acres is the rectangle in it. Altogether, 64 acres. And the figure will be as follows [figure 6.7.1]:

Problem 4. If there be given a scalene trapezoid such as the one drawn below, according to the conditions of the problem, the 13 squared is 169, and the 15 squared is 225. Subtract the 169; the remainder is 56. Subtract the 2 of the upper side from the 16 of the base; the remainder is 14. Take $\frac{1}{14}$ of 56; it is 4. This from the 14 of the base leaves 10; $\frac{1}{2}$ of this is 5. This squared is 25. Subtract this from the 169; the remainder is 144, of which the square root is 12, which is the length of the perpendicular. This by the 5 of the base is 60, $\frac{1}{2}$ of which is 30; of so many acres is each of the right-angled triangles. And the 12 by the 2 of the upper side is 24; of so many acres is the rectangle in it. And the 12 multiplied by the 4 of the base is 48, $\frac{1}{2}$ of which is 24; of so many acres is the obtuse-angled triangle in it. Altogether, 108 acres. And the figure will be as follows [figure 6.7.2].

Although the author here presents an algorithm that solves the problem, they do not indicate why they take any particular steps. The simplest explanation seems to be the following: The step beginning "subtract the 2 of the upper side" collapses the inner rectangle, thus turning the trapezoid into a 13-14-15 triangle. Using figure 6.7.3, we see that the altitude of this triangle is a leg in two separate right triangles. So by the Pythagorean theorem, $15^2-(14-q)^2=13^2-q^2$, which simplifies to $15^2-13^2=14^2-2\cdot 14q$ or $56=14(14-2q)$. Then, as detailed in the description above, $14-2q=4$, $2q=10$, and $q=5$. At this point, the author uses the Pythagorean theorem to calculate the perpendicular as $\sqrt{13^2-5^2}=\sqrt{144}=12$. The areas of the two right triangles, the rectangle, and the obtuse triangle are then easily calculated.

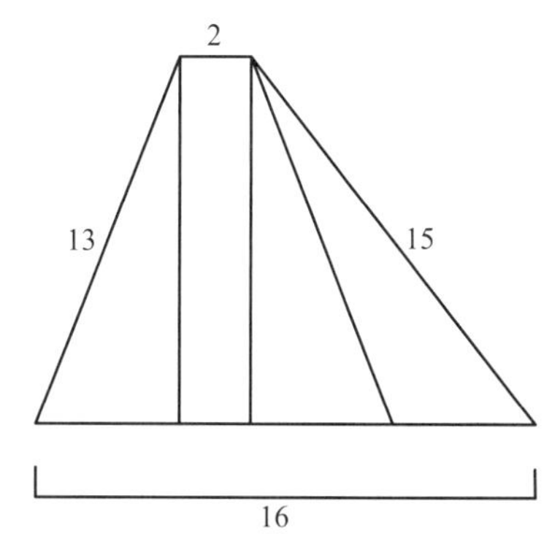

Figure 6.7.2.

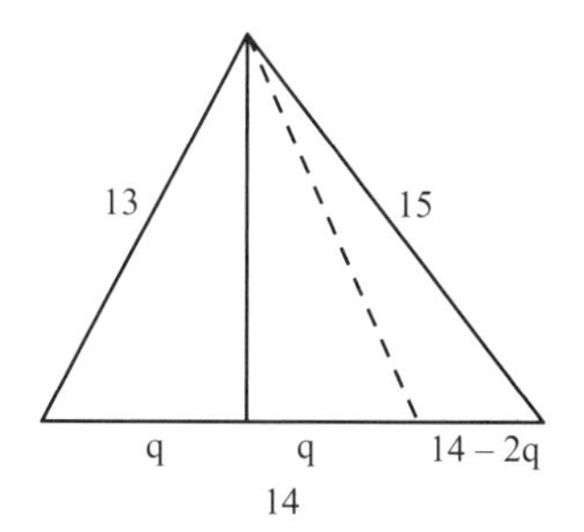

Figure 6.7.3.

Problem 6. If there be given a parallelogram such as the one drawn below, according to the conditions of the problem, the 13 of the side squared is 169, and the 15 of the side squared is 225. Subtract 169 from this; the remainder is 56. Subtract the 6 of the base from the 10 of the upper side; the remainder is 4. Take $\frac{1}{4}$ of 56; it is 14. Subtract the 4; the remainder is 10, $\frac{1}{2}$ of which is 5, which is the length of the base of the right-angled triangle. This squared is 25. And the 13 squared is 169. Subtract the 25; the remainder is 144, the square root of which is 12, which is the length of the perpendicular. And subtract the 5 from the 6 of the base; the remainder is 1. The 1 from the 10 of the upper side leaves 9, which is the length of the remainder of the upper base [which remainder is the base] of the right-angled triangle. And the 12 of the perpendicular by the 5 of the base is 60, $\frac{1}{2}$ of which is 30; of so many acres is the right-angled triangle in it. And the 12 by the 1 is 12; of so many acres is the rectangle in it. And 12 by the 9 of the base is 108, $\frac{1}{2}$ of which is 54; of so many acres is the other right-angled triangle. Altogether, 96 acres. And the figure will be as follows [figure 6.7.4].

First, note that the author uses the word *parallelogram* to mean a quadrilateral with two sides parallel, rather than our standard definition. As before, we do not know the reasoning behind the author's solution procedure, but we can explain it in a way similar to the previous solution. We can think of the step beginning "subtract the 6 of the base" as meaning one can draw a line parallel to the left side of the original

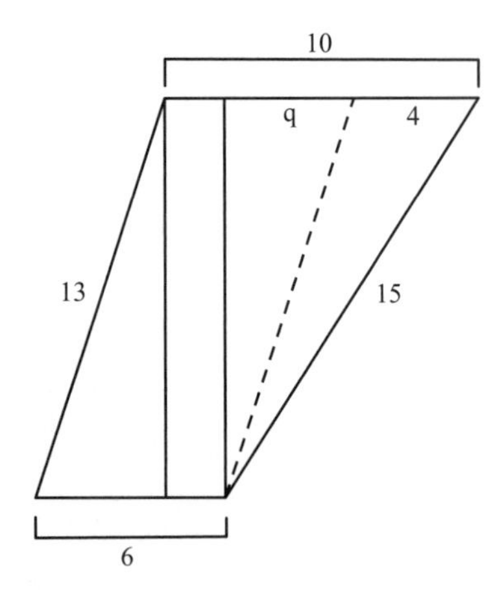

Figure 6.7.4.

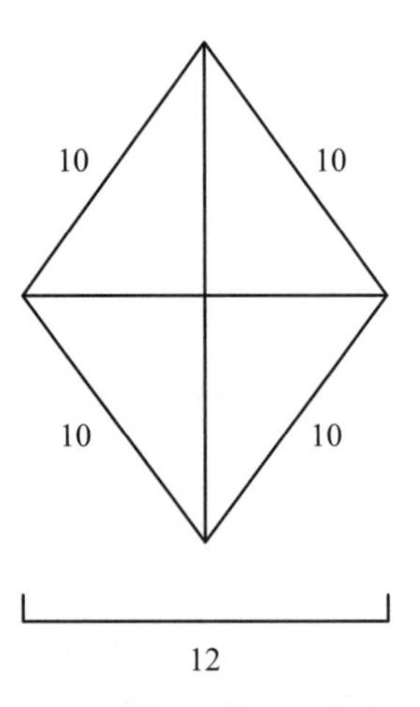

Figure 6.7.5.

diagram as shown in figure 6.7.4. This then creates an obtuse triangle at the right, with sides 13, 4, and 15. By again equating two expressions for the perpendicular, we get $15^2-(q+4)^2=13^2-q^2$. This becomes $15^2-13^2=8q+16$, or $56=4(2q+4)$. Then the author calculates that $4+2q=14$, so $2q=10$, and $q=5$. The perpendicular is calculated by the Pythagorean theorem, and then the areas of the rectangle and the two right triangles are easily found.

Problem 7. If there be given a rhombus such as the one drawn below, according to the conditions of the problem, the 10 squared is 100, and $\frac{1}{2}$ of the 12 of the base is 6. This squared is 36. Subtract this (from 100); the remainder is 64, of which the square root is 8, which is the length of the perpendicular. This by the 6 of the base is 48, $\frac{1}{2}$ of which is 24; of so many acres is each of the right-angled triangles. Altogether, 96 acres. And the figure will be as follows [figure 6.7.5].

6.8 *Papyrus Rainer*

Papyrus Rainer (also known as *Papyrus Graecus Vindobonensis*) is a collection of thirty-eight problems in metrical geometry written in Greek in the style of Heron's *Stereometrica*. That is, the problems are set out and the solution is found via an algorithm, with no discussion of the source of the algorithm. Hence, these are believed to be model problem-solving situations designed for students, very similar to more ancient Egyptian papyri or Mesopotamian tablets. The papyrus was found in Egypt and is dated to the first century CE at the earliest. But because the scribe here uses the value of 3 for π, it appears that he was unaware of Heron's work. We present here problems 13, 24, and 28: calculations of the volumes of a frustum of a triangular pyramid with base an equilateral triangle, a frustum of a cone, and a frustum of a square pyramid, respectively.

In problem 13, we are told that the base triangle has sides of length 18, the top triangle has sides of length 3, and the inclined edge has length 10.

Problem 13. If there is given another pyramid, half-complete, the top each is 3, the base each is 18, the slopes each are 10. Take away the top, the 3 from the 18, there is left 15. The 15 to itself—225, of which the third is 75, and the 10 to itself—100. From these take away the 75, there is left 25, of which a side is 5. So much is the height.

The first step is to calculate the height. The scribe calculates this by imagining a triangle at the base analogous to triangle $BH\Theta$ in book 2, problem 6 of Heron's *Metrica*, that is, an equilateral triangle of side $18-3=15$, the difference of the sides of the base triangle and the top triangle. They then use *Elements* 13:12, which asserts that the square on the circumradius of this triangle is one-third of the square on the side. The Pythagorean theorem then shows that the square on the height is the difference between the square on the inclined edge and the square on the circumradius. To calculate the volume, they use the same algorithm discussed in that problem from Heron, namely, that the volume is given by

$$\left[\frac{\sqrt{3}}{4}\left(\frac{a_1+a_2}{2}\right)^2+\frac{1}{3}\frac{\sqrt{3}}{4}\left(\frac{a_1-a_2}{2}\right)^2\right]h,$$

where a_1, a_2 are the lengths of the base and top triangles, respectively, and h is the height of the frustum. The scribe uses, without comment or explanation, the approximation $\frac{1}{3}+\frac{1}{10}$ for $\frac{\sqrt{3}}{4}$, an approximation used in other papyri as well. They also use some other approximations.

Then put together the 3 and the 18: 21, of which the half is $10\frac{1}{2}$. Measure a triangle, of which the sides are each $10\frac{1}{2}$: $47\frac{1}{2}\ \frac{1}{4}$.[21] Then take the 3 from the 18, there is left 15 of which the half is $7\frac{1}{2}$. Measure out another triangle, equilateral, of which the sides are

[21] The square of $10\frac{1}{2}$ is $110\frac{1}{4}$, of which one-third is $36\frac{1}{2}\ \frac{1}{4}$ and of which one-tenth is $11\frac{1}{40}$. The scribe neglects the $\frac{1}{40}$ in their calculation.

each $7\frac{1}{2}$: $25\frac{1}{5}$, of which the third is $8\frac{2}{5}$. These add to the first, elsewhere set out, to the $47\frac{1}{2}\ \frac{1}{4}$: $56\frac{3}{20}$. This to the height 5 gives $280\frac{1}{2}\ \frac{1}{4}$. Of so many feet is the pyramid.

In problem 24, to calculate the volume of the frustum of a cone, we are given circumferences of the base and top circles of 24 and 6, while the inclined edge is 5.

Problem 24. Suppose a half complete cone is given with the truncated surface, i.e., its top, in circumference 6 feet, its base 24 feet, and its inclinations each 5. Take one-third of the circumference which is of the base 24: 8, and one-third of 6 of the circumference of its top: 2. So much is the diameter where the altitudes fall. Subtract this [2] from 8, leaving 6. Take one-half, leaving 3. Each length between the circumference and the point to which the altitudes fall is so large. Multiply by themselves: 9. And the inclinations 5 multiply by themselves: 25. From this subtract 9, leaving 16, whose side is 4. The altitudes are so much.

Before the height can be calculated, the scribe needs the diameters of the base and top circles. They use the old value of 3 for π, finding the two diameters as 8 and 2. They then take half the difference of the diameters, $\frac{d_1-d_2}{2}$ to calculate the height using the Pythagorean theorem, that is, $h^2 = s^2 - (\frac{d_1-d_2}{2})^2$, in direct analogy with the previous problem.

To find the solid proceed as follows. Add the diameter of the top, 2, and the diameter of the base, 8, giving 10. Take one-half, giving 5. Multiply by itself: 25. Take one-fourth: $6\frac{1}{4}$. Subtract this from 25, leaving $18\frac{3}{4}$. Write it down. And suppose it is possible to find the diameter of the frustum which is the top. Then subtract 2 from the base 8, leaving 6. Take one-half: 3. Multiply by itself: 9, of which one-fourth is $2\frac{1}{4}$. Add this to $18\frac{3}{4}$: total is 21. Multiply by 4, the altitude, giving 84. Of so many feet is the solid.

The scribe here uses the second algorithm for calculating the volume of a conic frustum that we saw in Heron's *Stereometrica* (see section 6.3 in this volume), with the difference that they use 3 as the value of π rather than $\frac{22}{7}$. So the area of a circle of diameter d becomes $\frac{3}{4}d^2$.

Finally, in problem 28, to calculate the frustum of a square pyramid, the base square has length 10, the upper square has length 2, and the inclined edge is given as 6.

Problem 28. Given another square truncated pyramid, whose lower base has side 10, whose upper base is 2, and whose slant height is 6. To find the volume. Subtract 2 from 10, giving 8, and multiply it by itself, giving 64. Then $\frac{1}{2}$ of that is 32. Multiply the 6 by itself, giving 36. From the 36, subtract the 32, giving 4, of which the square root is 2. That is the height. Now add the 2 to the 10, making 12, of which $\frac{1}{2}$ is 6. Multiply it by itself, giving 36. Next subtract the 2 from the 10, giving a remainder 8, of which $\frac{1}{2}$ is 4. Multiply that by itself, giving 16, of which $\frac{1}{3}$ is $5\frac{1}{3}$. Add this to 36, giving $41\frac{1}{3}$. Multiply this by the height 2, giving $82\frac{2}{3}$ as the volume.

As in the earlier problems, the first task is to calculate the height of the frustum, and this is again done using the Pythagorean theorem. Here, the algorithm is $h = \sqrt{s^2 - \frac{1}{2}(a_1 - a_2)^2}$, where s is the inclined edge; a_1, a_2 are the sides of the lower and upper squares, respectively; and h is the height. Then, just as in an almost identical problem in Heron's *Stereometrica* (see section 6.3), the scribe calculates the volume by an algorithm expressed in the formula

$$V=\left[\left(\frac{a_1+a_2}{2}\right)^2+\frac{1}{3}\left(\frac{a_1-a_2}{2}\right)^2\right]h.$$

As noted earlier, this formula reduces to $V=\frac{h}{3}(a_1^2+a_1a_2+a_2^2)$, the algorithm used for a similar problem in the Egyptian *Moscow Mathematical Papyrus* nearly two millennia earlier.

6.9 *Papyrus Genevensis*

The *Papyrus Graecus Genevensis* 259 is a fragmentary Greek papyrus found in Egypt that has been dated to the second century CE. As is usual in ancient collections of problems, the scribe just writes the steps in the solution, without giving any theoretical justification. So we must speculate on the possible methodology from which the algorithm was developed.

Problem 1. Given a right triangle having a height of 3 feet and a hypotenuse of 5, to find the base. We proceed as follows. Multiply 5 by itself to give 25, and the 3 by itself to give 9. From 25, subtract 9. The remainder is 16, whose root is 4. The base is 4. We can solve similarly problems with other numbers.

The first problem is simply an application of the Pythagorean theorem. In the second problem, it appears that the scribe used both that theorem and the identity $d^2-h^2=(d-h)(d+h)$. Thus, when he divided $b^2=d^2-h^2$ by $d+h$, the result is $d-h$. Subtracting that value from $d+h$ gives $2h$, and thus h and d are found. The identity about the difference of squares was apparently used in ancient Mesopotamia and was probably discovered through a geometric argument.

Problem 2. If in a right triangle the sum of the height and the hypotenuse is 8 feet, and the base is 4 feet, we want to find the height and hypotenuse separately. We proceed as follows. Multiply the 4 by itself, giving 16. Divide this by the 8, which gives 2. Subtract the 2 from the 8 and there remains 6, of which half is 3. The height is 3. Then subtract the 3 from 8 and there remains 5. The hypotenuse is therefore 5 feet.

Problem 3. If in a right triangle the sum of the height and base is 17 feet and the hypotenuse is 13 feet, to find the height and base separately. We proceed as follows. Multiply the 13 by itself to give 169. And multiply the 17 by itself to give 289. Double the 169, giving 338. Subtract the 289 from the 338. There remains 49, whose root is 7. Subtract this from the 17; there remains 10, whose half is 5. The height is 5. Subtract this from 17; there remains 12. The base is thus 12 feet.

A virtually identical problem appears in the Seleucid Babylonian tablet BM 34568. The solution method is slightly different, but both solutions seem to rely on figure 6.4.2 above. A glance at the diagram shows that the square on the sum of the base and

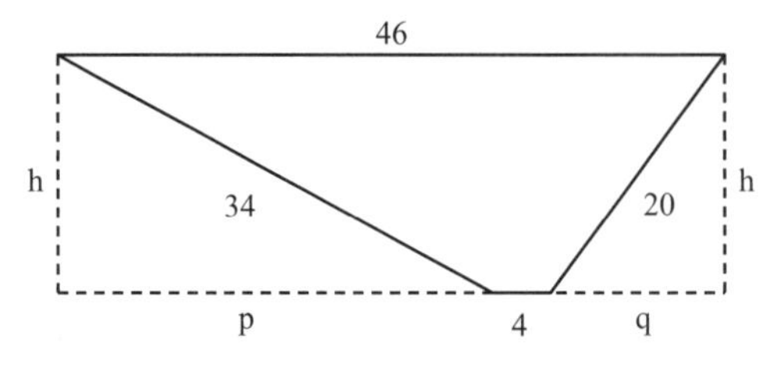

Figure 6.10.1.

height of a right triangle is equal to the square on the diagonal plus four times the right triangle, and also that the square on the diagonal equals four times the right triangle plus the square on the difference of the base and height. It is then clear that twice the square on the diagonal less the square on the sum of the base and height equals the square on the difference of the base and height. The scribe's procedure reflects that geometric idea, followed by the fact that if one subtracts the difference of the base and height from the sum of the base and height, the result is twice the height.

6.10 *Papyrus Cornell*

This Greek papyrus was probably written in the second century CE somewhere in Egypt. The parts of it that can be restored consist of three problems with their solutions, each of which shows how to calculate the area of a quadrilateral. The problems are similar to some found in the work of Heron. The first two problems are reproduced here. The first one asks to find the area of a trapezoidal field with all four sides given.

Problem 1. How one should do it. Make the 34 on the south by itself; there results 1156. And the 20 on the north by itself; there results 400. From 1156 there is left behind 756. Cast out the 4 on the east from the 46 on the west; there is left behind 42. Apply to 756; there results 18. Bring it to the 42; there results 60. Half, 30, which is the east of the triangle to the south. The 18 from the 30, there is left behind 12, which is the east of the triangle to the north. How one should find the upright base. Make the 34 on the south by itself; there results 1156. And the 30 on the east of the triangle to the south by itself; there results 900. From 1156, there is left behind 256. Side, 16, which is the upright base. Similarly the 20 on the east of the north by itself; there results 400. And the 12 on the east of the triangle to the north by itself; there results 144. From 400, there is left behind 256. Side, 16. [figure 6.10.1]

The problem is solved by circumscribing the field by a rectangle, as in figure 6.10.1, and then calculating the areas of various pieces of this rectangle. The first step is to calculate the east sides, p and q, of the two triangles outside of the main field. If h is the upright base of the trapezoid, the scribe evidently realized that $34^2 = h^2 + p^2$, while $20^2 = h^2 + q^2$, so $34^2 - 20^2 = p^2 - q^2 = (p+q)(p-q)$. But $p+q = 46-4 = 42$. It then follows that $(34^2 - 20^2) \div (46-4) = p-q$, or $p-q = 756 \div 42 = 18$. Then, since $p+q = 42$, we get $2p = 60$, $p = 30$, and $q = 12$. The upright base h is then found $\sqrt{34^2 - 30^2}$ and as $\sqrt{20^2 - 12^2}$, that is, $h = 16$. Although most of the remaining text

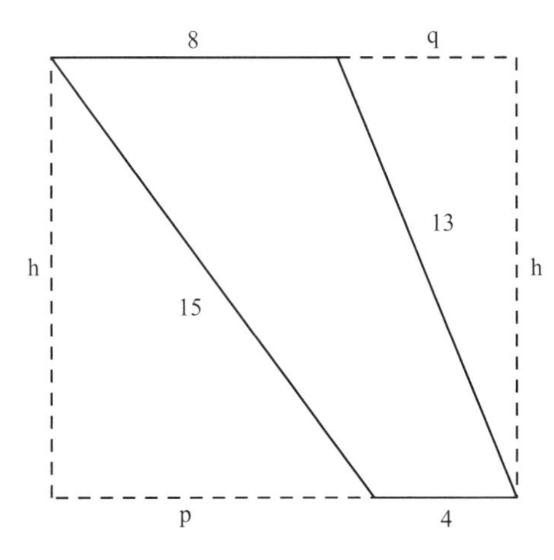

Figure 6.10.2.

of this problem is missing, the few words that can still be read seem to show two different calculations of the area, first as $4 \cdot 16 + (16 \cdot 15 + 16 \cdot 6) = 64 + 336 = 400$ and then as $16 \cdot 46 - (16 \cdot 15 + 16 \cdot 6) = 736 - 336 = 400$.

Problem 2. A different determination of *arourai* [an area measure]. The example of the shape; how one should do it. Make the 15 on the south by itself; there results 225. And the 13 on the north by itself; there results 169. From 225, there is left behind 56. The 4 on the west from the 8 on the east; there is left behind 4. Apply to 56; there results 14. Bring it to the 4; there results 18. Half, 9, which is the west of the plot to the south. From 14, there is left behind 5, which is the east of the plot to the north. How one should find the upright base. Make the 15 on the south by itself. There results 225. And the 9 on the west of the triangle to the south by itself; there results 81. From 225, there is left behind 144. Side 12, which is the upright base. Similarly the 13 on the north by itself; there results 169. And the 5 on the east of the triangle to the north by itself; there results 25. From 169, there is left behind 144. Side 12, which is the upright base. Surveying of the entire plot, the sum of the two triangles to the south and to the north, 84, from the 156, there is left behind 72. So many *arourai* does it have [figure 6.10.2].

Problem 2 is solved similarly to problem 1, except that one of the sides of the field makes an obtuse angle with the base. Thus, $15^2 - 13^2 = (p^2 + h^2) - (q^2 + h^2) = p^2 - q^2 = (p+q)(p-q)$, but since $p+4 = q+8$, we have $p-q=4$. Then $p+q = (225-169) \div 4 = 56 \div 4 = 14$, so $2p = 18$, $p=9$, and $q=5$. The upright base is again calculated twice, both as $\sqrt{15^2-9^2}$ and as $\sqrt{13^2-5^2}$, both calculations giving $h=12$. The total area is then calculated as $12 \cdot 13 - (6 \cdot 9 + 6 \cdot 5) = 156 - 84 = 72$.

6.11 *Papyrus Mathematics*

This Greek papyrus codex dates to the fourth century CE and contains mathematical problems, metrological texts, and model contracts. Here we reproduce five of the geometrical problems, with the identifiers given by the authors of the published version. The problems and the methods are very elementary; the scribe evidently is not aware of the more advanced work done by Heron and his followers. The first problem is to find the area of an isosceles trapezoid. The procedure is simply to divide the trapezoid into two right triangles and one rectangle.

Problem a5. An isosceles trapezoid whose legs are each 15 *schoinia*,[22] common base 30, top 6. Since from the common base I subtract the top, we subtract 6 from 30. The remainder is 24. Half of this is 12. This will be the base of the right-angled triangle. The 15 of each right-angled triangle times itself: 225. And the 12 times itself. The result is 144. The remainder is 81. The square root of this is 9. The base of the rectangle [formed by the two right triangles] will be 9. To find the area: 12 times 9. The result is 108. Half of this is 54 [i.e., each triangle has area 54]. To find the area of the rectangle: 9 times 6. The result is 54.

The next problem shows the method to approximate the area of a quadrilateral by taking the average of each pair of opposite sides and multiplying them together. The scribe here mixes the units, so we need to keep in mind that a *reed* is equal to 3 *paces* and a *pace* is equal to 2 cubits. Also, one *aroura* equals 9216 square cubits.

Problem f5. On the south 8 paces, on the north 6 paces, on the east 15 reeds, on the west 13. To find the *arourai*, we proceed as follows. I add the south and the north sides, 8 and 6. The result is 14. Half of this is 7. Times 2. Why times 2? Because one pace contains 2 cubits. The result is 14 cubits. And we add the east and west, 15 and 13. The result is 28. Half of this is 14. I convert into cubits. The result is 84. 84 times 14. The result is 1176. Divided by 9216, the result is $\frac{1}{8}\ \frac{1}{384}$. It will be this many *arourai*.

Here we are asked to find the area and side of an isosceles triangle, given the altitude and the base. Unfortunately, the scribe erred in their calculation of the area by using the wrong value in calculating the area of the triangle.

Problem c3. An isosceles triangle with altitude 18 *schoinia* and common base 48. To find the other sides. 18 *schoinia* times itself. The result is 324. And half of the base is 24. 24 times itself. The result is 576. I add 576 and 324. The result is 900. The square root of this is 30. Hence the side was 30. To find also the area, we proceed as follows. The base times each vertical, 48 times 30. The result is 1440. Half of this is 720. This way for similar cases.

In the next problem, the scribe finds the volume of a cylinder. The area of the circular base is found using the formula $A=\frac{3}{4}d^2$, where d is the diameter. Here 1 *naubion* equals 27 cubic cubits.

Problem e2. A circular excavation, whose upper diameter is 16 cubits, depth 3 cubits. To find the *naubia* [a volume unit], we proceed as follows. The 16 of the diameter by itself. The result is 256. From this we subtract one-quarter, 64. We subtract 64 from 256. The

[22] One *schoinion* is approximately 96 cubits or 144 feet.

remainder 192. Times the depth, 3 cubits. The result is 576. The *naubia*: I divide by 27 [see next problem]. The result is $21\frac{1}{3}$. It will be $21\frac{1}{3}$ *naubia*. This way for similar cases.

In the final problem, the scribe finds the volume of the frustum of a cone. The scribe calculates this by assuming that the volume of the frustum is equal to the volume of a cylinder with diameter the average of the diameters of the top and bottom bases. This is only an approximation.

Problem n1. Upper diameter 40 cubits, lower diameter 32 cubits, depth 80 cubits. To find how many *naubia* were excavated, I proceed as follows. I add the two diameters. that is, 40 and 32. The result is 72. Half of this is 36. 36 times itself. The result is 1296. I subtract $\frac{1}{4}$ from this. The remainder is 972. 972 times the depth, 80 cubits. The result is 77,760. I divide this by 27. Why? Because one *naubion* contains 27 [cubic] cubits. The result is 2880. Hence 2880 *naubia* were excavated.

6.12 *Corpus Agrimensorum Romanorum*

The *Corpus Agrimensorum* is a collection of surveyor's manuals that have come down to us from the Roman Empire, mostly in fragmentary texts. The earliest manuscripts of these works still extant date from the sixth to the ninth centuries. The works in this collection, which was put together by editors in medieval and early modern times, deal with various aspects of surveying and often include maps of regions in the empire. The surveyors' earliest jobs were to divide conquered lands into plots to be distributed to settlers in the new Roman colonies. The training of the surveyors included astronomy, the geometry of areas, orientation, sighting and leveling, land law, mapping, and recording. In the late empire, the surveyors sometimes served as judges in cases of land law. Unfortunately, since the monks who copied these manuals in the medieval period were often unfamiliar with some of the technical Latin vocabulary, it is not always easy to understand the extant documents. The excerpts we include here deal mostly with the measurement of land.

MARCUS JUNIUS NIPSUS

Little is known about Marcus Junius Nipsus, author of five extant texts that have been collected in the *Corpus Agrimensorum*, except that he probably lived in the second century CE. In this first selection, *Fluminis Varatio*, he shows a method of determining the distance across a river. His method just uses congruent triangles, probably about the simplest method that could be devised.

If as you are squaring land you have a river in the way that needs measuring, do thus [figure 6.12.1]:

From a line perpendicular to the river [BA] make a right angle, placing a crossroad sign. Move the *groma*[23] along the line [AD] at right angles, and make a turn to the right at right

[23] The *groma* was a Roman surveying instrument consisting of two crossed arms resting on a bracket and attached to a vertical staff.

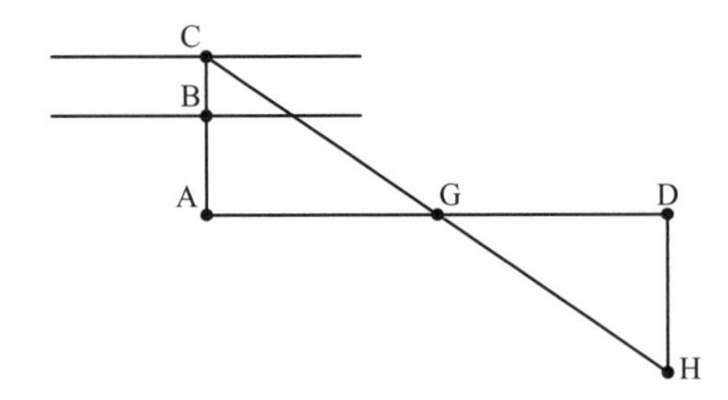

Figure 6.12.1.

angles [at D]. Bisect the line from this right angle to the original one, and place a pole at the midpoint [G]. With the *groma* here, make a line in the opposite direction from the pole you had placed on the other side of the river. Where this meets the line drawn at right angles [DH], place a pole [at H], and measure the distance from here to the crossroad sign [DH]. Since you have two [right] triangles with equal perpendiculars, their bases too will be equal [$CA = DH$]. So you now know the distance from the pole across the river to the first crossroad sign [CA]; subtract the distance from the latter to the river [AB], and you have the width [CB] of the river.

The following selection, from *Podismus* (*Areas*), is more mathematical and appears to be more of a teaching document than a description of methods one would actually use in surveying. Interestingly, the methods described are almost identical to methods found in some Greek and Demotic papyri found in Egypt. In particular, the first problem is very similar to problem 6 of the *Ayer Papyrus*, the second problem is similar to problem 34 of *Papyrus Cairo*, the third problem is similar to problem 3 of *Papyrus Graecus Genevensis*, and the fourth problem is very similar to problem 4 of the *Ayer Papyrus*. In addition, all of these problems bear a similarity to problems found in the Mesopotamian Seleucid tablet BM 34568. How the methods circulated from culture to culture is a matter for further study. But we also see that the solutions of problems 1 and 4 mirror the solutions to similar problems given in Heron's *Metrica*. It is furthermore curious that the fifth problem here uses Heron's formula for determining the area of a triangle, even though for a right triangle that is quite unnecessary.

Problem 1. In an obtuse-angled triangle, given the three sides to determine the perpendicular. We take such a triangle, whose major hypotenuse is 17, the base is 9, and the minor hypotenuse is 10. We multiply the major hypotenuse by itself, and from that subtract the sum of the two smaller sides multiplied by themselves [$289-(81+100)=108$]. Take half of that [54] and divide by the base [$54 \div 9 = 6$]. This will be [the extension of the base] on which the perpendicular falls. If you want to calculate the perpendicular, multiply the minor hypotenuse by itself and subtract the product of the extension with itself [$10^2 - 6^2 = 100 - 36 = 64$]. Take the square root of the result [8] and this will be the length of the perpendicular.

Problem 2. Given a right-angled triangle whose hypotenuse is 25 and whose area is 150, to find the other two sides separately. We multiply the hypotenuse by itself, giving 625, to which we add four times the area of 150, that is, 600, and we get 1225. The square root of that is 35, which we keep in mind. Then again multiply the hypotenuse by itself, giving 625, and subtract four times the area, giving 25, whose square root is 5. Add the 35 to the 5, giving 40, and take half, which is 20. This will be the base of the triangle. If we subtract 5 from the 20, we get 15, and this will be the third side of the triangle.

Problem 3. Given a right-angled triangle, whose perpendicular and base total 23 and whose area is 60 and hypotenuse 17, to find the perpendicular and base separately. Square the hypotenuse, 289. Then subtract four times the area, 240; the remainder is 49. Take the square root, 7. Add this to the sum of the two sides, 23. The result is 30, of which we take half, 15. This will be the base. Subtract this from the sum 23 to get 8 as the perpendicular.

Problem 4. Given an acute-angled triangle, whose three sides are given: the smallest side is 13, the base is 14, and the largest side is 15. To find the altitude of this triangle. Multiply 13 by itself, 169, and 14 by itself, 196. Add these to get 365. From this sum subtract the square of 15, 225. The remainder is 140. Take half of this, 70. Divide by the base, 14, and the result is 5. This is the shorter part of the base. Then from the square of 13 subtract the square of 5, 169 less 25, to get 144. The square root of that, 12, is the perpendicular.

Problem 5. Given a right-angled triangle with given sides: 6 is the altitude, 8 is the base, 10 is the hypotenuse. Add the three numbers together; the result is 24, of which you take half, 12. From this, subtract each of the sides individually. That is, subtract 6 from 12; subtract 8 from 12; and subtract 10 from 12. Then multiply the remainders: 6 times 4 is 24; then multiply by 2 to get 48; finally multiply by 12 to get 576. Take the square root; this is 24. This will be the area.

HYGINUS GROMATICUS

Hyginus Gromaticus probably lived during the second century CE. The excerpt here from *De Limitibus Constituendis* shows the method of determining the cardinal points of the compass. It is the same method recommended by Vitruvius.

First we shall draw a circle on a flat space on the ground, and in its center we shall place a sundial gnomon, whose shadow may at times fall inside the circle as well as outside [figure 6.12.2].... When the shadow touches the circle, we shall note the point on the circumference [E], and similarly when it leaves the circle [W]. Having marked these two points on the circumference, we shall draw a straight line [EW] through them and bisect it. Through this we shall plot our *cardo* [along line AN, going north-south], and *decumani* at right angles to it.[24]

DE IUGERIBUS METIUNDIS

This work, by an unknown author, presents algorithms for determining the areas of various shaped fields. It seems to be taken from the work of Columella, from the first century CE, although the units of measure have different names. In particular, the

[24] The *cardo* and *decumanus* were the two main perpendicular streets in any newly laid out Roman city.

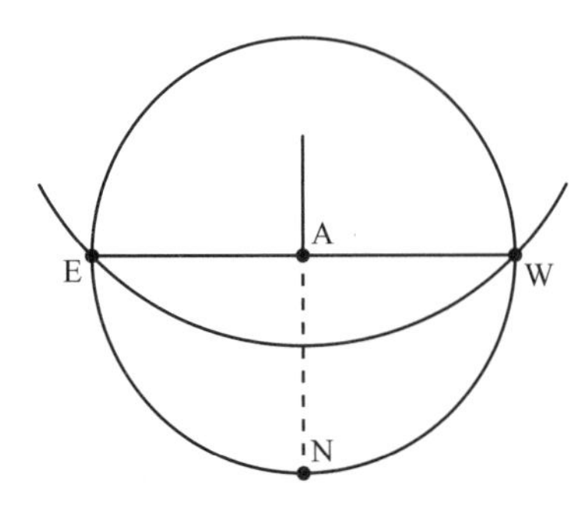

Figure 6.12.2.

author uses the square *pertica* instead of the *scrupulum* to designate 100 square feet, or $\frac{1}{288}$ of a *jugerum*. That is, a *pertica* equals 10 linear feet. They also use the *tabula* to designate 72 square *perticae*. Most of the algorithms are straightforward, but it is curious that the author did not understand Heron's method of calculating the area of a segment of a circle.

A *jugerum* is 288 square *perticae*, or 28,800 square feet; it is a rectangle with one side 18 *perticae* and the other 16 *perticae*. Four times a side gives 72 *perticae*, which, with a width of 1 *pertica*, makes a *tabula* of 72 square *perticae*. If there is a square field with one side equal to 50 *perticae*, to determine its measure, or how many *jugeri* it has, I multiply one side by another. So there are 2500 [square] *perticae*, which makes 8 *jugeri*, 2 *tabulae*, and 52 [square] *perticae*.

The following problem illustrates the ancient method of approximating the area of a quadrilateral by taking the average of each pair of opposite sides and then multiplying the results together.

If there is a field with four unequal sides, one side being 40, a second, 30, another, 20, and another 6, we add 40 and 30 to get 70; then divide this in half, which gives 35. Next add 6 to 20, giving 26, and divide that in half, which gives 13. We then multiply 35 by 13, which gives 455 [square] *perticae*, or 1 *jugerum*, 2 *tabulae*, and 23 [square] *perticae*.

Suppose there is a semicircular field with diameter 40 *perticae* and latitude equal to 20 *perticae*. We multiply the latitude by the diameter, giving 800 [square] *perticae*. Multiply this by 11, which gives 8800. Then divide this by 14, which gives 628 and a bit [square] *perticae*. This makes 2 *jugeri*, half of a *tabula*, and 16 [square] *perticae*.

Suppose there is a circular field whose diameter is 40 *perticae*. Multiply this by itself, which gives 1600. Multiply this by 11, which gives 17600, and then divide by 14, which gives 1257 [square] *perticae* and a bit, which makes 4 *jugeri*, 1 *tabula*, and 33 [square] *perticae*.

Suppose there is a field in the shape of a segment of a circle, but less than a semicircle, where the base is 20 *perticae* and the latitude 5 *perticae*. We add the latitude to the base, which gives 25 *perticae*, Multiply by 4, giving 100, of which half is 50. The same 20 *perticae* of the base, take half, which is 10, and then multiply this by itself giving 100. Take one-fourteenth of this, which is 7 [square] *perticae* and a bit. Add this to the 50 which I found earlier. The sum is then 57 [square] *perticae*, which is the area of the segment.

Heron's formula for a segment of base c and latitude h is $A=\frac{1}{2}(c+h)h+\frac{1}{14}(\frac{c}{2})^2$. But the author, instead of multiplying $c+h=25$ by $h=5$, multiplies by 4, which follows Columella's procedure exactly, except that in Columella's problem, $h=4$ but here $h=5$. Thus the author's answer is quite incorrect. The true area of the segment is 69.8906. If the author had used Heron's formula correctly, they would have found the answer to be about 69.6428.

Suppose there is a field in the shape of a regular hexagon, in which each side is 30 *perticae*. Multiply one side by itself, that is, thirty by thirty, which gives 900 [square] *perticae*. Take one third of this, which is 300 and then take one-tenth of this, which is 90. The sum of these is 390, which we multiply by 6. So the result is 2340. This is the number of [square] *perticae* in this field.

6.13 Polybius, *The Histories*

In the following two excerpts from Polybius's *Histories*, the author comments on two mathematical aspects of Roman civilization. In the first selection, from book 6, he describes a Roman military camp, emphasizing its strictly mathematical formation.

Book 6

The manner in which they form their camp is as follows. When the site for the camp has been chosen, the position in it giving the best general view and most suitable for issuing orders is assigned to the general's tent (*praetorium*). Fixing an ensign on the spot where they are about to pitch it, they measure off round this ensign a square plot of ground each side of which is one hundred feet distant, so that the total area measures four *plethra*. Along one side of this square in the direction which seems to give the greatest facilities for watering and foraging, the Roman legions are disposed as follows: As I have said, there are six tribunes in each legion; and since each consul has always two Roman legions with him, it is evident that there are twelve tribunes in the army of each. They place then the tents of these all in one line parallel to the side of the square selected and fifty feet distant from it, to give room for the horses, mules, and baggage of the tribunes. These tents are pitched with their backs turned to the *praetorium* and facing the outer side of the camp, a direction of which I will always speak as "the front." The tents of the tribunes are at an equal distance from each other, and at such a distance that they extend along the whole breadth of the space occupied by the legions.

They now measure a hundred feet from the front of all these tents, and starting from the line drawn at this distance parallel to the tents of the tribunes they begin to encamp the legions, managing matters as follows: Bisecting the above line, they start from this spot and along a line drawn at right angles to the first they encamp the cavalry of each legion facing each other and separated by a distance of fifty feet, the last-mentioned line being exactly halfway between them. The manner of encamping the cavalry and the infantry is very similar, the whole space occupied by the maniples and squadrons being a square. This square faces one of the streets or *viae* and is of a fixed length of one hundred feet, and they usually try to make the depth the same except in the case of the allies. When they employ the larger legions they add proportionately to the length and depth. The cavalry camp is thus something like a street running down from the middle of the tribunes' tents and at right angles to the line along which these tents are placed and to the space in front of them, the whole system of *viae* being in fact like a number of streets, as either companies of infantry or troops of horse are encamped facing each other all along each. Behind the cavalry, then, they place the *triarii*[25] of both legions in a similar arrangement, a company next to each troop, but with no space between, and facing in the contrary direction to the cavalry. They make the depth of each company half its length, because as a rule the *triarii* number only half the strength of the other classes. So that the maniples being often of unequal strength, the length of the encampments is always the same owing to the difference in depth. Next at a distance of 50 feet on each side they place the *principes* facing the *triarii*, and as they are turned towards the intervening space, two more streets are formed, both starting from the same base as that of the cavalry, i.e., the hundred-foot space in front of the tribunes' tents, and both issuing on the side of the camp which is opposite to the tribunes' tents and which we decided to call the front of the whole.

After the *principes*, and again back to back against them, with no interval they encamp the *hastati*. As each class by virtue of the original division consists of ten maniples, the streets are all equal in length, and they all break off on the front side of the camp in a straight line, the last maniples being here so placed as to face to the front. At a distance again of 50 feet from the *hastati*, and facing them, they encamp the allied cavalry, starting from the same line and ending on the same line. As I stated above, the number of the allied infantry is the same as that of the Roman legions, but from these the *extraordinarii* must be deducted; while that of the cavalry is double after deducting the third who serve as *extraordinarii*. In forming the camp, therefore, they proportionately increase the depth of the space assigned to the allied cavalry, in the endeavor to make their camp equal in length to that of the Romans. These five streets having been completed, they place the maniples of the allied infantry, increasing the depth in proportion to their numbers, with their faces turned away from the cavalry and facing the agger[26] and both the outer sides of the camp. In each maniple the first tent at either end is occupied by the centurions.

In laying the whole camp out in this manner they always leave a space of 50 feet between the fifth troop and the sixth, and similarly with the companies of foot, so that another passage traversing the whole camp is formed, at right angles to the streets, and parallel to the line of the tribunes' tents. This they called *quintana*, as it runs along the fifth troops and companies. The spaces behind the tents of the tribunes to the right and left of the *praetorium*, are used in the one case for the market and in the other for the office of the *quaestor* and the supplies of which he is in charge. Behind the last tent of the tribunes

[25] The *triarii*, along with the *hastati* and *principes* named below, were three different classes of soldiers making up a legion. They varied in their armor and weapons and generally had assigned places in the battle lines.

[26] An embankment along a street.

on either side, and more or less at right angles to these tents, are the quarters of the cavalry picked out from the *extraordinarii*, and a certain number of volunteers serving to oblige the consuls. These are all encamped parallel to the two sides of the agger, and facing in the one case the *quaestors'* depot and in the other the market. As a rule these troops are not only thus encamped near the consuls but on the march and on other occasions are in constant attendance on the consul and *quaestor*. Back to back with them, and looking towards the agger are the select infantry who perform the same service as the cavalry just described. Beyond these an empty space is left a hundred feet broad, parallel to the tents of the tribunes, and stretching along the whole face of the agger on the other side of the market, *praetorium*, and *quaestorium*, and on its further side the rest of the *equites extraordinarii* are encamped facing the market, *praetorium*, and *quaestorium*.

In the middle of this cavalry camp and exactly opposite the *praetorium* a passage, 50 feet wide, is left leading to the rear side of the camp and running at right angles to the broad passage behind the *praetorium*. Back to back with these cavalry and fronting the agger and the rearward face of the whole camp are placed the rest of the *pedites extraordinarii*. Finally the spaces remaining empty to right and left next to the agger on each side of the camp are assigned to foreign troops or to any allies who chance to come in. The whole camp thus forms a square, and the way in which the streets are laid out and its general arrangement give it the appearance of a town. The agger is on all sides at a distance of 200 feet from the tents, and this empty space is of important service in several respects. To begin with, it provides the proper facilities for marching the troops in and out, seeing that they all march out into this space by their own streets and thus do not come into one street in a mass and throw down or hustle each other. Again it is here that they collect the cattle brought into camp and all booty taken from the enemy, and keep them safe during the night. But most important thing of all is that in night attacks neither fire can reach them nor missiles except a very few, which are almost harmless owing to the distance and the space in front of the tents.

Given the numbers of cavalry and infantry, whether 4000 or 5000, in each legion, and given likewise the depth, length, and number of the troops and companies, the dimensions of the passages and open spaces, and all other details, anyone who gives his mind to it can calculate the area and total circumference of the camp. If there ever happen to be an extra number of allies, either of those originally forming part of the army or of others who have joined on a special occasion, accommodation is provided for the latter in the neighborhood of the *praetorium*, the market and *quaestorium* being reduced to the minimum size which meets pressing requirements, while for the former, if the excess is considerable, they add two streets, one at each side of the encampment of the Roman legions. Whenever the two consuls with all their four legions are united in one camp, we have only to imagine two camps like the above placed in juxtaposition back to back, the junction being formed at the encampments of the *extraordinarii* infantry of each camp whom we described as being stationed facing the rearward agger of the camp. The shape of the camp is now oblong, its area double what it was and its circumference half as much again. Whenever both consuls encamp together they adopt this arrangement; but when the two encamp apart the only difference is that the market, *quaestorium*, and *praetorium* are placed between the two camps.

In the following excerpt from book 9, Polybius notes that many in his time had a misconception of the relationship between area and circumference. In fact, he even criticized military leaders for their lack of mathematical knowledge, noting that they have "forgotten the lessons in geometry" they learned as children.

Book 9

Most people judge of the size of cities simply from their circumference. So that when one says that Megalopolis is fifty stades in circumference and Sparta forty-eight, but that Sparta is twice as large as Megalopolis, the statement seems incredible to me. And when in order to puzzle them still more, one tells them that a city or camp with a circumference of forty stades may be twice as large as one the circumference of which is one hundred stades, this statement seems to them absolutely astounding. The reason of this is that we have forgotten the lessons in geometry we learned as children. I was led to make these remarks by the fact that not only ordinary men but even some statesmen and commanders of armies are thus astounded, and wonder how it is possible for Sparta to be larger and even much larger than Megalopolis, although its circumference is smaller; or at other times attempt to estimate the number of men in a camp by taking into consideration its circumference alone. Another very similar error is due to the appearance of cities. Most people suppose that cities set upon broken and hilly ground can contain more houses than those set upon flat ground. This is not so, as the walls of the houses are not built at right angles to the slope, but to the flat ground at the foot on which the hill itself rests. The truth of this can be made manifest to the intelligence of a child. For if one supposes the houses on a slope to be raised to such a height that their roofs are all level with each other, it is evident that the flat space thus formed by the roofs will be equal in area and parallel to the flat space in which the hill and the foundations of the houses rest. So much for those who aspire to political power and the command of armies but are ignorant of such things and surprised by them.

6.14 Vitruvius, *On Architecture*

Vitruvius was a Roman architect and military engineer who lived during the first century BCE. He is best known for *On Architecture*, from which the following excerpts are taken. In book 1, chapter 1, he writes about the skills necessary for the successful architect.

Book 1: First Principles and the Layout of Cities

Chapter 1: The Education of the Architect

1. The architect should be equipped with knowledge of many branches of study and varied kinds of learning, for it is by his judgment that all work done by the other arts is put to test. This knowledge is the child of practice and theory. Practice is the continuous and regular exercise of employment where manual work is done with

any necessary material according to the design of a drawing. Theory, on the other hand, is the ability to demonstrate and explain the productions of dexterity on the principles of proportion.

2. It follows, therefore, that architects who have aimed at acquiring manual skill without scholarship have never been able to reach a position of authority to correspond to their pains, while those who relied only upon theories and scholarship were obviously hunting the shadow, not the substance. But those who have a thorough knowledge of both, like men armed at all points, have the sooner attained their object and carried authority with them.
3. In all matters, but particularly in architecture, there are these two points: the thing signified and that which gives it its significance. That which is signified is the subject of which we may be speaking; and that which gives significance is a demonstration on scientific principles. It appears, then, that one who professes himself an architect should be well versed in both directions. He ought, therefore, to be both naturally gifted and amenable to instruction. Neither natural ability without instruction nor instruction without natural ability can make the perfect artist. Let him be educated, skillful with the pencil, instructed in geometry, know much history, have followed the philosophers with attention, understand music, have some knowledge of medicine, know the opinions of the jurists, and be acquainted with astronomy and the theory of the heavens.
4. The reasons for all this are as follows. An architect ought to be an educated man so as to leave a more lasting remembrance in his treatises. Secondly, he must have a knowledge of drawing so that he can readily make sketches to show the appearance of the work which he proposes. Geometry, also, is of much assistance in architecture, and in particular it teaches us the use of the rule and compasses, by which especially we acquire readiness in making plans for buildings in their grounds, and rightly apply the square, the level, and the plummet. By means of optics, again, the light in buildings can be drawn from fixed quarters of the sky. It is true that it is by arithmetic that the total cost of buildings is calculated and measurements are computed, but difficult questions involving symmetry are solved by means of geometrical theories and methods.
8. Music, also, the architect ought to understand so that he may have knowledge of the canonical and mathematical theory, and besides be able to tune *ballistae*, *catapultae*, and *scorpiones* to the proper key. For to the right and left in the beams are the holes in the frames through which the strings of twisted sinew are stretched by means of windlasses and bars, and these strings must not be clamped and made fast until they give the same correct note to the ear of the skilled workman. For the arms thrust through those stretched strings must, on being let go, strike their blow together at the same moment; but if they are not in unison, they will prevent the course of projectiles from being straight.
9. In theaters, likewise, there are the bronze vessels (*echea*) which are placed in niches under the seats in accordance with the musical intervals on mathematical principles. These vessels are arranged with a view to musical concords or harmony, and apportioned in the compass of the fourth, the fifth, and the octave, and so on up to the double octave, in such a way that when the voice of an actor falls in unison

with any of them its power is increased, and it reaches the ears of the audience with greater clearness and sweetness. Water organs, too, and the other instruments that resemble them cannot be made by one who is without the principles of music.

16. Astronomers likewise have a common ground for discussion with musicians in the harmony of the stars and musical concords in tetrads and triads of the fourth and the fifth, and with geometricians in the subject of vision (*legos optikos*); and in all other sciences many points, perhaps all, are common so far as the discussion of them is concerned. But the actual undertaking of works which are brought to perfection by the hand and its manipulation is the function of those who have been specially trained to deal with a single art. It appears, therefore, that he has done enough and to spare who in each subject possesses a fairly good knowledge of those parts, with their principles, which are indispensable for architecture, so that if he is required to pass judgment and to express approval in the case of those things or arts, he may not be found wanting. As for men upon whom nature has bestowed so much ingenuity, acuteness, and memory that they are able to have a thorough knowledge of geometry, astronomy, music, and the other arts, they go beyond the functions of architects and become pure mathematicians. Hence they can readily take up positions against those arts because many are the artistic weapons with which they are armed. Such men, however, are rarely found, but there have been such at times: for example, Aristarchus of Samos, Philolaus and Archytas of Tarentum, Apollonius of Perga, Eratosthenes of Cyrene, and among Syracusans Archimedes and Scopinas,[27] who through mathematics and natural philosophy discovered, expounded, and left to posterity many things in connection with mechanics and with sundials.

In Chapter 6, Vitruvius shows how to determine due north. His method is the same as that given by Hyginus Gromaticus in the Roman *Corpus Agrimensorum*, written two centuries later.[28] But Vitruvius also names the winds coming from those directions, with Auster being the south wind and Septentrio that from the north. He then shows how to divide the circle so as to name the winds coming from various other directions. In any case, classical Roman city planning had streets running north-south as well as east-west.

Chapter 6: Orientation

6. In the middle of the city place a marble benchmark, laying it true by the level, or else let the spot be made so true by means of rule and level that no benchmark is necessary. In the very center of that spot set up a bronze gnomon or “shadow tracker.” At about the fifth hour in the morning, take the end of the shadow cast by this gnomon, and mark it with a point. Then, opening your compasses to this point which marks the length of the gnomon’s shadow, describe a circle from

[27] Although unattested in other literature, Scopinas is mentioned later by Vitruvius as having invented a type of sundial.

[28] See section 6.12.

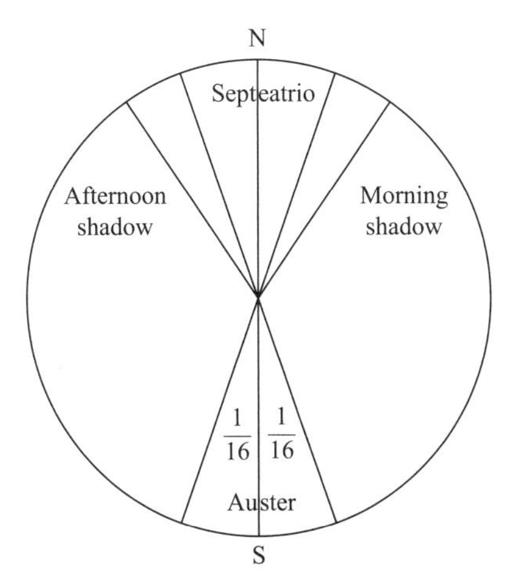

Figure 6.14.1.

the center. In the afternoon watch the shadow of your gnomon as it lengthens, and when it once more touches the circumference of this circle and the shadow in the afternoon is equal in length to that of the morning, mark it with a point [figure 6.14.1].

7. From these two points describe with your compasses intersecting arcs, and through their intersection and the center let a line be drawn to the circumference of the circle to give us the quarters of south and north. Then, using a sixteenth part of the entire circumference of the circle as a diameter, describe a circle with its center on the line to the south, at the point where it crosses the circumference, and put points to the right and left on the circumference on the south side, repeating the process on the north side. From the four points thus obtained draw lines intersecting the center from one side of the circumference to the other. Thus we shall have an eighth part of the circumference set out for Auster and another for Septentrio. The rest of the entire circumference is then to be divided into three equal parts on each side, and thus we have designed a figure equally apportioned among the eight winds. Then let the directions of your streets and alleys be laid down on the lines of division between the quarters of two winds.

In book 9 Vitruvius mentions some mathematical discoveries described by Plato, Pythagoras, and Archimedes. The first is the problem of doubling the square, as described in Plato's *Meno*. The second is a brief discussion of the use of the Pythagorean theorem to create a set square. Finally, Vitruvius gives his analysis of Archimedes's solution of Hieron's crown problem. Most likely, however, his story is not correct, since it does not reference the basic law of hydrostatics that Archimedes presented in his treatise *On Floating Bodies*, namely, that a solid heavier than a fluid will, when weighed in the fluid, be lighter than its true weight by the weight of the fluid displaced.

Book 9: Sundials and Clocks

Preface

1. The ancestors of the Greeks have appointed such great honors for the famous athletes who are victorious at the Olympian, Pythian, Isthmian, and Nemean games,[29] that they are not only greeted with applause as they stand with palm and crown at the meeting itself, but even on returning to their several states in the triumph of victory, they ride into their cities and to their fathers' houses in four-horse chariots, and enjoy fixed revenues for life at the public expense. When I think of this, I am amazed that the same honors and even greater are not bestowed upon those authors whose boundless services are performed for all time and for all nations. This would have been a practice all the more worth establishing, because in the case of athletes it is merely their own bodily frame that is strengthened by their training, whereas in the case of authors it is the mind, and not only their own but also man's in general, by the doctrines laid down in their books for the acquiring of knowledge and the sharpening of the intellect.
2. What does it signify to mankind that Milo of Croton[30] and other victors of his class were invincible? Nothing, save that in their lifetime they were famous among their countrymen. But the doctrines of Pythagoras, Democritus, Plato, and Aristotle, and the daily life of other learned men, spent in constant industry, yield fresh and rich fruit, not only to their own countrymen, but also to all nations. And they who from their tender years are filled with the plenteous learning which this fruit affords, attain to the highest capacity of knowledge, and can introduce into their states civilized ways, impartial justice, and laws, things without which no state can be sound.
3. Since, therefore, these great benefits to individuals and to communities are due to the wisdom of authors, I think that not only should palms and crowns be bestowed upon them, but that they should even be granted triumphs, and judged worthy of being consecrated in the dwellings of the gods. Of their many discoveries which have been useful for the development of human life, I will cite a few examples. On reviewing these, people will admit that honors ought of necessity to be bestowed upon them.
4. First of all, among the many very useful theorems of Plato, I will cite one as demonstrated by him. Suppose there is a place or a field in the form of a square and we are required to double it. This has to be effected by means of lines correctly drawn, for it will take a kind of calculation not to be made by means of mere multiplication. The following is the demonstration. A square place ten feet long and ten feet wide gives an area of one hundred feet. Now if it is required to double the square, and to make one of two hundred feet, we must ask how long will be the side of that square so as to get from this the two hundred feet corresponding to the doubling of the area. Nobody can find this by means of arithmetic. For if we take fourteen, multiplication will give one hundred and ninety-six feet; if fifteen, two hundred and twenty-five feet.

[29] These are the four Panhellenic Games of ancient Greece.

[30] He was a famous ancient Greek athlete of the late 6th c. BCE.

5. Therefore, since this is inexplicable by arithmetic, let a diagonal line be drawn from angle to angle of that square of ten feet in length and width, dividing it into two triangles of equal size, each fifty feet in area. Taking this diagonal line as the length, describe another square. Thus we shall have in the larger square four triangles of the same size and the same number of feet as the two of fifty feet each which were formed by the diagonal line in the smaller square. In this way Plato demonstrated the doubling by means of lines, as the figure appended at the bottom of the page will show.[31]
6. Then again, Pythagoras showed that a right angle can be formed without the contrivances of the artisan. Thus, the result which carpenters reach very laboriously, but scarcely to exactness, with their squares, can be demonstrated to perfection from the reasoning and methods of his teaching. If we take three rules, one three feet, the second four feet, and the third five feet in length, and join these rules together with their tips touching each other so as to make a triangular figure, they will form a right angle. Now if a square be described on the length of each one of these rules, the square on the side of three feet in length will have an area of nine feet; of four feet, sixteen; of five, twenty-five.
7. Thus the area in number of feet made up of the two squares on the sides three and four feet in length is equaled by that of the one square described on the side of five. When Pythagoras discovered this fact, he had no doubt that the Muses had guided him in the discovery, and it is said that he very gratefully offered sacrifice to them. This theorem affords a useful means of measuring many things, and it is particularly serviceable in the building of staircases in buildings, so that the steps may be at the proper levels.
8. Suppose the height of the story, from the flooring above to the ground below, to be divided into three parts. Five of these will give the right length for the stringers of the stairway. Let four parts, each equal to one of the three composing the height between the upper story and the ground, be set off from the perpendicular, and there fix the lower ends of the stringers. In this manner the steps and the stairway itself will be properly placed. A figure of this also will be found appended below.
9. In the case of Archimedes, although he made many wonderful discoveries of diverse kinds, yet of them all, the following, which I shall relate, seems to have been the result of a boundless ingenuity. Hieron, after gaining the royal power in Syracuse, resolved, as a consequence of his successful exploits, to place in a certain temple a golden crown which he had vowed to the immortal gods. He contracted for its making at a fixed price, and weighed out a precise amount of gold to the contractor. At the appointed time the latter delivered to the king's satisfaction an exquisitely finished piece of handiwork, and it appeared that in weight the crown corresponded precisely to what the gold had weighed.
10. But afterwards a charge was made that gold had been abstracted and an equivalent weight of silver had been added in the manufacture of the crown. Hieron, thinking it an outrage that he had been tricked, and yet not knowing how to detect the theft,

[31] This is the diagram described in Plato's *Meno* (see section 3.3).

requested Archimedes to consider the matter. The latter, while the case was still on his mind, happened to go to the bath, and on getting into a tub observed that the more his body sank into it, the more water ran out over the tub. As this pointed out the way to explain the case in question, without a moment's delay, and transported with joy, he jumped out of the tub and rushed home naked, crying with a loud voice that he had found what he was seeking; for as he ran he shouted repeatedly in Greek: "I found it! I found it! (*Eureka! Eureka!*)"

11. Taking this as the beginning of his discovery, it is said that he made two masses of the same weight as the crown, one of gold and the other of silver. After making them, he filled a large vessel with water to the very brim, and dropped the mass of silver into it. As much water ran out as was equal in bulk to that of the silver sunk in the vessel. Then, taking out the mass, he poured back the lost quantity of water, using a one-sextarius pitcher [= 1/2 liter], until it was level with the brim as it had been before. Thus he found the weight of silver corresponding to a definite quantity of water.
12. After this experiment, he likewise dropped the mass of gold into the full vessel and, on taking it out and measuring as before, found that not so much water was lost, but a smaller quantity: namely, as much less as a mass of gold lacks in bulk compared to a mass of silver of the same weight. Finally, filling the vessel again and dropping the crown itself into the same quantity of water, he found that more water ran over for the crown than for the mass of gold of the same weight. Hence, reasoning from the fact that more water was lost in the case of the crown than in that of the mass, he detected the mixing of silver with the gold and made the theft of the contractor perfectly clear.
13. Now let us turn our thoughts to the researches of Archytas of Tarentum and Eratosthenes of Cyrene. They made many discoveries from mathematics which are welcome to men, and so, though they deserve our thanks for other discoveries, they are particularly worthy of admiration for their ideas in that field. For example, each in a different way solved the problem enjoined upon Delos by Apollo in an oracle, the doubling of the number of cubic feet in his altars; this done, he said, the inhabitants of the island would be delivered from an offense against religion.
14. Archytas solved it by his figure of the semicylinders;[32] Eratosthenes, by means of the instrument called the mesolabe. . . .
15. These, then, were men whose researches are an everlasting possession, not only for the improvement of character but also for general utility. The fame of athletes, however, soon declines with their bodily powers. Neither when they are in the flower of their strength, nor afterwards with posterity, can they do for human life what is done by the researches of the learned.

SOURCES, CHAPTER 6

6.1 This translation is from E. W. Marsden, *Greek and Roman Artillery: Technical Treatises* (Oxford: Clarendon Press, 1971).

[32] See section 3.6.

6.2 This translation is from Marsden, *Greek and Roman Artillery*.

6.3 This material has been translated by Victor Katz from the French in Fabio Acerbi and Bernard Vitrac, *Metrica: Héron d'Alexandrie* (Pisa: Fabrizio Serra Editore, 2014), and from the German in Johan Ludvig Heiberg, ed., *Heronis Alexandrini opera quae supersunt omnia*, vol. 3, *Heronis quae feruntur Geometrica* (Leipzig: Teubner, 1912). The parts from the *Stereometrica* are from Heiberg, *Heronis Alexandrini*, vol. 5.

6.4 The translation is from Jens Høyrup, *Lengths, Widths, Surfaces: A Portrait of Old Babylonian Algebra and Its Kin* (New York: Springer, 2002), 392–99.

6.5 The translation of this papyrus is from Richard A. Parker, *Demotic Mathematical Papyri* (Providence, RI: Brown University Press, 1972).

6.6 This translation is from Lucius Junius Moderatus Columella, *L. Junius Moderatus Columella of Husbandry, in Twelve Books: And His Book, Concerning Trees*, trans. anon. (London: A. Millar, 1745).

6.7 This translation is from Edgar J. Goodspeed, "The Ayer Papyrus," *American Mathematical Monthly* 10, no. 5 (1903): 133–35.

6.8 The papyrus has been published in full in Hans Oellacher Gerstinger and Kurt Vogel, *Mitteilungen aus der Nationalbibliothek in Wien: Papyrus Erzherzog Rainer* (Vienna: Osterreichischen Staatsdruckerei, 1931). The translation here is adapted from that work.

6.9 This was translated by Victor Katz from the French in Jacques Sesiano, "Sur le Papyrus graecus genevensis 259," *Museum Helveticum* 56 (1999): 26–32.

6.10 This papyrus is analyzed and translated in Alexander Jones, "P. Cornell inv. 69 Revisited: A Collection of Geometrical Problems," in *Papyrological Texts in Honor of Roger S. Bagnall*, ed. Rodney Ast et al. (Durham, NC: American Society of Papyrologists, 2013), 159–73.

6.11 The translation of these problems comes from Roger S. Bagnall and Alexander Jones, eds., *Mathematics, Metrology, and Model Contracts: A Codex from Late Antique Business Education* (New York: New York University Press, 2019).

6.12 These selections were translated by Victor Katz from the Latin in F. Blume, K. Lachmann, and A. Rudorff, eds., *Die Schriften der Römischen Feldmesser* (Berlin: Georg Reimer, 1848).

6.13 This translation is from Polybius, *The Complete Histories of Polybius*, ed. and trans. W. R. Paton (Digireads.com, 2014).

6.14 The translation here is from Vitruvius, *The Ten Books on Architecture*, trans. Morris Hicky Morgan (Cambridge, MA: Harvard University Press, 1914). It can be accessed online at https://www.gutenberg.org/files/20239/20239-h/20239-h.htm#Page_251.

7

Astronomy

From the outset, geometry formed the heart of Greek astronomical reckoning. While observations were an important component of the discipline, astronomy was crucially advanced by the application of geometry to this field. Beginning in the fourth century BCE, Greek astronomers tackled fundamental cosmographical questions such as the shape of the celestial realm and the heavenly bodies therein with their sizes and relative positions. In time, aspirations to model the celestial motions introduced kinematic aspects to these geometric reckonings.

In particular, astronomers set themselves the task of accounting for the apparently irregular motions of the planets using circular, uniform, direct motions only, principles that had been established as foundational by philosophers. Eudoxus of Cnidus in the fourth century BCE was the first to propose a comprehensive model that adhered to these principles. He devised a system of nested spheres, concentric to the earth, with axes of rotation tilted with respect to one another. By adjusting the relative axes of rotation of certain spheres, Eudoxus was able to produce a figure-of-eight path—dubbed a *hippopede* (lit. "horse fetter")—designed to reproduce the occasional retrograde motions of the planets. But while his model, and the improvements suggested by later astronomers, could qualitatively capture aspects of planetary motion, they did so only crudely and, without quantitative data, had no predictive power. Although none of Eudoxus's own writings have survived, we begin this chapter with two excerpts from later authors describing his insights in section 7.1.

A century or so later (ca. 225 BCE), Apollonius proposed two equivalent geometric models that reproduced the back-and-forth motions of the planets but abandoned the requirement of concentricity. The epicyclic model was based on combinations of small circles (epicycles) carried around on deferent circles, and in the eccentric model, the planets advanced along large off-centered circular orbits; together these formed the basis of later Greek mathematical astronomy. But again, while these propositions could qualitatively account for planetary motion, it was another century before astronomers introduced numerical parameters onto these geometric models so that they had predictive power as well. Again, the works of Apollonius on this subject are lost, so in the excerpt from Ptolemy's *Almagest*, Book 3, we can read Ptolemy's discussion of these two models and their mathematical equivalence.

Hipparchus (ca. mid-2nd c. BCE) is often credited as the first to bring quantitative significance to bear on these geometrical models, inspired by numerical parameters that had been developed by astronomers in the ancient Near East. Before this time, numerical data remained in the realm of cosmographical speculation as seen in the work of Aristarchus on the measurements of the sizes and distances of the Sun and the Moon. We present an excerpt from this work in section 7.2.

Greek astronomy culminated with Ptolemy (2nd c. CE) and his comprehensive work on mathematical astronomy, the *Syntaxis*, later known as the *Almagest*. In this work, Ptolemy brought together and refined all aspects of Greek astronomy into a single definitive composition. Such was the success of this work that earlier astronomical treatises ceased to be copied, although some early Greek astronomy survives such as a composition by Hypsicles (early 2nd c. BCE), excerpted in section 7.3. Ptolemy's work became foundational for later astronomically active societies, such as found in the Islamic, Byzantine, and European traditions. His contribution was not substantially overturned until the early modern period.

One of the key mathematical tools that advanced quantitative astronomical reckoning was trigonometry, namely, techniques to determine chord lengths of given arcs/angles in a circle. Ptolemy's table of chords is the first extant table we have, along with a number of important trigonometric relations that were crucial to enhancing the numerical significance of the epicyclic/eccentric models. Related to this, the rising and setting of stars on the celestial sphere inspired the advancement of spherical geometry and trigonometry, emerging with the *Sphaerica* of Theodosius and that of Menelaus. We present excerpts from the *Almagest* showing how Ptolemy solved a number of mathematical problems arising in his astronomy. We follow this with a brief selection from Ptolemy's *Handy Tables*, which gave working astronomers easy access to the numerical parameters that he developed.

Ptolemy was also famous for his *Geography*, so we present some selections from this work as well as the *Planisphere*, which gave tools for representing the spherical earth on a flat surface. But evidently, Greek mathematicians and astronomers were able to translate some of their theoretical work into the construction of astronomical devices, so we share some articles dealing with the Antikythera Mechanism, discovered off the coast of Greece early in the twentieth century. We conclude the chapter with an excerpt from Cleomedes showing how to measure the size of the earth and finally look at some documents showing how practitioners applied astronomical tools to answer pressing questions.

7.1 Eudoxus of Cnidus (from Aristotle and Simplicius)

Eudoxus of Cnidus, whose *floruit* falls between Plato and Aristotle, studied mathematics under Archytas of Tarentum, as well as medicine and philosophy. Eudoxus is known for his writings and teaching in many subjects; in astronomy he made notable contributions. Most significant, he was the first to propose a geometrical kinematic model of planetary motion, expounded in the book *On Speeds*, which has not survived. His system involved nested homocentric spheres that rotated eastward or westward on different axes at various speeds and could thus reproduce the apparently irregular motions of the planets from a geocentric perspective.

While Eudoxus's proposals were ultimately abandoned by his successors, the general principle of modeling planetary motion via moving geometrical objects set the stage for future reckoning.

Eudoxus's system is preserved in Aristotle's *Metaphysics* and many centuries later by Simplicius (fl. ca. 6th c. CE) in his commentary on Aristotle's *On the Heavens*.

Aristotle, *Metaphysics* 12:8

In this passage Aristotle describes Eudoxus's proposals along with refinements to the system proposed by his contemporary Callippus, followed by his own. He begins by listing the number of spheres Eudoxus proposed to reproduce the motions of each planet—three each for the sun and the moon and four for each planet—along with the inclination of their axes and the individual aspect (daily rotation, latitudinal motion, etc.) each sphere accounted for. Aristotle then describes the additional spheres Callippus proposed to account for additional motions: two each for the sun and the moon, and an additional one for each of Mercury (Hermes), Venus (Aphrodite), and Mars.

Finally, Aristotle offers further modifications, by introducing "unrolling" spheres that were intended to counteract the effects of the moving spheres of each planet on other planets. Aristotle aspired to produce a singular coordinated working model that reproduced planetary phenomena. The result was a mechanistic system that is made up of 114 spheres!

Eudoxus's aim appeared to be a geometric model and broadly accounted for the motions of the planets. Indeed, his proposed system is entirely qualitative; the absence of any quantitative details meant it had no predictive power.

Eudoxus assumed that the Sun and Moon are moved by three spheres in each case; the first of these is that of the fixed stars, the second moves about the circle which passes through the middle of the signs of the zodiac, while the third moves about a circle latitudinally inclined to the zodiac circle; and, of the oblique circles, that in which the Moon moves has a greater latitudinal inclination than that in which the Sun moves. The planets are moved by four spheres in each case; the first and second of these are the same as for the Sun and Moon, the first being the sphere of the fixed stars which carries all the spheres with it, and the second, next in order to it, being the sphere about the circle through the middle of the signs of the zodiac which is common to all the planets; the third is, in all cases, a sphere with its poles on the circle through the middle of the signs; the fourth moves about a circle inclined to the middle circle [the equator] of the third sphere; the poles of the third sphere are different for all the planets except Aphrodite and Hermes, but for these two the poles are the same.

Callippus agreed with Eudoxus in the position he assigned to the spheres, that is to say, in their arrangement in respect of distances, and he also assigned the same number of spheres as Eudoxus did to Zeus and Kronos,[1] respectively, but he thought it necessary to add two more spheres in each case to the Sun and Moon, respectively, if one wishes to account for the phenomena, and one more to each of the other planets.

But it is necessary, if the phenomena are to be produced by all the spheres acting in combination, to assume in the case of each of the planets other spheres fewer by one [than the spheres assigned to it by Eudoxus and Callippus]; these latter spheres are those which unroll, or react on, the others in such a way as to replace the first sphere of the next lower planet in the same position [as if the spheres assigned to the respective planets above it did not exist], for only in this way is it possible for a combined system to produce the motions of the planets. Now the deferent spheres are, first, eight [for Saturn and Jupiter], then twenty-five more [for the Sun, the Moon, and the three other planets]; and of these, only the last set [of five] which carry the planet placed lowest [the Moon] do not require any reacting spheres. Thus the reacting spheres for the first two bodies will be six, and for the next four will be sixteen; and the total number of spheres, including the deferent spheres and those which react on them, will be fifty-five. If, however, we choose not to add to the Sun and Moon the [additional deferent] spheres we mentioned [i.e., the two added by Callippus], the total number of the spheres will be forty-seven. So much for the number of the 114 spheres.

Simplicius, *Commentary on Aristotle's On the Heavens*

Simplicius's account of Eudoxus's system highlights two important details. The first is the basic hypotheses of Greek geometric modeling of celestial motions: that planetary motion is ultimately uniform and circular and combinations of such uniform and circular motions should be devised to reproduce or "save" the phenomena, namely, the apparently irregular motions of the planets. The second is his explicit description of the *hippopede*, a special curve certain configurations of moving spheres could reproduce, that could account for retrograde motion.

And, as Eudemus related in the second book of his astronomical history, and Sosigenes also who herein drew upon Eudemus, Eudoxus of Cnidus was the first of the Greeks to concern himself with hypotheses of this sort, Plato having, as Sosigenes says, set it as a problem to all earnest students of this subject to find what are the uniform and ordered movements by the assumption of which the phenomena in relation to the movements of the planets can be saved.

. . .

The third sphere, which has its poles on the great circle of the second sphere passing through the middle of the signs of the zodiac, and which turns from south to north and from north to south, will carry round with it the fourth sphere which also has the planet attached to it, and will moreover be the cause of the planet's movement in latitude. But not the third sphere only; for, so far as it was on the third sphere (by itself), the planet would actually have arrived at the poles of the zodiac circle, and would have come near to the

[1] Jupiter and Saturn.

poles of the universe; but, as things are, the fourth sphere, which turns about the poles of the inclined circle carrying the planet and rotates in the opposite sense to the third, i.e., from east to west, but in the same period, will prevent any considerable divergence (on the part of the planet) from the zodiac circle, and will cause the planet to describe about this same zodiac circle the curve called by Eudoxus the *hippopede* (horse-fetter), so that the breadth of this curve will be the (maximum).

7.2 Aristarchus, *On the Sizes and Distances of the Sun and Moon*

Aristarchus of Samos flourished during the first half of the third century BCE. Thus, he was an older contemporary of Archimedes, who refers to him in his *Sand Reckoner* as the author of a book claiming that the earth revolves around the sun. Unfortunately, that book is no longer extant, but here we present a selection from another work. In *On the Sizes and Distances of the Sun and Moon*, Aristarchus presents several hypotheses about the sun, moon, and earth and then uses mathematics to develop and refine his results.

Hypotheses

1. That the moon receives its light from the sun.
2. That the earth is in the relation of a point and center to the sphere in which the moon moves.
3. That, when the moon appears to us halved, the great circle that divides the dark and the bright portions of the moon is in the direction of our eye.
4. That, when the moon appears to us halved, its distance from the sun is then less than a quadrant by one-thirtieth of a quadrant.
5. That the breadth of the earth's shadow is that of two moons.
6. That the moon subtends one-fifteenth part of a sign of the zodiac.

The first two hypotheses are natural, while the third simply means that the plane of the great circle passes through the eye of the observer. The fourth hypothesis uses the measure of angles standard before the adoption of degrees, namely, a right angle or quadrant. (Euclid uses this measure as well.) In astronomical contexts, a sign of the zodiac (30°) is sometimes used as a measure. So the statement means that at the time when the moon is exactly halved, that is, when the triangle formed by the earth, moon, and sun has a right angle at the moon [figure 7.2.1], the angular distance between the sun and the moon as observed from the earth is 87°.

There are two problems with this hypothesis. First, it is extremely difficult to determine the exact moment when the moon appears halved. But, more important, Aristarchus's value is much smaller than the accurate value of $89^\circ 51'$. Since Aristarchus's instruments were certainly accurate enough to notice that 87° was too small a value, it is curious that he stated this hypothesis. Perhaps he used it because it was an easier value with which to calculate than the true value, which he probably could not have measured to the accuracy stated. The sixth hypothesis that the diameter of the moon is one-fifteenth of an astronomical sign, that is, 2°, is

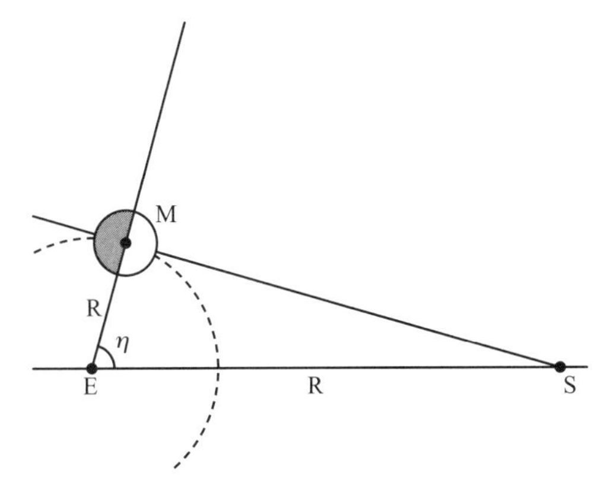

Figure 7.2.1.

also very curious. The actual value is closer to $\frac{1}{2}^{\circ}$ and again, Aristarchus must have known that his value was far too large. On the other hand, that hypothesis hardly plays any role in Aristarchus's argument. The fifth hypothesis presumably comes from a measurement during a lunar eclipse. However, since the apparent motion of the moon through the stars is about $\frac{1}{2}^{\circ}$ per hour, this hypothesis, combined with the sixth, would mean that the time of totality in a lunar eclipse would be about four hours, far longer than the normal length of such an eclipse. Since Aristarchus's hypotheses are not at all realistic, even to an astronomer at his time, perhaps all that he was trying to do was to demonstrate the power of a mathematical approach to solving astronomical problems. It is certainly clear that he understood how to present a mathematical argument, using his hypotheses and geometrical theorems, combined with a lettered diagram, precisely as Greek mathematicians had been doing for perhaps a century.

We are now in a position to prove the following propositions:

1. The distance of the sun from the earth is greater than eighteen times, but less than twenty times, the distance of the moon [from the earth]; this follows from the hypothesis about the halved moon.
2. The diameter of the sun has the same ratio (as aforesaid) to the diameter of the moon.
3. The diameter of the sun has to the diameter of the earth a ratio greater than that which 19 has to 3, but less than that which 43 has to 6; this follows from the ratio thus discovered between the distances, the hypothesis about the shadow, and the hypothesis that the moon subtends one-fifteenth part of a sign of the zodiac.

From figure 7.2.1, we see that, in modern terms, the ratio of the sun's distance to the earth to the moon's distance is equal to $\sec\eta$, where $\eta = 87^{\circ}$. Since trigonometry was not available to Aristarchus, he had to bound the possible values through some clever geometrical arguments.

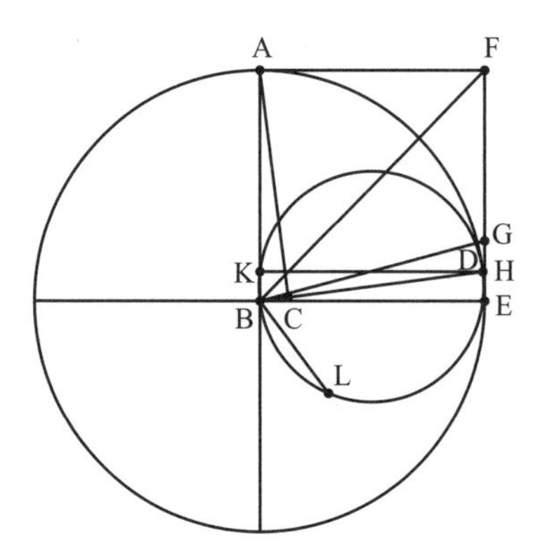

Figure 7.2.2.

Proposition 7. *The distance of the sun from the earth is greater than eighteen times, but less than twenty times, the distance of the moon from the earth.*

For let A be the center of the sun, B that of the earth. Let AB be joined and produced [figure 7.2.2]. Let C be the center of the moon when halved; let a plane be carried through AB and C, and let the section made by it in the sphere on which the center of the sun moves be the great circle ADE. Let AC, CB be joined, and let BC be produced to D. Then, because the point C is the center of the moon when halved, the angle ACB will be right.

Let BE be drawn from B at right angles to BA; then the circumference ED will be one-thirtieth of the circumference EDA; for, by hypothesis, when the moon appears to us halved, its distance from the sun is less than a quadrant by one-thirtieth of a quadrant. Thus the angle EBC is also one-thirtieth of a right angle. Let the parallelogram AE be completed, and let BF be joined. Then the angle FBE will be half a right angle. Let the angle FBE be bisected by the straight line BG; therefore the angle GBE is one fourth part of a right angle.

But the angle DBE is also one-thirtieth part of a right angle; therefore the ratio of the angle GBE to the angle DBE is that which 15 has to 2; for, if a right angle be regarded as divided into 60 equal parts, the angle GBE contains 15 of such parts, and the angle DBE contains 2.

Now, since GE has to EH a ratio greater than that which the angle GBE has to the angle DBE,[2] therefore GE has to EH a ratio greater than that which 15 had to 2. Next, since BE is equal to EF, and the angle at E is right, therefore the square on FB is double of the square on BE. But, as the square on FB is to the square on BE, so is the square on FG to the square on GE (*Elements* 6:3); therefore the square on FG is double of the square on GE. Now 49 is less than double of 25, so that the square on FG has to the square on GE a ratio greater than that which 49 has to 25; therefore FG also has to GE

[2] This is equivalent to the modern result that if $\alpha > \beta$, then $\frac{\tan\alpha}{\tan\beta} > \frac{\alpha}{\beta}$ and can easily be demonstrated by a geometric argument.

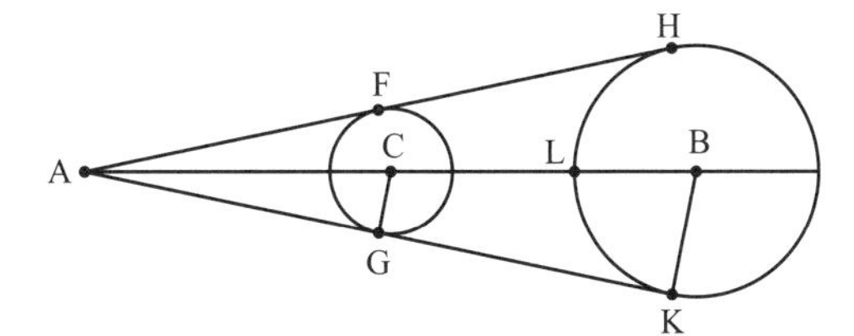

Figure 7.2.3.

a ratio greater than that which 7 has to 5. Therefore, *componendo*, FE has to EG a ratio greater than that which 12 has to 5, that is, than that which 36 has to 15.

But it was also proved that GE has to EH a ratio greater than that which 15 has to 2; therefore, *ex aequali*, FE has to EH a ratio greater than that which 36 has to 2, that is, than that which 18 has to 1; therefore FE is greater than 18 times EH. And FE is equal to BE; therefore BE is also greater than 18 times EH; therefore BH is much greater than 18 times HE. But, as BH is to HE, so is AB to BC, because of the similarity of the triangles; therefore AB is also greater than 18 times BC. And AB is the distance of the sun from the earth, while CB is the distance of the moon from the earth; therefore the distance of the sun from the earth is greater than 18 times the distance of the moon from the earth.

Again, I say that it is also less than 20 times that distance. For let DK be drawn through D parallel to EB, and about the triangle DKB let the circle DKB be described; then DB will be its diameter, because the angle at K is right. Let BL, the side of a hexagon, be fitted into the circle. Then, since the angle DBE is $\frac{1}{30}$th of a right angle, the angle BDK is also $\frac{1}{30}$th of a right angle; therefore the circumference BK is $\frac{1}{60}$th of the whole circle. But BL is also one-sixth part of the whole circle. Therefore the circumference BL is ten times the circumference BK.

And the circumference BL has to the circumference BK a ratio greater than that which the straight line BL has to the straight line BK;[3] therefore the straight line BL is less than ten times the straight line BK. And BD is double of BL; therefore BD is less than 20 times BK. But, as BD is to BK, so is AB to BC; therefore AB is also less than 20 times BC. And AB is the distance of the sun from the earth, while BC is the distance of the moon from the earth; therefore the distance of the sun from the earth is less than 20 times the distance of the moon from the earth. And it was before proved that it is greater than 18 times that distance.

Proposition 9. *The diameter of the sun is greater than 18 times, but less than 20 times, the diameter of the moon.*

For let our eye be at A, let B be the center of the sun, and C the center of the moon when the cone comprehending both the sun and the moon has its vertex at our eye, that is, when the points A, C, B are in a straight line [figure 7.2.3]. Let a plane be carried through ACB; this plane will cut the spheres in great circles and the cone in straight lines.

[3] This is equivalent to the modern result that if $\alpha > \beta$, then $\frac{\sin\alpha}{\sin\beta} < \frac{\alpha}{\beta}$. This can be demonstrated geometrically.

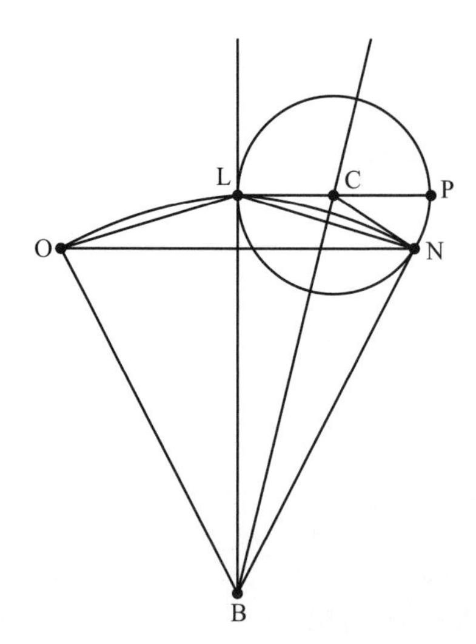

Figure 7.2.4.

Let it cut the spheres in the great circles FG, KLH, and the cone in the straight lines AFH, AGK, and let CG, BK be joined. Then, as BA is to AC, so will BK be to CG. But it was proved that BA is greater than 18 times, but less than 20 times, AC. Therefore, BK is also greater than 18 times, but less than 20 times, CG.

Proposition 10. *The sun has to the moon a ratio greater than that which 5832 has to 1, but less than that which 8000 has to 1.*

Let A be the diameter of the sun, B that of the moon. Then A has to B a ratio greater than that which 18 has to 1, but less than that which 20 has to 1. Now, since the cube on A has to the cube on B the ratio triplicate of that which A has to B, while the sphere about A as diameter also has to the sphere about B as diameter the ratio triplicate of that which A has to B; therefore, as the sphere about A as diameter is to the sphere about B as diameter, so is the cube on A to the cube on B.

But the cube on A has to the cube on B a ratio greater than that which 5832 has to 1, but less than that which 8000 has to 1, since A has to B a ratio greater than that which 18 has to 1, but less than that which 20 has to 1. Accordingly the sun has to the moon a ratio greater than that which 5832 has to 1, but less than that which 8000 has to 1.

Proposition 15. *The diameter of the sun has to the diameter of the earth a ratio greater than that which 19 has to 3, but less than that which 43 has to 6.*

We give only the proof of the first half of this proposition. To do this, we need first to show that the diameter of the earth's shadow at the moon is less than double the moon's diameter. But if we consider figure 7.2.4, where the moon has just entered total eclipse and assuming hypothesis 5, it is clear that $ON < 2LN$ and therefore that $ON < 2LP$, as desired. It then follows, since $LP < \frac{1}{18}$(sun's diameter), that the diameter of the earth's shadow is less than $\frac{1}{9}$(sun's diameter).

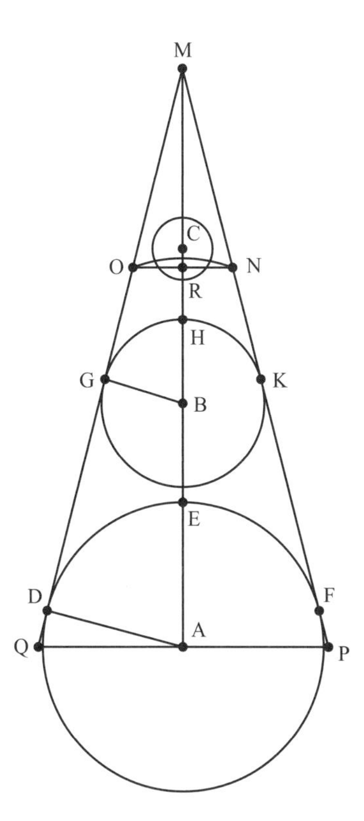

Figure 7.2.5.

For let A be the center of the sun, B the center of the earth, C the center of the moon when the eclipse [of the moon] is total, so as to secure that A, B, C may be in a straight line. Let a plane be carried through the axis, and let it cut the sun in the circle DEF, the earth in GHK, the shadow in the circumference NO, and the cone in the straight lines DM, FM. Let NO be joined, and from A let PAQ be drawn at right angles to AM [figure 7.2.5].

Then, since NO is less than one-ninth part of the diameter of the sun, therefore PQ has to NO a ratio much greater than that which 9 has to 1. Therefore AM also has to MR a ratio greater than that which 9 has to 1; and, *convertando*, MA has to AR a ratio less than that which 9 has to 8. Again, since AB is greater than 18 times BC, therefore it is much greater than 18 times BR; therefore AB has to BR a ratio greater than that which 18 has to 1; therefore, inversely, BR has to BA a ratio less than that which 1 has to 18; therefore, *componendo*, RA has to AB a ratio less than that which 19 has to 18.

But it was proved that MA also has to AR a ratio less than that which 9 has to 8; therefore *ex aequali*, MA will have to AB a ratio less than that which 171 has to 144, and therefore less than that which 19 has to 16; for parts have the same ratio as the same multiples of them; therefore, *convertendo*, AM has to BM a ratio greater than that which 19 has to 3. But as AM is to MB, so is the diameter of the circle DEF to the diameter of the circle GHK; therefore the diameter of the sun has to the diameter of the earth a ratio greater than that which 19 has to 3.

7.3 Hypsicles, *Little Work on Rising Times*

Hypsicles was a Greek astronomer and mathematician who flourished around 190 BCE in Alexandria. Only two of his works have survived to the present day: the so-called book 14 of Euclid's *Elements*, a continuation of the *Elements* covering mathematical relations of the dodecahedron and icosahedron inscribed in the same sphere over eight propositions and a number of lemmas (see section 3.16 in this volume), and the *Anaphoricus* (Little (Work) on Rising Times). The latter work sets out an arithmetical scheme to compute the rising times of zodiacal signs (also known as oblique ascensions), that is, the time it takes for arcs of the ecliptic to rise above the horizon. Rising times were a key astronomical phenomenon in ancient astronomical traditions—among other things, they were used to tell time during the night.

The *Anaphoricus* is historically important for a number of reasons. It is the earliest known source to explicitly define "degrees" as 360th parts of the ecliptic and the first surviving Greek text to use sexagesimal (base 60) numbers in its mathematical reckoning.

The *Anaphoricus* opens with three Euclidean-style propositions on arithmetical sequences that later on in the work are used to model these rising times. Employing such linear approaches to compute rising times was a clever approximation, as they approximated the circular arcs of the ecliptic—a great circle on the celestial sphere—by straight line segments. These linear approximations were theoretically eclipsed by the advent of trigonometric approaches that were probably developed at around the same time as the *Anaphoricus*. However, because these linear schemes involved only simple arithmetic, they remained popular among many practitioners for a long time.

The three propositions Hypsicles presents and proves invoke decreasing arithmetical sequences with either an odd $(2n-1)$ or an even $(2n)$ number of terms with given difference d that are equivalent to the following:

$$\sum_{k=1}^{n} a_k - \sum_{k=n+1}^{2n} a_k = n^2 d,$$

$$\sum_{k=1}^{2n-1} a_k = (2n-1)a_n,$$

and

$$\sum_{k=1}^{2n} a_k = n(a_1 + a_{2n}).$$

These results are then used to compute the rising times for particular latitudes, which are modeled as decreasing arithmetic sequences. We present the first proposition along with its proof.

The second half of this work applies this proposition and the other two to the task of computing the rising times for a particular instance: the local latitude of Alexandria. For this task, Hypsicles uses only a single empirically derived proportion, namely, the ratio of the longest to shortest daylight at this terrestrial latitude, which he gives as

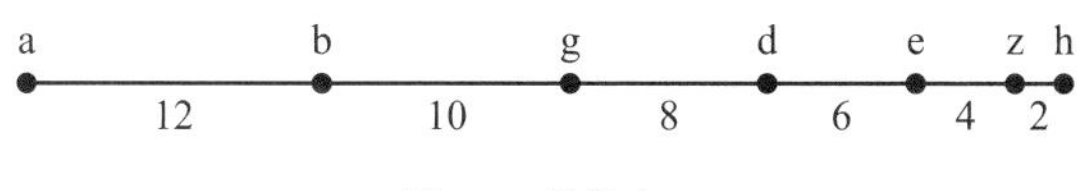

Figure 7.3.1.

7:5. Using this proportion, Hypsicles derives the rising times for each zodiacal sign and presents the results in a circular graphical table [see figure 7.3.2]. This worked example is presumably intended as a rubric from which astronomer-practitioners could compute the rising times for their own local latitude.

Proposition 1. *If there should be as many terms as we please, with a common difference, lying one after another, their number being even, the difference, by which the sum of half of the number (of terms) beginning from the greatest exceeds (the sum of) the remaining (terms), is the product of the common difference and the square of the (amount) of half of the number of terms (originally) set out.*

Let there be as many terms as we please, *ab*, *bg*, *gd*, *de*, *ez*, *zh*, with a common difference, lying one after another, their number being even, beginning from the greatest, *ab*, and let half of the number be *ad* [figure 7.3.1]. I say that the difference by which the sum of half of the number of terms exceeds the sum of the remaining terms, that is, the difference by which *ad* exceeds *dh*, is the product of the common difference and the square of half of the number of terms.

For since the difference between *ab* and *bg* is equal to the difference between *de* and *ez*, therefore, "alternately," the difference between *ab* and *de* is equal to the difference between *bg* and *ez*. Again, since the difference between *bg* and *gd* is equal to the difference between *ez* and *zh*, "alternately," the difference between *bg* and *ez* is equal to the difference between *gd* and *zh*. Consequently, the difference between *ab* and *de* and the difference between *bg* and *ez* and the difference between *gd* and *zh*, that is, the difference between *ad* and *dh*, is the product of the difference between *ab* and *de* and the number *ab*, *bg*, and *gd*. And the difference between *ab* and *de* is the product of the difference between *ab* and *bg* and the number of terms *ab*, *bg*, and *gd*. Consequently, the difference between *ad* and *dh* is the product of the difference between *ab* and *bg* and the square of the number of terms, *ab*, *bg*, and *gd*, that is, the square of half the number of terms originally set out.

. . .

[Astronomical section] With the circle of zodiacal (signs) being divided into 360 equal arcs, let each of the arcs be called spatial degrees. Indeed, in the same way, with the time in which the zodiacal circle returns from a point to the same point being divided into 360 equal time intervals, let each of these time intervals be called time degrees. With these things having been established, we will demonstrate, using the aforementioned theorems, that for a given locality, when the ratio of the longest day to the shortest is known, the time degrees in which each of the zodiacal signs rises will be known.

Let the *klime* through Alexandria in Egypt be assumed, where the longest day to the shortest day has a ratio of 7 to 5; for that it is thus, we demonstrated using the solsticial midday shadows taken from gnomons.

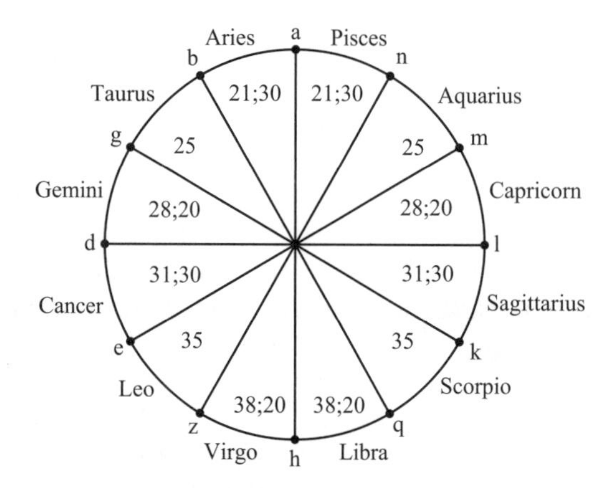

Figure 7.3.2.

Let the circle of zodiacal signs be laid out, in which the diameter of this circle intersecting with the celestial equator is *ah*. And let the circle be divided into the zodiacal signs at *a b g d e z h q k l m n* [figure 7.3.2]. And let the point *a* be the beginning of Aries, the point *b* the beginning of Taurus, *g* the beginning of Gemini, and let the successive points be understood as the beginnings of the successive zodiacal signs. Since the longest day with respect to the shortest day has a ratio of 7 to 5, and the time of the longest day is that in which the semicircle beginning with Cancer rises, that is to say, *dhl*, and the time of the shortest day is that in which the semicircle beginning from Capricorn rises, that is, *lad*, therefore the rising time of the semicircle *dhl* with respect to the rising time of the semicircle *lad* has a ratio of 7 to 5. Now, the whole circle rises in 360 time degrees: therefore the semicircle *dhl* rises in 210 time degrees and the semicircle *lad* in 150 time degrees. And the quadrant *dh* rises in the same time as the quadrant *hl*, and the quadrant *la* rises in the same time as *ad*. For each is offset an equal amount from the celestial equator. Therefore the quadrant *dh* will have risen in 105 time degrees, and the quadrant *da* will have risen in 75 time degrees. Therefore the rising time of the quadrant *hzed* will exceed the rising time of the quadrant *dgba* by 30 time degrees.

Since the six terms are the rising times of the arcs *hz*, *ze*, *ed*, *dg*, *gb*, *ba*, with common difference, lying one after another, beginning from the greatest, the one with respect to *h*, for this was proposed by the original assumptions concerning rising times, the difference, by which the sum from half of the number of terms beginning from the greatest exceeds the sum of the remaining terms, is the product of the difference of each term and the square of half of the number of terms. And the difference of rising times, by which the sum of the half of the number of remaining terms exceeds the sum of the remaining, is 30 time degrees; the square of half of the number of terms is 9; 30 divided by 9 is 3;20 time degrees.[4] Therefore, the difference of the rising times for the "twelfth parts" [henceforth zodiacal signs] *hz*, *ze*, *ed*, *dg*, *gb*, *ba* is 3;20.

[4] This notation means $3\frac{20}{60}$, or $3\frac{1}{3}$, and is the earliest recorded use of sexagesimal fraction notation in astronomical calculations in a Greek text.

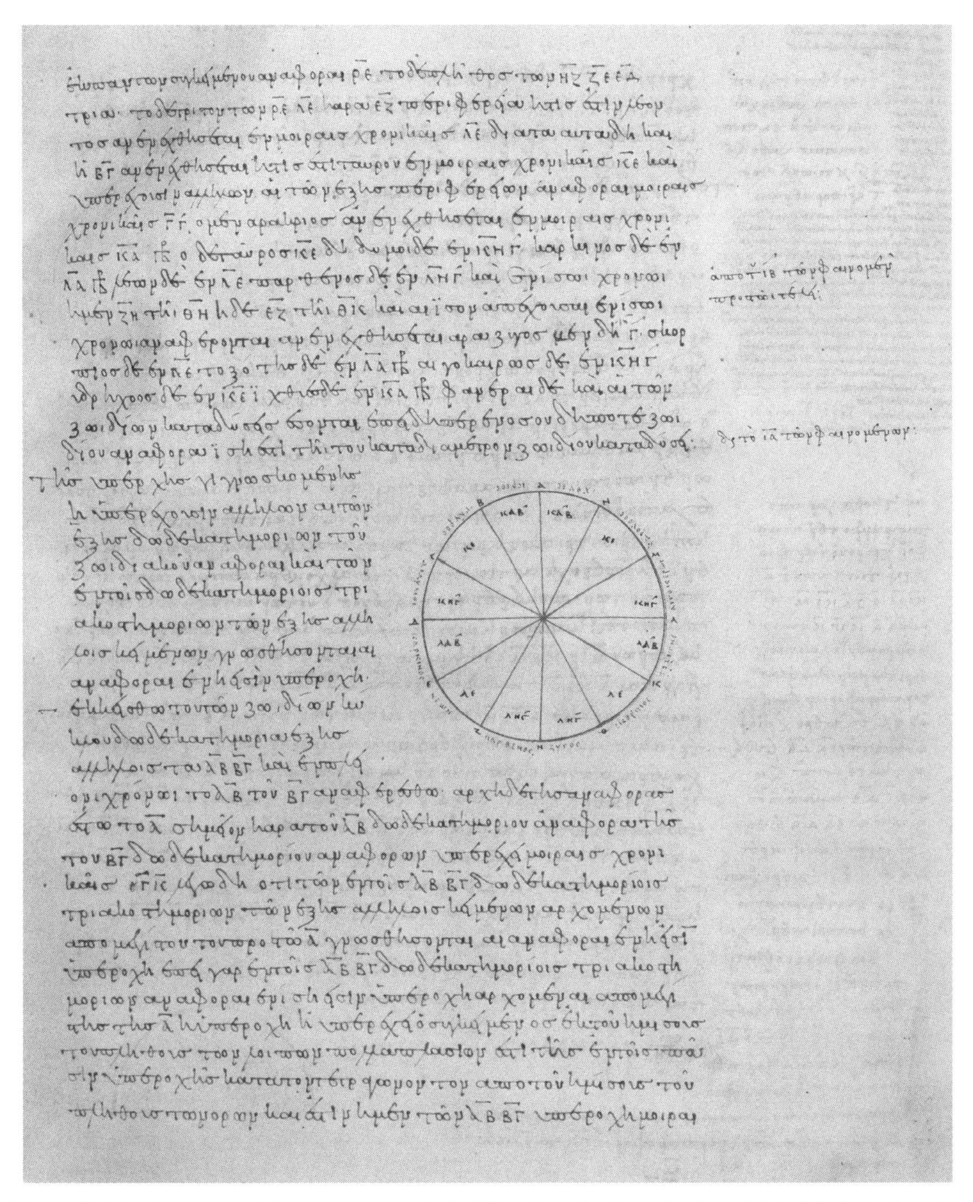

Figure 7.3.3. An excerpt from a manuscript of the *Anaphoricus* (Vat. gr. 204, f. 135r), showing the zodiacal circle displaying the numerical values associated with the rising time of each zodiacal sign for the locality of Alexandria. Image courtesy of the Vatican.

Again, since as many terms as we please are the rising times of the arcs *hz*, *ze*, *ed*, with common difference, lying one after another, their number being odd, beginning from the greatest, the sum of all of them is the product of the middle term and the number of them. Now, the rising time of the sum of them all is 105, the number of terms, *hz*, *ze*, *ed*, is three, a third of 105 is 35. Therefore the arc *ez*, which is Leo, will have risen in 35 time degrees. Through the same reasonings, indeed the arc *bg*, which is Taurus, will have risen in 25 time degrees.

And the rising times of the successive arcs will exceed each other by 3;20 time degrees. Therefore, Aries will rise in 21;40, Taurus in 25, Gemini in 28;20, Cancer in 31;40, Leo in 35, Virgo in 38;20, and *zh* in a time interval equal to *qh*, and *ez* to *qk*. And those offset an equal distance from the celestial equator rise in an equal time interval. Therefore, Libra will rise in 38;20, Scorpio in 35, Sagittarius in 31;40, Capricorn in 28;20, Aquarius in 25, Pisces in 21;40.

The setting times of the zodiacal signs will be obvious, as the rising time of any one zodiacal sign is equal to the setting time of the zodiacal sign diametrically opposite.

7.4 Theodosius, *Sphaerica*

Theodosius (2nd c. BCE) was born in Bithynia (now Turkey), but little is known of his life. From various references in Euclid's *Phaenomena* and in *On a Moving Sphere*, by Autolycus of Pitane (ca. 300 BCE), we know that much of the material in the *Sphaerica* was known earlier. Nevertheless, this work is the only surviving work from Greece containing the basic results on spherical geometry, results that were generally assumed by Menelaus in his own work on spherical trigonometry. In the *Elements*, Euclid dealt with the sphere only from the point of view of inscribing in it the five regular solids. Thus he defined a sphere as the surface created by the revolution of a semicircle around its diameter. On the other hand, Theodosius's work, in three books, starts with the definition of a sphere that is the generalization of Euclid's definition of a circle. And, in fact, many of the results in book 1 are closely related to Euclid's results on circles discussed in *Elements* 3. We present here many of Theodosius's propositions from book 1, with proofs for some of them, and then reproduce one important proposition from book 2.

Book 1

Definitions

1. A sphere is a solid figure contained by a unique surface such that all lines drawn to it from one particular point in its interior are equal.
2. The particular point in the previous definition is called the center of the sphere.
3. A diameter of the sphere is a line segment drawn through the center, ending in both directions on the surface of the sphere.
4. The poles of a sphere are the extremities of a diameter.
5. The pole of a circle drawn on the sphere is a point on the surface of the sphere, from which all the lines drawn to the circumference of the circle are equal.

Propositions

Proposition 1. *If a spherical surface is cut by a plane, the intersecting curve on the surface of the sphere is the circumference of a circle.*

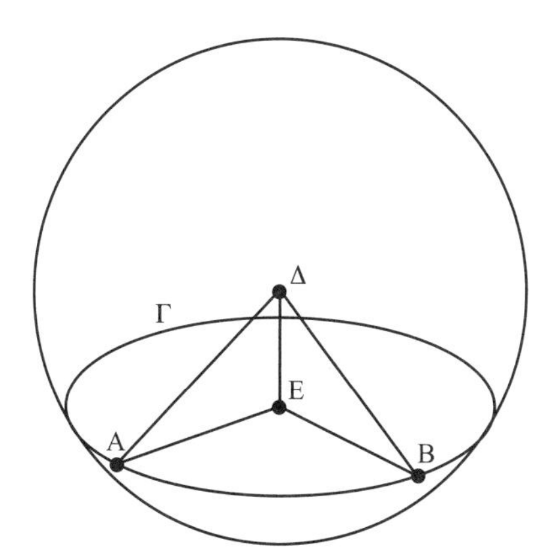

Figure 7.4.1.

Suppose a spherical surface is cut by a plane, and that the intersecting curve on the surface of the sphere is $AB\Gamma$; I say that $AB\Gamma$ is the circumference of a circle. If the cutting plane goes through the center of the sphere, it is clear that the curve $AB\Gamma$ is the circumference of a circle. . . . So now suppose that the cutting plane does not pass through the center of the sphere [figure 7.4.1].

Let the point Δ be the center of the sphere, and draw the perpendicular ΔE from the point Δ to the cutting plane. Connect the lines EA, EB from E to the curve $AB\Gamma$, and draw the lines ΔA, ΔB. Now ΔA is equal to ΔB. Also the sum of the squares on the lines AE, $E\Delta$ equals the square on the line $A\Delta$, and the sum of the squares on the lines BE, $E\Delta$ equals the square on the line ΔB. Therefore, the sum of the squares on AE, $E\Delta$ equals the sum of the squares on BE, $E\Delta$. Subtracting the square on $E\Delta$ shows that the line AE equals the line BE. It follows that all lines from the point E to points on the curve $AB\Gamma$ are equal. Therefore the curve $AB\Gamma$ is the circumference of a circle with center E.

Corollary. *Given a circle on a sphere, the perpendicular drawn from the center of the sphere to the circle passes through the center of the circle.*

Proposition 2. *To find the center of a given sphere.*

Corollary. *If there is a circle on a sphere and if from the center of the circle, one erects a perpendicular to its plane, that line passes through the center of the sphere.*

Proposition 3. *A sphere can not touch a plane that does not cut it in more than one point.*

Proposition 4. *If a sphere touches a plane that does not cut it, the line from the center of the sphere to the point of contact is perpendicular to the plane.*

Proposition 5. *If a sphere touches a plane that does not cut it, and if one draws a line from the point of contact at right angles to the plane, the line will pass through the center of the sphere.*

Proposition 6. *Among the circles on the sphere, those whose plane passes through the center of the sphere are the greatest, and, among the others, those whose planes are*

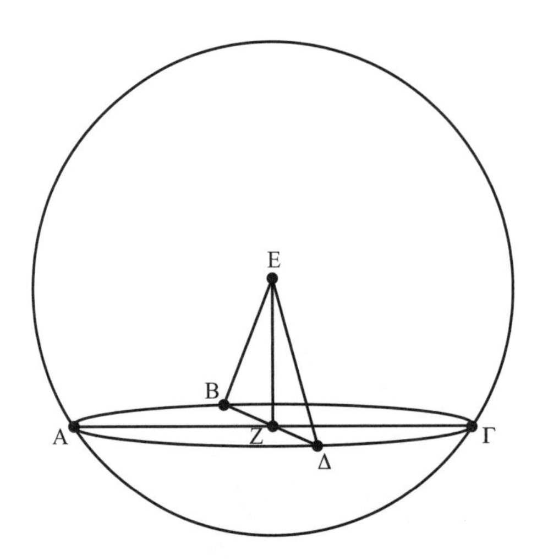

Figure 7.4.2.

equally distant from the center are equal, and of two circles whose planes are not equally distant from the center, the further one is smaller.

Proposition 7. *If one draws a line from the center of the sphere to the center of a circle on the sphere, this line is perpendicular to the plane of the circle.*

Let $AB\Gamma\Delta$ be a circle in the sphere. Let E be the center of the sphere, Z be the center of the circle, and connect EZ [figure 7.4.2]. I say that the line EZ is perpendicular to the plane of the circle $AB\Gamma\Delta$.

Draw two diameters of the circle $AZ\Gamma$ and $BZ\Delta$, and let BE, $E\Delta$ be joined. Since the line BZ is equal to the line $Z\Delta$, and since the line EZ is common, the two lines BZ, ZE are respectively equal to the two lines ΔZ, ZE, and the base BE is equal to the base $E\Delta$, since the point E is the center of the sphere and the points B, Δ are on the surface. Therefore the angle between the lines BZ, ZE is equal to the angle between the lines ΔZ, ZE. But since the line EZ erected on the line $B\Delta$ forms these two adjacent angles, each of those angles is right. Therefore, each of the angles formed by the lines BZ, ZE and by the lines ΔZ, ZE are right, and the line EZ is perpendicular to the line $B\Delta$. One demonstrates similarly that the line EZ is also perpendicular to the line $A\Gamma$. Therefore, the line EZ is at right angles to $A\Gamma$, $B\Delta$ at their point of intersection, so it is also perpendicular to the plane that passes through $A\Gamma$, $B\Delta$. But this plane is the plane of the circle $AB\Gamma\Delta$. So the line EZ is perpendicular to the plane of the circle $AB\Gamma\Delta$.

Proposition 8. *If one draws a line from the center of the sphere through the center of a circle on the sphere and perpendicular to the plane of the circle, the prolongation of the line will go through the poles of the circle.*

Proposition 9. *Given a circle on a sphere, if one draws a perpendicular from one of its poles to the plane of the circle, the line will pass through the center of the circle and, if prolonged, through the other pole of the circle.*

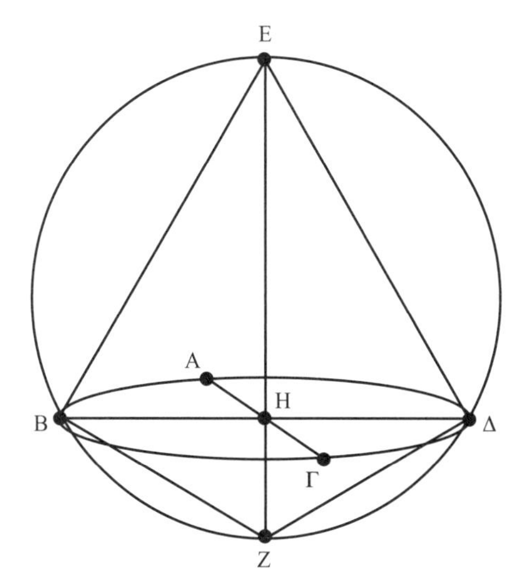

Figure 7.4.3.

Proposition 10. *Given a circle on a sphere, the line drawn through the two poles of the circle is perpendicular to the plane of the circle and passes through the center of the circle and the center of the sphere.*

Let $AB\Gamma\Delta$ be a circle on a sphere and E, Z its two poles. Let the straight line EZ, which passes through the two poles, be joined [figure 7.4.3]. I say that line EZ is perpendicular to the circle $AB\Gamma\Delta$, and passes through its center and the center of the sphere.

Let the line EZ intersect the plane of circle $AB\Gamma\Delta$ at point H. Draw two lines through point H, lines $AH\Gamma$ and $BH\Delta$, and join the lines BE, $E\Delta$, BZ, and $Z\Delta$. Line BE is equal to line $E\Delta$, and line EZ is common. Thus the two lines BE, EZ are respectively equal to the two lines $E\Delta$, EZ, and the base BZ is equal to the base $Z\Delta$; thus the angle between the lines BE, EZ is equal to the angle between the lines ΔE, EZ. Again, since line BE is equal to line $E\Delta$ and EH is common, the two lines BE, EH are equal respectively to the two lines ΔE, EH. Also, the angle between the lines BE, EH is also equal to the angle between the lines ΔE, EH. Consequently, the base BH is equal to the base $H\Delta$; the triangle EBH is equal to the triangle $E\Delta H$, and the remaining angles are equal to the remaining angles which are subtended by equal sides. Therefore, the angle between the lines BH, HE is equal to the angle between the lines ΔH, HE. Therefore, each of the angles formed by the lines BH, HE and the lines ΔH, HE are right, and the line EH is at right angles to the line $B\Delta$. Similarly, one demonstrates that the line EH is at right angles to the line $A\Gamma$ and therefore line EHZ is at right angles to the plane that passes through lines $B\Delta$, $A\Gamma$, that is to say, to the circle $AB\Gamma\Delta$. Thus the line EZ is perpendicular to the circle $AB\Gamma\Delta$.

I say also that this line passes through the center of the circle and of the sphere. In fact, since $AB\Gamma\Delta$ is a circle in a sphere and since the line EH is drawn from one of the poles of the circle, perpendicularly to the circle and meets the plane at point H, point H is the center of circle $AB\Gamma\Delta$ [proposition 9]. I say, also that this line passes through the center of the sphere. In fact, since $AB\Gamma\Delta$ is a circle in a sphere, and since the line EHZ

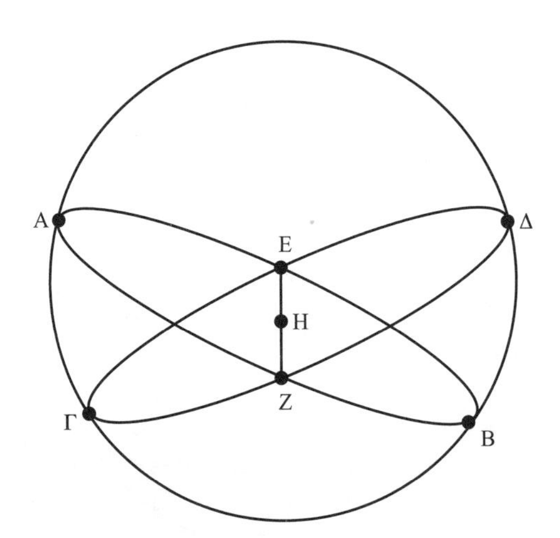

Figure 7.4.4.

has been drawn through the center of the circle, perpendicular to the plane of the circle, the center of the sphere lies on the line EHZ [proposition 2, corollary]. Consequently, the line EZ passes through the center of the sphere. Therefore, the line EZ is perpendicular to the circle $AB\Gamma\Delta$ and passes through the center of the circle and that of the sphere.

Proposition 11. *Two great circles on a sphere cut each other into two equal parts.*

On a sphere, let the two great circles AB, $\Gamma\Delta$ cut each other at the points E, Z [figure 7.4.4]. I say that the great circles AB, $\Gamma\Delta$ cut each other into two equal parts.

So let the center of the two circles be the point H, which is also the center of the sphere, and draw the lines EH, HZ. The points E, H, Z are in the plane of AB and also in the plane of $\Delta\Gamma$. Thus, the points E, H, Z are in the planes of both great circles AB, $\Gamma\Delta$ and are situated on the common section of the two planes. But the common section of two planes is always a straight line [*Elements* 11:3]. Thus EHZ is a straight line. And since the point H is the center of the circle AB, the line EZ is a diameter of the circle. So arcs EAZ and EBZ are semicircles. Similarly, since the point H is the center of circle $\Gamma\Delta$, the line EZ is a diameter of that circle, and arcs $E\Gamma Z$ and $E\Delta Z$ are semicircles. Therefore, the circles AB, $\Gamma\Delta$ cut each other into two equal parts.

Proposition 12. *Two circles on a sphere that cut each other into two equal parts are great circles.*

Proposition 13. *If a great circle on a sphere cuts another circle at right angles, then it cuts the circle into two equal parts and passes through its poles.*

Let great circle $AB\Gamma\Delta$ on a sphere cut some other circle on the sphere, circle $EBZ\Delta$, at right angles [figure 7.4.5]. I say that it cuts the circle into two equal parts and passes through its poles.

Let the common section of the two circles be joined, line $B\Delta$. Let H be the center of circle $AB\Gamma\Delta$. It is also the center of the sphere. Let there be drawn from point H the perpendicular $H\Theta$ to line BD and prolong this to meet the surface of the

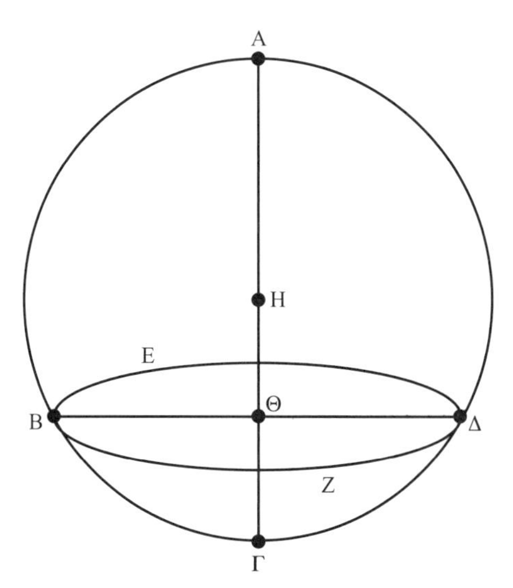

Figure 7.4.5.

sphere at points A, Γ. Since the two planes, that of the circle $AB\Gamma\Delta$ and that of the circle $EBZ\Delta$, are perpendicular, and since the line ΘA lying in one of the planes makes a right angle with their common section, the line $A\Gamma$ is at right angles to the plane $EBZ\Delta$. Consequently, since $EBZ\Delta$ is a circle in a sphere and the line $H\Theta$ has been drawn from the center of the sphere perpendicularly to the circle, and this line cuts the plane of circle $EBZ\Delta$ at point Θ, the point Θ is the center of circle $EBZ\Delta$ [proposition 1, corollary]. Consequently, arcs $BE\Delta$, $BZ\Delta$ are both semicircles, and the circle $AB\Gamma\Delta$ cuts circle $EBZ\Delta$ in two equal parts. I say that it also passes through its poles. Since $EBZ\Delta$ is a circle in a sphere and line $H\Theta$ is drawn from the center of the sphere perpendicularly to the circle, and its prolongation meets the surface of the sphere at points A, Γ, and since if a circle is on a sphere and if one draws a perpendicular from the center of the sphere to the circle and prolongs it on both sides, it passes through the poles of the circle [proposition 8], therefore the points A, Γ are the poles of the circle $EBZ\Delta$. Consequently, the circle $AB\Gamma\Delta$ cuts the circle $EBZ\Delta$ and passes through its poles. But since it also cuts it into two equal parts, the circle $AB\Gamma\Delta$ cuts the circle $EBZ\Delta$ into two equal parts and passes through its poles.

Proposition 14. *If a great circle on a sphere cuts into two equal parts another circle that is not a great circle, then it cuts that circle at right angles and passes through its poles.*

Proposition 15. *If a great circle on a sphere cuts another circle on the sphere and passes through its poles, then it cuts that circle into two equal parts and at right angles.*

Suppose that the great circle $AB\Gamma\Delta$ on a sphere cuts another circle $EBZ\Delta$ and passes through its poles [figure 7.4.5]. I say that it cuts that circle into two equal parts and at right angles. Since the points A, Γ are the poles of the circle $EBZ\Delta$, it is clear that the points are situated on the circle $AB\Gamma\Delta$, since the circle $AB\Gamma\Delta$ cuts the circle $EBZ\Delta$ and passes through its poles. Connect the line $A\Gamma$. Since $EBZ\Delta$ is a circle in a sphere and the line

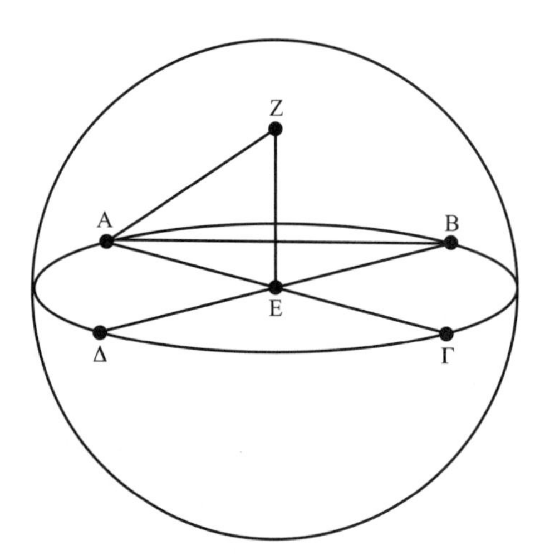

Figure 7.4.6.

$A\Gamma$ has been drawn through its poles, and since, in a circle on a sphere, the line drawn through its poles is perpendicular to the circle and passes through its center and that of the sphere [proposition 10], it follows that the line $A\Gamma$ is perpendicular to the circle $EBZ\Delta$, and all the planes that contain the line $A\Gamma$ are thus perpendicular to the circle $EBZ\Delta$ [*Elements* 11:18]. But the circle $AB\Gamma\Delta$ is one of the planes that contain the line $A\Gamma$. Therefore the circle $AB\Gamma\Delta$ is perpendicular to the circle $EBZ\Delta$. Thus the circle $AB\Gamma\Delta$ cuts the circle $EBZ\Delta$ at right angles. It also cuts the circle into two equal parts [proposition 13]. Thus the circle $AB\Gamma\Delta$, which cuts the circle $EBZ\Delta$ and passes through its poles, cuts it into two equal parts and at right angles.

Proposition 16. *If a great circle is on a sphere, the straight line that is drawn from its pole to its circumference is equal to the side of the square that is inscribed in the great circle.*

Let the great circle on the sphere be $AB\Gamma\Delta$. I say that the straight line that is drawn from its pole to its circumference is equal to the side of the square that is inscribed in the great circle. Let there be drawn two diameters of circle $AB\Gamma\Delta$ cutting each other at right angles, lines $A\Gamma$ and $B\Delta$ [figure 7.4.6]. Then, since circle $AB\Gamma\Delta$ is a great circle, its center and the center of the sphere are the same. Let that be point E. Let us draw from E a perpendicular to the plane of circle $AB\Gamma\Delta$, meeting the surface of the sphere at Z. Therefore, Z is a pole of circle $AB\Gamma\Delta$ [proposition 8]. Let the two lines ZA, AB be joined. Therefore, line AB is the side of the square which is inscribed in circle $AB\Gamma\Delta$ and line ZA is drawn from the pole to the circumference of the circle. I say that line ZA is equal to line AB.

For line ZE is a perpendicular to the circle $AB\Gamma\Delta$, and it makes a right angle with every straight line that is drawn from its endpoint in the plane of circle $AB\Gamma\Delta$. Therefore, line ZE is perpendicular to each of the lines AE, EB, $E\Delta$, $E\Gamma$. Since point E is the center of the sphere, EB is equal to EZ and line EA is common. Therefore, the two lines EA, EB are equal to the two lines EA, EZ respectively, and right angle BEA is equal to right angle AEZ, so base BA is equal to base AZ. Therefore, the line that is drawn from the pole of circle $AB\Gamma\Delta$ to its circumference is equal to the side of the square that is inscribed in the great circle.

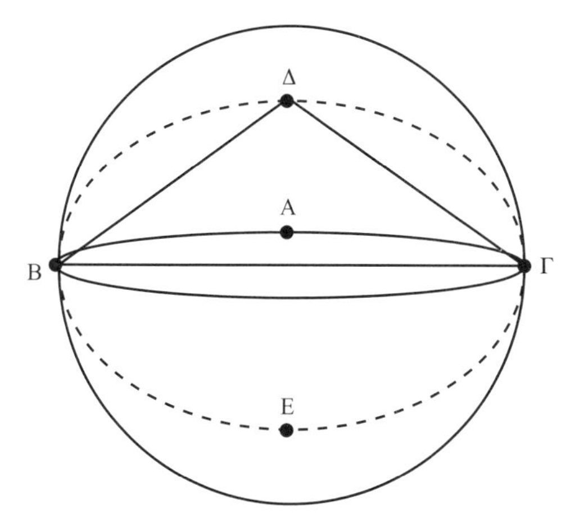

Figure 7.4.7.

Proposition 17. *If a circle is on a sphere, and the straight line that is drawn from its pole to its circumference is equal to the side of the square that is inscribed in a great circle on the sphere, then the circle is great.*

Let circle $AB\Gamma$ be on the sphere with pole Δ, and let the line $\Delta\Gamma$ that is drawn from its pole to its circumference be equal to the side of the square that is inscribed in a great circle on the sphere. I say that circle $AB\Gamma$ also is great. Let there be drawn a plane that passes through line $\Delta\Gamma$ and the center of the sphere. It makes a section on the surface of the sphere that is a great circle, circle $B\Delta\Gamma E$ [figure 7.4.7].

Let the common section of it and circle $AB\Gamma$ be line $B\Gamma$, and let line ΔB be joined. Therefore, line ΔB is equal to line $\Delta\Gamma$. But line $\Gamma\Delta$ is a side of the square, so line $B\Delta$ is also a side of the square, and each of the arcs $B\Delta$, $\Delta\Gamma$ is a quarter circle. Therefore, arc $B\Delta\Gamma$ is a semicircle, and line $B\Gamma$ is a diameter of circle $\Delta BE\Gamma$. Since point Δ is a pole of circle $AB\Gamma$, circle $\Delta BE\Gamma$ cuts circle $AB\Gamma$ and passes through its poles. Since great circle $\Delta BE\Gamma$ is on a sphere and it has cut a circle on the sphere, circle $AB\Gamma$, and passed through its two poles, it also bisects it and at right angles [proposition 15]. Therefore, the two circles $AB\Gamma$, $\Delta EB\Gamma$ bisect each other. Since circles that bisect one another on a sphere are great, therefore circle $AB\Gamma$ is great [proposition 12].

The following proposition gives a construction to find a great circle arc through two given points on a sphere, assuming they are not diametrically opposed. Theodosius does not explicitly note that the constructed arc is unique, but this follows immediately from proposition 11.

Proposition 20. *To describe a great circle passing through two given points on the surface of a sphere.*

Let the two given points on the surface of the sphere be A, B. We want to describe a great circle passing through them. If these two points are on a diameter of the sphere, it is clear that great circles without limit can be described on the two points. So let us assume that A, B are not on a diameter of the sphere, and let us describe circle $\Gamma E\Delta$ with

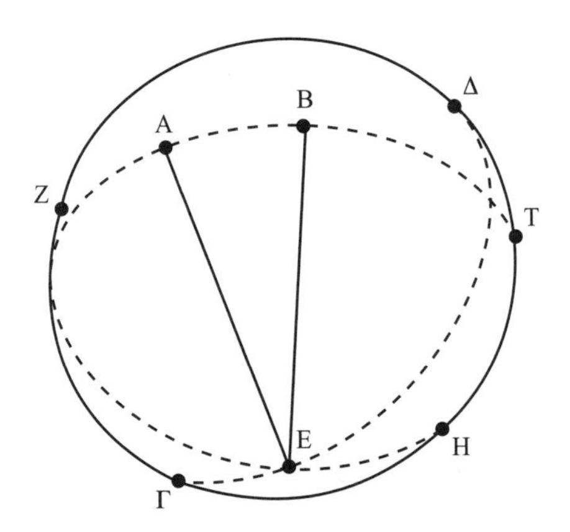

Figure 7.4.8.

pole A and a distance equal to the side of the square that is inscribed in a great circle [figure 7.4.8]. Therefore, circle $\Gamma E\Delta$ is great, for the straight line that is drawn from its pole to is circumference is equal to the side of the square that we inscribe in a great circle [proposition 17].

Again, let us describe circle ZEH with pole B and a distance equal to the side of the square that is inscribed in a great circle. Therefore, circle ZEH is also great. Let us join the two lines EA, EB. Each of the lines EA, EB is equal to the side of the square that is inscribed in a great circle. Therefore, line EA is equal to line EB, and the circle that is described with pole E and distance EB will pass through point A also. Let this circle be ABT. Therefore, circle ABT is great, for the straight line that is drawn from its pole to its circumference is equal to the side of the square that is inscribed in a great circle [proposition 17]. Therefore, a great circle ABT has been described, passing through the two given points A, B on the surface of the sphere.

From book 2 of the *Sphaerica*, we include just proposition 11. Although it is a very specialized result, similar to other such results in book 2, it is needed in the proof of proposition 2 in book 1 of Menelaus's *Sphaerica*, whose proof we include in the next section on Menelaus. Other results in books 2 and 3 deal with tangent circles on a sphere and with parallel circles and are often special cases of results needed for astronomy on the celestial sphere.

Book 2

Proposition 11. *We construct on the diameters of two equal circles equal and perpendicular arcs of circles. We cut off from the extremities of these arcs two equal arcs that are smaller than half of the entire arcs, and from the points so obtained, we draw equal lines onto the circumferences of the first two circles. Then these lines cut equal arcs on the first circles, from the extremities of their diameters.*

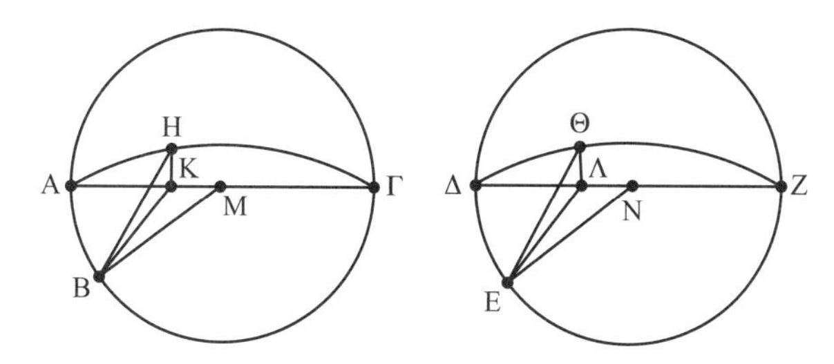

Figure 7.4.9.

Suppose we construct, on the equal circles $AB\Gamma$, ΔEZ, on the diameters $A\Gamma$, ΔZ, equal and perpendicular arcs of circles $AH\Gamma$, $\Delta\Theta Z$. Cut off from the extremities A, Δ of these arcs equal segments AH, $\Delta\Theta$ that are smaller than half of the arcs $AH\Gamma$, $\Delta\Theta Z$, and from the points H,Θ, draw equal lines HB, ΘE to the first circles $AB\Gamma$, ΔEZ [figure 7.4.9]. I say that the arc AB is equal to the arc ΔE.

Indeed, let fall from the points H, Θ perpendiculars on the planes of the circles $AB\Gamma$, ΔEZ, which therefore fall on the common sections, that is, on the lines $A\Gamma$, ΔZ. Let these be the lines HK, $\Theta\Lambda$. Let M, N be the centers of the circles $AB\Gamma$, ΔEZ. Connect the lines KB, BM, ΛE, EN. Then, since the line HK is perpendicular to the plane of the circle $AB\Gamma$, it forms also a right angle with all the lines that meet it and that are in the plane of the circle $AB\Gamma$. But the line KB, lying in the plane of the circle $AB\Gamma$, meets it; therefore the angle between the lines HK and KB is right. Similarly, the angle between the lines $\Theta\Lambda$ and ΛE is also right.

In addition, since the two segments $AH\Gamma$, $\Delta\Theta Z$ are equal, and since the arcs cut off, AH, $\Delta\Theta$, are equal, and since the lines HK, $\Theta\Lambda$ were made perpendicular, it follows that the line AK is equal to the line $\Delta\Lambda$ and the line HK is equal to the line $\Theta\Lambda$.[5] And since the line BH is equal to the line ΘE, the square on the line BH is also equal to the square on the line ΘE. But the squares on the lines HK, KB form the square on the line BH, and the squares on the lines $\Theta\Lambda$, ΛE form the square on the line ΘE. Therefore, the squares on the lines HK, KB are equal to the squares on the lines $\Theta\Lambda$, ΛE. But the square on the line HK is equal to the square on the line $\Theta\Lambda$. As a consequence, the remaining square constructed on the lines KB is equal to the remaining square constructed on the line ΛE. It follows that the line KB is equal to the line ΛE.

And since the line AM is equal to the line ΔN, lines on which the line AK is equal to the line $\Delta\Lambda$, it follows that the remaining line KM is equal to the remaining line ΛN. But the line BM is also equal to the line NE; thus the two lines KM, MB are respectively equal to the two lines ΛN, NE. Also, the base KB is equal to the base ΛE. Thus the angle formed by the lines KM, MB is equal to the angle formed by the lines ΛN, NE. But, in equal circles, equal angles from the centers to the circumferences subtend equal arcs. Thus the arc AB is equal to the arc ΔE.

[5] Angles HKA and $\Theta\Lambda\Delta$ are right. The chords AH and $\Delta\Theta$ are equal since they subtend equal arcs. Also, angle HAK equals angle $\Theta\Delta\Lambda$ since they are measured by equal arcs $H\Gamma$ and ΘZ. So the triangles AHK and $\Delta\Theta\Lambda$ are congruent. So $AK = \Delta\Lambda$ and $HK = \Theta\Lambda$.

7.5 Menelaus, *Sphaerica*

We know that Menelaus was active in the years surrounding 100 CE, because Ptolemy mentions an observation he made in 98 CE and Plutarch (45–120 CE) refers to him in one of his writings. The *Sphaerica* is his major work, although later authors mention several other mathematical works. In any case, the *Sphaerica* is the earliest work devoted to spherical triangles and, in particular, to spherical trigonometry. Although Ptolemy quotes some theorems of Menelaus, as does Pappu later on, it was not until the Islamic period that detailed studies of his work were accomplished. Evidently, a Greek manuscript of the *Sphaerica* reached Baghdad by the ninth century, for the initial translation into Arabic was made then by an unknown translator. Subsequently, it was translated again by Isḥāq ibn Ḥunayn (ca. 830–910) or perhaps his father, Ḥunayn ibn Isḥāq (ca. 809–873), whose work was revised and corrected by Muḥammad ibn 'Īsa ibn Aḥmad al-Māhānī (ca. 820–880) and further revised by Aḥmad ibn Abī Sa'd al-Harawī (mid-10th c.). A somewhat later translation, by Abū 'Uthmān al-Dimashqī (d. after 914), was also revised and amended by other mathematicians, including Abū Naṣr Manṣūr ibn 'Irāq (ca. 970–1036). The second of these translations was rendered into Hebrew by Jacob ben Makhir ibn Tibbon (1236–1304), while Gerard of Cremona (1114–1187) used both the second and third Arabic translations to produce a Latin version. Edmond Halley (1656–1741) retranslated the work into Latin, based mostly on the Hebrew version. With so many translations, each of which made modifications to the original Greek, which itself is lost, it is very difficult to determine exactly what Menelaus himself wrote, or even how he divided the work into books. Nevertheless, in what follows, we attempt to present a readable version in English of excerpts from this seminal work, using the Latin version's numbering of the propositions and division into three books.

Although Menelaus wrote his work in classical Greek style, beginning with some definitions, he clearly assumed some basic knowledge of spherical geometry. In fact, he often uses results from Theodosius's *Sphaerica*, many of which are included in our translation of that work (section 7.4). In general, book 1 is concerned with properties of spherical triangles. Many of the theorems are analogues of theorems on plane triangles given by Euclid in *Elements* 1. We give the statements of selected theorems, with proofs of only some of these. From book 3, we include just proposition 1, which has become known as Menelaus's theorem on the sector figure, and some lemmas necessary for its proof.

Book 1

Definitions

1. A spherical triangle is a region bounded by [three] great circle arcs on the surface of a sphere.
2. And each arc, which is always less than a semicircle, is called a side or leg of the triangle.

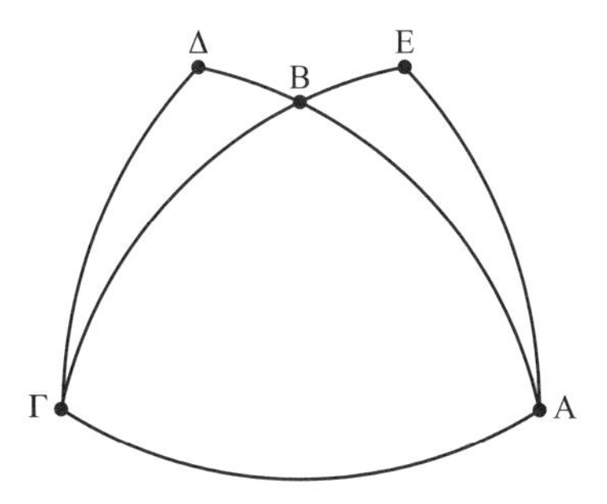

Figure 7.5.1.

3. The angles [of the triangle] are the angles contained by the arcs of the great circles on the sphere.
4. And two angles are said to be equal, when the planes containing the arcs forming the first angle are inclined at the same angle as the planes containing the arcs forming the second.
5. And if of these two angles, the planes containing the arcs of the first are inclined at a greater angle than the planes containing the arcs of the second, that angle is said to be greater.
6. And if the planes containing the arcs of an angle form a right angle, then the angle is said to be right.

Propositions

Proposition 1. *Given a point on the arc of a great circle on the surface of a sphere, and given an angle between two arcs of great circles on the surface, to construct an angle equal to the given angle at the given point.*

Proposition 2. *In any spherical triangle having two equal legs, the two angles adjoining the third side will be equal.*

Let $AB\Gamma$ be a spherical triangle with two equal legs, AB, $B\Gamma$. I say that the two angles at the base $A\Gamma$, namely, angles $BA\Gamma$ and $A\Gamma B$, are equal. With center A and radius $A\Gamma$, describe the arc $\Gamma\Delta$, and with center Γ and radius ΓA, the arc AE. Extend arc AB to $AB\Delta$ and arc ΓB to ΓBE [figure 7.5.1].[6]

Now arcs ΓBE and $AB\Delta$ are each equal to arc $A\Gamma$, and arc AB is equal to arc ΓB. Therefore, the remainder, arc $B\Delta$, is equal to the remainder BE. Since the arc $\Gamma\Delta$ is described around A at the same distance as the arc AE around Γ, therefore the circle of the arc $\Gamma\Delta$ is equal to the circle of the arc AE. And since the arc $AB\Delta$ passes through the pole of the circle $\Gamma\Delta$, it will be perpendicular to it. Likewise, the arc ΓBE will be perpendicular to the arc AE. So we have elevated perpendicularly equal segments of the equal circles

[6] We are assuming that the two drawn arcs lie outside the original triangle. If they are inside, the proof must be modified. And, of course, if the two arcs coincide with the original triangle, the proof is easy.

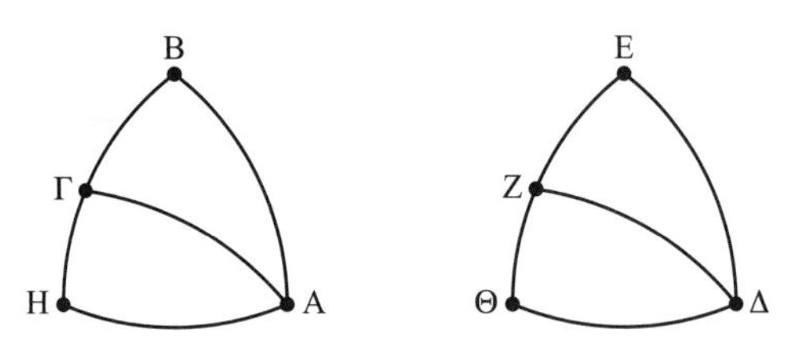

Figure 7.5.2.

AE, $\Gamma\Delta$ on the diameters from Δ, E, and which have arcs $A\Delta$, ΓE. We have elevated two equal arcs BE, $B\Delta$ which are not equal to half of the whole and which are set up in lines facing BA, $B\Gamma$. Therefore, the arc $\Gamma\Delta$ is equal to the arc AE.[7] Since on the sphere, two angles $BA\Gamma$, $A\Gamma B$, are contained under the arcs of great circles, and since there were drawn on the two points A, Γ and at the same distance two equal arcs $\Gamma\Delta$, AE, therefore, the angle $BA\Gamma$ is equal to the angle $A\Gamma B$.

Proposition 3. *In any spherical triangle having two equal angles, the two sides that subtend them are equal.*

Proposition 4. *If in two spherical triangles an angle of one is equal to an angle of the other, and if the two arcs containing the equal angles are respectively equal, then the remaining arcs are also equal. Also, if in the two triangles the remaining arcs are equal, then the angles opposite these equal arcs in the two triangles are also equal.*

Let the two spherical triangles be $AB\Gamma$ and ΔEZ. Assume that angle B of triangle $AB\Gamma$ equals angle E of triangle ΔEZ, that arc AB equals arc ΔE, and that arc $B\Gamma$ equals arc EZ [figure 7.5.2]. I say that arc $A\Gamma$ equals arc ΔZ.

Describe around the pole B with radius BA the arc AH and around the pole E with radius $E\Delta$ the arc $\Delta\Theta$. Since angle B is equal to angle E and the arc AH is described around the pole B with the same radius as the arc $\Delta\Theta$ is described around the pole E, it follows that arc AH is equal to arc $\Delta\Theta$. Since arc $B\Gamma$ passes through the pole of arc AH, it is perpendicular to it; similarly, arc EZ is perpendicular to arc $\Delta\Theta$. But arc $B\Gamma$ is equal to arc EZ, and arc BH is equal to arc $E\Theta$ (since both are equal to arc AB). Therefore the remaining arc $H\Gamma$ is equal to the remaining arc ΘZ. Since we elevated perpendicularly on two diameters of equal circles, whose segments are AH and $\Delta\Theta$, arcs of two equal circles, namely, the arcs $H\Gamma B$ and ΘZE, and [since we] have separated from them equal parts less than half, namely, the arcs ΓH and ΘZ, and we separated from the two equal circles, on the side of the extremities of the diameters, equal segments, AH and $\Theta\Delta$, therefore the line joining the points Z, Δ equals the line joining the points ΓA. Therefore the arc $Z\Delta$ is equal to the arc ΓA. [The proof of the second part of the theorem is similar and is omitted.]

Proposition 5. *In any spherical triangle, the sum of any two sides is greater than the third side.*

[7] Theodosius, *Sphaerica*, book 2, proposition 11. It is not entirely obvious that this theorem of Theodosius applies, since it is concerned with circles and lines in three-dimensional space while the current theorem deals with curves on the surface of a sphere. But a careful study of both results should convince the reader that one can in fact apply Theodosius's result in this situation. In particular, note that the diameters in question are diameters in three-dimensional space of the small circles drawn in the first step of the proof.

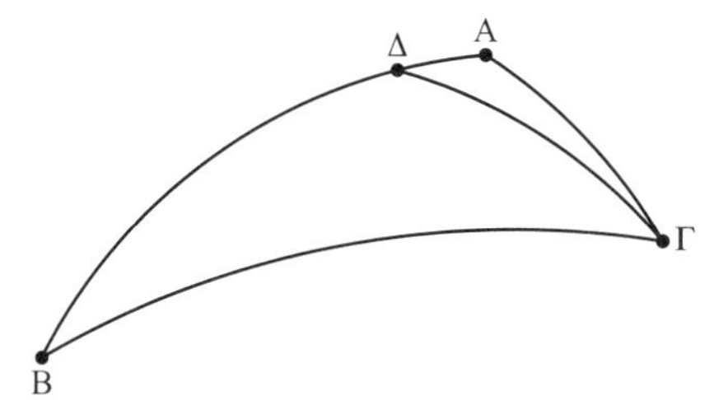

Figure 7.5.3.

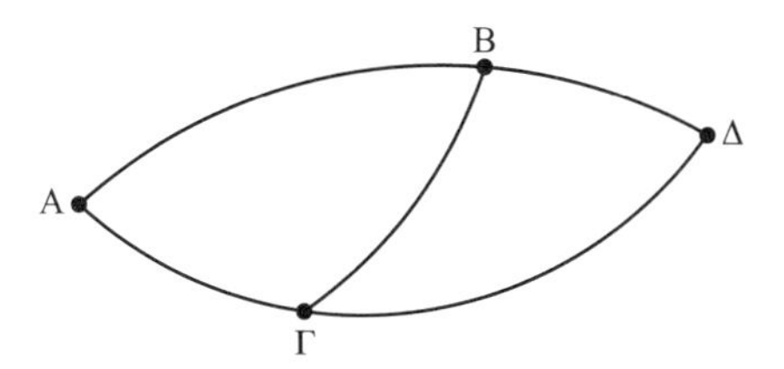

Figure 7.5.4.

Proposition 7. *In any spherical triangle, if one angle is greater than another, then the arc subtending the larger angle is greater than the arc subtending the smaller.*

Let $AB\Gamma$ be a spherical triangle, with angle Γ greater than angle B. I say that the arc AB is greater than the arc $A\Gamma$. At the point Γ, construct the angle $B\Gamma\Delta$ equal to angle B [figure 7.5.3]. Then arc $B\Delta$ is equal to arc $\Delta\Gamma$. Therefore the sum of the arcs $A\Delta$ and $\Delta\Gamma$ is equal to arc AB. But the sum of arcs $A\Delta$ and $\Delta\Gamma$ is greater than arc $A\Gamma$. Therefore arc AB is greater than arc $A\Gamma$.

Proposition 9. *In any spherical triangle, the greater arc subtends the greater angle.*

Proposition 10. *In a spherical triangle, if the sum of two sides is equal to a semicircle, then the exterior angle formed by extending the third side will be equal to the opposite interior angle. If the sum of the two sides is less than a semicircle, then the exterior angle produced as before will be greater than the opposite interior angle. Finally, if the sum of the two sides is greater than a semicircle, then the exterior angle will be less than the opposite interior angle.*

Let $AB\Gamma$ be a spherical triangle, and let the sum of the two sides AB, $B\Gamma$ be equal to a semicircle. Produce the arc $A\Gamma$ to Δ. I say that the exterior angle $B\Gamma\Delta$ is equal to the opposite interior angle A [figure 7.5.4]. And if the sum of the two arcs AB, $B\Gamma$ is less than a semicircle, then angle $B\Gamma\Delta$ will be greater than angle A. Finally, if the sum of the arcs AB, $B\Gamma$ is greater than a semicircle, then angle $B\Gamma\Delta$ will be smaller than angle A.

Produce arc AB so that it intersects with arc $A\Gamma$ at Δ. Since the sum of the arcs AB, $B\Delta$ is a semicircle,[8] and the sum of the arcs AB, $B\Gamma$ is also a semicircle, therefore $B\Delta$ is equal to

[8] The points of intersection of two great circles on a sphere are separated by a "semicircle"; that is, they are 180° apart.

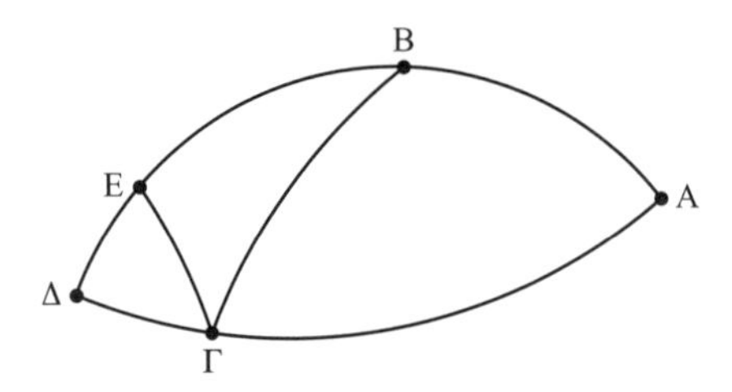

Figure 7.5.5.

$B\Gamma$, and angle $B\Gamma\Delta$ is equal to angle $B\Delta\Gamma$. However, angle $B\Delta\Gamma$ is equal to angle A.[9] Therefore angle $B\Gamma\Delta$ is also equal to A.

Now suppose that the sum of the two arcs AB and $B\Gamma$ is less than a semicircle. The sum of the two arcs AB, $B\Delta$ is equal to a semicircle. Therefore arc $B\Delta$ is greater than arc $B\Gamma$ and so angle $B\Gamma\Delta$ is greater than angle Δ. But angle Δ is equal to angle A. So the exterior angle $B\Gamma\Delta$ is greater than the interior angle A.

If the sum of the two arcs AB, $B\Gamma$ is greater than a semicircle, I say that the exterior angle $B\Gamma\Delta$ is less than the opposite interior angle A. For arc $AB\Delta$ is a semicircle and the sum of the two arcs AB, $B\Gamma$ is greater than a semicircle. Thus the arc $B\Gamma$ is greater than the arc $B\Delta$, and therefore angle Δ is greater than angle $B\Gamma\Delta$. But angle Δ is equal to angle A. So angle $B\Gamma\Delta$ is less than angle A.

The converse is manifest. That is, if in a spherical triangle with one side produced the exterior angle is equal to the interior and opposite angle, then the sum of the remaining two arcs is equal to a semicircle. If the exterior angle is greater than the interior and opposite angle, then the sum of the remaining arcs is less than a semicircle. And if the exterior angle is less than the interior and opposite angle, the sum of the remaining arcs is greater than a semicircle.

Proposition 11. *In a spherical triangle, if one side is extended, then the exterior angle so produced is less than the sum of the opposite interior angles. The sum of the three angles of the triangle is greater than two right angles.*

Let $AB\Gamma$ be a spherical triangle. I say that the exterior angle contained by the arcs $B\Gamma$, $\Gamma\Delta$ is less than the sum of the opposite angles A, B. Also, the sum of the three angles of the triangle, A, B, Γ is greater than two right angles.

Construct at the endpoint Γ of the arc $\Gamma\Delta$ an angle $\Delta\Gamma E$ equal to angle A, and produce arc AB until it meets arc $A\Gamma$ produced at the point Δ [figure 7.5.5]. Since angles Δ and Γ are equal, arcs ΔE and $E\Gamma$ are also equal. Therefore, the sum of arcs BE and $E\Gamma$ is the arc $B\Delta$, which is less than a semicircle. So the exterior angle ΓBA is greater than the interior angle $B\Gamma E$, and therefore the angle $B\Gamma\Delta$, the exterior angle of the triangle $AB\Gamma$, is less than the sum of angles ΓBA and $BA\Gamma$.[10] Adjoin the common angle $B\Gamma A$. Then the sum of the two angles $B\Gamma\Delta$, $B\Gamma A$ is less than the sum of the three angles A, B, Γ. But the

[9] The two angles of intersection of two great circles on a sphere are equal.

[10] Angle $B\Gamma\Delta$ equals the sum of angles $B\Gamma E$ and $E\Gamma\Delta$, so it is less than the sum of angles ΓBA and $BA\Gamma$.

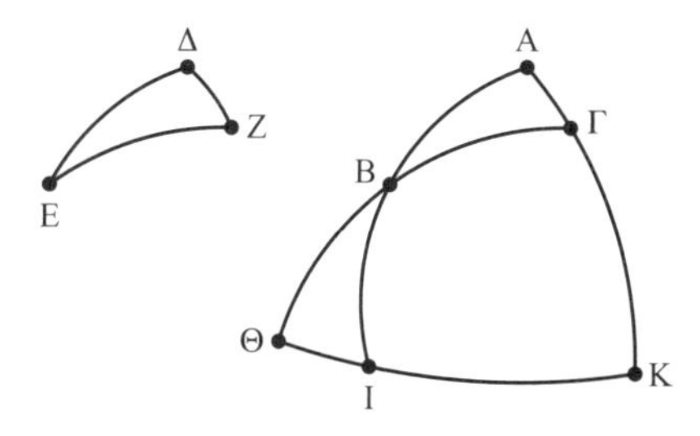

Figure 7.5.6.

two angles $B\Gamma\Delta$, $B\Gamma A$ are equal to two right angles. Therefore the three angles A, B, Γ exceed two right angles.

Proposition 14. *If in two spherical triangles, two angles of one are respectively equal to two angles of the other, and if the arcs adjoining the equal angles in the two triangles are also equal, then the two remaining arcs of one triangle are equal respectively to the two remaining arcs of the other, and the remaining angles are also equal.*

Proposition 17. *If two spherical triangles have the three angles of one respectively equal to the three angles of the other, then the arcs of the first triangle are equal to the arcs of the second connecting equal angles.*

Let $AB\Gamma$, ΔEZ be two spherical triangles in which angle A is equal to angle Δ, angle B is equal to angle E, and angle Γ is equal to angle Z. I say that arc AB is equal to arc ΔE, arc $B\Gamma$ to arc EZ, and arc $A\Gamma$ to arc ΔZ.

Produce arc AB to I, where BI is equal to ΔE. Produce $B\Gamma$ to Θ, where $B\Theta$ equals EZ. Draw a great circle arc through $I\Theta$ meeting arc $A\Gamma$ produced at point K [figure 7.5.6]. The two arcs ΘB and BI equal the two arcs ΔE and EZ, and angle E is equal to angle B. Therefore arc $I\Theta$ is equal to arc ΔZ, and angle I equals angle Δ, which is equal to angle A. So angle I equals angle A. Also, angle Θ is equal to angle Z, which is equal to angle Γ. Therefore angle Θ is equal to angle Γ. And since the exterior angle Γ of triangle $\Theta K\Gamma$ is equal to the interior and opposite angle Θ, the sum of the arcs ΘK, $K\Gamma$ is a semicircle. But the exterior angle I of triangle AIK is equal to its interior and opposite angle A, so the sum of the two arcs AK, KI is also a semicircle. Therefore the sum of the two arcs $K\Theta$, $K\Gamma$ is equal to the sum of the two arcs IK, KA. Subtracting the two common arcs IK, $K\Gamma$, we get that the remainders ΘI and $A\Gamma$ are equal. But arc ΘI is equal to arc $Z\Delta$. So arc $Z\Delta$ is equal to arc $A\Gamma$. But also angle A is equal to angle Δ and angle Γ is equal to angle Z. Therefore [proposition 14], arc AB is equal to arc ΔE and arc $B\Gamma$ is equal to arc EZ.

Book 3

Perhaps the most important result in Menelaus's *Sphaerica* is proposition 1 of book 3, the theorem on the spherical "sector figure." It was this result that enabled Ptolemy to solve spherical triangles in the *Almagest*. Then Islamic mathematicians used the theorem to provide simpler and more direct methods to solve spherical triangles as they developed spherical trigonometry beginning in the ninth century. However,

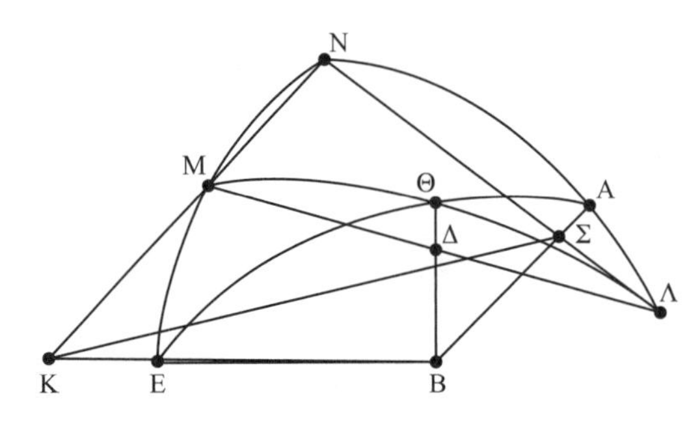

Figure 7.5.7.

Menelaus clearly assumed in his proof of this theorem that readers would know the analogous result for the plane sector figure. He also assumed that it would be clear how to convert that result into a result about chords. Thus, when Ptolemy included Menelaus's result in the *Almagest*, he provided lemmas to complete its proof. Similarly, many of the translations, into Arabic, Hebrew, and Latin, also included such lemmas. In addition, most of these translations converted Menelaus's chords into the more modern sines. We include these lemmas after this proposition, taking them directly from the *Almagest*. We then restate Menelaus's theorem in terms of sines.

Proposition 1. *Suppose there are two arcs of great circles on the surface of a sphere NME, $NA\Lambda$, between which are drawn two other great circle arcs $E\Theta A$, $\Lambda\Theta M$, meeting at the point Θ. I say that the ratio of the chord of double the arc AN to the chord of double the arc $A\Lambda$ is compounded of the ratio of the chord of double the arc NE to the chord of double the arc ME and of the ratio of the chord of double the arc $M\Theta$ to the chord of double the arc $\Theta\Lambda$.*

We suppose B is the center of the sphere and join the lines ΛN, ΛM, MN, EB, and ΘB. Let Δ be the intersection of $M\Lambda$ with ΘB, and Σ be the intersection of AB with $N\Lambda$, and extend $\Delta\Sigma$ to where it meets the extension of MN at K[11] [figure 7.5.7].

Then point K will be in the plane of the circles $A\Theta E$ and NME. But points E and B are also in that plane. So KEB is a straight line. Also, since the point Σ is the intersection of the lines AB, $N\Lambda$ and the point Δ of the lines ΘB, $M\Lambda$, therefore the point K is on the extension of the line $\Sigma\Delta$. Therefore the three points Σ, Δ, K are in the plane of the triangle $N\Lambda M$. Therefore [lemma 1], the ratio of $N\Sigma$ to $\Sigma\Lambda$ is compounded of the ratios of NK to KM and that of $M\Delta$ to $\Delta\Lambda$. And these are the ratios of the chords of the doubles of the arcs constructed on them [lemmas 2, 3]. Therefore, the chord of double the arc AN to the chord of double the arc $A\Lambda$ is compounded of the ratio of the chord of double the arc NE to the chord of double the arc ME and of the ratio of the chord of double the arc $M\Theta$ to the chord of double the arc $\Theta\Lambda$.

Lemma 1. *Let two straight lines, BE and GD, which are drawn to meet two other straight lines, AB and AG, cut each other at point Z. I say that the ratio of GE to AE is compounded*

[11] Menelaus also considers the case where $\Delta\Sigma$ is parallel to MN, but we skip its proof.

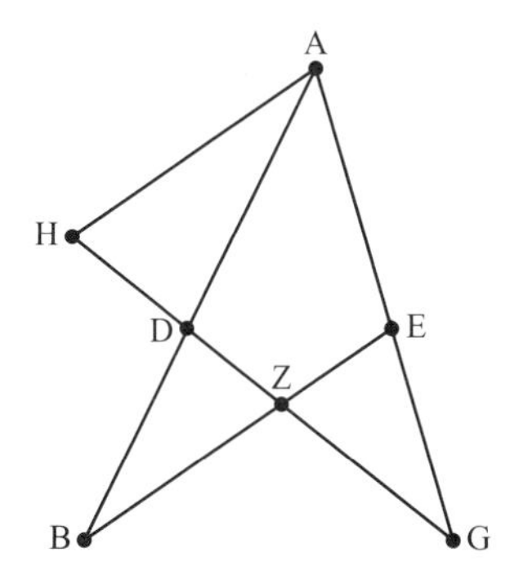

Figure 7.5.8.

of the ratio of GZ to DZ and the ratio of DB to BA. In addition, the ratio of GA to EA is compounded of the ratios of GD to DZ and ZB to BE.

Draw a line through A parallel to EB, and produce GD to cut it at H [figure 7.5.8]. Since AH is parallel to EZ, we have GE is to EA as GZ is to ZH. But, if we bring in ZD, the ratio of GZ to ZH is compounded of the ratios of GZ to ZD and DZ to ZH. But DZ is to ZH as DB is to BA (since BA and ZH meet the parallel lines AH and ZB). Therefore the ratio of GZ to ZH is compounded of the ratios of GZ to DZ and of DB to BA. But GZ is to ZH as GE is to EA. Therefore, the ratio of GE to EA is compounded of the ratio of GZ to DZ and the ratio of DB to BA. One can prove similarly that the ratio of GA to EA is compounded of the ratios of GD to DZ and ZB to BE.

The basic idea of the proof of proposition 1 is that the spherical sector figure is projected onto a planar sector figure on a plane that goes through the sphere. Thus line $N\Sigma\Lambda$ becomes a chord of the circular arc $NA\Lambda$. The same is true for line $M\Delta\Lambda$ in relation to arc $M\Theta\Lambda$. On the other hand, only part of line NMK is a chord of arc NME, that is, segment NM, while the remainder of the line is outside the circle. The same is true of line $\Sigma\Delta K$ with relation to arc $A\Theta E$. So we need two different results converting line segments to chords of arcs, one in which the entire line segment is a chord of a circle and one in which only part of the line segment is a chord.

Lemma 2. *On circle ABG, with center D, take any three points A, B, G, on the circumference, provided that each of the arcs AB and BG is less than a semicircle. Draw AG and DEB. I say that the chord of arc$2AB$ is to the chord of arc$2BG$ as AE is to EG.*

Drop perpendiculars AZ and GH from points A and G onto DB [figure 7.5.9]. Then, since AZ is parallel to GH, and they meet line AEG, we have AZ is to GH as AE is to EG. But AZ is to GH as the chord of arc$2AB$ is to the chord of arc$2BG$, for AZ is half of the chord of arc$2AB$ and GH is half of the chord of arc$2BG$. Therefore, AE is to EG as the chord of arc$2AB$ is to the chord of arc$2BG$.

Lemma 3. *On circle ABG with center D take three points on the circumference, A, B, G. Join DA and GB and produce them to meet at E. I say that the chord of arc$2GA$ is to the chord of arc$2AB$ as GE is to BE.*

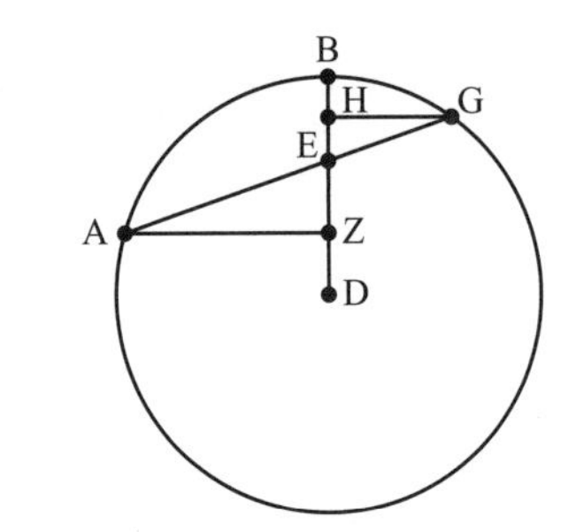

Figure 7.5.9.

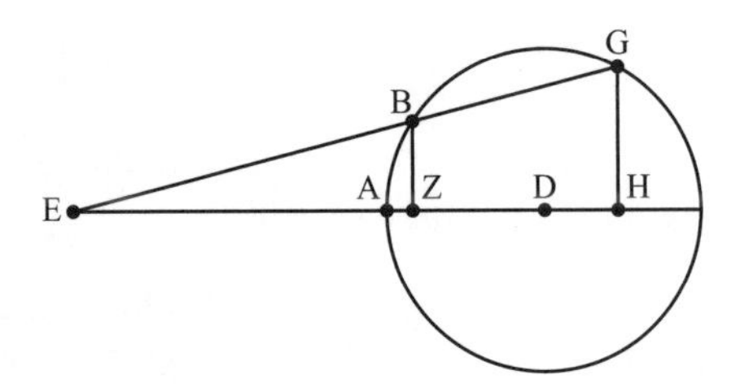

Figure 7.5.10.

Drop perpendiculars BZ and GH from B and G onto DA [figure 7.5.10]. Since they are parallel, GH is to BZ as GE is to EB. But GH is half of the chord of arc$2AG$ and BZ is half of the chord of $2AB$. Therefore, the chord of arc$2GA$ is to the chord of arc$2AB$ as GE is to BE.

In more modern terms, the conclusion of Menelaus's proposition 1 can be rewritten as

$$\sin AN : \sin A\Lambda = (\sin NE : \sin ME) \cdot (\sin M\Theta : \sin \Theta\Lambda).$$

There is a second closely related result that does not seem to have been mentioned by Menelaus, although Ptolemy includes it (without proof), as does the Latin translator (with proof):

$$\sin N\Lambda : \sin AN = (\sin M\Lambda : \sin M\Theta) \cdot (\sin E\Theta : \sin EA)$$

This one follows from the second conclusion in lemma 1. (Of course, for either of these results one can interchange the left and right sides of the diagram.) Ptolemy used these results to prove various results on spherical triangles, while Islamic mathematicians converted these results to somewhat simpler ones in their revision of the entire subject of spherical trigonometry.

7.6 Ptolemy, *Almagest*

Ptolemy of Alexandria was the most influential astronomer from Greek antiquity. Dated observations included in his works suggest his life span ranged from around 100 until 175 CE. Ptolemy lived and worked in Alexandria, which at that time boasted one of the best resourced libraries in the ancient world. Throughout his career Ptolemy wrote a number of books on mathematical astronomy, as well as geography and optics.

His most famous work, the *Mathematical Syntaxis* (Mathematical Systematic Treatise), or the *Almagest*, as it was later known, is thought to be the earliest of his works. It was founded on the principles of geometric kinematic models to reproduce celestial phenomena and was so influential that it became the standard treatise for astronomy for at least the next thousand years. Numerous commentaries were produced on the *Almagest*; some of the more notable ones in late antiquity were by Pappus of Alexandria (ca. 320 CE) and Theon of Alexandria (ca. 370 CE). In time, it was translated into Syriac and Arabic in the late eighth and ninth centuries and circulated widely across Eurasia in ensuing centuries.

Book 1

Ptolemy opens the *Almagest* with a philosophical preface that, among a discussion on the nature and configuration of the heavens, outlines the role of mathematics in astronomical inquiry.

1. *Philosophical Preface*

Hence we thought it fitting to guide our actions (under the impulse of our actual ideas [of what is to be done]) in such a way as never to forget, even in ordinary affairs, to strive for a noble and disciplined disposition, but to devote most of our time to intellectual matters, in order to teach theories, which are so many and beautiful, and especially those to which the epithet "mathematical" is particularly applied. For Aristotle divides theoretical philosophy too, very fittingly, into three primary categories, physics, mathematics, and theology.

. . .

From all this we concluded: that the first two divisions of theoretical philosophy should rather be called guesswork than knowledge, theology because of its completely invisible and ungraspable nature, physics because of the unstable and unclear nature of matter; hence there is no hope that philosophers will ever be agreed about them; and that only mathematics can provide sure and unshakable knowledge to its devotees, provided one approaches it rigorously. For its kind of proof proceeds by indisputable methods, namely, arithmetic and geometry.

. . .

As for physics, mathematics can make a significant contribution. For almost every peculiar attribute of material nature becomes apparent from the peculiarities of its motion from place to place. [Thus one can distinguish] the corruptible from the incorruptible by [whether it undergoes] motion in a straight line or in a circle, and heavy from light, and passive from active, by [whether it moves] towards the center or away from the center. With regard to virtuous conduct in practical actions and character, this science, above all things, could make men see clearly; from the constancy, order, symmetry and calm which are associated with the divine, it makes its followers lovers of this divine beauty, accustoming them and reforming their natures, as it were, to a similar spiritual state. It is this love of the contemplation of the eternal and unchanging which we constantly strive to increase, by studying those parts of these sciences which have already been mastered by those who approached them in a genuine spirit of enquiry, and by ourselves attempting to contribute as much advancement as has been made possible by the additional time between those people and ourselves.

After discussing the role of mathematics in astronomy and outlining the fundamental assumptions (e.g., the sphericity of the celestial realm and the earth, the centrality of the earth in the celestial sphere, and the principle celestial motions), Ptolemy addresses one of the most significant groups of mathematical identities: the theoretical bases from which to build a table of chords. The chord (Crd) of an arc (or a central angle) in a circle of radius R is related to the modern sine by the relation

$$\text{Crd}\,\theta = 2R\,\sin\left(\frac{\theta}{2}\right).$$

Similarly, given that $\text{Crd}^2\,(180-\theta)+\text{Crd}^2\,\theta=(2R)^2$, we also have

$$\text{Crd}(180-\theta)=2R\,\cos\left(\frac{\theta}{2}\right).$$

To enable him to calculate a detailed chord table, Ptolemy begins by dividing the circumference of the circle into 360 parts and setting the base circle radius to be $R=60$. First, the chord lengths Crd(60) and Crd(120) are easily determined using simple geometry. The chords Crd(36) and Crd(72) involve slightly more geometry and invoke Euclid's *Elements* 13:9 and 13:10. Ptolemy then derives the chord addition and subtraction and half-arc formulas for generating more chords. Using these established chord lengths and identities, Ptolemy is able to calculate the lengths of the chords of all arcs that are integral multiples of 3° (and indeed, to $\frac{3°}{2}$).

Finally, Ptolemy determines the chord of 1° to fill in the remaining missing chords. However, understanding the impossibility of trisecting an arbitrary arc, Ptolemy realizes that the chord of 1° is impossible to determine geometrically. So he uses a clever inequality, allowing him to approximate this quantity, and uses this value henceforth to complete his chord table. The table has numerical values for the chords from $\frac{1°}{2}$ to 180° with half-degree increments, precise to two fractional sexagesimal places, and a column of "sixtieths" from which to interpolate to produce the chords of finer divisions of an arc. This chord table forms the heart of numerical reckoning throughout the *Almagest*. In what follows, we have often, for the reader's convenience, replaced Ptolemy's words, such as "equals" with modern symbols, such as "=". The reader should remember, however, that these symbols are not in the original text.

9. *On the Individual Concepts*

Such, then, are the necessary preliminary concepts which must be summarily set out in our general introduction. We are now about to begin the individual demonstrations, the first of which, we think, should be to determine the size of the arc between the aforementioned poles [of the ecliptic and equator] along the great circle drawn through them. But we see that it is first necessary to explain the method of determining chords; we shall demonstrate the whole topic geometrically once and for all.

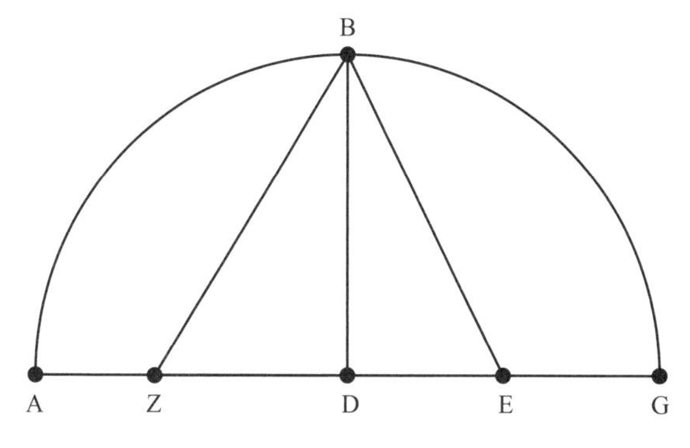

Figure 7.6.1.

10. *On the Size Of Chords*

For the user's convenience, then, we shall subsequently set out a table of their amounts, dividing the circumference into 360 parts, and tabulating the chords subtended by the arcs at intervals of half a degree, expressing each as a number of parts in a system where the diameter is divided into 120 parts. [We adopt this norm] because of its arithmetical convenience, which will become apparent from the actual calculations. But first we shall show how one can undertake the calculation of their amounts by a simple and rapid method, using as few theorems as possible, the same set for all. We do this so that we may not merely have the amounts of the chords tabulated unchecked, but may also readily undertake to verify them by computing them by a strict geometrical method. In general we shall use the sexagesimal system for our arithmetical computations, because of the awkwardness of the [conventional] fractional system. Since we always aim at a good approximation, we will carry out multiplications and divisions only as far as to achieve a result which differs from the precision achievable by the senses by a negligible amount.

First, then [figure 7.6.1], let there be a semicircle ABG about center D and on diameter ADG. Draw DB perpendicular to AG at D. Let DG be bisected at E, join EB, and let EZ be made equal to EB. Join ZB. I say that ZD is the side of the [regular] decagon, and BZ the side of the [regular] pentagon.

[Proof:] Since the straight line DG is bisected at E, and a straight line DZ is adjacent to it, $GZ \cdot ZD + ED^2 = EZ^2$. But $EZ^2 = BE^2 (EB = ZE)$, and $EB^2 = ED^2 + DB^2$. Therefore $GZ \cdot ZD + ED^2 = ED^2 + DB^2$. Therefore $GZ \cdot ZD = DB^2$ (subtracting ED^2, common). Therefore $GZ \cdot ZD = DG^2$.

So ZG has been cut in extreme and mean ratio at D. Now since the side of the hexagon and the side of the decagon, when both are inscribed in the same circle, make up the extreme and mean ratios of the same straight line,[12] and since GD, being a radius, represents the side of the hexagon, DZ is equal to the side of the decagon.

Similarly, since the square on the side of the pentagon equals the sums of the squares on the sides of the hexagon and decagon when all are inscribed in the same circle,[13] and,

[12] *Elements* 13:9.
[13] *Elements* 13:10.

in the right-angled triangle BDZ, the square on BZ equals the sum of the squares on BD, which is the side of the hexagon, and on DZ, which is the side of the decagon, it follows that BZ equals the side of the pentagon.

Since, then, as I said, we set the diameter of the circle as 120 parts, it follows from the above that $DE=30^p$ (DE half the radius) and $DE^2=900^p$; $BD=60^p$ (BD a radius) and $BD^2=3600^p$. And $EZ^2=EB^2=4500^p$, the sum [of DE^2 and BD^2]. Therefore $EZ\approx 67;4,55^p$,[14] and by subtraction [of DE from EZ], $DZ=37;4,55^p$. So the side of the decagon, which subtends 36°, has $37;4,55^p$ where the diameter has 120^p.

Again, since $DZ=37;4,55^p$, $DZ^2=1375;4,15^p$, and $DB^2=3600^p$, so $BZ^2=DZ^2+DB^2=4975;4,15^p$. Therefore $BZ\approx 70;32,3^p$.

Therefore the side of the pentagon, which subtends 72°, contains $70;32,3^p$ where the diameter has 120^p. It is immediately obvious that the side of the [inscribed] hexagon, which subtends 60° and is equal to the radius, contains 60^p. Similarly, since the side of the [inscribed] square, which subtends 90°, is equal, when squared, to twice the square on the radius, and since the side of the [inscribed] triangle, which subtends 120°, is equal, when squared, to three times the square on the radius, and the square on the radius is 3600^p, we compute that the square on the side of the square is 7200^p and the square on the side of the triangle is 10800^p. Therefore $\mathrm{Crd}(90^\circ)\approx 84;51,10^p$ and $\mathrm{Crd}(120^\circ)\approx 103;55,23^p$, where the diameter is 120^p.

We can, then, consider the above chords as established individually by the above straightforward procedures. It will immediately be obvious that if any chord be given, the chord of the supplementary arc is given in a simple fashion, since the sum of their squares equals the square on the diameter. For instance, since the chord of 36° was shown to be $37;4,55^p$, and the square of this is $1375;4,15^p$, and the square on the diameter is 14400^p, the square on the chord of the supplementary arc (which is 144°) will be the difference, namely, $13024;55,45$, and so $\mathrm{Crd}(144^\circ)\approx 114;7;37^p$. Similarly for the other chords [of the supplements].

We shall next show how the remaining individual chords can be derived from the above chords, first of all setting out a theorem which is extremely useful for the matter at hand.

Let there be a circle with an arbitrary quadrilateral $ABGD$ inscribed in it. Join AG and BD. We must prove that $AG\cdot BD=AB\cdot DG+AD\cdot BG$ [figure 7.6.2].[15]

[Proof:] Make $\angle ABE=\angle DBG$. Then, if we add $\angle EBD$ common, $\angle ABD=\angle EBG$. But $\angle BDA=\angle BGE$ also, since they subtend the same segment. Therefore triangle ABD is similar to triangle BGE. Therefore $BG:GE=BD:DA$. Therefore $BG\cdot AD=BD\cdot GE$. Again, since $\angle ABE=\angle DBG$, and $\angle BAE=\angle BDG$, triangle ABE is similar to triangle BGD. Therefore $BA:AE=BD:DG$. Therefore $BA\cdot DG=BD\cdot AE$. But it was shown that $BG\cdot AD=BD\cdot GE$. Therefore, by addition, $AG\cdot BD=AB\cdot DG+AD\cdot BG$.

Having established this preliminary theorem, we draw semicircle $ABGD$ on diameter AD, and draw from A two chords, AB, AG, each given in size in terms of a diameter of 120^p [figure 7.6.3]. Join BG. I say that BG too is given.

[14] This sexagesimal notation means 67 + 4/60 + 55/3600. Ptolemy generally uses sexagesimal notation in his numerical work.

[15] This result is known as Ptolemy's theorem.

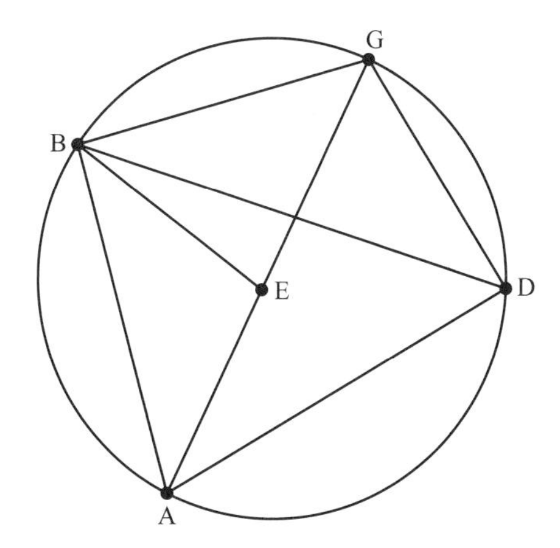

Figure 7.6.2.

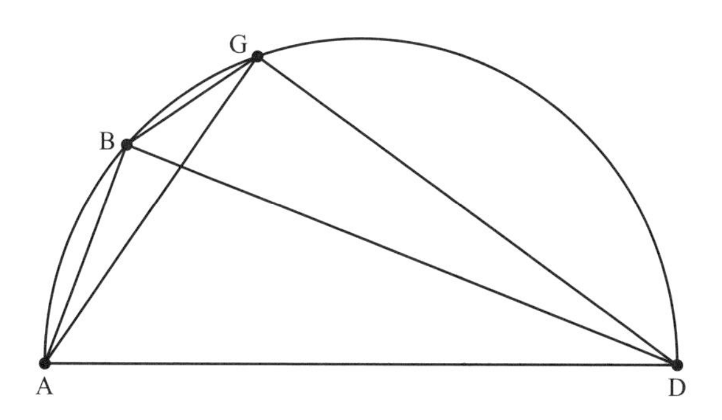

Figure 7.6.3.

[Proof:] Join BD, GD. Then, clearly, BD and GD too will be given, since they are chords of [arcs] supplementary [to the arcs of the given chords AB and AG]. Now since $ABGD$ is a cyclic quadrilateral, $AB \cdot GD + AD \cdot BG = AG \cdot BD$. But $AG \cdot BD$ and $AB \cdot GD$ are given. Therefore $AD \cdot BG$ is given by subtraction. And AD is a diameter. Therefore chord BG is given.

And we have shown that, if two arcs and the corresponding chords are given, the chord of the difference between the two arcs will also be given. It is obvious that by means of this theorem we shall be able to enter [in the table] quite a few chords derived from the difference between the individually calculated chords, and notably the chord of 12°, since we have those of 60° and 72°.

Let us now consider the problem of finding the chord of the arc which is half that of some given arc. Let ABG be a semicircle on diameter AG. Let GB be a given chord. Bisect arc GB at D, join AB, AD, BD, DG, and drop perpendicular DZ from D on to AG [figure 7.6.4]. I say that $ZG = \frac{1}{2}(AG - AB)$.

[Proof:] Let $AE = AB$, and join DE. Then since [in the triangles ABD, ADE] $AB = AE$, and AD is common, the two pairs of sides AB, AD and AE, AD are equal. Furthermore,

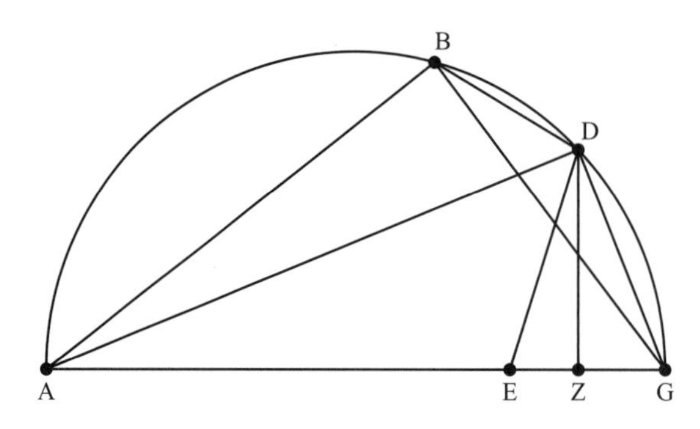

Figure 7.6.4.

angle BAD = angle EAD. Therefore base BD = base DE. But $BD = DG$ [by construction], therefore $DG = DE$. So, since, in the isosceles triangle DEG, perpendicular DZ has been drawn from apex to base, $EZ = ZG$. But $EG = [AG - AE =]AG - AB$. Therefore $ZG = \frac{1}{2}(AG - AB)$.

Now, if the chord of arc BG is given, the supplementary chord AB is immediately given. Therefore ZG, which is $\frac{1}{2}(AG - AB)$, is also given. But, since, in the right-angled triangle AGD, the perpendicular DZ has been drawn, triangle ADG is similar to triangle DGZ (both right-angled). Therefore, $AG : GD = GD : GZ$. Therefore $AG \cdot GZ = GD^2$. But $AG \cdot GZ$ is given. Therefore GD^2 is given, and so chord GD, which subtends an arc half of [the arc of the given chord] BG, is also given.

By means of this theorem too a large number of chords will be derived by halving [the arcs of] the previously determined chords, and notably, from the chord of $12°$, the chords of $6°$, $3°$, $1\frac{1}{2}°$, and $\frac{3}{4}°$. By calculation we find the chord of $12°$ to be approximately $1;34,15^p$ where the diameter is 120^p, and the chord of $\frac{3}{4}°$ to be approximately $0;47,8^p$ in the same units.

The two previous results are easily shown to be equivalent to the modern difference formula for the sine and the half-angle formula for the sine. Ptolemy next shows (although we omit it) that the chord of the sum of two arcs can be found, if the chords of each are known, a result that also can be easily shown to be equivalent to the sum formula for the cosine. He then continues:

It is obvious that by combining [in this way] the chord of $1\frac{1}{2}°$ with all the chords we have already obtained, and then computing successive chords, we will be able to enter [in the table] all chords [of arcs] which when doubled are divisible by three [i.e., multiples of $1\frac{1}{2}°$]. Then the only chords remaining to be determined will be those between the $1\frac{1}{2}°$ intervals, two in each interval, since our table is made at $\frac{1}{2}°$ intervals. If, therefore, we can find the chord of $\frac{1}{2}°$, this will enable us to complete [the table with] all the remaining intermediate chords, by finding the sum or difference [of $\frac{1}{2}°$] from the given chords at either end of the [$1\frac{1}{2}°$] intervals. Now, if a chord, e.g., the chord of $1\frac{1}{2}°$, is given, the chord corresponding to an arc which is one-third of the previous one cannot be found by geometrical methods.[16]

[16] The Greeks had not been able to trisect an angle or to show that this was impossible by "geometrical methods." However, Ptolemy clearly believed that trisection was impossible by such methods.

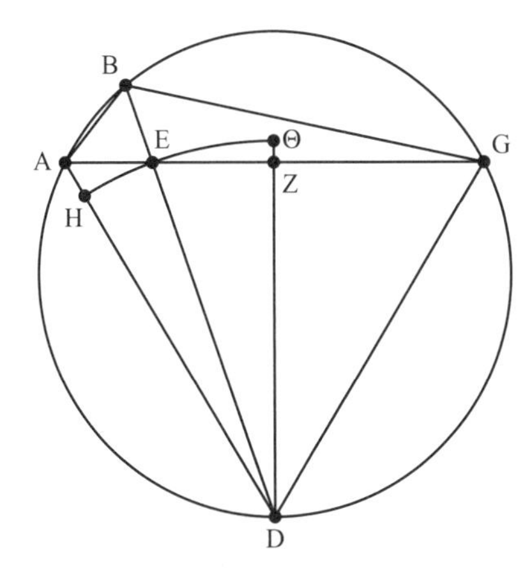

Figure 7.6.5.

(If this were possible, we should immediately have the chord of 1°). Therefore we shall first derive the chord of 1° from those of $1\frac{1}{2}^\circ$ and $\frac{3}{4}^\circ$. [We shall do this] by establishing a lemma which, though it cannot in general exactly determine the sizes [of chords], in the case of such very small quantities can determine them with a negligibly small error.

I say, then, that if two unequal chords be given, the ratio of the greater to the lesser is less than the ratio of the arc on the greater to the arc on the lesser [figure 7.6.5]. Let there be a circle $ABGD$, in which there are drawn two unequal chords, the lesser AB and the greater BG. I say that $GB : BA <$ arc BG : arc BA.

[Proof:] Let $\angle ABG$ be bisected by [chord] BD. Join AEG, AD, and GD. Then, since $\angle ABG$ is bisected by chord BED, $GD = AD$ and $GE > EA$. So drop perpendicular DZ from D on to AEG. Then, since $AD > ED$ and $ED > DZ$, a circle drawn on center D with radius DE will cut AD and pass beyond DZ. Let it be drawn as $HE\Theta$, and let DZ be produced to Θ. Now, since sector $DE\Theta$ is greater than triangle DEZ, and triangle DEA is greater than sector DEH, triangle DEZ : $\text{triangle } DEA < \text{sector } DE\Theta : \text{sector } DEH$.

But triangle $DEZ : \text{triangle } DEA = EZ : EA$, and sector $DE\Theta$: sector $DEH = \angle ZDE : \angle EDA$. Therefore, $ZE : EA < \angle ZDE : \angle EDA$. So, *componendo*, $ZA : EA < \angle ZDA : \angle ADE$. And, doubling the first members [of the ratios], $GA : AE < \angle GDA : \angle EDA$. Then *dividendo*, $GE : EA < \angle GDE : \angle EDA$. But $GE : EA = GB : BA$, and $\angle GDB : \angle BDA = \text{arc } GB : \text{arc } BA$. Therefore $GB : BA < \text{arc } GB : \text{arc } BA$.

Having established this, let us draw circle ABG, and in it two chords, AB and AG [figure 7.6.6].

Let us suppose, first, that AB is the chord of $\frac{3}{4}^\circ$ and AG the chord of 1°. Then, since $AG : BA < \text{arc } AG : \text{arc } AB$ and arc $AG = \frac{4}{3}$ arc AB, [therefore] $GA < \frac{4}{3}AB$. But, in units of which the diameter contains 120, we showed that $AB = 0;47,8^p$. Therefore $GA < 1;2,50^p$ (for $1;2,50 \approx \frac{4}{3} \cdot 0;47,8$).

Again, using the same figure, let us set AB as the chord of 1° and AG as the chord of $1\frac{1}{2}^\circ$. By the same argument, since arc $AG = \frac{3}{2}$ arc AB, $GA < \frac{3}{2}BA$. But, in units of which the diameter contains 120, we showed that $AG = 1;34,15^p$. Therefore $AB > 1;2,50^p$ (for $1;34,15 = \frac{3}{2} \cdot 1;2,50$).

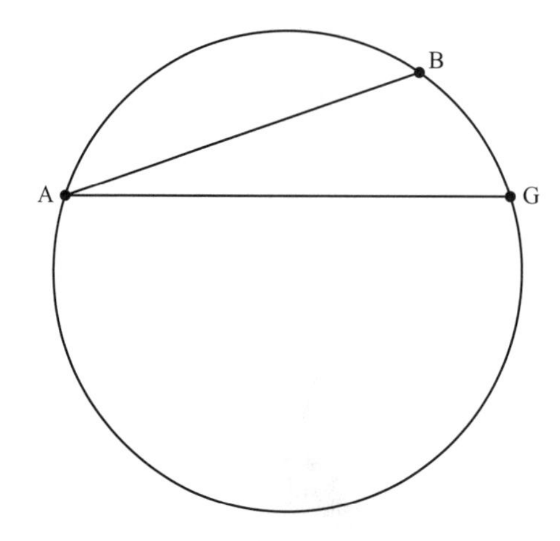

Figure 7.6.6.

Therefore, since the chord of 1° was shown to be both greater and less than the same amount, we can establish it as approximately $1;2,50^p$ where the diameter is 120^p. By the preceding propositions we can also establish the chord of $\frac{1}{2}^\circ$ which we find to be approximately $0;31,25^p$. The remaining intervals can [now] be completed, as we said. For example, in the first [$1\frac{1}{2}^\circ$] interval, we can calculate the chord of 2° by using the addition formula for the chord of $\frac{1}{2}^\circ$ applied to the chord of $1\frac{1}{2}^\circ$, while the chord of $2\frac{1}{2}^\circ$ is given by using the difference formula for [the chord of $\frac{1}{2}^\circ$] applied to the chord of 3°. Similarly for the remaining chords.

11. *Table of Chords*

Table 7.6.1 presents the first 90 entries of Ptolemy's chord table, followed by an illustration of the same table from an early manuscript [figure 7.6.7]. Modern reconstructions indicate that Ptolemy must have carried out his computations to five sexagesimal places to produce the precision given in the numerical entries! Note that the final column, labeled sixtieths, enables one to interpolate in the table to calculate the chord of an arc between any two given values by adding or subtracting the appropriate multiple of an entry in that column. As an example, to calculate the chord of 24;17,20°, as Ptolemy does in Book 2, section 5 below, we first find Crd(24°) in the table as 24;56,58. We then multiply the corresponding entry for sixtieths of 0;1,1,26 by 17,20. The product is 0;17,44,51, which, when added to 24;56,58, gives 25;14,42,51. Ptolemy rounds this to 25;14,43.

14. *On the Arcs between the Equator and the Ecliptic*

One of the main coordinate systems in astronomy uses two measurements, the declination and the right ascension. The former is the angular distance in degrees, measured on a great circle passing through the poles of the equator, between the equator and a point on the ecliptic, the oblique great circle tracing the apparent annual path of the sun. The latter is the angular distance along the equator of the point where that first great circle meets the equator. The other coordinate system uses longitude and latitude, where the first is the angular distance of a point along the ecliptic and

TABLE 7.6.1
Ptolemy's table of chords (first 90 entries)

Arcs	Chords			Sixtieths			Arcs	Chords			Sixtieths		
$\frac{1}{2}$	0	31	25	1	2	50	23	23	55	27	1	1	33
1	1	2	50	1	2	50	$23\frac{1}{2}$	24	26	13	1	1	30
$1\frac{1}{2}$	1	34	15	1	2	50	24	24	56	58	1	1	26
2	2	5	40	1	2	50	$24\frac{1}{2}$	25	27	41	1	1	22
$2\frac{1}{2}$	2	37	4	1	2	48	25	25	58	22	1	1	19
3	3	8	28	1	2	48	$25\frac{1}{2}$	26	29	1	1	1	15
$3\frac{1}{2}$	3	39	52	1	2	48	26	26	59	38	1	1	11
4	4	11	16	1	2	47	$26\frac{1}{2}$	27	30	14	1	1	8
$4\frac{1}{2}$	4	42	40	1	2	47	27	28	0	48	1	1	4
5	5	14	4	1	2	46	$27\frac{1}{2}$	28	31	20	1	1	0
$5\frac{1}{2}$	5	45	27	1	2	45	28	29	1	50	1	0	56
6	6	16	49	1	2	44	$28\frac{1}{2}$	29	32	18	1	0	52
$6\frac{1}{2}$	6	48	11	1	2	43	29	30	2	44	1	0	48
7	7	19	33	1	2	42	$29\frac{1}{2}$	30	33	8	1	0	44
$7\frac{1}{2}$	7	50	54	1	2	41	30	31	3	30	1	0	40
8	8	22	15	1	2	40	$30\frac{1}{2}$	31	33	50	1	0	35
$8\frac{1}{2}$	8	53	35	1	2	39	31	32	4	8	1	0	31
9	9	24	54	1	2	38	$31\frac{1}{2}$	32	34	22	1	0	27
$9\frac{1}{2}$	9	53	13	1	2	37	32	33	4	35	1	0	22
10	10	27	32	1	2	35	$32\frac{1}{2}$	33	34	46	1	0	17
$10\frac{1}{2}$	10	58	49	1	2	33	33	34	4	55	1	0	12
11	11	30	5	1	2	32	$33\frac{1}{2}$	34	36	1	1	0	8
$11\frac{1}{2}$	12	1	21	1	2	30	34	35	5	5	1	0	3
12	12	32	36	1	2	28	$34\frac{1}{2}$	35	36	6	0	59	57
$12\frac{1}{2}$	13	3	50	1	2	27	35	36	5	5	0	59	52
13	13	35	4	1	2	25	$35\frac{1}{2}$	36	35	1	0	59	48
$13\frac{1}{2}$	14	6	16	1	2	23	36	37	4	55	0	59	43
14	14	37	27	1	2	21	$36\frac{1}{2}$	37	34	47	0	59	38
$14\frac{1}{2}$	15	8	38	1	2	19	37	38	4	36	0	59	32
15	15	39	47	1	2	17	$37\frac{1}{2}$	38	34	22	0	59	27
$15\frac{1}{2}$	16	10	56	1	2	15	38	39	4	5	0	59	22
16	16	42	3	1	2	13	$38\frac{1}{2}$	39	33	46	0	59	16
$16\frac{1}{2}$	17	13	9	1	2	10	39	40	3	25	0	59	11
17	17	44	14	1	2	7	$39\frac{1}{2}$	40	33	0	0	59	5
$17\frac{1}{2}$	18	15	17	1	2	5	40	41	2	33	0	59	0
18	18	49	19	1	2	2	$40\frac{1}{2}$	41	32	3	0	58	42

TABLE 7.6.1
Ptolemy's table of chords (first 90 entries) (continued)

Arcs	Chords			Sixtieths			Arcs	Chords			Sixtieths		
$18\frac{1}{2}$	19	17	21	1	2	0	41	42	1	30	0	58	48
19	19	48	21	1	1	57	$41\frac{1}{2}$	42	30	54	0	58	42
$19\frac{1}{2}$	20	19	19	1	1	54	42	43	0	15	0	58	36
20	20	50	16	1	1	51	$42\frac{1}{2}$	43	29	33	0	58	31
$20\frac{1}{2}$	21	21	11	1	1	48	43	43	58	49	0	58	25
21	21	52	6	1	1	45	$43\frac{1}{2}$	44	28	1	0	58	18
$21\frac{1}{2}$	22	22	58	1	1	42	44	44	57	10	0	58	12
22	22	53	49	1	1	39	$44\frac{1}{2}$	45	26	16	0	58	6
$22\frac{1}{2}$	23	24	39	1	1	36	45	45	55	19	0	58	0

the second is measured in degrees north or south of the ecliptic. In figure 7.6.8, great circle *AEG* is the equator, while great circle *BED* is the ecliptic. They intersect at *E*, the vernal equinox. The maximum value of the declination, here arc *AB*, is known as the obliquity of the ecliptic. In what follows, Ptolemy describes how to compute the solar declination; the calculation requires the second version of Menelaus's theorem, which in the configuration of figure 7.6.8 can be written as

$$\mathrm{Crd}(2ZA):\mathrm{Crd}(2AB)=(\mathrm{Crd}(2\Theta Z):\mathrm{Crd}(2\Theta H))\cdot(\mathrm{Crd}(2HE):\mathrm{Crd}(2EB)).$$

Ptolemy's value for the obliquity of the ecliptic, stated earlier in the *Almagest*, is approximately 11 parts where the meridian is 83. This produces an angular value in degrees of the obliquity of the ecliptic of $\frac{11}{83}$ of $\frac{360°}{2} \approx 23;51,20$. Ptolemy then works through two examples: the first when the solar longitude is 30° and the second for 60°, the declinations of which he finds to be 11;40° and 20;30,9°, respectively. These values can be found below as part of Ptolemy's table of computed declinations for ecliptic longitudes 1° to 90° in argument increments of 1°. Ptolemy calls this the "table of inclination."

This passage explaining the calculation shows Ptolemy invoking the chord table, both finding the chord given an arc, and the inverse process, finding the arc given a chord.

Having set out this preliminary theorem, we shall first of all demonstrate the amounts of the arcs we set ourselves to determine, as follows [figure 7.6.8].

Let the circle through both poles, that of the equator and that of the ecliptic, be *ABGD*; let the semicircle representing the equator be *AEG*, and that representing the ecliptic *BED*, and let point *E* be the intersection of the two at the spring equinox, so that *B* is the winter solstice and *D* the summer solstice. On arc *ABG* take the pole of the equator *AEG*: let it be point *Z*. Cut off arc *EH* on the ecliptic; let us suppose it to be 30°, and draw through *Z* and *H* an arc of a great circle *ZHΘ*. Our problem, obviously, is to determine *HΘ*.[17] Let us

[17] *HΘ* is the declination of the sun when the sun is at *H*, a point on the ecliptic with longitude *HE*.

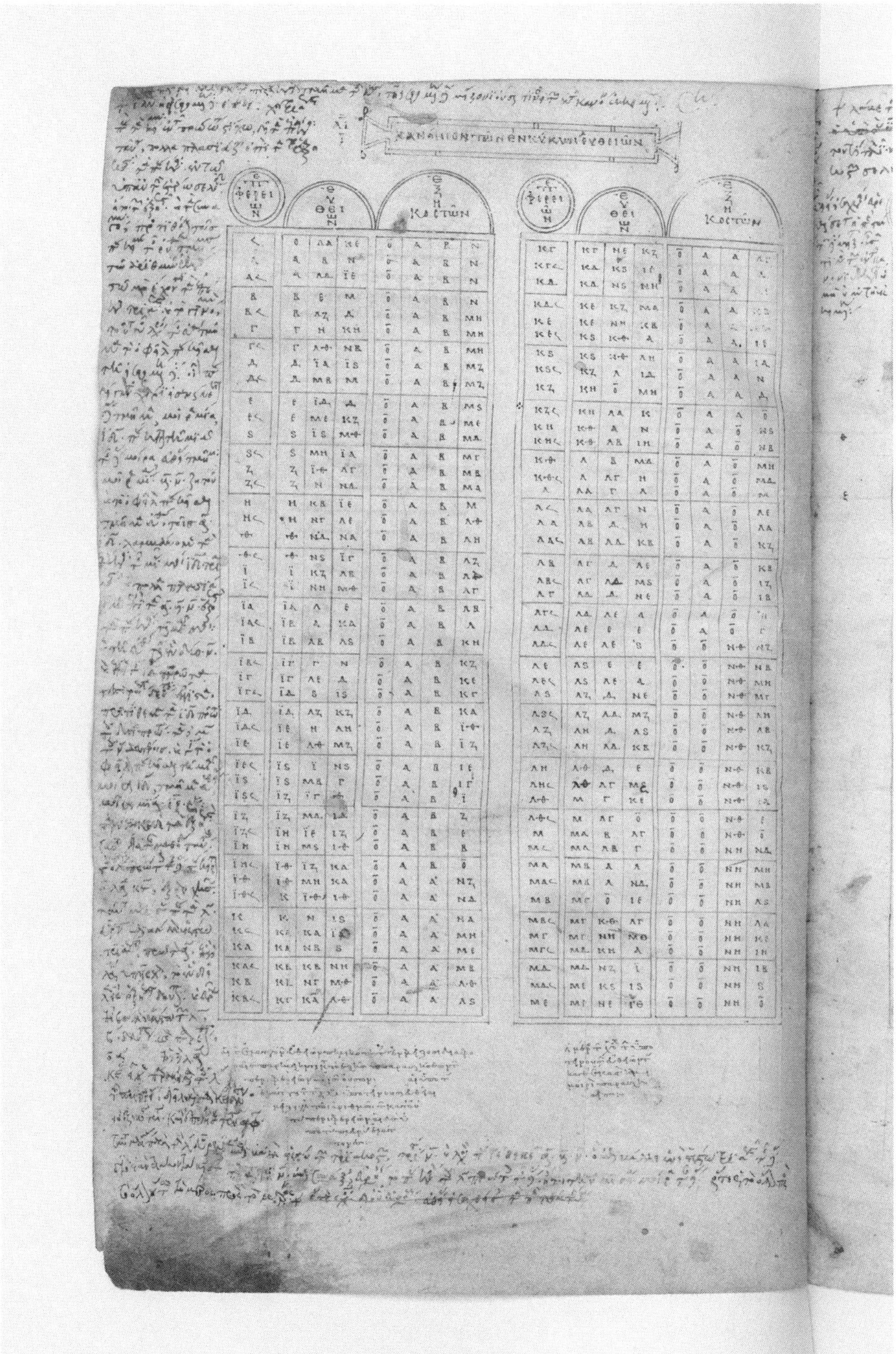

Figure 7.6.7. An excerpt of part of the chord table from a manuscript of the *Almagest* (MS Vat. Gr. 1594, f. 20v). Image courtesy of the Vatican.

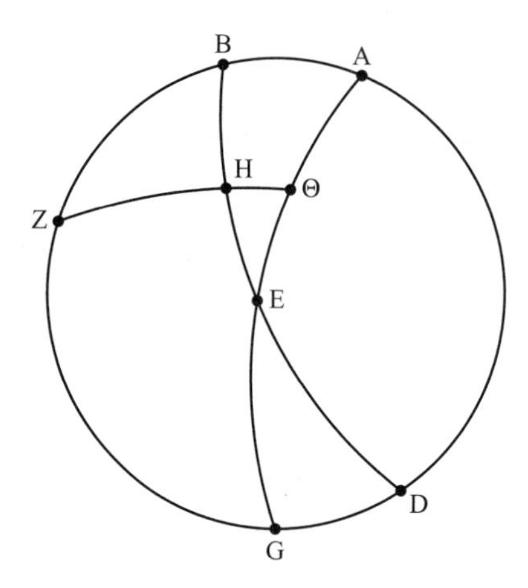

Figure 7.6.8.

take for granted both here and in general for all such demonstrations (to avoid repeating ourselves on each occasion), that when we speak of the sizes of arcs or chords in terms of "degrees" or "parts" we mean (for arcs) those degrees of which the circumference of a great circle contains 360, and (for chords) those parts of which the diameter of the circle contains 120.

Now since, in the figure, the two great circle arcs $Z\Theta$ and EB are drawn to meet the two great circle arcs AZ and AE, and intersect each other at H,

$$\text{Crd}(2ZA):\text{Crd}(2AB) = (\text{Crd}(2\Theta Z):\text{Crd}(2\Theta H)) \cdot (\text{Crd}(2HE):\text{Crd}(2EB)).$$

But arc$2ZA = 180°$. so Crd$(2ZA) = 120^p$, and arc$2AB = 47;42,40°$ (according to the ratio 11:83, with which we agreed), so Crd$(2AB) = 48;31,55^p$. Again, arc$2HE = 60°$, so Crd$(2HE) = 60^p$, and arc$2EB = 180°$, so Crd$(2EB) = 120^p$. Therefore Crd$(2Z\Theta)$: Crd$(2\Theta H) = (120 : 48;31,55)/(60 : 120) = 120 : 24;15,57$. And arc$2Z\Theta = 180°$, so Crd$(2Z\Theta) = 120^p$. Therefore Crd$(2\Theta H) = 24;15,57^p$. Therefore, arc$2\Theta H = 23;19,59°$, and arc$\Theta H \approx 11;40°$.

Again, let arc EH be taken as 60°. Then the other magnitudes will remain unchanged, but arc$2EH = 120°$, so Crd$(2EH) = 103;55,23^p$. Therefore, Crd$(2Z\Theta)$: Crd$(2\Theta H) = (120 : 48;31,55)/(103;55,23 : 120) = 120 : 42;1,48$. But Crd$(2Z\Theta) = 120°$. Therefore, Crd$(2\Theta H) = 42;1,48^p$. Therefore, arc$2\Theta H = 41;0,18°$, and arc$\Theta H = 20;30,9°$.

In the same way we shall compute the sizes of [the other] individual arcs, and set out a table giving for each degree of the quadrant the arc corresponding to those computed above.

15. *The Table of Inclination*

We include here the entries in the table just for selected solar longitudes.

TABLE 7.6.2
Table of inclination

Arcs				Arcs			
of the ecliptic	of the meridian			of the ecliptic	of the meridian		
3	1	12	46	48	17	29	27
6	2	25	22	51	18	19	15
9	3	37	37	54	19	5	57
12	4	49	24	57	19	49	42
15	6	0	31	60	20	30	9
18	7	10	45	63	21	7	21
21	8	20	0	66	21	41	0
24	9	28	5	69	22	11	1
27	10	34	57	72	22	37	17
30	11	39	59	75	22	59	41
33	12	42	28	78	23	18	11
36	13	45	6	81	23	32	30
39	14	44	39	84	23	43	2
42	15	42	2	87	23	49	16
45	16	37	1	90	23	51	20

Book 2

5. *Computing Right Triangles*

With his chord table in hand, Ptolemy could determine chords from arcs and the converse. However, unlike the modern sine function, the chord table gives the chords for a circle whose radius is 60. When Ptolemy wanted to compute the chords for circles of different sizes he had to adjust his values to the length of that particular radius. The following excerpt shows how Ptolemy carries out this conversion. The computation occurs in the process of determining the length of shadows cast when the rays of the sun (positioned at H, B, and L) fall on the gnomon GE to produce shadow lengths GK, GZ, and GN, respectively [figure 7.6.9]. These measurements are taken at latitude 36° at noon on the summer solstice (H), the equinox (B), and the winter solstice (L). Ptolemy's procedure is equivalent to the modern approach using the tangent.

Then, since arc GD, which is equal to the elevation of the north pole from the horizon, is 36° (where meridian ABG is 360°) at the latitude in question, and both arc ΘD and arc DM are 23;51,20°, by subtraction arc $G\Theta = 12;8,40°$, and by addition arc $GM = 59{:}51,20°$.

Therefore the corresponding angles are $\angle KEG = 12;8,40°$, $\angle ZEG = 36°$, and $\angle NEG = 59;51,20°$. . . .

Therefore in the circles about right-angled triangles KEG, ZEG, NEG, arc$GK = 24;17,20°$, arc$GE = 155;42,40°$ (supplement), arc$GZ = 72°$, arc$GE = 108°$, similarly [as supplement], arc$GN = 119;42,40°$, and arc$GE = 60;17,20°$ (again as supplement).

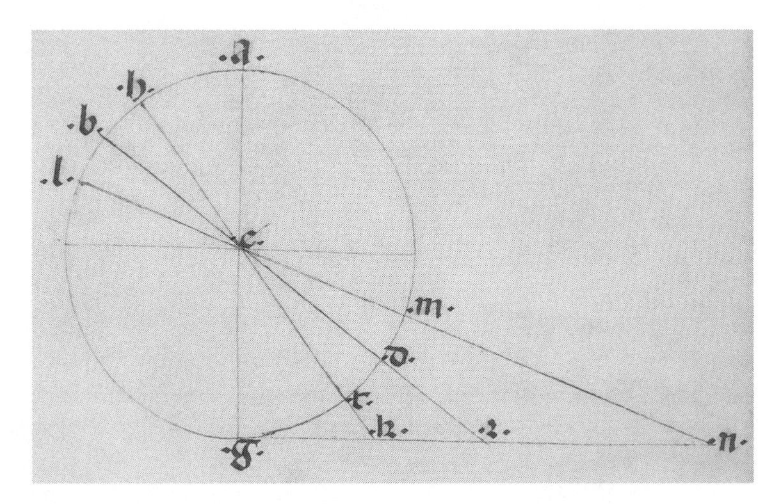

Figure 7.6.9. From a manuscript of Cremona's translation of Ptolemy's *Almagest*. Image courtesy of Bibliothéque nationale de France.

Therefore where Crd(arc GK) = $25;14,43^p$, Crd(arc GE) = $117;18,51^p$; where Crd(arc GZ) = $70;32,3^p$, Crd(arc GE) = $97;4,56^p$; and where Crd(arc GN) = $103;46,16^p$, Crd(arc GE) = $60;15,42^p$.

Therefore, where the gnomon GE has 60^p, in the same units, the summer [solsticial] shadow, $GK \approx 12;55^p$, the equinoctial shadow, $GZ \approx 43;36^p$, and the winter [solsticial] shadow, $GN \approx 103;20^p$.

Ptolemy's goal here is to find the lengths of GK, GZ, and GN. Now, for example, in right triangle GEZ, we know that $GE = 60$ and angle $ZEG = 72°$. To do computations using his chord table, Ptolemy needs GZ to be the chord of an arc in a circle. Thus, he imagines a circle through the points E, G, Z. In that circle, since angles at the center are double the angles at the circumference and angle $ZEG = 36°$, $GZ = \text{Crd}(72°) = 70;32,3$, while $EG = \text{Crd}(108°) = 97;4,56$. But, in fact, $GE = 60$. So Ptolemy reduces the calculated value of GZ by the ratio of 60 to 97;4,56, giving the final result that $GZ \approx 43;36$. Note that this calculation for finding the leg a of a right triangle, given α and b, can be rewritten as

$$a = b \cdot \frac{\text{Crd}(2\alpha)}{\text{Crd}(180 - 2\alpha)} = b \cdot \frac{2R\sin\alpha}{2R\cos\alpha} = b\tan\alpha,$$

in agreement with modern procedure.

Book 3

3. *The Mathematical Equivalence of the Epicycle and Eccentric Models*

One of the key challenges in Greek astronomy was to describe the apparently irregular motions of the planets in terms of geometric kinematic models involving combinations of uniform circular motions. Indeed, as viewed from the earth, the planets appear to advance sometimes faster and sometimes slower, and even go backward (retrograde) at certain points. Two distinct proposals to account for this using combinations of uniform circular motions were the eccentric and epicyclic models. The eccentric model involves a large concentric circle and another large circle whose center is offset with respect to the center of the first circle; the planet moves uniformly on this offset circle. The other, the epicyclic model, involves a smaller circle (an epicycle) on which the planet rotates, which is itself carried around a large, concentric circle, called the deferent.

In the following passage, Ptolemy demonstrates that these two models are in fact mathematically equivalent, and a remark later (*Almagest*, book 12, section 1) reveals that this equivalence was known to others, including his predecessor Apollonius of Perga (ca. 200 BCE; see chapter 5). The acknowledgment of two distinct mathematical models that both produce identical results and ancient debates about which model reflected better the physical reality of the configurations and motions of the celestial bodies is a fascinating issue that has prompted much speculation among historians of astronomy as to whether Greek astronomers such as Ptolemy were interested in uncovering the actual physical configurations of planetary motion or merely in "saving the phenomena" with theoretical apparatus disconnected from the physical reality.

Our next task is to demonstrate the apparent anomaly of the sun. But first we must make the general point that the rearward displacements of the planets with respect to the heavens are, in every case, just like the motion of the universe in advance, by nature uniform and circular. That is to say, if we imagine the bodies or their circles being carried around by straight lines, in absolutely every case the straight line in question describes equal angles at the center of its revolution in equal times. The apparent irregularity [anomaly] in their motions is the result of the position and order of those circles in the sphere of each by means of which they carry out their movements, and in reality there is in essence nothing alien to their eternal nature in the "disorder" which the phenomena are supposed to exhibit. The reason for the appearance of irregularity can be explained by two hypotheses, which are the most basic and simple. When their motion is viewed with respect to a circle imagined to be in the plane of the ecliptic, the center of which coincides with the center of the universe (thus its center can be considered to coincide with our point of view), then we can suppose, either that the uniform motion of each [body] takes place on a circle which is not concentric with the universe, or that they have such a concentric circle, but their uniform motion takes place, not actually on that circle, but on another circle, which is carried by the first circle, and [hence] is known as the "epicycle." It will be shown that either of these hypotheses will enable [the planets] to appear, to our eyes, to traverse unequal arcs of the ecliptic (which is concentric to the universe) in equal times.

In the eccentric hypothesis [see figure 7.6.10]: we imagine the eccentric circle, on which the body travels with uniform motion, to be $ABGD$ on center E, with diameter AED, on which point Z represents the observer. Thus A is the apogee, and D the perigee. We cut off equal arcs AB and DG, and join BE, BZ, GE, and GZ. Then it is immediately obvious that the body will traverse the arcs AB and GD in equal times, but will [in so doing] appear to have traversed unequal arcs of a circle drawn on center Z. For $\angle BEA = \angle GED$. But $\angle BZA < \angle BEA$ (or $\angle GED$), and $\angle GZD > \angle GED$ (or $\angle BEA$).

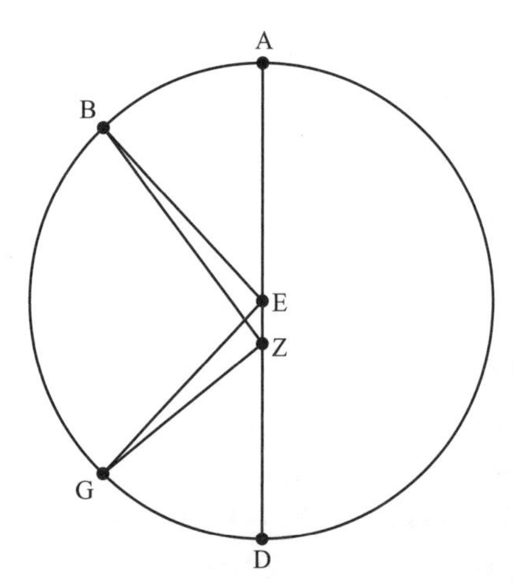

Figure 7.6.10.

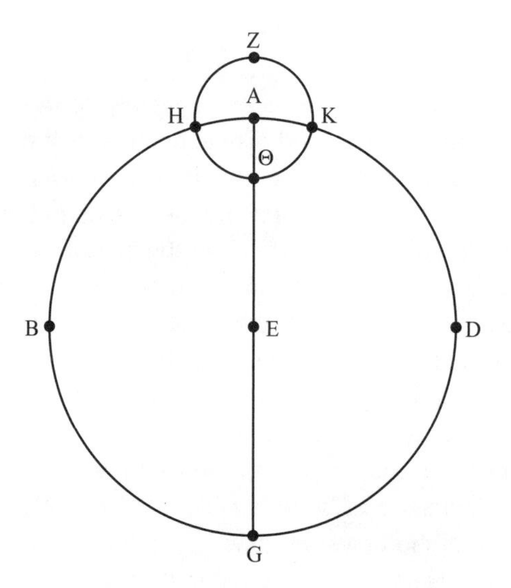

Figure 7.6.11.

In the epicyclic hypothesis [figure 7.6.11]: we imagine the circle concentric with the ecliptic as $ABGD$ on center E with diameter AEG, and the epicycle carried by it, on which the body moves, as $ZH\Theta K$ on center A.

Then here too it is immediately obvious that, as the epicycle traverses circle $ABGD$ with uniform motion, say from A towards B, and as the body traverses the epicycle with uniform motion, then when the body is at points Z and Θ, it will appear to coincide with A, the center of the epicycle, but when it is at other points it will not. Thus when it is, e.g., at H, its motion will appear greater than the uniform motion [of the epicycle] by arc AH, and similarly when it is at K its motion will appear less than the uniform by arc AK.

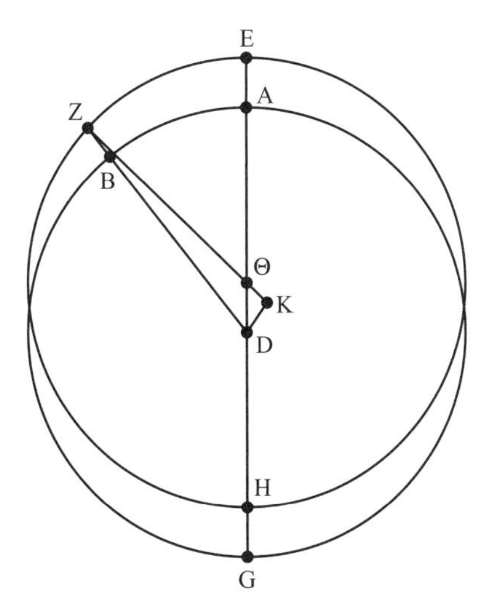

Figure 7.6.12.

Now in this kind of eccentric hypothesis the least speed always occurs at the apogee and the greatest at the perigee, since $\angle AZB$ [in figure 7.6.10] is always less than $\angle DZG$. But in the epicyclic hypothesis both this and the reverse are possible. For the motion of the epicycle is towards the rear with respect to the heavens, say from A towards B [in figure 7.6.11]. Now if the motion of the body on the epicycle is such that it too moves rearwards from the apogee, that is, from Z towards H, the greatest speed will occur at the apogee, since at that point both epicycle and body are moving in the same direction. But if the motion of the body from the apogee is in advance on the epicycle, that is, from Z towards K, then the reverse will occur: the least speed will occur at the apogee, since at that point the body is moving in the opposite direction to the epicycle.

If these conditions are fulfilled, the identical phenomena will result from either hypothesis.

5. *Calculating Oblique Triangles*

Ptolemy also needs to solve oblique triangles. In the following example involving the eccentric model for the sun, Ptolemy wants to determine the direction of the sun at a certain point in its orbit. His solution is equivalent to the modern law of cosines where two sides and an angle is known.

In order to enable one to determine the anomalistic motion over any subdivision [of the circle], we shall show, again for both hypotheses, how, given one of the arcs in question, we can compute the others.

First, let the circle concentric to the ecliptic be ABG on center D, the eccenter EZH on center Θ, and let the diameter through both centers and the apogee E be $EA\Theta DH$. [That is, the earth is at D and the center of the sun's orbit is at Θ.] Cut off arc EZ, and join ZD, $Z\Theta$ [figure 7.6.12]. First, let arc EZ be given, e.g., as 30°. Produce $Z\Theta$ and drop the perpendicular to it from D, DK.

Then, since arc EZ is, by hypothesis, $30°$, $\angle E\Theta Z = \angle D\Theta K = 30°$ where 4 right angles $= 360°$.... Therefore, in the circle about right-angled triangle $D\Theta K$, arc $DK = 60°$ and arc $K\Theta = 120°$ (supplement). Therefore the corresponding chords $DK = 60^p$ and $K\Theta = 103;55^p$, where hypotenuse $D\Theta = 120^p$.

Ptolemy had previously determined, using the fact that the four seasons are not exactly equal, that if one takes $Z\Theta = 60^p$, then $D\Theta = 2;30^p$.

Therefore, where $D\Theta = 2;30^p$ and radius $Z\Theta = 60^p$, $DK = 1;15^p$ and $\Theta K = 2;10^p$.[18] Therefore, by addition [of ΘK to radius $Z\Theta$], $K\Theta Z = 62;10^p$.

Now since $DK^2 + K\Theta Z^2 = ZD^2$, the hypotenuse $ZD \approx 62;11^p$. Therefore, where $ZD = 120^p$, $DK = 2;25^p$, and, in the circle about right-angled triangle ZDK, arc $DK = 2;18°$.[19] Therefore $\angle DZK = 1;9°$ where 4 right angles $= 360°$. That $[1;9°]$ will be the amount of the equation of anomaly at this position.

And $\angle E\Theta Z$ was taken as $30°$. Therefore, by subtraction, $\angle ADB$ (which equals arc AB of the ecliptic) equals $28;51°$.

To convert Ptolemy's calculation to modern terms, note that we are asked to solve triangle $Z\Theta D$, where $a = \Theta Z = 60$, $b = D\Theta = 2\frac{1}{2}$, and $\gamma = \angle Z\Theta D = 150°$. Ptolemy proceeds by dropping a perpendicular DK to $Z\Theta$ extended, so let us set $DK = h$ and $K\Theta = p$. It follows that

$$p = \frac{\text{Crd}(2\gamma - 180)\cdot b}{2R} = \frac{\text{Crd}(120)\cdot 2\frac{1}{2}}{120} \text{ and } h = \frac{\text{Crd}(360 - 2\gamma)\cdot b}{2R} = \frac{\text{Crd}(60)\cdot 2\frac{1}{2}}{120}.$$

Therefore $c^2 = ZD^2 = h^2 + (a+p)^2 = a^2 + b^2 + 2ab\frac{\text{Crd}(2\gamma - 180)}{2R}$, or $c^2 = a^2 + b^2 - 2ab\cos\gamma$, precisely the law of cosines. In addition, Ptolemy also noted that $\text{Crd}(2\beta) = \frac{h\cdot 2R}{c}$, where $b = \angle DZK$, and then found β from his chord table. That is, $\sin\beta = \frac{h}{c} = \frac{b\sin\gamma}{c}$, so Ptolemy has also used the equivalent of the modern law of sines.

7.7 Ptolemy, *Handy Tables*

Ptolemy's *Handy Tables* is a substantial collection of numerical tables that are adaptations of those included in the *Almagest*. The compilation of tables, along with Ptolemy's original introduction, is extant only in a version made by Theon of Alexandra (ca. 360 CE); it circulated widely throughout Eurasia in later antiquity and through the medieval period. As its name implies, this treatise offered the working astronomer "handy," easy-to-use sets of tabularized data from which to compute astronomical phenomena. Comparing the tables from the *Almagest* (see, e.g., table 7.6.2) with their counterparts in the *Handy Tables* [figure 7.7.1, table 7.7.1] shows the sorts of modifications Ptolemy made for this more practically oriented handbook.

[18] Calculated by multiplying 60 and 103;55, respectively, by the scale factor $\frac{2;30}{120}$.

[19] By use of the chord table in reverse.

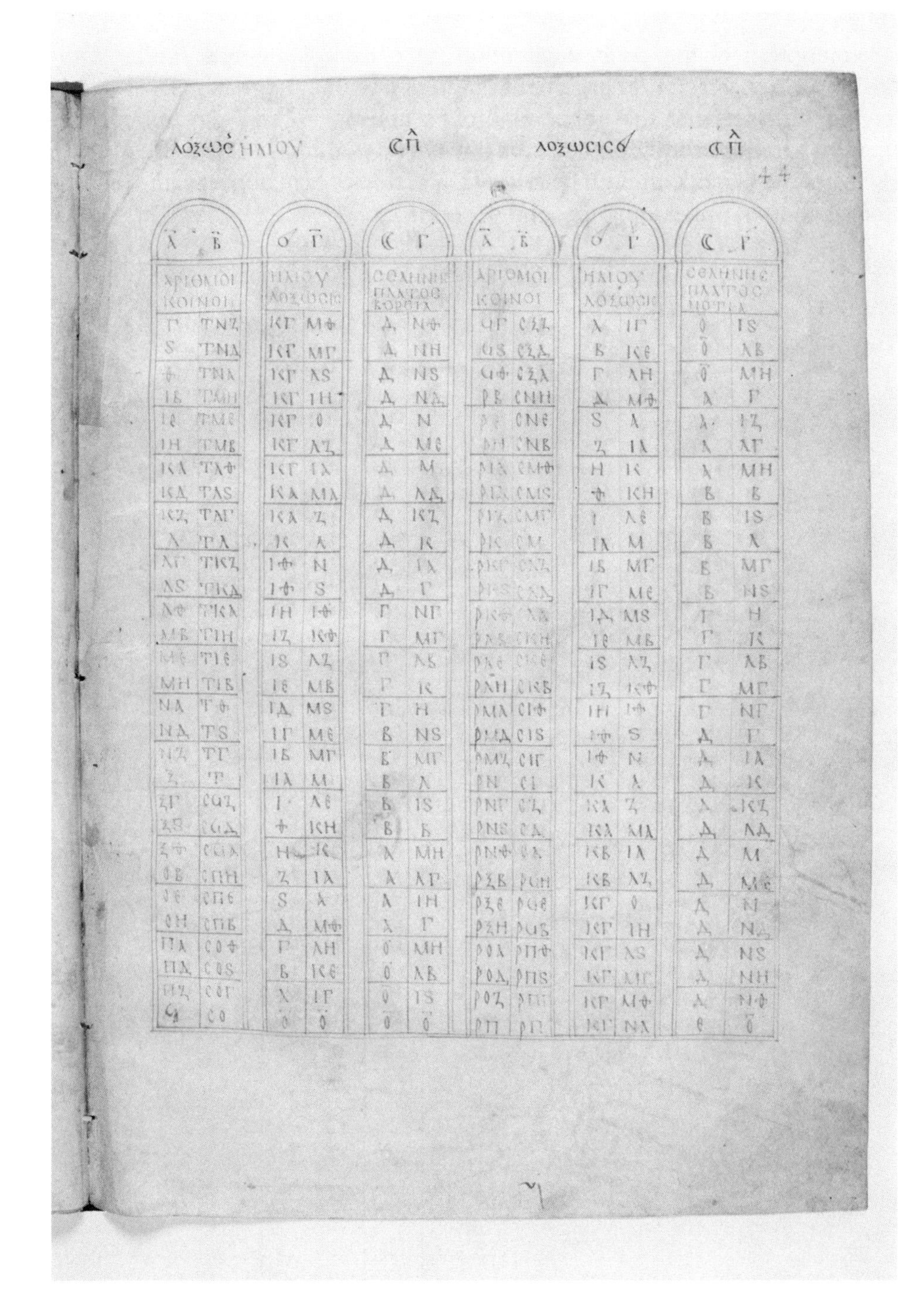

Figure 7.7.1. An excerpt from a manuscript of the *Handy Tables* (MS Vat. Gr. 1291, f. 44r) showing a solar declination table (cols. 2, 5) and lunar latitude (cols. 3, 6) for an argument range of 0 to 360° in 3° increments (cols. 1, 4). Image courtesy of the Vatican.

TABLE 7.7.1

A transcription of the solar declination table from Ptolemy's *Handy Tables* (Vat. Gr. 1291, f. 44r) for an argument range of 0 to 360 degrees in three-degree increments. Note that the solar longitudes are measured from the solstices rather than from the equinox, as in the table in the *Almagest*. The corresponding values are the same, although here they are stated only in degrees and minutes, rather than in degrees, minutes, and seconds.

Com. Nos		Obliquity		Com. Nos		Obliquity	
3	357	23	49	93	267	1	13
6	354	23	43	96	264	2	25
9	351	23	36	99	261	3	38
12	348	23	18	102	258	4	49
15	345	23	0	105	255	6	1
18	342	22	37	108	252	7	11
21	339	22	11	111	249	8	20
24	336	21	41	114	246	9	28
27	333	21	7	117	243	10	35
30	330	20	3	120	240	11	40
33	327	19	50	123	237	12	43
36	324	19	6	126	234	13	45
39	321	18	19	129	231	14	46
42	318	17	29	132	228	15	42
45	315	16	37	135	225	16	37
48	312	15	42	138	222	17	29
51	309	14	46	141	219	18	19
54	306	13	45	144	216	19	6
57	303	12	43	147	213	19	50
60	300	11	40	150	210	20	30
63	297	10	35	153	207	21	7
66	294	9	28	156	204	21	41
69	291	8	20	159	201	22	11
72	288	7	11	162	198	22	37
75	285	6	1	165	195	23	0
78	282	4	49	168	192	23	18
81	279	3	38	171	189	23	36
84	276	2	25	174	186	23	43
87	273	1	13	177	183	23	49
90	270	0	0	180	180	23	51

Introduction

The organization of these *Handy Tables* for the movements of the planets, my dear Syrus,[20] was made by us conformable to their mean and circular hypotheses, with a view to being able, by means of the eccentrics and epicycles drawn on a surface, according to the explanations demonstrated in the *Syntaxis*, to indicate their movements in longitude referred to the zodiac in accord with results obtained by calculations, while the deviations in latitude, which cannot result from such drawings, will be calculated methodically by the relevant tables.

The first tables have the positions in longitude and latitude of the famous towns of our inhabited earth; those that follow, the ascensions of the right sphere, the ecliptic and equator together, the sixtieths of an equinoctial hour being provided for each degree of the ecliptic: (the sixtieths) by which the mean *nychthemera*[21] differ from the origin up to that time, compared with apparent *nychthemera*; and again for the seven parallels in the interval of our habitable earth, are written, for each degree of the ecliptic, the seasonal times of the day, in order to be able to convert differences of times in a way that will be demonstrated in the following.

7.8 Ptolemy, *Planisphere*

Ptolemy's *Planisphere* is the first known treatise that develops a plane diagram of the celestial sphere using methods that are mathematically related to stereographic projection. There is no known Greek manuscript of this treatise; what exists today is an Arabic translation probably made in the ninth or tenth century in Baghdad, as well as a medieval Latin translation by Hermann of Carinthia based on a different Arabic version. It seems clear that Ptolemy wrote this for advanced students, given that he assumed many results from Euclid's *Elements* as well as trigonometric ideas from his own *Almagest*, and in general used ideas that a second-century mathematically knowledgeable reader would have understood. The basic idea of the treatise is to represent the important circles on the three-dimensional spherical universe on a plane through the celestial equator via stereographic projection from the south pole. Under this projection, circles on the sphere remain as circles and meridians become straight lines. We present here just the first seven sections of the treatise. In sections 1–3, Ptolemy constructs the objects in the plane that will stand for the equator, the meridians, and the declination circles on the celestial sphere. Then, in sections 4–7, he uses the trigonometric methods developed in the *Almagest* to calculate the correct ratios to use in actually drawing the plane diagram.

Section 1

Since it is possible, Oh Syrus, and useful in many subjects that there be, in a flat surface, the circles that occur on the solid sphere, as though spread out, I considered it necessary

[20] Syrus was the addressee of several of Ptolemy's works, but nothing is known about him.
[21] The length of a day and night.

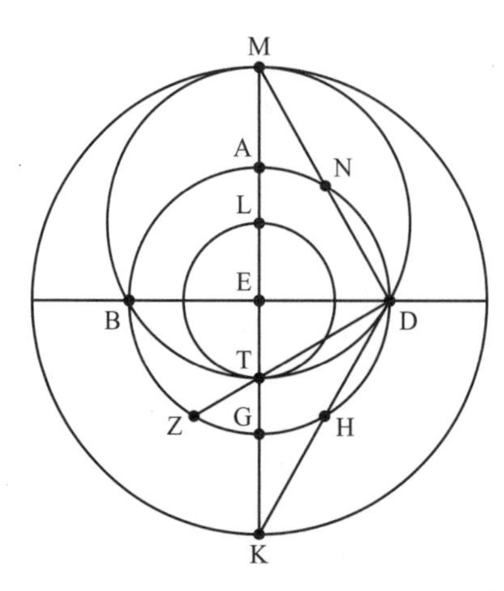

Figure 7.8.1.

with regards to the expert that I write a treatise for whoever desired knowledge of this, in which I briefly show how it is possible to draw the ecliptic, the circles parallel to the equator, and the circles known as the meridians, so that all of what occurs in this will be consistent with what is apparent in the solid sphere.

This aim we intend may be prepared for us when we use straight lines representing the meridians and arrange the circles parallel to the equator as a configuration, in which it is, firstly, proved that the drawn great circles, of the inclined circles tangent to circles parallel to the equator, which are the same distance from it in both directions, always bisect the equator. This is congruous for us in the following manner.

We assume the equator is circle $ABGD$ and that it is around center E [figure 7.8.1]. We draw in it two diameters intersecting at right angles, which are line AG and line BD. We imagine these lines representing meridians, and point E as the north pole, because it is not possible to place the other pole on a plane surface, since its plane extends without limit, as we shall show in what follows. Since the north pole is always visible in our countries, it is more appropriate that we specifically use it, in that we want a drawing of it.

Clearly, the circles parallel to the equator that are north of the equator should be drawn inside circle $ABGD$, while the parallel circles that are to the south must be drawn outside of it. We produce lines AG, BD and cut off two equal arcs of the circle on either side of point G, which are GZ and GH. We join line DTZ, and line DHK. We make point E a center, and we draw circle TL with a distance of line ET, and circle KM with a distance of line EK.

Then, I say that these circles are the correlates of two of the circles on the solid sphere that are the same distance from the equator on either side, and that the ecliptic, drawn about a center bisecting line TM such that it touches these circles at point T and at point M, bisects circle $ABGD$; that is, it passes through point B and point D. The proof of this is that we join line DNM. So, because arc AN is equal to arc GH, which is equal to arc GZ, arc NDZ is a semicircle. Hence, angle MDT is right and the circle drawn about diameter TM, of right triangle MDT, passes through point D. Hence, it bisects the equator.

So it is clear from this that, for all circles parallel to the equator, if we cut off arcs on both sides of point *G*, whose magnitude depends on the distance of each of these circles from the equator, and we join the endpoints of the arcs with straight lines to point *D*, and we make what the straight lines cut off from line *EK* distances, and we make point *E* a center, and we describe circles, then these arcs and lines are analogous to what we have set out earlier.

Clearly, if we assume both of arcs *ZG* and *GH* to be approximately $23;51^\circ$ (in the degrees in which the equator, circle *ABGD*, is 360°), which is the distance between the equator and both of the tropics along the circle drawn through the poles of the equator, then, of the two circles drawn through point *T* and point *M*, circle *TL* is the summer tropic and circle *KM* is the winter tropic. In this way, the circle drawn through point *M*, point *B*, point *T*, and point *D* (the circle through the signs) is tangent to the tropics at point *T* (the summer tropic) and at point *M* (the winter tropic); and it bisects the equator at points *B* and *D*. So, point *B* is the vernal point and point *D* the autumnal point, because the motion of the cosmos is indeed as though from point *B* toward point *A* and then to point *D*. It is neither possible for a division of the ecliptic into signs to take place through equal arcs, nor again for its division into four parts to take place through equal arcs. Rather, its division into what is required is strictly in this way: that is, the beginnings of the signs are put at the points at which circles parallel to the equator divide the ecliptic, which are drawn according to the explained method, with the distance consistent with the distance of each of the signs from the equator in the solid sphere. For, at this degree alone, all of the straight lines passing through pole *E*, representing the meridians, cross the ecliptic at parts that are the correlates of parts diametrically opposite on the solid sphere.[22]

Section 2

Every horizon circle, drawn in the same way as the ecliptic, not only bisects the equator but also functionally bisects the ecliptic. That is, it is also drawn through parts that are functionally the correlates of parts diametrically opposite on the solid sphere.

Let the equator be circle *ABGD* around center *E* [figure 7.8.2]. The circle through the signs is circle *ZBHD*, and it bisects the equator at point *B* and point *D*. We pass an arbitrary straight line through the pole *E* representing a meridian. Let it be line *ZAEHG*. I say that points *Z* and *H* are the correlates to diametrically opposite points on the solid sphere. That is, circles parallel to the equator that are drawn through these points will cut off equal arcs on both sides of the equator, in the way we described, just as occurs on the solid sphere as well.

The proof of this is that we produce a straight line, line *ET* from point *E* at right angles to line *AG*. We join line *AT*, line *GT*, line *ZKT*, and line *THL*. So, clearly, angle *ATG* is right, for arc *ATG* is a semicircle. Since the product of line *ZE* by line *EH* is equal to *ED* squared,[23] that is, equal to *ET* squared, the ratio of line *ZE* to line *ET* is as the ratio of line *ET* to line *EH*. So, triangle *ZTH* is also right angled, and angle *ZTH* is right. Hence,

[22] In other words, only when the signs of the zodiac are constructed in the manner described will the degrees determined as the beginnings of opposite signs be joined by straight lines that pass through the center of the equator.

[23] *Elements* 3:35.

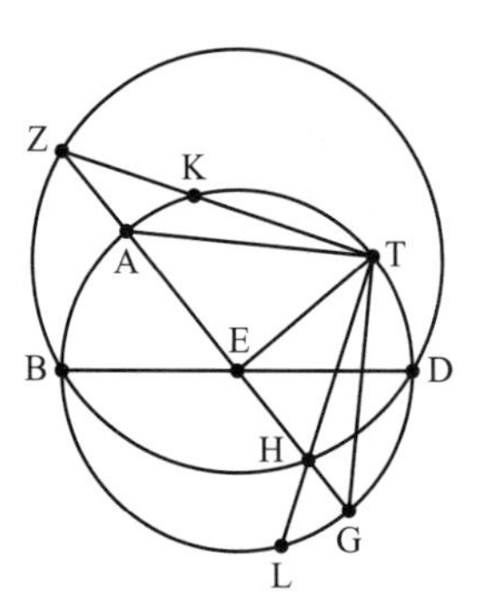

Figure 7.8.2.

angle ZTL is equal to angle ATG. Then, if we omit the common angle ATH, the remaining angle KTA will equal the remaining angle HTG. So, arc KA is also equal to arc GL. Now, we have shown that since lines TKZ and TL join the endpoints of arcs that are the same distance from the equator, and their origin is from the point, the distance of which from point A and point G is a quadrant, which is point T, then on line ZG we get point Z and point H, which are the points through which are drawn two circles parallel to the equator the same distance from it.[24] Therefore, line ZEH has passed through points that are functionally on the diameter of the ecliptic.

Section 3

I say that even if we draw another circle, inclined to the equator, representing the horizon circle, so that this circle bisects the equator alone, then the two places of the intersection of this circle and the circle through the signs are functionally diametrically opposite. That is, the line joining them passes through the center of the equator.

Again, let the equator be circle $ABGD$ around center E, and the ecliptic circle $HBTD$, and let it bisect the equator along diameter BED [figure 7.8.3]. The horizon circle is circle $HATG$, and this circle also bisects the equator along diameter AEG. Let the intersection common to the ecliptic and the horizon circle be point H and point T. Then, I say that if we join point H with center E by a straight line, representing a meridian, and we extend that line rectilinearly, it will arrive at point T.

The proof of this is that we join line HE and produce it recilinearly until it intersects the horizon circle, circle HAG, at point T.[25] Then, I say that point T is also common to the ecliptic, circle $HBTD$. So, because lines HT and AG have been produced in circle $HATG$ intersecting at point E, line HE by line ET is equal to line AE by line EG, and likewise, line AE by line EG is equal to line BE by line ED. Hence line BE by ED is equal to HE

[24] Ptolemy is imagining line *TE* as perpendicular to the plane of the equator. Thus arc *GL* represents the declination of the point *L* while arc *KA* represents the declination of the point *K*. The declinations are equal but in opposite directions; that is, circles parallel to the equator through these points cut off equal arcs on both sides of the equator.

[25] It would have been clearer if Ptolemy had initially differentiated between *T* as the intersection of *HE* and circle *HAG* and *T* as the intersection of circle *HAG* and circle *HBD* and then had proceeded to show that they were one and the same.

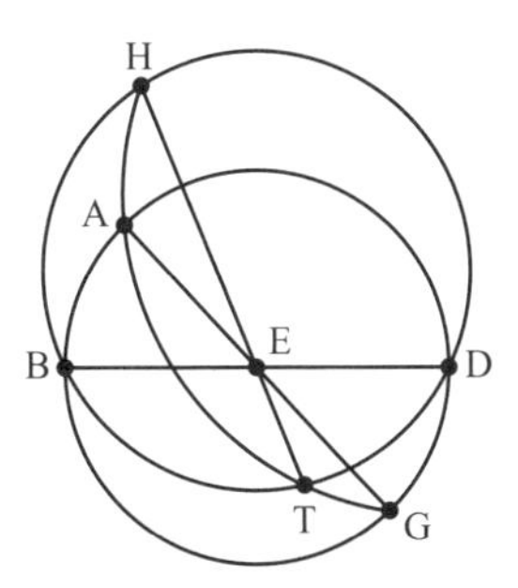

Figure 7.8.3.

by ET. Hence, line BD, TH are in a single circle.[26] From this it follows that point T is on the ecliptic, circle $HBTD$, and we had stated that it is on the horizon circle, circle $HATG$. So, the line joining the two places of the intersection of the ecliptic and the horizon is a line that passes through the center of the equator, point E. So, it is clear from this that the horizon circle and the ecliptic intersect at functionally diametrically opposite points.

Section 4

Then having previously demonstrated this, let us next consider the ratio of the radii of the parallel circles drawn according to the signs of the ecliptic to the radius of the equator, which we previously set out, so that we come to know that their rising times are also found numerically to be consistent with what is manifest with respect to the solid sphere.

Again, let the equator be circle $ABGD$ around center E [figure 7.8.4]. We produce two of its diameters, intersecting at right angles, AG and BD. We produce line AG rectilinearly to point Z. We cut off two equal arcs, GH and GT, on either side of point G. We join line DKH and line DTZ. We have previously explained, of circles parallel to the equator that are the same distance from it, the one of them to the north is indeed drawn about center E with distance EK, while that to the south [is drawn] about center E with distance EZ.

The ratio of line EZ to line EK is evident to us in this way. Because arc GH is equal to arc GT, arc BH and arc BGT together are a semicircle. So, the angles opposite them, that is, angle EDK and angle EDZ, are together equal to a right angle.[27] Also, angle EDK with angle EKD is right, so angle EDZ is equal to angle EKD. Hence, right triangle ZED is similar to right triangle DEK, so the ratio of line ZE to line ED is as the ratio of line DE to line EK. The ratio of arc BT to the supplement—that is, the arc equal to arc BH—is, however, as the ratio of angle EDZ to angle EZD, and as the ratio of the arc on line EZ, in the circle drawn around right triangle DEZ, to the arc on line ED, in the same circle.[28] So the ratio of the chord of arc BT to the chord of the supplement, that is, arc BH, is as the ratio of line ZE to line ED, and as the ratio of line DE to line EK.[29]

[26] Converse of *Elements* 3:35.
[27] Angle EDK is half of arc BH, and angle EDZ is half of arc BGT, by *Elements* 3:20.
[28] Equal arcs subtend equal angles, by *Elements* 3:26.
[29] This follows from the similarity of triangle BTD to triangle EDZ.

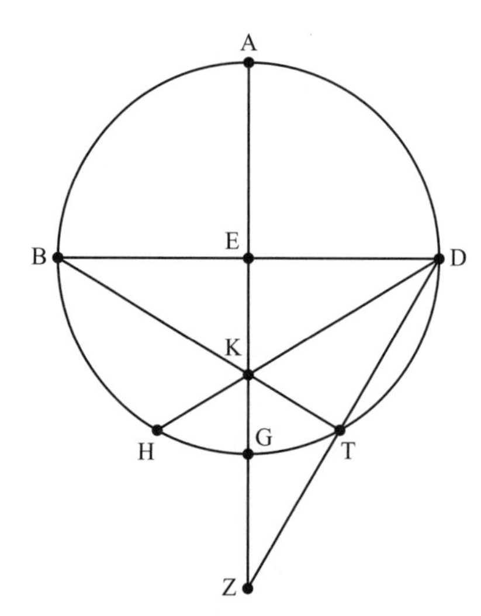

Figure 7.8.4.

Then having previously deduced that, in a similar diagram we first assume that both of the arcs GH and GT are $23;51,20°$ (in the degrees in which this circle is $360°$), which are the degrees that we assumed for the distance between the equator and both of the tropics in our discussion with respect to the solid sphere as well.

Then arc BT is $113;51,20°$ (in the degrees in which this circle is $360°$), and arc BH, the supplement, is $66;8,40°$. The chord of arc BT is $100;33,28^p$ (in the parts in which the diameter is 120^p, for we have assumed this in the *Almagest*), and chord BH is $65;29^p$ (of these parts).[30] So, the ratio of line ZE to line ED, and the ratio of line ED to line EK is the ratio of $100;33,28^p$ to $65;29^p$. Therefore, line EZ, the radius of the winter tropic, is $92;8,15^p$ (in those parts in which the radius of the equator, line ED, is 60^p), and the radius of the summer tropic is $39;4,19^p$.[31]

From this it is evident that the diameter of the ecliptic (since it is tangent to these two circles at the endpoints of its diameter) is the sum of their radii, $131;12,34^p$ (in the parts in which the radius of the equator is 60^p), and that the radius of the ecliptic is $65;36,17^p$. The line between its center and the center of the equator is $26;31,58^p$ (of these parts).[32]

Section 5

Again, we assume both arcs HG and GT to be $20;30,9°$—which is the distance between the equator and each of the two circles parallel to the equator that cut off 60° of the circle

[30] According to the table of chords in *Almagest*, book 1, section 11, the chord of an arc of $113;30°$ is $100;21,16^p$, with the value of a sixtieth being $0;0,34,20$. Thus, we need to multiply that latter value by $0;21,20$, giving a value of $0;12,12^p$ to add to $110;21,16^p$, giving the stated value of $100;33,28^p$. Similarly, the table's value of the chord of 66° is $65;21,24^p$ with the value of a sixtieth being $0;0,52,37^p$.

[31] The ratio of ZE to ED is, in decimals, 1.535623993. If $ED = 60$, then $EZ = 92.13743955$, which, in sexagesimal notation, is $92;8,15^p$, as stated. Similarly, $EK = 60 \div 1.535623993 = 39.07206469$, which, in sexagesimals, is $39;4,19^p$.

[32] This is the difference between the radius of the ecliptic and the radius of the summer tropic.

through the signs on both sides of the solstitial points[33]—so that arc BT is $110;30,9^{\circ}$ and its chord is $98;35,57^p$, and arc BH is $69;29,51^{\circ}$ and its chord is $68;23,51^p$. Hence, the ratio of line ZE to line ED, and also the ratio of line ED to line EK, is the ratio of $98;35,57^p$ to $68;23,51^p$. So, of the parts in which line ED is 60^p, line EZ is $86;29,42^p$ and line EK is $41;37,15^p$ (of these parts).[34]

Section 6

In this way, we assume both arcs HG and GT to be $11;39,59^{\circ}$, which is the distance, along the great circle drawn through the poles of the equator, between the equator and each of the two circles parallel to it that cut off 30° from the circle through the signs on both sides of the solstitial points.[35] So, the whole arc BT is $101;39,59^{\circ}$ and its chord is $93;2,14^p$, and arc BH is $78;20^{\circ}$ and its chord is $75;47,23^p$. So the ratio of line ZE to line ED and the ratio of line DE to line EK is the ratio of $93;2,14^p$ to $75;47,23^p$, and of the parts in which line DE is 60^p, line EZ is $73;39,7^p$ and line EK is $48;52^p$ (of these parts).[36]

Section 7

Likewise, if we make both arcs HG and GT 54°, which is the distance, on either side of the equator, of each of the circles parallel to the equator that are tangent to the horizon at the latitude of Rhodes, which is the horizon we used as an example on the solid sphere—then, in this case as well, arc BT is 144° and its chord is $114;7,37^p$, and arc BH is 36° and its chord is $37;4,55^p$. The ratio of line ZE to line ED, and line DE to line EK, is the ratio of $114;7,37^p$ to $37;4,55^p$. So, of the parts in which line ED is 60^p, line EZ again sums to $184;39,48^p$ and line EK is $19;29,42^p$ (of these parts).[37] Clearly, since it is these lines, when summed, that are the diameter of the horizon we previously assumed—just as the diameter of the ecliptic is the diameters of the tropics—this diameter will be $204;9,30^p$ (in the parts in which the diameter of the equator is 120^p). It follows from this that the radius of the horizon circle is $102;4,45^p$, and the line between the center of this circle and the equator is $82;35,3^p$ (of these parts).

7.9 Ptolemy, *Geography*

The main feature of Ptolemy's *Geography* is an enormous list of place names with their latitudes and longitudes, intended as the basis for drawing maps of the known world and its principal regions. But preliminary to this is a description of two methods of projecting the spherical globe onto flat paper so that one can in fact draw reasonable

[33] Ptolemy is here stating the declination of the sun when its longitude is 60°. This calculation is in *Almagest*, book 1, section 14, and is included above.

[34] Calculation here gives $EZ = 86;29,37^p$ and $EK = 41;37,18^p$.

[35] See *Almagest*, book 1, section 14.

[36] Calculation here gives $EZ = 73;39,15^p$ and $EK = 48;52,37^p$.

[37] Again, Ptolemy seems to be a bit incorrect in his calculation. The reader should attempt the calculation to check the values given, using Ptolemy's chord table.

maps, which we have excerpted below. Ptolemy was interested only in making maps of the *oikoumenē*, the known inhabited world. For him, the northern boundary of this region was the parallel of Thulē, near today's Shetland Islands, at 63° north, while the southern boundary was the parallel at 16°25′ south of the equator. In east-west extent, Ptolemy's world stretched close to 180°. The westernmost part was what he called the Islands of the Blest, identified with today's Canary Islands off the coast of Spain. The meridian through these islands was Ptolemy's prime meridian, that is, the meridian from which longitude was measured. The eastern boundary of Ptolemy's world was somewhere on the east coast of China. A glance at a modern map shows that the actual longitudinal distance between these two places is about 135° rather than Ptolemy's 180°. Ptolemy's erroneous value was crucial in Columbus's argument to the Spanish monarchs that he could reach China by sailing west. A second error of Ptolemy, also crucial to Columbus's argument, was his value for the size of the earth, namely, that the earth's circumference was 180,000 stades, a much lower estimate than Eratosthenes's value of 250,000. If the stade was approximately 185 meters, then Ptolemy's estimate is about 18 percent too small.

In section 23, Ptolemy begins by noting that the meridians through the Islands of the Blest and through a location just off the coast of China are 180° apart; that is, they enclose twelve hour-intervals. He then writes about the meridians themselves and the parallels (i.e., lines of latitude) that he will use. The parallels are described by their length of longest daylight, which, at least for those north of the equator, occurs on the summer solstice.

Section 23. We have decided it is appropriate [to the size of the map] to draw the meridians at intervals of a third of an equinoctial hour, that is, at intervals of five of the chosen units [i.e., degrees] of the equator, and [to draw] the parallels north of the equator as follows:

1. The first parallel differing [in length of longest daylight] from [the equator's twelve hours] by $\frac{1}{4}$ hour, and distant [from the equator] by $4\frac{1}{4}^\circ$, as established approximately by geometrical demonstrations.
2. The second, differing by $\frac{1}{2}$ hour, and distant $8\frac{5}{12}^\circ$.[38]
3. The third, differing by $\frac{3}{4}$ hour, and distant $12\frac{1}{2}^\circ$.
4. The fourth, differing by 1 hour, and distant $16\frac{5}{12}^\circ$, and drawn through Meroē.[39]
5. The fifth, differing by $1\frac{1}{4}$ hours, and distant $20\frac{1}{4}^\circ$.
6. The sixth, which is on the Summer Tropic, differing by $1\frac{1}{2}$ hours, and distant $23\frac{5}{6}^\circ$, and drawn through Soēnē.[40]
7. The seventh, differing by $1\frac{3}{4}$ hours, and distant $27\frac{1}{6}^\circ$.
8. The eighth, differing by 2 hours, and distant $30\frac{1}{3}^\circ$.
9. The ninth, differing by $2\frac{1}{4}$ hours, and distant $33\frac{1}{3}^\circ$.

[38] Recall that Ptolemy is using sexagesimal notation, so this would be expressed as 8;25.

[39] Meroē was an ancient city on the east bank of the Nile in what is now Sudan, approximately 200 km northeast of Khartoum. It was the capital of the Kingdom of Kush during Ptolemy's time.

[40] Soēnē is the modern city of Aswan in southern Egypt on the Nile.

10. The tenth, differing by $2\frac{1}{2}$ hours, and distant $36°$, and drawn through Rhodes.
11. The eleventh, differing by $2\frac{3}{4}$ hours, and distant $38\frac{7}{12}°$.
12. The twelfth, differing by 3 hours, and distant $40\frac{11}{12}°$.
13. The thirteenth, differing by $3\frac{1}{4}$ hours, and distant $43\frac{1}{12}°$.
14. The fourteenth, differing by $3\frac{1}{2}$ hours, and distant $45°$.
15. The fifteenth, differing by 4 hours, and distant $48\frac{1}{2}°$.
16. The sixteenth, differing by $4\frac{1}{2}$ hours, and distant $51\frac{1}{2}°$.
17. The seventeenth, differing by 5 hours, and distant $54°$.
18. The eighteenth, differing by $5\frac{1}{2}$ hours, and distant $56°$.
19. The nineteenth, differing by 6 hours, and distant $58°$.
20. The twentieth, differing by 7 hours, and distant $61°$.
21. The twenty-first, differing by 8 hours, and distant $63°$, which is drawn through Thulē.
22. And another parallel will be drawn south of the equator, containing a difference of $\frac{1}{2}$ hour, and it will pass through Cape Rhapton and Kattigara,[41] and be approximately the same number of degrees, $8\frac{5}{12}°$, from the equator as the oppositely situated [localities north of the equator with the same longest daylight].
23. And the parallel that marks the southern limit will also be drawn; it is as far south of the equator as the parallel through Meroē is north of it.

Section 24. Method of making a map of the *oikoumenē* in the plane in proper proportionality with its configuration on the globe

Ptolemy's first projection uses straight lines for the meridians and circular arcs for the parallels, although he does bend the meridian lines inward south of the equator.

[Ptolemy's first projection]

[Stage 1: Preparation of the rectangular surface on which the map is to be drawn; construction of the central meridian, the common intersection of all meridians, and the parallel through Rhodes]

Let us fashion a [planar] surface in the shape of a rectangular parallelogram $ABGD$, with side AB approximately twice AG. Let line AB be assumed to be in the top position; this is going to be at the north end of the map. Then we will bisect AB by the perpendicular straight line EZ, and attach a rule EH to AB, of suitable size and perpendicular to AB so that the line EH down the middle of its length is in a straight line with EZ [figure 7.9.1].

Let there be taken on it [a length] EH of 34 [units] such that straight line HZ is $131\frac{5}{12}$, and with a center H and radius [to reach] the point 79 units away on HZ we will describe a circle ΘKL, which will represent the parallel through Rhodes.[42]

[41] Although these place names appear on Ptolemy's maps—off the coast of Sri Lanka and in modern Vietnam, respectively—they are both in the northern hemisphere rather than south of the equator. Ptolemy's geographical knowledge about places in the southern hemisphere was quite imperfect.

[42] The length 34 of HE is arbitrary and probably chosen for convenience in drawing the map. But note that the arc of the equator, RST, has radius 115 $(=131\frac{5}{12}-16\frac{5}{12})$, since arc MZN is the parallel $16\frac{5}{12}°$ south of the equator.

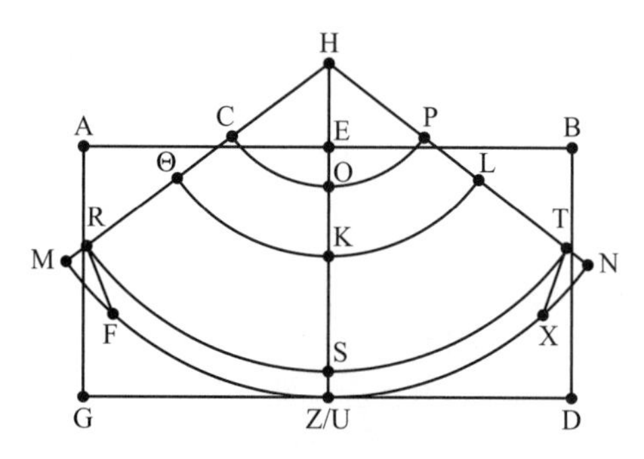

Figure 7.9.1.

[Stage 2: Construction of meridian lines at five-degree intervals]

For the limits of the longitude, which comprise six hour-intervals on each side of K, we take an interval of 4 units on the middle meridian HZ (i.e., five degrees on the parallel through Rhodes because of the approximate ratio of $5:4$ between the great circle and [the parallel]),[43] and we count off eighteen intervals of this size on each side of K along arc ΘKL. [In this way] we get the points through which the meridians that will enclose the intervals of one-third of an hour will have to be drawn from H, and consequently also the [meridians] marking off the limits [of longitude], namely, $H\Theta M$ and HLN.

[Stage 3: Construction of arcs representing the equator and limiting parallels, and the other parallels]

Next the parallel COP through Thulē will be drawn, with radius 52 units from H on HZ, and the equator RST, [with radius] 115 units from H, and the parallel MUN oppositely situated to the parallel through Meroē—this is the farthest south [of the parallels], [with radius] $131\frac{5}{12}$ units from H. Hence the ratio of RST to COP will amount to $115:52$, in agreement with the ratio of these parallels on the globe, since HO is 52 of such units as HS was assumed to be 115, and as HS is to HO, so is arc RST to [arc] COP.[44]

Also, the interval OK of the meridian, that is, the [interval] from the parallel through Thulē to that through Rhodes, will turn out to be 27 units; and KS, the [interval] from the parallel through Rhodes to the equator, will turn out to be 36 of the same [units]; and SU, the [interval] from the equator to the parallel oppositely situated to that through Meroē, will turn out to be $16\frac{5}{12}$ of the same [units].[45] Moreover, of such [units] as the latitudinal dimension OU of the known world is $79\frac{5}{12}$ (or as a round number, 80), ΘKL, the middle interval in longitude [measured along the parallel through Rhodes], will be 144, in agreement with the hypotheses derived from the demonstrations [earlier]—i.e., the 40,000 stades of latitude

[43] Since the latitude of Rhodes is 36° (see below), the length of a degree along the parallel through Rhodes is $\cos 36° = 0.81$ times that along the equator.

[44] Note that $115 - 52 = 63$, the number of degrees between the parallel through Thulē and the equator.

[45] Since $HK = 79$ and $HO = 52$, we have $OK = 27$; that is, the parallel through Rhodes is 27 degrees farther south than that through Thulē, so is at 36°.

have approximately the same ratio to the 72,000 stades of longitude on the parallel through Rhodes as [the ratio $79\frac{5}{12} : 144$].[46]

We will also draw the rest of the parallels, if we choose, again using H as center and radii [extending to the points] as many units away from S as the [numbers] set out in the [list of] distances from the equator.

[Stage 4: Inflection of meridian lines south of the equator]

Instead of having the lines representing the meridians straight as far as parallel MUN, we can have them [straight] just as far as the equator RST; then, dividing arc MUN into parts that are equal and equal in number to [the parts] established on the parallel through Meroē, [we can] draw between these divisions and the [divisions] on the equator the straight lines for the meridians that fall between [the parallel and the equator] (e.g., lines RF and TX), so that the bending away on the other, south side of the equator is in some way apparent from the inflection [of the meridian lines] incorporated [in the map].

[Stage 5: Drawing of the map]

Next, to make the labeling of the localities that are to be included convenient, we will also make a narrow little ruler, equal in length to HZ (or just to HS), and peg it to H so that as it is revolved along the whole longitudinal dimension of the map, one of its edges will exactly fit the straight lines of the meridians because it is cut away so as to be in line with the middle of the pole. We divide this edge into the $131\frac{5}{12}$ parts corresponding to HZ (or the 115 parts corresponding to just HS), and label the numbers starting from the division at the equator in order not to divide the middle meridian of the map into all the [115] parts and label them [all], thereby making a mess of the inscriptions of the localities that are to be made next to [the middle meridian]. It will also be possible to draw the parallels using these [marks].

Then we divide the equator, too, in the $180°$ of the twelve hour-intervals, and annex the numbers starting at the westernmost meridian. And we shift the edge of the ruler in each case to the indicated degree of longitude, and, using the divisions on the ruler, we arrive at the indicated position in latitude as required in each instance, and make a mark just as we have explained for the globe.

In his second projection, Ptolemy uses circular arcs for both the meridians and the parallels. As he notes, we are to imagine that we are viewing the globe from a point above the intersection of the central meridian ($90°$ east of the Islands of the Blest, or near the east coast of the Arabian Peninsula) and the central parallel, taken to be at latitude $23°50'$, the Tropic of Cancer, that is, the parallel through Soēnē. That parallel is roughly halfway between the southernmost and northernmost parallels on his map.

[46] Since a full great circle is assumed to be 180,000 stades, the 80 degrees of arc OU amounts to 40,000 stades. Also, since the size of a degree along the parallel of Rhodes is $\frac{4}{5}$ of that along the equator, the 180 degrees of arc ΘKL corresponds to $\frac{4}{5}$ of 90,000, or 72,000 stades as noted. Thus, Ptolemy has shown how distances are preserved by his map, at least near the central meridian and the parallel through Rhodes.

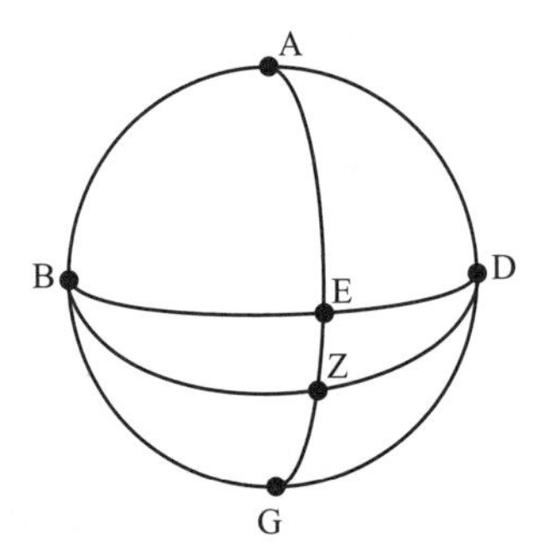

Figure 7.9.2.

[Ptolemy's second projection]

We could make the map of the *oikoumenē* on the [planar] surface still more similar and similarly proportioned [to the globe] if we took the meridian lines, too, in the likeness of the meridian lines on the globe, on the hypothesis that the globe is so placed that the axis of the visual rays passes through both [1] the intersection nearer the eye of the meridian that bisects the longitudinal dimension of the known world and the parallel that bisects its latitudinal dimension, and also [2] the globe's center. In this way the oppositely situated limits [of latitude and longitude] will be taken in and perceived by the visual rays at equal distances.

[Stage 1: Determination of an appropriate point to serve as the common center of the arcs representing the parallels]

First, [we want] to establish the magnitude of inclination of the parallel circles with respect to the plane that is perpendicular to the meridian in the middle of the longitude and [that passes] through both the stated intersection [of the bisecting meridian and parallel] and the sphere's center. Let us imagine the great circle $ABGD$ that delimits the visible hemisphere; the semicircle AEG of the meridian that bisects the hemisphere; and point E, which is the intersection nearer the eye of the [central meridian] and the parallel bisecting the latitudinal dimension [figure 7.9.2].

And let there be described through E another semicircle of a great circle, BED, perpendicular to AEG. Obviously the plane of BED lies along the axis of the visual rays. Let arc EZ be measured off as $23\frac{5}{6}^{\circ}$ (since the equator is this many [degrees] from the parallel through Soēnē, which is approximately the middle of the latitudinal breadth), and let the semicircle BZD of the equator be described through Z. Then the plane of the equator and the [planes] of the other parallels will appear inclined with respect to the [aforesaid plane] through the axis of the visual rays at [the angle of] arc EZ, which is $23\frac{5}{6}^{\circ}$.

Now let $AEZG$ and BED be imagined as straight lines representing arcs [figure 7.9.3], such that BE has a ratio to EZ of $90 : 23\frac{5}{6}$.[47] And let GZ be produced, and let the center

[47] Since BED is half of a great circle passing through the middle point of the parallel through Soēnē, BE itself is one-quarter of a great circle, therefore equal to 90°. Then EZ is taken to be $23\frac{5}{6}^{\circ}$, the distance from the central point Z on the equator to the parallel of Soēnē.

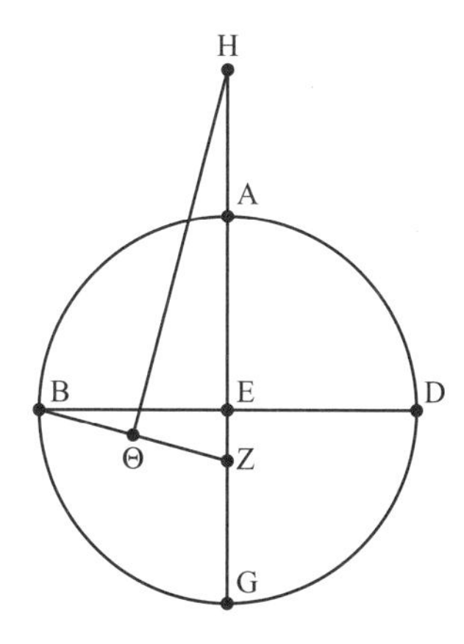

Figure 7.9.3.

about which the circular segment BZD is to be described be at H, and let it be required to find the ratio of HZ to EB.

Let straight line ZB be drawn, and bisected at Θ, and let ΘH (which is of course perpendicular to BZ) be drawn. Then since EZ was assumed to be $23\frac{5}{6}$ of such [units] as straight line BE is 90, the hypotenuse BZ will be $93\frac{1}{10}$ of the same [units]. And angle BZE will be $150\frac{1}{3}$ of such [units, i.e., half-degrees] as two right angles are 360,[48] and the remaining angle ΘHZ will be $29\frac{2}{3}$ of the same [half-degrees]. Consequently the ratio of HZ to $Z\Theta$ is $181\frac{5}{6}:46\frac{11}{20}$.[49] But of such [units] as ΘZ is $46\frac{11}{20}$, straight line BE is 90; so that also of such [units] as straight line BE is 90 (and ZE is $23\frac{5}{6}$ of the same), we will have straight line HZ too as $181\frac{5}{6}$. And [we will thus obtain] point H, about which all the parallels in the plane map are to be described.

[Stage 2: Construction of the arcs for the parallels]

Now that these things have been established, let the [plane] surface $ABGD$ be set out with AB again being twice AG, and AE equal to EB, and EZ at right angles to AEB [figure 7.9.4].[50] Also let some straight line equal to EZ be divided into the 90 units [corresponding to the degrees] of the quadrant. Let ZH be taken with length $16\frac{5}{12}$ units, and $H\Theta$ with length $23\frac{5}{6}$ units, and HK with length 63 units. If H is assumed to be on the equator, Θ will be the point through which the parallel through Soēnē (which is approximately in the middle of the latitudinal dimension) will be drawn; and Z will be the point through which the parallel will be drawn that marks the southern limit and is opposite to the parallel through Meroē, and K will be the point through which the parallel will be drawn that marks the northern limit and passes through the island of Thulē.

We now produce [line EZ's] extension HL with length $181\frac{5}{6}$ of the same units (or for that matter just 180 units, since the map will not be significantly different on this account). And

[48] Ptolemy could calculate this value by using his table of chords. Today we can calculate this as arcsin $\frac{90}{93.1} = 75\frac{1}{6}^{\circ}$, that is, $150\frac{1}{3}$ half degrees.

[49] By similar triangles, $HZ : Z\Theta = BZ : ZE$.

[50] Points L, Θ, H in figure 7.9.4 correspond to points H, E, Z in figure 7.9.3.

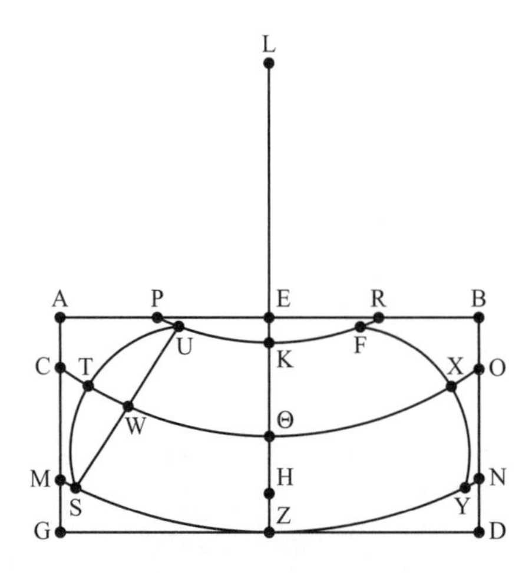

Figure 7.9.4.

with center L and radii [extending to] Z, Θ, and K, we describe arcs PKR, $C\Theta O$, and MZN. The proper pattern of inclination of the parallels with respect to the plane through the axis of the visual rays will thus have been preserved, since here, too, [as in the hypothetical view of the globe], the axis [of the visual rays] ought to point to Θ and be at right angles to the plane of the map, so that the oppositely situated limits of the map will again be perceived by the sight as equidistant [from the center].

[Stage 3: Construction of the arcs for the meridians]

The longitudinal dimension should be proportional to the latitudinal dimension. On the globe, of such [units] as the great circle is 5, the parallel through Thulē amounts to approximately $2\frac{1}{4}$, and that through Soēnē $4\frac{7}{12}$, and that through Meroē $4\frac{5}{6}$.[51] One has to place eighteen meridians at intervals of one-third hour on each side of the meridian line ZK to complete the semicircles [of the parallels of latitude] contained by the total longitudinal dimension, so that we will take, on each of the three parallels that have been set out, segments equivalent to $5°$, i.e., one-third of an hour-interval. [Thus] we will make the divisions from K at intervals of $2\frac{1}{4}$ units such as straight line EZ is 90, and from Θ at intervals of $4\frac{7}{12}$, and from Z at intervals of $4\frac{5}{6}$. Then we will draw the arcs to represent the remaining meridians through [each set of] three corresponding points, e.g., the [meridians] that are to mark the limits of the whole longitudinal dimension, STU and FXY. We shall then add the rest of the [arcs] representing the remaining parallels, with center again L and radii [extending to] the divisions on ZK according to their distances from the equator.

It is immediately obvious how such a map is more like the shape on the globe than the former map. For there [i.e., on the globe], too, when the globe is stationary and not turned about (which is necessarily the case with [the situation represented on] the [planar]

[51] Since Thulē is at latitude 63°, one degree along its parallel is $\cos 63°$ times one degree along the equator. But $\cos 63° = .454 \approx 2.25 : 5$. The ratios at the two other parallels are also approximately equal to the cosines of their latitude.

surface), since the sight is directed toward the middle of the map, a single meridian, [namely,] the one in the middle, would be in the plane through the axis of the visual rays and so would give the illusion of a straight line; whereas the [meridians] on either side of it all appear curved with their concavities toward it, and the more so the farther from it they are. Here, too, [in the present map] one will retain this [appearance] with the proper relative curvatures. Moreover, the proportionality of the parallel arcs with respect to each other preserves the proper ratio, not just for the equator and the parallel through Thulē (as in the former [map]), but also as very nearly as possible for the other [parallels], as anyone can discover who makes the experiment.

And [the ratio] of the total latitudinal dimension to the total longitudinal dimension [will be preserved] again not only for the parallel drawn through Rhodes (as in the former [map]), but [at least] roughly for absolutely all [the parallels]. For if here, too, as in the former drawing, we draw the straight line SWU, arc ΘW will obviously make a smaller ratio to [arcs] ZS and KU than the correct ratio in this map, which was obtained using all of [arc] ΘT (imagined as being along the equator). And if we make this [arc ΘW] in correct ratio to the latitudinal interval ZK, then ZS and KU will be greater than the [arcs] that are in correct ratio to ZK [on these parallels], just as ΘT is. Or if we keep ZS and KU in correct ratio to ZK, then ΘW will be less than the [arc] that is in correct ratio to KZ, just as it is less than ΘT.

In these respects, then, this method is superior to the former. But it might be inferior to the other with respect to the ease of making the map, since [in the former method] it was possible to inscribe each locality by revolving and moving the ruler from side to side, with just one of the parallels drawn and divided [into degrees]; whereas here such [a ruler] is of no advantage because of the bending of the meridian lines from the central [meridian], so that all the circles have to be drawn on the map, and positions falling between the grid lines have to be guessed at by calculating on the basis of the recorded fractional parts with reference to the [four] whole sides [of the grid] that surround [the place in question]. Even so, I think that, here as on all occasions, the superior and more troublesome [method] is to be preferred to the inferior and easier one; but all the same, one should hold on to descriptions of both methods, for the sake of those who will be attracted to the handier one of them because it is easy.

7.10 Antikythera Mechanism

In 1900, sponge divers discovered a sunken ship by Antikythera, a small island off the coast of mainland Greece. Among the haul of marvelous ancient treasures that were retrieved from the wreck—amphorae, bronze and marble statues, glassware, coins, jewelry—was an item initially described as a "slab," a wooden and metal object about the size of a shoebox inscribed with some legible Greek letters. After preliminary investigations in the years that followed, the significance of this object was eventually realized. The intricacy of its remaining mechanisms suggested an apparatus that was without parallel in antiquity, a remarkably complex device with interlocking gearwheels, some staggeringly small, that was some sort of working astronomical or calendric mechanism.

Dubbed a "scientific wonder of the ancient world," the item, known today as the Antikythera Mechanism, occupied teams of experts in various periods for the next hundred years up until the present day, in which studies are ongoing. Efforts to understand and contextualize this incredible device have yielded numerous speculations. As new technologies became available, among them radiography, computed tomographic (CT) scanning, and reflectance transformation imaging, details of internal workings and hidden inscriptions emerged that were beyond the naked eye observations of the initial investigations, eventually revealing a mechanism comprising around 30 intricate and interconnected gearwheels related to solar and lunar phenomena. Descriptions of the device, such as a mechanized planetarium, an astrolabe, an astronomical calculating machine, a calendar computer, a "cosmochronicon," all represent the evolving understanding of this astonishing device, culminating in a reconstruction and explanation of the Mechanism in 2006 that is now widely accepted, at least regarding application to solar and lunar motions. Further questions, such as its use and ownership, reconstructing the tools and techniques of construction, its connections to the mathematical and astronomical writings of its time, and its cultural significance, are the topics of ongoing studies.

We present here some analyses of the Mechanism by various scholars of the past century.

Derek J. de Solla Price, "An Ancient Greek Computer" (*Scientific American*, 1959)

Among the treasures of the Greek National Archaeological Museum in Athens are the remains of the most complex scientific object that has been preserved from antiquity. Corroded and crumbling from 2,000 years under the sea, its dials, gear wheels and inscribed plates present the historian with a tantalizing problem. Because of them we may have to revise many of our estimates of Greek science. By studying them we may find vital clues to the true origins of that high scientific technology which hitherto has seemed peculiar to our modern civilization, setting it apart from all cultures of the past.

. . .

Many of the Greek scientific devices known to us from written descriptions show much mathematical ingenuity, but in all cases the purely mechanical part of the design seems relatively crude. Gearing was clearly known to the Greeks, but it was used only in relatively simple applications. They employed pairs of gears to change angular speed or mechanical advantage or to apply power through a right angle, as in the water-driven mill. Even the most complex mechanical devices described by the ancient writers Hero of Alexandria and Vitruvius contained only simple gearing.

. . .

As soon as the fragments had been discovered they were examined by every available archaeologist; so began the long and difficult process of identifying the mechanism and determining its function. Some things were clear from the beginning. The unique importance of the object was obvious, and the gearing was impressively complex. From the inscriptions and the dials the mechanism was correctly identified as an astronomical device. The first conjecture was that it was some kind of navigating instrument—perhaps an astrolabe (a sort of circular star-finder map also used for simple observations). Some thought that it might be a small planetarium of the kind that Archimedes is said to have made. Unfortunately the fragments were covered by a thick curtain of calcified material and corrosion products, and these concealed so much detail that no one could be sure of his conjectures or reconstructions. There was nothing to do but wait for the slow and delicate work of the museum technicians in cleaning away this curtain.

. . .

The sun is mentioned several times, and the planet Venus once; terms are used that refer to the stations and retrogradations of planets; the ecliptic is named. Pointers, apparently those of the dials, are mentioned. A line of one inscription significantly records "76 years, 19 years." This refers to the well-known Calippic cycle of 76 years, which is four times the Metonic cycle of 19 years, or 235 synodic (lunar) months. The next line includes the number "223," which refers to the eclipse cycle of 223 lunar months.

Putting together the information gathered so far, it seems reasonable to suppose that the whole purpose of the Antikythera device was to mechanize just this sort of cyclical relation, which was a strong feature of ancient astronomy. Using the cycles that have been mentioned, one could easily design gearing that would operate from one dial having a wheel that revolved annually, and turn by this gearing a series of other wheels which would move pointers indicating the sidereal, synodic and draconitic months. Similar cycles were known for the planetary phenomena; in fact, this type of arithmetical theory is the central theme of Seleucid Babylonian astronomy, which was transmitted to the Hellenistic world in the last few centuries BCE. Such arithmetical schemes are quite distinct from the geometrical theory of circles and epicycles in astronomy, which seems to have been essentially Greek. The two types of theory were unified and brought to their peak in the second century CE by Claudius Ptolemy, whose labors marked the triumph of the new mathematical attitude toward geometrical models that still characterizes physics today.

The Antikythera Mechanism must therefore be an arithmetical counterpart of the much more familiar geometrical models of the solar system which were known to Plato and Archimedes and evolved into the orrery and the planetarium. The mechanism is like a great astronomical clock without an escapement, or like a modern analogue computer which uses mechanical parts to save tedious calculation.

Tony Freeth, "Decoding the Antikythera Mechanism" (*Nature*, 2006)

The 82 fragments that survive in the National Archaeological Museum in Athens are shown to scale [figure 7.10.1] . . . The major fragments A, B, C, D are across the top, starting at top left, with E, F, G immediately below them. Twenty-seven hand-cut bronze gears are in fragment A and one gear in each of fragments B, C, and D. Segments of display scales are in fragments B, C, E, and F . . . It is not certain that every one of the remaining fragments (numbered 1–75) belongs to the mechanism. The distinctive fragment A, which contains most of the gears, is approximately 180 x 150 mm in size. We have used three principal techniques to investigate the structure and inscriptions of the Antikythera Mechanism. (1) Three-dimensional X-ray microfocus computed tomography

Figure 7.10.1. The fragments of the Antikythera Mechanism that survive in the National Archaeological Museum in Athens. From "Decoding the Ancient Greek Astronomical Calculator Known as the Antikythera Mechanism" by T. Freeth, et al. (*Nature* 444: 587–591) © 2006 Nature Publishing Group. Reproduced with permission.

(CT), developed by X-Tek Systems Ltd. The use of CT has been crucial in making the text legible just beneath the current surfaces. (2) Digital optical imaging to reveal faint surface detail using polynomial texture mapping (PTM), developed by Hewlett-Packard Inc. (3) Digitized high-quality conventional film photography.

Alexander Jones, "The Antikythera Mechanism and the Public Face of Greek Science" (*Proceedings of Science, 2012*)

The remnants of the Hellenistic device known as the Antikythera Mechanism were salvaged in 1900–1901 from the site of a shipwreck dated to approximately 70–50 BCE, near the coast of the island of Antikythera. They lay unnoticed in the National Archaeological Museum in Athens among miscellaneous bronze fragments of statuary recovered from the wreck site until May 1902, when Spyridon Stais, the former Minister of Education who had commissioned the salvage operations on the part of the Greek government, visited the museum and chanced to observe fragments of corroded metal bearing gears and inscribed texts on some of their surfaces. Since 1902 there have been three periods of active research on the Mechanism: 1902–1910 (many archaeologists and other scholars), late 1920s–early 1930s (Ioannis Theofanidis), and 1953–present (Derek de Solla Price, Allan Bromley, M. T. Wright, the Antikythera Mechanism Research Project, and others).

From the outset two questions have dominated the study of the Mechanism: what did it do? and what was it for? The first question is essentially the problem of reconstructing the Mechanism as the integral, functioning object that it was when it was manufactured, and perhaps as it still was when it was taken on board the ill-fated vessel. The principal obstacle to such a reconstruction is the incomplete, damaged, and corroded state of the existing fragments. The Mechanism may well have suffered from impacts at the time of the wreck, for example from the rolling about of marble statuary and other heavy objects that formed part of the ship's cargo. But even if it escaped immediate damage, one could scarcely expect it to have held together after two thousand years of lying at the sea bottom.

. . .

The second question, what was the Mechanism for, remains open to dispute: purpose turns out to be more elusive than function. The question can be approached at more than one level, drawing on different categories of evidence. Imagine a situation in which we could study (or could reliably reconstruct) the interior mechanical components and their interconnections, while having no knowledge of the exterior whatsoever. From this kind of evidence alone, taken together with certain scientific facts—or still better, with information about ancient scientific theories—we would be able to establish the meaning of the Mechanism's functions in terms of astronomy and chronology. For example, a gear train that exactly translates nineteen revolutions of one gear into 254 revolutions of another would, all by itself, be compelling evidence that the first gear represents the Sun's longitudinal revolution while the second represents the Moon's longitudinal revolution, because the equation of nineteen years with 235 lunar months and 254 lunar revolutions is an accurate period relation that was well known in antiquity. On the other hand, one would need to see part of the exterior, in particular the scale of the dial on which the Sun's revolution was displayed, to establish that the Sun's position was expressed in terms of zodiacal signs and degrees or that certain solar longitudes were associated with the dates of first and last visibility of stars and constellations. Even a complete and perfectly preserved Antikythera Mechanism probably would not tell us explicitly who would have wanted a machine that displayed this astronomical information and why. For such questions, we need to look at context: the archaeological context of the find, but also the cultural and intellectual context of the Greco-Roman civilization that produced the Mechanism.

. . .

The most important developments with respect to the present understanding of what the Antikythera Mechanism was can be summarized as follows:

(1) A strong evidence-based consensus now exists about most aspects of the Mechanism's gearwork and displays relating to the Sun, Moon, calendars, and eclipses, through the researches of Wright and the AMRP. The upper back dial displayed uniform passage of time through the 235 lunar months composing a nineteen-year cycle, with scale inscriptions relating the cycle to a specific Greek regional calendar (the Corinthian calendar as known primarily from inscriptions in Epirus, Illyria, and Corcyra). Its extant subsidiary dial displayed a four-year cycle, with inscriptions relating it to several Greek athletic competitions held at 2-year and 4-year intervals; a second subsidiary dial is conjectured displaying a 76-year calendrical cycle. The lower back dial displayed uniform passage of time through the 223 lunar months of a so-called Saros eclipse cycle, with scale inscriptions marking the months in which lunar and solar eclipses might occur; its subsidiary dial displayed a triple-Saros or Exeligmos eclipse cycle. The front dial had pointers representing the apparent positions of the Sun and Moon, with a fixed scale representing the signs and degrees of the zodiac and a movable scale representing the

Egyptian calendar. In addition to a pointer, the Moon was represented by a revolving particolored ball showing the current lunar phase.

(2) The lunar gearwork is now known to have incorporated a pin-and-slot coupling that introduces a nonuniformity in the Moon's rate of longitudinal motion, conforming to the behavior of a simple epicyclic or eccentric model with a shifting lunar apogee such as is familiar from the theoretical work of Hipparchus and Ptolemy.

(3) The theory that the Mechanism's front had some kind of planetary display has won wide acceptance, and there is increasing support for the supposition (strongly urged by Wright since the early 2000s) that this took the form of pointers on the front dial representing the longitudinal motions of all five planets known in antiquity according to epicyclic or eccentric models, making the front dial a planetarium. Despite the absence of any surviving gearwork that can definitely be assigned to a planetary mechanism (only one extant gear, isolated in Fragment D, remains unaccounted for after the reconstruction of the lunisolar system), physical remains on A-1 indicate that a substantial part of the mechanism has been lost, and progress in reading the inscriptions on Fragments B and G has revealed detailed references to planetary pointers and planetary nonuniform motion.

The new picture we have of the Antikythera Mechanism is more complex than Price's, and his expression "calendar computer" clearly will no longer serve. On the other hand, we cannot follow Rehm in simply calling it a "planetarium." We would have to invent a new word to describe a device that simultaneously displays time cycles on one face and celestial motions on the other: perhaps a "cosmochronicon"? But the uncertainty remains whether the Mechanism existed for the sake of illustrating or calculating the phenomena of time and the heavens;

Tony Freeth, "The Antikythera Mechanism" (*World Archaeology*, 2021)

When I first studied the Antikythera Mechanism, many questions remained unresolved. So, in 2000, I proposed new investigations of the Antikythera Mechanism to get more data to tackle the outstanding issues. After years of struggle to get all the necessary permissions, this was finally carried out in 2005 by an Anglo-Greek team of researchers in collaboration with the National Archaeological Museum in Athens.

Our work used two high-tech, non-destructive techniques: Microfocus X-ray Computed Tomography (X-ray CT)—high-resolution 3D X-rays—and Polynomial Texture Mapping (PTM)—a digital imaging technique for looking at surface features, invented by Tom Malzbender (then at Hewlett-Packard). In 2005, an eight-ton X-ray machine from X-Tek Systems (now owned by Nikon Metrology) was coaxed into the basement of the National Archaeological Museum in Athens. Roger Hadland, then owner of X-Tek Systems, made a special prototype X-ray machine and brought a top-quality team of experts to scan all 82 fragments of the Antikythera Mechanism. The results have transformed our knowledge.

Two important discoveries resulted from the new X-ray CT. I established that a dial on the lower back predicted eclipses according to the 223-month cycle, which Rehm had recognized in the inscriptions. The gearing enabling this involved a disregarded 223-tooth gear at the back, which had found no role in any previous model. I then made a startling discovery about how the Mechanism calculated the variable motion of the Moon (caused by its elliptical orbit around the Earth). Wright had made an acute observation about how two of the epicyclic gears could interact to represent variable motion. Ultimately, he discarded this observation because it did not fit into his model, but I realized that this device could model the movement of the Moon. This idea also involved the 223-tooth gear—so this gear now had two essential functions. The system was utterly astonishing in its operation.

The X-ray CT delivered another crucial revelation: extensive new inscriptions were buried inside the fragments and could now be read for the first time in over 2,000 years. Critically for the UCL Antikythera Team, these texts referred to the planets on both the front and back covers. The back cover essentially acted as a user manual, describing the principles on which the Mechanism was based, which included the cycles that Rehm had observed. Even more important was a description of what the front display of the Mechanism looked like, deciphered by a professor of ancient astronomy, Alexander Jones, with whom I was working at the time. This inscription described a model of the Cosmos, with the Sun shown by a pointer, a ring system for the planets, and marker beads to indicate each planet. Now the UCL team knew that, to reconstruct the Mechanism, they must recreate this ring Cosmos system, where previous attempts had failed.

The X-ray CT of the front cover inscription disclosed extensive new information about the planets. There is a section for each planet, enumerating days between events in their synodic cycles as well as the planet's synodic period in years. For example, Venus takes 584 days to return to the same position relative to the Sun—a number that can be read in the Front Cover Inscription. In 2016, Jones discovered the period 462 years in the Venus section of the inscription and 442 years in the Saturn section. These were astounding numbers, unknown from previous studies of ancient astronomy. Clearly, the UCL team needed to incorporate these periods into the gearing for Venus and Saturn.

. . .

The UCL team created compact five-gear mechanisms for Mercury and Venus. For Venus, they discovered a way of designing the gearing that exactly matched a bearing on one of the spokes of the Main Drive Wheel and also included the mysterious 63-tooth gear in Fragment D. A comparable mechanism for Mercury also fitted all the surviving evidence. For Mars, Jupiter, and Saturn, the UCL team found ingenious seven-gear mechanisms, based on the concept behind the beautiful lunar mechanism established from previous research. Using shared gears, ensured by the choices of planetary periods, these could be shoe-horned into the tight spaces available.

Earlier models had wrongly used pointers to indicate the planets—an idea that contradicts the Back Cover description of a ring system. The ancient Greek text not only specified that the planets were arranged as a ring system, but also that they had to be shown in a particular cosmological order: Earth, Moon, Mercury, Venus, Sun, Mars, Jupiter, Saturn. For the planets to be displayed on concentric rings, the gearing systems for the planets needed to output in a very particular way. For example, the rotation of Mars is calculated by gearing that output on a tube, to which is attached a ring for the Mars display. Outside the Mars tube is another tube for Jupiter and outside this another tube

Figure 7.10.2. Model of Antikythera Mechanism as envisioned by Freeth et al., The Antikythera Mechanism. Courtesy of *World Archaeology* and Tony Freeth. © 2021 Tony Freeth, Images First Ltd.

for Saturn. In this way, the planetary positions can be shown on a system of nested tubes and displayed as concentric rings.

. . .

To complete the system, the team added a hypothetical Dragon Hand (a term from medieval astronomy), which is a long, double-ended pointer that indicates the Nodes of the Moon, showing when it is possible that eclipses may occur. Though no direct physical evidence for the Dragon Hand survives, its gearing explains a prominent bearing on one of the spokes of the Main Drive Wheel, and the idea thematically links both Front and Back Dials.

The UCL team set about assembling the whole system. All the gears for the Sun, planets, and the Nodes of the Moon were crammed into the constricted space defined by the pillars and attached plates. This culminates in a beautiful ring display for the Cosmos, with the added advantage that it considerably enhances the astronomical results calculated by the machine.

The positions of Sun and Moon are shown, as well as the phase of the Moon. The age of the Moon, in terms of the number of days from a new Moon, is read by the Moon pointer on the Sun ring. The team conjecture that the synodic events of the planets—such as conjunctions, stationary points, and maximum elongations—are marked on the planetary rings and indexed to the information in the front cover inscription. The Dragon Hand indicates eclipse possibilities when it is close enough to the Sun pointer [figure 7.10.2].

In March 2021, the UCL Antikythera Research Team published a radical paper in Nature's *Scientific Reports*, showing this new reconstruction.[52] Ours is the first model

[52] Tony Freeth et al., "A Model of the Cosmos in the ancient Greek Antikythera Mechanism," *Scientific Reports* 11 (2021): 5821.

that conforms to all the physical evidence and matches the descriptions given in the scientific inscriptions that were engraved on the Mechanism itself. This machine is an impressive tour de force of ancient Greek brilliance, which displays the Sun, Moon, and planets, allowing their future positions and events such as possible eclipses to be calculated.

7.11 Cleomedes, *On the Circular Motion of Heavenly Bodies*

Little is known about Cleomedes other than his authorship of an elementary astronomy textbook, *On the Circular Motions of the Celestial Bodies*, often referred to as *On the Heavens*. There is considerable scholarly debate even about when Cleomedes lived, but the first century CE seems to be the most likely. Much of Cleomedes's work is taken from an earlier work by Posidonius (ca. 135–51 BCE), but for our purposes, the most important part is his description, in chapter 7 of book 1, of the methods of both Posidonius and Eratosthenes (ca. 276–194 BCE) for calculating the circumference of the earth. It is interesting that both methods come up with values very close to each other (240,000 and 250,000 stades, respectively), even though their assumptions are not always accurate. But what we do not know, in either case, is the length of the stade that was being used. Various scholars have suggested values ranging from about 157 meters to about 191 meters. Thus, although modern texts usually credit Eratosthenes with calculating a rather accurate value for the earth's circumference, it is difficult to be sure. In any case, the astronomical/geometrical methods for the calculations, as described by Cleomedes, are accurate, given his assumptions.

Natural philosophers have held numerous doctrines about the size of the Earth, but two of these are superior to the rest. Eratosthenes's doctrine demonstrates its size by a geodesic procedure, while Posidonius's is less complicated. Each takes certain assumptions and then arrives at demonstrations via the implications of the assumptions. The first that we shall discuss is Posidonius's.

He states that [*P*1] Rhodes and Alexandria are located below the same meridian.[53] [Definition 1] Meridians are the circles drawn through the poles of the cosmos, and through a point that lies at the zenith of each of those [observers] who stands on the Earth. (Thus, while the poles are the same for everybody, the point at the zenith is different for different [observers], which is why infinitely numerous meridians can be drawn.) Rhodes and Alexandria, then, are located below the same meridian, and [*P*2] the distance between the cities is held to be 5,000 stades. Let it be assumed that this is so. [Definition 2] All the meridians are also included among the great circles in the cosmos, since they divide it into 2 equal parts by being drawn through its poles.

Now with this assumed to be the case, Posidonius next [*P*3] divides the zodiacal circle (which, since it too divides the cosmos into 2 equal parts, is [by definition 2] equal to the meridians) into 48 parts by dividing each of its *dōdekatēmoria* into quarters. Now if the meridian through Rhodes and Alexandria is also divided into the same 48 parts as the zodiacal circle, then its sections will be equal to the sections of the zodiacal circle just

[53] In fact, Rhodes is $1^{\circ}50'$ west of Alexandria.

identified. The reason is that [definition 3] when 2 equal magnitudes are divided into equal parts, their parts must also be equal to the parts of what has been divided.

Now with [$P1$-$P3$] assumed to be the case, Posidonius next says that the star called Canobus, which is located in the south at the rudder of Argo, is very bright. (This star is not seen at all in Greece; that is why Aratus does not mention it in his *Phaenomena*.) But for people going to the south from the north, the star starts to be seen at Rhodes, and once seen on the horizon immediately sets along with the revolution of the heavens. But when we reach Alexandria by sailing the 5,000 stades from Rhodes, this star, when precisely at the meridian, is determined as being elevated above the horizon $\frac{1}{4}$ of a zodiacal sign, that is, [by $P1$ and $P3$] $\frac{1}{48}$ of the meridian through Rhodes and Alexandria.[54]

Now it is necessary [by $P1$ and $P3$] that the section of the same meridian located above the distance separating Rhodes and Alexandria also be $\frac{1}{48}$ of that meridian, because the Rhodians' horizon is also distant by $\frac{1}{48}$ of the zodiacal circle from that of the Alexandrians. So since [by $P2$] the portion of the Earth located below this section is held to be 5,000 stades, the portions located below the other sections also consist of 5,000 stades. And in this way the circumference of the Earth is determined as 240,000 stades—if [by $P2$] there are 5,000 stades between Rhodes and Alexandria. Otherwise, [it will be determined] in proportion to the [true] distance. That, then, is Posidonius's procedure for dealing with the size of the earth.

Eratosthenes's [calculation], by contrast, involves a geodesic procedure, and is considered to possess a greater degree of obscurity. But the following [assumptions], when stated by us as presuppositions, will clarify his account.

Let us first assume here too that [$E1$] Syene and Alexandria are located below the same meridian; [second] that [$E2$] the distance between the two cities is 5,000 stades. Third, [assume] that [$E3$] the rays sent down from different parts of the Sun to different parts of the Earth are parallel, as geometers assume to be the case. Fourth, let the following assumption demonstrated by geometers be made: that [$E4$] straight lines intersecting with parallel lines make the alternate angles equal. Fifth, [assume] that [$E5$] the arcs [of a circle] standing on equal angles are similar, that is, have the same proportion (namely, the same ratio) to their own circles as is also demonstrated by geometers. (For example, when arcs stand on equal angles, then if one of them is one-tenth part of its own circle, all the remaining arcs will also be one-tenth parts of their own circles.)

Someone who has mastered these [assumptions] would have no difficulty in learning Eratosthenes's procedure, which is as follows. He says that [$E1$] Syene and Alexandria are located below the same meridian.[55] So since [definition 2] the meridians are included among the great circles in the cosmos, the circles of the Earth located below them are necessarily also great circles. Thus the size that this procedure demonstrates for the [arc of the] circle of the Earth through Syene and Alexandria will be in a ratio with the great circle of the Earth.

Eratosthenes says, and it is the case, that [$E6$] Syene is located below the summer tropical circle. So when the Sun, as it enters Cancer and produces the summer solstice, is precisely at this meridian, the pointers on the sundials are necessarily shadowless, since

[54] Rhodes is actually 5° north of Alexandria, that is, $\frac{1}{6}$ of a zodiacal sign or $\frac{1}{72}$ of the meridian.

[55] Syene is actually 3° east of Alexandria.

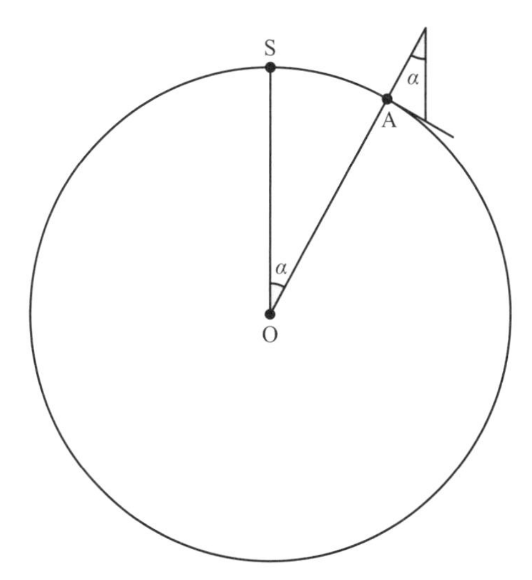

Figure 7.11.1.

the Sun is located vertically above them. (This shadowless area is reportedly 300 stades in diameter.) But in Alexandria at the same hour pointers on sundials do cast a shadow, since this city is located further north than Syene. Now since [by $E1$] the two cities are located below a meridian (a great circle [by definition 2]), if we draw an arc from the tip of the pointer's shadow on the sundial at Alexandria round to the base of the pointer, this arc will be a section of the great circle in the sundial's bowl, since the sundial's bowl is located below a great circle.

If we next conceive of straight lines produced through the Earth from each of the pointers, they will coincide at the center of the Earth. So since [by $E6$] the sundial at Syene is located directly below the Sun, then if we also conceive of a straight line going from the Sun to the tip of that sundial's pointer, the line going from the Sun to the center of the Earth will be a single straight line. If we conceive of a second straight line drawn from the bowl at Alexandria, that is, from the tip of the pointer's shadow up to the Sun, this line and the first one will be parallel [by $E3$], since they extend from different parts of the Sun to different parts of the Earth [figure 7.11.1].

Now a third straight line extending from the center of the Earth to the pointer at Alexandria meets these parallel lines, and as a consequence [of $E4$] makes the alternate angles equal. One of these [angles] is at the center of the Earth where the lines drawn from the sundials to the center of the Earth coincide. The other is where the tip of the pointer at Alexandria and the line drawn from the tip of the pointer's shadow up to the Sun through the point where the line touches [the tip] coincide. The arc drawn from the tip of that pointer's shadow round to its base stands on this second angle, while the arc extending from Syene to Alexandria stands on the angle at the center of the Earth.

Now the arcs are similar to one another, since [by $E5$] they stand on angles that are equal. Thus the ratio that the arc in the bowl [at Alexandria] has to its own circle is the same as the ratio of the [arc] from Syene to Alexandria [to its own circle]. The arc in the bowl is certainly determined as one-fiftieth part of its own circle. So the distance from Syene to

Alexandria is necessarily one-fiftieth part of the great circle of the Earth,[56] namely, [by $E2$] 5,000 stades. Therefore, the [great] circle [of the Earth] totals 250,000 stades. And that is Eratosthenes's procedure.

Also, at winter solstices sundials are positioned in each of these cities, and when each sundial casts shadows, the shadow at Alexandria is necessarily determined as the longer because this city is at a greater distance from the winter tropic. So by taking the amount by which the shadow at Syene is exceeded by that at Alexandria, they also determine this amount as one-fiftieth part of the great circle in the sundial. So it is evident from this [calculation] too that the great circle of the Earth is 250,000 stades. Thus the diameter of the Earth will exceed 80,000 stades, given that it must certainly be $\frac{1}{3}$ of the great circles of the Earth.[57]

Those who say that the Earth cannot be spherical because of the hollows occupied by the sea and the mountainous protrusions, express a quite irrational doctrine. For neither is there a mountain determined higher, nor a depth of sea [greater], than 15 stades. But 30 stades has no ratio to over 80,000 stades, but is just like a speck of dust would be on a ball. The protrusions on the rondures from plane trees also do not stop them from being rondures. Yet these protrusions have a ratio to the total sizes of the rondures greater than that of the hollows of the sea and the mountainous protrusions to the total size of the Earth.

7.12 A Hellenistic Horoscope

Our understanding of Greek astronomy derives largely from theoretical works, among the most significant are Ptolemy's *Almagest* and those texts related to it. Sources that provide insight into the actual mathematical practices and hand computations made by practitioners are harder to come by. Horoscopes from antiquity are one such historical source where computational practices can be detected, as they include data on localized planetary and stellar positions for a particular date and other mathematical astrological computed phenomena.

The following excerpt is taken from a Hellenistic astrological horoscope that is precisely dated to October 28, 497 CE. This date is confirmed by the computed planetary and stellar positions included in the text [figure 7.12.1]. As the text indicates, the parameters are ultimately derived from the *Almagest*, and planetary and stellar positions are interwoven with corresponding astrological schema.

At the end of the text an astrological horoscopic diagram is drawn [figure 7.12.2]. Progressing counterclockwise from the top left octant, the zodiacal signs, beginning with Pisces (12), are set out along with planetary positions and other significant astrological divisions, serving as a panoptic representation of the data included in the text.

[56] The true distance from Syene to Alexandria is just under one-fiftieth of a great circle of the earth, about 7.1°.

[57] Cleomedes is using the ancient approximation 3 for the ratio of circumference to diameter of a circle, although he certainly lived later than Archimedes, who gave a much better approximation of $\frac{22}{7}$.

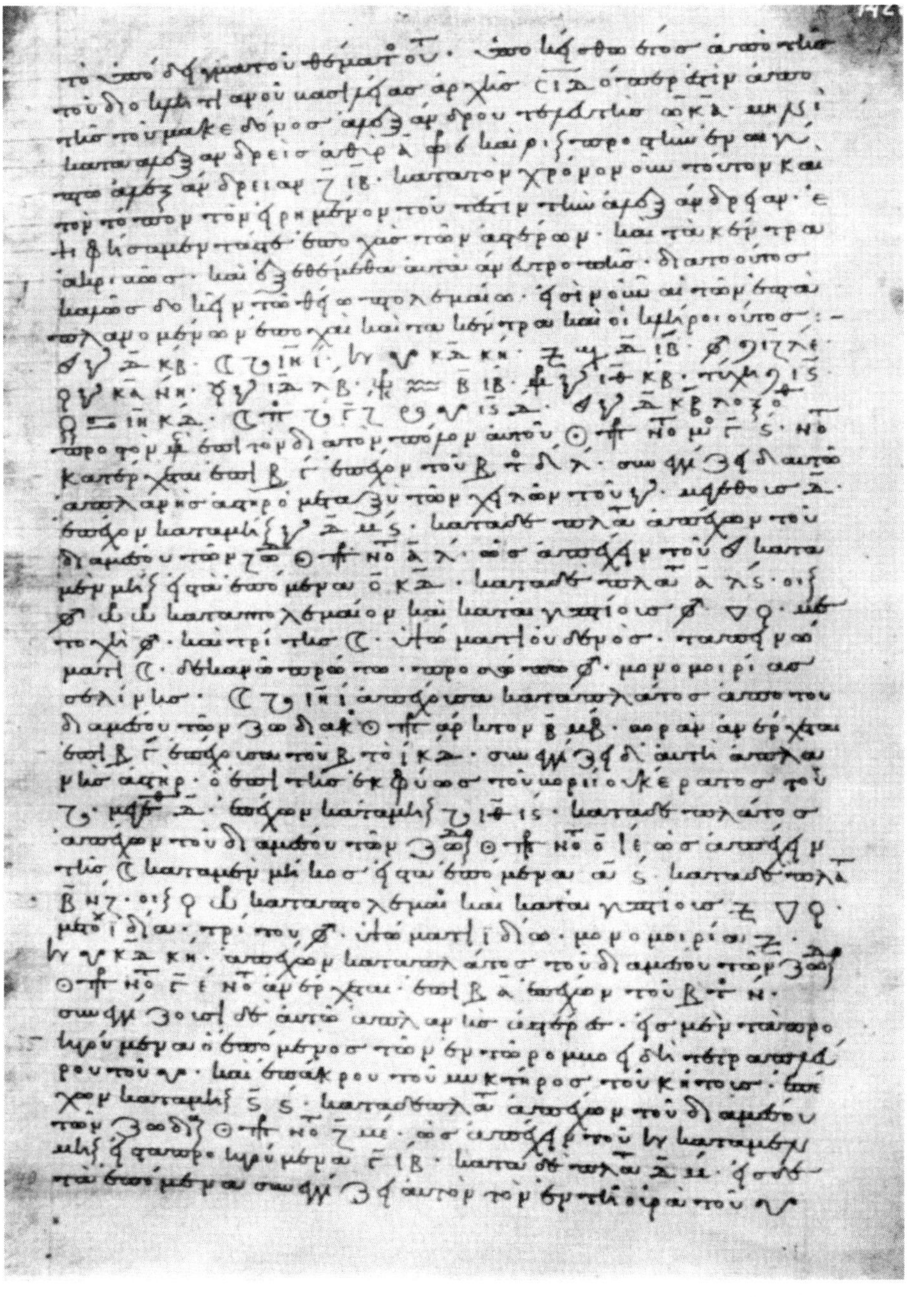

Figure 7.12.1. Hellenistic astrological horoscope dated to October 28, 497 CE. See Neugebauer and Van Hoesen, *Greek Horoscopes* (Philadelphia: American Philosophical Society, 1959), 231.

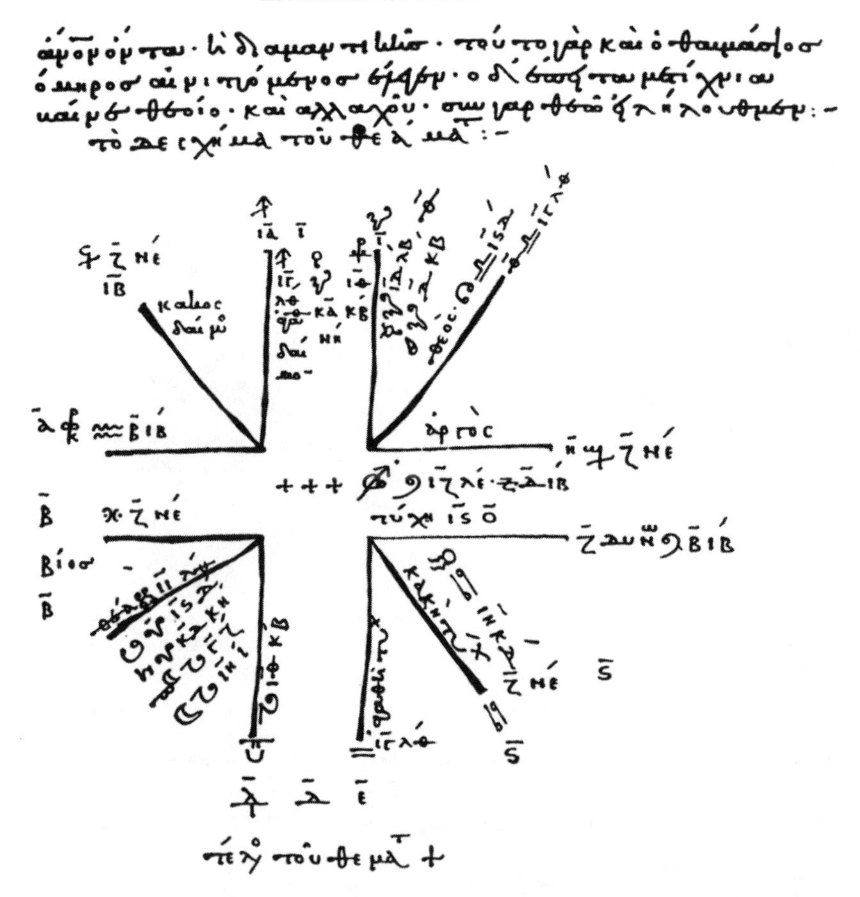

Figure 7.12.2. Horoscope diagram at the end of the horoscope text presenting planetary and stellar data and other astrological details for the particular date in question. See Neugebauer and Van Hoesen, *Greek Horoscopes* (Philadelphia: American Philosophical Society, 1959), 156.

Horoscope from the Reign of Diocletian (497 CE)

The example of the theme is as follows: let it be the year 214 from the beginning of the reign of Diocletian which is the year 821 from the death of Alexander the Macedonian, first (day) of the Alexandrian month Athyr (III), seasonal hour for Alexandria in Egypt 7;12 of the day. For this time and the said locality, that is, Alexandria, we computed accurately the positions of the stars and the centers and put them down without change because thus it seems right to the godly Ptolemaios. Thus the positions of the seven planets and the centers and the lots are as follows: ☉ ♏ 4;22, ☽ ♉ 18;10, ♄ ♈ 24;28, ♃ ♍ 4;12, ♂ ♌ 17;35, ♀ ♏ 21;58, ☿ ♏ 14;32, Horoscopos ♓ 2;12, Midheaven ♏ 19;22, (Lot of) Fortune ♌ 16, Daimon ♋ 18;24, full moon ♉ 3;7, ascending node ♈ 16;4. ☉ ♏ 4;22, inclined towards the equator on the circle which passes through its pole towards the south 13;6 degrees; it descends to the south in the third step, occupying of this (?) step $\frac{1}{4}\ \frac{1}{30}$. Near to it is the fixed star which is between the claws of ♏, of magnitude 4. It occupies longitude ♏ 4;46; in latitude it occupies in the circle through the middle of the zodiacal (signs) towards the south 1;30.

Figure 7.13.1. Fragment of a numerical table on Oxyrhynchus Papyrus 4167. It tabulates planetary latitudes for Venus and Mercury. Image from Alexander Jones, *Astronomical Papyri from Oxyrhynchus (P. Oxy. 4133-4300a)*, Vol. 1 (Philadelphia: American Philosophical Society, 1999).

Thus it is distant from the ⊙ in longitude towards following (longitudes) 0;24, in latitude 1;36. House of ♂, terms according to Ptolemy and the Egyptians (of) ♂. Triangle (of) ♀, partner ♂ (and third (partner) ☽. In the exaltation of none, in the depression (of) ☽. In the first decan, in the countenance (of) ♂; monomoiria of the moon. ☽ ♉ 18;10 being distant in latitude from the circle through the middle of the zodiacal (signs) towards north 2;42 north. . . .

7.13 Oxyrhynchus Astronomical Papyri

The Oxyrhynchus papyri are a huge collection of manuscripts that were discovered during the nineteenth and early twentieth centuries in the rubbish mounds of ancient Oxyrhynchus, Egypt. Of the estimated half a million papyri that were recovered, only a very small fraction (1–2%) have been studied; of those, among the administrative documents, inventories, public and private correspondence, and the like, are astronomical and astrological texts. These papyri—fragments of horoscopes, numerical tables, and astronomical prose texts—contain computed predictions of the

	]	0	5[4]	[				
[2]1	[33]9	0	52	0	[			
[2]4	336	0	50	0	[			
[2]7	333	0	49	0	[			
30	330	0	44	0	[			
33	327	0	41	0	[			
36	324	0	38	0	[			
39	321	0	34	[				
42	318	0	31	[				
45	315	0	27	0	[			
48	312	0	23	0	1[4]	[		
51	309	0	19	0	14	1	[	
54	306	0	15	0	14	1	[	
57	303	0	11	0	14	1	[	
60	300	0	7	0	14	1	[	
63	297	0	2	0	15	1	[	
		addi	tive					
66	294	0	2	0	15	2	[	
69	291	0	8	0	15	2	3	0
72	288	0	14	0	16	2	5	[
75	285	0	20	0	16	2	7	0
78	282	0	25	0	17	2	9	0
81	279	0	30	0	17	2	11	0
84	276	0	35	0	18	2	13	0
87	273	0	40	0	18	2	15	0
90	270	0	44	0	19	2	18	0

Figure 7.13.2. Translation of Oxyrhynchus Papyrus 4167.

positions of the sun, moon, and planets, eclipses, and various stellar phenomena. The underlying parameters and algorithms the practitioners used to do this can frequently be linked back to Ptolemaic astronomy.

The first example, from papyrus 4167, is a fragment of a numerical table tabulating planetary latitudes for Venus and Mercury [figures 7.13.1, 7.13.2]. These can be derived from Ptolemy's *Handy Tables*. The second example, from papyrus 4175, is an almanac ephemeris that dates to 24 BCE, making it the oldest astronomical papyrus in the corpus [figure 7.13.3]. In a complex set of ruled grids, dates and longitudes are given for planetary phases for the five planets, and underneath the daily lunar longitudes, truncated to degrees, are tabulated.

7.14 A Parapegma from Miletus

One of the many activities associated with the Greek mathematical astronomy tradition were *parapegmata*, originally physical devices designed to correlate astronomical cycles with weather events, using a peg that could be moved along a series of holes in the device as the phenomena or calendar days elapsed. (Later these devices transformed into literary texts and physical pegs were replaced with a calendar day count.) Such astronomical cycles might be equinoxes and solstices, or other solar phenomena, or phases of important fixed stars. Broadly, then, these parapegmata

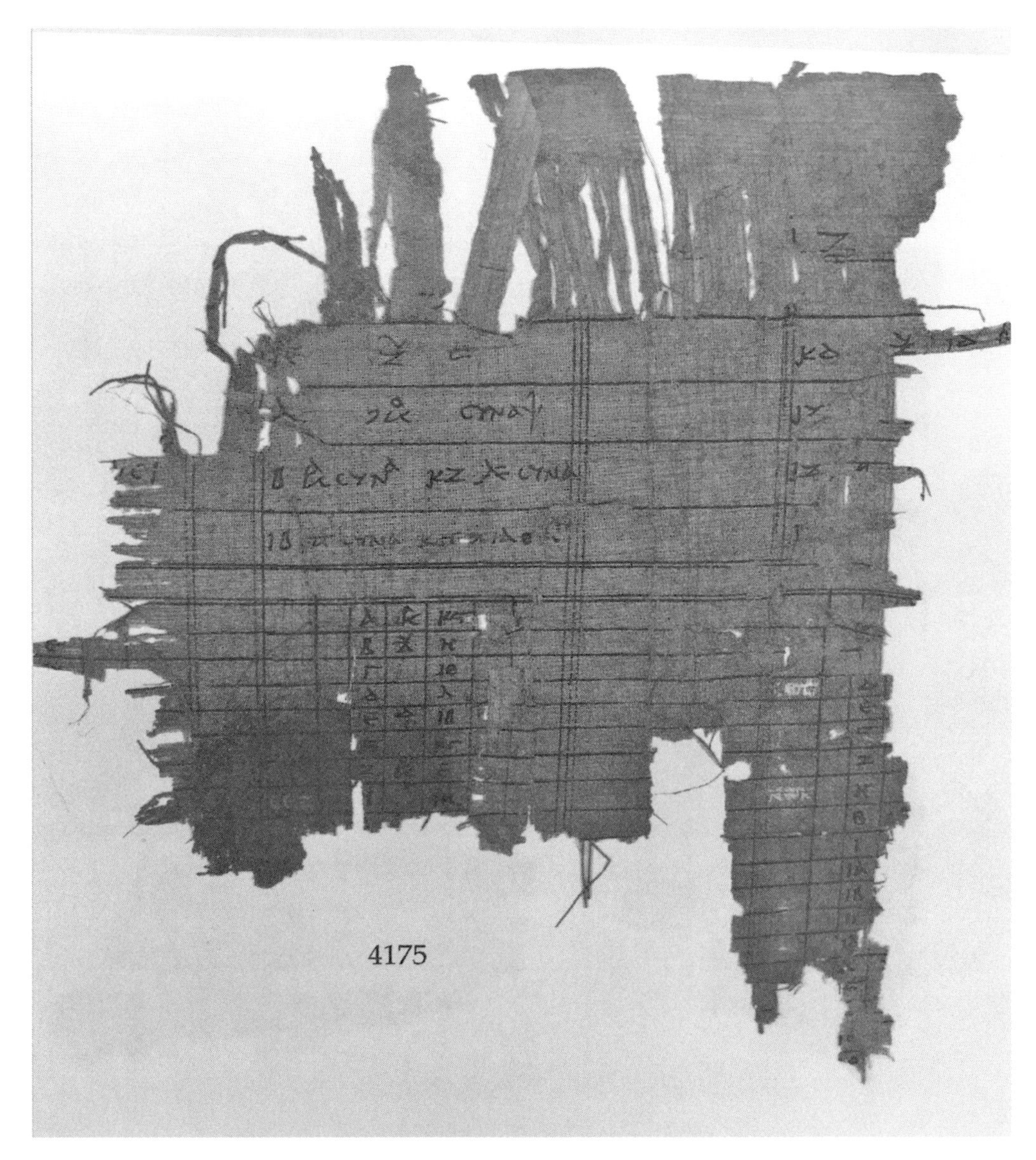

Figure 7.13.3. Oxyrhynchus Papyrus 4175. Image from Alexander Jones, *Astronomical Papyri from Oxyrhynchus (P. Oxy. 4133-4300a)*, Vol. 1 (Philadelphia: American Philosophical Society, 1999).

served as a type of tracking device for astronomical and meteorological cycles. Parapegmata generated their astronomical data from calculations and observations and thus reveal how practitioners produced computed data from theoretical astronomical models and how they recorded and used the results.

The following passages derive from the so-called Miletus *parapegmata*, six fragments carved in stone that had holes drilled alongside the inscribed text [figure 7.14.1].

figure IV: 456A

figure V: 456B

Figure 7.14.1. Miletus parapegmata, fragments 456A and 456B. Figures IV and V from "The Parapegma Fragments from Miletus" by D. Lehoux (*Zeitschrift für Papyrologie und Epigraphik* 152 (2005), p. 139). Reproduced with permission by Dr. Rudolf Habelt GmbH.

These holes correspond in number to the day count between astronomical events that were listed on the inscription.

Fragment 456A [figure 7.14.2] has two columns and is 54 cm high and 22 cm wide. It explicitly links astronomical and weather phenomena to actual named authorities. Fragment 456B [figure 7.14.3] is around 44 cm wide and 26 cm high. There are three

II.ii. 456 A, Translation

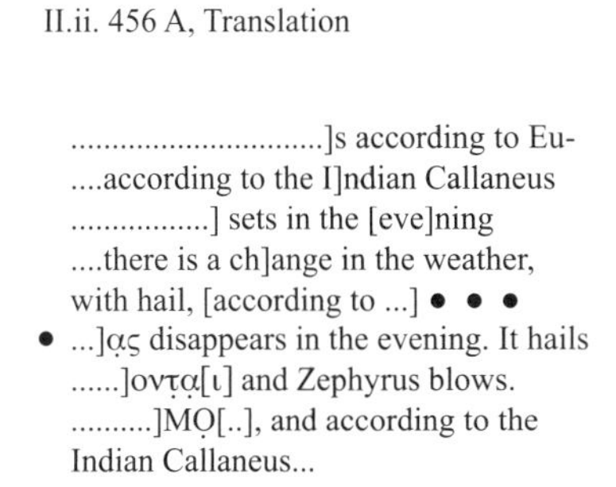
..............................]s according to Eu-
....according to the I]ndian Callaneus
.................] sets in the [eve]ning
....there is a ch]ange in the weather,
with hail, [according to ...] • • •
• ...]ας disappears in the evening. It hails
......]οντα[ι] and Zephyrus blows.
..........]ΜΟ[..], and according to the
Indian Callaneus...

...]ev[en]ing[...
[accordi]ng to Euctemon. •
• Capella sets acronychally ac[cording
to both Philippus and the Egypti[ans.
• Capella sets in the evening according to
the Ind[ian] Callaneus. •
• Aquila rises in the evenin[g
according to Euctemon.
• Arcturus sets in the morning and there is
a cha[nge in the] weather according to
Euctemon. On this day [Aqu-
ila raise in the evening also, ac[cording to
Philippus.

Figure 7.14.2. Translation of Miletus parapegmata Fragment 456A from "The Miletus Parapegma Fragments" by Daryn Lehoux (*Zeitschrift für Papyrologie und Epigraphik* 152 (2005), 125–140), p. 131. Reproduced with permission by Dr. Rudolf Habelt GmbH.

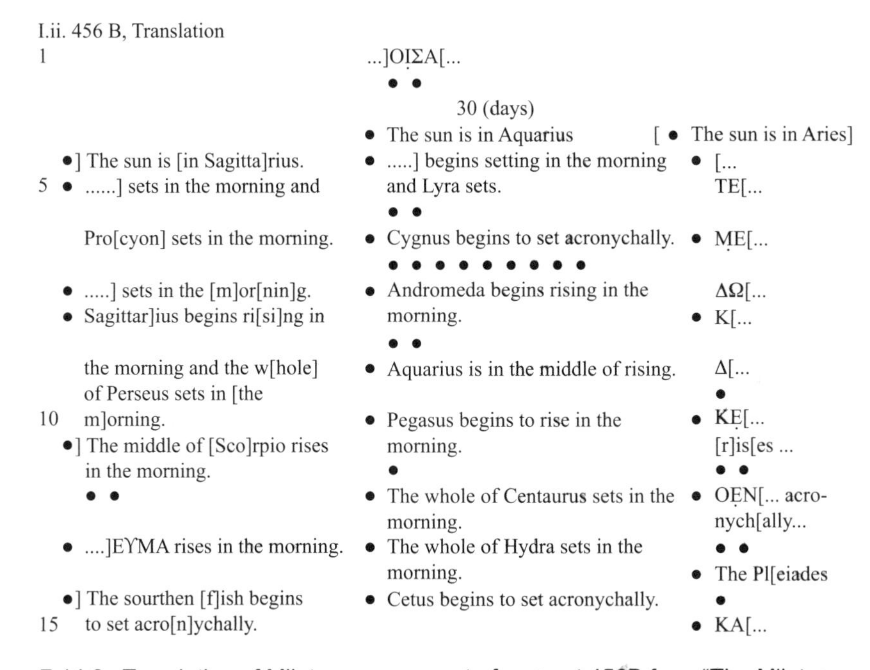
I.ii. 456 B, Translation

1		...]ΟΙΣΑ[...	
		• •	
		30 (days)	
		• The sun is in Aquarius	[• The sun is in Aries]
	•] The sun is [in Sagitta]rius.	• .....] begins setting in the morning	• [...
5	• ......] sets in the morning and	and Lyra sets.	TE[...
		• •	
	Pro[cyon] sets in the morning.	• Cygnus begins to set acronychally.	• ΜΕ[...
		• • • • • • • • • •	
	• .....] sets in the [m]or[nin]g.	• Andromeda begins rising in the	ΔΩ[...
	• Sagittar]ius begins ri[si]ng in	morning.	• Κ[...
		• •	
	the morning and the w[hole]	• Aquarius is in the middle of rising.	Δ[...
	of Perseus sets in [the		•
10	m]orning.	• Pegasus begins to rise in the	• ΚΕ[...
	•] The middle of [Sco]rpio rises	morning.	[r]is[es ...
	in the morning.	•	• •
	• •	• The whole of Centaurus sets in the	• ΟΕΝ[... acro-
		morning.	nych[ally...
	• ....]ΕΥΜΑ rises in the morning.	• The whole of Hydra sets in the	• •
		morning.	• The Pl[eiades
	•] The sourthen [f]ish begins	• Cetus begins to set acronychally.	•
15	to set acro[n]ychally.		• ΚΑ[...

Figure 7.14.3. Translation of Miletus parapegmata fragment 456B from "The Miletus Parapegma Fragments" by Daryn Lehoux (*Zeitschrift für Papyrologie und Epigraphik* 152 (2005), 125–140), p. 129. Reproduced with permission by Dr. Rudolf Habelt GmbH.

columns of text; the two outer columns have been damaged and so can only be partially read.

SOURCES, CHAPTER 7

7.1 These translations are found in Thomas L. Heath, ed., *Greek Astronomy* (London: J. M. Dent, 1932), 157–59 (Aristotle), 159–60 (Simplicius).

7.2 This translation is by Thomas L. Heath, *Aristarchus of Samos: The Ancient Copernicus* (New York: Dover, 1981), 353–409.

7.3 This translation is from Clemency Montelle, "The *Anaphoricus* of Hypsicles of Alexandria," in *The Circulation of Astronomical Knowledge in the Ancient World*, ed. John M. Steele (Leiden: Brill, 2016), 287–315.

7.4 The translation here was adapted by Victor Katz from the French translation of the Greek in Paul Ver Eecke, *Les Sphériques de Théodose de Tripoli* (Paris: Albert Blanchard, 1959), and from the English translation from the Arabic in Thomas J. Martin, "The Arabic Translation of Theodosius's Sphaerica," PhD diss., University of St Andrews, 1975.

7.5 This material was translated by Victor J. Katz from the Latin in Edmond Halley, ed. and trans., *Menelai Sphaericorum, Libri III* (Oxford, 1758), with assistance from Roshdi Rashed and Athanase Papadopoulos, *Menelaus' Spherics: Early Translation and al-Māhānī / al-Harawī's Version* (Berlin: De Gruyter, 2017). The material from Ptolemy's *Almagest* is from Gerald Toomer, *Ptolemy's Almagest* (Princeton, NJ: Princeton University Press, 1998), 64–67.

7.6 This translation is from Toomer, *Ptolemy's Almagest.*

7.7 This translation is from Raymond Mercier, ed., *Ptolemy's Handy Tables: Transcription and Commentary* (Louvain-La-Neuve: Université Catholique de Louvain, Institut Orientaliste, 2011).

7.8 This translation is from Nathan Sidoli and J. Lennart Berggren, "The Arabic Version of Ptolemy's *Planisphere* or *Flattening the Surface of the Sphere*: Text, Translation, Commentary," *SCIAMUS* 8 (2007): 37–139.

7.9 This translation is from J. Lennart Berggren and Alexander Jones, *Ptolemy's Geography: An Annotated Translation of the Theoretical Chapters* (Princeton, NJ: Princeton University Press, 2000), 84–93.

7.10 This section contains excerpts from four articles. The first is from Derek J. de Solla Price, "An Ancient Greek Computer," *Scientific American* 200, no. 6 (June 1959): 60–67. The second is from Tony Freeth, "Decoding the Antikythera Mechanism: Supplementary Notes 2," *Nature* 444 (2006): 587–91. The third is from Alexander Jones, "The Antikythera Mechanism and the Public Face of Greek Science," *Proceedings of Science* 038 (2012). https://pos.sissa.it/170/038/pdf. The fourth is from Tony Freeth, "The Antikythera Mechanism: An Ancient Greek Machine Rewriting the History of Technology," *World Archaeology*, July 22, 2021. It is used with permission of the author and the publisher.

7.11 This translation is taken from Alan C. Bowen and Robert B. Todd, ed. and trans., *Cleomedes' Lectures on Astronomy: A Translation of "The Heavens"* (Oakland: University of California Press, 2004), 78–85.

7.12 This material comes from Otto Neugebauer and Henry Bartlett Van Hoesen, *Greek Horoscopes* (Philadelphia: American Philosophical Society, 1959), p. 152ff.

7.13 This material comes from Alexander Jones, *Astronomical Papyri from Oxyrhynchus: (P. Oxy. 4133–4300a)*, vol. 1 (Philadelphia: American Philosophical Society, 1999), 160–62 and 118–29.

7.14 This material comes from Daryn Lehoux, "The Miletus Parapegma Fragments from Miletus," *Zeitschrift für Papyrologie und Epigraphik* 152 (2005): 125–40.

8

Music

Greek musical theory and harmonics encompass the systemic study of the building blocks of melody, how they lie in relation to one another, and the ways in which they can be organized into increasingly complex structures to produce more sophisticated compositions. This discipline finds its origins in the fifth century BCE. A survey of the important contributors on this topic—among them Archytas, Aristoxenus, Euclid, Nicomachus, and Ptolemy—reveal a range of approaches to harmonics from strictly rationalist descriptions or fully empiricist accounts to elaborate combinations of elements from each. Some authors were more oriented toward the perceptual qualities of musical theory as an organizing principle, and others appealed to the mathematical order that could be uncovered in music and structured their investigations around deeper principles and relations found in the quantitative sciences. In particular, the excerpts we present mainly deal with the mathematical relationships Greek scholars found in developing the principles of harmony.

We begin this chapter with some pieces from Plato and Aristotle, both of whom were certainly knowledgeable in the theory as it had been developed up to their day. Though there is little extant of the actual works of Archytas, there are several excerpts in this chapter that report about his work, since he is acknowledged by later writers as one of the most important of the early contributors to musical theory. We then illustrate the more empiricist approaches to harmonics with the works of Aristoxenus which contrasts with Euclid's account that analyzes musical structures through the mathematics of ratios. Passages from Nicomachus and Theon of Smyrna give us insight into the origins of harmonics and the Pythagorean accounts of this discipline. Next, Ptolemy's sophisticated approach to harmonics epitomizes the centrality of mathematical theory for analyzing harmonics while keeping central the critical importance of aural perception. We conclude with the work of Boethius in the sixth century CE, whose Latin text was central in passing on the major ideas of Greek musical theory to the medieval period.

8.1 Plato and Aristotle on Music Theory

While Plato and Aristotle did not write specific studies on music theory or harmonics per se, throughout their works they included passages exploring many aspects of the field more broadly. These were more philosophical than quantitative or mathematical but proved to be inspirational to later scholars in the discipline. The way in which they considered harmonics alongside other disciplines, such as astronomy and optics, meant that the field maintained its high status. We have already seen Plato's initial description of harmonics in the *Republic*. Here we consider his more detailed comments on the subject in the *Philebus* and the *Timaeus*.

The *Philebus* is a dialogue among Socrates, Protarchus, and Philebus in which they discuss the virtues of physical pleasures versus pleasures of the mind. In this passage, they are discussing the merits of music and the general relationship of musical understanding to mathematics.

From *Philebus*

Protarchus: I think I more or less understand some of what you say, Socrates, but I still need to have some of it put more clearly.

Socrates: Well, Protarchus, the letters give a clear exemplification of what I am saying. Since you have been taught them, try grasping it in them.

Protarchus: In what way?

Socrates: We can agree that vocal sound, that of people in general and that of each individual, is both a single thing that comes through the mouth, and again indefinite in number.

Protarchus: Certainly.

Socrates: And we are by no means yet experts by virtue of either of these things, because we grasp either the indefinite aspect of it or the unitary aspect. What makes each of us literate is knowing the number of the sounds, and their qualities.

Protarchus: Perfectly true.

Socrates: And again, what makes a man a musician is the very same thing.

Protarchus: How is that?

Socrates: Sound, as considered in this art too, is somehow one.

Protarchus: Of course.

Socrates: But we posit two things, low pitch and high, and a third, sameness of pitch.

Protarchus: Yes.

Socrates: But you would not yet be an expert in music by knowing only those things, though if you did not know them you would be virtually worthless in these matters.

Protarchus: So I would.

Socrates: But suppose, my good friend, that you grasp the number and the qualities of the intervals related to high and low pitch of sound, and the boundaries of the intervals, and the systems that have arisen out of them. These systems were identified by people in the past, and they handed down to us, their successors, the practice of calling them harmonies; and in the movement of the body they identified other, similar inherent features which, they say, we must measure by numbers, and call rhythms and measures, while being aware that this is how we should investigate every one and many. For when you grasp them in this way, then it is that you have become an expert; and when you have grasped any other one by investigating it in this way, you have by so doing understood it. But the indefinite plurality inherent in any kind of thing makes you, in each case, indefinite in your understanding, not numbered among persons of repute, since you have never turned your attention to number in anything.

In this passage from the *Timaeus*, Plato describes the dividing of the substance of the soul. However, the numerical calculation he provides undoubtedly comes from the study of the division of the musical octave, credited to the Pythagoreans, but more probably due to Plato's friend Archytas.

From *Timaeus*

Now God did not devise the soul [of the universe] later [than its body], in the way that we are setting off to speak of it later, for when he constructed them he would not have allowed an older to be ruled by a younger; but since ourselves partake greatly in chance and randomness, we are speaking also in this rather random way. But he made the soul prior and older in both birth and excellence, to be mistress and ruler of its subject, the body; and he put it together from the following things and in the following way. From the Being that is indivisible and always the same, and that which occurs divided among bodies, he mingled together a third, intermediate form of Being; and he did this also with the nature of the Same and that of the Different, putting together in the same way something intermediate between the indivisible form of each and that which is divided up among bodies. Taking the three things, he next blended them all into a single form, fitting the nature of the Different, which is resistant to mixing, by force into the Same. When he had mixed this with Being, and out of three had made one, he again divided the whole into as many parts as was proper, each part being a mixture of the Same, the Different, and Being.

This is how he began to divide. First, he took away one part of the whole, then another, double the size of the first, then a third, hemiolic [one and a half times] with respect to the second and triple the first, then a fourth, double the second, then a fifth, three times the third, then a sixth, eight times the first, then a seventh, twenty-seven times the first. Next he filled out the double and triple intervals, once again cutting off parts from the material and placing them in the intervening gaps, so that in each interval there were two means, the one exceeding [one extreme] and exceeded [by the other extreme] by the same part of the extremes themselves [e.g., 1, $\frac{4}{3}$, 2], the other exceeding [one extreme] and exceeded [by the other] by an equal number [e.g., 1, $\frac{3}{2}$, 2].[1] From these links within the previous intervals there arose hemiolic, epitritic [one and one-third] and epogdoic [one and one-eighth] intervals; and he filled up all the epitritics with the epogdoic kind of interval, leaving

[1] If we combine these two sequences, we get 1, 2, 3, 4, 8, 9, . . . Now, the harmonic mean between 1 and 2 is $\frac{4}{3}$, that between 2 and 4 is $\frac{8}{3}$, that between 4 and 8 is $\frac{16}{3}$, that between 1 and 3 is $\frac{3}{2}$, and that between 3 and 9 is $\frac{9}{2}$. Similarly, the arithmetic mean between 1 and 2 is $\frac{3}{2}$, between 2 and 4 is 3, between 4 and 8 is 6, between 1 and 3 is 2, and between 3 and 9 is 6. Putting everything together, we get the sequence $1, \frac{4}{3}, \frac{3}{2}, 2, \frac{8}{3}, 3, 4, \frac{9}{2}, \frac{16}{3}, 6, 8, 9$.

a part of each of them, where the interval of the remaining part had as its boundaries, the *leimma*, a ratio of 256 to 243.[2]

Aristotle refers to musical ideas in many passages in his works. Here we give two examples, the first from the *Posterior Analytics* and the second from the *Problemata*, a work compiled by students at the Lyceum during and after Aristotle's time.

From *Posterior Analytics*

In all these cases it is plain that the "What is it?" and the "Why is it?" are the same. What is an eclipse?—the deprivation of light from the moon by the interposition of the earth. Why is there an eclipse, or why is the moon eclipsed?— because the light fails when the earth is interposed. What is concord?—a ratio of numbers between the high-pitched and the low-pitched. Why does the high-pitched form a concord with the low-pitched?—because the high-pitched and the low-pitched stand in a ratio of numbers. Does there exist a concord between the high-pitched and the low-pitched?—Is their ratio in numbers? Granted that it is, what then is the ratio?

In *Problemata*, book 19, 4.21, the author discusses this idea of "concord" in a more mathematical way.

Why is the octave the finest concord? Is it because its ratios are between terms that are wholes, while those of the others are not between wholes? For *nete* is double *hypate*, so that if *nete* is 2, *hypate* is 1, if *hypate* is 2, *nete* is 4, and so on invariably. But *nete* is the hemiolic of *mese*: for the fifth, which is hemiolic, is not in whole numbers—if, for instance, the smaller term is 1, the greater is the same quantity and the half in addition. Hence wholes are not being compared with wholes, but parts are added. The case is similar with the fourth, for epitritic ratio is so much and one of the three in addition. Or is it because the completest concord is that constituted out of both, and because the measure of melody . . .

8.2 Archytas, from Porphyry and Ptolemy

Archytas of Tarentum was a prominent Pythagorean who flourished in the early fourth century BCE. Highly regarded as a mathematician and statesman, he was a close friend of Plato, who later adapted many of his ideas in his philosophical works. Archytas made some original contributions to harmonics, notably a detailed study of the ratios of scalar intervals, although later writers sometimes attribute this study to Pythagoras himself. Archytas's work has generally been preserved in later commentaries since his own books have been lost. We reproduce here two fragments from Archytas as reported by Porphyry (260–305 CE) in his commentary on Ptolemy's *Harmonics*.

[2] That is, the epitritic interval (the musical fourth) $4:3$ is composed of two epogdoic intervals (the musical tone) and one *leimma*: $4:3=(9:8)(9:8)(256:243)$. So, for example, the octave interval from 1 to 2 now is composed of a fourth, a tone, and another fourth. But it can be filled out by inserting tones as follows: $1, \frac{9}{8}, \frac{81}{64}, \frac{4}{3}, \frac{3}{2}, \frac{27}{16}, \frac{243}{128}, 2$. And the ratio of $\frac{4}{3}$ to $\frac{81}{64}$ is $\frac{256}{243}$ as is the ratio of 2 to $\frac{243}{128}$.

But now in addition let the words of Archytas the Pythagorean, whose writings most of all are said to be indeed genuine, be cited. In *On Mathematics*, just as he begins the discourse, he says the following:

Those concerned with the sciences seem to me to make distinctions well, and it is not at all surprising that they have correct understanding about individual things as they are. For, having made good distinctions concerning the nature of wholes they were likely also to see well how things are in their parts. Indeed concerning the speed of the stars and their risings and settings as well as concerning geometry and numbers and not least concerning music, they handed down to us a clear set of distinctions. For these sciences seem to be akin.

Well then, first they reflected that it is not possible that there be sound, if an impact of some things against one another does not occur; they said that an impact occurred whenever things in motion came upon and collided with one another. Some moving in opposing directions, when they meet, make a sound as each slows the other down, but others moving in the same direction but not with equal speed, being overtaken by the ones rushing upon them and being struck, make a sound. Indeed many of these sounds cannot be recognized because of our nature, some because of the weakness of the blow, others because of the distance of separation from us and some because of the excess of the magnitude. For the great sounds do not steal into our hearing, just as nothing is poured into narrow-mouthed vessels, whenever someone pours out a lot. Well then, of the sounds reaching our perception those which arrive quickly and strongly from impacts appear high in pitch, but those which arise slowly and weakly seem to be low in pitch.... It has become clear to us from many things that high notes move more quickly and low ones more slowly....

That ratio, then, differs from excess is clear. But that the ratio and the relation of the terms which are compared with one another is also called an interval, I will now show. But Demetrius too regards interval as the same as ratio, and many others of the ancients follow this usage, just as Dionysius of Halicarnassus (1st c. BCE), and Archytas in *On Music*, and the element man himself, Euclid, in *The Division of the Canon*, speak of intervals rather than ratios.... And Archytas speaking about the means writes these things:

There are three means in music: one is the arithmetic, the second geometric and the third sub-contrary [which they call "harmonic"]. The mean is arithmetic, whenever three terms are in proportion by exceeding one another in the following way: by that which the first exceeds the second, by this the second exceeds the third. And in this proportion it turns out that the interval of the greater terms is smaller and that of the smaller greater. The mean is geometric, whenever they [the terms] are such that as the first is to the second, so the second is to the third. Of these [terms] the greater and the lesser make an equal interval. The mean is sub-contrary, which we call harmonic, whenever they [the terms] are such that, by which part of itself the first term exceeds the second, by this part of the third the middle exceeds the third. It turns out that, in this proportion, the interval of the greater terms is greater and that of the lesser is less.[3]

[3] As examples, in the sequence (12, 9, 6), 9 is the arithmetic mean of 12 and 6; in the sequence (12, 6, 3), 6 is the geometric mean of 12 and 3; and in the sequence (12, 8, 6), 8 is the harmonic mean of 12 and 6. By interval, Archytas is thinking of the musical interval. Thus in the case of the arithmetic mean, $12:9<9:6$, for the geometric mean, $12:6=6:3$, and for the harmonic mean, $12:8>8:6$.

8.3 Aristoxenus, *Elementa Harmonica*

At the end of the fourth century BCE, the field of harmonics was impacted in new ways by the appearance of several important works. One of these was Aristoxenus's *Elementa Harmonica*, a comprehensive work of which only parts are extant. Aristoxenus's account was firmly grounded in musical perception as the foundation for musical theorizing, and his approach remained in the realms of experience and empirical data rather than in appeals to mathematics and physics to account for features and aspects in harmonics. Since Aristoxenus's approach is basically nonmathematical, we present here only a brief excerpt from book 2 explaining his position.

[Harmonics] is to be understood, speaking generally, as the science which deals with all melody, and inquires how the voice naturally places intervals as it is tensed and relaxed. For we assert that the voice has a natural way of moving, and does not place intervals haphazardly. We try to give these matters demonstrations which conform to the appearances, not in the manner of our predecessors, some whom used arguments quite extraneous to the subject, dismissing perception as inaccurate and inventing theoretical explanations, and saying that it is through ratios of numbers and relative speeds that the high and the low come about. Their accounts are altogether extraneous, and totally in conflict with the appearances. Others delivered oracular utterances on individual topics, without giving explanations or demonstrations, and without even properly enumerating the perceptual data. We, on the other hand, try to adopt initial principles which are all evident to anyone experienced in music, and to demonstrate what follows from them.

Taken as a whole, our science is concerned with all musical melody, both vocal and instrumental. Its pursuit depends ultimately on two things, hearing and reason. Through hearing we assess the magnitudes of intervals, and through reason we apprehend their functions. We must therefore become practiced in assessing particulars accurately. While it is usual in dealing with geometrical diagrams to say "let this be a straight line," we must not be satisfied with similar remarks in relation to intervals. The geometer makes no use of the faculty of perception: he does not train his eyesight to assess the straight or the circular or anything else of that kind either well or badly: it is rather the carpenter, the wood-turner, and some of the other crafts that concern themselves with this. But for the student of music, accuracy of perception stands just about first in order of importance, since if he perceives badly it is impossible for him to give a good account of the things which he does not perceive at all.

8.4 Euclid, *Sectio Canonis*

Although appearing at around the same time as Aristoxenus's work, Euclid's *Sectio Canonis* embodies a completely different approach to harmonics. After a short introduction, Euclid presents twenty distinct propositions with formal proofs. The first nine are distinct in their mathematical abstraction; while they refer to intervals, their scope is about ratios of any sort, not necessarily musical ones. It is not until proposition

10 that musical concepts are explicitly introduced and the work culminates in the division of a *kanon*, or string, according to the diatonic genus. We note that there has been considerable scholarly controversy as to the authorship of this work and even whether the introduction and the body of the text were written by the same author. From what is known about the development of the theory of harmonics, it does appear that the text can be dated to a time near to that of Euclid's life—which, of course, is not known precisely. So here we will keep the traditional attribution but refer the reader to André Barbera, *The Euclidean Division of the Canon* (Lincoln: University of Nebraska Press, 1991) for a detailed discussion of the dating and authorship of this work.

Introduction

If there were stillness and no movement, there would be silence: and if there were silence and if nothing moved, nothing would be heard. Then if anything is going to be heard, impact and movement must first occur. Thus since all sounds occur when some impact occurs, and since it is impossible for an impact to occur unless movement has occurred beforehand—and since of movements some are closer packed, others more widely spaced, those which are closer packed producing higher notes and those which are more widely spaced lower ones—it follows that some notes must be higher, since they are composed of closer packed and more numerous movements, and others lower, since they are composed of movements more widely spaced and less numerous. Hence notes that are higher than what is required are slackened by the subtraction of movement and so reach what is required, while those which are too low are tightened by the addition of movement, and so reach what is required. We must therefore assert that notes are composed of parts, since they attain what is required through addition and subtraction. Now all things that are composed of parts are spoken of in a ratio of number with respect to one another, so that notes, too, must be spoken of in a ratio of number to one another. Some numbers are spoken of in multiple ratio with respect to one another, some in epimoric ratio, and some in epimeric ratio,[4] so that notes must also be spoken of in these kinds of ratios to one another: and of these, the multiple and the epimoric are spoken of in relation to one another under a single name. Among notes we also recognize some as concordant, others as discordant, the concordant making a single blend out of the two, while the discordant do not. In view of this, it is to be expected that the concordant notes, since they make a single blend of sound out of the two, are among those numbers which are spoken of under a single name in relation to one another, being either multiple or epimoric.

Proposition 1. *If a multiple interval put together twice makes some interval, this interval too will be multiple.*

Let there be an interval BC, and let B be a multiple of C, and let B be to D as is C to B. I assert then that D is a multiple of C. For since B is a multiple of C, C therefore

[4] An epimeric ratio is one of the form $(n+m):n$, where $m>1$.

Figure 8.4.1.

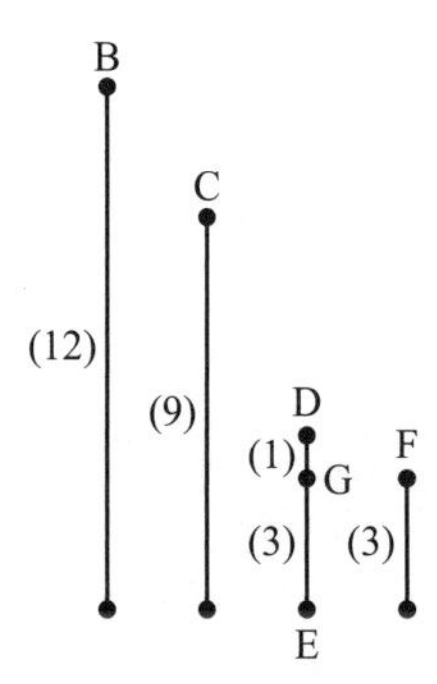

Figure 8.4.2.

measures B. But B was to D as C was to B, so that C measures D too. Therefore D is a multiple of C.

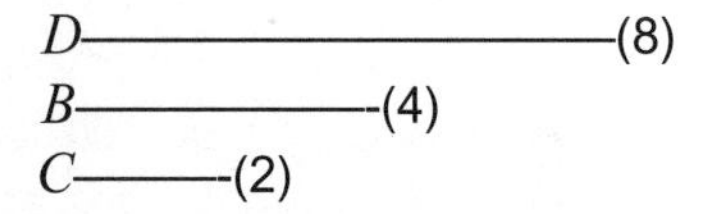

Proposition 2. *If an interval put together twice makes a whole that is multiple, then that interval will also be multiple.*

Let there be an interval BC, let B be to D as is C to B, and let D be a multiple of C [figure 8.4.1]. I assert that B is also a multiple of C. For since D is a multiple of C, C therefore measures D. But we have learned that where there are numbers in proportion—however many of them—and where the first measures the last, it will also measure those in between. Therefore C measures B, and B is therefore a multiple of C.

Proposition 3. *In the case of an epimoric interval,*[5] *no mean number, neither one nor more than one, will fall within it proportionally.*

Let BC be an epimoric interval. Let DE and F be the smallest numbers in the same ratio as are B and C [figure 8.4.2]. These then are measured only by the unit as a common measure. Take away GE, which is equal to F. Since DE is the epimoric of F, the remainder DG is a common measure of DE and F. DG is therefore a unit. Therefore no mean will fall between DE and F. For the intervening number will be less than DE and greater than F, and will thus divide the unit, which is impossible. Therefore no mean will fall between DE and F. And however many means fall in proportion between the smallest numbers,

[5] An epimoric ratio is what Nicomachus has called a superparticular ratio, one in modern terms expressible as $(a+1):a$, where a is an integer.

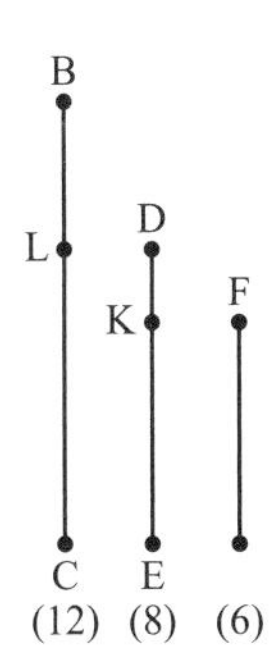

Figure 8.4.3.

there will fall in proportion exactly as many between any others which have the same ratio. But none will fall between DE and F; nor will one fall between B and C.

Proposition 4. *If an interval which is not multiple is put together twice, the whole will be neither multiple nor epimoric.*

Let BC be an interval which is not multiple, and let B be to D as C is to B. I say that D is neither a multiple nor an epimoric of C. First let D be a multiple of C. Now we have learned that if an interval put together twice makes a whole that is multiple, that interval is also multiple. Then B will be a multiple of C: but it was not. Hence it is impossible for D to be a multiple of C. Nor is it an epimoric: for within an epimoric interval there falls no mean in proportion. But B falls within DC. Therefore it is impossible for D to be either a multiple or an epimoric of C.

D———————————(9)
B——————————(6)
C————(4)

Proposition 5. *If an interval put together twice does not make a whole that is multiple, that interval itself will not be multiple either.*

Let BC be an interval; let B be to D as C is to B; and let D not be a multiple of C. I say that B will not be a multiple of C either. For if B is a multiple of C, D will therefore be a multiple of C. But it is not. Therefore B will not be a a multiple of C.

D————————————(9)
B——————————(6)
C—————(4)

Proposition 6. *The double interval is composed of the two greatest epimoric intervals, the hemiolic and the epitritic.*[6]

Let BC be the hemiolic of DE, and let DE be the epitritic of F. I say that BC is double F [figure 8.4.3]. I take away EK, equal to F, and CL, equal to DE. Then since BC is the hemiolic of DE, BL is a third part of BC, and half of DE. Again, since DE is the epitritic

[6] The hemiolic is also called the sesquialter, the ratio of 3 : 2. The epitritic is also called the sesquitertian, the ratio of 4 : 3. Both of these are epimoric.

of F, DK is a fourth part of DE, and a third part of F. Then since DK is a fourth part of DE, and BL is half of DE, DK will therefore be a half of BL. Now BL was a third part of BC: therefore DK is a sixth part of BC. But DK was a third part of F: therefore BC is double F.

Alternatively: let A be the hemiolic of B, and let B be the epitritic of C. I say that A is double C. Since A is the hemiolic of B, A contains B and half of B. Then two A's are equal to three B's. Again, since B is the epitritic of C, B contains C and a third of C. Therefore three B's are equal to four C's. Now three B's are equal to two A's. Therefore two A's are equal to four C's. Therefore A is equal to two C's: therefore A is double C.

Proposition 7. *From the double interval and the hemiolic, a triple interval is generated.*

Let A be double B, and let B be the hemiolic of C. I say that A is triple C. Since A is double B, A is therefore equal to two B's. Again, since B is the hemiolic of C, B therefore contains C and half of C. Hence two B's are equal to three C's. But two B's are equal to A. Therefore A is equal to three C's: therefore A is triple C.

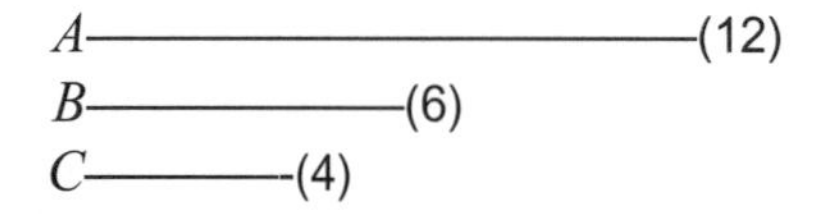

Proposition 8. *If from a hemiolic interval an epitritic interval is subtracted, the remainder left is epogdoic.*[7]

Let A be the hemiolic of B, and let C be the epitritic of B. I say that A is the epogdoic of C. Since A is the hemiolic of B, A therefore contains B and a half of B. Therefore eight A's are equal to twelve B's. Again, since C is the epitritic of B, C therefore contains B and a third of B. Therefore nine C's are equal to twelve B's. But twelve B's are equal to eight A's, and therefore eight A's are equal to nine C's. A is therefore equal to C and an eighth of C, and A is therefore the epogdoic of C.

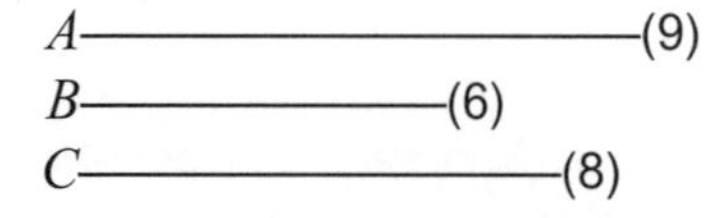

Proposition 9. *Six epogdoic intervals are greater than one double interval.*

Let A be one number. Let B be the epogdoic of A, let C be the epogdoic of B, let D be the epogdoic of C, let E be the epogdoic of D, let F be the epogdoic of E, and let G be the epogdoic of F. I say that G is more than double A. Since we have learned how to find seven numbers that are epogdoics of one another let the numbers A, B, C, D, E, F, G have been found. A is 262,144, B is 294,912, C is 331,776, D is 373,248, E is 419,904, F is 472,392, G is 531,441; and G is more than double A.

The Greek notes mentioned in the next few propositions are all listed in the table at the end, where the reader can compare them with modern notes over a two-octave interval.

[7] An epogdoic ratio is the ratio 9:8.

Proposition 10. *The octave interval is multiple.*

Let A be *nētē hyperbolaiōn*, let B be *mesē* and let C be *proslambanomenos*.[8] Then the interval AC, being a double octave, is concordant. It is therefore either epimoric or multiple. It is not epimoric, since no mean falls proportionally within an epimoric interval. Therefore it is multiple. Thus since the two equal intervals AB and BC put together make a whole that is multiple, AB is therefore multiple too.

C————————————————(16)
B————————-(8)
A————(4)

Proposition 11. *The interval of the fourth and that of the fifth are each epimoric.*

Let A be *nētē synēmmenōn*, let B be *mesē*, and let C be *hypatē mesōn*.[9] The interval AC, being a double fourth, is therefore discordant;[10] it is therefore not multiple. Thus since the two equal intervals AB and BC when put together make a whole which is not multiple, neither is AB multiple. And it is concordant; therefore it is epimoric. The same demonstration applies also to the fifth.

C——————————————-(16)
B——————————(12)
A——————-(9)

Proposition 12. *The octave interval is duple.*

We have shown that it is multiple; it is thus either duple or greater than duple. But since we showed that the duple interval is composed of the two greatest epimoric intervals, it follows that if the octave is greater than duple it will not be made up of just two epimoric intervals, but of more. But it is made up of two concordant intervals, the fifth and the fourth. Therefore the octave will not be greater than duple; therefore it is duple.

But since the octave is duple, and the duple is made up of the two greatest epimorics, it follows that the octave is made up of the hemiolic and the epitritic, since these are the greatest. But it is made up of the fifth and the fourth, and these are epimoric. The fifth, therefore, since it is greater, must be hemiolic and the fourth epitritic. It is clear, then, that the octave and a fifth is a triple interval. For we showed that the triple interval is generated from a duple and a hemiolic interval, so that the octave and a fifth is also a triple interval. The double octave is a quadruple interval. We have demonstrated, therefore, for each of the concords, in what ratios their bounding notes stand to one another.

Proposition 13. *It remains to consider the interval of a tone, to show that it is epogdoic.*

We have learned that if an epitritic interval is subtracted from a hemiolic interval, the remainder left is epogdoic. And if the fourth is taken from the fifth, the remainder is the interval of a tone. Therefore the interval of a tone is epogdoic.

[8] These are names of particular notes in the two-octave Greek musical scale. *Nētē hyperbolaiōn* is the highest note of that scale; *mesē* is one octave lower; and *proslambanomenos* is the lowest note, an octave below *mesē*.

[9] *Nētē synēmmenōn* is a fourth above *mesē*, which is in turn a fourth above *hypatē mesōn*.

[10] That this interval is discordant is taken from musical experience. It is not justified by any of the previous results in the treatise.

Proposition 14. *The octave is less than six tones.*

It has been shown that the octave is duple, and that the tone is epogdoic. Six epogdoic intervals are greater than a duple interval. Therefore the octave is less than six tones.

Proposition 15. *The fourth is less than two and a half tones, and the fifth is less than three and a half tones.*

Let B be *nētē diezeugmenōn*, let C be *paramesē*, let D be *mesē*, and let E be *hypatē mesōn*.[11] Then the interval CD is a tone, and BE, which is an octave, is less than six tones. Therefore the remainder, that is, BC and DE, which are equal, are less than five tones. Hence the interval in BC, which is a fourth, is less than two and a half tones, and BD, which is a fifth, less than three and a half tones.

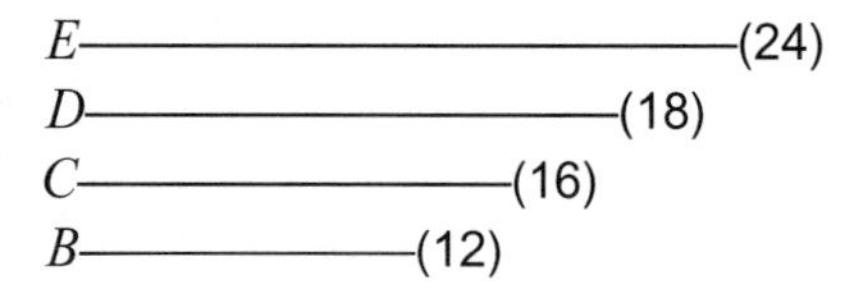

Proposition 16. *The tone will not be divided into two or more equal intervals.*

It has been shown that it is epimoric. Within an epimoric interval there falls neither one nor more than one mean in proportion. Therefore the tone will not be divided into equal intervals.

We conclude with Euclid's table giving the entire correspondence between the Greek and modern note names over two octaves in what is called the diatonic genus, one of three genera for tuning instruments in Greek times. For convenience, we repeat the same terms at the end of the first column and the beginning of the second. Note that modern notes in capital letters represent the first octave, lowercase letters represent the second octave, and the one note a′ is the first note of the next octave.

Greek notes	Modern notes	Greek notes	Modern notes
proslambanomenos	*A*	*mesē*	*a*
hypatē hypatōn	B	*paramesē*	*b*
parhypatē hypatōn	C	*tritē diezeugmenōn*	c
lichanos hypatōn	D	*paranētē diezeugmenōn*	*d*
hypatē mesōn	E	*nētē diezeugmenōn*	*e*
parhypatē mesōn	F	*tritē hyperbolaiōn*	*f*
lichanos mesōn	G	*paramētē hyperbolaiōn*	*g*
mesē	*a*	*nētē hyperbolaiōn*	*a′*

[11] *Nētē diezeugmenōn* is a fourth above *paramesē*; *paramesē* is a tone above *mesē*; *hypatē mesōn* is a fourth below *mesē* and therefore an octave below *nētē diezeugmenōn*.

8.5 Nicomachus, *The Manual of Harmonics*

Nicomachus's *Manual of Harmonics* is the earliest extant text giving us a full description of the Pythagorean origins of harmonics. Written probably around the turn of the second century CE, it presents in great detail what can only be a myth of Pythagoras discovering the ratios of musical notes by listening to smiths hammering iron on an anvil. We present here chapters 5 through 8 of the *Manual*. Note that chapter 5 gives the traditional Greek names for the various pitches in the musical scale. Nicomachus claims there that Pythagoras changed the basic scale from a seven-pitch one to our now standard eight-pitch one by inserting a new note in between two other ones. This does not seem possible. However, the eight-pitch scale that became standard can be represented by modern symbols as the scale *E, F, G, A, B, C, D, E*, where the intervals from *E* to *A* and from *B* to *E* are both fourths and the interval from *E* to *B* is a fifth, while that between *A* and *B* is a whole tone.

Chapter 5

Pythagoras is the first one who—in order that the middle note, when combined itself with both extremes through conjunction, might not produce the consonance of a fourth only, a fourth with *hypate* at one extreme, and with *nete* at the other; and in order that we might be able to envisage a more varied scheme, the extremes themselves producing with one another the most satisfying consonance, that is, the octave in a double ratio, which could not result from the two tetrachords—intercalated an eighth note, which he fitted between *mese* and *paramese* and separated it from *mese* by a whole tone and from *paramese* by a semi-tone. The result was that the string which was formerly *paramese* in the heptachord, still being the third string calculating from *nete*, is called *trite* and is situated in just this position; the intercalated string, on the other hand, is the fourth string calculating from *nete* and forms with it the consonance of a fourth, which is the consonance *mese* formed originally with *hypate*. The whole tone between both these two strings, *mese* and the intercalated string which is named after the former *paramese*, to whichever tetrachord it is added, whether to the one that is contingent on *hypate*, in which case it is relatively high in pitch, or to the one contingent on *nete*, in which case it is relatively deep in pitch, will produce the consonance of a fifth, which is a system of both the tetrachord itself and the added whole tone. Thus the hemiolic ratio of the fifth is found to be a system composed of the epitritic ratio and of the epogdoic ratio. The whole tone then is in an epogdoic ratio.

Chapter 6

The interval of strings comprising a fourth, that of a fifth, and that formed by the union of both, which is called an octave, as well as that of the whole tone lying between the two tetrachords, was confirmed by Pythagoras to have this numerical quantity by means of a certain method which he discovered. One day he was deep in thought and seriously considering whether it could be possible to devise some kind of instrumental aid for the ears which would be firm and unerring, such as vision obtains through the compass and the ruler or the dioptra, or touch obtains with the balance beam or the system of measures. While thus engaged, he walked by a smithy and, by divine chance, heard the hammers beating out iron on the anvil and giving off in combination sounds which were most harmonious with one another, except for one combination. He recognized in these sounds the consonance of the octave, the fifth and the fourth. But he perceived that the interval between the fourth and the fifth was dissonant in itself but was otherwise complementary to the greater of these two consonances. Elated, therefore, since it was as if his purpose was being divinely accomplished, he ran into the smithy and found by various experiments that the difference of sound arose from the weight of the hammers, but not from the force of the blows, nor from the shapes of the hammers, nor from the alteration of the iron being forged. After carefully examining the weights of the hammer and their impacts, which were identical, he went home.

He planted a single stake diagonally in the walls in order that no difference might arise from this procedure or, in short, that no variation might be detected from the use of several stakes, each with its own peculiar properties. From this stake he suspended four strings of the same material and made of an equal number of strands, equal in thickness and of equal torsion. He then attached a weight to the bottom of each string, having suspended each by each in succession. When he arranged that the lengths of the strings should be exactly equal, he alternately struck two strings simultaneously and found the aforementioned consonances being produced by a different pair of strings.

He found that the string stretched by the greatest weight produced, when compared with that stretched by the smallest, an octave. The weight on one string was twelve pounds, while that on the other was six pounds. Being therefore in a double ratio, it produced the octave, the ratio being evidenced by the weights themselves. Again, he found that the string under the greatest tension compared with that next to the string under the least tension (the string stretched by a weight of eight pounds), produced a fifth. Hence he discovered that this string was in a hemiolic ratio with the string under the greatest tension, the ratio in which the weights also stood to one another. Then he found that the string stretched by the greatest weight, when compared with that which was next to it in weight, being under a tension of nine pounds, produced a fourth, analogous to the weights. He concluded, therefore, that this string was undoubtedly in an epitritic ratio with the string under the greatest tension and that this same string was by nature in a hemiolic ratio with the string under the least tension (for this is the case with the ratio of 9 to 6). In a similar way, the string neighboring on that under the least tension, that is, the string stretched by a weight of eight pounds, compared with that stretched by a weight of six pounds, was in an epitritic ratio, but it was in a hemiolic ratio with the string stretched by a weight of twelve pounds.

Then that interval which is between the fifth and the fourth, that is, the interval by which the fifth is greater than the fourth, was confirmed to be in a epogdoic ratio, which is as 9 is to 8. And either way it was proved that the octave is a system consisting of the fifth and the fourth in conjunction, just as the double ratio consists of the hemiolic ratio and the epitritic, as for example, 12, 8, and 6; or conversely, it consists of the fourth and the fifth, just as the double ratio consists of the epitritic ratio and the hemiolic, as for example, 12, 9, and 6, in such order.

And having inured his hand and his hearing to the suspended weights and having established on their basis that ratio of their relations, he ingeniously transferred the bond, which fastened all the strings, from the diagonal stake to the bridge of the instrument, which he called *chordotonon* or string-stretcher, and he transferred the amount of tension on the strings analogous to the weights, to the commensurate turning of the tuning pegs set in the upper part of the instrument. Using this as a standard and as it were an infallible pointer, he extended the test henceforward to various instruments, namely, to the percussion on plates, to auloi and panpipes, to monochords and triangular harps, and the like. And in all of these he found consistent and unchanging, the determination by number.

He called the note partaking of the number 6, *hypate*, that of the number 8, *mese*, this number being in an epitritic proportion with the number 6; that of the number 9, he called *paramese*, which is higher than *mese* by a whole tone and what is more, stands in an epogdoic proportion with it; that of 12, he called *nete*. Filling out the intervening intervals in the diatonic genus with analogous notes, he thus subordinated the octachord to the consonant numbers, the double ratio, the hemiolic, the epitritic, and the difference between them, the epogdoic.

It is somewhat curious that the Greeks maintained this legend of Pythagoras discovering numerical ratios by listening to the hammers, since it is not true that the ratios between pitches correspond to those between weights of the hammers, and attaching weights to equal strings will not produce the musical notes. Anyone who tried to reproduce this experiment would surely have noticed that it did not produce Pythagoras's results. Still, this story is repeated over and over again through the centuries, including in the work of Boethius below. Most probably, then, these later Greek writers simply believed this was a folktale that did not need to be analyzed but merely accepted as showing that Pythagoras discovered the mathematical basis of music. Today, however, most scholars believe that, in fact, Pythagoras did not make this discovery and that most probably it was Archytas who worked out the basic mathematical details.

Chapter 7

Thus he discovered, on the basis of a certain natural necessity, the progression in this diatonic genus from the lowest note to the highest. (For from this procedure he also revealed the structure of the chromatic and enharmonic genera, as it will be possible for me to show you some time later.) This diatonic genus appears, however, to comprehend by nature the degrees and progressions such as follow: semi-tone, a whole tone, and then a whole tone. And this is the system of a fourth, consisting of two whole tones and the semi-tone, so-called. Then, by the addition of another whole-tone, namely, the intercalated whole-tone, the fifth results, being a system of three whole tones and a semi-tone. Then next in order to this come a semi-tone, a whole tone, and a whole tone, being another system of a fourth, that is, another epitritic proportion. So that in the more ancient heptachord all the notes four removed from one another, starting from the lowest, were consonant throughout with each other by a fourth, the semi-tone occupying by transference the first, the middle, and the third place in the tetrachord.

In the Pythagorean octachord, however, whether it be composed of a tetrachord and a pentachord by conjunction, or of two tetrachords separated by a whole tone from one another by disjunction, the progression by ascent will result in all the notes five removed from one another forming the consonance of a fifth with each other, the semi-tone as one advances, shifting into the four places, first, second, third, and fourth.

Using the same notes as above, the interval of a fourth from E to A consists of a semitone E to F, followed by two whole tones, F to G and G to A. Adding in the whole tone from A to B gives the interval of a fifth from E to B.

Chapter 8

It is useful, now that we have reached this point, to open up at this opportune moment the passage in the *Psychogony*[12] in which Plato expressed himself as follows: "so that within each interval there are two means, the one superior and inferior to the extremes by the same fraction, the other by the same number. He [the Demiurge] filled up the distance between the hemiolic interval and the epitritic with the remaining interval of the epogdoic."

For the double interval is as 12 is to 6, but there are two means, 9 and 8. The number 8, however, in the harmonic proportion is midway between 6 and 12, being greater than 6 by one-third of 6 and being less than 12 by one-third of 12. That is why Plato said that the mean, 8, inasmuch as it is of the harmonic proportion, is greater and lesser than the extremes by the same fraction. For the greatest term compared with the smallest is thus in a double proportion; and so it follows that the difference between the greatest term and the middle is 4, compared with the difference between the middle and the smallest, which is 2, for these differences are in a double proportion, 4 to 2. The peculiar property of such a mean is that, when the extremes are added to one another and multiplied by the middle term, a product is yielded which is the double of the product of the extremes; for 8 multiplied by the sum of the extremes, that is, 18, gives 144, which is double the product of the extremes, that is, 72.

The other mean, 9, which is fixed at the *paramese* degree, is observed to be at the arithmetic mean between the extremes, being less than 12 and greater than 6 by the same number. And the peculiar property of this mean is that the sum of the extremes is the double of the middle term itself, and the square of the middle term, 81, is greater than the product of the extremes, 72, by the whole square of the differences, that is, by 3 times 3, or 9, for this is the difference.

One can also point out the third mean, more properly called "proportion," in both the middle terms, 9 and 8. For 12 is in the same proportion to 8 as 9 is to 6; for both are in a hemiolic proportion. And the product of the extremes is equal to the product of the middle terms, 12 times 6 being equal to 9 times 8.

[12] This is another name for the *Timaeus*. This passage appears earlier in section 8.1.

8.6 Theon of Smyrna, *Mathematics Useful for Reading Plato*

Theon of Smyrna lived during the second century CE. In this passage of his work *Mathematics Useful for Reading Plato*, he is commenting on a treatise of Adrastus, who was born late in the first century CE and was known as the peripatetic because of his studies of Aristotle. Theon repeats the stories about Pythagoras and comments on how certain instruments are tuned to fit the basic concordant ratios.

It seems that Pythagoras was the first to have identified the concordant notes in their ratios to one another, those at the fourth in epitritic ratio [4:3], those at a fifth in hemiolic [3:2], those at an octave in duple [2:1]; those at an octave and a fourth in a ratio of 8 to 3, which is multiple-epimeric, since it is duple and double-epitritic; those at an octave and a fifth in triple ratio [3:1], those at a double octave in quadruple [4:1]; and of the other attuned notes, those bounding the tone in epogdoic ratio [9:8], those bounding what is now called the semitone but was then called the diesis in a ratio of number to number, that of 256 to 243. He investigated the ratios through both the lengths and the thicknesses of the strings, and again through the tension arising from the turning of the *kollaboi*, or in a more clearly discernible way from the attachment of weights; and also in wind instruments through the width of the bores or through the tension and relaxation of the breath; or again through solid bodies and weights such as discs and vessels. For whichever of these is taken according to one of the ratios mentioned, other factors being equal, it will produce the concord that corresponds to the ratio.

For the present let us be content to give a demonstration through the length of a string on what is called the *kanōn*. When the single string on the instrument is measured off into four equal parts, the note from the whole string, in relation to that from the three parts, will sound in concord at the fourth, being in epitritic ratio; in relation to that from the two parts, that is, from the half string, it will sound in concord at the octave, being in duple ratio; in relation to that from the fourth part it will sound in concord at the double octave, being in quadruple ratio. The note from the three parts in relation to that from the two parts will sound in concord at the fifth, being in hemiolic ratio; and in relation to the note from the fourth part it will sound in concord at the octave and a fifth, being in triple ratio. If the string is divided by measure into nine parts, the note from the whole string, in relation to that from the eight parts, will bound the interval of a tone, in epogdoic ratio.

All the concords are contained by the *tetraktys*, since the latter is composed of 1, 2, 3 and 4. In these numbers there is the concord of a fourth, the fifth and the octave, the epitritic ratio, the hemiolic, the duple, the triple and the quadruple.

Some people thought it proper to derive these concords from weights, some from magnitudes, some from movements and numbers, and some from vessels. Lasus of Hermione, so they say, and the followers of Hippasus of Metapontum, a Pythagorean, pursued the speeds and slownesses of the movements, through which the concords arise.... Thinking that ... in numbers, he constructed ratios of these sorts in vessels. All the vessels were equal and alike. Leaving one empty and filling the other up to halfway with liquid, he made a sound on each, and the concord of the octave was given out for him. Then, again leaving one of the vessels empty, he poured into the other one part out of the four, and when he struck it the concord of the fourth was given out for him, as was the fifth when he filled up one part out of the three. The one empty space stood to the other in the octave as 2 to 1, in the fifth as 3 to 2, and in the fourth as 4 to 3.

He studied the ratios in a similar way also in accordance with the divisions of strings, as was said above, not however on one string, as on the *kanōn*, but on two. For having made two strings equal in tension, when he divided one of them by pressing on it in the middle, the half of it made the concord of an octave in relation to the other string: and when he subtracted a third part, the remaining parts made the concord of a fifth in relation to the other string. Similarly too with the concord of a fourth: in this case he subtracted a quarter of one of the strings and related the remaining parts to the other string. This he did also on the *syrinx* according to the same method. Some people took the concordances from weights, attaching to two strings weights in the ratios specified; others did so from lengths and pressed down on various parts of the strings, so revealing the concordances in the strings.

8.7 Ptolemy, *Harmonics*

In addition to astronomy, optics, and geography, Ptolemy turned his attention to musical theory as well, composing a substantial treatise named *Harmonica*. Ptolemy's approach to harmonics was distinct from his predecessors in that he embraced the complementary roles of perception and more abstract reasoning in his framework rather than prioritizing one over the other. For him, principles of harmony and concordance could be generated by theoretical and mathematical analysis but needed to be confirmed by sense perception. To this end, he is critical of both the Pythagoreans and the Aristoxenians. In many ways, his conscious blending of the empirical with the theoretical echoes the methodology he employed in other disciplines such as astronomy.

Book 1

2. *What the aim of the student of Harmonics is*

The instrument of this kind of method is called the harmonic *kanōn*, a term adopted out of common usage, and from its straightening those things in sense perception that are inadequate to reveal the truth. The aim of the student of Harmonics must be to preserve in all respects the rational postulates of the *kanōn*, as never in any way conflicting with the perceptions that correspond to most people's estimation, just as the astronomer's aim is to preserve the postulates concerning the movements of the heavenly bodies in concord with their carefully observed courses, these postulates themselves having been taken from the obvious and rough and ready phenomena, but finding the points of detail as accurately as is possible through reason. For in everything it is the proper task of the theoretical scientist to show that the works of nature are crafted with reason and with an orderly cause, and that nothing is produced by nature at random or just anyhow, especially in its most beautiful constructions, the kinds that belong to the more rational of the senses, sight and hearing. To this aim some people seem to have given no thought at all, devoting themselves to nothing but the use of manual techniques and the unadorned and irrational exercise of perception, while others have approached the objective too theoretically. These are, in particular, the Pythagoreans and the Aristoxenians, and both are wrong. For the Pythagoreans did not follow the impressions of hearing even in those things where it is necessary for everyone to do so, and to the differences between sounds they attached ratios that were often inappropriate to the phenomena, so that they provided a slander to be directed at this sort of criterion by those whose opinions differed. The Aristoxenians, by contrast, gave most weight to things grasped by perception, and misused reason as if it were incidental to the route, contrary both to reason itself and to the perceptual evidence—contrary to reason in that it is not to the distinguishing features of sounds that they fit the numbers, that is, the images of the ratios, but to the intervals between them, and contrary to the perceptual evidence in that they also associate these numbers with divisions that are inconsistent with the submissions of the senses. Each of these things will become clear from the points I shall introduce, if the matters which contribute to an understanding of what follows are first given some analysis.

5. *Concerning the principles adopted by the Pythagoreans in their postulates about the concords*

Perception accepts as concords the fourth, as it is called, and the fifth, the difference between which is called a tone, and the octave; and also the octave and a fourth, the octave and a fifth, and the double octave. Let us ignore the concords greater than these for present purposes. The theory of the Pythagoreans rules out one of these, the octave and a fourth, by following its own special assumptions, ones which the leaders of the school put forward on the basis of ideas such as the following. They laid down a first principle of their method that was entirely appropriate, according to which equal numbers should be associated with equal-toned notes, and unequal numbers with unequal-toned; and from this they argued that just as there are two primary classes of unequal-toned notes, that of the concords and that of the discords, and that of the concords is finer, so there are also two primary distinct classes of ratio between unequal numbers, one being that of what are called "epimeric" or "number to number" ratios, the other being that of the epimorics and multiples; and of these the latter is better than the former on account of the simplicity of the comparison, since in this class the difference, in the case of epimorics, is a simple part, while in the multiples the smaller is a simple part of the greater. For this reason they fit the epimorics and multiple ratios to the concords, and link the octave to duple ratio [2:1], the fifth to hemiolic [3:2], the fourth to epitritic [4:3]. Their procedure here is very rational, since the octave is the finest of the concords, and the duple is the best of the ratios, the former because it is nearest to the equal-toned, the latter because it alone makes an excess equal to that which is exceeded; and again, because the octave consists of the two first concords taken successively, and the duple consists of the two first epimorics taken successively, the hemiolic and the epitritic; and while in the latter case the hemiolic ratio is greater than the epitritic, in the former the concord of the fifth is greater than that of the fourth, so that the difference between them—that is, the tone—is assigned to the epogdoic ratio [9:8], by which the hemiolic is greater than the epitritic; and in accordance with these points they also adopt among the concords the magnitude put together from the octave and the fifth, and again that put together from two octaves—that is, the double

Figure 8.7.1.

octave—since it follows that the ratio of the latter is constituted as quadruple, and that of the former as triple. But they do not adopt the magnitude put together from the octave and the fourth, because it makes the ratio of 8 to 3, which is neither epimoric nor multiple.

They argue to the same conclusion in a more geometrical way, as follows. Let AB, they say, be a fifth, and let BC be another fifth, continuous with the first, so that AC is a double fifth [figure 8.7.1]. Since the double fifth is not concordant, it follows that AC is not multiple, so that neither is AB multiple: but it is concordant, and hence the fifth is superparticular. In the same way they show that the fourth is also an epimoric, a smaller one than the fifth. Again, they say, let AB be an octave and let BC be another octave continuous with the first, so that AC is a double octave. Then since the double octave is concordant, it follows that AC is either epimoric or multiple: but it is not epimoric—since then no mean would fall within it proportionately—and hence AC is multiple, so that AB is also multiple: therefore the octave is multiple. From these things it is plain to them that the octave is duple, and that of the others the fifth is hemiolic and the fourth epitritic, since of the multiples only the duple ratio is constituted by the two greatest epimorics, so that ratios put together from two of the other epimorics are together smaller than the duple, and there is no multiple smaller than the duple. Since the tone is accordingly shown to be epogdoic, they reveal that the half-tone is unmelodic, because no epimoric ratio divides another proportionately as a mean, and melodic magnitudes must be in epimoric ratios.

11. *On the tension, density, and thickness of strings*

That the strings do not differ, even when there is more than one of them, if they are made to be of equal pitch in equal lengths, will be clear from the following. Since in strings there are three causes of difference in respect of high and low, of which one lies in the density of the strings, one in their thickness, and one in their length, and the sound made by the denser string, the thinner string, and that with smaller length is higher; and since in strings tension is substituted for increased density—for it tenses and stiffens, and for that reason is more uniform in strings with lesser lengths—it is clear that if the other factors are the same, then as the greater tension is to the smaller, so is the sound based on the greater to that based on the smaller; and as the greater thickness is to the smaller, so is the sound

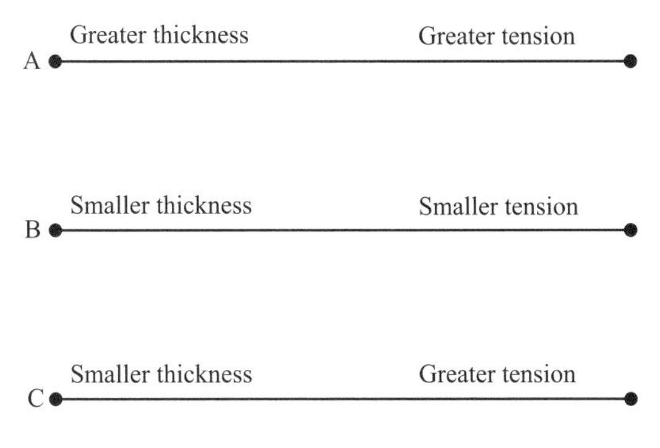

Figure 8.7.2.

based on the smaller to that based on the greater. I say, then, since these things are so, that in dissimilar strings, when they are made to be of equal pitch in equal lengths, the deficiency of the sound arising from the greater thickness is made up for in exchange by the excess of sound arising from the greater tension. And the ratio of the greater thickness to the lesser is always the same as that of the greater tension to the lesser.

Thus let there be two notes, A and B, of equal pitch in equal lengths, and let the thickness of A—and therefore its tension too, of course—be greater than that of B. Let another, C, be taken in an equal length, having a thickness equal to B and a tension equal to A [figure 8.7.2]. Since then C differs from B only in tension, then as is the tension of C to the tension of B, so will be the sound of C to the sound of B. Again, since C differs from A only in thickness, then as is the thickness of A to the thickness of C, so will be the sound of C to the sound of A, while the sound of C has the same ratio to each [of the others], both to the sound of A and to that of B, for the sounds of A and B are equal. Therefore as is the tension of C to that of B, so is the thickness of A to that of C: and as is the tension of C to that of B, so is the tension of A to that of B, for the tensions of A and C are equal: and as is the thickness of A to that of C, so is the thickness of A to that of B, for the thicknesses of B and C are equal. Hence as is the tension of A to the tension of B, so is the thickness of A to the thickness of B. This would be true of them even if they were in all respects unvarying and indistinguishable from a single string.

But again, if in strings that are like that, AB and CD, we make the lengths unequal by diminishing the latter to CE, then as is the distance AB to the distance CE, so will be the sound of CE to the sound of AB [figure 8.7.3]. For since as the length CD is to the length CE, so is the sound of CE to the sound of CD, and since the length AB is equal to CD and the sound of AB is equal to the sound of CD, then as is the length AB to the length CE, so is the sound of CE to the sound of AB.

We skip many of Ptolemy's criticisms of the Pythagoreans and the Aristoxenians and consider his own mathematical approach to the division of the tetrachord, the musical fourth. He assumes that the octave is always divided into a tetrachord, a whole tone, and then another tetrachord. So to fill in the remaining notes in the octave, it is the tetrachord that must be divided into three parts. He shows that there are different genera of musical scales and, for each of these, several different ways of dividing the

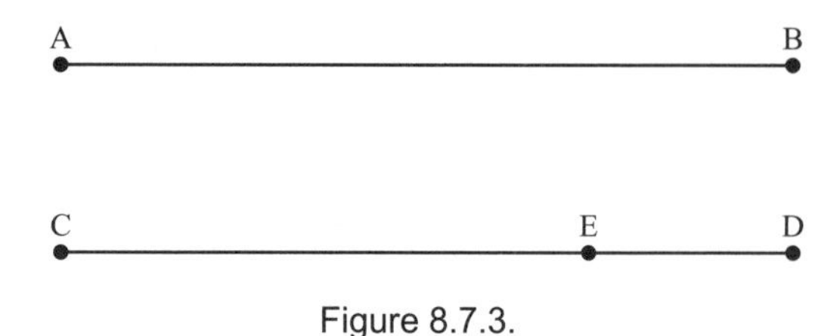

Figure 8.7.3.

tetrachord, each of which can be used to tune stringed instruments. Evidently, many of these methods were in fact used in Greek music. He begins with the methods due to Archytas, who was probably the first one to apply mathematics to musical harmonies.

13. *Concerning the division of the genera and the tetrachords according to Archytas*

Archytas of Tarentum, of all the Pythagoreans the most dedicated to the study of music, tried to preserve what follows the principles of reason not only in the concords but also in the divisions of the tetrachord, believing that a commensurable relation between the differences is a characteristic of the nature of melodic intervals. But though he sets off from this presupposition, at several points he seems to fall hopelessly short of it; and though in most cases he is well in control of this sort of thing, he is patently out of tune with what has already been straightforwardly accepted by the senses, as will be seen at once from the division of the tetrachords that he proposes. He posits three genera, the enharmonic, the chromatic, and the diatonic, and makes his division of each of them in the following way. He makes the "following" ratio the same in all three genera, 28:27; the middle one in the enharmonic 36:35 and in the diatonic 8:7, so that the "leading" interval in the enharmonic turns out to be 5:4, in the diatonic 9:8.

In the chromatic genus he locates the note second from the highest by reference to that which has the same position in the diatonic. For he says that the second note from the highest in the chromatic stands to the equivalent note in diatonic in the ratio of 256 to 243. Such tetrachords, on the basis of the ratios set out, are constituted in their lowest terms by the following numbers. If we postulate that the highest note of each tetrachord is 1512, and the lowest, in epitritic ratio with this, is 2016, this latter term will make the ratio 28:27 with 1944, and that will be the quantity of the second note from the lowest in all three genera. As to the second note from the highest, that in the enharmonic genus will be 1890, since that makes with 1944 the ratio 36:35, and with 1512 the ratio 5:4. The equivalent note in the diatonic genus will be 1701, since that makes with 1944 the ratio 8:7, and with 1512 the ratio 9:8. The equivalent note in the chromatic genus will be 1792, since that has a ratio to 1701 as is 256 to 243. The table of these numbers is set out below [figure 8.7.4]:

14. *A demonstration that [the division by Archytas does not] preserve what is truly melodic*

Now the chromatic tetrachord, as we said, was put together by [Archytas] in a way contrary to his own premises (for the number 1792 makes an epimoric ratio neither with 1522 nor with 1944); while both the chromatic and the enharmonic were put together in a way contrary to the plain evidence of the senses. For we grasp the "following" [i.e., the lowest] ratio of the familiar chromatic as greater than 28:27, while the "following" ratio in

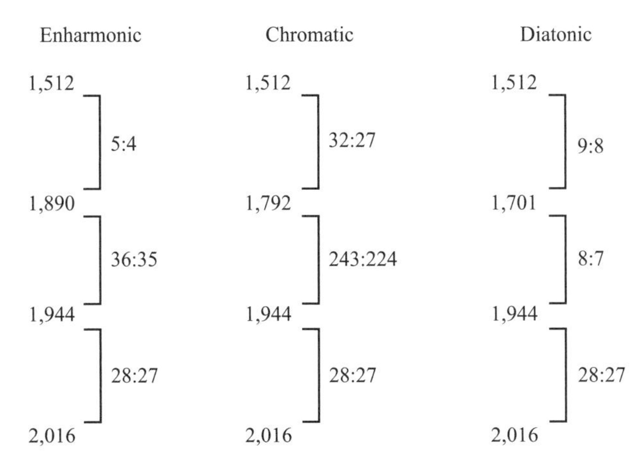

Figure 8.7.4.

enharmonic, once again, which appears much smaller than its equivalents in the other genera, he supposes to be equal to them; and further, he makes the middle ratio smaller than it, setting it in the ratio 36:35, though wherever it occurs such a thing is always unmelodic, that is, where the magnitude next to the lowest note is made greater than the middle one. . . .

15. *Concerning the division of the tetrachords by genus, according to what is rational and evident to perception*

. . . To find the positions and orders of the quantities, we adopt as our primary postulate and rational principle the thesis that all the genera have the following feature in common: that in the tetrachords too, the successive notes always make those epimoric ratios in relation to one another which amount to divisions into two or three that are nearly equal. . . . Secondly, on the basis of agreed perception, we adopt similarly, as common to all the genera, the thesis that the "following" magnitudes of the three are smaller than each of the remaining ones; as peculiar to the genera that have *pykna* the thesis that the two magnitudes next to the lowest note are together less than the one next to the highest note; and as peculiar to the *apykna* the thesis that none of the magnitudes is greater than the remaining two together.[13]

With these principles laid down, then, we first divide the epitritic ratio of the concord of the fourth, as many times as is possible, into two epimoric ratios: such a thing, once again, occurs only three times, when we adopt in addition the three epimoric ratios in succession below it, the ratios 5:4, 6:5, and 7:6. For the ratio 16:15 added to the ratio 5:4 fills out the epitritic, as does the ratio 10:9 added to the ratio 6:5, and the ratio 8:7 with the ratio 7:6; and after these we cannot find the ratio 4:3 put together from just two other epimorics.

Now in the genera that contain the *pyknon*, since in them the "leading" ratios are greater than the remaining ones together, they fitted the greater ratios in the pairings

[13] The terms *pykna* and *apykna* simply refer to the fact that in the first, the product of the two lower ratios is less than the first ratio, while in the second, this is not the case.

set out—that is, the ratios 5:4, 6:5, and 7:6—to their leading ratios, and the remaining and smaller ones—that is, the ratios 16:15, 10:9, and 8:7—to the remaining ones taken together. The division of each of these in respect of the two "following" ratios is achieved when it is divided into three sections (because by these means the three ratios of the tetrachord are at once produced), the differences being kept equal, and the ratios nearly equal, since it is not possible for them to be equal. For if we take the first numbers making the ratio 15:16, I mean 15 and 16, and triple them, we shall get 45 and 48, and their mean numbers in equal excesses are 46 and 47. Now since 47 does not make an epimoric ratio with both the extremes, but only 46 does so, making with 48 the ratio 24:23, and with 45 the ratio 46:45, the greater, the ratio 24:23, because of our initial postulates, will be conjoined with the ratio 5:4 , and the remaining ratio, 46:45, will fill up the "following" ratio. Again, if we take the first numbers that make the ratio 10:9, that is, 9 and 10, and triple them, we shall get 27 and 30, and their means in equal excesses are 28 and 29. But 29 does not make an epimoric ratio with both the extremes, whereas 28 makes with 30 the ratio 15:14, and with 27 the ratio 28:27, so that here too the ratio 15:14 is conjoined with the ratio 6:5, and the ratio 28:27 is left in the "following" position. Similarly, if we take the first numbers that make the ratio 8:7, which are 7 and 8, and triple them, we shall get 21 and 24, and their means in equal excesses are 22 and 23. Since the latter does not make an epimoric ratio with both the extremes, but only 22 does so, making with 24 the ratio 12:11, and with 21 the ratio 22:21, here too the ratio 12:11 will be conjoined with the ratio 7:6, and the ratio 22:21 will possess the "following" position.

Now since of all the genera the enharmonic is softest, there is as it were a road from it towards the more tense, by a process of increase through first the softer chromatic, then the tenser, towards the succeeding genera that are *apykna* and diatonic. In general those appear softer that have the larger leading ratio, and those appear tenser that have the smaller one. We have thus attached the tetrachord put together from the ratios 5:4, 24:23, and 46:45 to the enharmonic genus; that put together from the ratios 6:5, 15:14, and 28:27 to the softer of the chromatics; and that put together from the ratios 7:6, 12:11, and 22:21 to the more tense of the chromatics. The first numbers that contain these three tetrachords are these: common to all three genera, those of their extremes, 106,260 and 141,680; peculiar to each genus individually, those the ones second from the leaders [the highest], 132,825 and 127,512 and 121,970; and those of the ones that come third, 138,600 and 136,620 and 135,240.

As to the *apykna* genera, it follows from our previous definitions that the smaller ratios of those arising from the first division of the epitritic ratios, the division into two, should here, by contrast, be placed in the leading positions, and that the greater ratios coupled with them should be divided in the same way into the two "following" ratios. . . . When the ratio 8:7 is arranged in the leading position, the first numbers bounding the remaining ratio, 7:6, which are 6 and 7, when tripled in the same way will make 18 and 21, whose means, taken in equal excesses, are 19 and 20. Then 19, once again, will not make an epimoric ratio with both the extremes, but 20 will make with 18 the ratio 10:9, and with 21 the ratio 21:20, of which the greater, 10:9, will be conjoined as before with the ratio 8:7, while the lesser, 21:20, will fill out the "following" ratio. In the same way, when the ratio 10:9 is arranged in the leading position, if the numbers bounding the remaining ratio, 6:5, which are 5 and 6, are tripled, they will make 15 and 18, whose means falling in equal excesses are 16 and

17. Now 17 will not make an epimoric ratio with both the extremes, but 16 will make with 18 the ratio 9:8, and with 15 the ratio 16:15, so that the greater, 9:8, is conjoined with the ratio 10:9, while the remaining one, the ratio 16:15, is attuned to the "following" position.

But prior to all these ratios, the ratio 9:8 was found in its own right to contain the tone arising from the difference between the first two concords; and this, according to what is both rational and necessary, ought also to occupy the leading position, those closest to it being conjoined with it, since none of the epimoric ratios fills out with it the epitritic ratio. The ratio 10:9 has already been conjoined with it in the division set out above, but the ratio 8:7 has not yet. Hence we shall conjoin this with it, in the middle position, and allocate the remainder making up the epitritic ratio, which is the ratio 28:27, to the "following" position.

Here, once again, in correspondence with the magnitude of the leading ratios, we shall attach the tetrachord put together from the ratios 8:7, 10:9, and 21:20 to the soft diatonic, that put together from the ratios 10:9, 9:8, and 16:15 to the tense diatonic, and that put together from the ratios 9:8, 8:7, and 28:27 to the one lying somehow between the soft and the tense, which is called "tonic" reasonably enough, because that is the size of its leading position. The first numbers that contain these three tetrachords are these: common to all three genera, those of the extremes, 504 and 672; peculiar to each genus individually, those of the ones second from the leader, 576 and 567 and 562; and those of the ones that come third, 640 and 648 and 630.

The fact that the divisions of the genera set out above do not contain only what is rational but also what is concordant with the senses can be grasped, once again, from the eight-stringed *kanōn* that spans an octave, once the notes are made accurate, as we have said, in respect of the evenness of the strings and their equality of pitch.

16. *The genera more familiar to the hearing—how many there are and which ones they are*

Now of the genera that have been set out, we would find all the diatonic ones familiar to our ears, but not to the same extent the enharmonic, nor the soft one of the chromatics, because people do not altogether enjoy those of the characters that are exceedingly slackened, but it is enough for them in the movement towards the soft to stop at the tense chromatic. . . .

[There is] another genus, when we set out from the melodic-ness that is constituted in accordance with equalities, and investigate the question whether there is any appropriate ordering of the tetrachord when it is initally divided into the three nearly equal ratios, again in equal excesses. The ratios comprising this sort of genus are 10:9, 11:10, and 12:11. . . . When a division is taken in strings of equal pitch on the basis of these numbers, the character that becomes apparent is rather foreign and rustic, but exceptionally gentle, and the more so as our hearing becomes trained to it, so that it would not be proper to overlook it. . . . So let us call this genus the "even diatonic," from the characteristic it has.

[Certain musicians] actually tune to another genus, close to [the tonic diatonic], but plainly different; for they make the two leading intervals tones [9:8] and the remainder, as they think, a half-tone, but as reason implies, what is called the *leimma* [256:243]. This works for them well enough, since there is no noticeable difference either in the leading positions between the ratios 9:8 and 10:9, nor in the "following" positions between the ratio 16:15 and the *leimma*. . . . Let us then accept this genus too, both because of the ease of the modulations to it from the tonic diatonic genus, in the case of its mixture with this one, and also because the ratio of the *leimma* has a certain affinity with the fourth and the tone, marking it out from the other ratios that are not epimoric, since it follows inevitably when two epogdoics have been inserted into the epitritic. For the *leimma*, too, can be constructed by itself by means of concords, just as can the tone, the latter from the difference between the first two concords, the former from the difference between the ditone and the concord of a fourth. The first numbers that make this genus are 192, 216, 243, and 256. It may reasonably be called "ditonic," since it has tones as its two leading ratios.

In book 2, Ptolemy constructs the entire octave based on each of the various genera he discussed, crediting each possibility to either an earlier mathematician or to himself. For example, the diatonic according to Archytas gives the values for each note of the scale as $60, 67\frac{1}{2}, 77\frac{1}{7}, 80, 90, 101\frac{1}{4}, 115\frac{5}{7}, 120$, where in each of the fourths, the division is by the ratios 9:8, 8:7, and 28:27. Similarly, the ditonic diatonic according to Ptolemy has the notes $60, 67\frac{1}{2}, 75\frac{15}{16}, 80, 90, 101\frac{1}{4}, 113\frac{29}{32}, 120$, where here in each of the fourths, the division is by the ratios 9:8, 9:8, and 256:243. Our modern "even-tempered" scale has divisions very close to this latter scale. Note that in the modern scale, the ratios of each of the twelve notes (including the "black keys") to the previous note is $\sqrt[12]{2}$.

We skip Ptolemy's book 3, in which he attempts, without giving much detail, to show the relationship of the mathematics of music to the mathematics of the heavens.

8.8 Boethius, *Fundamentals of Music*

In medieval Europe, music, or harmonics, was one of the four core subjects of the quadrivium, along with arithmetic, geometry, and astronomy. Thus, it was a subject of considerable importance to Boethius. His work *De Institutione Musica* was dedicated to this topic and was part of his broader project to translate and circulate key works from Greek antiquity into Latin. Boethius drew inspiration from ancient accounts of the discipline, reformulating early ideas on consonance and scales. His work proved to be highly influential and circulated widely throughout the medieval period. The extant manuscripts of this work include intricate diagrams and figures alongside, which are visually stunning [figures 8.8.1, 8.8.2].

Book 1

3. *Concerning pitches and concerning the basic principles of music*

Consonance, which governs all setting out of pitches, cannot be made without sound; sound is not produced without some pulsation and percussion; and pulsation and percussion cannot exist by any means unless motion precedes them. If all things were immobile, one thing could not run into another, so one thing should not be moved by another; but if all things remained still and motion was absent, it would be a necessary consequence that no sound would be made. For this reason, sound is defined as a percussion of air remaining undissolved all the way to the hearing.

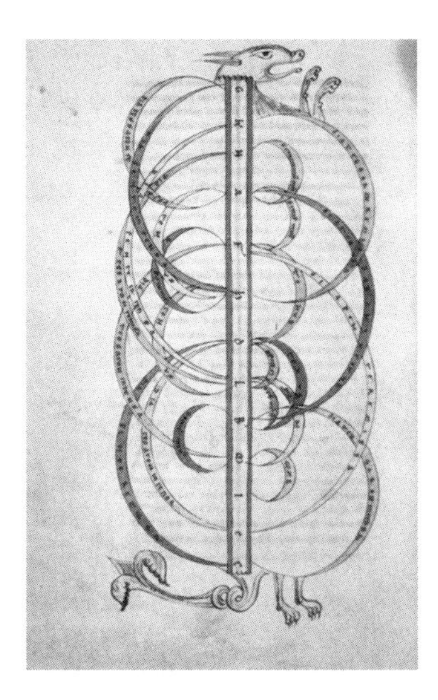

Figure 8.8.1. Illustration from a medieval manuscript of Boethius's *Fundamentals of Music*. Ref: MSR-05-f043v. Alexander Turnbull Library, Wellington, New Zealand. /records/23194763.

Figure 8.8.2. Illustration from a medieval manuscript of Boethius's *Fundamentals of Music*. Fragment of MS Ii.3.12; f. 61v. Reproduced by kind permission of the Syndics of Cambridge University Library.

Some motions are faster, others slower; some motions are less frequent, others more frequent. If someone regards an uninterrupted motion, he will necessarily observe in it either speed or slowness; moreover, if someone moves his hand, he will move it in either a frequent or less frequent motion. If motion is slow and less frequent, low sounds are necessarily produced by the very slowness and infrequency of striking. But if motions are fast and more frequent, high sounds are necessarily produced. For this reason, if the same string is made tighter, it sounds high, if loosened, low. For when it is tighter, it produces faster pulsation, recurs more quickly, and strikes the air more frequently and densely. The string that is looser brings about lax and slow pulsations, and, being less frequent because of this very weakness of striking, does not vibrate very long.

Since high pitches are incited by more frequent and faster motions, whereas low ones are incited by slower and less frequent motions, it is evident that high pitch is intensified from low through some addition of motions, while low pitch is relaxed from high through lessening of motions. For high pitch consists of more motions than low. Plurality makes the difference in these matters, and plurality necessarily consists in a kind of numerical quantity. Every smaller quantity is considered in relation to a larger quantity as number compared to number. Of those things which are compared according to number, some are equal, while others stand at an interval from each other by virtue of an inequality. In those pitches which do not harmonize through any inequality, there is no consonance at all. For consonance is the concord of mutually dissimilar pitches brought together into one.

4. *Concerning the species of inequality*

Things which are unequal hold within themselves five criteria relating to degrees of inequality. One is surpassed by another either by a multiple or by a singular part or by several parts or by a multiple and a part or by a multiple and parts.

The first class of inequality is called "multiple." The multiple is such that the larger number contains the whole smaller number within itself twice, three times, or four times, and so forth, nothing is either lacking or superfluous. It is called either "duple," "triple," or "quadruple," and the multiple class proceeds into infinity according to this series.

The second class of inequality is that which is called "superparticular"; it is such that the larger number contains within itself the whole smaller number plus some single part of it: either a half, as three to two (and this is called the "sesquialter" ratio), or a third, as four to three (and this is called the "sesquitertian"). According to this manner in subsequent numbers, some single part in addition to the smaller numbers is contained by the larger numbers.

The third class of inequality is such that the larger number contains within itself the whole lesser number plus several of its parts besides. If it contains two parts more, it will be called the "superbipartient" ratio, as five is to three; whereas if it contains three parts more, it will be called "supertripartient," as seven is to four. The pattern can be the same in other numbers.

The fourth class of inequality, which is combined from the multiple and the superparticular, is such that the larger number has within itself the lesser number either twice, or three times, or some other number of times, plus one other part of it. If the larger contains the smaller twice plus a half part, the ratio will be called "duple-sesquialter," as five is to two; whereas if the lesser number is contained twice plus a third part of it, it will be called "duple-sesquitertian," as seven is to three. But if the lesser number is contained three times plus its half part, it will be called "triple-sesquialter," as seven is to two. In this same way, by the names of multiplicity and superparticularity are varied in other numbers.

The fifth class of inequality, which is called "multiple-superpartient," is such that the larger number has the whole lesser number within itself more than once, plus more than one single part of it. If the larger number contains the smaller number twice plus two parts of it besides, it will be called "duple-superbipartient" as three is to eight; or again a ratio of this class may be called "triple-superbipartient."

We explain these things cursorily and briefly now, since we elucidated them carefully in the books we wrote concerning the fundamentals of arithmetic.

5. *What species of inequality pertain to consonance*

Of these classes of inequality, the last two may be set aside, since they are a mixture of others; theorizing ought to be carried out within the first three. The multiple seems to hold the greater authority for consonances, whereas the superparticular seems to occupy the next place. The superpartient is excluded from consonance of harmony—as acknowledged by various theorists, with the exception of Ptolemy.

7. *Which ratios should be fitted to which musical consonances*

Meanwhile this should be known: all musical consonances consist of a duple, triple, quadruple, sesquialter, or sesquitertian ratio. Moreover, that which is sesquitertian in number will be called "diatessaron" in sounds; that which is sesquialter in number is called "diapente" in pitches; that which is duple in ratios, "diapason" among consonances; the triple, "diapason-plus-diapente," the quadruple, "bis-diapason." For the present, let the above be stated generally and without particulars; later, the complete theory of ratios will be brought to light.

10. *In what manner Pythagoras investigated the ratios of consonances*

This, then, was primarily the reason why Pythagoras, having abandoned the judgment of hearing, had turned to the weights of rules. He put no credence in human ears, which are subject to change, in part through nature, in part by external circumstance, and undergo changes caused by age. Nor did he devote himself to instruments, in conjunction with which much inconstancy and uncertainty often arise. When you wish to examine strings, for example, more humid air may deaden the pulsation, or drier air may excite it, or the thickness of a string may render a sound lower, or thinness may make it higher, or, by some other means, one alters a state of previous stability. Moreover, the same would be true of other instruments.

Assessing all the instruments as unreliable and granting them a minimum of trust, yet remaining curious for some time, Pythagoras was seeking a way to acquire through reason, unfalteringly and consistently, a full knowledge of the criteria for consonances. In the meantime, by a kind of divine will, while passing the workshop of blacksmiths, he overheard the beating of hammers somehow emit a single consonance from differing sounds. Thus in the presence of what he had long sought, he approached the activity spellbound. Reflecting for a time, he decided that the strength of the men hammering caused the diversity of sounds, and in order to prove this more clearly, he commanded them to exchange hammers among themselves. But the property of sounds did not rest in the muscles of the men; rather, it followed the exchanged hammers. When he had observed this, he examined the weight of the hammers. There happened to be five hammers, and those which sounded together the consonance of the diapason were found to be double in weight. Pythagoras determined further that the same one, the one that was the double of the second, was the sesquitertian of another, with which it sounded a diatessaron. Then he found that this same one, the duple of the above pair, formed the sesquialter ratio of still another, and that it joined with it in the consonance of the diapente. These two, to which the first double proved to be sesquitertian and sesquialter, were discovered in turn to hold the sesquioctave ratio between themselves. The fifth hammer, which was discordant with all, was discarded.

Although some musical consonances were called "diapason," some "diapente," and some "diatessaron" (which is the smallest consonance) before Pythagoras, Pythagoras was the first to ascertain through this means by what ratio the concord of sounds was joined together. So that what has been said might be clearer, for sake of illustration, let the weights of the four hammers be contained in the numbers written next: 12, 9, 8, 6.

Thus the hammers which bring together 12 with 6 pounds sounded the consonance of the diapason in duple ratio. The hammer of 12 pounds with that of 9 (and the hammer of 8 with that of 6) joined in the consonance of the diatessaron according to the epitritic ratio. The one of 9 pounds with that of 6 (as well as those of 12 and 8) commingled the consonance of the diapente. The one of 9 with that of 8 sounded the tone according to the sesquioctave ratio.

11. *By what differing means Pythagoras weighed the ratios of consonances*

Upon returning home, Pythagoras weighed carefully by means of different observations whether the complete theory of consonances might consist of these ratios. First, he attached corresponding weights to strings and discerned by ear their consonances; then, he applied the double and mean and fitted other ratios to lengths of pipes. He came to enjoy a most complete assurance through the various experiments. By way of measurement, he poured ladles of corresponding weights into glasses, and he struck these glasses—set in order according to various weights—with a rod of copper or iron, and he was glad to have found nothing at variance. Thus led, he turned to length and thickness of strings, that he might text further. And in this way he found the rule, about which we shall speak later. It came to be called by that name, not because the rule with which we measure the sizes of strings is wooden, but because this kind of rule is tantamount to such fixed and enduring inquiry that no researcher would be misled by dubious evidence.

15. *Concerning the sequence of subjects, that is, of speculations*

Having set forth these matters, it seems that we should discuss the number of genera within which all song is composed and which the discipline of harmonic theory contemplates. They are these: diatonic, chromatic, and enharmonic. These must be explained, but first we must discuss tetrachords and in what manner the augmented number of strings came into existence (to which more are added now). This will be done after we have recalled in what ratios musical consonances are combined.

16. *Concerning the consonances and the tone and the semitone*

The consonance of the diapason is that which is made in the duple ratio such as this: 1 to 2.

The diapente is that which consists of these numbers: 2 to 3.

The diatessaron is that which occurs in this ratio: 3 to 4.

The tone is comprised of the sesquioctave ratio, but there is no consonance in this: 8 to 9.

The diapason-plus-diapente is brought together through the triple ratio in this manner: 2 to 4 to 6.

The bis-diapason is brought about through the quadruple comparison: 2 to 4 to 8.

The diapente plus the diatessaron produce one diapason, in this manner: 2 to 3 to 4.

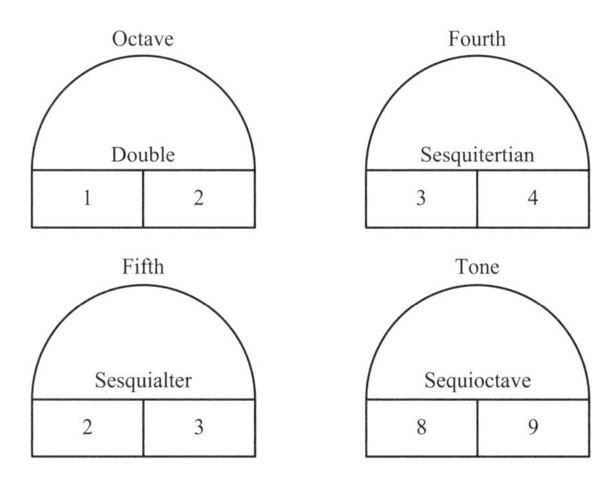

Figure 8.8.3. The basic harmonies and the ratios from which they are made.

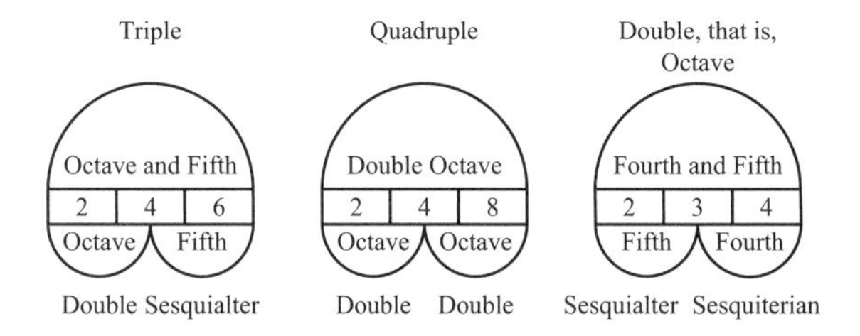

Figure 8.8.4. More harmonies and ratios.

If a pitch is higher or lower than another pitch by a duple, a consonance of the diapason will be made. If a pitch is higher or lower than another pitch by a sesquialter, sesquitertian, or sesquioctave ratio, then it will yield consonances of the diapente or of the diatessaron or a tone respectively. Likewise, if a diapason such as 4 to 2, and a diapente, such as 6 to 4, are joined together, they will make the triple consonance, which is the diapason-plus-diapente. If two diapasons are combined, such as 2 to 4 and 4 to 8, then the quadruple consonance will be made, which is the bis-diapason. If a sesquialter and a sesquitertian—that is, a diapente and a diatessaron, such as 2 to 3 and 3 to 4—are joined, the duple consonance, obviously the diapason, is formed; 4 to 3 brings about a sesquitertian ratio, while 3 to 2 is joined by a sesquialter relation, and the same 4 related to the 2 unites within a duple comparison. But the sesquialter ratio creates a consonance of the diapente, and the sesquitertian that of the diatessaron, whereas the duple ratio produces a consonance of the diapason. Therefore, a diatessaron plus a diapente forms a consonance of the diapason [figures 8.8.3, 8.8.4].

Moreover, a tone cannot be divided into equal parts; why, however, will be explained later. For now it is sufficient to know only that a tone is never divided into two equal parts. So that this might be very easily demonstrated, let 8 and 9 represent the sesquioctave ratio. No mediating number falls naturally between these, so let us multiply them by 2: twice 8 makes 16; twice 9, 18. However, a number naturally falls between 16 and 18—namely, 17. Let these numbers be set out in order: 16, 17, 18. Now when 16 and 18 are

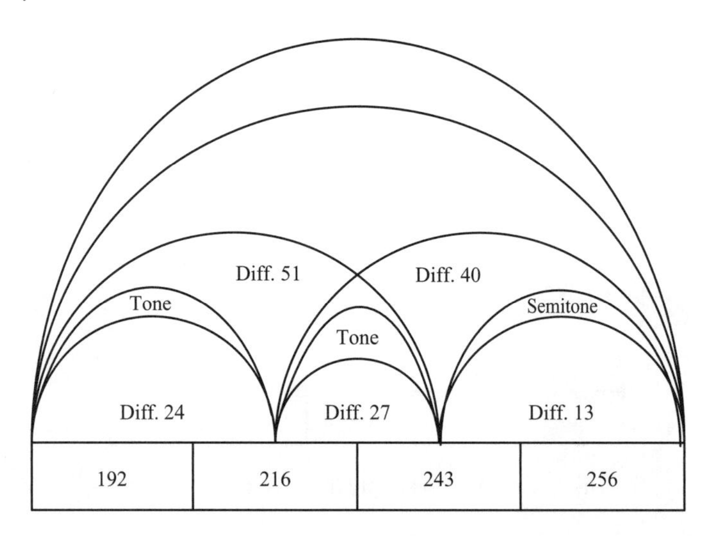

Figure 8.8.5. Diagram of the diatessaron.

compared, they yield the sesquioctave ratio and thus a tone. But the middle number, 17, does not divide this ratio equally. For compared to 16, 17 contains in itself the total 16, plus 1/16 part of it—that is, unity. Now if the third number—that is, 18—is compared to this number—that is, to 17—it contains the total 17 and 1/17 part of it. Therefore 17 does not surpass the smaller number and is not surpassed by the larger number by the same parts; the smaller part is 1/17, the larger 1/16. Both of these are called "semitones," not because these intermediate semitones are equal at all, but because something that does not come to a whole is usually called "semi." In the case of these, one semitone is called "major," and the other "minor."

17. *In what smallest integers the semitone is ascertained*

At this time I shall explain more clearly what an integral semitone is, or in what smallest integers it is ascertained. For what has been said about the division of the tone has nothing to do with our wanting to show the measurements of semitones; it applies rather to the fact that we said a tone cannot be divided into twin equal parts.

The diatessaron, which is a consonance of four pitches and of three intervals, consists of two tones and an integral semitone. A diagram of this is set out below [figure 8.8.5].

If the number 192 is compared to 256, a sesquitertian ratio will result, and it will sound a consonance of the diatessaron. But if 216 is compared with 192, the ratio is a sesquioctave, for the difference between them is 24, which is an eighth part of 192. Therefore it is a tone. Moreover, if 243 is compared to 216, the ratio will be another sesquioctave, for the difference between them is 27, which is an eighth part of 216. Therefore it is a tone. The comparison of 256 with 243 remains: the difference between these is 13, which, multiplied by 8, does not seem to arrive at a mean of 243. It is therefore not a semitone, but less than a semitone. For it might be reckoned to be a whole semitone, rigorously speaking, if the difference of these numbers, which is 13, multiplied by 8, could have equaled a mean of the number 243. Thus the comparison of 243 with 256 yields less than a true semitone.

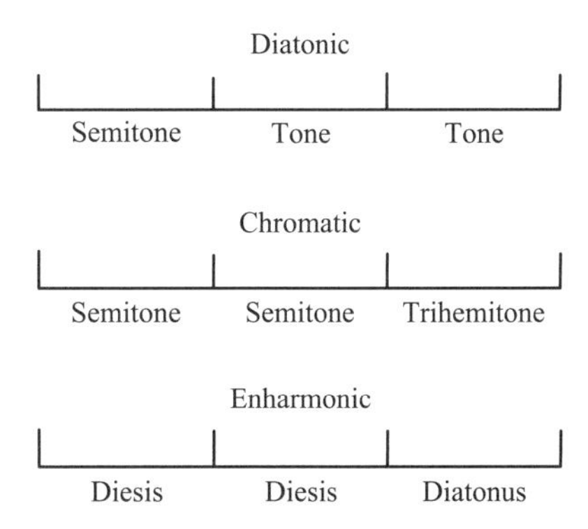

Figure 8.8.6. Diagram of the three genera and their tetrachords.

21. *Concerning the genera of song*

Now that these things have been explained, the genera of melodies should be discussed. There are three genera: the diatonic, the chromatic, and the enharmonic. The diatonic is somewhat more austere and natural, while the chromatic departs from that natural intonation and becomes softer; the enharmonic is very rightly and closely joined together.

There are five tetrachords: the *hypatōn*, *mesōn*, *synemmenōn*, *diezeugmenōn*, and *hyperboleōn*. In all these, according to the diatonic genus of song, the pitch progresses through semitone, tone, and tone in one tetrachord, again through semitone, tone and tone in the second, and thus in succession. For this reason it is called "diatonic," because it progresses, as it were, through tone and through tone.[14]

The chromatic, on the other hand, is named from "color," inasmuch as it is the first mutation from the above intonation. It is sung through semitone, semitone, and three semitones. For the whole consonance of the diatessaron contains two tones and a semitone (although not a full semitone). This word chroma, as has been said, is derived from surfaces which are transformed into another color when they are turned.

The enharmonic genus is that which is most closely joined together, for it is sung in all tetrachords through diesis and diesis and ditone (a diesis is half a semitone).

A diagram of the three genera progressing through all tetrachords is as follows [figure 8.8.6]:

Book 2

28. *Concerning the semitone: in what smallest numbers it is found*

It seems that semitones were so named not because they are truly halves of tones, but because they are not complete tones. The size of the interval which we now call "semitone," but which was called *leimma* or *diesis* by the ancients, is determined as follows: when two

[14] For example, the *hypatōn* tetrachord is, in modern musical notation, the fourth from *B* to *E*, which consists of the semitone from *B* to *C* followed by the two tones, *C* to *D* and *D* to *E*.

sesquioctave relations (which are tones) are subtracted from a sesquitertian ratio (which is a diatessaron), an interval called a "semitone" remains. Let us try to write the two tones in a continuous arrangement. As was said, since these consist of the sesquioctave ratio, and we cannot join two continuous sesquioctave ratios unless a multiple from which these can be derived is found, let unity be set out and also its first octuple, 8. From this I can derive one sesquioctave. But since I am seeking two, 8 should be multiplied by 8, making 64. This will be the second octuple, from which we can extract two sesquioctave ratios, for 8 (an eighth part of 64 unities), when added to the same, makes the total sum of 72. Similarly, let an eighth part of this be taken, which is 9, and it gives 81. And these two first continuous tones are written down in this arrangement: 64:72:81.

Now we should seek out sesquitertian of 64 unities. But since 64 proves not to have a third part, then, if all these numbers are multiplied by 3, the third part forthwith comes into being, and all remain in the same ratio as they were before the multiplier 3 was applied to them. Thus, let 64 be multiplied by 3, which makes 192. A third of this (64) added to it produces 256. Then 256:192 is the sesquitertian ratio, which holds the consonance of the diatessaron. Now let us assemble in appropriate series the two sesquioctave ratios to 192, ratios that will be contained in two numbers: let 72 be multiplied by 3, making 216, and 81 by 3, making 243. Let these be arranged between the two terms cited above in this manner: 192:216:243:256. In this arrangement of ratios, the ratio of the first number to the last constitutes the consonance of the diatessaron, and those of the first to the second and the second to the third contain two identical tones. The interval that remains consists of the ratio 243:256, which constitutes in smallest integers the form of the semitone.

29. *Demonstrations that 243:256 is not half a tone*

I am showing, then, that the interval of 243:256 is not the full magnitude of half a tone. The difference between 243 and 256 is contained in only 13 unities, which cover less than an eighteenth part of the smaler term, but more than a nineteenth part (for if you multiply 13 by 18, you will make 234, which by no means is equal to 243, and if you multiply 13 by 19, it surpasses 243). Every semitone, if it holds a full half of a tone, ought to be placed between the sixteenth part and the seventeenth—which will be demonstrated later.[15]

Now it will become clear that such an interval of a semitone, if doubled, cannot complete one interval of a tone. So let us, without further delay, arrange two such ratios continous with each other that contain the same relation as 256:243, according to the rule presented above. So let us multiply 256 by itself, and put the result as the largest term: 65,536; likewise 243 is increased by its own quantity, and the result is the smallest term: 59,049; again 256 is increased by the number 243, and this gives the number 62,208. Let the mean term be set down in this manner: 65,536 : 62,208 : 69,049. Therefore 256 and 243 are in the same ratio as 65,536 and 62,208, as well as 62,208 and 59,049.

But the largest term of these to the smallest does not produce one whole tone. But if the ratio of the first to the second, which is equal to the ratio of the second to the third, should

[15] In modern terms, what Boethius is conveying is that the true semitone ratio is that whose square is $\frac{9}{8}$. But $(\frac{17}{16})^2 > \frac{9}{8}$, while $(\frac{18}{17})^2 < \frac{9}{8}$. That is, the full half of a tone is between $1\frac{1}{17}$ and $1\frac{1}{16}$.

prove to be whole semitones, the two halves joined together would necessarily produce one tone. Since the ratio of the extreme terms is not sesquioctave, it is clear that these two intervals do not represent true halves of tones, for whatever is half of something, if it is doubled, makes that of which it is said to be the half. If it cannot fill that, then the part that is doubled is less than a half part; whereas if it exceeds it, it is more than a half part. Furthermore, it will be proved that 65,536 does not make a sesquioctave ratio with 59,049 unities if an eighth part of 59,049 is taken according to the rules that were given in the arithmetic books. Since this eighth part does not consist of a whole number, we leave the computing of the eighth part to the diligence of readers. It is thus evident that the ratio consisting of 256:243 is not a whole half of a tone. This which is truly called "semitone" is, then, less than a half part of the tone.

Book 3

11. *Archytas's demonstration that a superparticular ratio cannot be divided into equal parts.*

A superparticular ratio cannot be split exactly in half by a number proportionally interposed. This will be demonstrated conclusively later.[16] The demonstration that Archytas suggests is very weak. It is of this nature.

Let $A:B$ be a superparticular ratio, he says. I take the smallest integers in that same ratio, $C:DE$. Since $C:DE$ is the same ratio and the ratio is superparticular, the number DE exceeds the number C by one of its—that is, DE's—own parts. Let this part be D. I say then that D will not be a number, but unity. For if D is a number and is part of DE, the number D measures the number DE, and thus it will also measure the number E. Whence it follows that it should also measure C. Thus the number D would measure both the numbers C and DE, which is impossible. For these are the smallest integers in the same ratio as some other numbers, the first numbers so related, and they maintain the difference of unity alone. Therefore D is unity. So the number DE exceeds the number C by unity. For that reason no mean number comes between them that splits the ratio equally. It follows that between those greater numbers that maintain the same ratio as these, a mean number cannot be interposed that splits the same ratio equally.

SOURCES, CHAPTER 8

8.1 These translations are taken from Andrew Barker, ed., *Greek Musical Writings*: vol. 2, *Harmonic and Acoustic Theory* (Cambridge: Cambridge University Press, 1989), 53–97. The original translations are from the Oxford Classical Texts and from Aristotle, *Problems*, Vol. 1, Books I–XXI, ed. and trans. W. S. Hett, (Cambridge, MA: Harvard University Press, 1961).

[16] Boethius's later proof of this is essentially the same as Archytas's proof. Of course, this result is a special case of *Elements* 8:8.

8.2 The two translations are taken from Carl A. Huffman, *Archytas of Tarentum: Pythagorean, Philosopher and Mathematician King* (Cambridge: Cambridge University Press, 2005), 105–107, 163.

8.3 This translation is taken from Andrew Barker, ed., *Greek Musical Writings: Harmonic and Acoustic Theory* (Cambridge: Cambridge University Press, 1989), 149–50.

8.4 This translation is taken from Andrew Barker, ed., *Greek Musical Writings: Harmonic and Acoustic Theory* (Cambridge: Cambridge University Press, 1989), 191–208.

8.5 This translation is taken from Flora R. Levin, ed. and trans., *The Manual of Harmonics of Nicomachus the Pythagorean* (Grand Rapids, MI: Phanes, 1994).

8.6 This translation is taken from Andrew Barker, ed., *Greek Musical Writings: Harmonic and Acoustic Theory* (Cambridge: Cambridge University Press, 1989), 217–19.

8.7 This translation is taken from Andrew Barker, ed., *Greek Musical Writings: Harmonic and Acoustic Theory* (Cambridge: Cambridge University Press, 1989).

8.8 This translation is taken from Calvin M. Bower, *Fundamentals of Music: Anicius Manlius Severinus Boethius* (New Haven, CT: Yale University Press, 1989).

9

Optics

Optics is not one of the subjects of the classical quadrivium. However, it is noted as one of the parts of mathematics by Geminus, as reported by Proclus in section 1.11. So although we will not include any material on mechanics and geodesy, two other subjects mentioned by Geminus, we include here a brief survey of optics to conclude the *Sourcebook*.

The beginning of optics as a mathematical discipline probably dates back to at least the fourth century BCE, when it is thought that Archytas was one of its originators. But visual perception was a topic of interest among Greek thinkers even earlier. Its very name, derived from the ancient Greek word for "eye" (*opsis*), suggests that the primary orientation for the discipline in ancient times was the eye and the ray emanating from it to produce a line of sight. This ray concept and the geometry arising from it entailed the use of mathematical frameworks to study and explore visual phenomena.

Mathematical optics was advanced considerably over the following centuries. Four works are considered fundamental for this study: Euclid's *Optics*, a work called *Catoptrics* insecurely attributed to Euclid, Heron of Alexandria's *Catoptrics*, and Ptolemy's *Optics*. Among the important results in these works are those on reflection and refraction, some of which are included in the excerpts below.

Parallel to this theoretical development of the discipline, more perceptually based accounts of optics were developed. Issues of visual perception and the physiological aspects of sight were also significant themes in natural philosophy and were explored by authors such as Aristotle, Theophrastus, Diocles, Lucretius, and Galen, to name a few, but we cannot consider their works here.

9.1 Euclid, *Optics*

Euclid's *Optics* survives in two editions. One is the edited version by Theon that was first translated into Latin in 1505. The other, probably the original edition, was only discovered in the late nineteenth century by Heiberg in two manuscripts. This translation is from the original edition.

The style of the *Optics* is very much like that of the *Elements* in that Euclid opens his work with a collection of postulates, in which a number of important fundamentals are set out. Among them, he establishes the notion of the visual cone (postulate 2), a collection of visual rays or straight lines that emanate out from the eye. He also postulates a mathematical explanation for visual clarity in postulate 7. These postulates are then used in the fifty-eight propositions that follow, which, like in a mathematical text, are deliberately arranged in a deductive sequence. These propositions cover many different scenarios arising from how an object might appear as related to its spatial orientation with respect to the point of observation. Among these, Euclid includes the geometric analysis of objects at a distance, elevation with or without the sun casting shadows, depth and length, viewing arcs, spheres, chariot wheels, and moving objects.

Postulates

Let it be assumed

1. That rectilinear rays proceeding from the eye diverge indefinitely;
2. That the figure contained by a set of visual rays is a cone of which the vertex is at the eye and the base at the surface of the objects seen;
3. That those things are seen upon which visual rays fall and those things are not seen upon which visual rays do not fall;
4. That things seen under a larger angle appear larger, those under a smaller angle appear smaller, and those under equal angles appear equal;
5. That things seen by higher visual rays appear higher, and things seen by lower visual rays appear lower;
6. That, similarly, things seen by rays further to the right appear further to the right, and things seen by rays further to the left appear further to the left;
7. That things seen under more angles are seen more clearly.

Propositions

Proposition 1. *Nothing that is seen is seen at once in its entirety.*

For let the thing seen be AD, and let the eye be B, from which let the rays of vision fall, BA, BG, BK, and BD [figure 9.1.1]. So, since the rays of vision, as they fall, diverge from one another, they could not fall in continuous line upon AD; so that there would be spaces also in AD upon which the rays of vision would not fall. So AD will not be seen in its entirety at the same time. But it seems to be seen all at once because the rays of vision shift rapidly.

Proposition 2. *Objects located nearby are seen more clearly than objects of equal size located at a distance.*

Let B represent the eye and let GD and KL represent the objects seen; and we must understand that they are equal and parallel, and let GD be nearer to the eye; and let the rays of vision fall, BG, BD, BK, and BL [figure 9.1.2]. For we could not say that the rays falling from the eye upon KL will pass through the points G and D. For in the triangle

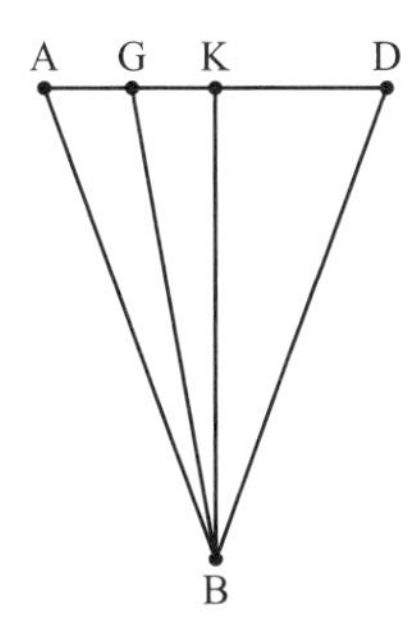

Figure 9.1.1.

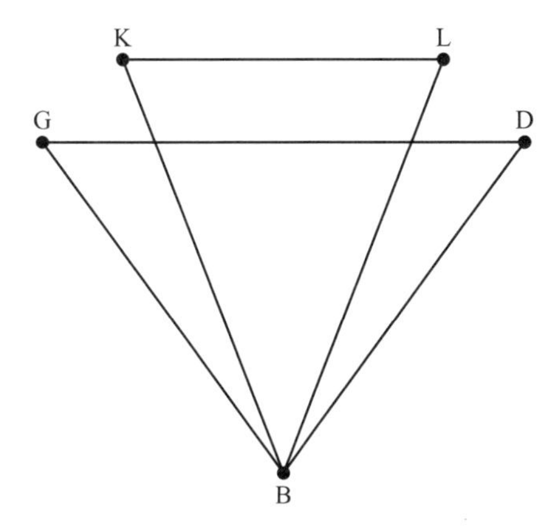

Figure 9.1.2.

$BDLKGB$ the line KL would be longer than the line GD; but they are supposed to be of equal length. So GD is seen by more rays of the eye than KL. So GD will appear more clear than KL; for objects seen within more angles appear more clear.

Proposition 3. *Every object seen has a certain limit of distance, and when this is reached it is seen no longer.*

For let the eye be B, and let the object seen be GD [figure 9.1.3]. I say that GD, placed at a certain distance, will be seen no longer. For let GD lie midway in the divergence of the rays, at the limit of which is K. So, none of the rays from B will fall upon K. And the thing upon which rays do not fall is not seen. Therefore, every object seen has a certain limit of distance, and, when this is reached, the object is seen no longer.

Proposition 6. *The intervals between parallels appear unequal when seen from an [increasing] distance.*

Let AB and GD be two parallel magnitudes, and let E be the eye [figure 9.1.4]. I say that AB and GD do not appear [to maintain] the same distance apart, and that a nearer interval between them will always appear larger than a farther one. Let visual rays EB,

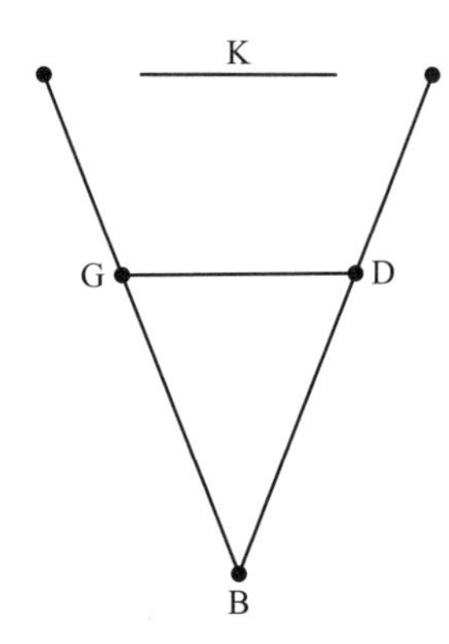

Figure 9.1.3.

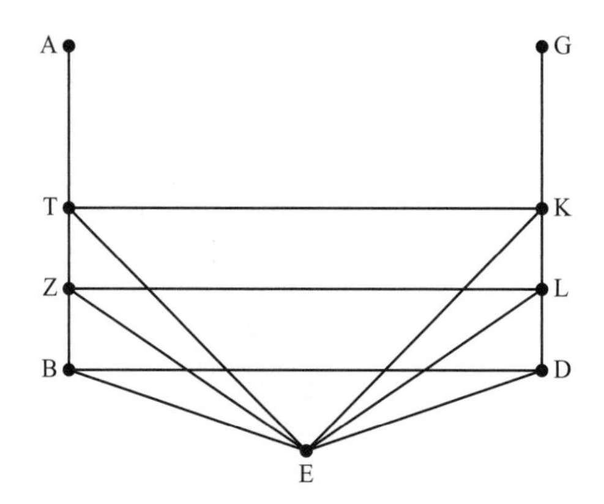

Figure 9.1.4.

EZ, ET, EK, EL, and ED be dropped [to the respective parallels], and let BD, ZL, and TK be joined. Thus, since angle BED is greater than angle ZEL, BD appears longer than ZL. Moreover, since angle ZEL is greater than angle TEK, ZL appears longer than TK. Therefore, BD appears longer than ZL, and ZL appears longer than TK. Consequently, parallels will not seem to lie the same distance apart [throughout] but will seem unequally separated.

Proposition 7. *Objects of equal size upon the same straight line, if not placed next to each other and if unequally distant from the eye, appear unequal.*

Let there be two objects of equal size, AB and GD, upon the same straight line, but not next to each other, and unequally distant from the eye, E, and let the rays EA and ED fall upon them and let EA be longer than ED [figure 9.1.5]. I say that GD will appear larger than AB. Let the rays EB and EG fall upon them, and let the circle, AED, be circumscribed about the triangle, AED. And let the straight lines BZ and GL be added as a continuation of the straight lines EB and EG, and from the points B and G let the equal straight lines BT and GK be drawn at right angles. And AB is equal to GD, but also the angle ABT is

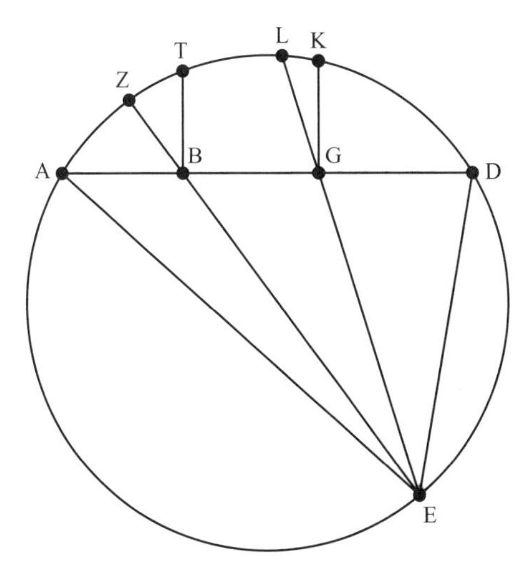

Figure 9.1.5.

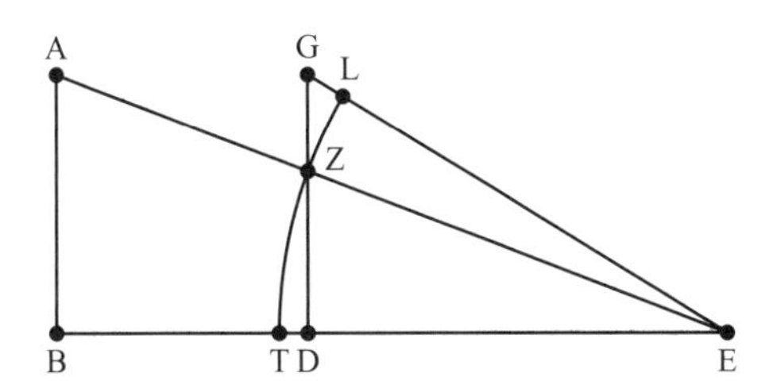

Figure 9.1.6.

equal to the angle DGK. And so the arc AT is equal to the arc DK. Thus, the arc KD is greater than the arc ZA. So the arc LD is much greater than the arc ZA. But upon the arc ZA rests the angle AEZ, and upon the arc LD rests the angle LED. So the angle LED is greater than the angle AEZ. But within the angle AEZ, AB is seen, and within the angle LED, GD is seen. Therefore, GD appears larger than AB.

Proposition 8. *Lines of equal length and parallel, if placed at unequal distances from the eye, are not seen in proportion to the distances.*

Let there be two lines, AB and GD, unequally distant from the eye E. I say that, as it appears, BE is not in the same relation to ED as GD is to AB [figure 9.1.6]. For let the rays fall, AE and EG, and with E as the central point and at the distance EZ let an arc be drawn, LZT. Now, since the triangle EZG is greater than the section EZL, and since the triangle EZD is less than the section EZT, the triangle EZG, compared with the section EZL, has a greater ratio than the triangle EZD, when compared with the section EZT. And, alternately, the triangle EZG, compared with the triangle EZD, has a greater ratio than the section EZL as, when compared with the section EZT, and, when put together, the triangle EGD, compared with the triangle EZD, has a greater ratio than the section

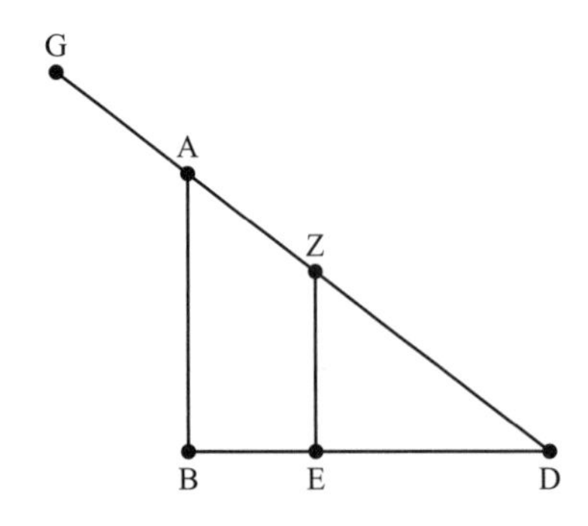

Figure 9.1.7.

ELT, compared with the section *EZT*. But as the triangle *EDG* is to the triangle *EZD*, so is *GD* to *DZ*. But *GD* is equal to *AB*, and, as *AB* is to *DZ*, so is *BE* to *ED*. Now, *BE* in comparison with *ED* has a greater ratio than the section *ELT* in comparison with the section *EZT*. And as one section is to the other section, so the angle *LET* is to the angle *ZET*. So the straight line *BE*, compared with the straight line *ED*, has a greater ratio than the angle *LET* compared with the angle *ZET*. And from the angle *LET*, *GD* is seen, and from the angle *ZET*, *AB* is seen. So lines of equal length are not seen in proportion to the distances.

The next two propositions are the earliest written methods for determining the height of an object. They are followed by methods of determining a depth and a length when these quantities cannot be measured directly. These methods became standard and were used well into the medieval period to determine heights and distances. This may seem particularly surprising once trigonometry was available, but it seems that trigonometry was, in general, only to be used in astronomical contexts. For measurements on earth, these methods using similarity were always preferred.

Proposition 18. *To know how great is a given elevation when the sun is shining.*

Let the given elevation be *AB*, and we have to know how great it is. Let the eye be *D*, and let *GA* be a ray of the sun falling upon the end of the line *AB*, and let it be prolonged as far as the eye *D* [figure 9.1.7]. And let *DB* be the shadow of *AB*. And let there be a second line, *EZ*, meeting the ray, but not at all illuminated by it below the end of the line at *Z*. So, into the triangle *ABD* has been fitted a second triangle, *EZD*. Thus, as *DE* is to *ZE*, so is *DB* to *BA*. But the ratio of *DE* to *EZ* is known; and, therefore, the ratio of *DB* to *BA* is known. Moreover, *DB* is known; so, *AB* also is known.

Proposition 19. *To know how great is a given elevation when the sun is not shining.*

Let there be a certain elevation, *AB*, and let the eye be *G*, and let it be necessary to know how high is *AB* when the sun is not shining. Let a mirror be placed, *DZ*, and let *DB* be prolonged in a straight line continuous with *ED*, to the point where it touches *B*, the end of *AB*, and let a ray fall, *GL*, from the eye *G*, and let it be reflected to the point where it touches *A*, the end of *AB*, and let *ET* be a prolongation of *DE*, and from *G* let the perpendicular *GT* be drawn upon *ET* [figure 9.1.8]. Now, since the ray *GL* has fallen

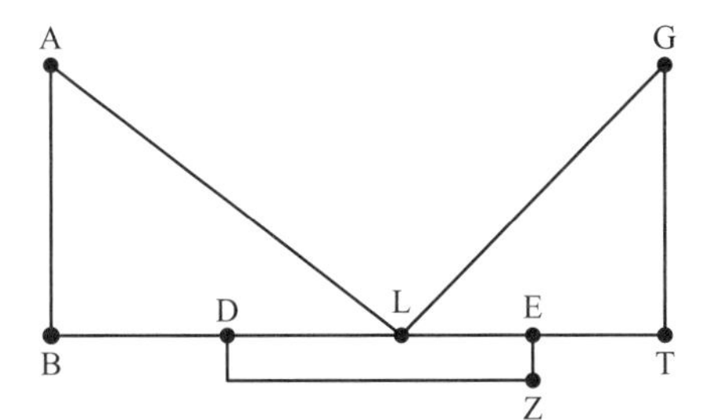

Figure 9.1.8.

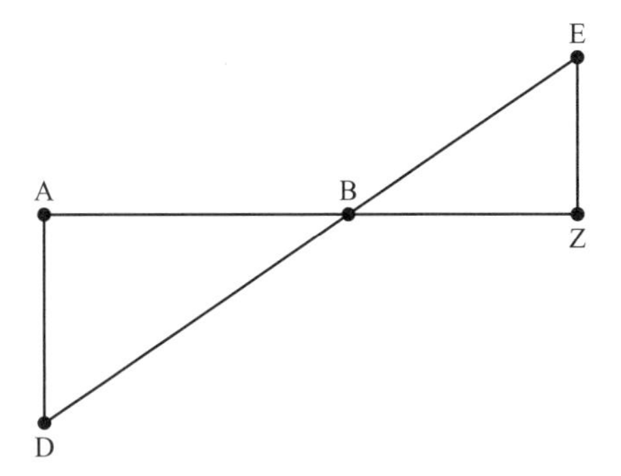

Figure 9.1.9.

and the ray LA has been reflected, they have been reflected at equal angles, as is said in the *Catoptrics*;[1] thus, the angle GLT is equal to the angle ALB. But also the angle ABL is equal to the angle GTL; and the remaining angle LGT is equal to the remaining angle LAB. So, the triangle ALB is equiangular with the triangle GLT. But the sides of equiangular triangles are proportional. Thus, as GT is to TL, so is AB to BL. But the ratio of GT to TL is known; and the ratio of BA to BL is known. But LB is known. And so, AB is known.

Proposition 20. *To know how great is a given depth.*

Let the given depth be AD, and let the eye be E, and let it be necessary to know how great is the depth. Let a ray of the sun fall before the eye, meeting the plane at the point B, and the depth at the point D. And let BZ be continued from B in a straight line, and let the perpendicular EZ be drawn from E to the horizontal line BZ [figure 9.1.9]. Now, since the angle EZB is equal to the angle BAD and the angle ABD is equal to the angle EBZ, the third angle also, BEZ, is equal to the angle ADB. So, the triangle ADB is equiangular with the triangle BEZ. And thus the sides will be proportional. Then, as EZ is to ZB, so is DA to AB. But the ratio of EZ to ZB is known; and the ratio of DA to AB is known. And also AB is known. And thus AD is known.

[1] See sections 9.2 and 9.3.

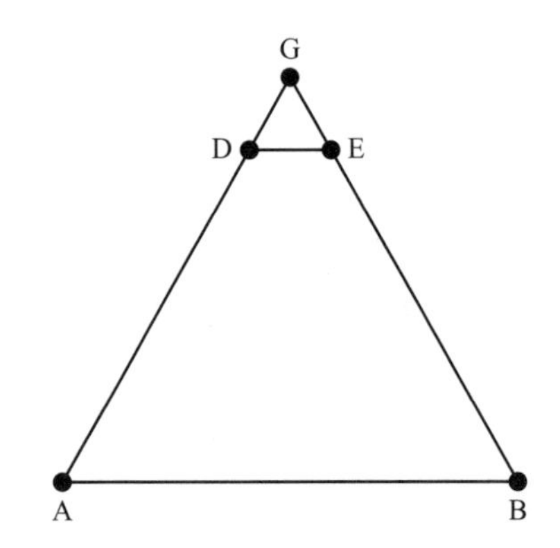

Figure 9.1.10.

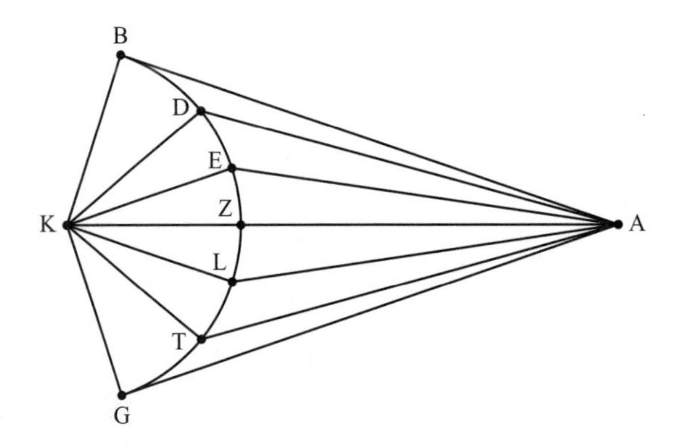

Figure 9.1.11.

Proposition 21. *To know how great is a given length.*

Let the given length be AB, and let the eye be G, and let it be necessary to know how great is the length of AB. Let the rays fall, GA and GB, and near the eye, G, let the point D be taken somewhere upon the ray (GA), and through the point D let the horizontal line DE be drawn, parallel to AB [figure 9.1.10]. Now, since DE has been drawn parallel to BA, one of the sides of the triangle ABG, as GD is to DE, so is GA to AB. But the ratio of GD to DE is known; and the ratio of AG to AB is known. And AG is known. So, AB also is known.

Proposition 22. *If an arc of a circle is placed on the same plane as the eye, the arc appears to be a straight line.*

Let BG be an arc lying on the same plane as the eye A, from which let the rays fall, AB, AD, AE, AZ, AL, AT, and AG [figure 9.1.11]. I say that the arc BG appears to be a straight line. Let K be the center of the arc, and let the straight lines KB, KD, KE, KZ, KL, KT, and KG be drawn. Now, since KB is seen within the angle KAB, and KD within the angle KAD, KB will appear longer than KD, and KD longer than KE, and KE longer than KZ, and, on the other side, KG will appear longer than KT, and KT longer than KL, and KL

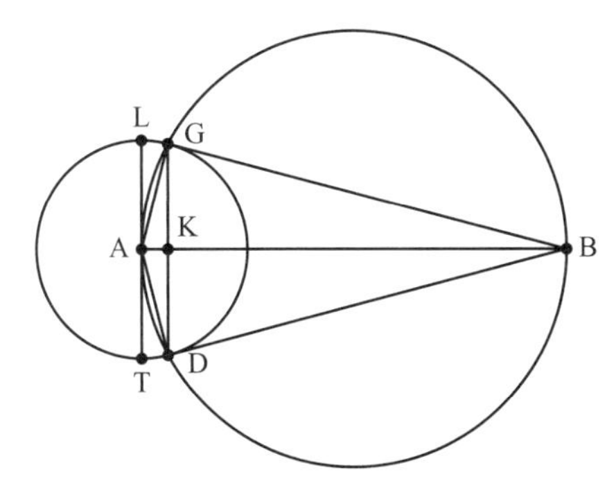

Figure 9.1.12.

longer than KZ. On this account, KA remaining a straight line, BG is always perpendicular. And the same things will happen also in the case of a concave arc.

An addition. It is possible to say these things also with reference to the visual rays themselves, that the ray between the eye A and the diameter is shortest, and that the ray nearer to it is always shorter than the one farther away. And the same things happen when AZ is a perpendicular upon the diameter. On this account the arc gives an impression of a straight line, and especially if it should be seen from a greater distance, so that we do not perceive the curve. On this account ropes not stretched tightly, when seen from the side, appear to sag, but seen from below, seem to be straight, and also the shadows of rings lying on the same plane with the source of light become straight.

Proposition 23. *Of a sphere seen in whatever way by one eye, less than a hemisphere is always seen, and the part of the sphere that is seen itself appears as an arc.*

Let there be a sphere, of which A is the center, and let B be the eye. And let A and B be joined, and let the plane be continued along the line BA [figure 9.1.12]. So it will make a circular section. Let it make the circle $GDTL$, and around the diameter AB let the circle GBD be inscribed, and let the straight lines be drawn, GB, BD, AD, and AG. Now, since AGB is a semicircle, the angle AGB is a right angle; similarly, also the angle BDA. So, the lines GB and BD touch the sphere. Now, let G and D be joined, and let LT be drawn through the point A parallel to GD. So, the angles at K are right angles. Now, with AB remaining in its place, if the triangle BGK is revolved about the right angle K, and is restored to the same position from which it started, the line BG will touch the sphere at one point and the line KG will make a circular section. So, an arc will be seen in the sphere. And I say that it is less than a hemisphere. For, since LT is a semicircle, GD is less than a semicircle. And the same part of the sphere is seen by the rays BG and BD. So, GD is less than a hemisphere; and it is seen by the rays BG and BD.

Proposition 36. *The wheels of the chariots appear sometimes circular, sometimes distorted.*

Let there be the wheel $AGBD$, and let the diameters be drawn, BA and GD, cutting each other at right angles at the point E, and let the eye not be on the plane of the circle [figure 9.1.13]. Now, if the line drawn from the eye to the center is at right angles to the plane

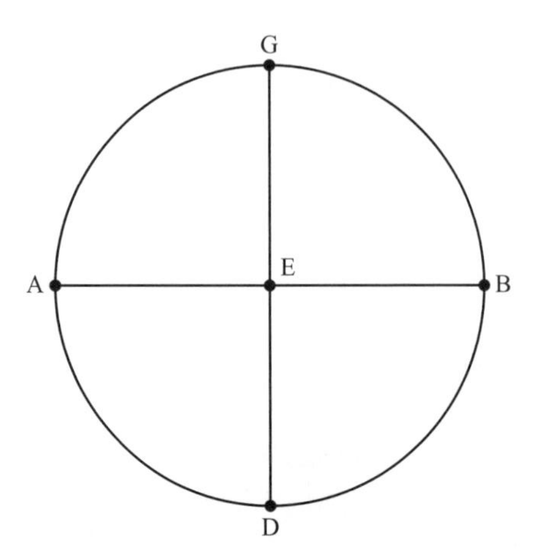

Figure 9.1.13.

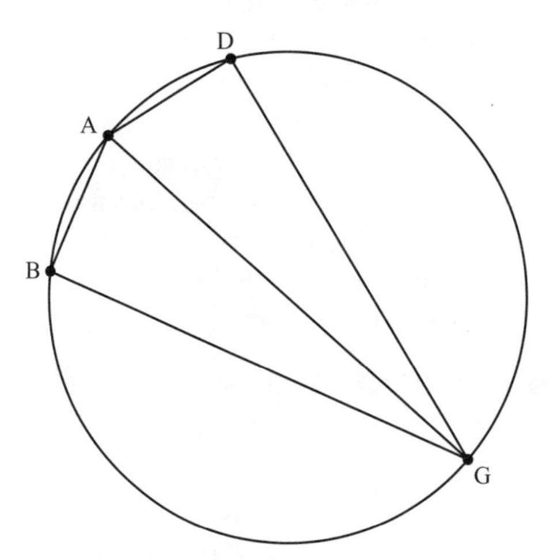

Figure 9.1.14.

or is equal to the radius, all the diameters will appear equal; so that the wheel appears circular. But if the line drawn from the eye to the center is not at right angles to the plane or equal to the radius, the diameters will appear unequal, one longer and one shorter, and to every other one which is drawn between the longer and shorter diameters only one other will appear equal, the one drawn to the opposite side; so that the wheel appears distorted.

Proposition 37. *There is a place where, if the eye remains in the same position, while the thing seen is moved, the thing seen always appears of the same size.*

Let the eye be A, and the size of the thing seen be indicated by BG, and from the eye let rays fall, AB and AG, and about ABG let a circle, ABG, be inscribed [figure 9.1.14]. I say that there is a place where, if the eye remains in the same position, while the thing seen is moved, the thing seen always appears of the same size. For let the object be moved and let it be indicated by the line DG, and let AD be equal to AB. Now, since BA is equal to AD and BG to GD, also angle BAG is equal to angle DAG. For they are upon equal arcs;

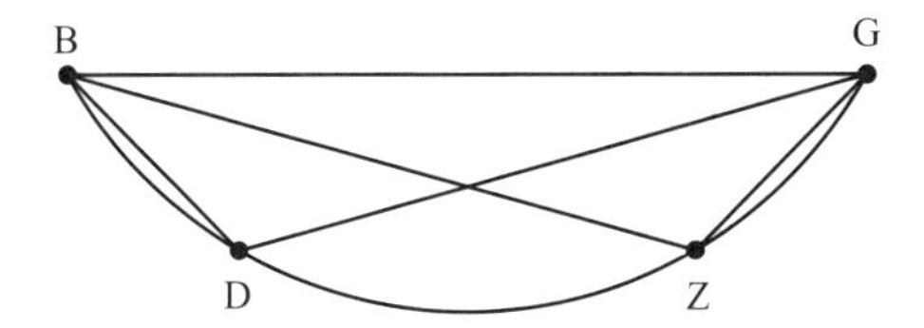

Figure 9.1.15.

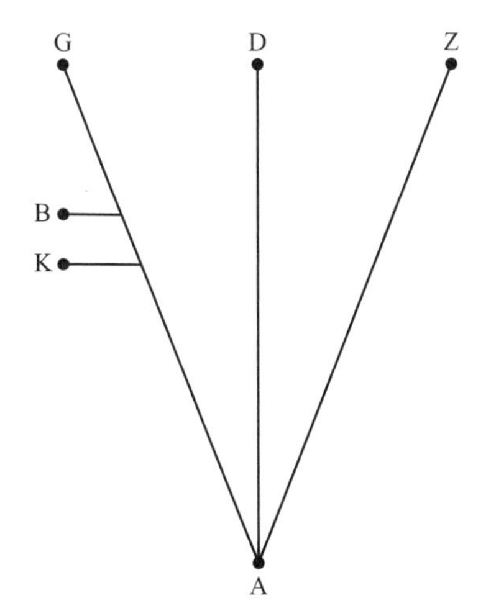

Figure 9.1.16.

so that they are equal. So the thing seen will appear of the same size. And the same thing will happen also if the eye should remain at the center of the circle, and the thing seen should move upon the arc.

Proposition 38. *There is a place where, if the position of the eye is changed while the thing seen remains in the same place, the thing seen always appears of the same size.*

For let BG be the thing seen and let Z be the eye, from which let rays fall, ZB and ZG, and about the triangle BZG let the arc of a circle, BZG, be inscribed, and let the eye, Z, be shifted to D, and let the rays fall, DB and DG [figure 9.1.15]. Now the angle D is equal to the angle Z; for they are on the same arc. And things seen at the same angle appear equal. So BG will always appear of the same size, when the eye is shifted upon the arc BGD.

Proposition 54. *When objects move at equal speed, those more remote seem to move more slowly.*

For let B and K move at equal speed, and from the eye, A, let rays be drawn, AG, AD, and AZ [figure 9.1.16]. So, B has longer rays than K extending from the eye. Therefore, it will cover a greater distance, and, later, passing the line of vision, AZ, will seem to move more slowly.

An addition. For let two points, A and B, move on parallel straight lines, and let Z be the eye, from which let the rays fall, ZA, ZB, ZE, and ZD [figure 9.1.17]. I say that the more remote object, A, seems to move more slowly than B. For, since AZ and ZD form a smaller

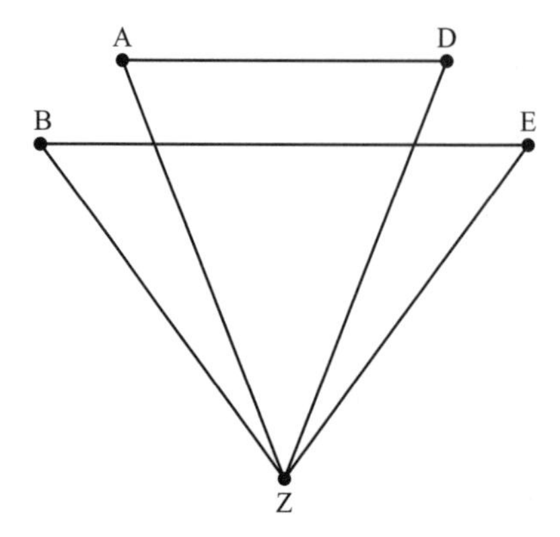

Figure 9.1.17.

angle than ZB and ZE, BE appears greater than AD. (*The following sentence is corrupt in the Greek text, and no satisfactory translation is possible.*) Therefore, if we extend the ray ZE in a straight line, since in the case of objects moving at the same speed B reaches the ray ZE later than the things moving at the same speed, the more remote objects seem to move more slowly.

9.2 (Pseudo-[?])Euclid, *Catoptrics*

Catoptrics is the study of reflection, as understood in the context of image distortion caused by mirrors. These mirrors could be plane, spherical convex, or spherical concave. The key to theories of reflection was the law of equal angles. The following proposition on reflection from the "pseudo?-Euclidean" *Catoptrics*[2] invokes horn-angles in its demonstration, a special sort of angular displacement treated in *Elements* 3:16.

Proposition 1. *Visual rays are reflected at equal angles by plane, convex, and concave mirrors.*

Let B be the eye and AC a plane mirror [figure 9.2.1]. Let visual ray BK emitted from the eye be reflected to D. I say that angle E is equal to angle Z. Let perpendiculars BC and DA be dropped. Therefore, BC is to CK as DA is to AK, for this was postulated in the definitions.[3] Therefore, triangles BCK and DAK are similar. Accordingly, angle E is equal to angle Z, because similar triangles are equiangular.

Now let mirror AKC be [spherical] convex, and ray BK be reflected to D [figure 9.2.2]. I say that angle E together with [horn] angle T equals angle Z together with [horn] angle L. I have applied plane mirror NM [to convex mirror AKC]. Accordingly, angle E is equal to angle Z. But [horn] angle T is equal to [horn] angle L, because MN is

[2] There has been some debate over the authorship of this work. See Smith (1999):17, who treats it as authentically Euclidean.

[3] Definition 3 of the *Catoptrics* says that "when a given perpendicular magnitude is viewed in a plane mirror, the straight distance between the mirror and the viewer is to the straight distance between the mirror and the perpendicular magnitude as the altitude of the viewer is to the altitude of the perpendicular magnitude."

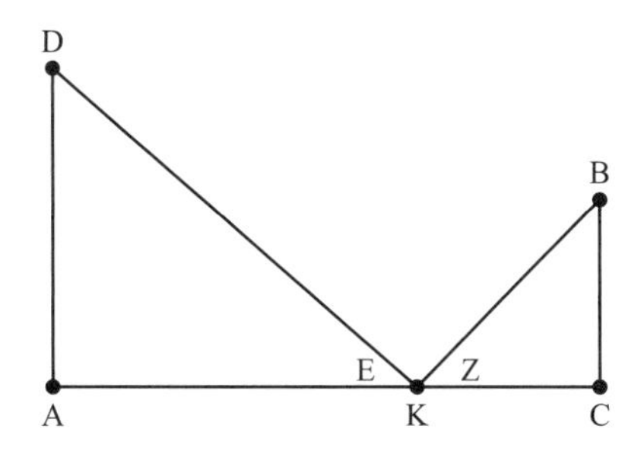

Figure 9.2.1.

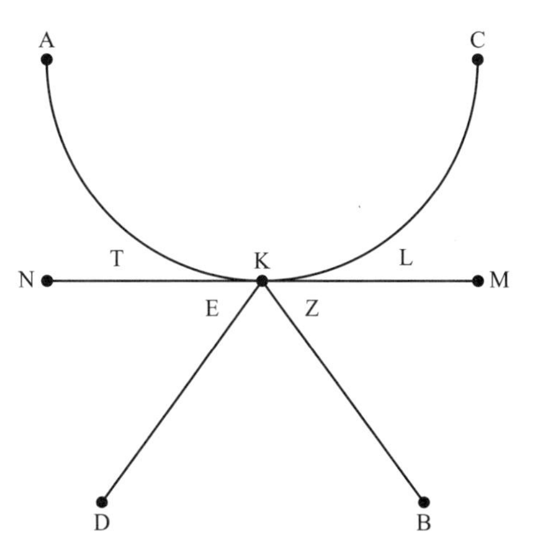

Figure 9.2.2.

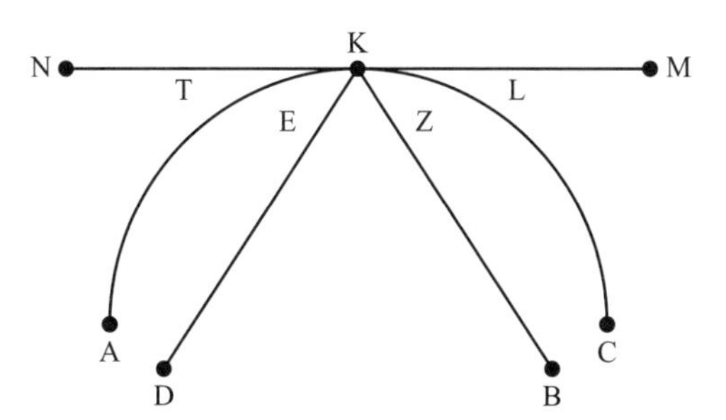

Figure 9.2.3.

tangent. Thus, angle E together with [horn] angle T equals angle Z together with [horn] angle L.

Once again, let mirror AKC be [spherical] concave, and let the visual ray BK be reflected to D [figure 9.2.3]. I say that angle E is equal to angle Z, for, with the plane mirror [NM] applied, [horn] angle T together with angle E is equal to [horn] angle L together with angle Z. Therefore, the remaining angles E and Z are equal.

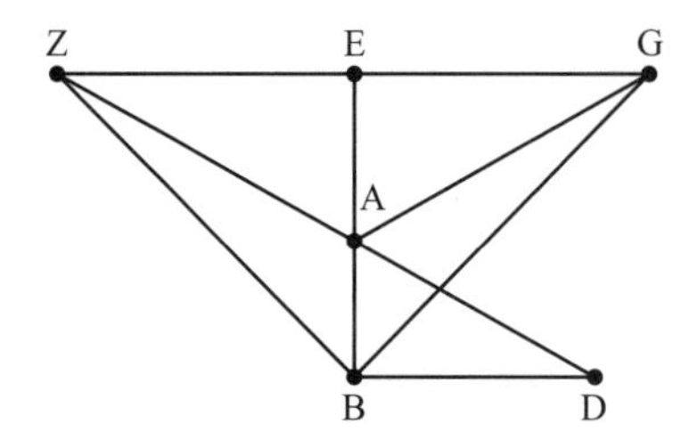

Figure 9.3.1.

9.3 Heron, *Catoptrics*

Heron of Alexandria wrote the *Catoptrics* to elucidate his theories of reflection. Knowledge of this work is through a Latin translation produced in the mid-thirteenth century; the original Greek text is lost. We present here his study of the basic law of reflection, prefaced by two comments about the trajectory of visual rays.

Chapter 2

The arrows we see shot from bows may serve as an example. For, because of the impelling force the object in motion strives to move over the shortest possible distance, since it does not have the time for slower motion, that is, for motion over a longer trajectory. And so, because of its speed, the object tends to move over the shortest path. Moreover, that the rays emitted by us travel at an immeasurable velocity is evidenced by the fact that, when we have closed our eyes and then reopened them to look at the heavens, it takes no perceptible time for the visual rays to reach the heavens, for just as our eyes are opened, we see the stars, even though the distance is, so to speak, infinite.

Chapter 3

Just as a rock hurled with great force against a compact body rebounds, . . . so also the rays sent forth by us at enormous velocity, as has been shown, are reflected by compact bodies when they strike them.

Chapter 4

On the same grounds—that is, according to the speed of incidence and reflection—we will demonstrate that reflections occur at equal angles in plane and [spherical] convex mirrors. For again it is necessary to reason according to the actual minimum lines. I say, then, that of all incident and reflected rays reaching the same point from a plane or [spherical] convex mirror, those that form equal angles [with the reflecting surface] are the shortest possible. Moreover, if such is the case, they are reflected at equal angles according to an [underlying] rationale.

Accordingly, let AB be a plane mirror, point G the eye, and D the visible [point-object] [figure 9.3.1]. Let GA be the incident ray, and let it be connected with [reflected ray] AD. And let the angle EAG be equal to the angle BAD. Then let a similar ray GB be the incident

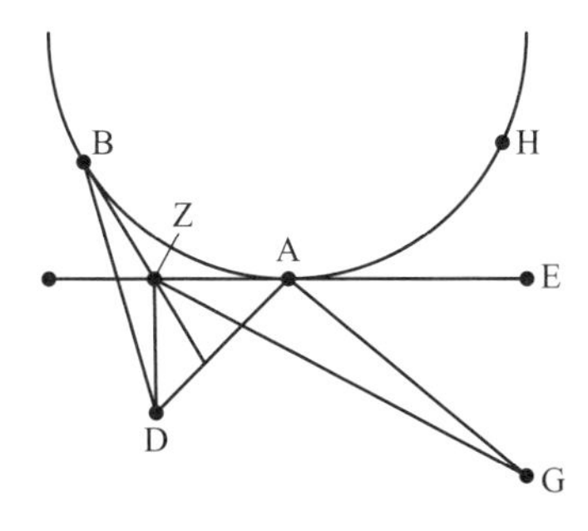

Figure 9.3.2.

ray, and let it be connected with [reflected ray] BD. I say that GA, AD together are less than GB, BD together. For let GE be dropped from G perpendicular to AB, let GE and DA be extended to Z, and let ZB be joined. Angle BAD is equal to angle ZAE, because they are vertical angles, and angle BAD is equal to angle EAG [which therefore equals ZAE]. But the angles at E are right, so ZA equals AG. Furthermore, ZB equals BG, so that ZD is less than ZB, BD taken together. But ZA equals AG, and ZB equals BG. Hence, GA, AD together are less than GB, BD together.

Chapter 5

Let AB be the surface of a [spherical] convex mirror, let G be the eye, and let D be the visible [object-point] [figure 9.3.2]. Let GA and AD form equal angles [with the reflecting surface], and let GB and BD form unequal angles. I say that GA, AD together are less than GB, BD. For let tangent EAZ be drawn. Accordingly, [horn] angle HAE equals [horn] angle BAZ, and the remaining [angle] EAG is equal to ZAD. If, then, ZD is joined, it follows from the previous demonstration that GA, AD together are less than GZ, ZD together. But GZ, ZD together are less than GB, BD together, so GA, AD together are less than GB, BD together.

9.4 Ptolemy, *Optics*

Ptolemy's *Optics* is the pinnacle of Greek theories of vision and mathematical optics, appearing over six hundred years after the very first speculations on visual perception. The only remaining evidence we have of this text, however, is a poorly preserved incomplete manuscript containing a twelfth-century Latin translation made in Byzantium on the basis of an Arabic translation of the original Greek text, both of which are lost.

Ptolemy was obviously influenced by Euclid's *Optics*, but his exposition encompasses so much more than just mathematical optics. He covers a spectrum of topics relating more broadly to visual perception, including illumination, color, clarity, size, shape, movement, and binocular vision, as well as detailed and original accounts of reflection and refraction. He approaches the latter two topics both empirically as well as theoretically, and he includes the results of experiments, observations, and measurements he made to study these visual phenomena. Of note are his experiments involving a bronze plaque to gauge reflection properties and the observation of rays traveling between air, water, and glass to study refraction. For the latter trio of refraction experiments, Ptolemy reported physical measurements of the angle of refraction against the angle of incidence at 10-degree intervals. The numerical data he tabulated roughly correspond to what would be predicted by the modern sine-law relation. Ptolemy's professional expertise in astronomy also comes through in his approach to optics; aspects related to celestial observation, such as atmospheric refraction, are featured throughout his work.

Book 2

2. Accordingly, we say that the visual faculty apprehends corporeity, size, color, shape, place, activity, and rest. Yet it apprehends none of these without some illumination and something [opaque] to block the passage [of the visual flux]. We need say no more on this score but must instead specify what characterizes each of the visible properties.

20. And since [each] visual ray terminates at its own unique point, what is seen by the central ray, i.e., the one that lies upon the axis [of the visual cone], should be seen more clearly than what is viewed to the sides [of the visual axis] by lateral rays. The reason is that those rays lie nearer to [the edge of the visual cone where there is an increasing] absence [of rays], whereas those rays that approach the [visual axis] lie farther from [such an area of] absence. The same holds for objects that lie toward the middle of spherical sections whose center point is the apex of the visual cone, because the generating point of the sphere itself and powers that approach their generating sources are more effective. The farther such powers extend from their sources, then, the weaker they become as, e.g., [the power of] projection [in relation to] the thrower, or of heat in relation to the heater, or of illumination in relation to the light-source. Therefore, since the visual ray within the cone has two primary referents, one being the center point of the [ocular] sphere where the vertex of the visual cone lies and the other being the straight line that originates at this point and extends the whole length [of the cone] to form its axis, it necessarily follows that the visual perception of what lies far from the vertex of the cone is carried out by a more weakly-acting ray than the visual perception of something lying at a moderate distance. The same holds for objects that lie far from the visual axis in comparison to those that lie near it.

26. The visual faculty also discerns the place of bodies and apprehends it by reference to the location of its own source-points [i.e., the vertices of the visual cones] . . . as well as by the arrangements of the visual ray falling from the eye upon those bodies. That is, longitudinal distance [is determined] by how far the rays extend outward from the vertex of the cone, whereas breadth and height [are determined] by the symmetrical displacement of the rays away from the visual axis. That is how differences in location are determined, for whatever is seen with a longer ray appears farther away, as long as the increase in [the ray's] length is sensible.

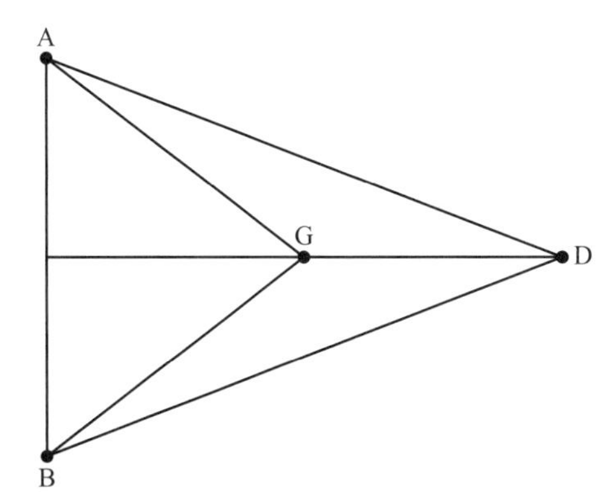

Figure 9.4.1.

33. [Experiment 2.1] Let points A and B [figure 9.4.1] be the vertices of the visual cones, and let B lie at the right eye and A at the left. Let two pegs, G and D, be erected vertically upon line [GD, which is] perpendicular to AB, and from each vertex of the two visual cones let rays GA, GB, DA, and DB be extended to the two pegs. Then let us first focus on G, which is nearer.

34. AG and BG will therefore lie upon the axes themselves. Of the remaining two rays, however, AD will be one of the left-hand rays [in the cone whose vertex is at A], and it is obvious that BD is one of the right-hand rays [in the cone whose vertex is at B]. It necessarily follows, then, that G is seen at one location, insofar as each of the axes corresponds with the other. On the other hand, D must be seen at two locations, since AD is one of the left-hand rays of the left eye, while ray BD is one of the right-hand rays of the right eye. Thus, when we cover the left eye, the left-hand [member of the doubled image] will disappear, and when we cover the right eye, the right-hand [member] will disappear.

35. Now if we focus on D, the opposite will happen. This is demonstrated from the fact that AD and BD will lie on the [visual] axes. Thus, D will be seen as one, and G will be seen double, since AG happens to be one of the right-hand rays of the left eye, while BG is one of the left-hand rays of the right eye. So the opposite of what happened before will occur: that is, if we cover the left eye, the image seen on the right-hand side by ray AG will disappear, and if we cover the right eye, the image seen on the left-hand side by ray BG will disappear.

54. Now the aforementioned two variables [i.e., distance and orientation] by which the apparent size of objects is determined make no difference in the sensible impression, but the remaining one does, for, if there is a difference in that third variable, i.e., in the visual angles, then the object will appear larger when the angle subtended by it is larger. For instance, if there are two magnitudes, such as AB and GD [figure 9.4.2], and if they lie the same distance away at the same orientation but subtend unequal angles, then AB, which subtends the larger angle at point E, will appear larger.

55. And if there is a difference in [either] of the two remaining features alone, then the object will never appear larger, no matter whether it faces us more directly or whether it is

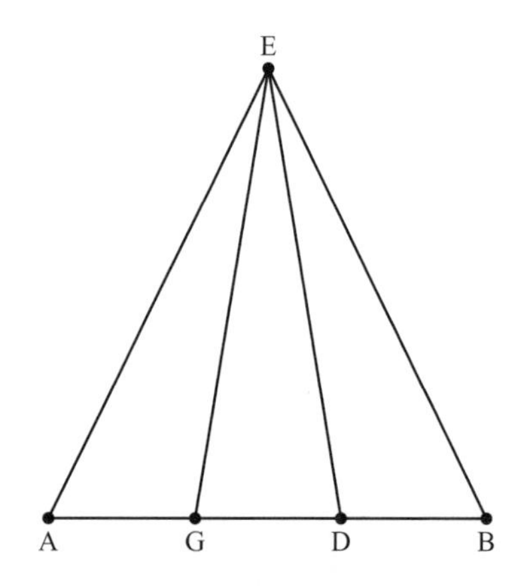

Figure 9.4.2.

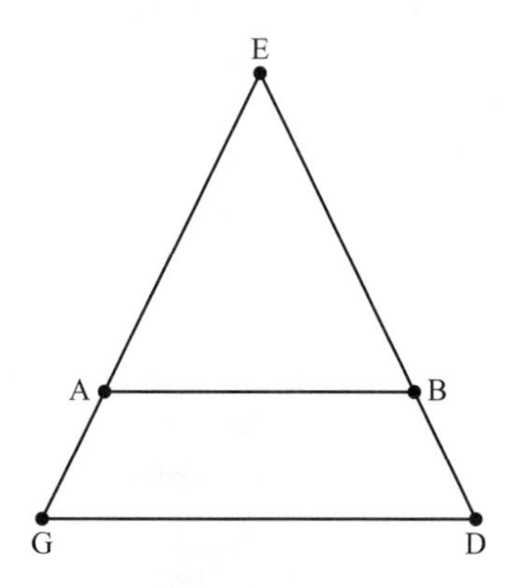

Figure 9.4.3.

closer. Thus, it will appear either smaller or equal and in each case the apparent size will depend on relative differences in actual size.

56. [Example 2.1] For instance, if two magnitudes, AB and GD [figure 9.4.3], have the same orientation and subtend the same angle at E, then, since AB does not lie the same distance as GD [from point E] but is closer to it, AB will never appear larger than GD, as seems appropriate from its proximity [to E]. Instead, it will either appear smaller (which happens when the distance of one from the other is perceptible), or it will appear equal (which happens when the difference in [relative] distance is imperceptible).

57. [Example 2.2] Likewise, if there are two magnitudes, such as AB and GD [figure 9.4.4], that subtend the same angle at E and lie the same distance from it, and if their orientation is different, [so that] one of them, AB, faces E directly and the other obliquely, then AB will never appear larger than GD on account of its facing orientation. Instead, AB will either appear smaller than GD (which happens when the difference in orientation between

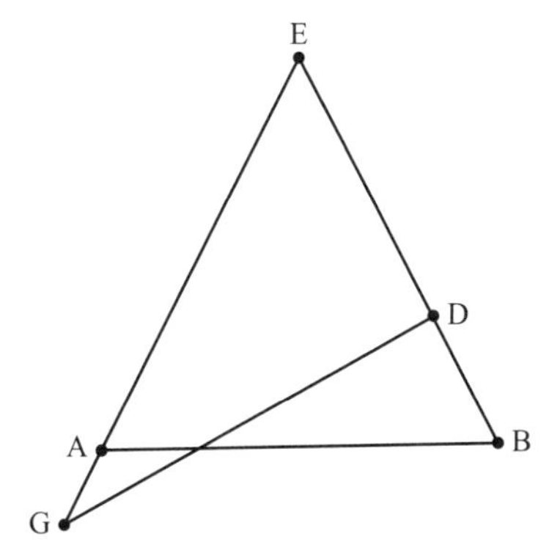

Figure 9.4.4.

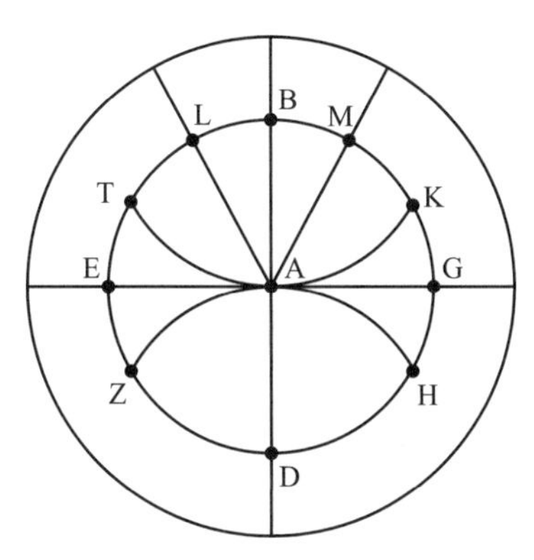

Figure 9.4.5.

the two magnitudes is perceptible), or it will appear equal to GD when the difference in orientation is imperceptible.

The four sections excerpted from book 3 present Ptolemy's experimental confirmation of the equal-angles law of reflection.

Book 3

8. [Experiment 3.1] Let a round, bronze plaque of moderate size, such as the one below [figure 9.4.5], be set up, and let A be its center. Let both faces be planed down as carefully as possible, and let its edges be rounded and polished. Then let a small circle be inscribed at center point A on either of its faces, and let it be $BGDE$. On this same face let two diameters, BD and GE, be inscribed to intersect at right angles; and let each quarter circle be divided into 90 degrees. Finally, let the two points B and D be taken as center points, and, using BA and DA as radii, let the two arcs ZAH and TAK be inscribed.

9. Now let three thin, small, square, straight sheets of iron be formed. Let one of them remain straight, and let one of its sides be polished so that it appears as a clear mirror. Let the remaining two sheets be curved in such a way that the convex surface of the one and the concave surface of the other [taken together] form a circular section equal to circle *BGDE*, and let the two [respective convex and concave] surfaces of these sheets be polished so that they are made into two mirrors.

10. Let us cut arcs from each of the [above] two sheets, and let them be represented by *ZAH* and *TAK*. Let line *BA* be drawn in white and *AL* in some other color. Then let a small diopter be mounted upon *AL*, and let the aforementioned plaque be disposed in such a way that [the sight-line of] the diopter passes easily through point *L* and along line *AL*. Now let the aforementioned plaque be placed with the side upon which the mirrors are [to be mounted] facing up. Of these mirrors, let the plane one lie on *GAE*, the convex one on *ZAH*, and the concave one on *TAK*. Finally, in the middle of the upper edge of each of the mirrors let a pin be attached axially to the mirror so as to keep it in place on point *A*.

11. Assume, then, that we view with either of our eyes through the diopter, which is placed at point *L* on *AL*, and that we direct our line of sight toward the axial pin of [each of] the mirrors. Accordingly, if we slide a small colored marker [along arc *TBK*] on the plaque's surface until it appears to us to lie on the same line of sight with *A*, then point *L*, point *A*, and the image that is seen in the three mirrors will appear [to lie] upon a single line [of sight]. If, therefore, we mark the point at which the colored marker stands on the plaque's surface—i.e., the place from which the marker's image appears in those mirrors (e. g., the place represented by point *M*)—and if we draw out straight line *AM*, we will find that arc *BM* is always equal to arc *BL*. Since that is the case, angle *LAB* will be equal to angle *MAB*, and line *BD* will be normal to all of those mirrors. Line *AL* [thus] delineates the [branch of the] visual ray incident to the mirror's surface from the eye, whereas line *AM* delineates the [branch of the] visual ray reflected from the mirror's surface to the visible object.

In book 5, Ptolemy discusses refraction. His table of values for refraction between air and water is reasonably close to the values calculated from the modern sine law of refraction.

Book 5

1. There are two ways in which the visual ray is broken. One involves rebound and caused by reflection from bodies that block the visual ray's passage and that are included under the heading of "mirrors." The other way, however, involves penetration and is caused by a deflection in media that do not completely block the visual ray's passage, and those media are included under the single heading "transparent." In the preceding books we have discussed mirrors; and, insofar as it is possible for it to be demonstrated, we have explained not only variations in the images of visible objects according to the principles laid out for the science of optics, but also what happens with each of the visible properties

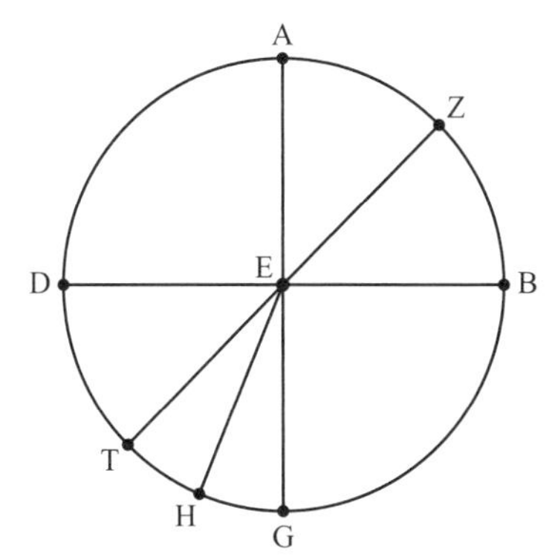

Figure 9.4.6.

in reflection. It thus remains for us at this point to analyze what sorts of variations occur in such objects when we look at them through transparent media.

7. The amount that the ray is refracted in water below the [original line of] sight is determined according to the following experiment, which is conducted by means of the bronze plaque that we constructed for analyzing the phenomena of mirrors.

8. [Experiment V.1] Let circle $ABGD$ be described on that plaque about center point E, and let the two diameters AEG and BED intersect one another at right angles [figure 9.4.6]. Let each of the [resulting] quadrants be divided into 90 equal increments. At the center point let a small marker of some color or other be attached, and let the plaque be stood upright in the small vessel [discussed in the previous experiment]. Then let a suitable amount of water that is clear enough to be seen through be poured into that vessel, and let the graduated plaque be placed erect at right angles to the surface of the water. Let all of semicircle BGD of the plaque, but nothing beyond that, lie under water, so that diameter AEG is normal to the water's surface. From point A, let a given arc AZ be marked off on either of the two quadrants that lie above the water. Furthermore, let a small, colored marker be placed at Z.

9. Now, if we line up both markers at Z and E along a line of sight from either eye so that they appear to coincide, and if we then move a small, thin peg along the opposite arc GD under water until the end of the peg, which lies upon that opposite arc, appears to lie directly in line with the two previous markers, and if we mark off the portion of the arc GH that lies between G and the point at which the object would appear unrefracted, the resulting arc will always turn out to be smaller than AZ. Moreover, if we join lines ZE and EH, angle AEZ is greater than angle GEH, which cannot be the case unless there is refraction—that is, unless ray ZE is refracted toward H according to the excess of one of the opposite angles over the other.

10. Furthermore, if we place our line of sight along normal AE, we will find the image directly opposite along its rectilinear continuation, which will extend to G; and this [radial line] undergoes no refraction.

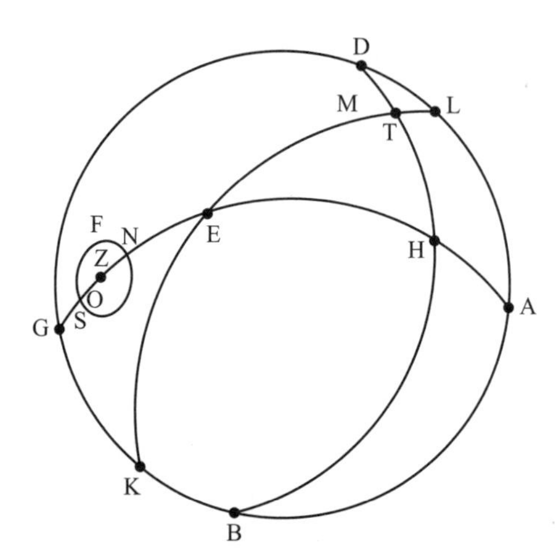

Figure 9.4.7.

11. In the case of all the remaining positions, when arc AZ is increased, arc GH in turn will be increased, and the refraction will be greater. When arc AZ is 10 degrees out of the 90 into which quadrant [AB] is divided, then arc GH will be around 8 degrees. When AZ is 20 degrees, then GH will be 15.5. When AZ is 30 [degrees], then GH will be 22.5. When AZ is 40 [degrees], then GH will be 29. When AZ is 50 [degrees], then GH will be 35. When AZ is 60 [degrees], then GH will be 40.5. When AZ is 70 [degrees], then GH will be 45.5. And when AZ is 80 [degrees], then GH will be 50.[4]

27. [Theorem V.2] With these points established, let ABG represent the circle of the horizon and $AEZG$ the semicircular arc of the meridian that lies above the earth [figure 9.4.7]. Let point E represent the zenith and point Z the apparent pole of the heavenly sphere [of fixed stars]. Let BHD be the arc of a line above the earth's surface that is parallel to the equator and that passes through certain stars. Let T be a star that lies on this line near the horizon, and let $KETL$ be the semicircular arc of the circle that passes above the earth through the zenith and through star T.

28. Accordingly, since the star appears to lie nearer the zenith than it truly is when it is near the horizon, and since the divergence in its apparent location from its true position is measured on the great circles passing through the points on the horizon, the point where the star that is [actually] at T appears will lie between E and T, such that it appears at point M. Moreover, the line parallel to the equator and passing through point M, will lie higher to the north than the line parallel to the equator and passing through point T, which at our particular latitude is inclined toward the north. And when the star rises to position

[4] It is not clear that Ptolemy is reporting on the results of the experiment here. It seems, in fact, that he is using an algorithm to calculate these values, since the angles of refraction increase by 7.5, 7, 6.5, 6, 5.5, 5, and 4.5 degrees, respectively, each time the angle of incidence increases by 10 degrees. In any case, a modern calculation by Snell's law, given that the index of refraction of water is 1.333, shows that Ptolemy's values are reasonably good. For example, by Snell's law, the ratio of sin(AZ) to sin(GH) should be 1.333. For Ptolemy, when $AZ = 30°$, $GH = 22.5°$; thus, the ratio of the sines is 0.5 : 0.3827 = 1.3066. Calculations for the other values are similar.

H, it reaches a point where the visual ray is refracted without any perceptible difference between apparent and true location.

29. Likewise, if we suppose *Z* to represent the north pole and if we draw one of the circles parallel to the equator that is always visible [at our latitude], e.g., circle *NSF*, then, when the star lies at point *S* on this circle, it will appear closer to point *E*, which lies at the zenith, and it will seem to lie at point *O*. But when the star is at point *N*, then there is no difference, or only an imperceptible one, between apparent and true location. And therefore, when a star approaches the horizon in its revolution, its distance from the north pole of the [celestial] sphere appears to be smaller [than it really is]; but when it approaches the zenith in the course of its revolution, that apparent distance seems larger, for arc *ZN* will be larger than arc *OZ*.

30. It has thus been demonstrated how stellar observation must be affected by the refraction of the visual ray. It would also be possible for us not only to examine the degree of such refractions, but also to analyze such refraction in the case of certain [celestial] bodies whose distance is given—e.g., the sun and the moon—and to determine the degrees [of refraction] toward the horizon as well as the amount by which the refraction of the visual ray shifts the apparent position upward if the distance of the interface between the two media [i.e., air and ether] were known. But, although this distance lies nearer than the earth to the lunar sphere, where the ether stops, it is not known whether the [refractive] interface lies at the same distance as the aforesaid surface, or whether it lies nearer the earth, or whether it lies farther from [the aforesaid] surface. Therefore, it is impossible to provide a method for determining the size of the angles of deviation that occur in this sort of refraction.

SOURCES, CHAPTER 9

9.1 This translation is taken from Harry Edwin Burton, "The Optics of Euclid," *Journal of the Optical Society of America* 35, no. 5 (1945): 357–72. Proposition 6 is translated by A. Mark Smith in "Ptolemy and the Foundations of Ancient Mathematical Optics: A Source Based Guided Study," *Transactions of the American Philosophical Society*, new ser., 89, no. 3 (1999): 56–57.

9.2 This translation is taken from A. Mark Smith, "Ptolemy and the Foundations of Ancient Mathematical Optics: A Source Based Guided Study," *Transactions of the American Philosophical Society*, new ser., 89, no. 3 (1999): 80.

9.3 This translation is taken from A. Mark Smith, "Ptolemy and the Foundations of Ancient Mathematical Optics: A Source Based Guided Study," *Transactions of the American Philosophical Society*, new ser., 89, no. 3 (1999): 54, 80–81.

9.4 This translation is taken from A. Mark Smith, "Ptolemy's Theory of Visual Perception: An English Translation of the *Optics* with Introduction and Commentary," *Transactions of the American Philosophical Society*, new ser., 86, no. 2 (1996).

Bibliography

Acerbi, Fabio, and Bernard Vitrac. *Metrica: Héron d'Alexandrie*. Pisa: Fabrizio Serra, 2014.

Apollonius of Perga. *Conics, Books I–III*. Translated by R. Catesby Taliaferro. Santa Fe, NM: Green Lion, 2000.

———. *Conics, Book IV*. Translated by Michael Fried. Santa Fe, NM: Green Lion, 2002.

———. *Conics, Books V to VII: The Arabic Translation of the Lost Greek Original in the Version of the Banū Mūsā*. Edited and translated by G. J. Toomer. New York: Springer, 1990.

Archibald, Raymond Claire. *Euclid's Book on Divisions of Figures: With a Restoration Based on Woepcke's Text and on the Practica Geometriae of Leonardo Pisano*. Cambridge: Cambridge University Press, 1915.

Archimedes. *The Sand Reckoner*. Edited and translated by Gerard Michon. Bures-sur-Yvettes: Institut des Hautes Études Scientifiques. www.numericana.com/answer/archimedes.htm.

Aristophanes. *The Birds of Aristophanes*. Edited and translated by Benjamin Bickley Rogers. London: G. Bell, 1920. *Great Books of the Western World*, vol. 5. Edited by Robert Maynard Hutchins. Chicago: Encyclopedia Britannica, 1952.

———. *The Clouds*. Translated by the Athenian Society, 1912. https://en.wikisource.org/wiki/Aristophanes:_The_Eleven_Comedies/Clouds

———. *The Frogs*. Translated by the Athenian Society, 1912. https://en.wikisource.org/wiki/Aristophanes:_The_Eleven_Comedies/Frogs

Aristotle. *Metaphysics*. Edited and translated by W. D. Ross. In *The Works of Aristotle*, vol. 1, 495–626. *Great Books of the Western World*, vol. 8. Edited by Robert Maynard Hutchins. Chicago: Encyclopedia Britannica, 1952.

———. *Physics*. Edited and translated by R. P. Hardie and R. K. Gaye. In *The Works of Aristotle*, vol. 1, 259–355. *Great Books of the Western World*, vol. 8. Edited by Robert Maynard Hutchins. Chicago: Encyclopedia Britannica, 1952.

———. *Posterior Analytics*. Edited and translated by G. R. G. Mure. In *The Works of Aristotle*, vol. 1, 97–137. *Great Books of the Western World*, vol. 8. Edited by Robert Maynard Hutchins. Chicago: Encyclopedia Britannica, 1952.

———. *Prior Analytics*. Edited and translated by A. J. Jenkinson. In *The Works of Aristotle*, vol. 1, 39–93. *Great Books of the Western World*, vol. 8. Edited by Robert Maynard Hutchins. Chicago: Encyclopedia Britannica, 1952.

———. *Problems*. Vol. 1, *Books I–XXI*. Edited and translated by W. S. Hett. Cambridge, MA: Harvard University Press, 1961.

Bagnall, Roger S., and Alexander Jones. *Mathematics, Metrology, and Model Contracts: A Codex from Late Antique Business Education*. New York: New York University Press, 2019.

Baillet, Jules. "Le papyrus mathématique d'Akhmîm." *Mémoires publiés par les membres de la Mission Archéologique Française au Caire* 9, no. 1 (1892).

Barker, Andrew, ed. *Greek Musical Writings: Volume 2, Harmonic and Acoustic Theory*. Cambridge: Cambridge University Press, 1989.

Berggren, J. Lennart. "Abū al-Wafā' on the Geometry of Artisans." In *The Mathematics of Egypt, Mesopotamia, China, India, and Islam: A Sourcebook*, edited by Victor J. Katz, 612–16. Princeton, NJ: Princeton University Press, 2007.

Berggren, J. Lennart, and Alexander Jones. *Ptolemy's Geography: An Annotated Translation of the Theoretical Chapters*. (Princeton, NJ: Princeton University Press, 2000).

Blume, F., K. Lachmann, and A. Rudorff, eds. *Die Schriften der Römischen Feldmesser*. Berlin: Georg Reimer, 1848.

Bowen, Alan C., and Robert B. Todd, eds. and trans. *Cleomedes' Lectures on Astronomy: A Translation of "The Heavens."* Berkeley: University of California Press, 2004.

Bower, Calvin M. *Fundamentals of Music: Anicius Manlius Severinus Boethius*. New Haven, CT: Yale University Press, 1989.

Burton, Harry Edwin. "The Optics of Euclid." *Journal of the Optical Society of America* 35, no. 5 (1945): 357–72.

Christianidis, Jean and Jeffrey Oaks. *The Arithmetica of Diophantus: A Complete Translation and Cjommentary*. London: Routledge, 2023.

Columella, Lucius Junius Moderatus. *L. Junius Moderatus Columella of Husbandry, in Twelve Books: And His Book, Concerning Trees*. London: A. Millar, 1745.

Cuomo, Serafina. *Ancient Mathematics*. London: Routledge, 2001.

Diocles. *On Burning Mirrors: The Arabic Translation of the Lost Greek Original*. Edited and translated by G. J. Toomer. New York: Springer, 1976.

Fowler, D. H. *The Mathematics of Plato's Academy: A New Reconstruction*. Oxford: Clarendon Press, 1987.

Freeth, Tony. "The Antikythera Mechanism: An Ancient Greek Machine Rewriting the History of Technology" *World Archaeology*" July 22, 2021.

———. "Decoding the Antikythera Mechanism: Supplementary Notes 2." *Nature* 444, no. 7119 (2006): 587–91.

Gerstinger, Hans Oellacher, and Kurt Vogel, eds. *Mitteilungen aus der Nationalbibliothek in Wien: Papyrus Erzherzog Rainer*. Vienna: Osterreichischen Staatsdruckerei, 1931.

Goodspeed, Edgar J. "The Ayer Papyrus." *American Mathematical Monthly* 10, no. 5 (1903): 133–35.

Halley, Edmond, ed. and trans. *Menelai Sphaericorum, Libri III*. Oxford, 1758.

Heath, Thomas L. *Aristarchus of Samos: The Ancient Copernicus*. New York: Dover, 1981.

———, ed. *Greek Astronomy*. London: J. M. Dent, 1932.

———, ed. and trans. *The Thirteen Books of Euclid's Elements*. 3 vols. New York: Dover, 1956.

———. *The Works of Archimedes with the Method of Archimedes*. New York: Dover, 1953.

Heiberg, Johan Ludvig, ed. *Apollonii Pergaei quae Graece exstant cum commentariis antiquis*. Stuttgart: Teubner, 1974.

———. *Archimedis Opera*. 2nd ed. Leipzig: Tuebner, 1910–1915.

———. *Euclid's Elements*. Vol. 5. Leipzig: Teubner, 1888.

———. *Heronis Alexandrini opera quae supersunt omnia*. Leipzig: Teubner, 1912.

———. *Sereni Antinoensis opuscula*. Leipzig: Teubner, 1896.

Hogendijk, Jan P. "The Arabic Version of Euclid's *On Divisions*." In *Vestigia Mathematica: Studies in Medieval and Early Modern Mathematics in Honour of H. L. L. Busard*, edited by M. Folkerts and J. P. Hogendijk, 143—62. Amsterdam: Rodopi, 1993.

Høyrup, Jens. *Lengths, Widths, Surfaces: A Portrait of Old Babylonian Algebra and Its Kin*. New York: Springer, 2002.

Huffman, Carl A. *Archytas of Tarentum: Pythagorean, Philosopher and Mathematician King*. Cambridge: Cambridge University Press, 2005.

Hughes, Barnabas. "Leonardo of Pisa, *De practica geometrie* (*practical geometry*)." In *Sourcebook in the Mathematics of Medieval Europe and North Africa*, edited by Victor J. Katz et al., 130–39. Princeton, NJ: Princeton University Press, 2016.

———, ed. *Fibonacci's De practica geometrie.* New York: Springer, 2008.

Hultsch, Frederick. *Metrologicorum Scriptorum Reliquae*. Vol. 2. Leipzig: Teubner, 1866.

Hutchins, Robert Maynard and Mortimer Adler, eds. *Great Books of the Western World.* 54 vols. Chicago: Encyclopedia Britannica, 1952.

Iamblichus. *On the General Science of Mathematics*. Translated by J. O. Urmson, John Dillon, and Sebastian Gertz. London: Bloomsbury Academic, 2020.

Jones, Alexander. "The Antikythera Mechanism and the Public Face of Greek Science." *Proceedings of Science* 038 (2012). https://pos.sissa.it/170/038/pdf.

———. "P. Cornell inv. 69 Revisited: A Collection of Geometrical Problems." In *Papyrological Texts in Honor of Roger S. Bagnall*, edited by Rodney Ast, Hélène Cuvigny, T. M. Hickey, and Julia Lougovaya, 159–73. Durham, NC: American Society of Papyrologists, 2013.

———, ed. and trans. *Astronomical Papyri from Oxyrhynchus (P. Oxy. 4133-4300a)*. Vol. 1. Philadelphia: American Philosophical Society, 1999.

Karpinski, Louis C., and Frank E. Robbins. "Michigan Papyrus 620: The Introduction of Algebraic Equations in Greece." *Science* 70, no. 1813 (1929): 311–14.

Katz, Victor J., ed. *The Mathematics of Egypt, Mesopotamia, China, India, and Islam: A Sourcebook.* Princeton, NJ: Princeton University Press, 2007.

Katz, Victor J., Menso Folkerts, Barnabas Hughes, Roi Wagner, and J. Lennart Berggren, eds. *Sourcebook in the Mathematics of Medieval Europe and North Africa.* Princeton, NJ: Princeton University Press, 2016.

Knorr, Wilbur. "Techniques of Fractions in Ancient Egypt and Greece." *Historia Mathematica* 9 (1982): 133–71.

Knudtzon, Erik, J., and Otto Neugebauer. "Zwei astronomische Texte." *Humanistiska Vetenskapssamfundet i Lund, Ärsberättelse, 1946–47* 2 (1947): 77–88.

Lehoux, Daryn. "The Parapegma Fragments from Miletus." *Zeitschrift für Papyrologie und Epigraphik* 152 (2005): 125–40.

Levin, Flora R., ed. and trans. *The Manual of Harmonics of Nicomachus the Pythagorean*. Grand Rapids, MI: Phanes, 1994.

Macierowski, E. M., trans. *Apollonius of Perga, On Cutting Off a Ratio: An Attempt to Recover the Original Argumentation through a Critical Translation of the Two Extant Medieval Arabic Manuscripts*. Edited by Robert H. Schmidt. Fairfield, CT: Golden Hind, 1987.

Marsden, E. W. *Greek and Roman Artillery: Technical Treatises.* Oxford: Clarendon Press, 1971.

Martin, Thomas J. 'The Arabic Translation of Theodosius's Sphaerica." PhD diss., University of St Andrews, 1975.

Masi, Michael. *Boethian Number Theory: A Translation of the De Institutione Arithmetica*. Amsterdam: Rodopi, 2006.

McDowell, George L., and Merle A. Sokolik, trans. *The Data of Euclid.* Baltimore, MD: Union Square, 1993.

Mercier, Raymond, ed. *Ptolemy's Handy Tables: Transcription and Commentary.* Louvain-La-Neuve: Université Catholique de Louvain, Institut Orientaliste, 2011.

Montelle, Clemency. "The *Anaphoricus* of Hypsicles of Alexandria." In *The Circulation of Astronomical Knowledge in the Ancient World*, edited by John. M. Steele, 287–315. Leiden: Brill, 2016.

Netz, Reviel. *A New History of Greek Mathematics.* Cambridge: Cambridge University Press, 2022.

———. *The Transformation of Mathematics in the Early Mediterranean World: From Problems to Equations*. Cambridge: Cambridge University Press, 2004.

———. *The Works of Archimedes, Translation and Commentary*. Vol. 1, *The Two Books On the Sphere and the Cylinder*. Cambridge: Cambridge University Press, 2004.

———. *The Works of Archimedes, Translation and Commentary*. Vol. 2, *On Spirals*. Cambridge: Cambridge University Press, 2017.

———. *The Works of Archimedes, Translation and Commentary*. Vol. 3. Cambridge: Cambridge University Press, forthcoming.

Neugebauer, Otto, and Henry Bartlett Van Hoesen. *Greek Horoscopes.* Philadelphia: American Philosophical Society, 1959.

Nicomachus of Gerasa, *Introduction to Arithmetic*. Edited and translated by Martin Luther D'Ooge. New York: Macmillan, 1926. 0

Pappus of Alexandria. *Book 4 of the Collection*. Edited and translated by Heike Sefrin-Weis. New York: Springer, 2010.

———. *Book 7 of the Collection: Part 1. Introduction, Text, and Translation*. Edited and translated by Alexander Jones. New York: Springer, 1986.

Parker, Richard A. *Demotic Mathematical Papyri*. Providence, RI: Brown University Press, 1972.

———. "A Demotic Mathematical Papyrus Fragment." *Journal of Near Eastern Studies* 18, no. 4 (1959): 275–79.

Paton, W. R., ed. *The Greek Anthology*. Vol. 5. Cambridge, MA: Harvard University Press, 1979.

Philo. *The Works of Philo*. Translated by C. D. Yonge. Peabody, MA: Hendrickson, 1993.

Plato. *The Dialogues of Plato*. Edited and translated by Benjamin Jowett. *Great Books of the Western World*, vol. 7. Edited by Robert Maynard Hutchins. Chicago: Encyclopedia Britannica, 1952.

———. *Republic*. Translated and edited by C. D. C. Reeve. Indianapolis: Hackett, 2004.

Polybius. *The Complete Histories of Polybius*. Edited and translated by W. R. Paton. Digireads.Com, 2009.

Price, J. D. de Solla. "An Ancient Greek Computer." *Scientific American* 200, no. 6 (1959): 60–67.

Proclus. *A Commentary on the First Book of Euclid's Elements*. Translated by Glenn R. Morrow. Princeton, NJ: Princeton University Press, 1970.

Rashed, Roshdi, and Athanase Papadopoulos, eds. and trans. *Menelaus' Spherics: Early Translation and al-Māhānī / al-Harawī's Version*. Berlin: De Gruyter, 2017.

Rideout, Bronwyn. "Pappus Reborn: Pappus of Alexandria and the Changing Face of Analysis and Synthesis in Late Antiquity." MA thesis, University of Canterbury, New Zealand, 2008.

Sabra, A. I. "Simplicius's Proof of Euclid's Parallels Postulate." *Journal of the Warburg and Courtauld Institutes* 32 (1969): 1–24.

Sesiano, Jacques. "Greek Multiplication Tables." *SCIAMVS* 21 (2020–2021): 83–140.

———. "Sur le Papyrus graecus genevensis 259." *Museum Helveticum* 56, no. 1 (1999): 26–32.

———, ed. *Books IV to VII of Diophantus' Arithmetica in the Arabic Translation Attributed to Qustā ibn Lūqā*. New York: Springer, 1982.

Sidoli, Nathan, and J. Lennart. Berggren. "The Arabic Version of Ptolemy's *Planisphere* or *Flattening the Surface of the Sphere*: Text, Translation, Commentary." *SCIAMUS* 8 (2007): 37–139.

Smith, A. Mark. "Ptolemy and the Foundations of Ancient Mathematical Optics: A Source Based Guided Study." *Transactions of the American Philosophical Society, new ser.*, 89, no. 3 (1999): 3–172.

———. "Ptolemy's Theory of Visual Perception: An English Translation of the Optics with Introduction and Commentary." *Transactions of the American Philosophical Society*, new ser., 86, no. 2 (1996): 1–300.

Taisbak, Christian Marinus. *Euclid's Data or The Importance of Being Given*. Copenhagen: Museum Tusculanum Press, 2003.

Tannery, Paul, ed. *Diophanti Alexandrini Opera Omnia*. Leibniz: Teubner, 1893.

Thomas, Ivor, ed. and trans. *Greek Mathematical Works*. Vol. 1, *Thales to Euclid*. Cambridge, MA: Harvard University Press, 1980.

———. *Greek Mathematical Works*. Vol. 2, *Aristarchus to Pappus of Alexandria*. Cambridge, MA: Harvard University Press, 1980.

———. *Selections Illustrating the History of Greek Mathematics*. 2 vols. Cambridge, MA: Harvard University Press, 1980.

Thomson, William, ed. and trans. *The Commentary of Pappus on Book X of Euclid's Elements*. Cambridge, MA: Harvard University Press, 1930.

Toomer, Gerald J., ed. and trans. *Ptolemy's Almagest*. Princeton, NJ: Princeton University Press, 1998.

Ver Eecke, Paul, ed. and trans. *Pappus D'Alexandrie: Collection Mathématique*. Vol. 1. Paris: Desclée de Brouwer, 1933.

———. *Serenus d'Antinoë: Le Livre de la Section du Cylindre et le Livre de la Section du Cône*. Paris: Desclée de Brouwer, 1929.

———. *Les Sphériques de Théodose de Tripoli.* Paris: Albert Blanchard, 1959.
Vitruvius. *The Ten Books on Architecture*. Translated by Morris Hickey Morgan. Cambridge, MA: Harvard University Press, 1914. https://www.gutenberg.org/files/20239/20239-h/20239-h.htm#Page_251.
Wagner, Roi. "Abraham bar Ḥiyya: Ḥibur Hameshiḥa Vehatishboret (The Treatise on Measuring Areas and Volumes)." In *Sourcebook in the Mathematics of Medieval Europe and North Africa*, edited by Victor J. Katz et al., 296–313. Princeton, NJ: Princeton University Press, 2016.
Waterfield, Robin. *The Theology of Arithmetic: On the Mystical, Mathematical and Cosmological Symbolism of the First Ten Numbers; Attributed to Iamblichus.* Grand Rapids, MI: Phanes, 1988.
Woepcke, Franz. "Notice sur des traductions arabes de deux ouvrages perdus d'Euclide," *Journal Asiatique* 18, no. 4 (1851): 217–247.

Index